R 670
P9-CFA-676

SEATTLE PREP LIBRARY
2400 - 11th AVE. EAST
SEATTLE, WA 98102

# How PRODUCTS Are MADE

# How PRODUCTS Are MADE

*An Illustrated Guide to Product Manufacturing*

**Volume 5**

*Jacqueline L. Longe, Editor*

Detroit
San Francisco
London
Boston
Woodbridge, CT

**STAFF**



ISBN 07876-2444-6
ISSN 1072-5091

Printed in the United States of America
10 9 8 7 6 5 4 3 2 1

# Contents

# Introduction

## About the Series

Welcome to *How Products Are Made: An Illustrated Guide to Product Manufacturing*. This series provides information on the manufacture of a variety of items, from everyday household products to heavy machinery to sophisticated electronic equipment. You will find step-by-step descriptions of processes, simple explanations of technical terms and concepts, and clear, easy-to-follow illustrations.

Each volume of *How Products Are Made* covers a broad range of manufacturing areas: food, clothing, electronics, transportation, machinery, instruments, sporting goods, and more. Some are intermediate goods sold to manufacturers of other products, while others are retail goods sold directly to consumers. You will find items made from a variety of materials, including products such as precious metals and minerals that are not "made" so much as they are extracted and refined.

## Organization

Every volume in this series is comprised of many individual entries, each covering a single product. Although each entry focuses on the product's manufacturing process, it also provides a wealth of other information: who invented the product or how it has developed, how it works, what materials are used, how it is designed, quality control procedures, byproducts generated during its manufacture, future applications, and books and periodical articles containing more information.

To make it easier for you to find what you're looking for, the entries are broken up into standard sections. Among the sections you will find are the following:

- Background
- History
- Raw Materials
- Design
- The Manufacturing Process
- Quality Control
- Byproducts/Waste
- The Future
- Where To Learn More

Every entry is accompanied by illustrations. Uncomplicated and easy to understand, these illustrations generally follow the step-by-step description of the manufacturing process found in the text.

Every entry is accompanied by illustrations. Uncomplicated and easy to understand, these illustrations generally follow the step-by-step description of the manufacturing process found in the text.

A general subject index of important terms, processes, materials, and people is found at the end of the book. Bold faced items in the index refer to main entries.

Main entries from previous volumes are also included in the index. They are listed along with their corresponding volume and page numbers.

## About this Volume

This volume contains essays on over 100 products, arranged alphabetically, and 16 special boxed sections, describing interesting historical developments related to a product. Photographs are also included.

## Contributors/Advisor

The entries in this volume were written by a skilled team of technical writers and engineers, often in cooperation with manufacturers and industry associations. The advisor for this volume was David L. Wells, PhD, CMfgE, a long time member of the Society of Manufacturing Engineers (SME) and the Academic Dean at Focus: HOPE, a nonprofit civil and human rights organization dedicated to the technical training and education of the multicultural community of Detroit, Michigan.

## Suggestions

Your questions, comments, and suggestions for future products are welcome. Please send all such correspondence to:

*How Products Are Made*
Gale Group, Inc.
27500 Drake Rd.
Farmington Hills, MI 48331-3535

# Contributors

*Nancy EV Bryk*

*Chris Cavette*

*Loretta Hall*

*Gillian S. Holmes*

*Perry Romanowski*

*Randy Schueller*

*Rose Secrest*

*Laurel Sheppard*

*David L. Wells*

*Angela Woodward*

# Acknowledgments

The editor would like to thank the following individuals, companies, and associations for providing assistance with Volume 5 of *How Products Are Made*:

**Artificial Flower:** Ardith Beveridge, AAF, AIFD, PFCI, Director/Instructor, Koehler & Dramm's Institute of Floristry, Minneapolis, MN; Trice Whitaker, Owner, American Prestige Silks Inc., Nocona, TX. **Bean Bag Plush Toy:** Peggy Gallagher, Beanie Baby expert and authenticator of rare Beanie Baby collectibles; Ray Bolhouse, Co-owner, SWIBCO Inc., Lisle, IL. **Bed Sheet:** Jeff Day, Sheeting Supevisor, Fieldcrest Cannon Corp., Kanapolis, NC. **Brassiere:** Bernadette Chavez, Playtex Inc. **Castanets:** Morca Foundation, Bellingham, WA. **Cheese Curl:** Kent Hunold, Planning Manager, Frito-Lay Production Plant, IN; Liz Doyle, Snack Food Association, Chicago, IL. **Child Safety Seat:** Carol M. Dingledy, COSCO, Inc., Columbus, Indiana. **Computer Mouse:** Debra Reich, Kensington Technology Group, a division of ACCO Brands Inc., San Mateo, California. **Doughnut:** Larry Jabro, General Manager, Krispy Kreme, Dearborn, MI. **Eggs:** Lehman's Egg Service, Greencastle, PA. **Electric Automobile:** General Motors , Lansing Craft Centre, Lansing, MI. **Fishing Fly:** John Herzer, Owner, Blackfoot River Outfitters, Missoula, MT; Bob Knapp, hand-tier, Missoula, MT. **Fishing Lure:** David Nichols, President and CEO, Nichols Lures Inc., San Antonio, TX. **Fruit Leather:** Keith Barton, Technical Department, Favorite Brands International, Inc. **Galoshes:** Scott Hardy, Co-Owner, N.E.O.S.; John A. Greene, Kensington Cobblers, Kensington, CA. **Greeting Card:** Susan Millichamp, Assistant Production Technician, Avanti Press Inc., Detroit, MI; Matt Burckhardt, Customer Service Representative, Northwestern Printing, Grosse Pointe, MI. **High Heel:** John A. Greene, Kensington Cobblers, Kensington, CA. **Holiday Lights:** Sandy Kinderman, Chief Executive Officer, Brite Star Company, Inc. **Hourglass:** David W. Hood, The Hourglass Connection. **Incense Stick:** Mark Radlinski, Product Coordinator, Wild Berry Incense Inc., Oxford, OH; Deepak Roy, Senior Representative, Excelsior Incense Works, San Francisco, CA. **Lock:** John Crocco, Illinois Lock Co. **Lyocell:** Mike Finlen, Accordis Fibers. **Moustrap:** Harry Knuppel, Engineering Manager, Kness Manufacturing, Albia, IA. **Olives:** Craig A. Makela, President, Santa Barbara Olive Co., Santa Barbara, CA. **Paintbrush:** Scott Routledge, Marketing Manager, The Wooster Brush Co., Wooster, OH. **Parachute:** Strong Enterprises, Orlando, FL. **Pepper:** Carlo Busceme, III, Vice President of Operations, Texas Coffee Co. **Popcorn:** Jon Tiefenthaler, Snappy Popcorn Co.; Tom Elsen, American Pop Corn Co. **Sheet Music:** Don Zegel, Alafia Publishing and Music Sales; Robert Loughrige, Owner, Bob's Music Notation Service. **Statuary:** Design Toscano Inc., Arlington Heights, IL. **Thread:** Viola Hechinger, Gütermann AG, Germany. **Voting Machine:** Bill Carson, Carson Manufacturing. **Wind Chime:** Stacey Bowers, President, Woodstock Percussion Inc., West Hurley, NY.

Photographs appearing in Volume 5 of *How Products Are Made* were received from the following sources:

AP/Wide World Photos. Reproduced by permission: **Suspension Bridge.** Deegan, Patrick (standing on Tacoma Narrows Bridge, tying cable strands for the suspension cable of the bridge), photograph.

Archive Photos Inc. Reproduced by permission: **Cork.** Assorted vintage bottle caps, photograph.

Corbis-Bettmann. Reproduced by permission: **Molasses.** Boston Molasses Disaster, photograph.

Henry Ford Museum and Greenfield Village. Reproduced by permission: **Toilet.** Ceramic chamber pot, photograph.

The Library of Congress: **Artificial Blood.** Landsteiner, Karl, photograph; **Concrete Dam.** Bayless Dam on Freemans Run, photograph; **Frozen Vegetable.** Birdseye, Clarence (reading book at desk), photograph; **Galoshes.** Goodyear, Charles, sketch; **Glue.** Cooper, Peter, portrait; **Hang Glider.** Langley, Samuel, painting; **Hourglass.** Harrison, John, portrait; **LP Record.** Edison, Thomas Alva (pouring substance), photograph; **Nuclear Submarine.** Fulton, Robert (seated at drawing table), painting; **Telephone.** Gray, Elisha, photograph; **Vodka.** Pasteur, Louis (holding bottle in right hand, piece of paper in left hand), photograph.

U.S. National Aeronautics and Space Administration (NASA): **Spacesuit.** Ride, Sally (wearing a dark NASA jumpsuit), 1978, photograph.

All line art illustrations in this volume were created by **Electronic Illustrators Group (EIG)** of Fountain Hills, Arizona.

# Aluminum

The metallic element aluminum is the third most plentiful element in the earth's crust, comprising 8% of the planet's soil and rocks (oxygen and silicon make up 47% and 28%, respectively). In nature, aluminum is found only in chemical compounds with other elements such as sulphur, silicon, and oxygen. Pure, metallic aluminum can be economically produced only from aluminum oxide ore.

Metallic aluminum has many properties that make it useful in a wide range of applications. It is lightweight, strong, nonmagnetic, and nontoxic. It conducts heat and electricity and reflects heat and light. It is strong but easily workable, and it retains its strength under extreme cold without becoming brittle. The surface of aluminum quickly oxidizes to form an invisible barrier to corrosion. Furthermore, aluminum can easily and economically be recycled into new products.

## Background

Aluminum compounds have proven useful for thousands of years. Around 5000 B.C., Persian potters made their strongest vessels from clay that contained aluminum oxide. Ancient Egyptians and Babylonians used aluminum compounds in fabric dyes, cosmetics, and medicines. However, it was not until the early nineteenth century that aluminum was identified as an element and isolated as a pure metal. The difficulty of extracting aluminum from its natural compounds kept the metal rare for many years; half a century after its discovery, it was still as rare and valuable as silver.

In 1886, two 22-year-old scientists independently developed a smelting process that made economical mass production of aluminum possible. Known as the Hall-Heroult process after its American and French inventors, the process is still the primary method of aluminum production today. The Bayer process for refining aluminum ore, developed in 1888 by an Austrian chemist, also contributed significantly to the economical mass production of aluminum.

In 1884, 125 lb (60 kg) of aluminum was produced in the United States, and it sold for about the same unit price as silver. In 1995, U.S. plants produced 7.8 billion lb (3.6 million metric tons) of aluminum, and the price of silver was seventy-five times as much as the price of aluminum.

## Raw Materials

Aluminum compounds occur in all types of clay, but the ore that is most useful for producing pure aluminum is bauxite. Bauxite consists of 45-60% aluminum oxide, along with various impurities such as sand, iron, and other metals. Although some bauxite deposits are hard rock, most consist of relatively soft dirt that is easily dug from open-pit mines. Australia produces more than one-third of the world's supply of bauxite. It takes about 4 lb (2 kg) of bauxite to produce 1 lb (0.5 kg) of aluminum metal.

Caustic soda (sodium hydroxide) is used to dissolve the aluminum compounds found in the bauxite, separating them from the impurities. Depending on the composition of the bauxite ore, relatively small amounts of other chemicals may be used in the extrac-

*Aluminum is the third most plentiful element in the earth's crust, comprising 8% of the planet's soil and rocks (oxygen and silicon make up 47% and 28%, respectively).*

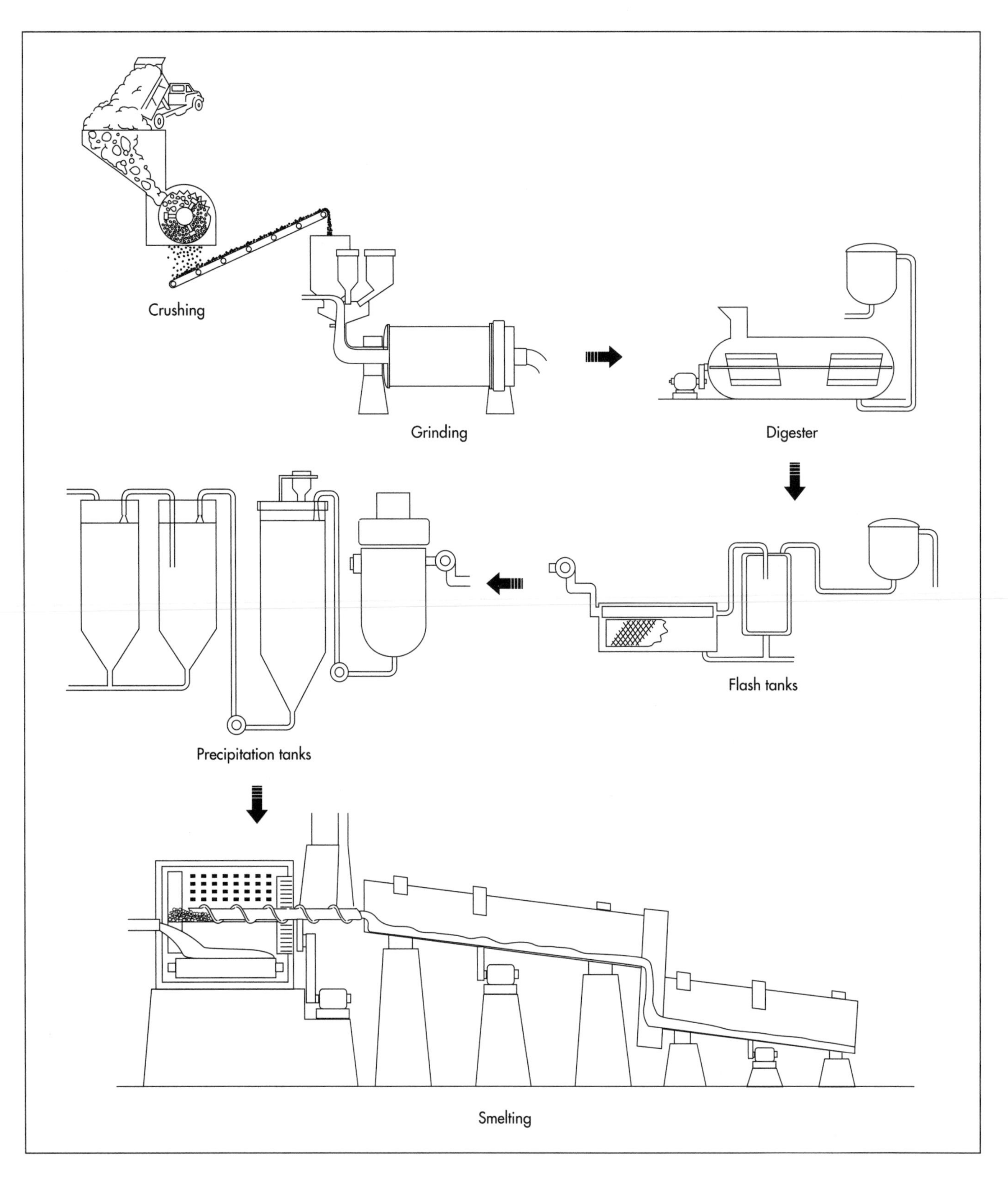


Aluminum is manufactured in two phases: the Bayer process of refining the bauxite ore to obtain aluminum oxide, and the Hall-Heroult process of smelting the aluminum oxide to release pure aluminum.

tion of aluminum. Starch, lime, and sodium sulphide are some examples.

Cryolite, a chemical compound composed of sodium, aluminum, and fluorine, is used as the electrolyte (current-conducting medium) in the smelting operation. Naturally occurring cryolite was once mined in Greenland, but the compound is now produced synthetically for use in the production of aluminum.

Aluminum fluoride is added to lower the melting point of the electrolyte solution.

The other major ingredient used in the smelting operation is carbon. Carbon electrodes transmit the electric current through the electrolyte. During the smelting operation, some of the carbon is consumed as it combines with oxygen to form carbon dioxide. In fact, about half a pound (0.2 kg) of carbon is used for every pound (2.2 kg) of aluminum produced. Some of the carbon used in aluminum smelting is a byproduct of oil refining; additional carbon is obtained from coal.

Because aluminum smelting involves passing an electric current through a molten electrolyte, it requires large amounts of electrical energy. On average, production of 2 lb (1 kg) of aluminum requires 15 kilowatt-hours (kWh) of energy. The cost of electricity represents about one-third of the cost of smelting aluminum.

## The Manufacturing Process

Aluminum manufacture is accomplished in two phases: the Bayer process of refining the bauxite ore to obtain aluminum oxide, and the Hall-Heroult process of smelting the aluminum oxide to release pure aluminum.

### *The Bayer process*

1 First, the bauxite ore is mechanically crushed. Then, the crushed ore is mixed with caustic soda and processed in a grinding mill to produce a slurry (a watery suspension) containing very fine particles of ore.

2 The slurry is pumped into a digester, a tank that functions like a pressure cooker. The slurry is heated to 230-520°F (110-270°C) under a pressure of 50 lb/in$^2$ (340 kPa). These conditions are maintained for a time ranging from half an hour to several hours. Additional caustic soda may be added to ensure that all aluminum-containing compounds are dissolved.

3 The hot slurry, which is now a sodium aluminate solution, passes through a series of flash tanks that reduce the pressure and recover heat that can be reused in the refining process.

4 The slurry is pumped into a settling tank. As the slurry rests in this tank, impurities that will not dissolve in the caustic soda settle to the bottom of the vessel. One manufacturer compares this process to fine sand settling to the bottom of a glass of sugar water; the sugar does not settle out because it is dissolved in the water, just as the aluminum in the settling tank remains dissolved in the caustic soda. The residue (called "red mud") that accumulates in the bottom of the tank consists of fine sand, iron oxide, and oxides of trace elements like titanium.

5 After the impurities have settled out, the remaining liquid, which looks somewhat like coffee, is pumped through a series of cloth filters. Any fine particles of impurities that remain in the solution are trapped by the filters. This material is washed to recover alumina and caustic soda that can be reused.

6 The filtered liquid is pumped through a series of six-story-tall precipitation tanks. Seed crystals of alumina hydrate (alumina bonded to water molecules) are added through the top of each tank. The seed crystals grow as they settle through the liquid and dissolved alumina attaches to them.

7 The crystals precipitate (settle to the bottom of the tank) and are removed. After washing, they are transferred to a kiln for calcining (heating to release the water molecules that are chemically bonded to the alumina molecules). A screw conveyor moves a continuous stream of crystals into a rotating, cylindrical kiln that is tilted to allow gravity to move the material through it. A temperature of 2,000° F (1,100° C) drives off the water molecules, leaving anhydrous (waterless) alumina crystals. After leaving the kiln, the crystals pass through a cooler.

### *The Hall-Heroult process*

Smelting of alumina into metallic aluminum takes place in a steel vat called a reduction pot. The bottom of the pot is lined with carbon, which acts as one electrode (conductor of electric current) of the system. The opposite electrodes consist of a set of carbon rods suspended above the pot; they are lowered into an electrolyte solution and held about 1.5 in (3.8 cm) above the surface of the molten aluminum that accumulates on the floor of the pot. Reduction pots are arranged in rows (potlines) consisting of 50-200 pots that are connected in series to form an electric circuit. Each potline can produce 66,000-110,000 tons (60,000-100,000 metric tons) of aluminum per year. A typical smelting plant consists of two or three potlines.

8 Within the reduction pot, alumina crystals are dissolved in molten cryolite at a temperature of 1,760-1,780° F (960-970° C) to form an electrolyte solution that will conduct electricity from the carbon rods to the carbon-lined bed of the pot. A direct current (4-6 volts and 100,000-230,000 amperes) is passed through the solution. The resulting reaction breaks the bonds between the aluminum and oxygen atoms in the alumina molecules. The oxygen that is released is attracted to the carbon rods, where it forms carbon dioxide. The freed aluminum atoms settle to the bottom of the pot as molten metal.

The smelting process is a continuous one, with more alumina being added to the cryolite solution to replace the decomposed compound. A constant electric current is maintained. Heat generated by the flow of electricity at the bottom electrode keeps the contents of the pot in a liquid state, but a crust tends to form atop the molten electrolyte. Periodically, the crust is broken to allow more alumina to be added for processing. The pure molten aluminum accumulates at the bottom of the pot and is siphoned off. The pots are operated 24 hours a day, seven days a week.

9 A crucible is moved down the potline, collecting 9,000 lb (4,000 kg) of molten aluminum, which is 99.8% pure. The metal is transferred to a holding furnace and then cast (poured into molds) as ingots. One common technique is to pour the molten aluminum into a long, horizontal mold. As the metal moves through the mold, the exterior is cooled with water, causing the aluminum to solidify. The solid shaft emerges from the far end of the mold, where it is sawed at appropriate intervals to form ingots of the desired length. Like the smelting process itself, this casting process is also continuous.

## Byproducts/Waste

Alumina, the intermediate substance that is produced by the Bayer process and that constitutes the raw material for the Hall-Heroult process, is also a useful final product. It is a white, powdery substance with a consistency that ranges from that of talcum powder to that of granulated sugar. It can be used in a wide range of products such as laundry detergents, toothpaste, and fluorescent light bulbs. It is an important ingredient in ceramic materials; for example, it is used to make false teeth, spark plugs, and clear ceramic windshields for military airplanes. An effective polishing compound, it is used to finish computer hard drives, among other products. Its chemical properties make it effective in many other applications, including catalytic converters and explosives. It is even used in rocket fuel—400,000 lb (180,000 kg) is consumed in every space shuttle launch. Approximately 10% of the alumina produced each year is used for applications other than making aluminum.

The largest waste product generated in bauxite refining is the tailings (ore refuse) called "red mud." A refinery produces about the same amount of red mud as it does alumina (in terms of dry weight). It contains some useful substances, like iron, titanium, soda, and alumina, but no one has been able to develop an economical process for recovering them. Other than a small amount of red mud that is used commercially for coloring masonry, this is truly a waste product. Most refineries simply collect the red mud in an open pond that allows some of its moisture to evaporate; when the mud has dried to a solid enough consistency, which may take several years, it is covered with dirt or mixed with soil.

Several types of waste products are generated by decomposition of carbon electrodes during the smelting operation. Aluminum plants in the United States create significant amounts of greenhouse gases, generating about 5.5 million tons (5 million metric tons) of carbon dioxide and 3,300 tons (3,000 metric tons) of perfluorocarbons (compounds of carbon and fluorine) each year.

Approximately 120,000 tons (110,000 metric tons) of spent potlining (SPL) material is removed from aluminum reduction pots each year. Designated a hazardous material by the Environmental Protection Agency (EPA), SPL has posed a significant disposal problem for the industry. In 1996, the first in a planned series of recycling plants opened; these plants transform SPL into glass frit, an intermediate product from which glass and ceramics can be manufactured. Ultimately, the recycled SPL appears in such products as ceramic tile, glass fibers, and asphalt shingle granules.

## The Future

Virtually all of the aluminum producers in the United States are members of the Voluntary Aluminum Industrial Partnership (VAIP), an organization that works closely with the EPA to find solutions to the pollution problems facing the industry. A major focus of research is the effort to develop an inert (chemically inactive) electrode material for aluminum reduction pots. A titanium-diboride-graphite compound shows significant promise. Among the benefits expected to come when this new technology is perfected are elimination of the greenhouse gas emissions and a 25% reduction in energy use during the smelting operation.

## Where to Learn More

### Books

Altenpohl, Dietrich. *Aluminum Viewed from Within: An Introduction into the Metallurgy of Aluminum Fabrication* (English translation). Dusseldorf: Aluminium-Verlag, 1982.

Russell, Allen S. "Aluminum." *McGraw-Hill Encyclopedia of Science & Technology.* New York: McGraw-Hill, 1997.

### Periodicals

Thompson, James V. "Alumina: Simple Chemistry—Complex Plants." *Engineering & Mining Journal* (February 1, 1995): 42 ff.

### Other

Alcoa Aluminum. http://www.alcoa.com/ (March 1999).

Reynolds Metals Company. http://www.reynoldswrap.com/gbu/bauxitealumina/ (April 1999).

*—Loretta Hall*

# Ambulance

*The first motorized ambulance went into operation in Chicago in 1899.*

An ambulance is a self-propelled vehicle specifically designed to transport critically sick or injured people to a medical facility. Most ambulances are motor vehicles, although helicopters, airplanes, and boats are also used. The interior of an ambulance has room for one or more patients plus several emergency medical personnel. It also contains a variety of supplies and equipment that are used to stabilize the patient's condition while en route.

## Background

The earliest ambulances were simple two-wheeled carts used to carry sick or wounded soldiers who were unable to walk by themselves. The word ambulance comes from the Latin word *ambulare*, meaning to walk or move about. The first ambulances specifically used to transport patients to a medical facility were developed in the late 1700s in France by Dominique-Jean Larrey, surgeon-in-chief in Napoleon's army. Larrey noted that it took almost a full day for wounded soldiers to be carried to field hospitals, and that most of them died in that time "from want of assistance." To render more immediate aid and provide faster transportation, he designed a horse-drawn carriage staffed by a medical officer and assistant with room for several patients on stretchers.

The first military ambulance corps in the United States was organized in 1862 during the Civil War as part of the Union army. The first civilian ambulance service in the United States was organized three years later by the Cincinnati Commercial Hospital. By the turn of the century, most major hospitals had their own private ambulances. The first motorized ambulance went into operation in Chicago in 1899.

In areas where there were no major hospitals, the local undertaker's hearse was often the only vehicle capable of carrying a patient on a stretcher, and many funeral homes also provided an ambulance service. As a result, the design and construction of ambulances and hearses remained closely related for many years.

Most early ambulances were simply intended to transport patients. After the doctor or fire department rescue squad applied first aid, the patient was loaded into the back of the ambulance for a quick ride to the hospital. In some cases, the doctor rode along, but most of the time the patient rode alone and unattended. In the United States that changed dramatically when the federal government passed the Highway Safety Act in 1966. Among its many standards, the new act set requirements for ambulance design and emergency medical care. Ambulances with low-slung, hearse-like bodies were replaced by high-bodied vans to accommodate additional personnel and equipment. Radios were installed. Many ambulances carried advanced equipment like cardiac defibrillators, along with an arsenal of life-saving medicines and drugs.

Today, ambulances come in a wide variety of shapes and sizes. The simplest designs are equipped to provide basic life support, or BLS, while larger, more sophisticated designs are equipped to provide advanced life support, or ALS. Ambulances may be operated by private companies, hospitals, the local fire or police department, or a separate city-run organization.

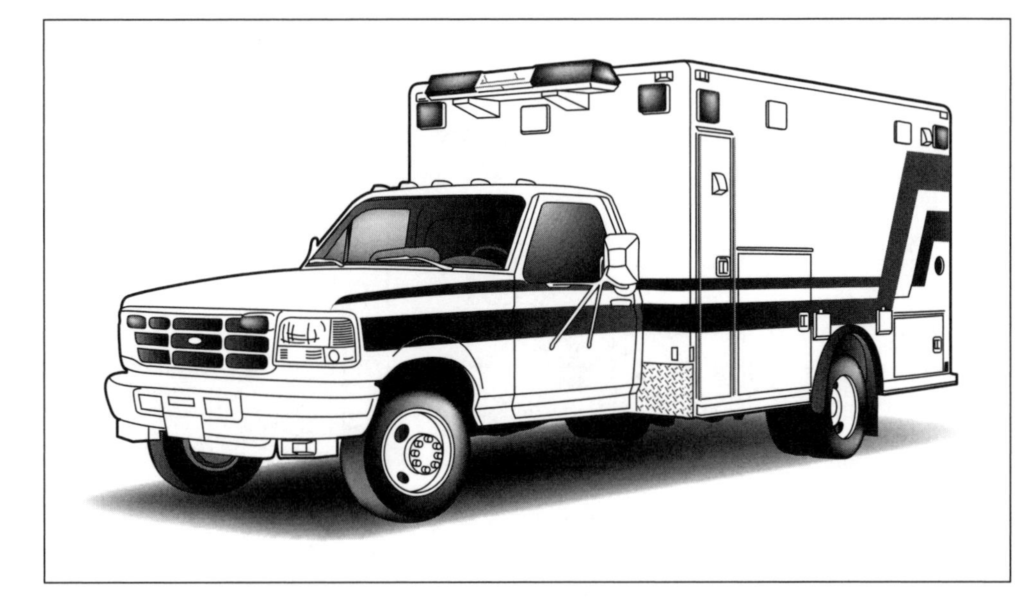

Ambulance manufacturers purchase many components from other suppliers rather than fabricate them themselves. The body framework is usually made of formed or extruded aluminum. The outer walls are painted aluminum sheet, and the interior walls are usually aluminum sheet covered with a vinyl coating or a laminated plastic. The subfloor may be made of plywood or may use an open-cored plastic honeycomb laminated to aluminum sheet. The interior floor covering is usually a seamless, industrial-grade vinyl that extends partially up each side for easy cleaning.

## Raw Materials

Ambulance manufacturers purchase many components from other suppliers rather than fabricate them themselves. These include the vehicle cab and chassis, warning lights and sirens, radios, most electrical system components, the heating and air conditioning components, the oxygen system components, and various body trim pieces like windows, latches, handles, and hinges.

If the ambulance has a separate body, the body framework is usually made of formed or extruded aluminum. The outer walls are painted aluminum sheet, and the interior walls are usually aluminum sheet covered with a vinyl coating or a laminated plastic. The subfloor may be made of plywood or may use an open-cored plastic honeycomb laminated to aluminum sheet. The interior floor covering is usually a seamless, industrial-grade vinyl that extends partially up each side for easy cleaning.

Interior cabinets in the patient compartment are usually made of aluminum with transparent, shatter-resistant plastic panels in the doors. The counter and wall surfaces in the "action area," the area immediately opposite the patient's head and torso in the left-hand forward portion of the ambulance body, are usually covered with a seamless sheet of stainless steel to resist the effects of blood and other body fluids. Interior seating and other upholstered areas have a flame-retardant foam padding with a vinyl covering. Interior grab handles and grab rails are made of stainless steel. Other interior trim pieces may be made of various rubber or plastic materials.

## Design

Ambulance designs fall into three categories. Type I ambulances have a modular, or detachable, body built on a truck chassis. The truck cab is connected to the body through a small window, but the occupants of the cab must go outside the vehicle to enter the ambulance body. Type II ambulances use a van with a raised roof. Because of the van construction, the occupants of the cab can easily enter the body from the inside, although the interior space is limited. Type III ambulances have a modular body built on a cut-away van chassis. This design combines the capacity of the larger modular body with the walk-through accessibility of a van.

The federal requirements for ambulances are defined by General Services Administration Standard KKK-A-1822: Federal Specifications for Ambulances. It covers overall construction, electrical systems, emergency warning lights, and many other aspects of ambulance design. Some states have adopted this federal standard, while others have their own design requirements. Because an ambulance is a motor vehicle, the Federal Motor Vehicle Safety Standards (FMVSS) apply to the vehicle portion. Certain Occupational Safety and Health Administration

Interior cabinets are usually made of aluminum with transparent, shatter-resistant plastic panels in the doors. The counter and wall surfaces in the "action area," are usually covered with a seamless sheet of stainless steel to resist the effects of blood and other body fluids. Interior seating and other upholstered areas have a flame-retardant foam padding with a vinyl covering. Interior grab handles and grab rails are made of stainless steel. Other interior trim pieces may be made of various rubber or plastic materials.

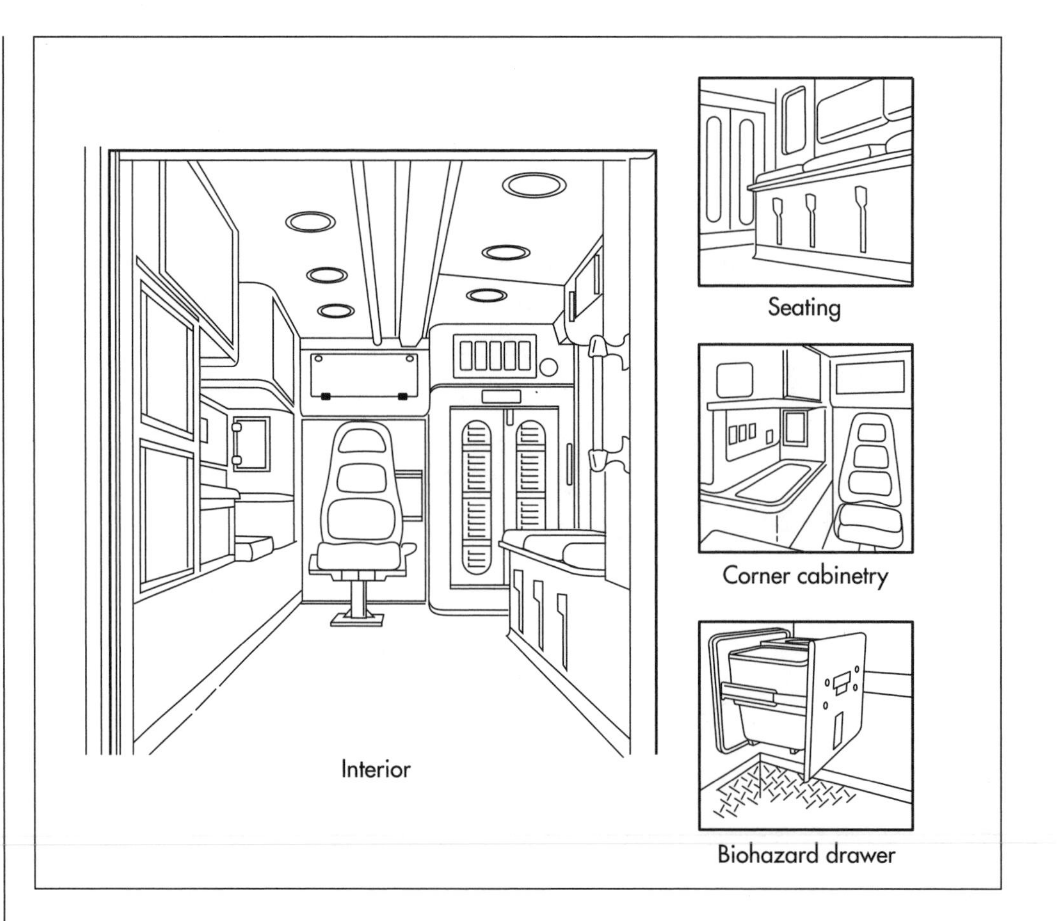


(OSHA) standards regarding blood-borne and airborne pathogens also apply. Within the framework of these standards, manufacturers may specify specific features and materials to provide their products with unique advantages in the marketplace.

## The Manufacturing Process

Ambulances are usually manufactured in a modified assembly line process, where the vehicle or body moves from one fixed area of a plant to another, rather than being pulled along an assembly line. Specific parts are brought to each area for installation or assembly. Different manufacturers may use slightly different processes. The following is a typical sequence of operations for the manufacture of a Type I ambulance with a modular body.

### *Building the body shell*

1 The structural components of the ambulance body—the supporting struts, braces, and brackets for the floor, sides, and roof—are either bent to shape using standard machine shop tools, or are cut from specially shaped aluminum extrusions that have been purchased from suppliers. The components are held in the proper position with a device called a jig and are welded together to form the body framework.

2 The exterior skin pieces are fabricated using standard sheet metal shop tools and are fastened to the outside of the framework using either mechanical fasteners or adhesive bonding. The external compartments are fabricated and welded in place. Finally, the external body doors are fabricated and are fastened in place on hinges.

3 The outside of the body shell is then cleaned, sanded, and spray painted with a primer. Next, a sealer is applied. This is followed by a base coat of paint, usually white, and then a clear coat of paint to protect the base color and give the surface a shiny appearance. Between each coat, the body is placed in an oven to dry.

### *Preparing the cab and chassis*

4 Additional wiring is added to the cab, chassis, and engine electrical system to accommodate the warning lights and sirens and to bring power to the body. Additional switches and controls are added to the dash as required. The heating and air conditioning system may also be modified.

5 Holes are drilled in the vehicle frame rails and mounting brackets are installed to support the ambulance body. The frame rails may be cut to the proper length for the body.

### *Mounting the body*

6 The painted body shell is lowered onto the chassis mounting brackets and is bolted in place.

7 The cab is usually ordered with the same background color as the body, and does not require priming or base/clear painting. Most ambulances are specified with one or more colored stripes that extend along the sides and rear of the cab and body. The areas around the stripes are masked off with paper and tape so that the position of the stripes on the cab and the body match. The stripes are then painted and dried, and the masking removed.

8 The front and rear bumpers, which are not painted, are then installed. If the mirrors have been removed to paint the stripes, they are reinstalled.

### *Finishing the body*

9 The electrical wiring in the body walls and ceiling is installed from the inside, and foam panels are bonded in place to provide thermal and noise insulation. With the wiring in place, the exterior lights are mounted and connected, and the exterior latches, grab handles, windows, and other trim pieces are installed.

10 The oxygen piping and outlets, which are part of the patient life-support system, are installed in the body walls. The vacuum system, which removes blood, saliva, and other body fluids is also installed. If the ambulance body requires an auxiliary heating and air-conditioning system, it is installed at this time.

11 With all the systems in place, the interior cabinets are installed and the walls, floors, and ceilings are covered. The electrical power distribution board is installed in a forward compartment of the body and the panel is connected to the cab and chassis electrical wiring. If the ambulance is specified with an inverter, which converts 12 volts direct current from the vehicle batteries into 120 volts alternating current for use with certain medical equipment, it is also installed at this time.

12 The seats and upholstery pieces, which are either purchased or assembled in a separate area, are fastened in place. The interior grab handles, containers, and trim pieces are installed as the final step.

## Quality Control

The design of ambulances is regulated by several standards, and the manufacturer must take appropriate steps to ensure compliance with those standards. Each system is inspected and tested for proper installation and operation as part of the manufacturing process. In addition, every material, from the aluminum in the body to the foam in the head rests, is certified by the manufacturer to meet the required specifications.

## The Future

Many fire departments are finding that approximately 80-90% of their calls are for medical emergencies, while only 10-20% are for fires. In the case of medical emergencies, an ambulance has to be called in addition to the fire engine. Instead of responding to all calls with large pumpers or ladder trucks, some fire departments are starting to use smaller, lower-cost first-response vehicles that combine the equipment and patient transport capabilities of a rescue truck and ambulance with the fire suppression capabilities of a small pumper. These combination vehicles are able to handle a variety of emergency situations, including those involving small fires such as might occur in vehicle accidents. This saves wear on the larger firefighting vehicles, and eliminates the need to dispatch two vehicles to the same incident. In the future, an increase in traffic congestion and an increase in the average age of the population in the United

States are expected to increase the number of medical emergency calls. When this happens, it is expected that the single-function ambulance may be replaced by a multi-function combination vehicle in many areas.

## Where to Learn More


### Books

Barkley, Katherine Traver. *Ambulance: The Story of Emergency Transportation of Sick and Wounded Through the Centuries.* ______, 1990.

Haller, John S. Jr. *Farmcarts to Fords: A History of the Military Ambulance, 1790-1925.* Southern Illinois University Press, 1992.

McCall, Walt and Tom McPherson.*Classic American Ambulances.* Iconografix, 1999.

### Periodicals

Kelly, Jack. "Rescue Squad."*American Heritage* (May/June 1996): 91-99.

Sachs, Gordon M. "The Evolution of the Fire Service EMS Vehicle."*Fire Engineering* (July 1998): 22, 24, 26.

### Other

Federal Specifications for Ambulances. General Services Administration Standard KKK-A-1822.

American Emergency Vehicles. September 28, 1998. http://www.ambulance.com/ (June 29, 1999).

McCoy Miller Corporation website. May 26, 1999. http://www.mccoymiller.com/ (June 29, 1999).

Road Rescue, Inc. http://www.roadrescue.com/ (June 29, 1999).


*—Chris Cavette*

# Antiperspirant/ Deodorant Stick

## Background

Antiperspirant/deodorant (APD) sticks are used to reduce underarm wetness and control body odor. These products are made by blending active ingredients with waxes, oils, and silicones and molding the mixture into stick form.

Body odor is primarily generated in the area under the arms where there is a high concentration of sweat glands. While sweat from these glands is initially odorless, it contains natural oils, called lipids, that provide a growth medium for bacteria living on the skin. These bacteria interact with the lipids, converting them into compounds that have a characteristic sweaty odor. Isovaleric acid, for example, is one chemical compound that gives sweat its smell.

There are two primary types of products used to control body odor. The first, deodorants, reduce body odor by killing the odor-causing bacteria. These products do not affect the amount of perspiration the body produces. Antiperspirants, on the other hand, inhibit the activity of sweat glands so less moisture is produced. In addition to avoiding unpleasant wetness, these products also decrease odor because there is less sweat for the bacteria to act upon. While deodorants are considered to be cosmetic products because they only control odor, antiperspirants are actually drugs because they affect the physiology of the body. Although the exact mechanism of this physiological interaction is not fully understood, theory has it that antiperspirant salts form temporary plugs in some of the sweat gland openings so that moisture is not secreted. While this moisture reduction is not severe enough to interfere with normal body metabolism, it does noticeably lessen underarm wetness.

## History

Products to control body odor and wetness have been used for centuries. Before bathing became commonplace, people used heavy colognes to mask body odor. In the late nineteenth century, chemists developed products that were able to prevent the formation of these odors. Early antiperspirants were pastes that were applied to the underarm area; the first such product to be trademarked in the United States was Mum in 1888. It was a waxy cream that was difficult to apply and extremely messy. A few years later, Everdry, the first antiperspirant to use aluminum chloride was developed. Within 15 years, a variety of products were marketed in a number of different forms including creams, solids, pads, dabbers, roll-ons, and powders.

In the late 1950s, manufacturers began using aerosol technology to dispense personal care products such as perfumes and shaving creams. In the early 1960s, Gillette introduced Right Guard, the first aerosol antiperspirant. Aerosols became a popular way to dispense antiperspirants because they allowed the user to apply without having to touch the underarm area. By 1967, half the antiperspirants sold in the United States were in aerosol form, and by the early 1970s, they accounted for 82% of all sales.

However, later that decade two technical issues arose which greatly impacted the popularity of these products. First, in 1977, the Food and Drug Administration (FDA) banned the primary active ingredient used in aerosols, aluminum zirconium complexes,

*Body odor is primarily generated in the area under the arms where there is a high concentration of sweat glands. While sweat from these glands is initially odorless, it contains natural oils, called lipids, that provide a growth medium for bacteria living on the skin. These bacteria interact with the lipids, converting them into compounds that have a characteristic sweaty odor.*

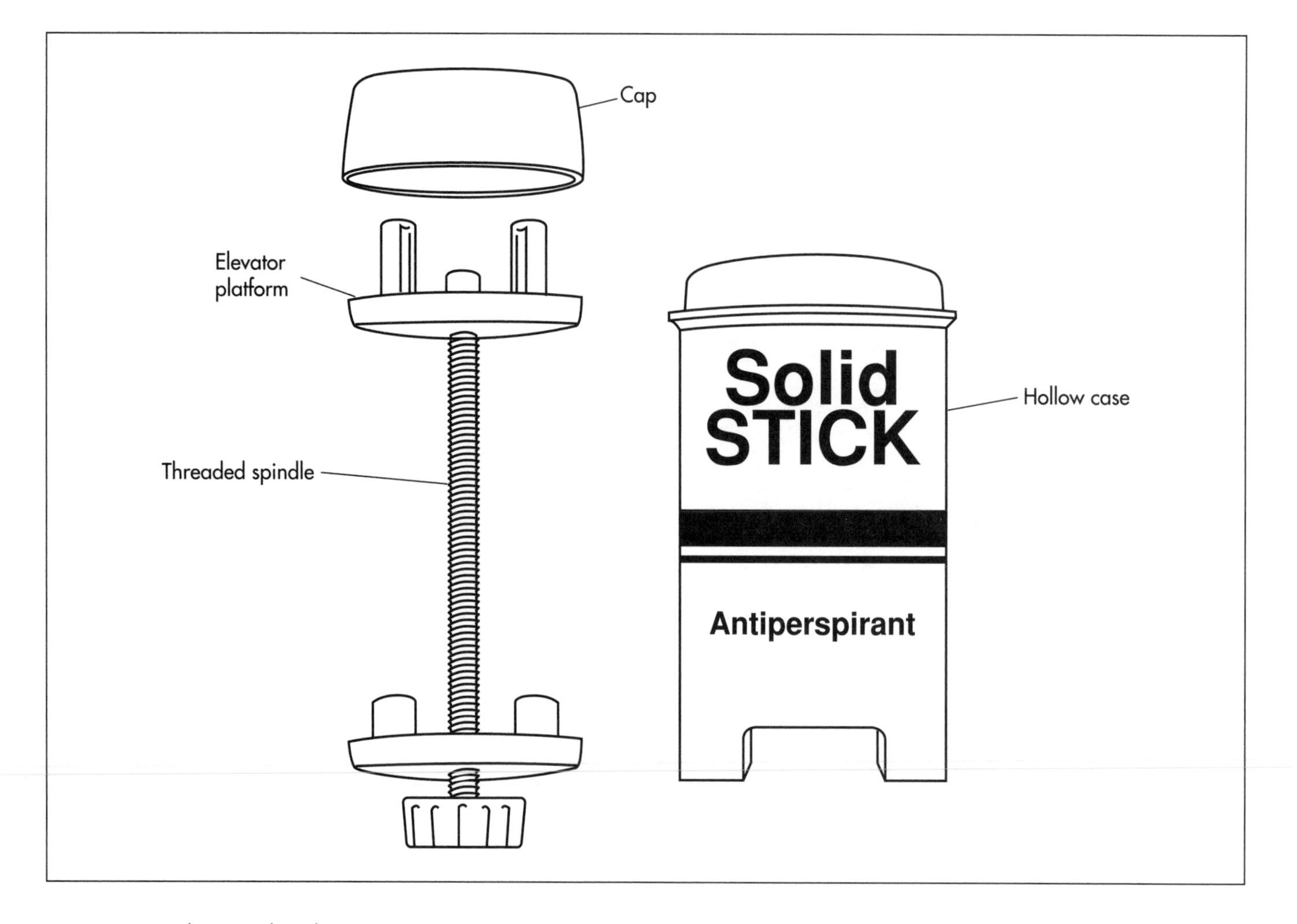


Antiperspirant sticks are packaged in hollow tubes with an elevator platform inside that moves up and down to dispense the product. In some packages, this platform can be pushed up by hand, in others it is elevated by turning a screw that causes it to travel up along a central threaded post.

due to concerns about long term inhalation safety. (This ingredient remains safe for use in stick form.) Next, the Environmental Protection Agency (EPA) strictly limited the use of chlorofluorocarbon (CFC) propellants used in aerosols due to growing concerns that these gases may contribute to the depletion of the ozone layer. CFCs were preferred as propellants for antiperspirants because they gave a soft dry spray. Although the industry reformulated their products to be safe and efficacious, it was too late. Consumers had lost confidence in aerosol antiperspirants. By 1977, sales of the reformulated versions dropped to only 50% of the market, and by 1982, they dipped below 32%. While some brands still offer antiperspirants in aerosol form, today these account for a very small percentage of the total market.

As the popularity of aerosols waned, antiperspirants in stick form became increasingly popular. In 1974, sticks held only about 4% of the market and they were considered to be wet and aesthetically unpleasing. Such products were generally associated with deodorants for men. Because of breakthroughs in ingredient technology that allowed for drier, more efficacious products, sticks gained acceptance between 1974-1978. Consumers embraced sticks as an alternative to aerosols and their market share swelled to over 35% by the mid 1980s. Today, sticks are the single most popular antiperspirant form.

## Raw Materials

Antiperspirants consist of the active drug ingredients that control perspiration; gelling agents that form the stick matrix; and other ingredients, such as fragrance or colorants, that make the product aesthetically pleasing.

### *Active ingredients*

The Food and Drug Administration (FDA) controls the active ingredients used in antiperspirants because they are legally classified as drugs. The FDA publishes an Over the Counter (OTC) Drug monograph that lists which ingredients are approved for use. The ingredients on this list are limited to aluminum chlorohydrate, aluminum chloride, aluminum sulfate, and aluminum zirconium complexes. Of these compounds, the most commonly used is aluminum zirconium tetrachlorohydrex glycine. Most of these materials are supplied as powders, and they are typically used at levels of 8-25% based on the weight of the finished product.

### *Gelling agents*

The bulk of the formulation consists of waxy or fatty materials that are gelled to form a solid stick. Common examples include stearyl alcohol, cetyl alcohol, hydrogenated castor oil, and glyceryl stearate. These waxy materials are blended with lubricating oils and emollients such as cyclomethicone, which is a volatile silicone compound. These silicones are liquids at room temperature, but they quickly evaporate and are used because they leave the skin feeling smooth and dry. In addition, talc, starches, or other powders may be added to control stick consistency and to give the product a dry feel and a smooth payoff.

### *Other ingredients*

Fragrance and colorants may be added to the formula to improve its odor or appearance. Some brands have fragrances that are time released. Other brands may add featured ingredients that contribute little functionality but are designed to increase consumer appeal.

## The Manufacturing Process

### *Batching*

1 In the batching process, ingredients are combined in a jacketed stainless steel kettle. Steam heat is applied to melt the ingredients while the batch is being mixed. During the blending process, the temperature must be carefully controlled to avoid scorching the waxy ingredients. Once all the ingredients have been added to the batch, it is blended until uniform.

### *Filling*

2 Stick packages are typically hollow tubes with an elevator platform inside that moves up and down to dispense the product. In some packages, this platform can be pushed up by hand, in others it is elevated by turning a screw that causes it to travel up along a central threaded post. These empty containers move along a conveyor belt where the molten product is dispensed through a filling nozzle. The exact process varies depending on whether the package is designed to be filled from the top or bottom. In general, the product is filled slightly above its congealing temperature so that it flows easily. If it is filled too hot, the dispersed solids may settle to the bottom; if it is filled too cold, air bubbles will be trapped in the stick.

### *Finishing operations*

3 Sticks may then go through subsequent finishing operations to ensure the surface is smooth and that they are free from trapped pockets of air. These operations usually involve heating the tops of the sticks slightly by passing them under an infrared lamp.

4 A probe is then stuck into the center of the stick to allow air to escape and the surface is heated again to remelt the product, allowing it to flow into the void.

5 At the next station, the sticks pass through a refrigeration tunnel that rapidly lowers the temperature and forces them to solidify. Depending on the package design, a top or bottom piece is put into place to seal the container.

6 Finally, the sticks may pass through cleaning stations before they are placed in cartons for shipping.

## Quality Control

### *Safety testing*

Safety testing guidelines are recommended by the Cosmetics, Toiletries, and Fragrance Association (CTFA), the primary trade organization for the cosmetic industry. While these guidelines are not absolute rules, they do give manufacturers an indication of the minimal level of testing that should be done to ensure their products are safe. These tests include evaluation of the irritation potential (for skin and eyes), contact sensitization (where contact with the product can result in a chemical delayed reaction), photodermatitis (where light interacts with the product to cause a reaction), as well as toxicity (both ingested and inhaled.)

### *Efficacy testing*

According to the OTC monograph, antiperspirants must reduce the amount of perspiration by at least 20% and a variety of test methods are used to ensure formulations meet this requirement. One method, known as the visualization technique, shows the action of the sweat glands via a color change. This is done by first painting the skin with a mixture of iodine castor oil and alcohol. After drying, the skin is then whitened with a layer of powdered starch. When sweat droplets are exuded, they appear as very dark spots against the white background. Another method involves painting a silicone polymer painted onto the skin to form a film. The subject is made to sweat by exposure to elevated temperature or by physical exertion and the film is peeled off and examined for tiny holes formed by the sweat drops. A relative measure of the amount of sweat produced by the body can be obtained by counting the number of holes in the film. Sweat production can also be measured using infrared gas sensors that detect moisture loss. In this process, a constant stream of gas is passed over the subject's armpits and is subsequently analyzed for moisture content. Gravimetric techniques are also used to measure the amount of sweat collected on cotton balls.

## Byproducts/Waste

During the filling process, overfilling or spillage may occur, resulting in scrap product. This can usually be returned to the batch tank and remelted. Depending on the quantity of material involved and the degree of reheating, the batch may have to be assayed to ensure it still meets quality specifications. Additional solvent or fragrance may be added to replace that which was driven off during the reheating operation. The product can then be filled into the packages. Any waste material that is contaminated or otherwise unsuited for refilling must be disposed of in accordance with local regulations.

## The Future

Clear APD sticks have gained popularity in the 1990s. Although the first clear stick products appeared as early as 1979, these early products had stability problems and did not have a significant impact on the market. Since the late 1970s, chemists for various companies have struggled to advance clear APD stick technology. It wasn't until 1993, when Bristol Myers introduced Ban for Man, that clear stick products achieved significant commercial success. It is also interesting to note earlier that year Gillette began spending millions of dollars in advertising for the launch of its Cool Wave clear APD gel-stick. This product was actually a gel, but it was dispensed from a stick-type package. The long term market impact of clear APD sticks and gel-sticks remains to be seen.

## Where to Learn More

### *Books*

Laden, Karl. *Antiperspirants and Deodorants* New York: Marcel Dekker, 1988.

### *Periodicals*

"Flexibility is the hallmark of fluid packaging."*Drug & Cosmetic Industry* (June 1996): 98.

"Gillette deodorants are fit to fill upside-down."*Packaging Digest* (September 1993): 54.

Kintish, Lisa."A Clear Advantage. The emergence of clear technology has given a much-needed boost to the antiperspirant and deodorant market."*Soap-Cosmetics-Chemical Specialties* (July 1997): 29.

Springer, Neil and Helga Tilton. "Staying Power."*Chemical Marketing Reporter* (August 9, 1993): SR12.

Strandberg, Keith. "Antiperspirant & Deodorant Update."*Soap-Cosmetics-Chemical Specialties* (April 1993): 30.

*—Randy Schueller*

# Artificial Blood

Artificial blood is a product made to act as a substitute for red blood cells. While true blood serves many different functions, artificial blood is designed for the sole purpose of transporting oxygen and carbon dioxide throughout the body. Depending on the type of artificial blood, it can be produced in different ways using synthetic production, chemical isolation, or recombinant biochemical technology. Development of the first blood substitutes dates back to the early 1600s, and the search for the ideal blood substitute continues. Various manufacturers have products in clinical trials; however, no truly safe and effective artificial blood product is currently marketed. It is anticipated that when an artificial blood product is available, it will have annual sales of over $7.6 billion in the United States alone.

## Background

Blood is a special type of connective tissue that is composed of white cells, red cells, platelets, and plasma. It has a variety of functions in the body. Plasma is the extracellular material made up of water, salts, and various proteins that, along with platelets, encourages blood to clot. Proteins in the plasma react with air and harden to prevent further bleeding. The white blood cells are responsible for the immune defense. They seek out invading organisms or materials and minimize their effect in the body.

The red cells in blood create the bright red color. As little as two drops of blood contains about one billion red blood cells. These cells are responsible for the transportation of oxygen and carbon dioxide throughout the body. They are also responsible for the "typing" phenomena. On the membranes of these cells are proteins that the body recognizes as its own. For this reason, a person can use only blood that is compatible with her type. Currently, artificial blood products are only designed to replace the function of red blood cells. It might even be better to call the products being developed now, oxygen carriers instead of artificial blood.

## History

There has been a need for blood replacements for as long as patients have been bleeding to death because of a serious injury. According to medical folklore, the ancient Incas were responsible for the first recorded blood transfusions. No real progress was made in the development of a blood substitute until 1616, when William Harvey described how blood is circulated throughout the body. In the years to follow, medical practitioners tried numerous substances such as beer, urine, milk, plant resins, and sheep blood as a substitute for blood. They had hoped that changing a person's blood could have different beneficial effects such as curing diseases or even changing a personality. The first successful human blood transfusions were done in 1667. Unfortunately, the practice was halted because patients who received subsequent transfusions died.

Of the different materials that were tried as blood substitutes over the years, only a few met with minimal success. Milk was one of the first of these materials. In 1854, patients were injected with milk to treat Asiatic cholera. Physicians believed that the milk helped regenerate white blood cells. In fact, enough of the patients given milk as a blood

*It is anticipated that when an artificial blood product is available, it will have annual sales of over $7.6 billion in the United States alone.*

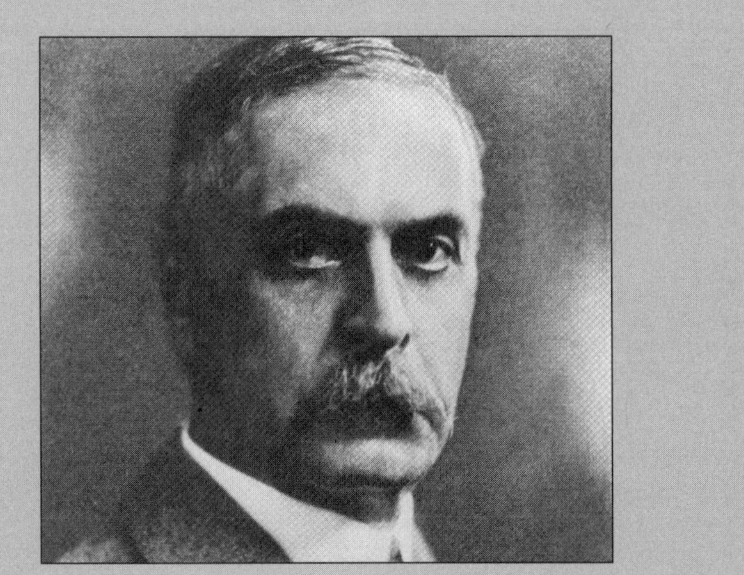

*Karl Landsteiner*

Karl Landsteiner, who has been called the father of immunology, was the only child of Leopold Landsteiner, a prominent Austrian journalist and editor, and Fanny Hess Landsteiner. Landsteiner was educated at the University of Vienna, where he received his medical degree in 1891. While in medical school, Landsteiner began experimental work in chemistry, as he was greatly inspired by Ernst Ludwig, one of his professors. After receiving his medical degree, Landsteiner spent the next five years doing advanced research in organic chemistry for Emil Fischer, although medicine remained his chief interest. During 1886-1897, he combined these interests at the Institute of Hygiene at the University of Vienna where he researched immunology and serology. Immunology and serology then became Landsteiner's lifelong focus. Landsteiner was primarily interested in the lack of safety and effectiveness of blood transfusions. Prior to his work, blood transfusions were dangerous and underutilized because the donor's blood frequently clotted in the patient. Landsteiner was intrigued by the fact that when blood from different subjects was mixed, the blood did not always clot. He believed there were intrinsic biochemical similarities and dissimilarities in blood.

Using blood samples from his colleagues, he separated the blood's cells from its serum, and suspended the red blood cells in a saline solution. He then mixed each individual's serum with a sample from every cell suspension. Clotting occurred in some cases; in others there was no clotting. Landsteiner determined that human beings could be separated into blood groups according to the capacity of their red cells to clot in the presence of different serums. He named his blood classification groups A, B, and O. A fourth group AB, was discovered the following year. The result of this work was that patient and donor could be blood-typed beforehand, making blood transfusion a safe and routine medical practice. This discovery ultimately earned Landsteiner the 1930 Nobel Prize in physiology or medicine.

substitute seemed to improve that it was concluded to be a safe and legitimate blood replacement procedure. However, many practitioners remained skeptical so milk injections never found widespread appeal. It was soon discarded and forgotten as a blood replacement.

Another potential substitute was salt or saline solutions. In experiments done on frogs, scientists found that they could keep frogs alive for some time if they removed all their blood and replaced it with a saline solution. These results were a little misleading, however, because it was later determined that frogs could survive for a short time without any blood circulation at all. After much research, saline was developed as a plasma volume expander.

Other materials that were tried during the 1800s include hemoglobin and animal plasma. In 1868, researchers found that solutions containing hemoglobin isolated from red blood cells could be used as blood replacements. In 1871, they also examined the use of animal plasma and blood as a substitute for human blood. Both of these approaches were hampered by significant technological problems. First, scientists found it difficult to isolate a large volume of hemoglobin. Second, animal products contained many materials that were toxic to humans. Removing these toxins was a challenge during the nineteenth century.

A significant breakthrough in the development of artificial blood came in 1883 with the creation of Ringer's solution—a solution composed of sodium, potassium, and calcium salts. In research using part of a frog's heart, scientists found that the heart could be kept beating by applying the solution. This eventually led to findings that the reduction in blood pressure caused by a loss of blood volume could be restored by using Ringer's solution. This product evolved into a human product when lactate was added. While it is still used today as a blood-volume expander, Ringer's solution does not replace the action of red blood cells so it is not a true blood substitute.

Blood transfusion research did not move forward until scientists developed a better understanding of the role of blood and the issues surrounding its function in the body. During World War I, a gum-saline solution containing galactoso-gluconic acid was used to extend plasma. If the concentration, pH, and temperature were adjusted, this material could be designed to match the viscosity of whole blood, allowing physicians to use less plasma. In the 1920s, studies suggested that this gum solution had some negative health effects. By the 1930s, the use of this material had significantly diminished. World War II reignited an interest in the research of blood and blood substitutes. Plasma donated from humans was commonly used to replace blood and to save soldiers from hemorrhagic shock. Eventually, this led to the establishment of blood banks by the American Red Cross in 1947.

In 1966, experiments with mice suggested a new type of blood substitute, perfluorochemicals (PFC). These are long chain polymers similar to Teflon. It was found that mice could survive even after being immersed in PFC. This gave scientists the idea to use PFC as a blood thinner. In 1968, the idea was tested on rats. The rat's blood was completely removed and replaced with a PFC emulsion. The animals lived for a few hours and recovered fully after their blood was replaced.

However, the established blood bank system worked so well research on blood substitutes waned. It received renewed interest when the shortcomings of the blood bank system were discovered during the Vietnam conflict. This prompted some researchers to begin looking for hemoglobin solutions and other synthetic oxygen carriers. Research in this area was further fueled in 1986 when it was discovered that HIV and hepatitis could be transmitted via blood transfusions.

## Design

The ideal artificial blood product has the following characteristics. First, it must be safe to use and compatible within the human body. This means that different blood types should not matter when an artificial blood is used. It also means that artificial blood can be processed to remove all disease-causing agents such as viruses and microorganisms. Second, it must be able to transport oxygen throughout the body and release it where it is needed. Third, it must be shelf stable. Unlike donated blood, artificial blood can be stored for over a year or more. This is in contrast to natural blood which can only be stored for one month before it breaks down. There are two significantly different products that are under development as blood substitutes. They differ primarily in the way that they carry oxygen. One is based on PFC, while the other is a hemoglobin-based product.

### *Perfluorocarbons (PFC)*

As suggested, PFC are biologically inert materials that can dissolve about 50 times more oxygen than blood plasma. They are relatively inexpensive to produce and can be made devoid of any biological materials. This eliminates the real possibility of spreading an infectious disease via a blood transfusion. From a technological standpoint, they have two significant hurdles to overcome before they can be utilized as artificial blood. First, they are not soluble in water, which means to get them to work they must be combined with emulsifiers—fatty compounds called lipids that are able to suspend tiny particles of perfluorochemicals in the blood. Second, they have the ability to carry much less oxygen than hemoglobin-based products. This means that significantly more PFC must be used. One product of this type has been approved for use by the Federal Drug Administration (FDA), but it has not been commercially successful because the amount needed to provide a benefit is too high. Improved PFC emulsions are being developed but have yet to reach the market.

### *Hemoglobin-based products*

Hemoglobin carries oxygen from the lungs to the other tissues in the body. Artificial blood based on hemoglobin takes advantage of this natural function. Unlike PFC products where dissolving is the key mechanism, oxygen covalently bonds to hemoglobin. These hemoglobin products are different than whole blood in that they are not contained in a membrane so the problem of blood typing is eliminated. However, raw hemoglobin cannot be used because it would break down into smaller, toxic compounds within the body. There are also problems with the stability of hemoglobin in a solution. The challenge in creating a hemoglobin-based artificial blood is to modify the hemoglobin molecule so these problems are resolved. Various strategies are employed to stabilize hemoglobin. This involves either chemically

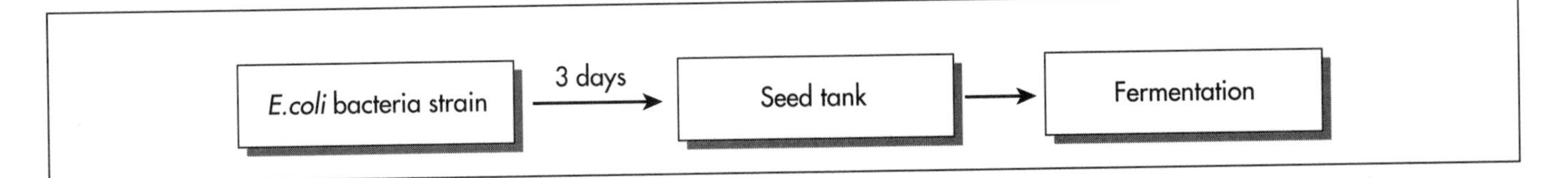


Artificial blood can be produced in different ways using synthetic production, chemical isolation, or recombinant biochemical technology. Synthetic hemoglobin-based products are produced from hemoglobin harvested from an *E. coli* bacteria strain. The hemoglobin is grown in a seed tank and then fermented.

cross-linking molecules or using recombinant DNA technology to produce modified proteins. These modified hemoglobins are stable and soluble in solutions. Theoretically, these modifications should result in products that have a greater ability to carry oxygen than our own red blood cells. It is anticipated that the first of these products will be available within one to two years.

## Raw Materials

Depending on the type of artificial blood that is made, various raw materials are used. Hemoglobin-based products can use either isolated hemoglobin or synthetically produced hemoglobin.

To produce hemoglobin synthetically, manufacturers use compounds known as amino acids. These are chemicals that plants and animals use to create the proteins that are essential for life. There are 20 naturally occurring amino acids that may be used to produce hemoglobin. All of the amino acid molecules share certain chemical characteristics. They are made up of an amino group, a carboxyl group, and a side chain. The nature of the side chain differentiates the various amino acids. Hemoglobin synthesis also requires a specific type of bacteria and all of the materials needed to incubate it. This includes warm water, molasses, glucose, acetic acid, alcohols, urea, and liquid ammonia.

For other types of hemoglobin-based artificial blood products, the hemoglobin is isolated from human blood. It is typically obtained from donated blood that has expired before it is used. Other sources of hemoglobin come from spent animal blood. This hemoglobin is slightly different from human hemoglobin and must be modified before being used.

## The Manufacturing Process

The production of artificial blood can be done in a variety of ways. For hemoglobin-based products, this involves isolation or synthesization of hemoglobin, molecular modification then reconstitution in an artificial blood formula. PFC products involve a polymerization reaction. A method for the production of a synthetic hemoglobin-based product is outlined below.

### *Hemoglobin synthesis*

1 To obtain hemoglobin, a strain of *E. coli* bacteria that has the ability to produce human hemoglobin is used. Over the course of about three days, the protein is harvested and the bacteria are destroyed. To start the fermentation process, a sample of the pure bacteria culture is transferred to a test tube that contains all the nutrients necessary for growth. This initial inoculation causes the bacteria to multiply. When the population is great enough, they are transferred to a seed tank.

2 A seed tank is a large stainless steel kettle that provides an ideal environment for growing bacteria. It is filled with warm water, food, and an ammonia source which are all required for the production of hemoglobin. Other growth factors such as vitamins, amino acids, and minor nutrients are also added. The bacterial solution inside the seed tank is constantly bathed with compressed air and mixed to keep it moving. When enough time has passed, the contents of the seed tank is pumped to the fermentation tank.

3 The fermentation tank is a larger version of the seed tank. It is also filled with a growth media needed for the bacteria to grow and produce hemoglobin. Since pH control is vital for optimal growth, ammonia water is added to the tank as necessary. When enough hemoglobin has been produced, the tank is emptied so isolation can begin.

4 Isolation begins with a centrifugal separator that isolates much of the hemoglobin. It can be further segregated and purified using fractional distillation. This standard

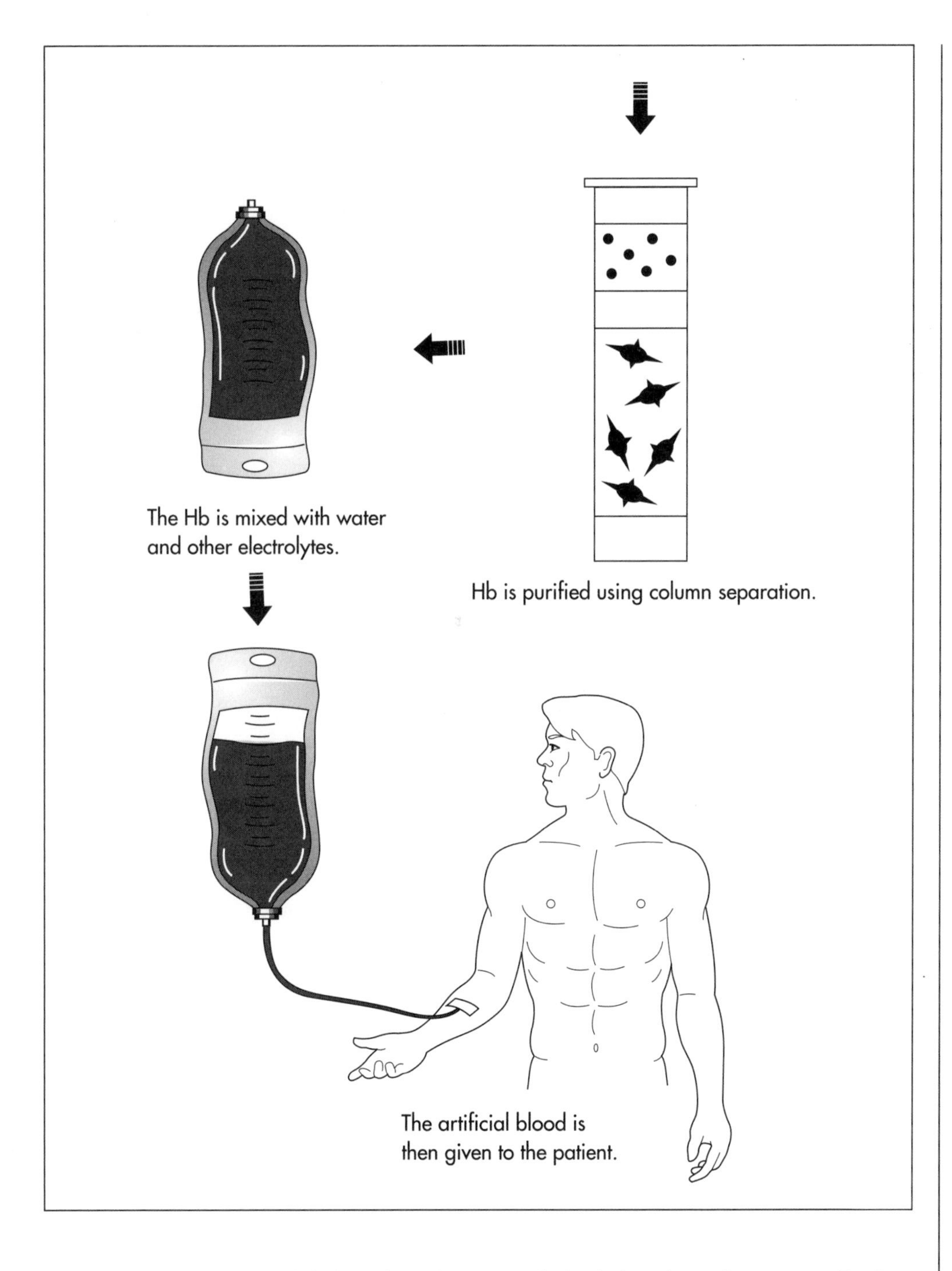


Once fermented, the hemoglobin is purified and then mixed with water and other electrolytes to create useable artificial blood.

column separation method is based on the principle of boiling a liquid to separate one or more components and utilizes vertical structures called fractionating columns. From this column, the hemoglobin is transferred to a final processing tank.

### *Final processing*

5 Here, it is mixed with water and other electrolytes to produce the artificial blood. The artificial blood can then be pasteurized and put into an appropriate packaging. The quality of compounds is checked regularly during the entire process. Particularly important are frequent checks made on the bacterial culture. Also, various physical and chemical properties of the finished product are checked such as pH, melting point, moisture content, etc. This method of production has been shown to be able to produce batches as large as 2,640 gal (10,000 L).

## The Future

Currently, there are several companies working on the production of a safe and effective artificial blood substitute. The various blood substitutes all suffer from certain limitations. For example, most of the hemoglobin-based products last no more than 20-30 hours in the body. This compares to transfusions of whole blood that lasts 34 days. Also, these blood substitutes do not mimic the blood's ability to fight diseases and clot. Consequently, the current artificial blood technology will be limited to short-term blood replacement applications. In the future, it is anticipated that new materials to carry oxygen in the body will be found. Additionally, longer lasting products should be developed, as well as products that perform the other functions of blood.

## Where to Learn More

### Books

Winslow, R. *Hemoglobin-Based Red Cell Substitutes.* Johns Hopkins University Press, 1992.

### Periodicals

Fricker, Janet. "Artificial blood - bad news for vampires?"*The Lancet* (May 11, 1996).

Maclean Hunter, "The Quest for Blood: Blood substitutes may ease chronic shortages."*Maclean's* (August 24, 1998).

Robb, W. J. "Searching for an ideal blood substitute."*RN* (August 1998).

Ross, Philip. "Brewing blood: can Somatogen bring its artificial blood to market?"*Forbes* (November 17, 1997).

Winslow, R. "Blood Substitutes - A Moving Target."*Nature Medicine* (1995).

*—Perry Romanowski*

# Artificial Flower

## Background

Silk and other artificial flowers manufactured today are breathtakingly real and must be touched if they are to be distinguished from nature's own. Silk trees bring the outdoors into sterile offices, and flower arrangements change the color and feel of a room for a relatively small investment. Hobbyists find them a joy to work with and take pleasure in completing arrangements that make beautiful, lasting gifts and ornaments.

The vast improvements in the quality of artificial flowers as well as lifestyles that demand carefree home decorating accessories have caused a flowering of the artificial flower industry into a multi-billion-dollar business. Many of the individual flowers, stems, and foliage are now imported from Thailand, China, and Honduras where the intensive hand labor can be acquired more readily.

Faux flowers allow home decorators to defy the seasons, not only by having summer blooms in the dead of winter but by mixing flowers from several seasons in a single display. Some manufacturers use real materials to enhance silk flowers, such as inserting artificial branches in real tree trunks. Real touches are also added to the false flora; leaves may have holes that look like insect damage, silk roses are complete to the thorns, and some fabulous fakes are even fragrant. Their ultimate attraction may be their least natural aspects; these plants don't need water, fertilizer, sunlight, or tender care.

## History

Florists call silk and other artificial flowers "permanent botanicals," and for many years, they looked down on both dried flowers and artificial flowers as inferior. Today, silk flowers are prized for their versatility and are used by florists to enhance live plants and mingle with cut blossoms. This tradition is hundreds of years old and is believed to have been started by the Chinese who mastered the skills of working with silk as well as creating elaborate floral replicas. The Chinese used artificial flowers for artistic expression, but they were not responsible for turning silk flower-making into a business.

As early as the twelfth century, the Italians began making artificial florals from the cocoons of silkworms, assembling the dyed, velvety blooms, and selling them. The French began to rival their European neighbors, and, by the fourteenth century, French silk flowers were the top of the craft. The French continued to improve both fabrics and the quality of flowers made from them. In 1775, Marie Antoinette was presented with a silk rosebud, and it was said to be so perfect that it caused her to faint. The Revolution that ended Marie Antoinette's reign also dispatched many French flower artisans to England, and, by the early 1800s, English settlers had taken the craft with them to America.

The Victorian Age was the setting for a true explosion in floral arts, including both living and artificial varieties. The Victorians favored an overdone style of decor in which every table and mantelpiece bore flowers or other ornaments. Flowers were so adored that "the language of flowers" grew to cult status in which floral bouquets carried messages and meanings. During the mid- to late-1800s, artificial flowers were made of a wider variety of materials than any time be-

*As early as the twelfth century, the Italians began making artificial florals from the cocoons of silkworms, assembling the dyed, velvety blooms, and selling them.*

The manufacture of high-quality artificial flowers made of silk, rayon, or cotton involves die-cutting each petal size from the chosen fabric, hand dyeing the petals, and then molding the petals to create a life-like effect. Wires are inserted by hand after the petals are pressed. Each flower is assembled individually, and once complete, the flowers are wrapped in florist's paper, and the stems are placed in boxes as if they are to be delivered like a bouquet of real flowers.

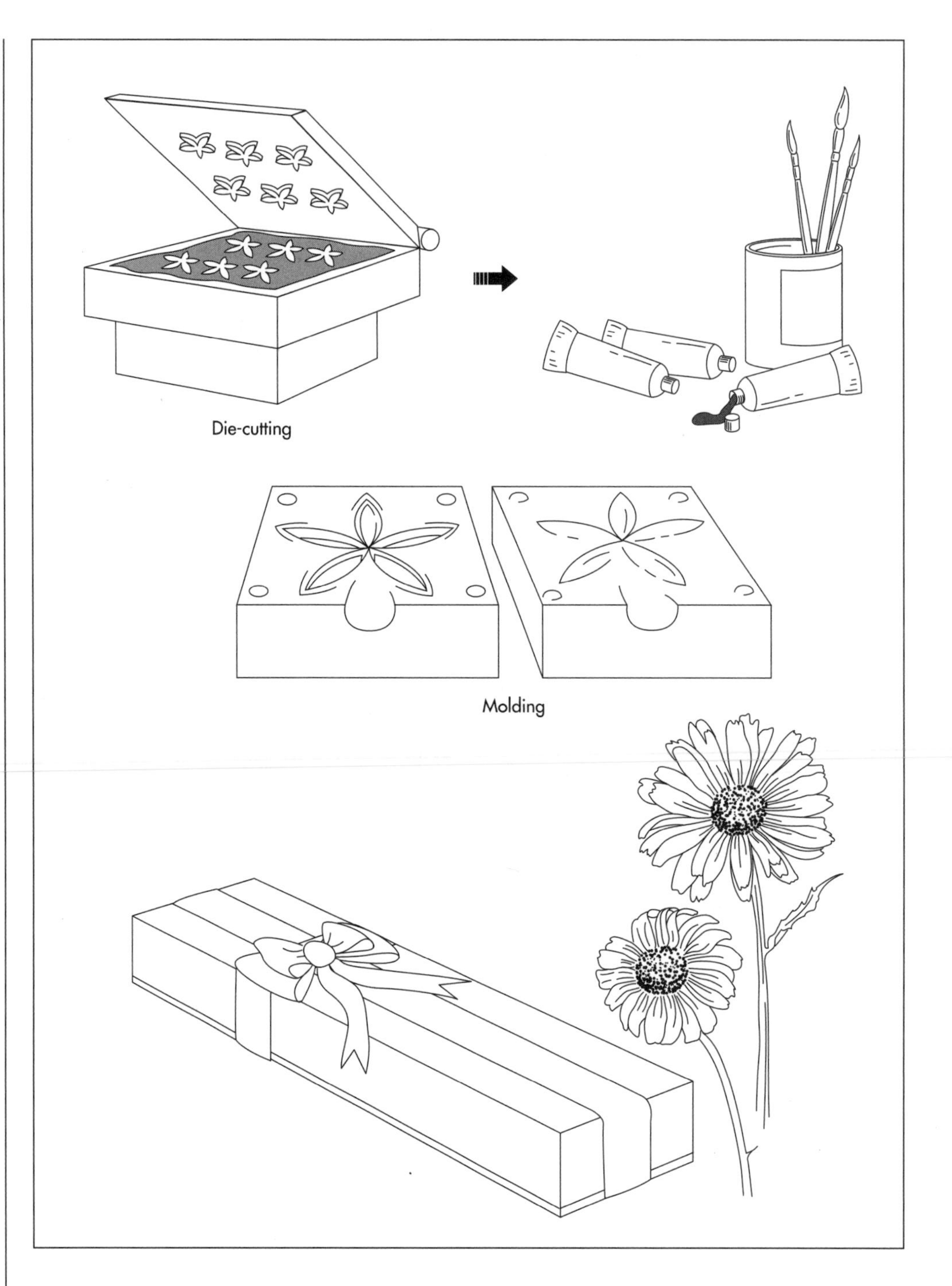


fore or since. Fabrics included satin, velvet, calico, muslin, cambric, crepe, and gauze. Other materials included wood, porcelain, palm leaves, and metal. Wax flowers were popular and became their own art form, and flowers were even made of human hair especially to commemorate deceased loved ones.

In the United States, lavish arrangements and apparel made use of permanent botanicals. The Parisian Flower Company, which had offices in both New York and Paris, supplied silk flowers and other artificial florals to milliners, makers of bridal and ball gowns, and other dressmakers, as well as for room decoration. They sold separate stems and arrangements that were either pre-made or commissioned. By 1920, florists began to add them to their products and services to cover those times when cut blossoms were in short supply.

The trend toward wreaths and ornaments using false fruit in the Italian *della Rob-*

*bia* style flourished in the 1920s and 1930s and waned by 1940. Celluloid became a popular material for flowers in the 1940s, but the highly flammable flowers were banned from importation from Japan after several disastrous fires. Plastic soon overwhelmed the industry, however, and is still responsible for its versatility in the 1990s. Inexpensive plastics to realistic silk blossoms offer something for everyone.

## Raw Materials

Artificial flowers are made in a wide variety of materials depending on the market the manufacturer is reaching. In quantity, polyester has become the fabric of choice by flower makers and purchasers because of lower cost, ability of the fabric to accept dyes and glues, and durability. Plastic is also the material used most often for the stems, berries, and other parts of flowers for the market that includes picks—small clusters of artificial flowers on short plastic and wire stems that can be inserted into forms to make quick, inexpensive floral decorations—and bulk sales of longer stems of flowers that are also less expensive. Artificial flowers are made of paper, cotton, parchment, latex, rubber, sateen (for large, bold-colored flowers and arrangements), and dried materials, including flowers and plant parts, berries, feathers and fruits.

For more upscale silk flowers, silk, rayon, and cotton are the fibers of choice. Wire in a wide range of gauges or diameters is used for firmness in creating the stems (and in stiffening some flower petals and parts), but the wire is wrapped with specially dyed, tear-resistant, durable paper. No plastic is used. Other natural materials such as dried flowers, feathers, and berries are also significant in the upper end market. To make fruit and some berries, specialty suppliers manufacture forms that are precisely sized and shaped to look like the real fruit from mixtures of tapioca or flour base. The forms are sold to the flower manufacturer who dyes them and mounts them on paper-wrapped stems or stalks. All dyes and glues are also derived from natural materials.

## Design

Most silk flowers are sold by the stem. Their designs begin with nature. When a silk flower manufacturer plans to make a new design of a magnolia, for example, the designer takes a magnolia fresh from the tree and dissects it to use the actual parts as models. Dies called tools must be made to cut the silk petals. The exact petals are used to design these tools, and three or four are required to make the different sizes of petals that comprise the flower. The leaves also require several tools. The cutting dies are expensive to machine, so the manufacturer makes a significant financial commitment when investing in a new design.

Silk flower design is also heavily influenced by trends in interior design and fashion. Manufacturers attend trade shows to learn about colors and styles in wallpaper and furniture or summer dresses and hats that are forecast for one to two years ahead.

## The Manufacturing Process

The manufacturing process described below features high-quality silk flowers that are sold by the stem and are made for custom decorating, millinery, other fashion accessories, displays, package ornamentation, candy companies, and floristry.

1 White silk, rayon, or cotton fabric are used for all petals, regardless of their finished color. The fabric is die-cut using the tools described above into the many petal sizes and shapes that go into a single type of flower. The petals are dyed in the first step of a detailed hand assembly process. The dyer uses cotton balls and paintbrushes to touch color onto the petals beginning with the edges of the petal and working in toward the center. Dyeing a single petal can take an hour of concentrated work.

2 To give them their distinctive curves, wrinkles, and other shapes, the petals are inserted in molds to which heat is applied to press the petals into individual shapes. After they are pressed, some petals and leaves are stiffened with thin wires. The wires are inserted by hand, and glue is touched on to fix the wire in place.

3 The separate flowers and sprays of leaves are assembled individually, but several of each may be used to construct a single stem. Another skilled worker has taken wire precut to specified lengths and covered it with floral paper or tape that has a waxy coating to make it self-sticking. Finally, assemblers add the individual flowers and sprays of leaves to the stem.

4 The finished stems are taken to the packing department. Each stem is wrapped in florist's paper, and the stems are placed in boxes as if they are to be delivered like a bouquet of real flowers. The boxes are sealed and stored for shipment.

## Quality Control

As with most hand-assembled products, silk flowers are inspected by workers at each step of the process. The assemblers are responsible for rejecting imperfect flower parts; for example, if the presser receives petals that have dye spots on them, the presser rejects the petals rather than proceeding with inserting them in molds for pressing.

Before the finished stems are wrapped and packed for shipping, they are subjected to three separate inspections. The finish inspectors work independently, but all three must approve the silk flower before it is hand wrapped and taped for boxing.

## Byproducts/Waste

There are no byproducts from the manufacture of silk flowers, but the manufacturer's line may include hundreds of different varieties. Waste is very limited and includes wire and fabric scraps that are disposed. Dyes are all natural and can be recycled. The materials also do not subject workers to any hazards. The die-cutting machines are enclosed to protect the operator's hands, and other metals like the florists' wire arrive at the factory in pre-cut lengths. Both glues and dyes are non-toxic, and assemblers wear latex gloves as an additional safeguard.

## The Future

New technologies like the permastem or permasilk processes that fuse flowers to their stems and makes them more durable continue to improve the functionality and beauty of faux flowers. Technology is also used to produce dried-look and soft-touch (velvet touch) plants; foliage especially has benefited from soft-touch processing that varies the sizes of leaves on a single branch and gives them a warm, gentle feel.

The future of artificial flowers is likely to imitate its long past. People like to be surrounded by beautiful representations from nature, but they also want the convenience of low-maintenance, everlasting flowers. Our homes and fashions benefit from the addition of artificial flowers, and many other businesses from millinery to confectionery rely on silk flowers to add the finishing touch to their products.

## Where to Learn More

### *Books*

Beveridge, Ardith and Shelly Urban. "Permanent Botanicals" In *A Centennial History of the American Florist.* Topeka, KS: Florist Review Enterprises, 1998.

Blacklock, Judith. *Silk Flowers: Complete Color and Style Guide for the Creative Crafter* Radnor, PA: Chilton Book Company, 1995.

Miller, Bruce W. and Mary C. Donnelly. *Handmade Silk Flowers.* New York: Prentice Hall Press, 1986.

### *Periodicals*

Caldera, Norman J. "How Honduras developed exports of artificial flowers."*International Trade Forum* (January-March 1990): 4+.

Kelly, Mary Ellen. "Fake flowers evolve further."*Discount Store News* (November 4, 1991): 21+.

Mastropoalo, Dominick. "Artificial flowers: looking better, selling more."*Home Improvement Market* (June 1997): GSR10.

### *Other*

American Prestige Silks, Inc. http://www.americanprestige.com/.

Koehler & Dramm, Inc., and the Institute of Floristry. http://www.kdfloral.com/.

*—Gillian S. Holmes*

# Bean Bag Plush Toy

## Background

Investors who worry about bull and bear markets should consider the alternatives—the moose, lobster, pink pig, platypus, and dolphin markets, just for starters. These stars in the investment firmament "Chocolate the Moose," "Pinchers the Lobster," "Squealer the Pink Pig," "Raspberry Patti the Platypus," and "Flash the Dolphin" are among the original nine Beanie Babies produced in 1993 by Ty Incorporated. The cute critters are more generically known as bean bag plush toys, and, not only have they shaken kids and the toy domain to its roots, they have forced many adults to rethink their retirement plans, storage space, and sanity.

Is the craze ridiculous? History will tell, but, in March 1999, "Peanut the Royal Blue Elephant" who originally sold for $5.95 in June 1995 was "experiencing a strong secondary market" at the price of $4,500.

## History

The bean bag plush toy exploded on the scene in 1993, but its origins are ages old. Bean bags are among the oldest toys and have been made in geometric, animal, and doll shapes and filled with beans, peas, rice, and pebbles for centuries. Rag dolls are another predecessor and are literally as old as fabrics themselves that could be tied in knots or shapes. Rags are the stuffing in many old dolls with cloth bodies and china or bisque porcelain heads. Bears—the most popular bean bag plush toy—are a merging of the bear shape of the classic teddy bear (born in 1903) and the ancient bean bag.

The phenomenon known as Beanie Babies is the brain child of one man. H. Ty Warner worked for Dakin, a major plush toy maker, before founding Ty Incorporated, in Oak Brook, Illinois, in 1986. He designed and manufactured larger plush toys (typically 12-20 in [30.5-51 cm] long) in the United States, England, Germany, Mexico, and Canada before inventing Beanie Babies. Mr. Warner developed Beanie Babies with the idea of creating small plush toys that fit children's hands easily and that also were priced to fit their allowances. He had previously used pellets of polyvinylchloride (PVC) to fill the feet of his larger stuffed animals, so a combination of understuffing with polyester filler and PVC pellets was used to make the little toys soft.

In November 1993, Mr. Warner debuted the first nine Beanie Babies, modeled after designs used for his larger stuffed toys, at a toy exposition. The first nine Beanie Babies found their way to store shelves in 1994, and Ty Inc., began introducing nine to 12 new designs every six months. By 1995, Beanie Babies had become a phenomenon, and Ty factories were unable to match supply to demand. So-called "beanie baby mania" began in Chicago near Ty's Oak Brook headquarters but was soon experienced as far afield as Canada and England.

Apart from endearing designs, Ty Inc. employed several strategic marketing tactics. When demand began to increase, production was limited to spur on that demand. The toys were and still are not sold in major stores; instead, small shops that sell cards, other types of small toys, candy, and other items attractive to children became the major sellers of beanie babies. The designs themselves included color and clever detailing in eyes, whiskers, feet, tails, and other parts of the small animals.

*In March 1999, the "Peanut the Royal Blue Elephant" Beanie Baby, who originally sold for $5.95 in June 1995, was "experiencing a strong secondary market" at the price of $4,500.*

Each also bears two tags. One printed paper tag is suspended from the ear of each animal and is termed a hang tag, swing tag, or heart tag because of its heart shape. The second tag called a tush or butt tag carries the manufacturing location, toy contents, and company insignia and date and is folded into a seam in the toy's tail area. The hang tag identifies the animal by name, birth date, and after 1996, it is inscribed with a short poem describing the animal's habits or most endearing characteristics.

All of these features attract kids and enhance the collectible value of the toys. In addition, Ty Inc., began retiring the toys routinely; the fact that a particular toy may become an endangered species also adds to its appeal and limited availability. Collectors are pushed to snap them up while they are available. These features and intelligent marketing campaigns have made Beanie Babies a colossal retail success, creating a strong secondary market and numerous offshoots as well.

The success of Beanie Babies can also be attributed to the Internet. In August 1996, Ty Inc., debuted on the Internet and, courtesy of a guest book, Beanie Baby collectors could exchange information and buy and sell toys. A host of web sites followed with every related opportunity from auctions of Beanie Babies to web pages created by children to show off photos of their toys. Another boost to this success came with McDonald's April 11, 1997, launch of its first Teenie Beanie Baby promotion, featuring 10 miniature versions of existing Beanie Baby designs that were sold in McDonald's Happy Meals. The promotion intended to last for five weeks was terminated in less than two weeks when the supply of 100 million toys was exhausted. Two months after the death of Britain's Princess Diana, Ty Inc., released its first special-issue Beanie, a purple bear with a white rose stitched over its heart and named "Princess." All of Ty's profits from sale of this Beanie Baby were dedicated to the Diana, Princess of Wales, Memorial Fund. Sports teams giveaways, charity auctions, the opening of the Beanie Baby Official Club on January 1, 1998, and other promotions further boosted public interest.

Changes in the hang tags and tush tags produced "generations" of Beanie Babies and again enhanced interest. Inevitably, errors in mass production of the toys were made, and color errors in materials, missing accessories, or incorrect tagging also created rarities among collectors. Some notable mistakes have been made. Spotted dogs have been produced without the right spots, legs are sometimes stitched in place backwards, fabrics have been mismatched to the wrong animals, and hang tags and tush tags don't always match. For collectors, these errors may add to the thrill of the hunt because the faulty animals may be valuable in their own right.

Demand has also caused a significant counterfeiting industry to grow. Legitimate manufacturers fight this with unique fabrics, accessories, and tags. Holograms on tush tags are an example of the manufacturers' attempts to prevent copying. Planet Plush issued its "Windy, the Chicago Bear" designed by famed plush artist Sally Winey in a limited edition of 36,000 with serial numbers, and Limited Treasures also released production figures to increase demand. For retired toys that are commanding high prices on the secondary (resale) market, manufacturers recommend that potential buyers have experts toys authenticate the little animals before investments are made.

Most soft toy makers and many other toy and novelty producers began generating their own designs and pitching unique takes on the plush bean bag. From Meanie Beanies to baseball and NBA bears to remakes of classic bunnies and bears in miniature, the market has responded with something for every taste—all based on small size, small price, plastic pellets, and polyester fiber.

## Raw Materials

Bean bag plush toys do not contain beans. Their characteristic soft stuffing consists of two materials, which are plastic pellets and polyester fiber fill. The plastic pellets are made of either polyvinyl chloride (PVC) or polyethylene (PE), and they are produced by specialty suppliers. The polyester fiber is the filling commonly used for decorative pillows, comforters, some furniture, and many other products.

The bean bag toy's outer fabric skin is made of synthetic plush. Manufacturers try to use unique fabrics to distinguish their products; for example, SWIBCO, which makes the

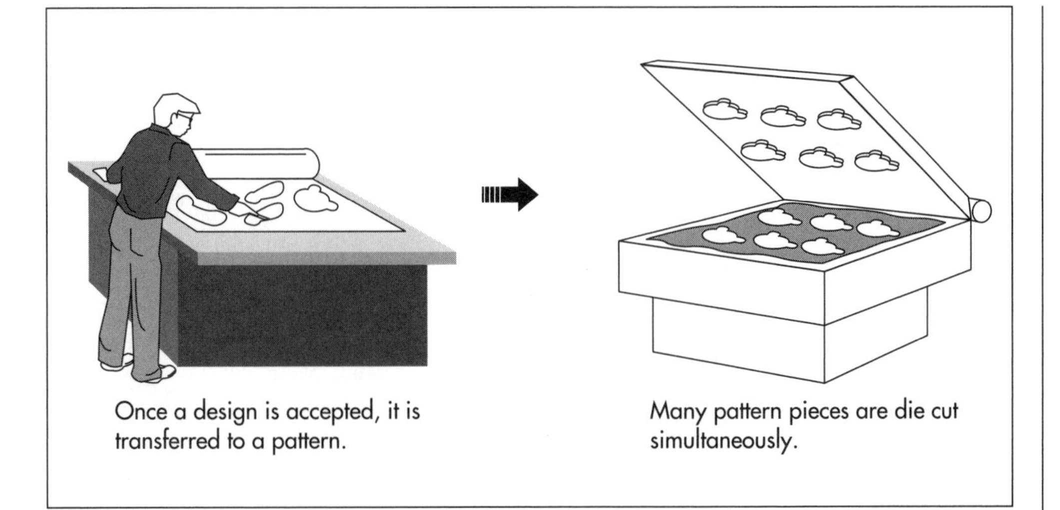

Once a design is accepted, it is transferred to a pattern.

Many pattern pieces are die cut simultaneously.

Plush toy patterns and their corresponding cutting dies are computer-generated to maximize efficiency and minimize the amount of waste fabric. Multiple layers of plush fabric are simultaneously cut with dies.

Puffkins Collection, uses high-pile fabric for a fluffy, furry appearance. Ty Inc., recently created its own fabric called Tylon to add shimmer and color variations to its line.

The eyes, noses, and other hard plastic features of bean bag plush toys are designed to suit the animal and are made by specialty subcontractors. All are child-proof parts that became an accepted industry standard in the 1960s because they can't easily be pulled off the toys. The eyes are mounted on plastic stems and fixed to the back side of the fabric with washers, collars, or grommets. Some manufacturers use felt eyes and other features that are prefabricated and stitched securely in place or overstitched features that are made of many layers or wraps of sewing thread by machine. Yarn and thread are used for insect antennae and cat whiskers. Larger appendages like legs, feet, beaks, wings, and ears are made of plush or other fabric and are also stuffed. Ribbons are high-quality double-sided satin.

The toy's other prominent attachments are the tags. Hang tags are printed on paper and bear the manufacturer's identification and information about the character of the animal toy. The hang tag is attached to the toy with a plastic strip fastener; these are made of either red or clear plastic, and each one is 0.5-0.74 in (1.3-1.9 cm) long. On the animals' hindquarters, a fabric safety tag tells the contents of the critter and its place of manufacture (usually China, Korea, Malaysia, or Indonesia), as well as its name, company, and registration and trademark data. Because of counterfeiting, holograms that are more difficult to copy have been added to tush tags.

## Design

The process of designing a bean bag plush toy begins with a prototype and may take several years to finalize. For Beanie Babies, Ty Warner himself designs the toys by making several prototypes of the same design. Shapes, colors, materials, features, and accessories are varied on the prototypes. Mr. Warner then polls friends and employees to help select the best design. Further evolutions are involved in the toy design itself but also in its name, tags, and the poem on the tag. Some designs have been reissued with color changes and other variations to improve the products.

Puffkins go through a similar process. These plush toys all have rounded shapes so some types of animals like snakes, worms, long-beaked or -legged birds are not suited to the Puffkins style. Employees often suggest new ideas—an employee contest resulted in the Puffkins name—and public opinion is recorded through e-mail, collectors' input, and suggestions from children. SWIBCO's art department produces up to six designs, and the sketches are reviewed by the firm's owners. The art work is sent to the factory where handmade prototypes are constructed from different fabrics and color combinations. These may be approved for production or new art boards may be requested, and the process repeats. Of the original six, four may be ordered. The two that are not selected may be revised and kept for future

Miscellaneous parts and accessories are attached to the fabric with a special hand tool. Next, pattern pieces are sewn together by seamstresses. Once sewn, the plush body is turned right side out and stuffed with fiber fill and PVC pellets. Stuffing is accomplished both by machine and by hand.

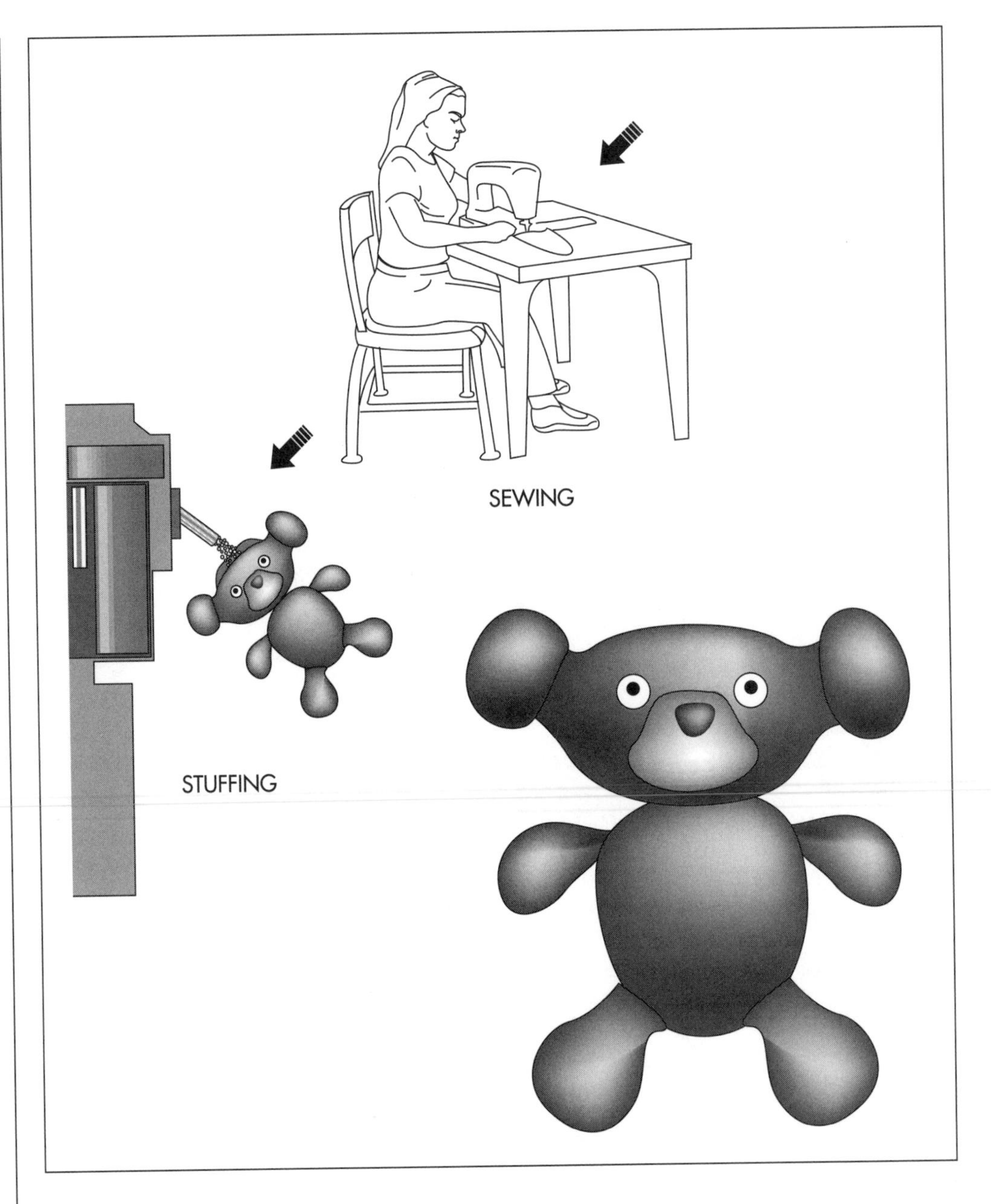


use. Filed designs are often studied later and may prompt new ideas.

## The Manufacturing Process

1 The patterns made for the selected prototype are computer-generated to fit a given length and width of fabric and are laid out for optimal use of the fabric. Cutting dies are also computer-generated from the pattern data, and pieces of the toy are stamped out of multiple layers of plush fabric with the dies. Hand-cutting is also done.

2 The animal's face and other parts with accessory attachments are assembled first. The grommeted eyes and nose are snapped into place with a special hand tool, and whiskers or other thread and yarn features are stitched into seams.

3 At long rows of sewing stations, seamstresses stitch segments of the animal together. One station may produce ears only or wings, paws, heads, or bodies. Industrial sewing machines are used, but the machines' access and attachments are specially made for the small pieces to be sewn. At other stations farther along the assembly line, arms and legs and tush tags are attached to bodies until construction of the toy

is nearly complete. The whole animal is turned right side out.

4 Depending on the manufacturer, fiber fill may be added to some pieces like legs before they are stitched to the body. Stuffing is added to the body after careful measuring of both the bean-like pellets and the polyester fiber. Measurements ensure a uniform weight and understuffed feel to each tiger or penguin, and assemblers also subject the creature to a touch and squeeze test to make sure it will sit in the hand, bend at the legs, and otherwise be appropriately cuddly. The last of the stuffing is forced in by hand, and the final opening in the head or side seam is stitched by hand.

5 Final details like neck ribbons are tied in place, and the hang tags are clipped on with plastic fasteners. The toys are sent to the packaging department where they are bagged and boxed 60 to a carton for shipment.

## Quality Control

Seamstresses and assemblers are responsible for the quality of their work. A final quality control review is done prior to packing at the factory, and when the boxed toys reach their distribution centers in the United States or elsewhere, they are inspected again when they are repackaged for shipment to retail stores.

## Byproducts/Waste

Makers of bean bag plush toys produce lines of toys with similar design characteristics but no true byproducts. They may use their designs to make other companion products. SWIBCO, for example, has adapted its Puffkins to smaller versions for key rings and magnets. Wastes are minimized to be able to keep the price of the toys within a child's affordability. Polyester fiber fill can be recycled.

## The Future

Naysayers claimed that the bean bag plush toy market was about to burst in 1998,but others including the manufacturers themselves say it has at least two to five more years to run its course. The toy market is very volatile, and new fads and interests tempt children and their parents every day. Still, these toys are easy to collect and store, given their small size, and they have something for everyone in color, type of animal, seasonal characters, and charm. Bean bag addicts claim the demand will last for many more years on the secondary market alone. Whether a toy stalwart or a fad, bean bag plush toys have the perennial attraction of bean bags and cloth toys behind them and future generations of kids to enrapture with their names, birth dates, bright eyes, welcoming price tags, and cuddly feel.

## Where to Learn More

### Books

*Collector's Value Guide.* Ty Beanie Babies. Meriden, CT: Collectors' Publishing Co., Inc., 1998.

Fox, Les and Sue.*The Beanie Baby Handbook.* Midland Park, NJ: West Highland Publishing Company, 1998.

King, Constance Eileen.*The Encyclopedia of Toys.* Crown Publishers Inc., 1978.

Phillips, Becky and Becky Estenssoro. *Beanie Mania II: The Complete Collector's Guide.* Sun Prairie, WS: Royale Communications Group Inc., 1998.

### Periodicals

Bryant, Adam. "Time to Short Beanies? Lessons about Investing from Peanut the Elephant."*Newsweek* (March 29, 1999): 46.

Chen, Kathy. "Modern Marco Polos head east in search of Peanut and Garcia."*The Wall Street Journal* (June 19, 1998): B1.

Dunne, Claudia and Mary Beth Sobolewski. "How to Protect Yourself from Counterfeits: Part II."*Beanie World Monthly* supplement (Fall/Winter 1998).

### Other

Beanie Mom's Newsletter. http://beaniemom.com/.

Beanie Nation. http://www.BeanieNation.com/ .

Mary Beth's Beanie World Monthly http://www.beanieworld.net/.

Peggy Gallagher Enterprises, Inc. http://www.beaniephenomenon.com/.

Planet Plush http://www.planetplush.com/.

SWIBCO, Inc. http://www.swibco.com/.

Ty Inc. http://www.ty.com/.

*—Gillian S. Holmes*

# Bed Sheet

## Background

A bed sheet is a flat-woven textile that is used on a bed between the occupant of a bed and the warm blanket above. It is generally a rectangle of broadloomed fabric, meaning it is made without a center seam. Bed sheets have hems at top and bottom. The selvages, or finished edges of the woven sheet as it is made on the loom are used as side seams and thus there is no need for hemming on the sides. Today, the bed sheet comes as part of a set of bed linens that match in color, fabric, and detail and includes the fitted sheet (to cover the mattress), the flat sheet and at least one pillow case.

The bed sheet may be made of a variety of fibers, including linen, cotton, synthetics (often blended with natural fibers such as cotton) and occasionally silk. Bed sheets are made of a wide variety of fabrics. Particularly popular is percale, a closely-woven plain weave of all cotton or cotton-polyester blend that is smooth, cool, and comfortable against the skin. Also of plain weave but more coarsely woven than percale is muslin. In winter months flannel sheets, which are woven with nappy cotton fibers, provides additional warmth. Silky, satiny bed sheets, generally woven of synthetics (silk is very expensive) are a novelty. Linen is also occasionally used for bed sheeting but is not generally commercially available in this country as linen is not processed in the United States. Linen sheeting is either imported from Eastern Europe or Britain.

## History

Beds of some sort have been around for millennia. It is unknown when sheeting was first used to keep the sleeper comfortable but it is likely that the first true bed sheets were linen. Linen, derived from the flax plant, has been cultivated for centuries and was expertly cultivated, spun, and woven by the Egyptians. It is a laborious plant to cultivate but the finished fabric is perfect for bed sheeting because it is more soft to the touch than cotton and becomes more lustrous with use. Linen sheeting was made on conventional looms that were between 30-40 in (76.2-101.6 cm) wide, resulting in bed sheets that had to be seamed down the center in order to be large enough for use. Europeans brought linen culture to the New World; linen processing flourished in the Northeast and Middle Colonies for two centuries. However, because of the painstaking cultivation process, linens were difficult and time-consuming to make. Nevertheless, many seventeenth, eighteenth, and early nineteenth century American women worked relentlessly producing linen goods—pillow cases, bed sheets, napkins, towels—for family use upon their marriage.

By about 1830 in the United States, cotton cultivation and processing was becoming well-established. Previously, it was difficult to remove the tenacious seeds found in short-staple cotton which grows easily in the American South. Eli Whitney's development of the cotton gin enabled the seeds to be stripped from the cotton wool easily and quickly; southern plantations immediately began growing the now-lucrative plant using enslaved labor. At the same time, New England textile mills were quickly adapting British cotton manufacturing technologies and were able to spin, weave, dye, and print cotton in huge quantities. By about 1860, few bothered to make bed sheets from linen anymore—why spend the time when cotton sheeting was cheap and easy to obtain?

*Typically, 8,000 yd (7,312 m) of sheeting is woven on a loom and wound up in rolls and shipped for further processing.*

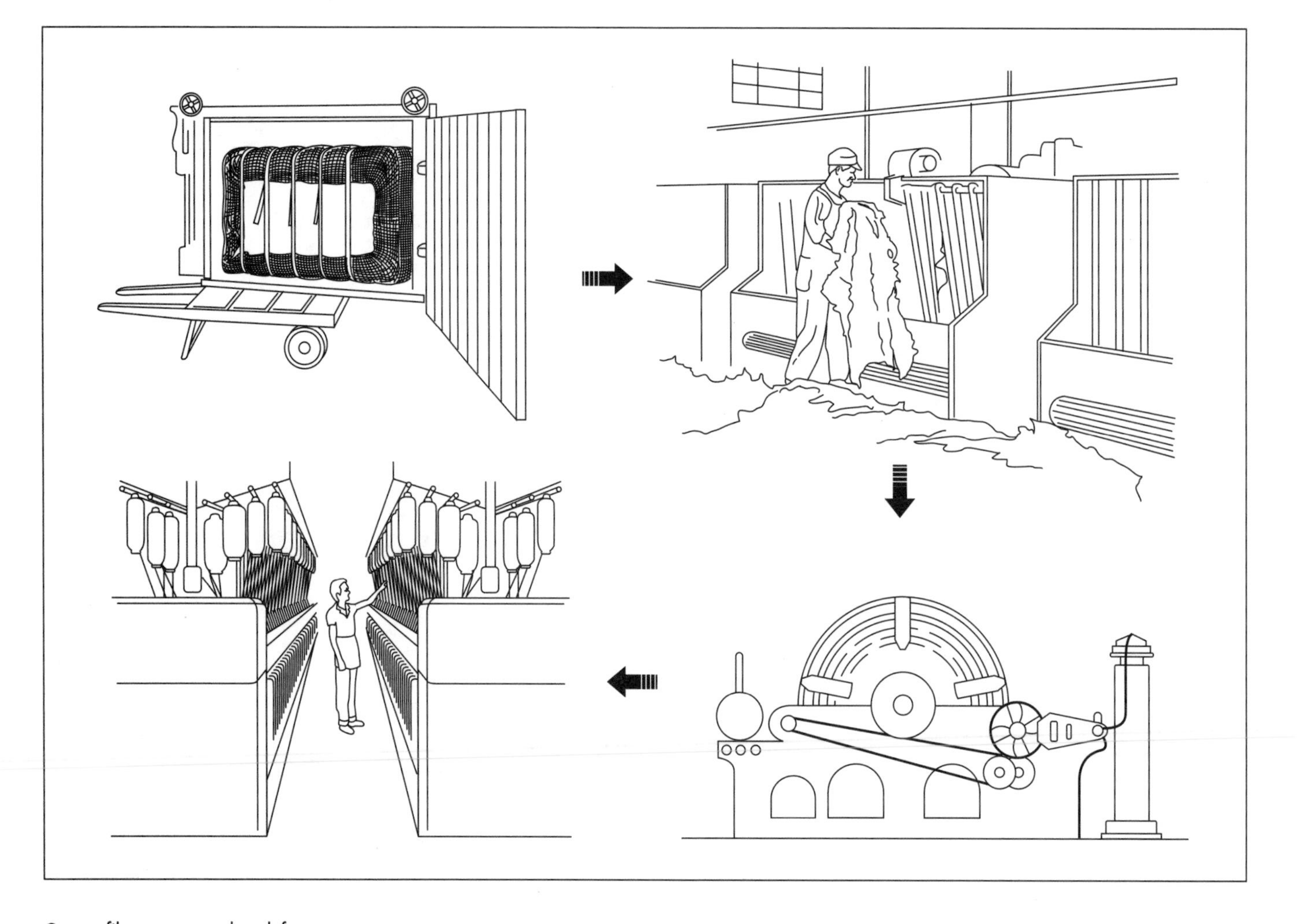

Cotton fibers are produced from bales of raw cotton that are cleaned, carded, blended, and spun. Once loaded onto a section beam, the bobbins are coated with sizing to make weaving easier. Several section beams are loaded onto a single large loom beam. As many as 6,000 yarns are automatically tied onto old yarns by a machine called a knotter in just a few minutes.

Looms became more mechanized with human hands barely touching the products and bed sheets have been made on such looms since the later nineteenth century. Recent innovations in the product include the introduction of blended fibers, particularly the blending of cotton with polyester (which keeps the sheet relatively wrinkle-free). Other recent developments include the use of bright colors and elaborate decoration. Furthermore, labor is cheaper outside the United States and a great many bed sheets are made in other countries and are imported here for sale. Today, the southern states, particularly the state of Georgia, includes a number of cotton processors and weavers. Many of our American cotton bed sheets are produced in the South.

## Raw Materials

If cotton is to be spun into yarn in the bed sheet manufactory, 480 lb (217.9 kg) bales are purchased from a cotton producer. This cotton is often referred to as cotton wool because it is fuzzy like wool. It is still dirty and includes twigs, leaves, some seeds, and other debris from harvesting. Other materials used in the weaving process include starches or sizing of some sort that is applied to the cotton threads to make them easier to weave. During the cleaning and bleaching process after the sheet has been woven, caustic chemicals and bleaches including chlorine and/or hydrogen peroxide solutions are used to remove all color before dyeing. Dyeing includes chemically-derived dyes (meaning they are not natural and not found in plants or trees but are created in laboratories) are used for standard coloration and color-fastness.

## The Manufacturing Process

Some manufacturers spin the bales of cotton delivered to the manufacturer. Others purchase the yarn already spun on spools.

This section will describe the process of making 100% sheeting from bales of cotton delivered to the plant which are not yet spun.

### *Procuring the cotton*

1 Bales of cotton weighing about 480 lb (217.9 kg) are purchased and shipped to the sheeting manufacturer.

### *Blending*

2 Bales are laid out side by side in a blending area. The bales are opened by a Uniflock machine that removes a portion of cotton from the top of each bale. Next, the machine beats the cotton together, removing impurities and initiating the blending process. The fibers are then blown through tubes to a mixing unit where the blending continues.

### *Carding*

3 Once blended, the fibers move through tubes to a carding machine, which aligns and orients the fibers in the same direction. Cylinders with millions of teeth pull and straighten the fibers and continue to remove impurities.

### *Drawing, testing, and roving*

4 Here, the cotton fibers are further blended together and straightened as many strands of fibers are drawn together into one strand by a roving frame. The frame twists the fibers slightly and winds a cotton roving onto bobbins.

### *Spinning*

5 The rovings are spun on a ring spinner, drawing the cotton into a single small strand and twisting it as it spins. The yarn is then wound onto bobbins and the bobbins are placed onto winders that wind the thread onto section beams that will eventually fit onto a loom for weaving.

### *Warping a section beam*

6 It takes between 2,000-5,000 warp (lengthwise yarns) to make up a single width of sheet. Thus, the warping beam, which holds all of the yarns, is very large and cannot be loaded at once. So 500-600 ends of yarn from spools are pulled onto a single section beam, thus warping it. Later, several section beams will be loaded onto the large warping beam, each contributing a portion of the warp.

### *Slashing*

7 Each section beam goes through a slasher—a machine that coats the yarn with starch or sizing to protect the ends and makes the yarn easier to weave.

### *Warping the beam*

8 Once coated with sizing, several section beams are loaded onto a single large loom beam. As many as 6,000 yarns are automatically tied onto old yarns by a machine called a knotter in just a few minutes. The knots are pulled through the machine and the weaving can begin.

### *Weaving*

9 The weaving, in which the weft or filler threads interlock with the warp or vertical threads, is done on high-speed automatic air jet looms. The filler threads are transported across the warp threads at a rate of 500 insertions per minute, meaning that a filler thread runs across the warp thread about every one-tenth of a second. It takes about 90 insertions to weave an inch of sheeting. Thus, about 5.5 in (14 cm) of sheeting is woven per minute—10 yd (9.14 m) per hour are woven. Typically, 8,000 yd (7,312 m) of sheeting is woven on a loom and wound up in rolls and shipped for further processing.

### *Cleaning and bleaching*

10 The fabric, called greige, is gray in color. It is further finished by singeing—a process in which bits of yarn are burned off of the surface. Then, the sheeting is ready to be bleached. This is done in three steps. First, it is de-sized by bathing it in water and soaps that removes contaminants. Next, caustic chemicals are applied to get rid of dirt and remnants of debris found in cotton yarn. The caustic is washed out and concentrated bleaches (chlorine and/or hydrogen peroxide) are applied to dissipate the gray color. Now whitened, the sheeting is rolled into a rope and put into a dryer which takes the moisture out prior to dyeing.

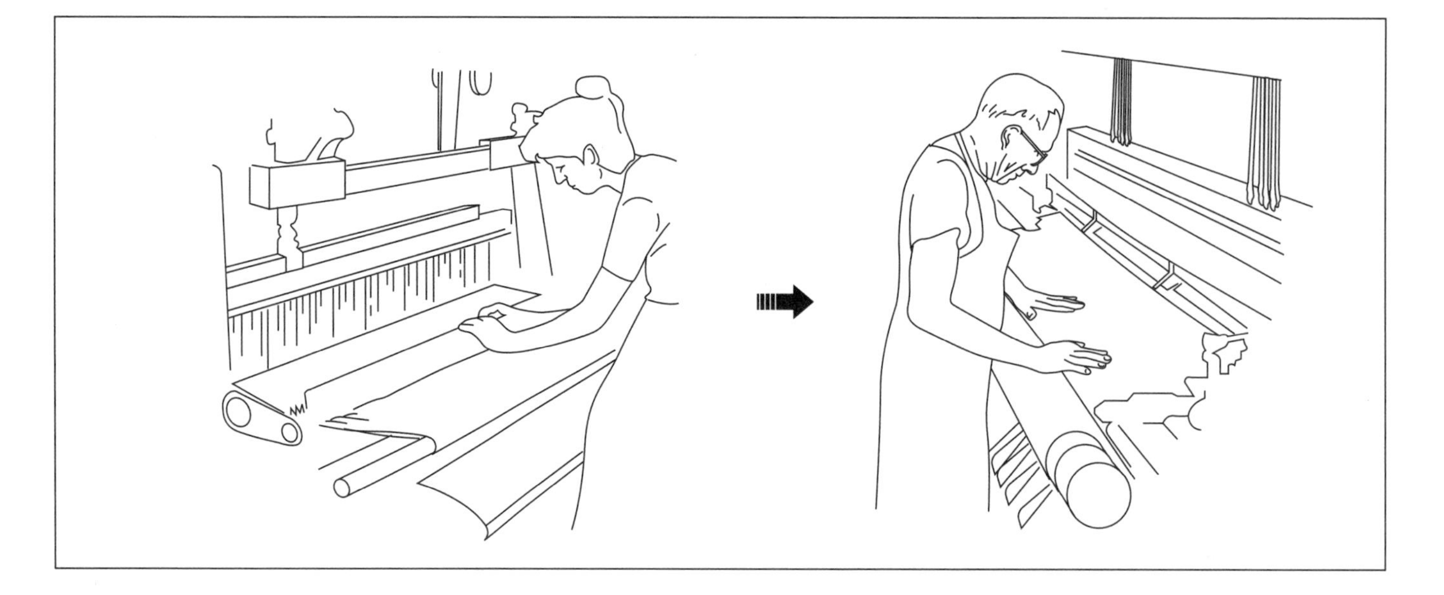

Weaving is done on high-speed automatic air jet looms. Typically, 8,000 yd (7,312 m) of sheeting is woven on a loom and wound up in rolls and shipped for further processing. Once woven, the sheeting is bleached, rolled into a rope and dried, dyed, and rolled. Automatic cutting equipment cuts the roll into standard sheet lengths and the sheet hems are sewn.

### Dyeing

11 All sheeting is dyed. Even sheeting sold as white must be dyed to become a truly white sheet. In order to give the gray-colored sheets color, pigments are applied to the sheeting in color vats that use large rollers to press the dyestuff into the material. Once dyed, the sheeting is steamed to set the color. Next, a resin is applied to the sheeting to control shrinkage. The sheeting is rolled onto huge rolls and is ready to be cut and sewn.

### Cutting and sewing

12 Automatic cutting equipment pulls the cloth off the rolls as it automatically cuts the sheeting to the requisite length. The rolls are transferred to a sewing machine that sews top and bottom hems.

### Packaging

13 The sewn sheet is either folded by hand or machine. Machine-folded sheets are ejected, shrink wrapped, and individually packaged for sale.

## Quality Control

Sheeting manufacturers carefully choose cotton bales. Cotton is classified by length (staple) and by quality (grade). Shorter staples are used for batting, while longer staples are used to make higher quality products. Egyptian cotton is made from longer staples. Medium staples is considered standard. There are nine grades used to classify cotton from middling to good. Cotton with much debris and residue would be of a lower grade than that with less impurities. The lower grade bales tend to slow down the processing of the cotton into spool yarn and may never render a quality product. Thus, many plants will purchase bales based on test data received from the U.S. Department of Agriculture to ensure the bales are fairly clean. Many weaving facilities perform their own tests on bales to be purchased to assess quality and cleanliness.

Rovings—the rope-like strand that is spun into yarn—generally undergoes quality control inspection prior to spinning. At major points in the production of yarn and sheeting, statistical samples are taken and tested in the laboratory. Physical tests are run on the completed products. Because the bleaching and dyeing processes include a number of chemicals that must be mixed exactly, the chemical solutions are monitored. Furthermore, employees within the plant carefully monitor the process and visually inspect the product at each manufacturing stage.

## Byproducts/Waste

Cotton weavers have worked diligently in recent years to reduce polluting effluvia and cotton lint. Occasionally, fiber wastes resulting from spinning can be recycled and used for other cotton products. In the past, cotton lint generated inside factories was hazardous to the employees; however, now the air-jet looms generate little cotton dust. At the point in which greige goods are handled, there are automatic sweeping and cleaning machines to rid the rooms of ambient dust.

Of greatest concern to the federal government are the chemicals used in the cleaning, bleaching, and dyeing of the goods. Federal regulations require that resulting liquids emitted from the factory (which may contain chlorine, hydrogen peroxide, and other miscellaneous caustics) meet state and federal clean water regulations, and mills are required to have National Pollutant Discharge Elimination System permits. The larger mills have invested in building their own wastewater treatment plants. Emissions are now governed by the Clean Air Act and must be within acceptable guidelines.

## Where to Learn More

### Books

Walker, Sandra Rambo. *Country Cloth to Coverlets*. Lewisburg, PA: Union County Historical Society, 1981.

### Other

National Council on Cotton. 1996-1999. http://www.cotton.org/ (June 22, 1999).

Linen Association. 1996. http://www.lin.asso.fr/ (June 22, 1999).

*—Nancy EV Bryk*

# Billboard

*The Federal Highway Administration estimates that in 1996 there were over 400,000 billboards on federally controlled roads, which generated revenues in excess of $1.96 billion.*

## Background

Billboard is the common term used to describe a type of outdoor advertising found along major highways. This name is most frequently given to large steel-framed signs, which are mounted on poles 20-100 ft (6.1-30.5 m) above the ground. Most often, the sign is printed on large poster sheets, which are affixed to the face of the sign. These signs may also be equipped with a variety of special lighting and display effects. This type of sign is one component of a unique advertising medium that communicates to audiences on the go. Such promotions are also referred to as "out of home" advertising because the intended audience is usually in transit, and is always away from their homes. As recently as 25 years ago, 90% of outdoor advertising consisted of billboards. Today, the industry has expanded to include smaller signs on bus shelters, kiosks, and malls. There are over 500 companies nationwide that specialize in this type of advertising (although not all of them construct large roadside billboards.) The Federal Highway Administration estimates that in 1996 there were over 400,000 billboards on federally controlled roads, which generated revenues in excess of $1.96 billion. According to Competitive Media Reporting the top 10 billboard revenue categories for 1996 included: entertainment and amusements; tobacco; retail; business and consumer services; automotive; travel, hotels and resorts; publishing and media; beer and wine; insurance and real estate; and drugs and remedies.

## History

Born out of necessity, billboards were probably first used to convey a message to the majority of individuals who were illiterate. The oldest known billboard ad was posted in the Egyptian city of Thebes over 3,000 years ago and offered a reward for a runaway slave. Prior to the late 1700s, the predecessor to the modern day billboard—billposting— was prevalent throughout Europe, but only as an informal source of information. It wasn't until the invention of lithography in the late eighteenth century that billboards as a medium expanded into an art form. The first art poster was created in 1871 by Englishman Frederick Walker, who was commissioned to create the playbill for the play "Lady in White" in London. By the early 1900s, schools for poster art were being formed and artists like Talouse Lautrec were making names for themselves.

The first large scale use of the billboard as an advertising tool was as circus posters printed or secured on horse-drawn trucks that would precede a show to town in order to increase interest and attendance. At this time, billboards were not standardized or controlled by any laws. During 1872-1912, organizations in the United States met to create billboard standards. Originally, the standard set was 24-sheet poster panels with a total size of 19.5 x 8.7 ft (6 x 2.6 m) . Today, that size remains the same, while technology has reduced 24 sheets to 10.

It was also during the early 1900s that electric billboards were used to light up cities. Prior to the electric billboard, cities were dark, foreboding places. The electric billboard brought the cities to life at night, creating a more hospitable atmosphere that induced people to stay on the streets. Hence, the birth of nightlife.

By the late 1920s, more people were purchasing automobiles and traveling beyond the city. Billboard advertising expanded as well, and for the first time, billboard advertising had to consider a wider range of demographic audiences. Billboard art and design changed with the times, reflecting new technologies and the mood of a generation. With the use of photography and comics, billboards portrayed a world without problems during the depression of the 1930s. The 1950s gave rise to the hand-painted billboard and use of sexual innuendo in campaigns. Billboards were extensively used in China to promote Red Army politics. It was also during this time that billboard companies utilized the boom truck with a crane to move billboards and place them in more prominent positions. During the 1960s, celebrity endorsements became essential and the advent of the superstar was born. Focus shifted from the family to the singles lifestyle and the medium itself was emulated in the Pop Art movement. As interest in environmentalism increased during the late 1960s and early 1970s, billboard ads borrowed images from nature. It was at this time that the Marlboro man on horseback was born. In the 1970s and 1980s, campaigns used sexually explicit rather than implied themes. Objects were omnipotent and were created larger than life with little or no accompanying text.

## Design

Billboard design depends on such factors as location of the sign, the advertising budget, and the type of product being promoted. The industry uses market research firms to aid in the design process. These firms supply detailed information on the number of people in vehicles in different metropolitan regions, even projecting traffic patterns 10-15 years into the future. They can estimate the frequency and number of exposures the advertising will have upon its target audience. Using data generated by Global Positioning Systems (GPS), billboard location data can be merged with other geographic and demographic business information to create customized marketing solutions for outdoor advertisers. Computerized data analysis is available that incorporates census data, traffic origins, travel patterns, trading zones, competitor locations, and other key facts to help optimize the use and location of billboards.

The location of the billboard also helps determine the type of sign selected by the advertiser. The term billboard is actually a generic classification, referring to several types of signs. The most common forms are known as bulletins and poster panels. Bulletins, the largest sign style, may be as large as 20 x 60 ft (6.1 x 18.3 m) and are found in high-density traffic locations. They use computer or hand-painted messages as advertising artwork and are usually purchased for multi-month contract periods. Poster panels are somewhat smaller and are designated by the number of sheets employed on the sign. Thirty-sheet poster panels are approximately 12 x 25 ft (3.6 x 7.6 m) and are found on primary and secondary traffic ways. They are lithograph- or silkscreen-printed and are usually displayed for 30 days. Eight sheets are somewhat smaller (6 x 12 ft [1.8 x 3.6 m]) and are designed more for pedestrian and some vehicular traffic. They are placed in high-density urban neighborhoods and suburban shopping malls. The larger bulletin style is the most challenging type to construct.

## Components

Large billboards have three main components: steel used to construct and support the frame, artwork that conveys the advertising message, and electrical equipment for lighting and other special effects.

### *Steel structure*

Modern billboards, also known as monopoles, are supported by steel poles ranging from 36-72 in (91.4-183 cm) in diameter and up to 100 ft (30.5 m) tall. At the top of the mounting pole is a frame constructed from steel I-beams. This frame supports the artwork and lighting equipment. Standard sizes for a large steel frame assembly are 20 x 60 ft (6.1 x 18.3 m), 20 x 48 ft (6.1 x 14.6 m), or 10 x 36 ft (3 x 11 m). When a company is interested in constructing one of these signs, they contact a steel erection firm with expertise in billboard construction. Typically, a customer will seek bids from three or four competitive vendors. These bids estimate the cost of designing the sign as well as the materials, transportation, and labor used to construct it. Some smaller billboards may be available from the steel company as

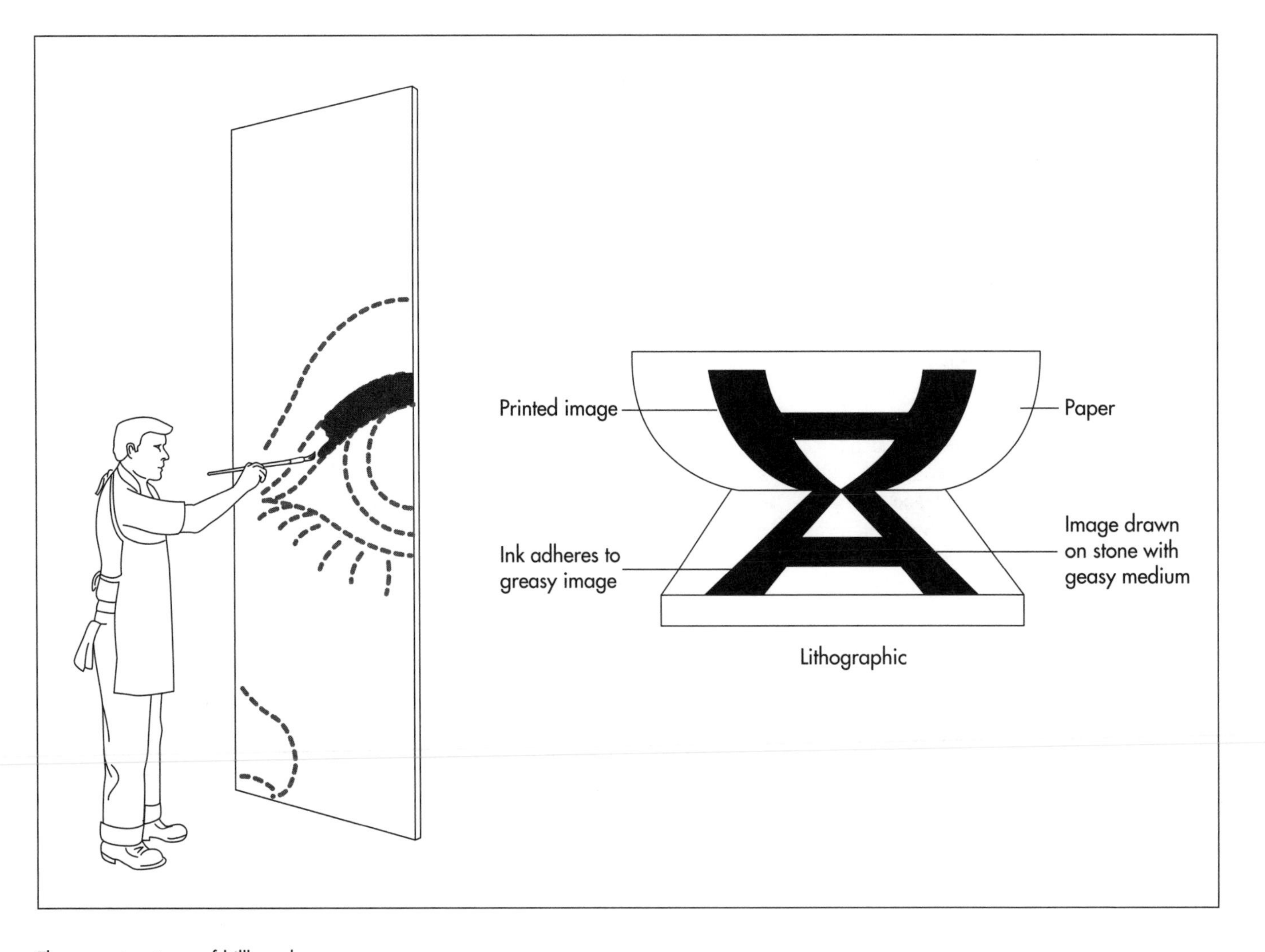


There are two types of billboards. Hand-painted billboards are usually used for small campaigns that want to create a special look. More often, billboards are produced from poster panels that are mass-produced copies of the original artwork.

stock items that are pre-made and stored in their warehouse awaiting a customer's order. Larger signs, or signs that have special design requirements like the ability to withstand severe weather or moving parts, must be custom ordered. After the design is finalized, the erection company orders the appropriate steel. Some common sizes and grades of steel are kept in stock and are quickly accessible. Unusual steel components, such as those designed for use in high wind conditions, must be specially ordered from the steel fabricators.

### *Artwork*

The steel frame is covered with a backing material, known as a facing. The artwork is affixed to the facing. The art is either preprinted on paper or vinyl sheets that are pasted onto the facing, or in some cases, the art is painted directly onto a plywood or canvas facing.

### *Electrical systems*

Most billboards are electrically lit and therefore require appropriate lighting and power systems along with a significant number of high wattage bulbs. Activation of these lighting systems is no simple matter. While many billboards are located in major metropolitan areas, others can be found in remote areas along interstates. In both cases, it is very impractical to have to travel to each sign every night to turn on the lights. Therefore, automatic switches have been developed to turn on the lights at specified times. Other systems use photosensitive cells to turn on the lights when dusk sets. Still other more advanced systems turn lights on and off electronically with a signal from a satellite system.

## The Manufacturing Process

Billboard manufacture requires three separate types of contractors. First, a steel erec-

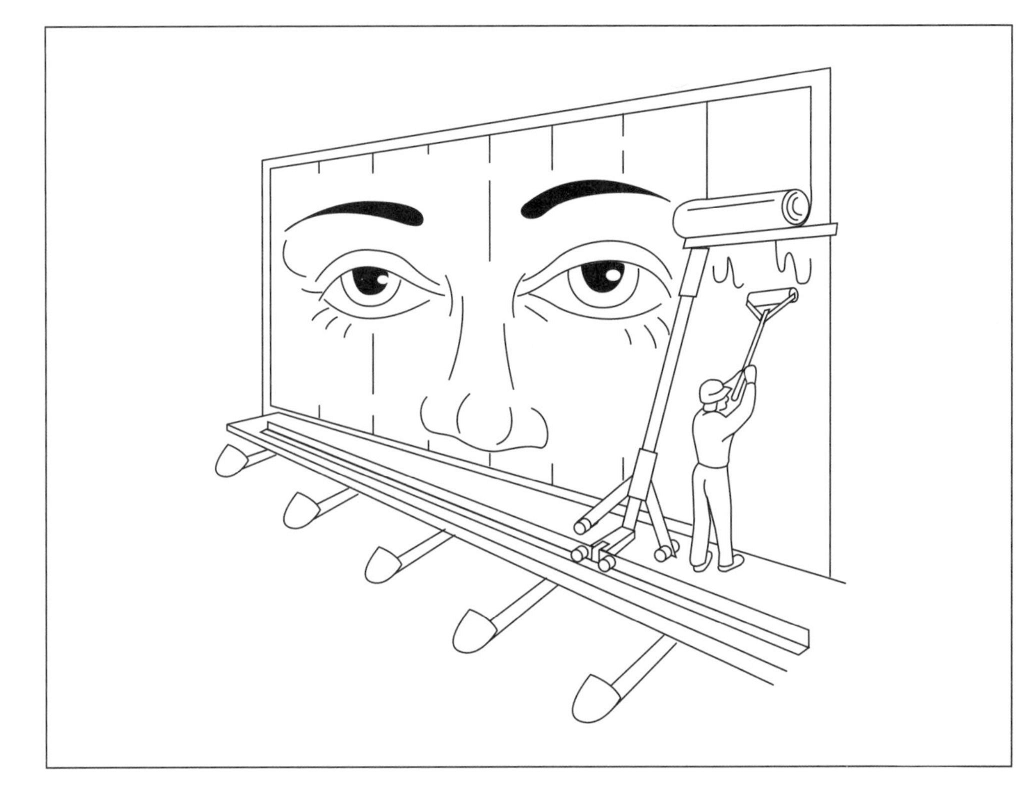

Poster panels are pasted onto the billboard frame. Once the campaign is finished, the poster panels are covered with the next advertisement's image. Hand-painted billboards are painted on plywood panels that are secured to the frame. Once the campaign is complete, the plywood panels are whitewashed in preparation for the next job.

tion firm is hired to install support pole and frame. Next, a company specializing in graphics creates and mounts the artwork, and finally electricians install the power and lighting.

### *Pre-assembly of structural components*

1 The steel erection company orders all the support components required for the job. Upon receipt of the components, they bolt and weld together as many pieces as possible before shipping them to the job site. Some of the longer steel pipes may be connected together as telescoping sections. Pre-assembly saves time on the job site and improves ease of shipping. The pre-assembled parts are then transported to the job site by truck.

### *Steel erection*

2 The job site must be properly prepared prior to installation of the steel. A subcontractor is typically assigned to drill a hole in the ground in which to place the support pole. Approximately 20-30 ft (6.1-9.1 m) deep, the hole is filled with concrete and the support pole is put into place.

3 Next, the frame is bolted into place on top of the pole which may be as little as 20 ft (6.1 m) or as much as 100 ft (30.5 m) above the ground. The term haggle is used to describe the distance from the sign face to ground level. The frame is equipped with catwalk-style walkways to allow access to the surfaces where the advertising elements are mounted. These catwalks may run along the front as well as the back of the sign with an access ladder located in the rear. The walkways are also built with attachments for safety cables used by the workers. The entire installation process takes a crew of three or four men approximately one week to complete.

### *Artwork fabrication*

The artwork is added after the structural elements are in place. The method of application depends on the design of the sign. Some advertisements are hand painted directly onto plywood sections that are directly attached to the billboard frame; others use lithographic prints prepared on large vinyl sheets, which are pasted onto the sign face. Usually, hand-painted billboards are used for small campaigns that want to achieve a higher quality look.

4 The pounce pattern technique is used to create a billboard-sized stencil of an original artwork. Using this technique, the artwork is projected onto a billboard-size sheet that is placed over a grounded copper mesh screen. Charcoal pencils with 500 volts of electricity passing through them transfer the enlarged image to the paper by creating pounce points.

5 Once attached to the plywood sections, the initial pattern of dots is defined further by charcoal dust. The charcoal is spread over the rough image, and it adheres to the points that were made by the electrically charged charcoal.

6 Artists work using a reducing glass to paint the enlarged version of the original. Oftentimes, several artists work on one billboard, each focusing on either the detailed images or the background. Mechanical scaffolding increases the artists' maneuverability.

7 Poster panels are printed on lithographic printers that mass produces the original artwork onto sheets of poster paper.

8 Once the campaign is completed, hand-painted panels are disassembled and whitewashed. Poster panels remain on the billboard where they are covered by the next ad campaign.

### *Electrical connection*

9 Typically, local electrical contractors are hired to install the power and lighting systems. In many cases, both sides of the sign are used for artwork, so lighting assemblies are required for both sides. In addition, the sign may require special wiring to operate moving parts or other special effects. All wiring must be done in accordance with relevant electrical codes.

## Quality Control

There are no universal guidelines for billboard construction. Each company has its own proprietary standards. However, firms engaged in this type of work expend significant effort in repairing and maintaining the quality of the signs. These efforts are necessary due to the effects of weathering which causes deterioration of the sign components, particularly the paper and vinyl used to post the artwork. In addition, steel components can rust after extended exposure to the elements. Severe weather conditions can cause damage to signs that require more serious repairs.

In 1965, the Highway Beautification Act was passed in the United States. It governs the amount, spacing, and quality of billboards placed along highways. As a result, many dilapidated frames were removed. In the late 1990s, cigarette manufacturers, who traditionally used billboards to advertise their products, made an agreement with the federal government in order to prevent total regulation of the industry. One of their concessions was to replace their ads with anti-smoking campaigns.

## Byproducts/Waste

The processes used in billboard manufacture generate little usable byproducts or waste material. However, after a billboard has outlived its usefulness a steel firm may be required to cut the sign down. In some cases, the components may be recycled for use in other jobs.

## The Future

Billboard manufacture is becoming increasingly sophisticated. More and more sign companies are taking advantage of computerized market research data to optimize the placement of their billboards. This trend is likely to continue as marketers continue to seek out the most effective advertising mediums. Bar coding technology is becoming a popular way of tracking information related to billboards. Some companies even videotape their signs and put together a computer display for their clients so they can just click on a location and see a picture or video of the billboard that is there. Satellite technology will also play a larger role in the billboard market of the future. Already satellite systems are being used to control lighting and to track sign locations.

However, perhaps the most interesting innovations in the industry are occurring in the area of advertising artwork. For example, one new type of sign uses a multi-faceted prismatic facing to actually deliver two different advertising messages. As the viewer approaches the sign, they see one picture, but as they pass the sign, their angle of view changes, revealing a different picture. This type of clever innovation continues to make billboards a popular and economically viable method of advertising.

## Where To Learn More

### *Books*

Henderson, Sally and Robert Landau. *Billboard Art.* San Francisco: Chronicle Books, 1986.

### *Other*

Erected Steel Products. Steve McDowell, VP Operations. PO Box 360347, Birmingham, AL 35236. (205) 481-3700.

*—Randy Schueller*

# Bioceramics

*Bioceramics, made from a calcium phosphate material containing tiny pores, have been used to coat metal joint implants or used as unloaded space fillers for bone ingrowth. Ingrowth of tissue into the pores occurs, with an increase in interfacial area between the implant and the tissues and a resulting increase in resistance to movement of the device in the tissue. As in natural bone, proteins adsorb to the calcium phosphate surface to provide the critical intervening layer through which the bone cells interact with the implanted biomaterial.*

## Background

Over the last several decades, bioceramics have helped improve the quality of life for millions of people. These specially designed materials—polycrystalline aluminum oxide, hydroxyapatite (a mineral of calcium phosphate that is also the major component of vertebrate bone), partially stabilized zirconium oxide, bioactive glass or glass-ceramics, and polyethylene-hydroxyapatite composites—have been successfully used for the repair, reconstruction, and replacement of diseased or damaged parts of the body, especially bone. For instance, aluminum oxide has been used in orthopedic surgery for more than 20 years as the joint surface in total hip prostheses because of its exceptionally low coefficient of friction and minimal wear rates.

Clinical success requires the simultaneous achievement of a stable interface with connective tissue and a match of the mechanical behavior of the implant with the tissue to be replaced. Bioceramics, made from a calcium phosphate material containing tiny pores, have been used to coat metal joint implants or used as unloaded space fillers for bone ingrowth. Ingrowth of tissue into the pores occurs, with an increase in interfacial area between the implant and the tissues and a resulting increase in resistance to movement of the device in the tissue. As in natural bone, proteins adsorb to the calcium phosphate surface to provide the critical intervening layer through which the bone cells interact with the implanted biomaterial.

Resorbable biomaterials have also been designed to degrade gradually over time to be replaced by the natural host tissue. Porous or particulate calcium phosphate ceramic materials (such as tricalcium phosphate) have been successfully used as resorbable materials for low mechanical strength applications, such as repairs of the jaw or head. Resorbable bioactive glasses are also replaced rapidly with regenerated bone.

Bioactive materials form a biologically active layer on the surface of the implant, which result in the formation of a bond between the natural tissues and the material. A wide range of bonding rates and thickness of interfacial bonding layers are possible by changing the composition of the bioactive material.

Bioactive materials include glass and glass-ceramics based on silicon dioxide-phosphate systems containing apatite (a natural calcium phosphate containing some fluorine or chlorine), dense synthetic hydroxyapatite, and polyethylene-hydroxyapatite composites. Applications include orthopedic implants (vertebral prostheses, intervertebral spacers, bone grafting), middle-ear bone replacements, and jawbone repair. Bioactive glass and glass-ceramic implants have been used for more than 10 years in the middle-ear application. Bioactive glass particles have also been used as fillers around teeth that have had gum disease, preventing the teeth from falling out.

## Design

The performance of the artificial bone depends on its composition and end use application. Careful selection of the right material with suitable properties is thus important. Computer-aided design software is also used for optimizing the shape and for simulating the mechanical behavior of the implant with the surrounding bone tissue. A mathematical technique called finite element analysis is used to determine the stress distribution on both the implant and biological structure. Prototypes are then fabricated that undergo testing of properties, as well as clinical tests, before final production.

## Raw Materials

The major raw material is usually a ceramic powder of specific composition and high purity. Additives include binders, lubricants, and other chemicals to assist in the shape forming process. The powder may also contain a sintering aid, which helps the ceramic material to densify properly during firing and sometimes at a lower temperature. If a chemical-based process is used, organic precursors and solvents are combined into a solution to make the final product.

## The Manufacturing Process

Depending on its composition, artificial bone is made using two processes, the traditional ceramic process and a chemical-based method called sol gel. In the sol gel method, two approaches can be used. In one, a suspension of extremely small particles is allowed to gel inside a mold, followed by aging at 77-176° F (25-80° C) for several hours, drying, and several thermal treatments to chemically stabilize and densify the material. The other approach uses a solution of chemical precursors as the starting material followed by the same process. Since the ceramic process is more common, it will be discussed in more detail here.

### *Raw material preparation*

1 The ceramic powder is manufactured elsewhere from mined or processed raw materials. Additional crushing and grinding steps may be necessary to achieve the desired particle size. The ceramic powder plus additives are carefully weighed in the appropriate amounts and then mixed in some type of mixing machine equipped with blades or revolving rolls. Sometimes mixing and particle size reduction takes place at the same time, using a milling machine. A ball mill uses rotating cylinders filled with the mixture and spherical media to disperse the material and reduce its particle size. An attrition mill uses tiny beads and rotating agitators to accomplish the same thing.

### *Forming*

2 After mixing, the ceramic material is of plastic consistency and now ready for forming into the desired shape. A variety of methods can be used, including injection molding, extrusion, or pressing. In injection molding, the mix is loaded into a heated cylinder, where it softens. A steel piston forces the hot mixture into a cooled metal mold. Extrusion compacts the material in a high-pressure cylinder and then forces the material out through a specially shaped die orifice. Pressing involves compaction of the material in steel dies or the material is placed in a rubber mold inside a high-pressure oil or water cylinder, with uniform pressure applied. Another variation of pressing called hot pressing combines forming and firing in one step using heated dies.

### *Drying and firing*

3 After forming, the ceramic bone must undergo several thermal treatments. The first dries the material to remove moisture using a drying oven or chamber. After drying, a kiln or furnace is used to heat the material at high temperatures in order to remove organics and densify the material. The firing cycle will depend on the material composition and must be designed at the appropriate heating rates to prevent cracking.

### *Finishing*

4 After firing, one or more finishing processes may be required depending on application. To achieve the desired dimensional and surface finish specifications, grinding and/or polishing is conducted. Grinding and polishing of the harder materials usually requires diamond tooling or abrasives. Drilling may be needed to form holes of various shapes. If the application requires joining of two or more components, a brazing or cementing method is used.

## Quality Control

During the manufacture of the artificial bone material or component, control of each processing step is required to control the properties that affect performance. The

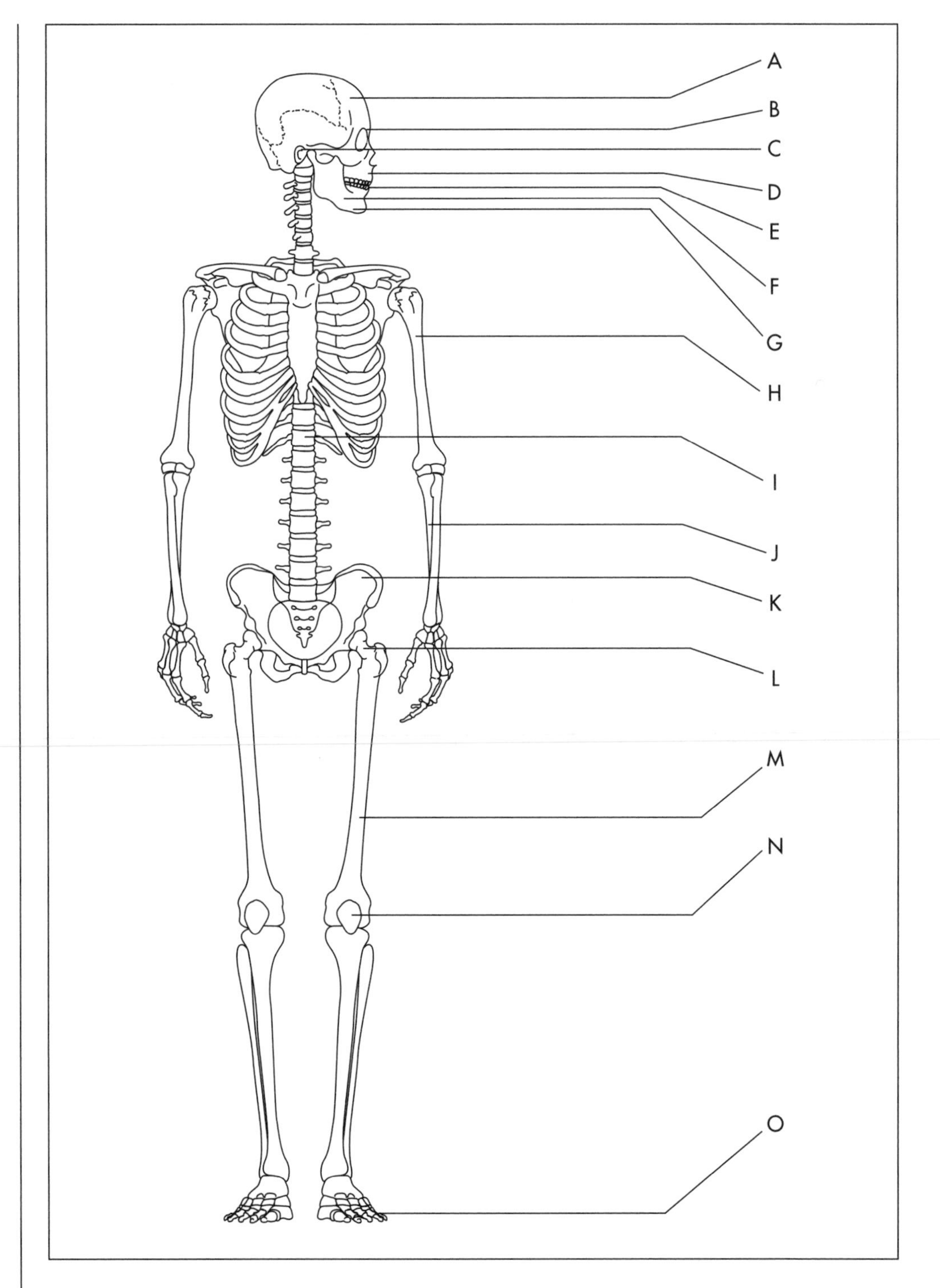


Bioceramic applications. **A.** Cranial repair. **B.** Eye lens. **C.** Ear implants. **D.** Facial reconstruction. **E.** Dental implants. **F.** Jaw augmentation. **G.** Periodontal pockets. **H.** Percutaneous devices. **I.** Spinal surgery. **J.** Iliac crest repair. **K.** Space fillers. **L.** Orthopedic support purposes. **M.** Orthopedic fillers **N.** Artificial tendons. **O.** Joints.

properties of interest for most implant applications are mechanical performance and surface chemical behavior. These in turn depend on the chemical composition (type and amount of impurities), the particle size, shape and surface characteristics of the starting powder, crystalline structure, microstructure (grain size, type and content of each phase), and surface behavior (measured by comparing the chemical composition of the surface before and after it is tested in a simulated environment relevant to the application). Some of these properties may be more important than others, depending on the type of artificial bone material and its application.

Since artificial bone can sometimes be considered a medical device or at least part of a medical device, it must meet national and international standards for such devices and materials, as well as regulations estab-

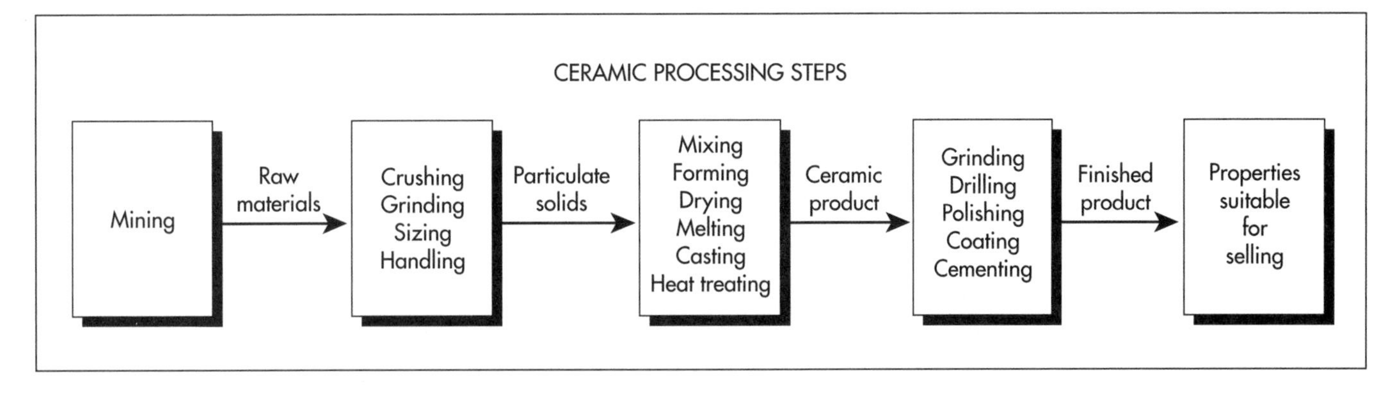


The ceramic powder is manufactured elsewhere from mined or processed raw materials. Additional crushing and grinding steps may be necessary to achieve the desired particle size. After mixing, the ceramic material is ready for forming into the desired shape. Once formed, the ceramic bone must undergo several thermal treatments in order to remove organics and densify the material. One or more finishing processes may be required depending on application. To achieve the desired dimensional and surface finish specifications, grinding and/or polishing is conducted. Drilling may be needed to form holes of various shapes. If the application requires joining of two or more components, a brazing or cementing method is used.

lished by the Food and Drug Administration (FDA). The American Society for Testing and Materials has developed a number of specifications (over 130 standards total) for certain materials used as surgical implants that cover chemical, physical and mechanical properties, as well as characterization methods. The International Organization for Standardization has two committees that have developed standards for surgical devices and biocompatibility of materials.

The FDA has the authority to regulate medical devices during most phases of their development, testing, production, distribution and use, with a focus on the pre- and post-market phases to ensure safety and effectiveness. The level of regulation or control is based on how the device is classified (I, II, or III). The higher the class, the more regulation—Class III devices must have an approved Pre-market Approval Application.

All classes are subject to General Controls, which involves registering each manufacturing location, listing marketed medical devices, submitting a Premarket Notification for a new device, and manufacturing the device according to the Good Manufacturing Practices regulation. This regulation includes requirements for the quality assurance program used by the manufacturer.

## Byproducts/Waste

Since careful control of the manufacturing process is so important, waste is minimal. Since contamination must be avoided, any waste produced can only be recycled if the properties match the starting material. Sometimes waste material can be used to make other ceramic products of lower quality. Byproducts that must be controlled throughout the process include dust and organic emissions from firing.

## The Future

In the next century—as a better understanding of the interactions of artificial bone with organic components is achieved on the molecular level—it will be possible to tailor the physical and chemical properties of the material with the specific biological and metabolic requirements of bone tissues or disease states. Since the population is continuing to grow older, artificial bone will play an even more important role in improving the health of many people around the world.

## Where to Learn More

### Books

Fischman, Gary and Clare, Alexis, ed. "Bioceramics: Materials and Applications." In *Ceramic Transactions.* The American Ceramic Society, 1995.

Hench, Larry and June Wilson. *An Introduction to Bioceramics.* World Scientific Publishing Col. Ltd., 1993.

Ravaglioli, A. and A. Krajewski, ed. *Bioceramics and the Human Body.* Elsevier Applied Science, 1992.

### Periodicals

Hench, Larry. "Bioceramics, 'A Clinical Success.'"*Ceramic Bulletin.* The American Ceramic Society (July 1998): 67-74.

*—Laurel Sheppard*

# Bisque Porcelain Figurine

*Bisque porcelain was called* fan ts'u *or turned porcelain by the Chinese, but elsewhere, it is also called biscuit ware, parian ware, or unglazed ware.*

## Background

Bisque porcelain is unglazed, white ceramic ware that is hard-fired, non-poreous, and translucent. Today's bisque porcelain industry has arisen out of hundreds of years of experimentation with clay products and untold sources of artistic inspiration. Manufacturers of bisque porcelain collectibles are unanimous in their approach to each product as a work of art. Quality begins with the design and is controlled throughout the process, which can take several months from the time a collectible is first sketched by an artist until it reaches a collector's hands. In the factory alone, a single figurine may be more than a week in production, with its manufacture scrutinized intensely every step of the way.

## History

The Chinese were the creators and first masters of the art of producing porcelain. Chinese mastery of the art form made them virtually the only porcelain producers for hundreds of years. Bisque porcelain was called *fan ts'u* or turned porcelain by the Chinese, but elsewhere, it is also called biscuit ware, parian ware, or unglazed ware. All porcelain is fired at least once. Originally, the biscuit stage referred to porcelain after its initial firing when it was so brittle it could be broken by finger pressure and it remained porous. Dipping it in glaze that was absorbed by the porous material preserved the porcelain. The second firing caused the glaze to melt or fuse with the clay and become vitrified or glass-like. This differs from modern production of bisque porcelain, which is hard and durable without the addition of glaze.

In Europe, the production of bisque porcelain wares rose to prominence in the mid-1700s. The French made busts and medallion-like portraits at the factories of Sévres, Mennency-Villeroy, and Vincennes. The Frenchman Desoches and the German artist Rombrich modeled portrait plaques from life in bisque and represented Greek subjects in frames of laurel leaves in the style of the Englishman, Josiah Wedgwood, who succeeded in adding colors to clay that were retained through firing in his unglazed Jasperware. By the end of the century, a number of sculptors were modeling figurines (usually of classical figures or ordinary characters including idealized children, street sweepers, and peasant girls) in biscuit ware. The popularity of bisque seems to have been due to the vulgarity of glazed porcelain. The colors made at this time were raucous and garish, and the bisque effect was softer and warmer. By Victorian times, bisque porcelain was used to make the heads and arms of dolls, and these dolls (both antiques and modern forms) form another branch of the bisque collectibles industry. Figurines made from both glazed and unglazed porcelain have remained highly collectible since the eighteenth century throughout changes in fashion and style and with improvements in processing.

## Raw Materials

The raw materials required for making figurines include plaster for molds, porcelain clay, pumice and water for polishing the fired pieces, paints or pigments specially created to suit the designer's intent, and packaging materials. Porcelain clay is a mixture of kaolin, feldspar, and ground flint. Kaolin is a naturally-occurring, fine clay that primarily consists of aluminum silicate. Feldspar is a crystalline mineral that also contains aluminum silicate as well as potassium, sodium, calcium, or barium. Flint is hard quartz.

## Design

The creation of a porcelain figurine begins with an artist's conception. Perhaps the best-known examples are the doe-eyed children (each with a teardrop near one eye) drawn by artist Sam Butcher and featured in the Precious Moments figurines.

After the artist has sketched or painted a design, master sculptors use moist clay to make models from the artwork. A rough model is made first, then the sculptor works it by adding and subtracting subtle pieces of clay until the model is complete. Ideally, the sculpture has not only the correct shape but matches the original intent or feeling expressed in the artwork. The completed sculpture is then reviewed and approved for mass production.

## The Manufacturing Process

### *Making molds*

Casts of the original sculpture are used to make plaster molds for production of the figurines. The details transferred from the artist's original conception to the original sculpture are sometimes tiny and complex, so the original sculpture is subdivided into multiple parts to make a set of molds for reproduction. Sometimes a dozen sets of molds are required to produce a single figurine. The process of making the molds is very carefully done so the original sculpture is duplicated in the molded porcelain figurines. The mold-making process may go through several steps, including the making of sample molds and case molds, before a production mold is finally produced. The plaster production molds can be reused as many as 50 times, but each use robs the mold of a tiny bit of detail. Porcelain factories limit use of the molds to preserve the quality from figurine to figurine, so they are often destroyed after 30 or so castings.

1 The plaster used to make the molds is highly refined to produce extremely fine powder that will capture the most detail and make finished surfaces smooth. Water and plaster powder is mixed. To eliminate bubbles from the soup, a vacuum blender is used to draw off the entrained air.

2 The liquefied plaster is poured into a case mold to create the production mold. The plaster solidifies in about 20 minutes, but its solid appearance is deceptive because it still contains a lot of moisture.

3 The production molds are fire-dried for about 48 hours at a temperature of about 90° F (32.2° C). The dried plaster molds are fastened tightly together with rubber straps and are ready for production.

4 The porcelain clay is prepared while the molds are dried. The clay is also mixed with water until a slurry called slip is formed. The slip resembles thick cream, and its color is usually very different from the finished color of the figurine because the processes of firing the clay alter its color.

5 The mold filled with slip is allowed to rest while the plaster mold absorbs liquid from the clay. After about 30 minutes, the surfaces of the mold bear deposits of clay that are thick enough to form the parts of the figurines. The slip that remains in the middle of this part of the figurine is poured off.

6 The mold rests again until the figurine cast can be safely removed from the mold by gentle tapping. At this stage, the figurine is a collection of pieces of greenware that have not yet been assembled or fired.

7 The components of the figurine are still somewhat pliable. They are assembled by using more of the liquid slip like glue that is applied by brush. The object now resembles the finished product.

### *Finishing the greenware*

8 In the next process called finishing, all seams are gently removed, any traces of the mold are smoothed away, and artistic detail is added to the greenware. Artisans complete the finishing work quickly because the greenware begins to dry as soon as it is exposed to air.

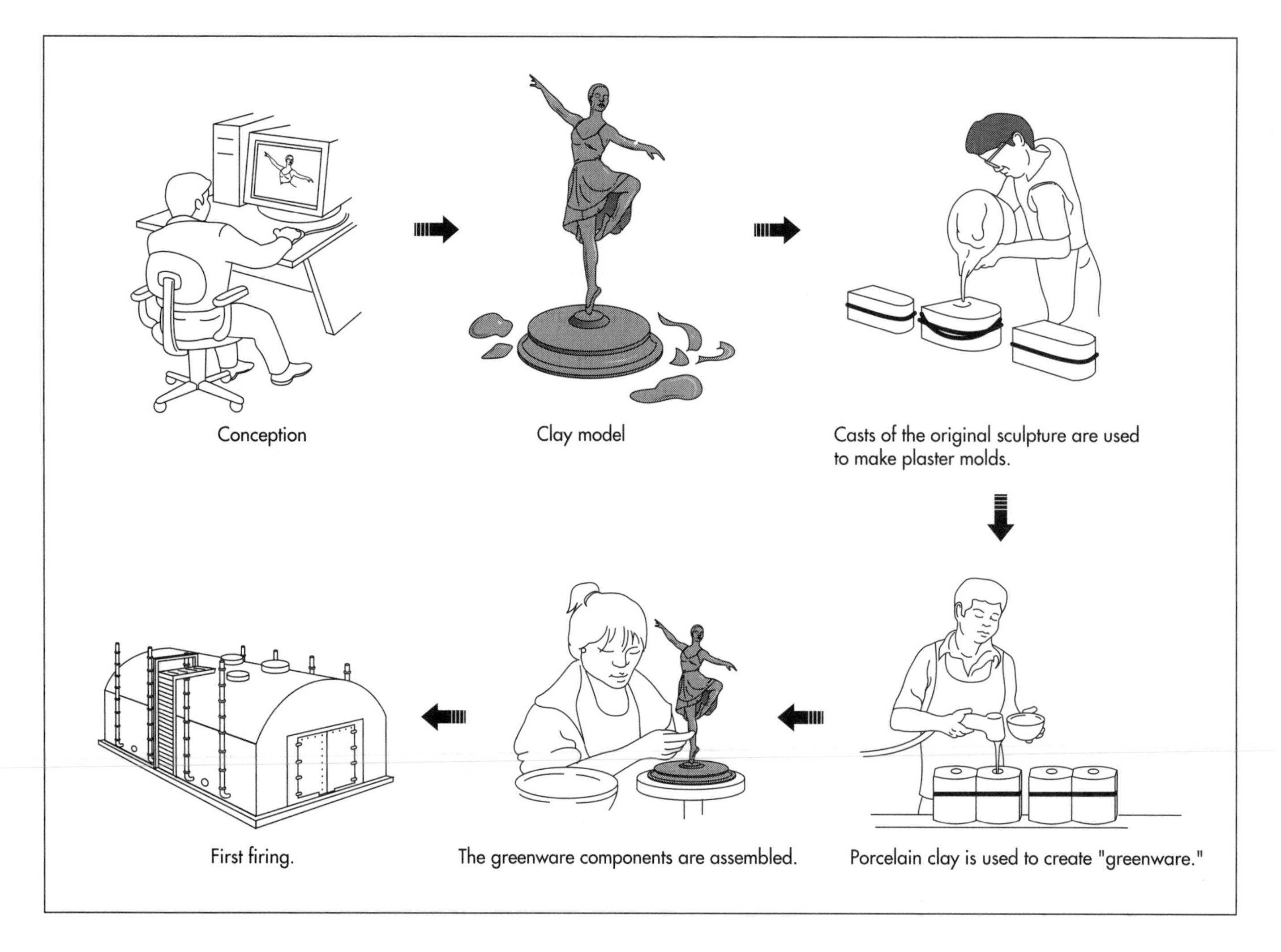


The manufacture of bisque figurines begins as an artist's conception that is modeled in clay. Once the model is perfected, casts are used to make plaster molds. Greenware is molded and then fired.

9 The figurines are then air-dried thoroughly to allow moisture to evaporate before the pieces are fired. Any moisture remaining in the porcelain will cause it to shatter in the kiln.

### *Firing*

10 The bisque kiln is used to fire molded wares. The collectibles are subjected to a first firing in a gas-fired kiln in which they are baked at a temperature or about 2,300° F (1,260° C) for 14 hours. Other types of bisque pottery may be fired for as much as 70 hours to produce a hardened piece with an unglazed finish. Finished porcelain is non-absorbent, translucent, vitreous, and can be as hard as steel. During the first firing, the color of the greenware also transforms to the finished color of the porcelain. Any impurities present in the clay will appear during firing and cause discoloration, requiring that the pieces be destroyed. During firing, vitrification occurs, changing the color and consistency of the clay. Physically, the figurine may have shrunk in size by as much as 15%.

### *Polishing and painting*

11 If the particular collectible is to have a satiny smooth finish, the next step in its production is polishing. The figurines are placed in a tumbler much like the device used to polish precious stones. In this case, the tumbling action is very gentle as water and fine pumice polish the surfaces of the figurines.

12 The polishing process has an added advantage in that the painters who will work on the figurines next have beautiful, primed surfaces for their artistry. The painters are highly skilled, not only in technique but also in capturing the spirit of the figurine. They hold as many as six brushes at a time and apply specially formulated pig-

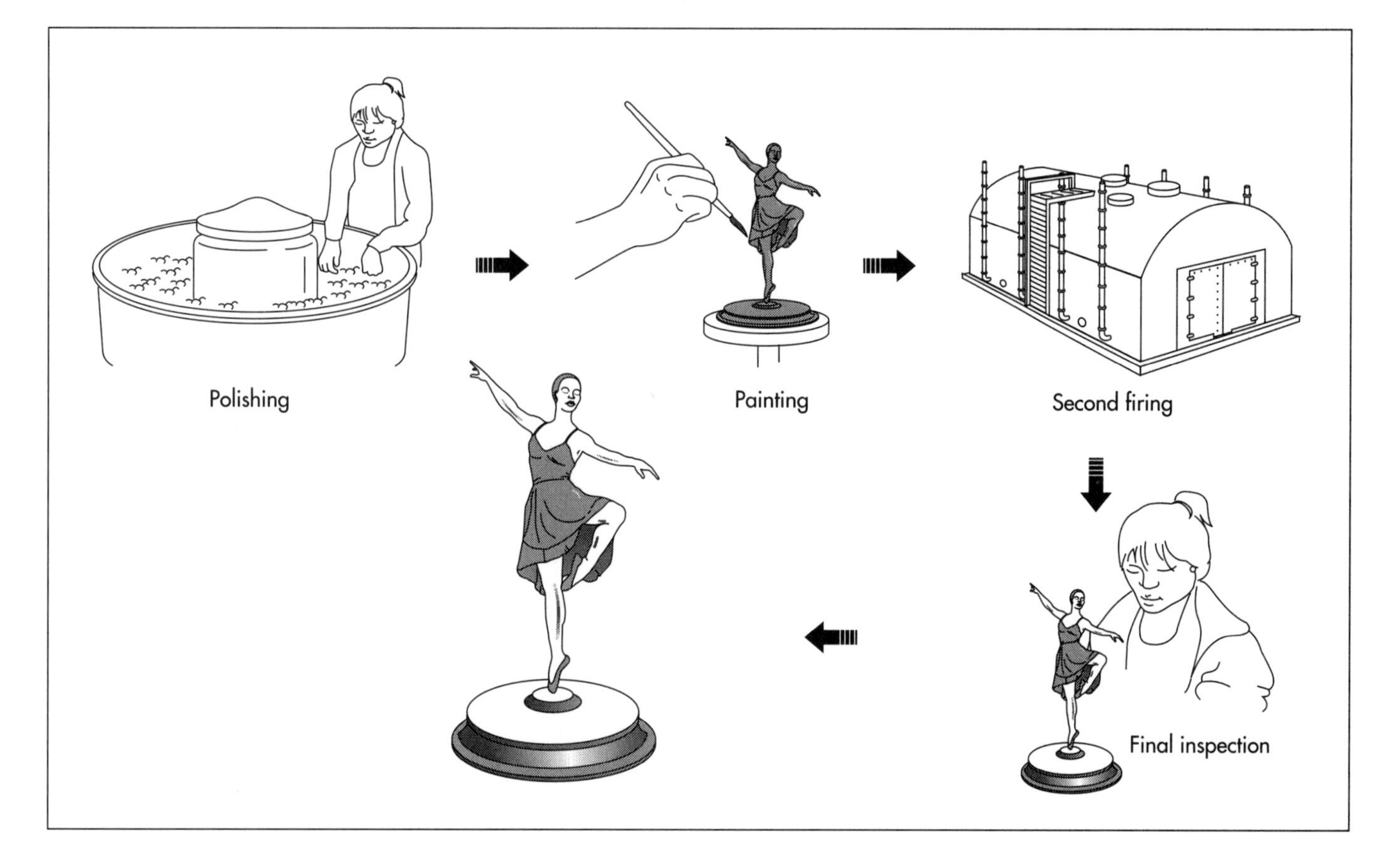


The unglazed pottery is polished, painted, and fired for a second time.

ments to the figurines. Painting is a multistep process in which various artists use different skills from fine line work to painting using airbrushes. The painted characters are inspected before decoration firing.

### Second firing

13 The second, or decoration, firing uses an electrically-heated tunnel kiln to affix the pigments permanently to the porcelain. Although it does not apply for most collectibles that are left unglazed, some porcelain products are fired in third or even fourth firings to harden glaze (in a gloss kiln) or harden added relief-type decorations. During decoration firing, the figurines are fired for four hours at a temperature of about 1,300° F (704.4° C).

## Quality Control

Attentive quality control is essential to the production of detailed collectibles. Materials, particularly the plaster and clay, are selected, processed, mixed, and used with great care. A porcelain clay containing impurities may color a fired ornament bright pink instead of pure white, wasting the entire batch of slip and fired figurines. Figurines themselves typically pass six or more inspections before they are shipped. Air-dried greenware is inspected to verify detailing, assembly, and smoothness of seams and mold imprints. After firing, the pieces are again inspected for flaws. Polishing can be a damaging process especially when tiny birds or butterflies are part of the figurine. Another detailed inspection occurs following polishing. Pigments are then applied, and the figures are scrutinized immediately after painting and again after decoration firing to ensure that color quality is true. If painted details are omitted, the figurine can be repainted and refired, but any detailing errors result in destruction of the piece.

## Byproducts/Waste

Byproducts do not result from manufacture of bisque porcelain figurines, but a single manufacturer may produce any number of product lines using the same basic processes. Figurines are far from the only products. Bells, ornaments, photo frames, music boxes, plates, and accent pieces for decorating are often made in the same style as a leading line of figurines and in the same materials. Some waste is the result of stringent quality control because figurines that are even slightly flawed are destroyed. Inevitably, there is also some breakage. As new collectibles are added to a product line, old products are suspended or retired. Molds for retired collectibles are destroyed, and the item is never produced again.

## The Future

Bisque porcelain collectibles are highly prized as gifts, collectibles, and decor for the home. Many lines of figurines inspire extraordinary loyalty among their fans. For example, the Precious Moments Collectors' Club was established in 1981 and has become the largest such organization in the world with over 400,000 members. The gift and collectibles industry, which grossed $9.1billion in 1997, prides itself on adding individual and artistic touches of warmth to a busy world. Ever-changing fashion and taste, combined with the loyalty of confirmed collectors, guarantee the future of bisque porcelain figurines, among many other types of collectibles.

## Where to Learn More

### Books

Cox, Warren E. *The Book of Pottery and Porcelain.* New York: Crown Publishers Inc., 1970.

Ketchum, William C. *The Pottery & Porcelain Collector's Handbook: A Guide To Early American Ceramics From Maine To California.* New York: Funk & Wagnalls, 1971.

### Other

Ceramitech, Inc. http://www.ceramitech.com/.

Enesco Corporation. http://www.enesco.com.

International House. http://user.mc.net/~intlhse/.

Lucy & Me Collecting Network. http://www.lucyandme.com/.

"The Making of A Precious Moments Figurine." Itasca, IL: Enesco Corp., 1989.

*—Gillian S. Holmes*

# Bow and Arrow

In simplest terms, a bow is a long, flexible staff; a shorter string is attached to the staff's two ends, causing the staff to bend. An arrow is a shaft with feather-type vanes near one end, which is notched, and a pointed head on the other end. The notched end of the arrow is mounted against the bowstring, with the pointed head extending just beyond the bow. With one hand braced against the bow and the other gripping the string, an archer pulls back on the string, storing potential energy in the bow. When the archer releases the string, that potential energy is converted to kinetic energy, which is imparted to the arrow, propelling it forward suddenly and swiftly. Bows are used primarily for hunting and for target shooting.

## Background

Archaeologists believe hunters used bows and arrows as early as 50,000 years ago. Indigenous people used such weapons in every part of the world except Australia. In addition to hunting and warfare, bows and arrows were used for sport in ancient cultures of Egypt, China, and India.

Originally, bows were made of any springy material, including bamboo as well as various types of wood, and the bowstrings were made of animal gut. Native American and Asian bow makers independently made an important innovation when they reinforced the weapon by gluing animal sinew (tendon) to the back of the bow (the side facing the target). The composite bow (one made of three or more layers of dissimilar substances) was invented by several cultures in Central, North, and Southwest Asia as much as 4,500 years ago. The technique included reinforcing the bow's back with up to three layers of shredded sinew mixed with glue, and strengthening the face of the bow (the side facing the archer) with a glued-on layer of animal horn. Northern Europeans used a different method to strengthen bows; by the beginning of the fourth century A.D., they were bonding a back of sapwood to a face of heartwood (dense wood taken from the non-living core of a tree).

Arrows, which were normally made of wood shafts, were tipped with arrowheads shaped from hardwood, bone, horn, flint, bronze, or (eventually) steel. In India, weapon makers experimented widely with iron and steel, and they produced an all-metal arrow during the third century B.C. Although it is likely that they made metal bows at that time as well, it was not until the seventeenth century that steel bows truly became popular in India.

Archery (using a bow to shoot arrows) was a dominant means of warfare (with standard bows proving to be generally superior to mechanically assisted crossbows) until the late sixteenth century, when firearms became practical. Since then, hunting and target shooting have developed as the main activities in archery.

From 1929-1946, seven archers who were also scientists or engineers studied the performance of equipment designs and materials using techniques like high-speed photography. They published their findings in various journals, and in 1947, three of them edited a collection of these articles, calling the book *Archery: The Technical Side.* These experimental and mathematical analyses of bow dynamics laid the groundwork for the first significant improvements in archery equipment design since the Mid-

*Archaeologists believe hunters used bows and arrows as early as 50,000 years ago. Indigenous people used such weapons in every part of the world except Australia.*

dle Ages. Among the innovations that appeared after World War II were the use of new materials like plastics and fiberglass, and modification of the bow's grip section to resemble a pistol handle.

## Design

The most basic type of bow, called a longbow, is formed from an essentially straight shaft. Additional power and stability are achieved by recurved bows, which have permanent curves that make the bow's back concave at each end. Even more power can be achieved with a compound bow, a mechanically assisted device that attaches the bowstring to a system of pulleys rather than to the tips of the bow.

A recurved bow consists of three parts—two flexible limbs extending from opposite ends of a rigid riser. The bow's total length may be 50-70 in (125-175 cm). The riser, which is about 20 in (50 cm) long, provides a comfortable handgrip and a ledge on which the arrow rests prior to release. The limbs may be permanently attached to the riser, or they may be removable, allowing the archer to take the bow apart for ease of transportation and storage or to interchange limbs with different operating characteristics.

## Raw Materials

When made of a single piece of wood, a bow can warp from moisture or become brittle in cold weather. It can also permanently deform into the curved shape attained when the bow is strung (the bowstring is attached to both ends, bending the bow). When this happens, the bow's springiness is decreased and it loses power. Making bows from fiberglass solves some of these problems, but with reduced performance characteristics. The best results are obtained with composite materials that are formed by gluing together layers of various woods, fiberglass, or carbon fiber. Among the woods commonly used for bows are red elm, maple, cedar, bamboo, and exotic woods such as bubinga.

Historically, bowstrings have been made from sinew, twisted rawhide, gut, hemp, flax, or silk. Today, strings for wooden longbows are often made of linen thread. Compound bows may be strung with steel wire. Bowstrings for popular recurved bows are usually made of Dacron, which stretches very little and wears well. Nylon thread is wrapped around the bowstring to reinforce it at the ends and in the middle where the arrow and the archer's fingers contact the string during shooting.

Arrows have traditionally been made of solid shafts of wood such as ash, elm, willow, oak, cedar, or Sitka spruce. Hollow arrow shafts may be formed of modern materials like aluminum, fiberglass, graphite, or carbon fiber. Feathers (commonly from turkey wings) mounted on the shaft near one end cause the arrow to spin during flight, steadying its path. Because of better durability and moisture resistance, vanes made of plastic or molded rubber have become more popular than natural feathers for this purpose. A nock (a plastic piece that is grooved to fit around the bowstring) is attached to the back end of the arrow. Arrowheads, which were historically made of flint, bone, horn, bronze, or hardwood, are now commonly made of steel. They may have two to six protruding blades, or they may simply bring the shaft to a rounded or pointed end.

## The Manufacturing Process

### *The bow*

The following paragraphs describe the construction of a recurved bow with permanently attached limbs.

1 Various materials are cut into rectangles for the layers of the limbs. Wood layers are dyed the desired color. Glue is applied, and the layers are stacked in the proper sequence.

2 The multi-layer limb section is mounted on a form that will determine its final curvature. While attached to the form, the limb is cured in an oven at 180° F (80° C) for six hours.

3 The riser is made from a solid block of aluminum or a block formed by laminating various layers of wood. After cutting the block down to a basic outline of its final shape, pins are inserted near the riser's ends to allow attachment of the limbs.

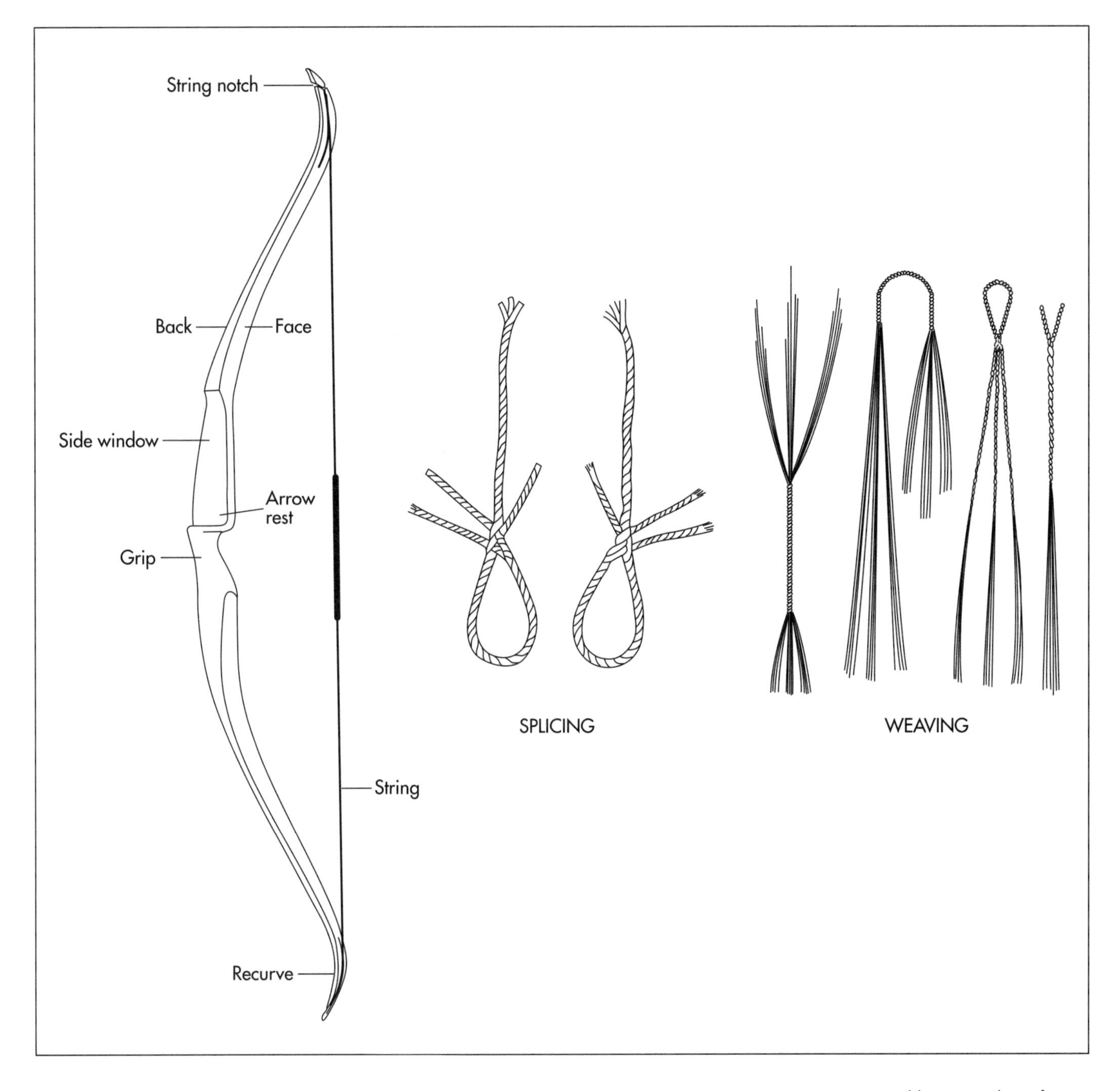


A typical bow. In order to form a bowstring loop, the string can either be spliced or woven.

4 Holes are drilled in the limbs to match the position of the pins in the riser, and the limbs are temporarily attached to the riser. After the joints are sanded smooth, the limbs are removed from the riser.

5 Using a template, the bowyer (bow maker) marks the limbs for cutting. Using a power saw and a sander, the craftsman tapers and shapes the ends of the limbs from their originally rectangular shape. The ends of the limbs are filed to make grooves where the bowstring can be mounted.

6 The bowyer begins to shape the riser by cutting out sections to form a shelf on which the arrow can rest and to provide a sighting window. Using a power saw, a sander, and a hand rasp (wood file), the bowyer contours the riser into a shape that will be comfortable to grip.

7 The limbs are attached to the finished riser and glued into place. Final shaping is done on the limb tips. The entire bow is sanded by hand and then finished with a protective coating of clear epoxy.

During bow manufacture, the limb is mounted on a form that will determine its final curvature. While attached to the form, the limb is cured at a high temperature and the riser is then attached to the bow with pins.

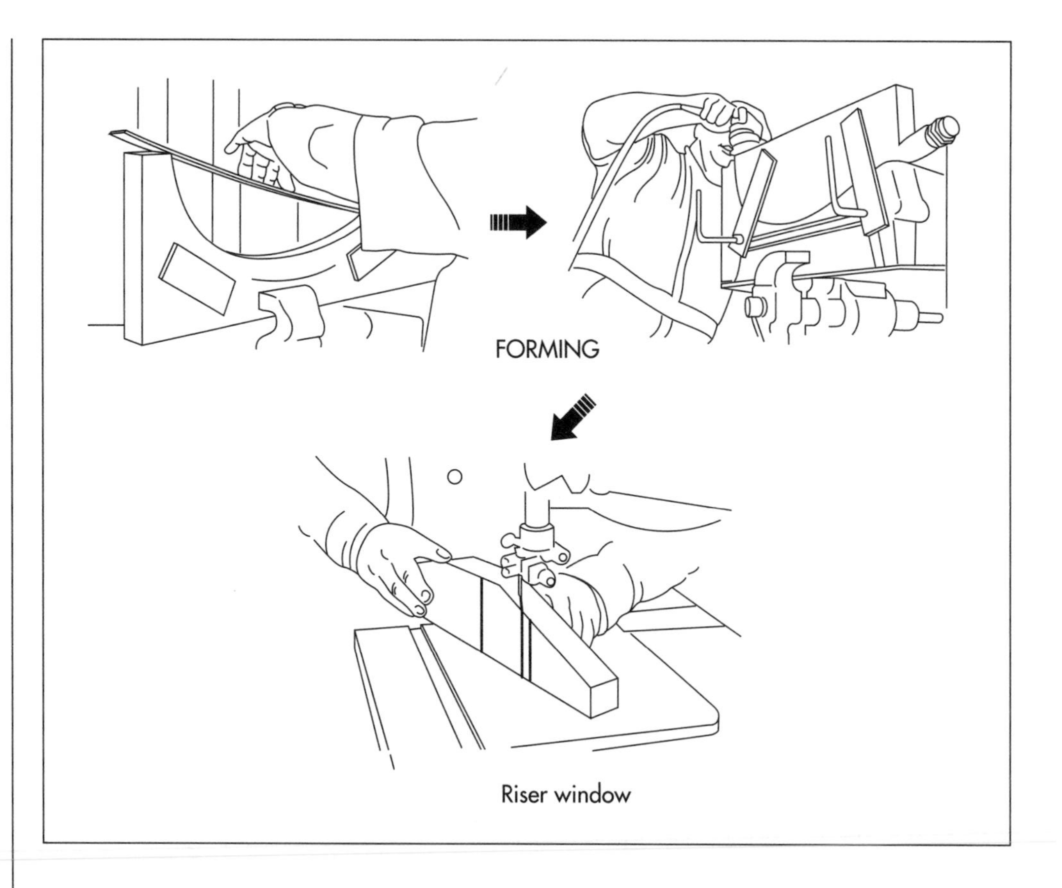


### *The bowstring*

Although manufactured bowstrings are available, some archery enthusiasts prefer to make their own.

8 The number of strands of thread needed is determined. This depends on the strength of the thread being used and the draw weight (strength) of the bow. The bundle of strands is divided into three equal sets, and each set is coated with beeswax (perhaps with added resin). The sets of strands are then formed into a cord by twisting and weaving them together.

9 When enough cord has been formed, a loop is formed by bringing the cord's end around and splicing or weaving it into the new section that is being corded. When the bowstring's desired length is nearly achieved, the string is pre-stretched by hanging it from the initial loop while attaching a weight to the free end. The length is then reevaluated, and cording continues until the desired length is attained. Forming another loop finishes off the string.

10 "Serving" is applied by wrapping nylon thread around a 10-in (25-cm) section in the center of the bowstring and a 5-in (13-cm) section near each end loop. A reinforcement called a nocking point, which is made of rubber or plastic, is attached at the point where arrows will be mounted against the string.

### *The arrow*

The following steps describe how wooden arrows are made.

11 A "two by four" (2 in [5 cm] thick and 4 in [10 cm] wide) of appropriate wood is selected, making sure the grain of the wood runs as close as possible to the length of the board. A section is cut that is about 3 in (7.5 cm) longer than the planned arrow length. Using a heavy knife or an axe, the board is split down one side to form an edge that truly runs along the grain of the wood.

12 Following the split edge, square blanks are sawed that are slightly larger than the desired shaft diameter. If necessary, the blanks can be straightened by heating them and bending them.

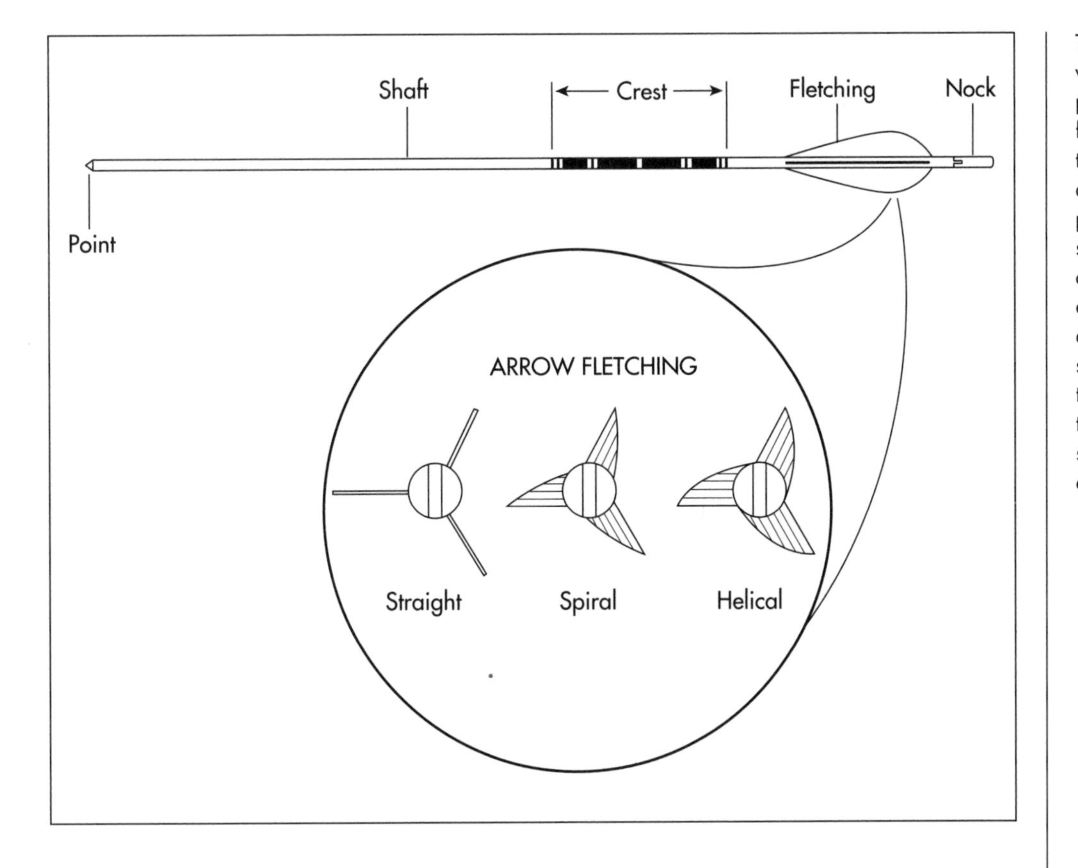


The arrow is typically made of wood and coated with polyurethane and paint. Trimmed feathers or plastic vanes are glued to the shaft between the cresting and the nock in a pattern that is parallel to the shaft, spiral(in a straight-line diagonal to the shaft), or helical (in a curve that begins and ends parallel to the shaft). An arrowhead is mounted on the shaft. The shape of the head is determined by the purpose for which the arrow will be used—target shooting or hunting specific types of animals.

13 Each side of the blank is planed to ensure its smoothness and straightness. Then the four corners are planed off to form an octagonal rod. Again, the corners are planed off. Finally, the shaft is sanded to form a round dowel.

14 A nock, or slot, is cut into one end of the arrow shaft. Alternatively, the end of the shaft can be inserted into a plastic nock.

15 The shaft is coated with polyurethane or varnish. Cresting (bands of color that identify the arrow's owner or manufacturer) is applied around the shaft.

16 The arrow is fletched by gluing trimmed feathers or plastic vanes to the shaft between the cresting and the nock. These real or artificial feathers may be applied parallel to the shaft, spirally (in a straight-line diagonal to the shaft), or helically (in a curve that begins and ends parallel to the shaft). Usually three feathers are applied, one of which will face directly away from the bow when the arrow is mounted for shooting. This is called the cock feather, and the other two are called shaft feathers.

17 An arrowhead is mounted on the shaft. The shape of the head is determined by the purpose for which the arrow will be used—target shooting or hunting specific types of animals.

## The Future

Building on the analytical approach begun in the 1930s, modern researchers are refining mathematical models that describe the performance of bows, in order to evaluate possible design changes. In addition to varying the size and shape of bow components, bowyers also experiment with new materials. For example, at least one manufacturer now offers limbs made with a core layer of syntactic foam (a high-strength, low-density material, composed of epoxy resin and microscopic glass beads that can be cast and machined).

Some archers use attachments on their bows to improve their performance, and manufacturers are developing increasingly sophisticated models of such accessories. For example, an electronic sighting device is now available that not only helps archers fix their aim on a target, but also acts as a digital-display rangefinder. New designs are also being developed for stabilizers that are mounted on rods extending outward from the back of the bow. These stabilizers consist of weights or hydraulic damping devices (movable weights encased in a fluid-filled cylinder) that help prevent twisting of the bow during shooting by absorbing some of the shock when the bowstring is released.

## Where to Learn More

### Books

Paterson, W. F. *Encyclopaedia of Archery*. New York: St. Martin's Press, 1984.

Williams, John C. with Glenn Helgel. *Archery for Beginners*. Chicago: Contemporary Books Inc., 1985.

### Other

McNeur, Rob. *Arrow Making FAQ*. http://snt.student.utwente.nl/~sagi/artikel/faq/arrwmake.shtml#question (December 7, 1998).

"Steps in Handcrafting Our Traditional Bows." Harrelson Traditional Archery Inc. http://www.mindspring.com/~bowyer/ (October 25, 1998).

*—Loretta Hall*

# Brassiere

## Background

Derived from the french word meaning upper arm, the brassiere is a mass-produced support undergarment worn by women that consists of two fabric cups attached to two side panels, a back panel, and shoulder straps (unless strapless) that fits snugly. They are sized according to a universal grading system first introduced by Ida Rosenthal, the founder of Maidenform, in 1928. Two measurements are crucial to determining bra size: the chest circumference below the underarm and the fullest part of the breast. Cup size is calculated from the difference between the two measurements. The greater the difference the larger the cup size. Brassieres support breasts, separate them, and give them a shape or form.

These undergarments are made of many different materials including cotton, rayon, silk, spandex, polyester, and lace. They are available in many styles from cups that come without any padding (and are quite sheer) to those that add significantly to the size and shape of the cup. A woman can alter her silhouette by simply purchasing a brassiere with cups that are designed to render a specific shape.

## History

Prior to the advent of the modern bra, a term coined in 1937, corsets were the only support garments available. Originally fashioned with whalebones, the one-piece corset was made popular by Catherine de Médici's demand for slim-waisted court attendants during her husband's—King Henri II—reign in France in the 1550s. The corset's popularity was withstanding and lasted over 350 years, with whalebone being replaced by steel rods. The corset design changed to accommodate the reigning ideal figure, pushing bust and hips around according to the fashionable silhouette.

In the late nineteenth century, several precursors to the modern bra were developed. In 1875, a loose, unionsuit was manufactured by George Frost and George Phelps. During this period, corsets were lengthened to produce the fashionable figure type, the top of the corset dropped low, often not supporting or covering the breasts. As added support, fabric undergarments called bust bodices were worn over the corset to cover and shape the breasts (by pushing them together but not separating them), somewhat similar to the modern brassiere. In 1889, a Frenchwoman named Mme. Herminie Cadolle devised the a garment called the *Bien-Étre* (meaning well-being), which connected with sashes over the shoulders to the corset in back.

Early in the twentieth century, the need for a less obtrusive undergarment became necessary as the fashions changed. In 1913, the modern brassiere was born out of necessity when New York socialite Mary Phelps Jacobs' whalebone corset poked up above her low cut gown. Fashioned from silk handkerchiefs and ribbons, the mechanism proved useful and Jacobs filed the first patent for a brassiere and began producing brassieres under the name Caresse Crosby. Jacobs sold the patent and business to Warner Brothers Corset Company for $1,500.

## Raw Materials

The raw materials gathered for the production of brassieres vary tremendously depending on the product. Some are all cotton,

*In 1913, the modern brassiere was born out of necessity when New York socialite Mary Phelps Jacobs' whalebone corset poked up above her low cut gown.*

some are all polyester, some are combinations of natural and synthetics, and so forth. Most brassieres include an elastic material of some sort on the back panel that allows some expansion and movement of back muscles. Spandex, a modern synthetic fiber extensively processed from Malaysian tree sap, must be processed prior to the assembling of the brassiere because it is, in some products, the most important material in the brassiere. A closure of some sort (most often metal hooks and eyes) must be included on the brassiere unless it is an elastic sports brassiere which can be put on over the head. Cups, padding, and straps vary not only from manufacturer to manufacturer but by style.

## Design

The design process for developing a new brassiere style is an important part of the manufacturing process. Brassiere manufacturers, like other clothing manufacturers, must supply not just a functional item but one that appeals to a large enough segment of women that the products can be sold with a profit. Before a new product or product line is designed, the marketing and sales departments review data on the current line of products. They examine comments from retailers as to what they feel might sell well, female consumer attitudes in general and trends in women's purchasing habits. They may also talk to focus groups who offer their opinions on products and needs.

By the time this review is complete, the marketers and designers have decided on the next season's collection. Decisions are based on how the new styles will be positioned within the collection, special features, cut, sizing, production costs, market pricing, quality specifications, and when the new product will be launched publicly. These general specifications are essential for the designers and design engineers to use as guidelines once they leave that meeting.

Prototype drawings are made, pattern pieces are designed, and often the pattern pieces are devised using computerized programs. Components of the brassiere—cup top and bottom and side, central and back panels—render the shape. These components are cut out of cardboard using a computerized cutter. This prototype is assembled and is subject to important fine-tuning and modification. It is important to note that more styles and prototypes are created than the company intends to produce. After modifications, the appropriate prototypes are selected. Computer production of pattern is useful to size the pattern in order to fit different sizes of women.

Final selections are tested by laboratories to ensure quality, fit, sizing, etc. Then, the prototype is manufactured in the factory in some quantity and tested once again by everyone from designers to shop foremen to marketers. When all agree in the quality, fit, and market appeal, the brassiere is ready to be produced in quantity.

## The Manufacturing Process

The methods for constructing brassieres vary from one company to the next. It is a product that is still pieced out in some plants, meaning that the sewing that connects all the components may be contracted out of the plant to smaller sewing operations. In addition, materials utilized in the construction of the brassiere affects the manufacturing method. For example, if an undergarment company utilizes spandex within the product, they may manufacture the material on premises. If a company uses cotton, it may be supplied from a manufacturer who makes the material based on their specifications.

### *Cutting out the components*

1 The components of the brassiere—the cup top and bottom (if seamed), the straps, and the central, side and back panels—must be cut out according to the pattern specifications residing in the computerized specifications. Many layers of fabric are cut out at a time using either a bandsaw-like shearing device or a more contemporary computer-controlled laser. The cups, panels, and straps are cut; kept together in stacks via style; and sent out to various locations to be sewn.

### *Sewing*

2 The stacks may be sent to different parts of the factory or even off premises to piece workers who assemble the brassieres using industrial grade sewing machines. However, large operations send the pieces

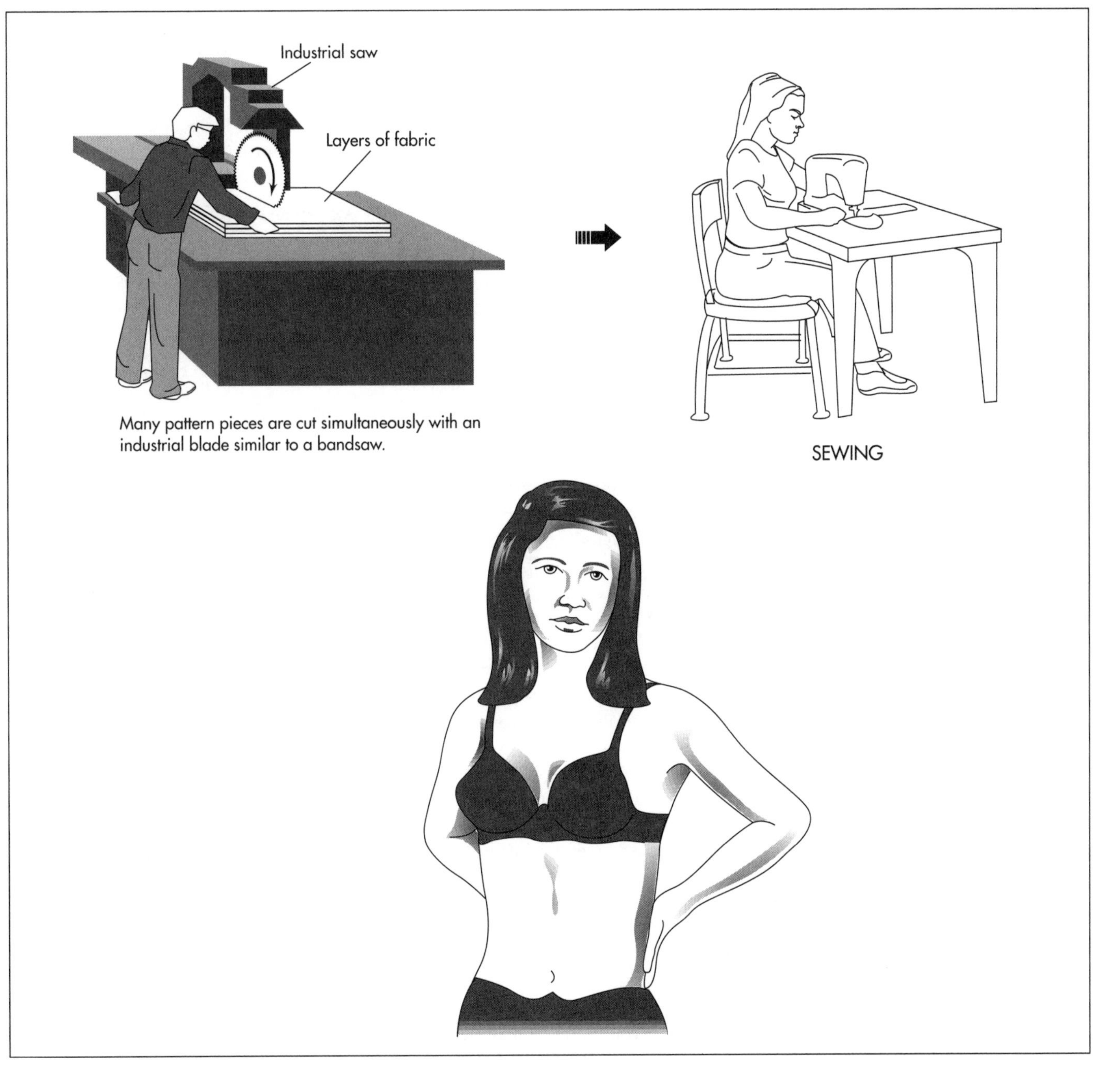


Many pattern pieces are cut simultaneously with an industrial blade similar to a bandsaw.

through on a line in which machines sew parts in quick succession. (Cups might be sewn onto a side panel, the parts move along and another piece is sewn on, etc.) In larger facilities, humans rarely sew anything onto the brassiere unless it is a peculiar or unusual design.

### Closures and labels

3 The brassiere, assembled a bit at a time as it moves through the machinery, is ready for the closures. Coated metal hooks and eyes are both sewn in by machine and heat processed or ironed into the two halves of the back panel. The label is usually machine-stitched into the back or side panel as well at this time.

### Packaging

4 The completed brassieres are sent (either transported in a bin or on a line) to another location and sorted by style and folded (either by hand or machine depending on the size of the operation). Boxes into which many brassieres come arrive at the manufacturer completely flat. Machines must crease and

The manufacture of brassieres involves first cutting many layers of fabric at one time using either a bandsaw-like shearing device or a more contemporary computer-controlled laser. Once cut, the pattern pieces are assembled at the factory by workers, off site by piece workers, or by automated machinery. Hooks and eyes are both sewn in by machine and heat processed or ironed into the two halves of the back panel.

fold the packages that are fed into the machine and a rectangular box is created. A worker called a picker puts a brassiere into the box, the box is closed, and then sent down a chute. A laser reads that the box is fully packed and ready to go to the holding area, awaiting transportation to the wholesaler.

## Quality Control

Quality is controlled in all phases of the design and manufacture of the brassiere. First, experienced designers and design engineers understand the requirements of the wearer as well as the marketers and design with activities and cleaning requirements in mind. Second, a very important part is procuring fabrics and components (underwire, hooks and eyes, or buckles) that are durable. Testing of materials include assessing shrink-resistance, color-fastness and durability, shape-retention, stretch, manufacturing stability, and comfort. Companies work with suppliers in order to acquire new materials that provide service as well as value. In fact, some manufacturers have developed their own fabrics or underwire because all other similar support materials on the market were inferior. Third, prototypes are extensively examined by many members of the company and problems are discovered and solved when many are involved in the assessment of new products. An essential part of this is when the prototype moves from a single example to early manufacturing. Those involved in the manufacturing assist in solving the problems that can occur in the initial stages of manufacturing. Finally, manufacturers must offer consumers brassieres that fit well. In prototyping and in manufacturing, the brassieres are inspected and expected to be within 0.125 in (0.3175 cm) of the desired measurements (one French company requires that the brassiere must not deviate from the standard pattern more than 1 mm[0.0394 in]). If not, the brassiere is rejected as an inferior or second.

## Byproducts/Waste

Fabric wastes are the primary byproducts of this manufacturing process. They may be recycled or discarded.

## Where to Learn More

### Books

Ewing, Elizabeth. *Dress and Undress: A History of Women's Underwear.* London: Anchor Press, Ltd: 1987.

Fontanel, Beatrice. *Support and Seduction: A History of Corsets and Bras.* Translated by Willard Wood. Harry N. Abrams, 1997.

Hawthorne, Rosemary. *Bras: A Private View.* Souvenir Press Ltd., 1993.

### Periodicals

Dowling, Claudia Glenn. "Ooh-la-la! The Bra."*Life* (June 1989): 88.

Wadyka, Sally. "Bosom Buddies"*Vogue* (August 1994): 122.

### Other

Bali Company. http://www.balinet.com/ (June 7, 1999).

*—Nancy EV Bryk*

# Castanets

Castanets are pairs of shell-shaped clappers that are hinged together with string. A Spanish dancer holds a pair in each hand, clicking the clappers together rapidly to produce rhythmic patterns of sound to accompany the dance movements. Castanets are not used in flamenco dancing, however, as the rhythmic accompaniment is produced by stomping the feet.

## Background

The word castanet comes from *castaña*, the Spanish word for chestnut. Besides *castañuelas*, there are several other Spanish words for castanets, including *pulgaretes* (because some dancers attach them to their thumb, or *pulgar*) and *platillos* (saucers).

The classical technique for playing castanets is to let one clapper rest in the palm, with the string looped around the thumb. Striking the other clapper with the fingertips knocks it against its mate, producing a tone. Rapidly striking the clapper with a succession of different fingers on the dominant hand produces trills that embellish the sound and provide counter rhythms. The pair in the other hand (e.g., the left hand of a right-handed person) is played with single strokes to mark the basic rhythm of the music. An alternative technique, used by folkloric dancers, consists of looping the string around one or more fingers in the middle of the hand and flicking the wrist to throw the two clappers toward the palm, where they strike each other.

A pair of castanets should fit comfortably in the dancer's hand, so the diameter is about 1.5-2.75 in (4-7 cm). The smaller sizes, usually used by women, produce a higher tone that is crisper in quality; the larger sizes, usually used by men, produce a lower tone that is richer and more mellow in quality.

Mass-produced castanets made of poor-quality wood or plastic cost less than $10 a set. Custom-made sets that are handcrafted from high-quality material, such as hardwoods and composites, to suit an individual performer cost $100-400.

## History

Musical instruments similar to castanets have been developed in many parts of the world. Ancient versions of small, wood or metal clappers were used by Egyptian, Greek, Roman, Arab, Moorish, and Chinese dancers, for example. It is not known whether such instruments were brought to Iberia (the region now known as Spain and Portugal), perhaps by the Greeks, or whether they developed independently in that region. Archaeological evidence indicates that Iberians made small clappers from sticks, shells, flat stones, and bone.

## Raw Materials

Novelty castanets have been made from ivory, marble, crystal, gold, silver, bronze, and aluminum, but few of these are musically useful. The traditional material used for good castanets has been very hard (but not brittle) wood such as granadillo, rosewood, ebony, pomegranate, or oak. The best hardwoods come from equatorial forests, and they are becoming quite expensive; some people prefer not to use them out of a sense of environmental consciousness.

*The word castanet comes from castaña, the Spanish word for chestnut.*

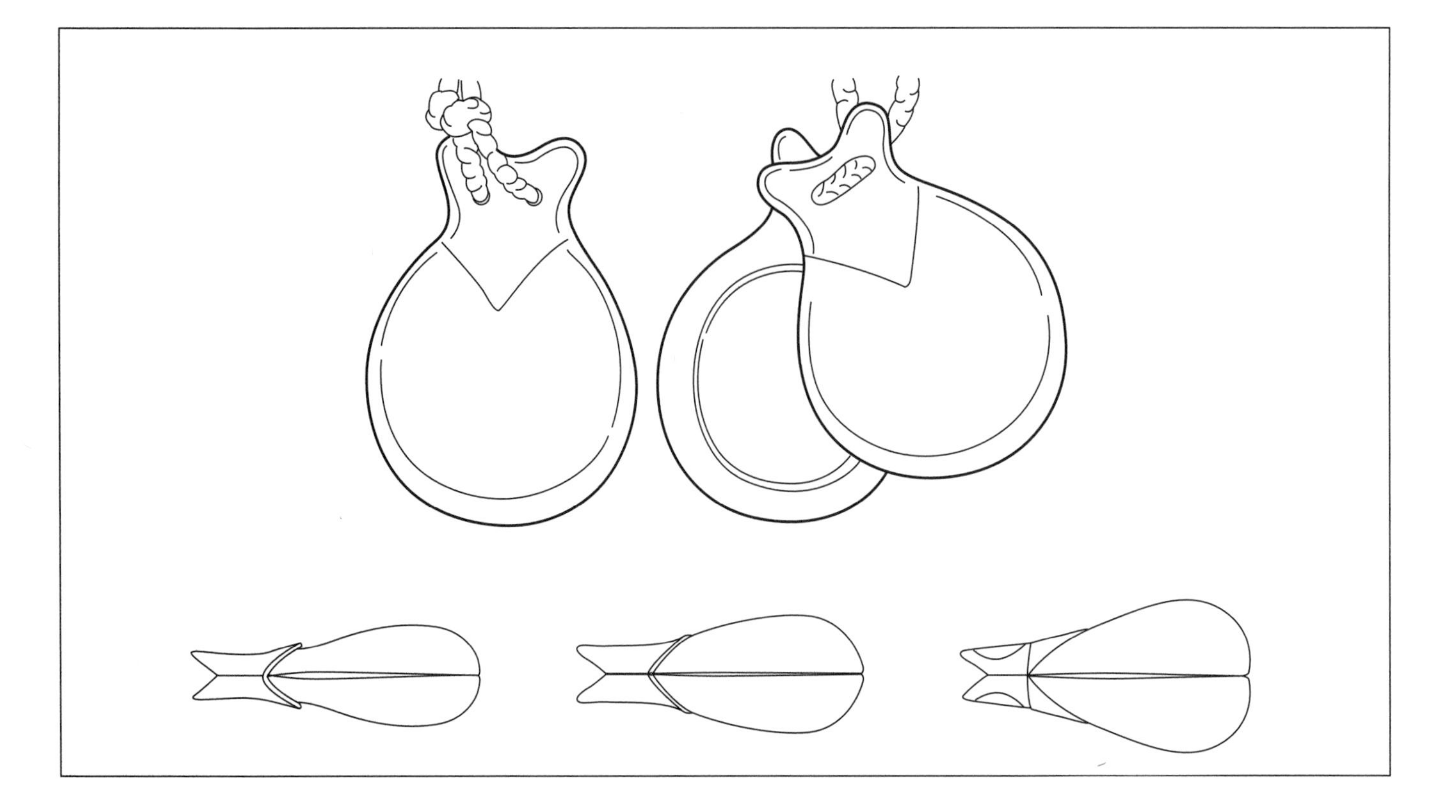

Castanets are commonly shaped like a clam shell that is circular or slightly oval, with an extension on one side for the hinge holes.

Most professional-quality castanets are currently made from a synthetic, laminated material such as Micarta. Called *tela de música* (cloth of music) by castanet makers, this material is made by applying heat and pressure to many layers of paper, cotton cloth, or glass-fiber cloth that have been impregnated with a phenolic resin. For the purposes of castanet making, this material is quite similar to high-quality ebony or granadillo.

Cotton strings are generally used to hinge together the two clappers of a pair of castanets. Nylon string can be used, especially by an orchestra musician; however, a dancer who keeps his or her arms up in the air while playing the castanets may find nylon strings too slippery.

## Design

Castanets are commonly shaped like a clam shell that is circular or slightly oval, with an extension on one side for the hinge holes. Occasionally, however, castanets are made in more novel shapes like squares, rectangles, or triangles.

Besides the overall size and the material from which they are made, another factor that influences the tone and sound quality of a pair of castanets is the size and depth of the hollows on the insides of the clappers. Also important is a good area of contact at the tips of the castanets, where the two clappers strike each other. Although some dancers prefer the two pairs of a set of castanets to have the same tone, it is traditional for one pair to produce a tone one-third lower than the other. This bass pair, called the *macho* or male pair, is played with the left hand to mark the beat of the music. The treble pair, called the *hembra* or female pair, is played with the right hand in a way that embellishes the music.

Several factors influence the degree of control a dancer has over the castanets when playing them in the classical manner. One is the exterior slope of the clappers—a steeper slope makes proper finger action easier. Other factors are the angle of the string holes and the curvature of the wood between the holes, where the bases of the two clappers rotate against each other when being played.

Castanets are sometimes used as a percussion instrument by orchestras rather than dancers. In this case, castanets may be mounted on wooden handles rather than being attached to the player's fingers or

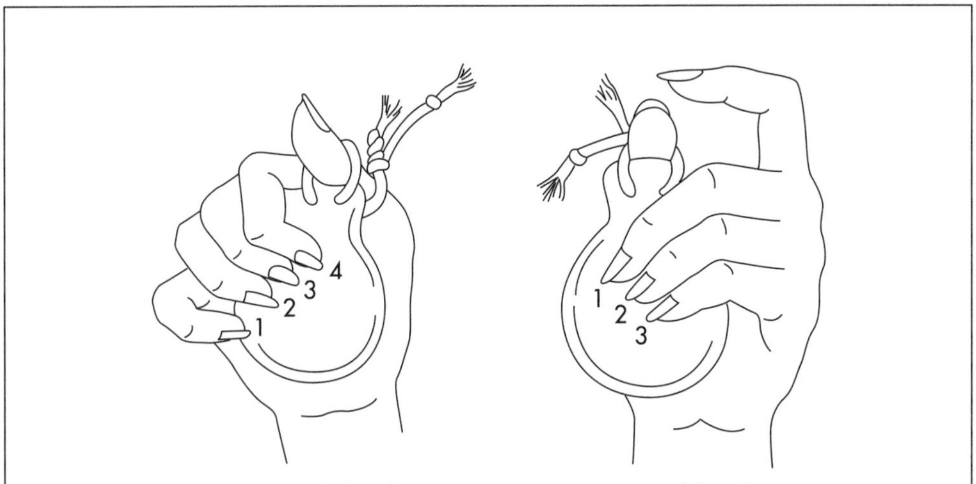

Right hand

Left hand

Castanets must be attached to the thumb of each hand by a string and the fingers must be lined up at certain points on the castanets in order to properly play the instrument.

thumbs. The musician can pick them up quickly by the handle and flick them in the air or strike them against his or her knees. Elastic cords are used to connect the clappers so they remain open while at rest. Sets of castanets can also be mounted on a base; the musician plays them by tapping on the upper clappers, sending them down against the lower clappers.

### The Manufacturing Process

Castanets may be either mass produced or individually hand crafted. The following steps describe the manual method, although a few comments compare this to mass production techniques.

1 A block of material approximately the size of the intended pair of castanets is sawed in half lengthwise to produce a blank for each shell. After placing a sheet of paper between centers of the halves, the pieces are glued back together.

2 Using either a standard pattern or one derived from a tracing of the intended user's hand, an outline of the shell-shaped clapper is traced on the block.

3 Using a bandsaw, the clapper is cut to the desired shape.

4 Two holes are drilled to accommodate the cord that will join the finished clappers together.

5 The two halves are separated, somewhat like opening an oyster.

6 Using hand tools and sandpaper, the outside of each shell is smoothed to its final shape.

7 The hollow is cut out of the inside of each shell. Depending on the preferences of the castanet maker, various tools may be used, such as a Carborundum wheel or a sanding ball.

8 Additional shaping involves sloping the faces of the clappers so that they contact each other only at the base and the lips. The base of the hinge area must also be rounded to allow the clappers to pivot properly.

9 When the two shells are practically finished, the maker tests them for sound quality. If necessary, they can be tuned by making the hollow deeper or broader.

In mass production, steps one through nine are largely automated. For example, a mechanical duplicator or a computer-programmed cutting tool may be used to create uniform exterior and interior curvatures according to a master design. Individual tuning is not necessary.

10 When the shaping has been finalized, the castanets are polished with jeweler's rouge. Depending on the preferences of the maker and the user, wood castanets may be treated with olive oil. Future oiling will not be necessary, as the castanets will be

continually conditioned by the natural oils of the user's hands.

11 If the set of castanets consists of a bass pair and a treble pair, it is customary to mark the treble pair. The mark is characteristic of the castanet maker—some use a notch, and some use a particular color and shape of paint marking.

12 A pair of clappers is joined with a 12 in (30.48 cm) length of cord thick enough to barely fit through the holes. The cord is threaded through the four holes so that both ends of the cord are on the outside of one of the clappers. Both ends of the cord are knotted separately to prevent fraying. A slipknot is formed in one end of the cord, and the other end is drawn through it. The user is able to adjust the tightness of the cord for proper operation of the castanets.

## Where to Learn More

### Books

Lalagia. *Spanish Dancing: A Practical Handbook*. London: Dance Books, 1985.

La Meri [Russell Meriwether Hughes]. *Spanish Dancing*. Pittsfield, MA: Eagle Print and Binding Co., 1967.

### Other

LP Music Group. http:/www.bongo.com/ (June 7, 1999).

*—Loretta Hall*

# Ceramic Filter

## Background

During many industrial processes, a filtering step may be required to remove impurities and improve quality of the final product. Depending on the process, the filter may be subjected to high temperatures and a corrosive environment. A filter material with good temperature and chemical resistance is therefore needed.

Ceramic filters meet these requirements and are finding use in a wide range of applications. One major application is filtration of molten metal during casting of various components. Another is diesel engine exhaust filters. The world market for molten metal filters exceeds $200 million per year.

The metal casting industry is the sixth largest in North America, contributing over $20 billion to the U.S. economy. About 13 million tons of metal castings are shipped every year, with 85% made from ferrous (iron) metals. Castings are used in over 80% of all durable goods.

In the casting process, a solid metal is melted, heated to proper temperature (and sometimes treated to modify its chemical composition), and is then poured into a cavity or mold, which contains it in the proper shape during solidification. Thus, in a single step, simple or complex shapes can be made from any metal that can be melted. Cast parts range in size from a fraction of an inch and a fraction of an ounce (such as the individual teeth on a zipper), to over 30 ft (9.14 m) and many tons (such as the huge propellers and stern frames of ocean liners).

Though there are a number of different casting processes, die casting is used for over one-third of all metal castings and contributes over $7.3 billion to the U.S. economy every year. This process involves injecting molten metal into a steel die under high pressure. The metal—either aluminum, zinc, magnesium, and sometimes copper—is held under pressure until it solidifies into the desired shape. Parts range from automobile engine and transmission parts; to intricate components for computers and medical devices; or to simple desk staplers.

The various casting processes differ primarily in the mold material (whether sand, metal, or other material) and the pouring method (gravity, vacuum, low pressure, or high pressure). All of the processes share the requirement that the materials solidify in a manner that would maximize the properties, while simultaneously preventing potential defects, such as shrinkage voids, gas porosity, and trapped inclusions.

These inclusions can be removed by placing ceramic filters in the gating system leading to the mold. Such filters must resist attack at high temperature by a variety of molten metals. These metals can contain such reactive elements as aluminum, titanium, hafnium, and carbon. Using these filters can reduce scrap rates by 40% and increase yields by 10% for manufacturing a wide range of parts made out of iron alloys, stainless steel, super alloys, aluminum, or other nonferrous alloys.

Molten metal filters generally come in two forms: a porous foam-like structure with interconnected pores that vary in direction or cross section, or an extruded porous cellular or honeycomb structure with cells of various shapes (square or triangular) and constant cross section. Though globally the most popular type of filter is foam, cellular filters

*The world market for molten metal filters exceeds $200 million per year.*

are used in 75% of applications in North America.

Filters can have either open cells or closed cells. Open cell (reticulate) filters consist of a network of interconnected voids surrounded by a web of ceramic and are widely used for molten metal filtration. Closed cell filters (foams) consist of a similar network but the beams are bridged by thin faces which isolate the individual cell. The open porosity in an open cell structure is critical in filter applications. The properties of a filter depend on both the cellular geometry (density, cell size) and the properties of the material. Advantages include high temperature stability and low weight.

The pore size of these filters are defined as cells or pores per linear inch (ppi). For honeycomb filters, this ranges from 64-121 ppi or 240 ppi. For foam filters, pore size is much more difficult to measure but generally ranges from 10-30 ppi.

Foam filters, which were first introduced over 20 years ago for nonferrous casting, are also used in direct pour units for casting steel. Inclusions that range from 0.125-2 in (0.3175-5.1 cm). or more in length and up to 0.25 in (0.635 cm) in depth can be removed. These inclusions come from molding materials, ladle refractories, and reoxidation during the pouring process.

Filtration occurs by mechanical interference, with large inclusions separated at the filter face and smaller inclusions trapped within the filter. Foam filters are able to trap inclusions significantly smaller than their open pore areas and can also remove liquid inclusions.

Thermal shock behavior (the resistance to sudden changes in temperature) for foam filters is dependent on their cell size, increasing with larger cells. Strength is initially retained after thermal shock and then gradually decreases with increasing quench temperature. A higher density may also improve thermal shock resistance.

## Raw Materials

The filter material is usually a metal oxide powder of various compositions. These include aluminum oxide, zirconium oxide, spinel (a combination of magnesium and aluminum oxides), mullite (a combination of aluminum and silicon oxides), silicon carbide, and combinations thereof. Ceramic fibers of various compositions may also be added to improve certain properties. Other additives include binders (alumina hydrate, sodium silicate), antifoaming agents (silicone or alcohol), and other chemicals to improve slurry properties. Water is usually used to make the ceramic slurry.

## Design

For optimal filter performance, a filter must be designed with the proper composition, pore size, and properties that match the specific application. Size and shape must be tailored to fit the mold system of the part being cast. Sufficient port area must be allowed so the filter does not choke the gating system during filtration. Filter area should be between three to five times the total choke area that the filter is feeding.

The major performance criteria when designing a filter are flow rate, filtering efficiency, hot/cold strength, slag resistance, thermal shock resistance, quality level, and cost. Each design is better at some than others, with significant design tradeoffs required in many cases.

## The Manufacturing Process

There are several methods used to make ceramic filters. The polymeric-sponge method, which will be described in more detail here, produces open-cell structures by impregnating a polymeric sponge with a ceramic slurry, which is then burned out to leave a porous ceramic. The direct foaming method can produce both open-cell and closed-cell structures, with the foam structure more common. In this method, a chemical mixture containing the desired ceramic component and organic materials is treated to evolve a gas. Bubbles are then produced in the material, causing it to foam. The resulting porous ceramic material is then dried and fired. For the honeycomb or cellular structure, a plastic-forming method called extrusion is used, where a mixture of ceramic powder plus additives is forced through a shaped die (like play dough). The cellular struc-

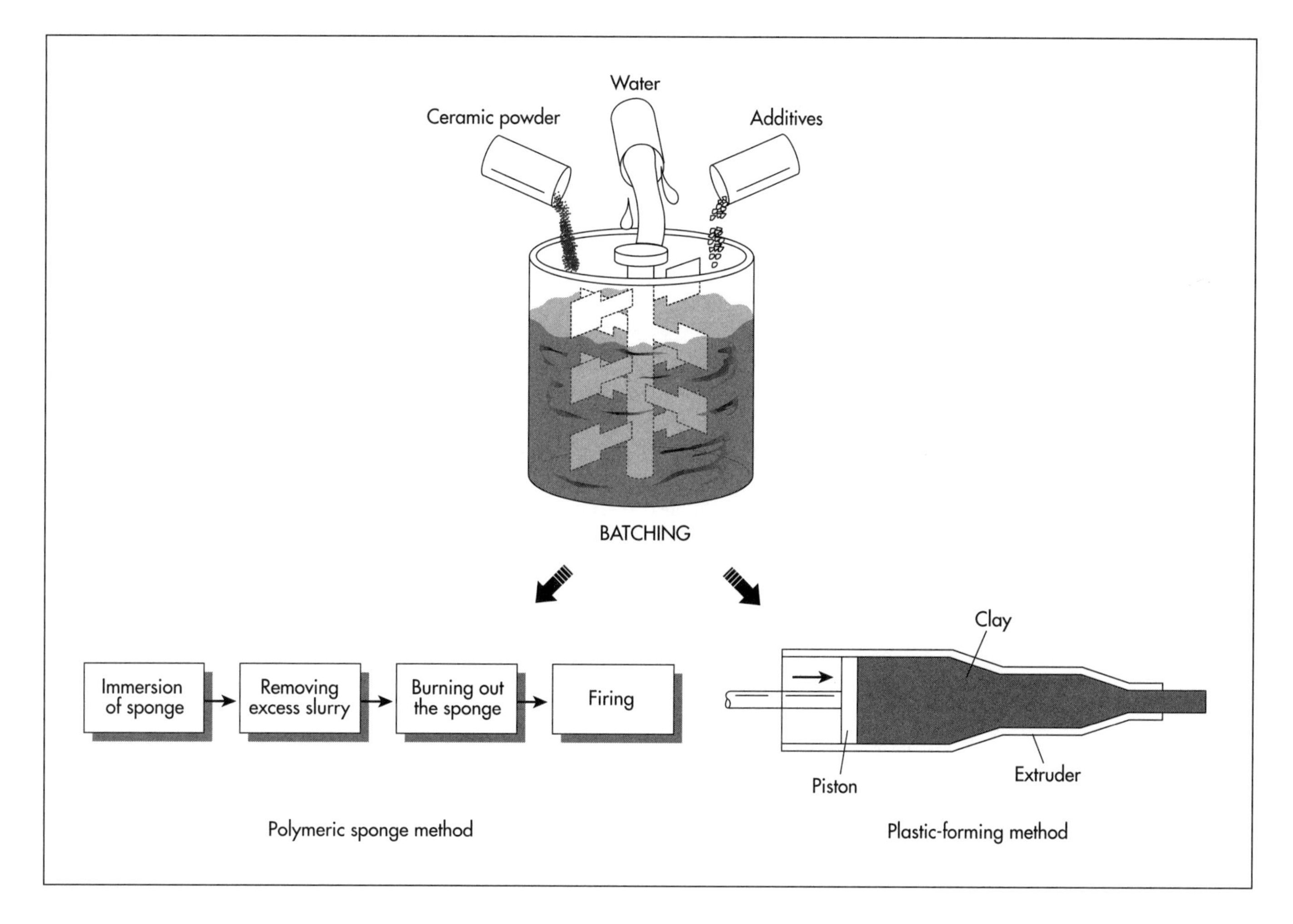


ture can also be produced using a pressing method.

### *Selecting the sponge*

1 First, a polymeric sponge must be selected with suitable properties. The pore size of the sponge determines the pore size of the final ceramic after firing. It must also be able to recover its original shape and convert into a gas at a temperature below that required to fire the ceramic. Polymers that can satisfy these requirements include polyurethane, cellulose, polyvinyl chloride, polystyrene, and latex. Typical polymeric sponges range in size from 3.94-39.4 in (10-100 cm) in width and 0.394-3.94 in (1-10 cm) in thickness.

### *Preparing the slurry*

2 After the sponge is selected, the slurry is made by mixing the ceramic powder and additives in water. The ceramic powder usually consists of particles less than 45 microns in size. The amount of water can range from 10-40% of the total slurry weight.

### *Immersing the sponge*

3 Before immersion, the sponge is usually compressed to remove air, sometimes using a mechanical plunger several times. Once it is immersed in the slurry, the sponge is allowed to expand and the slurry fills the open cells. The compression/expansion step may be repeated to achieve the desired density.

### *Removing excess slurry*

4 After infiltration, between 25-75% of the slurry must be removed from the sponge. This is done by compressing the sponge between wooden boards, centrifuging, or passing through preset rollers. The gap between rollers determines the amount removed. Sometimes the impregnated foam goes through another shaping step since it is still flexible.

Ceramic filters are manufactured in several different ways. The polymeric-sponge method produces open-cell structures by impregnating a polymeric sponge with a ceramic slurry, which is then burned out to leave a porous ceramic. In order to create a honeycomb or cellular structure, a plastic-forming method called extrusion is used, where a mixture of ceramic powder plus additives is forced through a shaped die (like play dough).

### *Drying*

5 The infiltrated sponge is then dried using one of several methods—air drying, oven drying, or microwave heating. Air drying takes from eight to 24 hours. Oven drying takes place between 212-1,292° F (100-700° C) and is completed in 15 minutes to six hours.

### *Burning out the sponge*

6 Another heating step is required to drive off the organics from the slurry and burn out the sponge. This takes place in air or inert atmosphere between 662-1,472° F (350-800° C) for 15 minutes to six hours at a slow and controlled heating rate to avoid blowing apart the ceramic structure. The temperature depends on the temperature at which the sponge material decomposes.

### *Firing the ceramic*

7 The ceramic structure must be heated to temperatures between 1,832-3,092° F (1,000- 1,700° C) to densify the material at a controlled rate to avoid damage. The firing cycle depends on the specific ceramic composition and the desired final properties. For instance, an aluminum oxide material may require firing at 2,462° F (1,350° C) for five hours.

## Quality Control

Raw materials usually must meet requirements regarding composition, purity, particle size, and other properties. Properties monitored and controlled during manufacturing are usually dimensional and then design specific. For foam filters, the weight of the filter must be measured to determine coating efficiency. Extruded filters are measured for density. Both parameters relate to strength properties.

## Byproducts/Waste

The manufacturing process is carefully controlled to minimize waste. In general, excess slurry cannot be recycled since it could change the purity and solid loadings of the original slurry, thereby affecting final properties.

## The Future

The metal casting market is expected to decline by 2.7% in 1999, mainly because of the weakening global economy, with total shipments expected to reach 14.5 million tons. Sales will increase slightly to $28.8 billion. Though casting shipments will continue to decline slightly in 2000 and 2001, over the long term, shipments are expected to reach almost 18 million tons in 2008, with sales of $45 billion. Shipments and sales will see 10-year growth rates of 1.7% and 4.75%, respectively.

The increased use of lighter-weight metal components, such as aluminum die castings, has spurred growth in the automotive sector. Today, there is an average of 150 lb (68.1kg) of aluminum castings per vehicle, an amount projected to grow to 200 lb (90.8 kg) per year by the year 2000.

Ceramic filters will continue to play an important role in producing quality castings and will follow the growth of the casting market. Dollar volume may decrease due to continued price reductions. Quality and productivity demands for metal castings are increasing the need for filters since they provide a fast and reliable way to obtain good castings. Thus, casting buyers are specifying "filtered" more and more often.

## Where to Learn More

### *Books*

Ishizaki, Kozo et al., ed.*Porous Materials, Ceramic Transactions*. The American Ceramic Society, 1993.

### *Periodicals*

"Metal pouring/filtering."*Foundry Management & Technology* (January 1996): C2-C6.

Outen, John. "Reduce defects with direct pour technology." *Foundry Management & Technology* (August 1996): 108-111.

Saggio-Woyansky, J., C. Scott, and W. Minnear. "Processing of Porous Ceramics."*American Ceramic Society Bulletin* (November 1992): 1674-1682.

*Other*

American Foundryman's Society. 505 State Street Des Plaines, IL 60016-8399. (800) 537-4237. (847) 824-0181. Fax: (847) 824-7848. http://www.afsinc.org/.

Hamilton Porcelains Ltd. Hamilton Technical Ceramics. 25 Campbell Street, Box 594, Brantford, Ontario, Canada N3t 5N9. (519) 753-8454. Fax: (519) 753-5014. http://www.hamiltonporcelains.com/.

Kirgin, Kenneth. "1999 Contraction to cause demand to dip to 14.5 million tons."*Modern Casting Online* (January 1999). Http://www.moderncasting.com/archive/feature_026_01.html/.

North American Die Casting Association. 9701 West Higgins Road, Suite 880, Rosemont, Illinois 60018-4721. (847) 292-3600. Fax: 847-292-3620. twarog@diecasting.org. http://www.diecasting.org/.


*—Laurel Sheppard*

# Cheese Curl

*An estimated 82% of American family households have eaten cheese curls.*

## Background

Cheese curls, sometimes referred to as corn curls or cheese puffs, have been a popular American snack food since the 1950s. These crispy cheese snacks are formed from cornmeal, water, oil, and flavored coatings. Cheese curls are an extruded snack, meaning they are cooked, pressurized, and pushed out of a die that forms the particular snack shape. They are then baked (or fried, depending on the product) and flavored with oil and seasonings. Today, we eat extruded corn snacks that are ball-shaped, curly, straight, or irregularly shaped depending on the shape of the die. Flavorings added after baking or frying vary greatly and different brands have distinctive flavors. These snacks are extremely popular snacks with children. The cheese curl industry packages them so that they are an easy choice to drop into school lunches.

## History

The invention of the cheese curl was quite serendipitous. During the 1930s, the Flakall Company that produced corn-based feed for livestock sought a way to produce feed that did not contain sharp hulls and grain dust and eventually produced a machine that broke the grain into small pieces by flaking it. The Flakall Company became successful manufacturers of flaked feed. One day as Edward Wilson was working as a flake operator at the Flakall Company, he noticed that workers poured moistened corn kernels into the machine to reduce clogging. He found that when the flaking machine ran continuously it made parts of it quite hot. The moistened cornmeal came out of the machine in puffy ribbons, hardened as it hit the air, and fell to the ground. Wilson took the ribbons home, added oil and flavor and made the first cheese curls. The company ran another flaker just for the production of Korn Kurls. By 1950, the Adams Corporation was mass-producing the Korn Kurl. There were dozens of small snack companies that followed the Adams Corporation and produced cheese curls, with many devising their special shape using innovative dies for their extruded snacks. Today, perhaps the most popular cheese snacks are produced by Frito-Lay although they did not offer any such snacks until 1980. This company offers the product in a variety of shapes and flavorings (including one that is flavored with cayenne pepper). Despite their minimal nutritional value (they are high in calories and fat and offer little fiber or protein), they are quite popular. It is estimated that 82% of Americans with families have eaten cheese curls at some point.

## Raw Materials

The cheese extrusion itself is generally made from two primary ingredients: cornmeal and water. All other ingredients are sprayed or applied to the corn curl after it leaves the extruder and is dried. The coatings vary greatly according to flavor and the manufacturer. Soybean and/or cottonseed and/or coconut oil may be sprayed on the extrusions. Then, powders such as cheddar cheese powders, acid whey powder, artificial cheese flavor, salt, and other spices are often applied over the oil.

## The Manufacturing Process

This section will focus on the production of baked cheese curls. The process for making

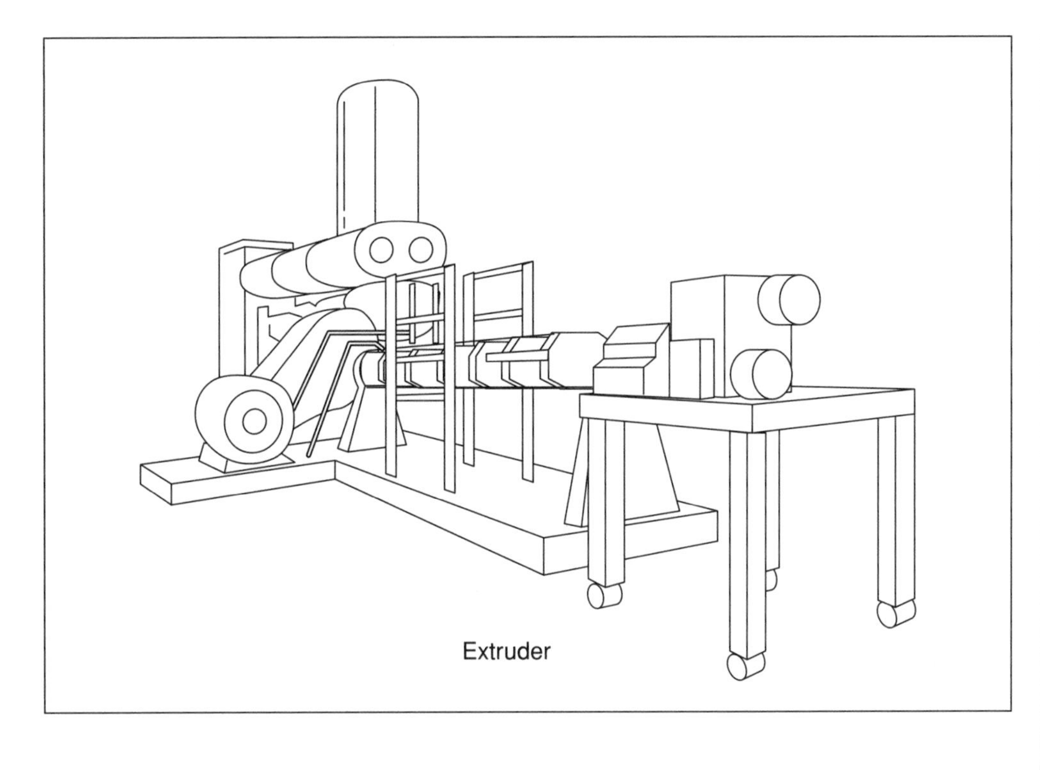


The primary step in the manufacture of cheese curls is the extrusion of the cornmeal mixture. Using a tapered screw, the extruder forces the mixture against the inside of the extrusion chamber, creating a shearing effect when pressure is increased. Steam jackets line the extrusion chamber to assist in cooking the meal mixture. When the cornmeal reaches the die it is hot, elastic, and viscous. The moisture is liquid under high pressure but changes to steam as it reaches lower pressure on the other side of the extrusion process. The result is that the cornmeal dough expands and puffs up as it moves through the die.

fried cheese curls differs only in that the product is fried after drying. In general, the manufacturing process for any extruded puff is relatively standard. It includes the heating of kernels of grain and then subjecting the meal to extremely high temperature and pressure. Starch in the mixture is gelatinized prior to its extrusion, and as the gelatinous mix is extruded, it is shaped and puffed.

### *Mixing*

1 Cornmeal is carefully chosen for use in puffed corn curls. The cornmeal must be of fine texture so that it produces a pleasant feel as it is eaten. Also, cornmeal with a low moisture content of 6-10% is used in this process. Too much moisture in the meal will render the product soggy. Cornmeal inspected for moisture content is placed into a mixing bin and sprayed with a fine mist of water as it is stirred. Water poured directly into the meal will result in gluten formation, uneven distribution of moisture, and clumping of wet grain. When the desired consistency is achieved, the mixture is immediately extruded. Holding the mixture too long often results in additional water absorption and the mixture cannot be used. If the cornmeal slurry cannot be used immediately, it must be stored in an airtight container.

### *Extruding the collettes*

2 The cornmeal mixture is loaded into the top of the extruding machine. The batch becomes rather gelatinous as it is exposed to heat, moisture, and pressure. The mixture is then propelled through the extruder using an auger or tapered screw. This screw forces the mixture against the inside of the extrusion chamber, creating a shearing effect when pressure is increased. Steam jackets line the extrusion chamber to assist in cooking the meal mixture. When the cornmeal reaches the die it should be hot, elastic, and viscous. The moisture is liquid under high pressure but changes to steam as it reaches lower pressure on the other side of the extrusion process. The result is that the cornmeal dough expands and puffs up as it moves through the extrusion dies.

The shape of the die is a critical aspect of its product because it gives the product its distinctive shape. As the dough is pushed through the dies, it looks like puffy snakes. These extruded snakes called collettes are cut to the desired length by a rotating knife.

### *Drying the collettes*

3 The collettes still contain between 6-10% water at this point (depending on the recipe and the manufacturer) and must be dried out. So, the collettes are convey-

The extruded cheese curls can be seasoned using a flavor reel. In this process, the oils, flavors, spices, and color are mixed together in a tank and sprayed on the curls as they are tumbled in a barrel. Once flavored, the curls are dried and packaged.

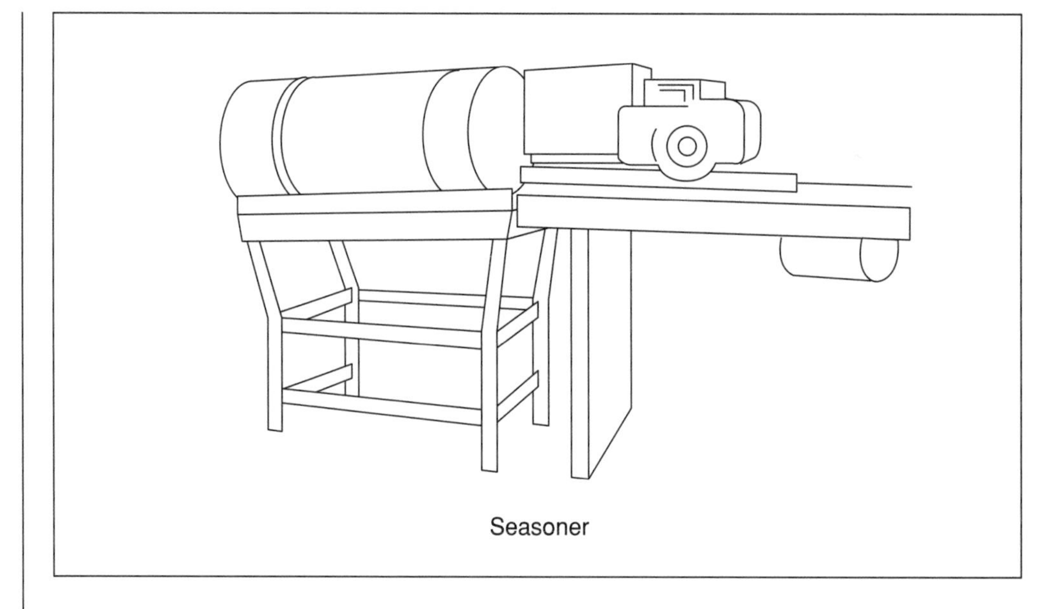


ored to a large oven called a dryer. They move continuously through the dryer until they emerge at the other end crisped up considerably. The dryer generally is about 140° F (59.9° C)—hot enough to get the moisture out but not to darken the collette. The collettes require about five minutes to dry out and are reduced to between 1-2% moisture.

### *Separating the fines*

4 The fines or small particles that are produced as the collettes are conveyed to the dryers must be separated from the nicely-formed collettes. (The fines tend to absorb oil and flavorings and are undesirable to include in a package of cheese curls. Some modern dryers are self-cleaning and automatically deposit the fines in a cross conveyor for removal from the machine.

### *Flavoring the collettes*

5 As the collettes move through the oven, they are sent to the flavor coating station. They are still bland cornmeal crisps and must receive a coating in order to be palatable. They may be coated using one of two methods. In the first method, the extruded corn curls are first sprayed with vegetable oil and then dusted with a variety of dry flavors, seasonings, and color. In the second method, the oils, flavors, spices, and color may be mixed together in a tank and sprayed on the collettes as they are tumbled in what is called a flavor reel. In either process, it is essential that oil is placed on the surface of the collette in order for the flavors and spices to stick to the snack. The flavored snacks now sit on a conveyor and dry before they are packaged.

### *Packaging the cheese curls*

6 The cheese curls move along the conveyor belt and are dried. A vibrating conveyor belt moves the snacks along until they fall into the weigher. The weighing machine weighs out just the right amount of curls to go into a bag and puts them into a chute. (Cheese curl bags may be made of polypropylene which are resistant to moisture and keep the product crisp.) Huge rolls of flattened bags are loaded into the weighing machine, formed, and prepared to receive the corn curls that drop into the bags. The bags are then heat sealed. Cartons of bags are packed and shipped to a warehouse.

## Quality Control

One of the most important quality control steps in the entire process is selecting cornmeal for use. First, it must be finely ground because big grains or gritty meal result in an undesirable feel as the finished product is consumed. Cornmeal is tested for grit size by weighing 1.75 oz (50 g) of a well-mixed, representative sample of meal. Then, the test sample is transferred to the top sieve of a series of different grades of sieves. The grain is poured on top and agitated as it is put through the sieves. The grain remaining on the sieves is considered too coarse to use;

this amount is weighed and the approximate usable grain is calculated and compared to specifications sent from the supplier. Next, moisture content of the grain is assessed because too-wet or too-dry cornmeal will create inferior product. (Excessive moisture will cause formation of small, heavy, hard puffs, while too-dry cornmeal will create light, long, straight puffs. Puffs that are too dry will burn in the dryer.) Moisture content is determined by weighing a 0.07-0.105 oz (2-3 g) portion of cornmeal, placing it a uncovered in the self of an oven for exactly 60 seconds, then re-weighing the samples and calculating the amount of moisture loss. The meal must not lose more than 0.2% of weight (moisture) of the cornmeal may be considered too wet for use at that point.

The equipment used in the manufacture of the extruded snacks is calibrated and checked very carefully throughout the production process. Industry and food technology manuals indicate the proper rate of feed into and through the extruder, the number of revolutions per minute at which the extruder must run, the temperature of the extruder, the pressure (measured in Atmospheres) that must be maintained in the extruder, the speed of the extruder, and the rotating speed of the knife that cuts the collettes. Human operators constantly check the extruding dies to ensure they have not plugged up. Extremely important is determining the amount of water that must be added to the cornmeal in order for it to become gelatinous. The moisture content of the product as it enters the extruder, as well as when it exits (just before the baking that crisps it up), is carefully assessed. Machinery is programmed so that the duration of the processes described above are perfectly timed in order to ensure each step is thoroughly completed.

## The Future

The puffed corn snack is a recently developed manufactured product. As such, it has seen many interesting variations and improvements in the past two decades in particular. Companies are constantly seeking new flavors to spray onto the extruded cornmeal collettes. Recently, very spicy coatings have been added to one company's product and they are selling well. One company that sells extruding machines and dies advertises that they look forward to working with food manufacturers in order to develop new, unique product. Furthermore, it is possible to extrude other grains in the extruder and perhaps the snack food industry will see how well wheat or rye puffs may sell in the near future.

## Where to Learn More


### Books

Herbst, Sharon Tyler. "Cornmeal," In *The Food Lover's Second Edition*. Barron's Educational Services, Inc.

Snack Food Association. *Corn Quality Assurance Manual*. Alexandria, VA: Snack Food Association, 1992.

Snack Food Association. *Corn Quality Assurance Manual*. Alexandria, VA: Snack Food Association, 1994.

### Other

Frito-Lay. 1998. http://www.fritolay.com/ (June 7, 1999).

Snack Food Association. http://www.snax.com/ (June 7, 1999).

Wenger Extrusion Systems.1999. http://www.wenger.com/ (June 7, 1999).


*—Nancy EV Bryk*

# Chicken

*In the 1980s, 50% of all chicken in the United States was purported to be infected with salmonella. According to the USDA, the industry altered its quality control procedures, and brought the incidence down to 16% in 1996, and to below 10% in 1998.*

Chicken in the United States is a cheap and readily available meat. It is packed in a variety of formats, from whole roasting chickens to selections of one particular cut, such as thighs or wings. Highly automated, large-scale chicken farming and processing complexes run by large corporations fuel the American chicken market. The development of so-called factory farming sharply reduced the price and increased the availability of chicken, when this method was introduced in the1920s.

## Background

The ancestor of today's domestic chicken is the wild red jungle fowl *Gallus gallus*, native to India and Southeast Asia. The red jungle fowl was first domesticated apparently for use in religious rituals involving cockfighting. The domesticated bird spread westward from India to Greece, and was later introduced to Western Europe by invading Roman armies. By the Roman era, chickens were used as food, both for their meat and for their eggs. Romans commonly carried them on their ships, as a convenient source of fresh food.

The first European settlers in North America brought chickens with them. But until the twentieth century, there was no chicken industry as such in this country. Care of the chicken flock was for the most part considered work for women and children. At that time, a typical hen laid only 30 eggs a year, and farm wives sold their excess at market as supplemental income. Chicken meat was usually only plentiful in the early summer, when chickens that had hatched in the spring were big enough to eat. Because chicken husbandry was primarily women's work, only as an adjunct to the major farm production, distribution channels were limited. Whereas railroads were built to bring cattle from the West to waiting urban markets, no such effort was put into chicken production, and chicken was available in cities more or less sporadically, with large seasonal jumps in prices and amount of supply.

Several inventors perfected chicken incubators in the late nineteenth century. These machines could keep hundreds of eggs at a time warm, and so made possible commercial breeding of chicks. In the nineteenth century, breeding of chickens was mostly a hobby, with many poultry enthusiasts raising fabulously feathered chickens. Showy and colorful exotic breeds were the most popular; however, with the advent of mechanical incubators, poultry breeders began to breed birds with good egg-laying and meat production potential.

The first person in the United States to raise broiler chickens (chickens for meat) on a large scale strictly for profit was a Mrs. Wilmer Steele, of Ocean View, Delaware. In 1923, Mrs. Steele bought 500 chicks and sold the surviving 387 of them when they matured to 2 lb (0.9 kg). Her profit was enormous, and within just a few years, Delaware became the center of a thriving chicken industry. In 1926, the state produced around one million broiler chickens.

By 1934, it was raising about seven million chickens annually. In the 1930s, the National Poultry Improvement Plan, a federal-state cooperative mission, helped chicken farmers use scientific breeding principles to produce superior strains of birds. At this time, birds were first bred specifically for meat production. The important qualities of broiler chickens were rapid growth, white feathers (dark feathers left unsightly stubs), and meaty breasts and thighs. The advances in breeding made quite an impact: in 1900, a typical chick took 16 weeks to reach 2 lb (0.9 kg), which was considered frying weight. Today, a commercial broiler chicken lives only about six weeks, and weighs about 4 lb (1.8 kg) at slaughter.

Advances in nutrition were also important to the development of a commercial chicken industry. Chicken nutrition has actually been studied more, and is better understood, than human nutrition. The combined efforts of the feed industry, the U.S. Department of Agriculture, and agricultural scientists led to optimum feed. The ratio of feed necessary per pound of chicken meat has fallen through this century, making chicken ever cheaper to produce. By the 1950s, several large companies had integrated feed production with chicken farming and meat processing, so that only a few large corporations controlled a high percentage of the chicken produced in this country. These major producers each slaughter millions of chickens a week.

## Commercial chicken production

### The production complex

Chicken production is typically carried out at so-called complexes. Each complex contains a feed mill, a hatchery, a processing plant, and chicken farms where the chicks are raised, usually in a 30-40 mi (48.3-64.4 km) radius from the processing plant. Contract farmers receive chicks from the hatchery, and house them in climate-controlled chicken houses. The houses are typically 400 x 50 ft (122 x 15.24 m), and hold up to 20,000 chickens. The interior is open, with no cages or partitions. When the chickens are old enough for slaughter, they are collected and shipped to the processing plant.

### The hatchery

1 Broiler chickens are bred especially for meatiness, quick growth, and weight gain. Most chickens used for meat in this country are a hybrid of Cornish males and White Rock females. The hatchery houses a flock of thousands of chickens. The hatchery building is a large open space similar to the house where broilers are raised, except it contains many small houses set inside it, which look like miniature versions of the traditional chicken coop. When the hens are ready to lay, they seek shelter in the coop. The eggs are collected from the coops and taken to incubate. The breeder hens live for about 45 weeks, after which they are no longer considered productive. These "spent" hens are slaughtered and their meat is usually used for pet food or bought by food companies that use cooked, diced meat (such as in soups).

### Incubation

2 The eggs are placed in large walk-in incubators. The eggs are kept warm and periodically rotated by machine. They begin to hatch in about 20 days. Shortly before hatching, the eggs are transferred to drawers. Many processors now inoculate chicks for diseases in ovo, that is, in the shell before they hatch. This is usually done three days before hatching. The chicks peck their way out of their shells when they are ready. For their first several days of life, the chicks are still absorbing nutrients from their yolk sacs, so they do not need food at this time. Trays of newly hatched chicks are wheeled on carts to an inoculation area, where they are sprayed with a mist of vaccine against common diseases. Some producers "debeak" the chicks at this point, which actually means clipping the sharp tip off the beak. This prevents the birds from damaging each other by pecking. This practice was discontinued at some large producers in the late 1990s, as for the most part the growing chicks are not overly aggressive, and debeaking was deemed costly and unnecessary. Next, the chicks are shipped to the nearby "grow-out" farms.

### Growing out

3 The chicks live in large houses which hold as many as 20,000 birds. These grow-out houses are kept at about 85° F (29.4° C) through heating and ventilation controls. The birds are not caged, and typically they are provided with approximately 0.8 sq ft per bird. The floor of the house is covered with a dry bedding material such as wood chips, rice hulls, or peanut shells. The birds are fed a diet of chicken feed, which is typically 70% corn, 20% soy, and 10% other ingredients such as vitamins and minerals. Broiler chickens in the United States are not

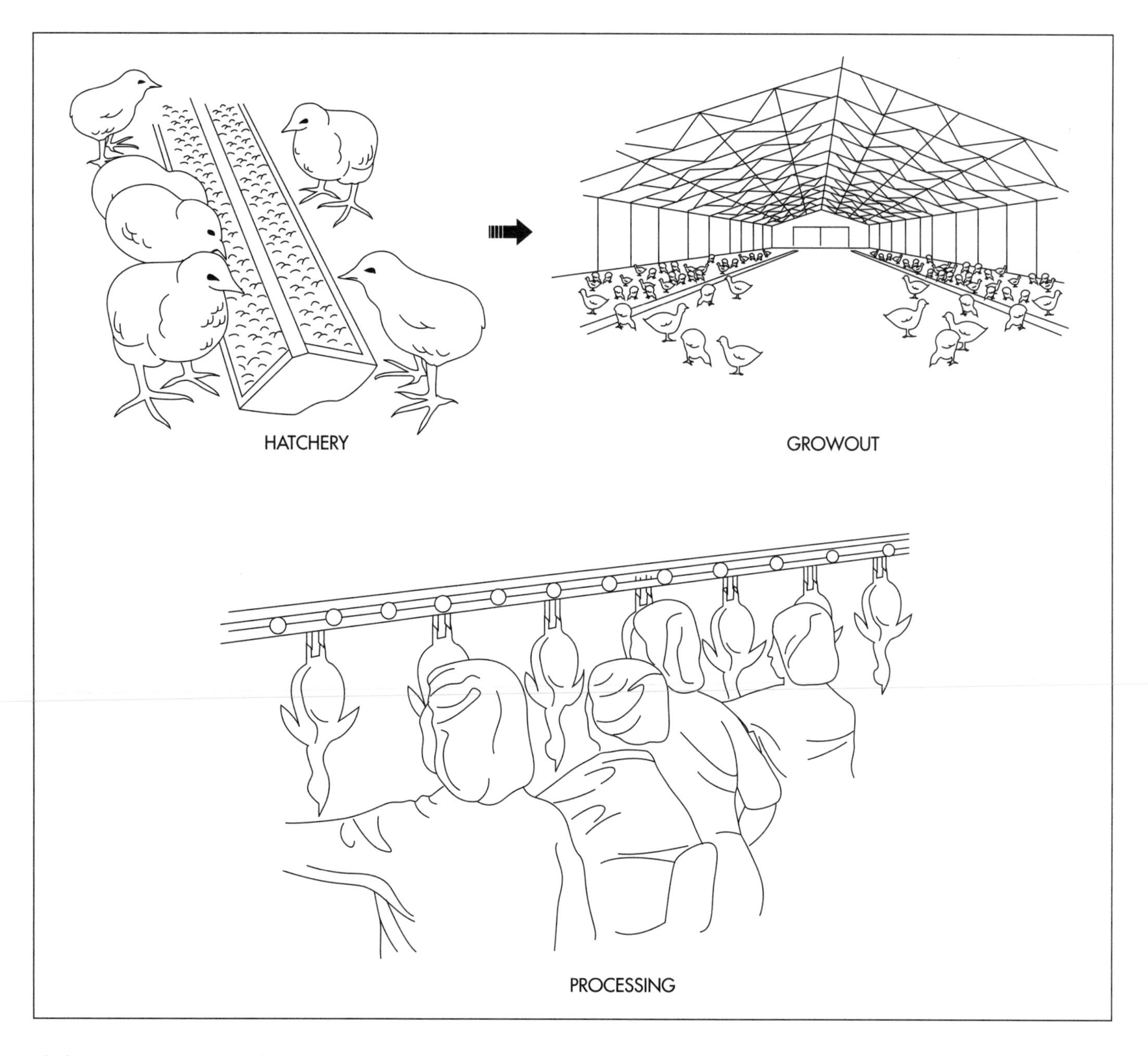


Chicken processing begins at the hatchery where hens lay eggs. The eggs are collected and incubated until they begin to hatch in about 20 days. The chicks live in large, growout houses where they are fed a diet of chicken feed. After growout, the birds are conveyored through a stun cabinet. The mild electrical current in the water stuns or paralyses the birds. Next, the birds are conveyed to an automatic neck cutter. The carcasses hang until all the blood has drained and then they are defeathered. Next, they are washed, cleaned, and immersed in cooled, chlorinated water for 40-50 minutes.

fed any steroids or hormones. Sick birds are treated with antibiotics or other medications. These birds then go through a withdrawal period before slaughter, to make sure no medication residue remains in their meat. The birds are usually watered through nipple drinkers, so that they don't spill and wet their bedding.

### Collecting

4 The chicks live in the growing-out houses for about six weeks. Broiler chickens have been bred for excessive weight gain, especially in their breasts and thighs. At six weeks, the chicks usually weigh about 4 lb (1.8 kg), and are ready for slaughter. Collecting of the chickens is usually done at night. Though a variety of mechanical collectors have been developed, such as vacuum devices and plow-like chicken pushers, the simplest and most effective way to get the chickens crated for transport to the processing plant is to have farm employees enter the house and gather the birds by hand. The workers catch the birds and stuff them into cabinet-like boxes. The boxes are stacked, and a driver with a forklift picks them up and loads them onto a waiting truck. The boxed chickens are stacked in the truck and driven to the processing plant. The processing center of the chicken complex is generally no

more than 30-40 mi (48.3-64.4 km) from the grow-out farm, so that the birds do not have to be driven an excessive distance.

### *Slaughter*

5 At the processing plant, workers take the birds from their boxes and hang them by their feet on a conveyor belt. In a typical process, the birds on the conveyor are first passed through a vat of electrified salt water called a stun cabinet. About 20 birds occupy the stun cabinet at one time, and they remain in the water for about seven seconds. The mild electrical current in the water stuns or paralyses the birds. Next, the birds are conveyed to an automatic neck cutter—rotating blades that sever the two carotid arteries. The birds' carcasses hang until all the blood has drained.

### *Defeathering and evisceration*

6 The carcasses are then briefly immersed in hot water to scald the skins. This makes removal of the feathers easier. The carcasses move to automatic feather pickers, which are moving rubber fingers that rub off most of the feathers. Then the carcasses are scalded a second time and run through another feather picker. Lastly, a specialized machine removes the wing feathers. The defeathered carcasses next pass to a washer, which scrubs the outside of the body. The feet and head are cut off, and the carcass is conveyed to the evisceration area. Next, the carcass is suspended in shackles by the feet and neck, cut open, and the viscera (internal organs) are removed. When the carcass is empty, it is washed again inside and out by a multiple-nozzled sprayer.

### *Chilling and cutting*

7 The cleaned carcasses are sent down a chute and immersed in a "chiller" of cooled, chlorinated water for 40-50 minutes. The entire slaughter process takes only about an hour, and the bulk of that time is taken up by the chilling. The internal temperature of the chicken must be brought down to 40° F (4.4° C) or lower before further processing. The chilled carcasses are then passed to a cutting room, where workers cut them into parts, unless they are to be packaged whole. Some carcasses may be cooked and the cooked meat removed and diced for foods such as chicken pot pie or soups. Meat from backs, necks, and wings may be processed separately for sale in other meat products such as hot dogs or cold cuts. In whatever format, the meat is packaged by workers at the processing plant, loaded into cases, and stored in a temperature-controlled warehouse.

## Quality Control

Quality control is a particularly important issue in poultry farming because the end product is raw meat, which has the potential to carry disease-causing microorganisms. To prevent diseases in the chickens themselves, the chicks are vaccinated for common avian diseases. Veterinarians visit the growing-out farms and tend to any sick birds. Corporations that contract with the growing-out farms also typically send a service technician out on a weekly visit to each farm to monitor conditions.

Quality control at chicken-processing plants is done by the company and also by inspectors from the U.S. Department of Agriculture. A USDA inspector is required to be in the plant whenever chickens are being slaughtered. The government inspector examines the birds both before and after slaughter for obvious signs of disease and for injury, such as broken wings. The meat from injured parts is not usable.

In a typical process, there are two critical control points where the company continually monitors conditions. There may be additional control points as well. The first critical control point is just before the cleaned carcass goes to the chiller. An inspector pulls carcasses at random and visually inspects them under bright light. No fecal matter is allowed on the carcass at this point. If any is found in the random check, the production line must be stopped and all the birds that have gone through the chiller since the last inspection must be rewashed and chilled. The second critical control point is when the birds come out of the chiller. The internal temperature of the carcass must be 40° F (4.4° C) or lower at this stage. Inspectors make random sample checks to verify internal temperatures. Though these are the most important control points, each plant designs its own quality control program, and inspectors may also periodically verify the temperature of the scalding water, check the automatic equipment, and whatever else the company deems necessary.

Until 1998, USDA inspectors at chicken processing plants were required to do only what is called an organaleptic test of the chickens before and after slaughter. This translates to looking and smelling; that is, inspectors verified that the birds were disease-free and healthy by looking them over and perhaps giving the carcass a quick sniff.

In 1998, the USDA instituted a new quality control program for all meat processors known as Hazard Analysis Critical Control Points, or HACCP. Under HACCP, in addition to the organaleptic method, inspectors are also required to take periodic microbiological tests to look for dangerous bacteria. The most problematic bacteria in chicken meat are salmonella. Though this organism is killed with proper cooking of the meat, it can cause illness if the consumer does not handle the meat properly. In the 1980s, 50% of all chicken in the United States was purported to be infected with salmonella. The industry altered its quality control procedures, and brought the incidence down to 16% in 1996, and to below 10% in 1998, according to the USDA. Under HACCP, chicken must be randomly tested for salmonella at the production plant, and the rate of infection must be lower than 20%. Also under HACCP, USDA inspectors have the authority to shut down plants that they deem dirty or unsafe. The plant is not allowed to re-open until it comes up with a plan for remedying the situation. Some incidents that caused chicken processing plants to be shut in 1998 included carcasses falling on the floor, rodent infestation of the facility, and most commonly, failure to prevent fecal contamination.

## Byproducts/Waste

Many of the byproducts of chicken slaughter can be used. Chicken feet are removed at the processing plant because they are not considered edible in the United States. However, chicken feet are a delicacy in Asia, and so large amounts of them are exported. The feathers can be ground up and used as a protein supplement in animal feed. Substandard meat is also commonly sold to pet food makers. However, many chickens die before slaughter, either at the growing-out farm or en route to the processing plant. These birds are disposed of in landfills. Sick or deformed chicks are culled—taken from the flock and killed (usually by wringing the neck)—after hatching, and these bodies must also be disposed. Unused viscera and parts also produce waste in chicken processing.

A significant waste produced in chicken farming is the feces of the birds. Because the flocks are so large, with 20,000 birds typical for a broiler growing-out farm, the amount of feces is enormous. Decomposing poultry manure produces ammonia, an irritating gas that can cause disease and distress in poultry workers and in the chickens themselves if chicken houses are not adequately cleaned and ventilated. Flies are attracted to chicken manure, and large-scale broiler farming may cause an unwelcome increase in the fly population in surrounding areas. The odor associated with large-scale chicken farming can also be a problem for neighbors. Of more concern than odor is the threat to water quality by run-off from chicken farming. Some chicken manure is used as fertilizer for crops, and when it rains, excessive nitrogen and phosphorus are washed into nearby bodies of water. Outbreaks of a harmful bacteria in the Chesapeake Bay area in 1997 were blamed on water conditions caused by run-off from chicken farms. To control run-off, chicken producers may opt to alter the feed they give their broilers, adding enzymes that help breakdown some of the nutrients in the waste.

## Where to Learn More

### *Books*

Davis, Karen. *Prisoned Chickens, Poisoned Eggs* Summertown, Tennessee: Book Publishing Company, 1996.

Smith, Page, and Daniel, Charles. *The Chicken Book.* Boston: Little, Brown, 1975.

### *Periodicals*

Gordon, John Steele. "The Chicken Story." *American Heritage* (September 1996).

"Poultry Growers Unite to Address Waste Issue." *New York Times* (August 25, 1998).

Sharpe, Rochelle. "U.S. Shut 34 Meat and Poultry Plants in Quarter Due to Sanitation or Safety." *Wall Street Journal* (May 8, 1998).

*—Angela Woodward*

# Child Safety Seat

*In the United States, more than 2,000 children under 14 years of age die each year in vehicle crashes.*

## Background

In the United States, more than 2,000 children under 14 years of age die each year in vehicle crashes. Not only are vehicle crashes the leading killers of children, in 1997 they also injured nearly 320,000 youngsters. Most of the fatalities happen because the children aren't secured in specially designed car seats. Instead, they are strapped in the wrong kinds of seats, such as adult seats that neither position the children's bodies properly nor cushion them against impact, or they aren't wearing any restraints at all. Children aged four to 14 are least likely to be restrained properly.

Most of the 50 states require that children under the age of four be secured in child safety seats or seat belts. Many states also mandate booster seats (or seat belts) for kids between four and 14 depending on their age, weight, and height. Up to 95% of the safety seats that are installed in vehicles may not be the right seat for the child, may be hooked so loosely with an incompatible belt in the car that the seat rotates or pitches forward, or may have harnesses incorrectly fastened in some way. Child seats are also often incorrectly placed rear-facing in front of air bags. In 1997, six out of 10 children who were killed in vehicle crashes were unbelted. Enormous energies and talents have produced child safety seats that, when used properly, have vastly improved the likelihood that a child can survive a serious vehicle crash without injury.

## History

Car seats for children have been manufactured since 1933. The Bunny Bear Company made several designs of children's car seats, but their purpose was not to protect the child in the event of an accident. Instead, these seats confined the children, raised them above the level of the passenger seat, and made them more visible to adults from the front seat. The true safety seat for children was invented in England by Jean Ames in 1962. The Ames design had straps that held the padded seat against the rear passenger seat. Within the seat, the child was restrained by a Y-shaped harness that slipped over its head and both shoulders and fastened between the legs. Other designs to accommodate growing children followed quickly over the next several years.

From 1956-1970, lap-type seat belts were developed and became standard equipment for adults. Even though crash-test results proved that seat belts saved lives, in the 1960s seat belts were met with resistance. In 1966, Congress passed the Twin Highway Acts that empowered the Department of Transportation (DOT) to set standards for vehicle design; the separate states have the authority to enforce driver laws, which vary among the states. Shoulder harnesses, self-applying belts, and front- and side-impact air bags increased the level of protection for adults; these types of safety equipment became available from 1966-1995. During this period, child safety seats grew to include rear-facing seats for infants under 22 lb (10 kg), convertible seats that start as rear-facing infant seats and convert to face forward for toddlers weighing less than 40 lb (18.2 kgs), and booster seats that elevate a growing child weighing between 30-70 lb (13.6-31.8 kg) so the car's seat belt can be fastened around child and booster seat. Child safety seats are in the news almost daily, yet more than 30 years after their development, many of these news items focus on the inconvenience of the seats rather than the lives they save.

## Raw Materials

The child safety seat is made of polypropylene, a tough plastic that flexes under pressure and doesn't crack easily the way some other plastics do. The plastic is transported to the factory in the form of pebble-sized pellets; a major car seat maker receives train cars full of the pellets for its production. Coloring, which varies among the seat models, is added to the plastic.

Several of the smaller components like buckle latch plates, harness adjusters, locking clips, and the buckles themselves are usually made by specialty manufacturers. Indiana Mills makes buckle and adjustment mechanisms for almost all American manufacturers of child safety seats.

Fabrics and vinyls are used to make safety seat covers and harnesses. Both covers and harnesses have to be able to withstand flammability tests, but they also have to be washable. Only mild soap and water can be used because detergents or chemicals break down the flame resistant fabric treatments. The thread that sews these materials together has to meet the same requirements. Color, durability, and fashion are other considerations in selecting the cover fabrics. The harnesses meet the same strength requirements as those for adult seat belts.

The seats are padded with foam. Types of foams are heavily regulated to meet standards for flame resistance and energy absorption. Pads and covers are sewn by the child safety seat manufacturer or by outside suppliers.

Printed paper components are among the most important. Labels are designed by the manufacturer in accordance with Federal standards. Locations of labels on the device, precise wording, and paper that withstands tears (so missing information is evident) are among the specifications. Instructions are also prepared to meet exacting requirements, and the child safety seat must include permanent storage for the instructions. Physical printing of labels and instructions is usually done by printers subcontracted by car seat manufacturers.

Assembly also includes small parts such as rivets and fabric fasteners. These parts are all manufactured and supplied by specialty firms.

## Design

A wide variety of design issues must be considered for each type of child safety seat. The four most important design issues are safety (including meeting government regulations), ease of use (and this includes the child's comfort), style or appearance, and manufacturing feasibility. A strict set of government regulations (Federal Motor Vehicle Safety Standard [FMVSS] 213) establishes the seat back height, buckle release pressure, type of impact-absorbing foam, the wording of some labels, and much more. The seat also must be able to withstand a crash test of either two cars each traveling at 30 mph (48 kph) hitting each other head-on or a car crashing into a parked car at 60 mph (96 kph). After the crash test, the seat must still meet certain performance criteria like buckle release pressure.

Two major changes in regulations are radically altering safety seat design. The distance the child's head is allowed to move forward in a crash (called the "head excursion") has been reduced from 32 in (81 cm) to 28.5 in (72 cm). To meet this requirement, most seats will probably need a tether or strap on the top of the child seat that will be anchored behind the car seat. The shells of many designs of child safety seats will also have to be reinforced to meet this standard (and vehicles will also require anchor points for the tethers).

Another new regulation becomes effective in 2002. Special anchor attachments will be added to safety seats that will secure them to new anchors in the vehicles and reduce the problem of using the safety belt to secure the child seat to the car seat. This "universal attachment system" uses attachment belts in different locations on the safety seat than current designs, so the safety seat will require considerable redesigning and strengthening.

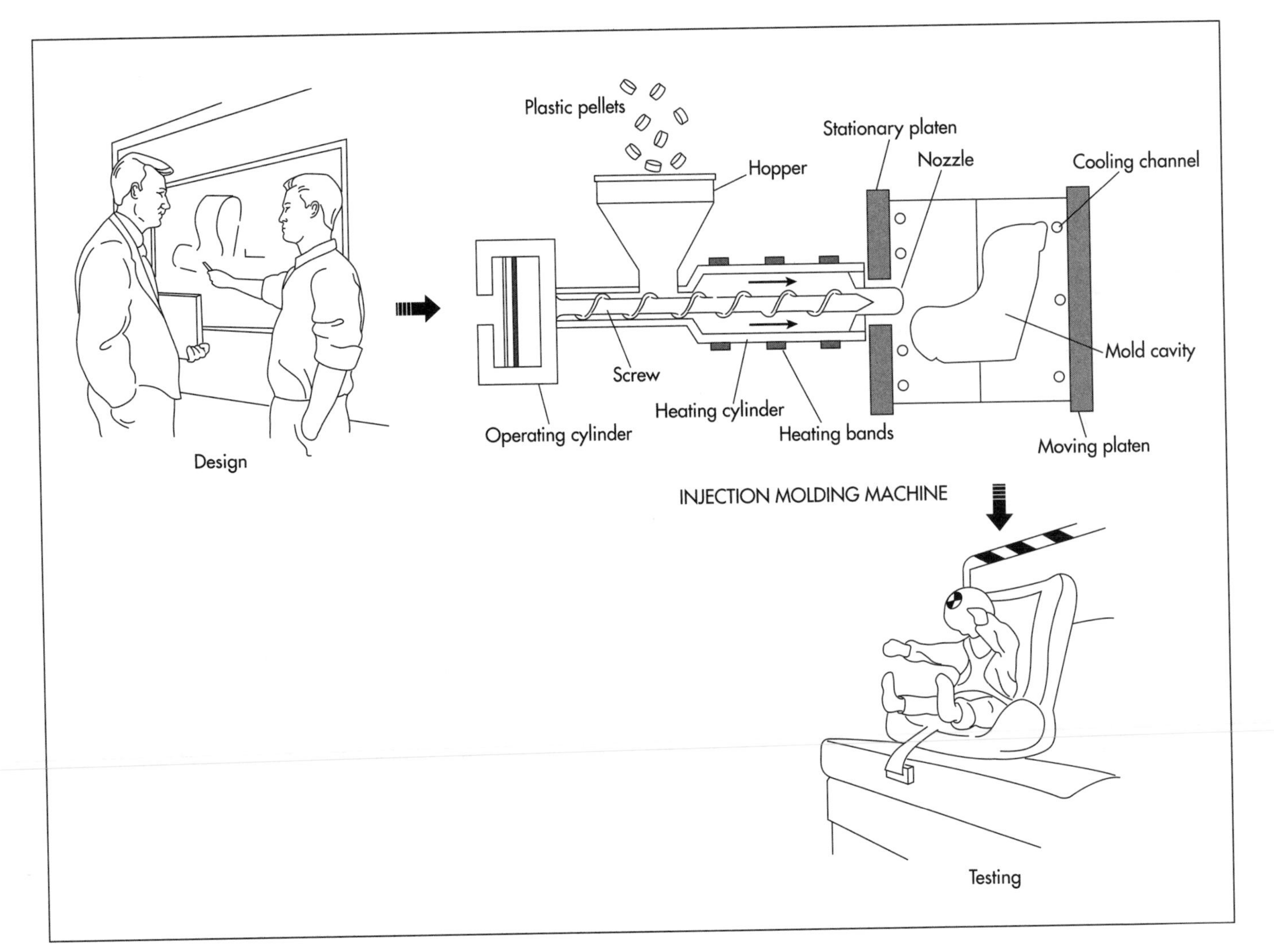


The manufacture of a child safety seat takes one to three years from concept through production. The plastic seat shell is injection-molded from plastic pellets. All parts are hand assembled at work stations. Crash testing is done to ensure the effectiveness and quality of the child safety seat design.

The child's comfort and ease of use of the seat are much more subjective. Charts of average child sizes within particular age and weight ranges help designers fit the seat to the child and make it comfortable. To understand the parents' perspective on using the seats, manufacturers consult with focus groups; analyze sales trends as indicators of ease of operations, ease of inserting and removing infants, and popularity of features such as pillows and seat protectors; and listen to consumer feedback through vehicles like consumer e-mail.

Consumer opinion also includes fashion and appearance of the safety seats. Purchasers prefer infant seats with lighter colors of fabrics and child-like patterns. Seats for older children tend to match trends in vehicle design, including darker colors and more sophisticated patterns that are coordinated with vehicle interiors.

Design must also consider constructability. The method of molding the seat, the materials used, the method of assembly, and other characteristics may simply not be compatible with other desirable design features. The process of design conception through production startup takes from one to three years, depending on the complexity of the design. Design time is expected to be slower as the first seats with the new anchorage system are designed, constructed, and tested, although it will quickly resume speed as designers become accustomed to working with the new system.

## The Manufacturing Process

1 Manufacture of the child safety seat begins with molding the shell. The plastic pellets are melted and injection-molded into forms for the shell. The molded forms are

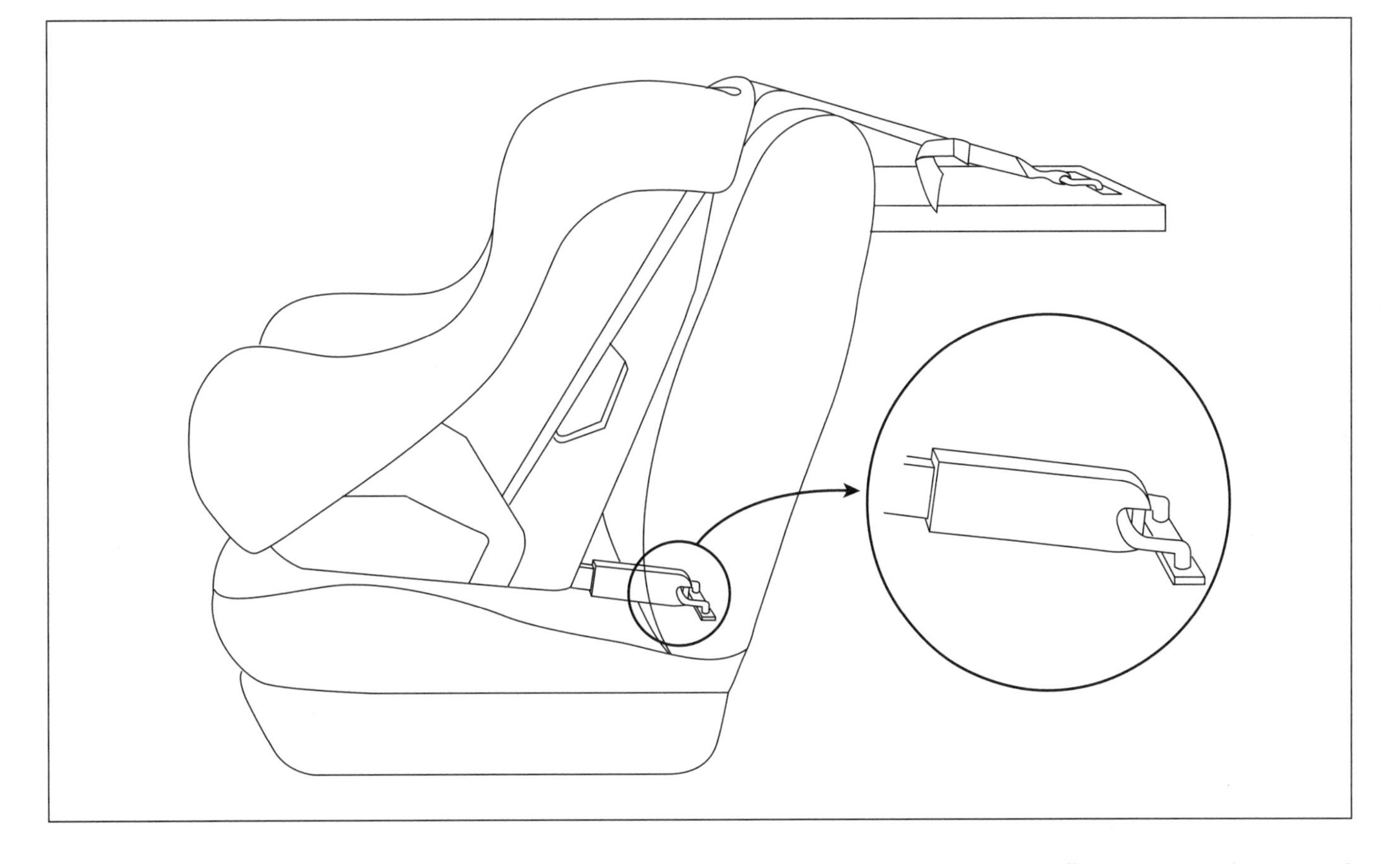

trimmed and cleaned. As soon as they are cooled, they are delivered to the assembly line.

2 The parts made by outside suppliers are distributed to work stations along the assembly line. These include the foam padding, cover, harness, buckle, labels, and instruction. Usually, the assembly line does not use a conveyor belt; workers simply complete their portion of the work and hand the seat to the next person along the line. This enables personnel to work at their own pace and check their own work on the product.

3 The padded cover is placed on the shell and attached. The buckle assembly is secured to the shell, and the harness is threaded through the buckle, adjuster, and harness retainer.

4 The labels are secured on the safety seat, and instructions are packed in the storage compartment that is a mandatory part of design of the seat.

5 If the product is also to be sold with point-of-purchase pieces (hang tags) related to marketing or advertising, these are added before the seats are packed in cartons. Before packing, some seats may be selected for quality and performance reviews including crash testing.

6 In the packing department, the seats are packed in cartons that carry information and designs developed by marketing and advertising. Generally, the cartons are stacked and wrapped in plastic so the cartons are kept clean until they are ordered and shipped. The wrapped batches of cartons are stored on pallets and moved by forklift. Some are loaded directly into trucks for shipment or taken to inventory.

## Quality Control

Manufacturers maintain a quality control department and an established inspection system. At one manufacturer, for example, every person on the assembly line is expected and encouraged to report errors, and all seats are inspected on the line for visually detectable problems. Individual parts are typically compared to masters for correctness, and each product has a bill of materials that lists the part numbers of every part in

Effective in 2002, the "universal attachment system" will secure child safety seats to anchors located in every automobile. The switch from seat belt attachment to an anchoring system will reduce the problems that occur with the use of seat belts.

the product. Product managers may also pull products off the line for review.

Crash testing is also done to test child safety seat models. Cosco, Inc., is the only domestic car seat manufacturer with its own dynamic crash test sled for assuring quality and performance. Quality can be aided by the sharing of safety-related information among manufacturers. The Juvenile Products Manufacturers' Association car seat committee assists with distributing information and collaborating on labeling and education programs. Industry representatives participate on committees, such as the Blue Ribbon Panel and the Society of Automotive Engineering (SAE), that develop recommendations for car seat makers, vehicle manufacturers, and government agencies.

## Byproducts/Waste

Manufacturers usually produce several lines of child safety seats. For example, Cosco makes a car bed/car seat, three kinds of infant-only seats, four kinds of convertible seats, a line with three car seats in one to adapt to a growing child, a high backed booster seat, a travel vest, and an auto booster.

Shells that are rejected by quality control or that have been used in crash testing are reground and combined with new plastic for remolding. Only a very small percentage of reground plastic is allowed in remoldings. Covers may not fit or may have been sewn incorrectly; if so, they are returned to the supplier and resewn. Metal parts like clips that may not have been plated properly can be replated or recycled. Other parts like incorrectly made buckles are discarded. Very little waste results.

The assembly process is also environmentally and worker friendly. Mold operators wear protective gloves. Power screwdrivers and riveters are the only other equipment used in the assembly.

## The Future

Computer systems like Cosco's Tattle Tales system will allow car seats to give verbal warnings to caregivers if the child is climbing out of the car seat or the buckle is unlocked, for example. These systems perform several checks per second, and, when the driver hears the warning, the vehicle can be stopped and the child can be resecured in the seat.

The universal attachment system (mandatory by 2002) will standardize the way in which child safety seats are attached in all vehicles. Air bags remain a concern, but "smart" air bags are in design that will recognize the sizes of vehicle occupants and whether they are correctly seated or are out of position. Ultimately, automated highway systems will reduce opportunities for driver error and other crash-causing circumstances on major highways, but seat belts and child safety seats are likely to be necessary.

## Where to Learn More

### Periodicals

"Crash-test favorites."*Daily Review* (Hayward, California), 9 May 1999, p. BAL-2.

Fix, Janet L. "U.S. to promote safe use of child safety seats, seat belts."*Knight-Ridder/Tribune News Service*, 18 November 1998.

"How to keep traveling children safe."*Childhood Education* (Spring 1997): 174+.

Shelness, Annemarie and Charles, Seymour. "Children as Passengers in Automobiles: The Neglected Minority on the Nation's Highways."*Pediatrics* Vol. 56, 1975, pp. 271-284.

Stevens, Liz. "Car seats the bane of every parent's existence."*Daily Review* (Hayward, California), 9 May 1999, p. BAL-2.

### Other

American Academy of Pediatrics (AAP). http://www.aap.org/family/mncrseat.htm/.

Cosco, Inc. http://www.coscoinc.com/.

Family Health and Safety. http://www.saferidenews.com/.

National Highway Traffic Safety Association. http://www.nhtsa.dot.gov/people/injury/childps/.

National Safe Kids Campaign Online. http://www.safekids.org/.

Safety Belt Safe U.S.A. http://www.carseat.org/.

Safe Within. http://safewithin.com/childsafe/child.seat.cgi/.

Society of Automotive Engineers. http://www.sae.org/.

*—Gillian S. Holmes*

# Compost

*By 1992, almost 1,500 cities had yard waste composting facilities.*

Compost is a finely divided, loose material consisting of decomposed organic matter. It is primarily used as a plant nutrient and soil conditioner to stimulate crop growth. Although many people associate compost production with small garden compost piles that are tended with a shovel, most compost is produced in large municipal, industrial, or agricultural facilities using mechanized equipment.

## Background

The expression "older than dirt" certainly applies to compost. Nature has been producing compost for millions of years as part of the cycle of life and death on Earth. The first human use of animal manure, a raw form of compost, was in about 3,000B.C. in Egypt when it was spread directly on the fields as a fertilizer. Later, manure was mixed with dirty stable straw and other refuse and allowed to sit in piles until it was needed. Rain kept the piles wet and aided the decomposition process, producing a rich compost.

The Greeks and Romans knew the value of compost to boost crop production and even used the warmth of decomposing compost to produce summer vegetables in winter. Christian monasteries kept the art of composting alive in Europe after the fall of the Roman Empire, and by about 1200 compost was again being used by many farmers. Shakespeare mentions it in several of his plays written in the early 1600s.

In the United States, Presidents George Washington and Thomas Jefferson were prominent landowners during the late-1700s and early-1800s. When they were not involved with affairs of state, they both spent much of their time trying innovative farming practices, including experiments with various composting methods and materials. As years of successive crops depleted the nutrients in the soil on the East Coast, the practice of composting became widespread. This trend continued until the early 1900s when it was estimated that 90% of the fertilizer used in the United States came from compost.

That all changed in 1913, when a German company began producing synthetic nitrogen compounds, including fertilizers. These new chemical fertilizers could be produced less expensively than messy animal manure compost, and the farmyard compost pile quickly became a thing of the past. By 1950, it was estimated that only 1% of the fertilizer used in the United States was derived from compost.

One notable exception to this trend was the work started in 1942 by J.I. Rodale, a noted pioneer in the development of the organic method of farming. Rodale was one of the first to see the hazards of relying on synthetic fertilizers and the benefits of using compost derived from natural sources. Composting got a short-lived boost during the environmentally conscious era of the 1960s, but it wasn't until the 1980s when it became a big business. This surge wasn't the result of a renewed awareness of the positive aspects of compost, but rather a growing concern over the negative aspects of refuse. In short, in our efforts to get rid of our refuse, we were polluting our air, poisoning our rivers, and quite literally burying ourselves in it with our landfills.

In order to divert some of the municipal refuse away from landfills, several cities established recycling centers in the early

1970s where people could bring cans, bottles, and newspaper rather than throw them in the trash. This was followed by curbside recycling, where people could place these recyclable materials in separate containers for pickup in front of their houses. Finally, many cities added additional curbside containers for yard wastes to be composted. By 1992, almost 1,500 cities had yard waste composting facilities.

At the same time, tough new environmental laws mandated that industries could no longer simply dump their waste products onto the surrounding land or discharge them into nearby rivers. To meet these laws, many industries began their own recycling and composting programs. Environmental concerns also affected farmers, who were being blamed for the negative health effects that chemical fertilizers and pesticides had on humans and wildlife. As a result, many farmers decided to cut back or eliminate chemicals in favor of using compost.

Today, most compost is processed in large facilities designed to handle a specific type of raw material. Agricultural compost is usually produced and used on the same farm that generated the raw materials. Industrial compost may be bagged and sold to individual buyers, or the raw materials may be sold in bulk to other composting facilities. Municipal yard waste compost is usually produced in facilities operated by the city or the refuse collection company and is sold to local landscaping companies and garden centers.

## Raw Materials

Technically, compost may be made from any organic material. That is, it may be made from any part of an organism, plant or animal, that contains carbon. Compost also requires a source of nitrogen, oxygen, and water, plus small amounts of a variety of elements usually found in organic material, including phosphorus, copper, potassium, calcium, and others.

In order for the organic materials to combine with the other materials and decompose into compost, several living organisms and microorganisms are needed. These include sowbugs, which help digest the materials and transport bacteria; earthworms, which aerate the materials with their tunnels; a variety of fungi, which help digest decay-resistant cellulose; mold-like bacteria called actinomycetes, which attack raw plant tissues; and many others.

The most common raw materials used to make compost are yard wastes such as grass clippings, leaves, weeds, and small prunings from shrubs and trees. Most home garden compost piles and municipal compost facilities use yard wastes exclusively because of the large volume of materials available.

Industrial compost facilities tend to use waste materials generated within a particular plant or region. For example, sugar beet pulp is mixed with other materials to make compost in an area where sugar refineries operate. Spent hops and grain from breweries also make excellent compost materials. Other materials include sawdust and wood chips from lumber mills, fish waste from canneries, and dried blood and pulverized animal bones from slaughterhouses.

Agricultural compost facilities use materials readily available on nearby farms. These include animal manure, used stable straw, spoiled fruits and vegetables, field refuse, vineyard and orchard prunings, rotted hay, and other agricultural waste products.

Some of the more unusual raw materials used to make compost include seaweed, chicken feathers, peanut shells, and hair clippings.

## The Manufacturing Process

The production of compost is both a mechanical and a biological process. The raw materials must first be separated, collected, and shredded by mechanical means before the biological decomposition process can begin. In some cases, the decomposition process itself is aided by mechanical agitation or aeration of the materials. After decomposition, the finished compost is mechanically screened and bagged for distribution.

There are several methods for producing compost on a large scale. The methane digester method places the raw materials in a large, sealed container to exclude oxygen. The resulting oxygen-starved decomposition not only produces compost, but also methane gas, which can be used for cooking or heating. The aerated pile method places the raw materials in piles or trenches containing perforated pipes that circulate air. The resulting oxygen-rich decomposition produces a great amount of heat, which kills most harmful bacteria. The windrow method places the raw materials in long piles, called windrows, where they are allowed to decompose naturally over a period of several weeks or months. It is the least expensive method of all. Here is a typical sequence of operations used to convert municipal yard wastes into compost using the windrow method.

### Separating

1 Yard wastes are deposited in separate containers by homeowners, and the containers are placed at the curb for pickup on the regular refuse collection day. Homeowners are instructed that only certain yard wastes are acceptable for collection. These include grass clippings, leaves, weeds, and small prunings from shrubs and trees. Short pieces of tree limbs up to about 6 in (15 cm) in diameter are also acceptable. Homeowners are also instructed that certain other yard wastes are not acceptable. These include rocks, sod, animal excrement, and excessive amounts of dirt. Palm fronds are prohibited because the frond spikes do not decompose and carry a poison. Food scraps, fruits, and vegetables are also prohibited because they can attract rodents, carry unwanted seeds, and contribute to odors.

2 The yard wastes are collected by separate refuse trucks and are transported to the processing center where they are dumped in piles. The piles are visually inspected, and any oversized or unacceptable materials are manually removed.

### Grinding

3 A large, wheeled machine called a front loader picks up material from the piles and dumps it into a tub grinder. The tub grinder has a stationary vertical cylindrical outer shell with a rotating cylindrical inner shell. As the material passes between the two shells, it is ground into smaller pieces and thoroughly mixed. The ground material falls out the bottom and through a screen where the larger pieces are screened out. The remaining material is transported by a conveyor belt to a holding pile.

4 The larger pieces are sold to landscaping companies for use as mulch or groundcover without further processing. The rest is loaded into large dump trucks and transported to the composting area where it is dumped in long rows, called windrows. Each row is about 6-10 ft (2-3 m) high and several hundred feet (m) long with a triangular cross section. A flat space about 10 ft (3 m) wide is left between each row to allow vehicles to move along the length.

### Composting

5 The composting area may cover several acres (hectares). After a windrow is laid in place, the material is dampened by a tank truck that moves along the row spraying water. The water aids in the composting process and helps minimize wind-blown dust.

6 Every few weeks, a special machine straddles each windrow and moves along its length to turn and agitate the material. This breaks down the material into even smaller pieces and exposes it to oxygen, which aids in the decomposition process. After the windrow is turned, it is sprayed with water again. This process continues for two or three months. In hot, dry weather, the windrows may have to be watered more often. During decomposition, the internal temperature of the pile may reach 130° F (54° C), which helps kill many of the weed seeds that might be present.

### Curing

7 The raw compost is scooped up with a front loader and moved to a large conical pile where it is allowed to finish the decomposition process over a period of several weeks. This process is called curing and it allows the carbon and nitrogen in the compost to adjust to their final levels.

### Screening

8 After the compost has cured, it is scooped up with a front loader and dumped into the hopper of a rotary screen. This device consists of a large cylindrical screen rotating on an axis that is slightly inclined above the horizontal. The openings in

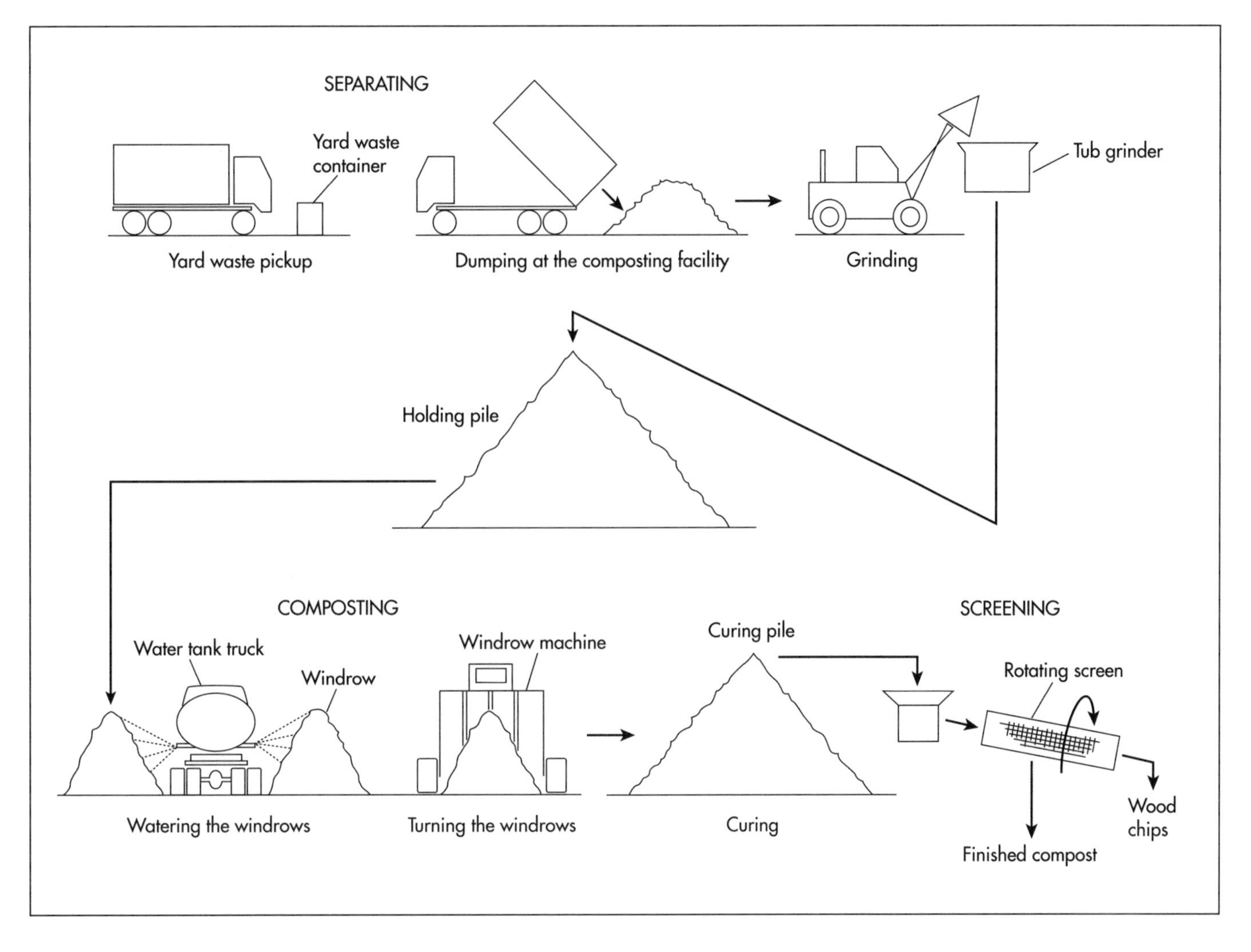


Diagram depicting the commercial processing of yard waste into compost.

the screen are about 0.5 in (1 cm) in diameter. The compost is fed into the raised end of the rotating screen from the hopper by a conveyor belt. As the compost tumbles its way down the length of the rotating screen, the smaller material falls through the screen and is moved to a storage pile by a conveyor belt. The larger material that cannot pass through the screen falls out the lower end of the cylinder and is either returned to the compost piles for further decomposition or is sold as wood chips.

### *Distributing*

9 Much of the finished compost is loaded into large dump trucks and sold in bulk to landscaping companies, municipalities, nurseries, and other commercial customers. Some of it is sealed in 40 lb (18 kg) plastic bags for retail sale to homeowners. Using the windrow method, a typical suburban yard waste processing facility can produce as much as 100,000 tons (91,000 metric tons) of compost a year.

## Quality Control

Composting companies regularly have their finished compost tested to ensure it is free of harmful materials and contains the proper amounts of plant nutrients. The tests measure the size of the particles, moisture level, mineral content, carbon-to-nitrogen ratio, acidity, nutrient content, weed seed germination rate, and many other factors. For example, waste particles should be between 0.5-2 in (1.2-5 cm) in diameter in order to encourage the flow of oxygen within the compost. Likewise, the level of moisture should be above 40% to facilitate the compost process. Moisture levels that dip below 40% slow the process and present the risk of spontaneous combustion. Also, the ideal ratio of carbon to nitrogen should average 30 parts carbon to one part nitrogen by weight. The ideal balance maintains a healthy microbial population that speeds decomposition and minimizes odor.

## Harmful Materials

Compost made from yard wastes, such as leaves and grass clippings, rarely contains any harmful materials. Problems can occur, however, when compost is made from partially sorted municipal refuse, certain industrial wastes, or sewage sludge. In those cases, unacceptable levels of toxic metals, chemicals, or harmful bacteria may be present.

To protect the public, the federal Environmental Protection Agency (EPA) sets acceptable levels for thousands of materials that might be present in compost. Each state may have its own standards as well. For municipal refuse, source separation—that is, having homeowners sort their yard wastes into separate containers rather than throw them away with the rest of their trash—is felt to be one of the most effective way to produce clean, safe compost.

## The Future

By separating home yard wastes and turning them into compost, it is estimated that municipalities can reduce the amount of trash going to landfills by about 20%. While that is a significant reduction, it is expected that even more trash will have to be diverted from landfills in the future. Materials such as soiled food packaging, disposable diaper padding, food scraps, natural fiber rags, pieces of wood, and other organic materials could all be composted. To do this, municipalities may have to establish municipal solid waste (MSW) treatment facilities to separate the compostible materials from the harmful materials, such as discarded batteries, motor oil, asbestos, and many household chemicals.

Eventually composting may also provide a means for handling and neutralizing even the harmful materials. For example, at several older military ammunition factories and storage facilities the surrounding soil is contaminated with the explosive material trinitrotoluene, also known as TNT. Researchers are using a specially formulated compost mix of vegetable wastes and buffalo manure to neutralize the soil through a simple biological composting process that converts the explosive organic components of TNT into less harmful compounds.

## Where to Learn More

### Books

Christopher, Tom and Marty Asher.*Compost This Book!* Sierra Club Books, 1994.

Hansen, Beth, editor.*Easy Compost*. Brooklyn Botanic Gardens, Inc., 1997.

Martin, Deborah L. and Grace Gershuny, ed.*The Rodale Book of Composting*. Rodale Press, Inc., 1992.

### Periodicals

Raloff, Janet. "Cleaning Up Compost: Municipal waste managers see hot prospects in rot"*Science News* (January 23, 1993): 56-58.

### Other

The Compost Resource Page. http://www.oldgrowth.org/compost/ (June 7, 1999).

Composting Council. May 1999. http://www.compostingcouncil.org/ (June 7, 1999).

*—Chris Cavette*

# Computer Mouse

## Background

Designers in the computer industry seek not only to "build the better mousetrap" but to build the best mouse. The computer mouse is an accessory to the personal computer that has become an essential part of operation of the computer. The small device fits neatly in the curve of the user's hand and enables the user, through very limited movements of the hand and fingers to "point and click" instructions to the computer. A rolling ball on the underside of the mouse gives directions on where to move to the cursor (pointer) on the monitor or screen, and one to three buttons (depending on design) allow the user to say yes by clicking the buttons on the right instruction for the computer's next operation.

## History

Dr. Douglas Engelbart, a professor with the Stanford Research Institute in Menlo Park , California, developed the first device that came to be known as the mouse in 1964. At that time, the arrow keys on the keyboard were the only way of moving the cursor around on a computer screen, and the keys were inefficient and awkward. Dr. Engelbart made a small, brick-like mechanism with one button on top and two wheels on the underside. The two wheels detected horizontal and vertical movement, and the unit was somewhat difficult to maneuver. The unit was linked to the computer by a cable so the motion signals could be electrically transmitted to the computer for viewing on the monitor. One of Dr. Engelbart's co-workers thought the device with its long cable tail looked something like a mouse, and the name stuck.

Other scientists, notably those at the National Aeronautics and Space Administration (NASA), had also been seeking methods of moving cursors and pointing to objects on the computer screen. They tried steering wheels, knee switches, and light pens, but, in tests of these devices versus Engelbart's mouse, it was the mouse that roared. NASA's engineers were concerned, however, about the spacewalks the mouse would take from its work surface in the weightlessness of space.

By 1973, the wheels on the mouse's undercarriage had been replaced by a single, free-rolling ball; and two more buttons (for a total of three) had been added to the top. The creature was called both a mouse and a pointing device, and Xerox combined it with its Alto computer, one of the first personal computers. The Alto had a graphical user interface (GUI); that is, the user pointed to icons, or picture symbols, and lists of operations called menus and clicked on them to cause the computer to open a file, print, and perform other functions. This method of operating the computer was later adapted by Macintosh and Windows operating systems.

The development of the personal computer stimulated an explosion of applications for the device that was small enough to be used at a number of work stations. Engineers could develop computer-aided designs at their own desks, and the mouse was perfect for drawing and drafting. The mouse also began to generate offspring, collectively called input/output devices, such as the trackball, which is essentially a mouse lying on its back so the user can roll the ball instead of moving the entire unit over a surface. The military, air traffic controllers, and video game players now had a pet of their

*Dr. Douglas Engelbart, a professor with the Stanford Research Institute in Menlo Park , California, developed the first device that came to be known as the mouse in 1964. One of Dr. Engelbart's co-workers thought the device with its long cable tail looked something like a mouse, and the name stuck.*

own. Mechanical sensors in both types of devices were replaced by optical-electronic sensor systems patented by Mouse Systems; these were more efficient and lower in cost. An optical mouse with no moving parts was developed for use on a special mouse pad with grid lines; light from inside the mouse illuminates the grid, a photodetector counts the number and orientation of the grid lines crossed, and the directional data are translated into cursor movements on screen.

The mouse began to multiply rapidly. Apple Computers introduced the Macintosh in 1984, and its operating system used a mouse. Other operating systems like Commodore's Amiga, Microsoft Windows, Visicorp's Vision, and many more incorporated graphical user interfaces and mice. Improvements were added to make sensors less prone to collecting dust, to make scrolling easier through an added wheel on the top, and to make the mouse cordless by using radio-frequency signals (borrowed from garage door openers) or infrared signals (adapted from television or remote controls).

## Mouse Anatomy

### *Body*

The mouse's "skin" is the outer, hard plastic body that the user guides across a flat surface. It's "tail" is the electrical cable leading out of one end of the mouse and finishing at the connection with the Central Processing Unit (CPU). At the tail end, one to three buttons are the external contacts to small electrical switches. The press of a button closes the switch with a click; electrically, the circuit is closed, and the computer has received a command.

On the underside of the mouse, a plastic hatch fits over a rubberized ball, exposing part of the ball. Inside, the ball is held in place by a support wheel and two shafts. As the ball rolls on a surface, one shaft turns with horizontal motion and the second responds to vertical motion. At one end of each of the two shafts, a spoked wheel also turns. As these spokes rotate, infrared light signals from a light-emitting diode (LED) flicker through the spokes and are intercepted by a light detector. The dark and light are translated by phototransistors into electrical pulses that go to the interface integrated circuit (IC) in the mouse. The pulses tell the IC that the ball has tracked left-right and up-down, and the IC instructs the cursor to move accordingly on the screen.

The interface integrated circuit is mounted on the printed circuit board (PCB) that is the skeleton to which all the internal workings of the mouse are attached. The integrated circuit, or computer chip, collects the information from the switches and the signals from the phototransistors and sends a data stream to the computer.

### *Brain*

Each mouse design also has its own software called a driver. The driver is an external brain that enables the computer to understand the mouse's signals. The driver tells the computer how to interpret the mouse's IC data stream including speed, direction, and clicked commands. Some mouse drivers allow the user to assign specific actions to the buttons and to adjust the mouse's resolution (the relative distances the mouse and the cursor travel). Mice that are purchased as part of computer packages have the drivers built in or preprogrammed in the computers.

## Raw Materials

The mouse's outer shell and most of its internal mechanical parts, including the shafts and spoked wheels, are made of acrylonitrile butadiene styrene (ABS) plastic that is injection-molded. The ball is metal that is coated in rubber; it is made by a specialty supplier. The electrical micro-switches (made of plastic and metal) are also off-the-shelf items supplied by subcontractors although mouse designers can specify force requirements for the switches to make them easier or firmer to click. Integrated circuits or chips can be standard items, although each manufacturer may have proprietary chips made for use in its complete line of products. Electrical cables and overmolds (end connectors) are also supplied by outside sources.

The printed circuit board (PCB) on which the electrical and mechanical components are mounted is custom-made to suit the mouse design. It is a flat, resin-coated sheet. Electrical resistors, capacitors, oscillators, integrated circuits (ICs), and other components are made of various types of metal, plastic, and silicon.

## Design

Design of a new mouse begins with meetings among a product development manager, designer, marketing representative, and consulting ergonomist (a specialist in human motion and the effects various movements have on body parts). A list of human factors guidelines is developed specifying size range of hands, touch sensitivity, amount of work, support of the hand in a neutral position, the user's posture while operating the mouse, finger extension required to reach the buttons, use by both left- and right-handed individuals, no prolonged static electricity, and other comfort and safety requirements; these can differ widely, depending on whether the mouse is to be used in offices or with home computers, for example. A design brief for the proposed mouse is written to describe the purpose of the product and what it achieves; a look is also proposed in keeping with the anticipated market.

The design team returns to the table with foam models; scores of different shapes may be made for a single mouse design. User testing is done on these models; the engineers may do this preliminary testing themselves, or they may employ focus groups as typical users or observe one-on-one testing with sample users. When the selection of models is narrowed down, wooden models that are more refined and are painted are made of the winning designs. Input is gathered again on the feel, shape, and look of the models; the ergonomist also reviews the likely designs and confirms that the human factors guidelines have been achieved.

When the optimal model is chosen, the engineering team begins to design the internal components. A three-dimensional rendering is computer-generated, and the same data are used to machine-cut the shapes of the exterior shell with all its details. The mechanical and electronics engineers fit the printed circuit board (and its electronics) and the encoder mechanism (the ball, shafts, wheels and LED source and detector) inside the structure. The process of fitting the workings to the shell is iterative; changes are made, and the design-and-fit process is repeated until the mouse meets its design objectives and the design team is pleased with the results. Custom chips are designed, produced on a trial basis, and tested; custom electronics will help the design meet performance objectives and give it unique, competitive, and marketable characteristics.

The completed design diagrams are turned over to the project tooler who begins the process of modifying machines to produce the mouse. Tooling diagrams are generated for injection-molding the shell, for example. The size, shape, volume of the cavity, the number of gates through which the plastic will be injected into the mold, and the flow of the plastic through the mold are all diagramed and studied. After the final tooling plan is reviewed, tools are cut using the computer-generated data. Sample plastic shells are made as "try shots" to examine actual flow lines and confirm that voids aren't induced. Changes are made until the process is perfect. Texture is added to the external appearance of the shell by acid etching or by sand blasting.

In the meantime, the engineering team has set up the assembly line for the new mouse design and conducted trial assemblies. When the design details are finalized, tools have been produced, and test results have met the design team's objectives and standards, the mouse is ready for mass production.

## The Manufacturing Process

To make the computer mouse, several manufacturing processes are performed simultaneously to make different pieces of the unit. These processes are described in the first three steps below. The pieces are then brought together for final assembly, as described in steps 4 through 7.

1 In one of the sets of manufacturing and assembling steps, the printed circuit board (PCB) is cut and prepared. It is a flat, resin-coated sheet that can be of surface-mount design or through-hole design. The surface-mount version is assembled almost entirely by machine. A computer-controlled automatic sequencer places the electrical components in the proper order onto the board in a prescribed pattern.

For through-hole PCB assembly, attachment wires of the electronic components are inserted in holes in the PCB. Each assembly line worker has a drawing for part of the board and specific units to add. After all the components are mounted on the board, the bottom surface of the board is passed through molten lead solder in a wave soldering machine. This machine washes the board with flux to remove contaminants, then heats the board and the components it carries by infrared heat to lessen the possibility of thermal shock. As the underside of the board flows over the completely smooth, thin liquid sheet of molten solder, the solder moves up each wire by capillary action, seals the perforations, and fixes the components in place. The soldered boards are cooled. The PCB is visually inspected at this stage, and imperfect boards are rejected before the encoder mechanism is attached.

2 The encoder mechanism (including the rubber-covered ball, the support wheel, both spoked wheels and their axles, the LED, and its detector) is assembled as a separate unit. The plastic parts were also manufactured by injection-molding in accordance with proprietary specifications and trimmed of scrap plastic. After the mechanism is assembled, the unit is fastened to the PCB using either clips or screws. The board is now completely assembled and is subjected to an electronics quality control test.

3 The mouse's tail—its electrical cable—has also been manufactured using a set of wires, shielding, and the rubber cover. The cable has two additional pieces of molded rubber called overmolds. These are strain relief devices that prevent the cable from detaching from the mouse or its connector plug if the cable is tugged. Mouse makers typically design their own shapes for overmolds. The near-mouse overmold is hooked to the housing, and, at the opposite end of the tail, the connector is soldered to the wires and the connector overmold is popped into place.

4 The pieces of the outer shell are visually inspected after molding, trimming, and surface (finish) treatment and prior to assembly. The outer shell is assembled in four steps. The completed PCB and encoder assembly is inserted into the bottom of the shell. The buttons are snapped into the top part of the housing, the cable is attached, and the top and bottom are screwed together using automated screwdrivers.

5 The final electronics and performance quality check is performed when assembly is essentially complete. Rubber or neoprene feet with adhesive sheeting pre-applied to one side are added to the underside of the mouse.

6 While the tooling designs and physical assembly described above have been in progress, a programming team has been developing, testing, and reproducing the mouse driver firmware. The firmware so-called because it lies in the realm between software and hardware consists of a combination of codes in the integrated circuit and the translation of the mouse's directional movements and micro-switch signals that the receiving computer needs to understand when the mouse is attached. When the driver has been developed, the manufacturer's own testers run it through rigorous trials, and both the Federal Communications Commission (FCC) and the European Commission (CE—an organization that governs radio emissions and electrostatic discharge) also approve the electronics. Approved driver data is encoded and mass-produced on diskettes.

7 The FCC requires that signaling or communications devices including the mouse bear labels identifying the company and certain product specifications. The labels are preprinted on durable paper with strong adhesive so they cannot easily be removed. A label is pasted on the mouse bottom, and the mouse is bagged in plastic. The device, its driver diskette, and an instruction booklet with registration and warrantee information are boxed and prepared for shipment and sale.

## Quality Control

Use of computer-generated designs builds quality and time savings into the product. Data can be stored and modified quickly, so experiments with shapes, component layouts, and overall look can be attempted and iterative adjustments can be made. Computer-aided design data also speeds review of

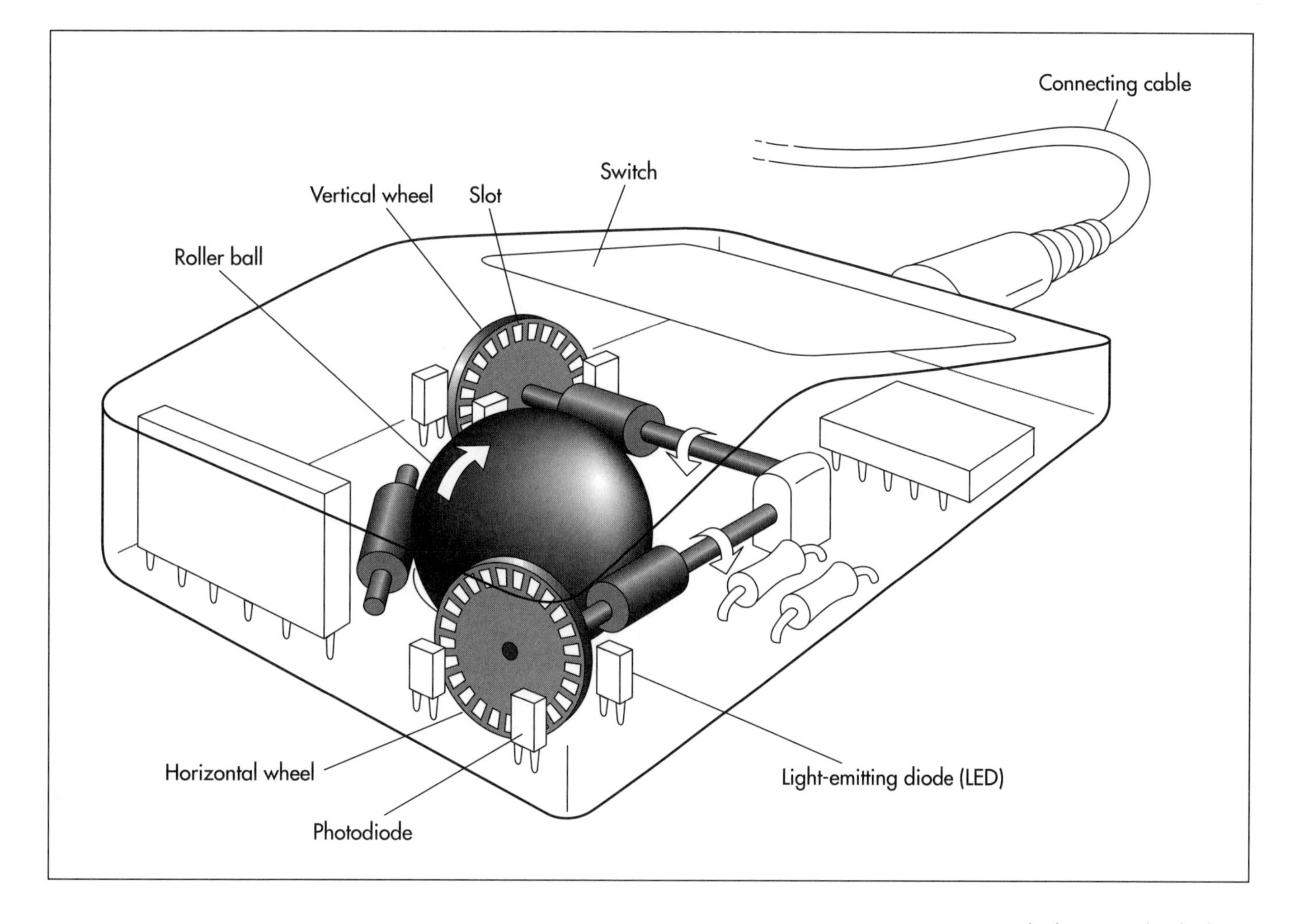


Beneath the outer, hard plastic body that the user maneuvers across a mouse pad is a rubberized ball that turns as the mouse moves. The ball is held in place by a support wheel and two shafts. As it rolls, one shaft turns with horizontal motion and the second responds to vertical motion. At one end of each of the two shafts, a spoked wheel also turns. As these spokes rotate, infrared light signals from a light-emitting diode (LED) flicker through the spokes and are intercepted by a light detector. The dark and light are translated by phototransistors into electrical pulses that go to the interface integrated circuit (IC) in the mouse. The pulses tell the IC that the ball has tracked left-right and up-down, transmits the command through the cable to the Central Processing Unit (CPU), and instructs the cursor to move accordingly on the screen.

parts specifications, the tooling process, and design of assembly procedures so the opportunity for conflicts is small.

At least three quality control steps are performed during assembly. An electronics check is carried out on the PCB after its components are attached (and soldered into place if through-hole assembly methods are used) and before any of the plastic mechanism is attached. The plastic parts (the encoder mechanism and the outer shell) are visually inspected when they are complete but before they are connected to the board and electronics; this prevents disassembly or wasting electronics due to a defective shell, for example. Finally, the completely assembled device is subjected to another electronics and performance check; 100% of the mice manufactured by Kensington Technology Group are plugged into operating computers and tested before they are packaged. As noted above, both the FCC and CE regulate aspects of mouse operations, so they also test and approve driver data.

## Byproducts/Waste

Computer mice makers do not generate byproducts from mouse manufacture, but most offer a range of similar devices for different applications. Compatible or interchangeable parts are incorporated in new designs or multiple designs whenever possible to avoid design, tooling, and assembly modification costs.

Waste is minimal. The mouse's ABS plastic skin is highly recyclable and can be ground, molded, and reground many times. Other plastic and metal scrap is produced in minute quantities and can be recycled or disposed.

## The Future

Devices that are modifications of mice are currently on the market. The Internet mouse inserts a scrolling wheel between the two buttons to make scrolling of web pages easier; a still more sophisticated version adds buttons that can be programmed by the user to perform Internet functions, like moving back or forward, returning to the home page, or starting a new search. One mouse version has returned to the floor where two foot pads or pedals replace the ball and buttons; one pedal is pushed to relocate the cursor and the second clicks. Cordless mice that communicate with radio signals are available, and the mouse has been disposed of altogether by the touchpad. The user runs a finger across the touchpad to reposition the cursor, and web pages can be scrolled and advanced by other, specific moves. Many of these adaptations are designed to eliminate repetitive stress ailments and save forearm strain.

The mouse's inventor, Dr. Engelbart, never believed the mouse would reach thirty-something or retain its nontechnical name. In fact, both the mouse and its trackball offspring are increasingly popular as shapes become more comfortable, less cleaning and maintenance are required, and reliability and longevity improve. Future developments in mice will follow the evolution of the Internet and include more options for programmability, such as switching hands to double the number of available functions. The mouse may become extinct someday, and the most likely candidate to replace it is a device that tracks the eye movement of the computer user and follows it with appropriate cursor motions and function signals.

## Where to Learn More

### Books

Ed., Time-Life Books.*Input/Output: Understanding Computers.* Alexandria, VA: Time-Life Books, 1990.

### Periodicals

Alexander, Howard. "Behold the Lowly Mouse: Clever Technology Close at Hand."*New York Times* (October 1, 1998): D9.

"The Mouse." *Newsweek* (Winter 1997): 30.

Randall, Neil. *PC Magazine* (January 5, 1997): 217.

Terrell, Kenneth. "A new clique of mice: designers turn the computer mouse on its head; some cut its tail."*U.S. News & World Report* (March 23, 1998): 60+.

### Other

Kensington Technology Group. http://www.kensington.com/ (June 7, 1999).

Logitech. http://www.logitech.com/ (June 7, 1999).

Microsoft Corporation. http://www.microsoft.com/ (June 7, 1999).

*—Gillian S. Holmes*

# Concrete Dam

## Background

Concrete dams are built in four basic shapes. The concrete gravity dam has weight as its strength. A cross section of this dam looks like a triangle, and the wide base is about three-fourths of the height of the dam. Water in the reservoir upstream of the dam pushes horizontally against the dam, and the weight of the gravity dam pushes downward to counteract the water pressure. The concrete buttress dam also uses its weight to resist the water force. However, it is narrower and has buttresses at the base or toe of the dam on the downstream side. These buttresses may be narrow walls extending out from the face of the dam, much like the "flying buttresses" supporting cathedral walls or a single buttress rather like a short dam may be built along the width of the toe of the dam.

The arch dam is one of the most elegant of civil engineering structures. In cross section, the dam is narrow in width, but, when viewed from above, it is curved so the arch faces the water and the bowl of the curve looks downstream. This design uses the properties of concrete as its strength. Concrete is not strong in tension (when it is pulled or stretched), but it is very strong in compression (when it is pushed or weighed down). The arch dam uses the weight of the water behind it to push against the concrete and close any joints; the force of the water is part of the design of the dam. The arch-gravity dam is a combination of the arch type and gravity type, as the name suggests; it is a wider arch shape. Multiple-arch dams combine the technology of arch and buttress designs with a number of single arches supported by buttresses.

Concrete dams are used more often than fill dams to produce hydroelectric power because gates (also called sluices) or other kinds of outlet structures can be built into the concrete to allow for water to be released from the reservoir in a controlled manner. When water for power, drinking water, or irrigation is needed downstream, the gates can be opened to release the amount needed over a specified time. Water can be kept flowing in the river downstream so fish and other wildlife can survive. Both concrete and fill dams are required to have emergency spillways so that flood waters can be safely released downstream before the water flows over the top or crest of the dam and potentially erodes it. Spillways channel the water downstream and well below the base or toe of the dam so the dam and its foundation are not eroded.

Most dams built in the twentieth century and those being designed today have several purposes. Over 40,000 dams higher than 45 ft (15 m) and classified as large dams exist, and more than half of these have been built since 1960. Of these dams, 16% of them are in the United States and 52% are in China; 83% are fill dams used primarily for water storage, and the remaining 17% are concrete or masonry dams with multiple purposes. Dams that generate hydroelectric power produce 20% of the electricity in the world.

## History

Fill dams may be a far older construction technique than concrete or masonry dams, but the oldest surviving dam is Sadd el Kafara about 20 mi (32 km) south of Cairo, Egypt. This dam is actually a composite

*Over 40,000 dams higher than 45 ft (15 m) and classified as large dams exist, and more than half of these have been built since 1960.*

*Reminants of the Austin, Pennsylvania, dam after its failure on September 30, 1911.*

On September 30, 1911, the town of Austin (population 3,200) in the mountainous country of north central Pennsylvania was ravaged by a torrent of water roaring through the valley, channeled by its narrowness and rugged walls. The force ripped gas mains from beneath the streets; and as soon as the wall of water passed, an errant flame lit the gas and the fire jumped from gas pipe to gas pipe and house to building throughout the standing remains of the 30-year-old town of Austin. Initial reports claimed that 1,000 people had perished, although later information placed the death toll at between 50 and 149. The source of this grief was also the source of livelihood for Austin. The Bayless Pulp and Paper Mill owned the concrete dam, which it had constructed in 1909 to provide a water storage reservoir for its water-intensive pulp- and papermaking business. There had been a precursor to this disaster in January 1910 when, following intense winter rain and snowmelt, cracks had been observed in the dam. The cracks were repaired, but they were not recognized as indications of problems related to the foundation, design, and construction of the structure.

The dam was still under construction as the winter of 1909-10 approached. Temperatures were below freezing when some of the concrete was placed, and the final stages of construction were finished hurriedly. The dam was completed on about December 1, 1909, and a crack running from the crest of the dam vertically to the ground was visible when construction was finished. By the end of the month, a second crack had appeared. Both cracks seemed to have resulted from contraction of the concrete. On January 17, 1910, a warm spell brought heavy rains and caused rapid snowmelt, and four days later, flood water was pouring over the spillway.

All technical aspects of Austin Dam were poor. The construction failures were obvious and included use of weak, oversized aggregate placed in improperly cured concrete in freezing weather. When the January 1910 failure occurred, it showed that the dam structure and the founding bedrock had failed. The owner/operator's disregard of the engineer's recommended repairs was the fatal seal.

consisting of two masonry walls with the space between filled with gravel; it was built between 2,950 and 2,750B.C.

The ancient Romans developed superior techniques for building with masonry, but, curiously, they did not often use their masonry skills in dam construction. An exception was the Proserpina Dam in Mérida, Spain, that is still standing today. Developments by the Romans were not overlooked by others. In about 550A.D., the Byzantines on the eastern fringes of the Roman Empire used the shape of the Roman masonry arch to build what history believes was the world's first arch-gravity dam. Dam building came to America with the conquistadors. In Mexico, they saw dry land in need of irrigation and imitated the dams built by the Romans, Muslims, and Spanish Christians in their homeland; the Catholic Church financed construction, and many of the missionaries were skilled engineers.

Dam building was rare in Europe until the Industrial Revolution. The northern climates produced more rainfall, so water power was naturally occurring and water supply was plentiful. In the eighteenth century, however, the rise of industry necessitated constant, reliable supplies of water power delivered at greater force, so masonry and concrete dam construction became popular in Europe. The Industrial Revolution also powered developments in science and engineering, and the civil engineering specialty, which includes designing and building structures to better the quality of life, came into existence in the 1850s. Early civil engineers began to study Sir Isaac Newton's physics and other scientific theories and apply them to practical structures including dams.

## Raw Materials

The key raw materials for concrete dams are the concrete itself and steel reinforcement. A number of other materials and components made by specialty contractors may be used in dam building and include steel gates and tunnel liners, rubber waterstops, plastic joint-filling compounds to prohibit the movement of water, electrical controls and wiring, siphons, valves, power generators, a wide assortment of instruments, and even Teflon sheeting to line water outlet structures to prevent turbulence and cavitation (damage due to swirling water).

Concrete itself is made of cement, water, and materials collectively called aggregate that consist of sand or gravel. Cement has unique properties that must be considered in selecting the cement, designing the dam, and timing construction. Mixing of cement and water causes a chemical reaction that makes concrete hard but that also releases heat. This causes a distinct rise in the temperature inside a mass of concrete, and, when the concrete begins to cool, it shrinks and cracks, potentially causing leaks. To limit these effects, concrete can be placed when the air temperature is low, low-heat cement can be used, and water can be circulated through pipes in the concrete. Furthermore, the concrete has to be placed in shallow lifts (i.e., only a few feet or meters are added at a time) and in narrow blocks; then it has to be allowed to cure over a specified minimum time so the heat dissipates. Depending on the design of the dam, engineers will choose the concrete mix (including the cement and type of aggregate) very carefully; a thin arch dam is designed with a different concrete mix than a massive gravity dam.

## Design

Design of a concrete dam depends on the purpose of the dam and the configuration of the site where it will be built. Dams are of two general types. Overflow dams block flow in a stream and harness the water for generating power or to improve navigation and provide irrigation water. The components of an overflow dam are designed so the water can be released and the level of the water in the reservoir regulated by a series of sluice gates, spillways, or outlet tunnels. Non-overflow dams store water for drinking water supply, irrigation, or power; they also have a spillway, but its use is restricted for emergencies to lower the water level quickly during floods. Methods for releasing the stored water are much more limited than in overflow dams, and the dam itself may not contain any outlet structures. Instead, water may be pumped out for irrigation, for example, from part of the reservoir.

Some sites are best suited to particular types of dams. An arch dam is most appropriate for construction in a high, narrow gorge where the arch of the structural shape provides strength. But an arch can also be built across a wider canyon where other effects like friction on the base of the dam add strength and resistance to movement. Similarly, a gravity dam is the typical choice for a shallow, wide canyon, but if it is built with some curvature, arching action will also strengthen a gravity dam in a narrower and higher gorge. Where the riverbed is exceptionally wide, the dam may be designed to have several spans, each with different engineering properties depending on the variation of foundation materials. The separate spans are usually supported on the downstream (air) side by buttresses or the extended curves of multiple arches. Sometimes, the spans of multiple span dams are constructed of concrete slabs or steel plates supported on piers.

Like fill dams, concrete dams go through extensive rounds of preliminary design and feasibility studies to choose and explore the site, to evaluate the quantity of water retained and its value (as a power source or source of supply) versus the cost of the project over the anticipated years of operation, to consider a wide range of other effects such as changes to the environment, and to choose a dam of the optimal size and configuration. Hundreds of factors enter into these studies, and the process is usually iterative. A design is chosen and tested against all these factors until it fails to satisfy one or more factors, and the next variation in design is chosen and studied until it fails—or passes.

The design process for a concrete dam typically involves professionals from a more extensive range of disciplines than design of a fill dam. The technical professionals who contribute their expertise to design of a concrete dam may include geologists, seismologists, environmental scientists, geotechnical (soil) engineers, civil engineers, structural engineers, computer analysts (specialists in software applications that examine the dam's strength and safety), hydrologists and hydraulic engineers, mechanical engineers, and electrical engineers if the dam is to be used for power generation. Still more specialists may study aspects like corrosion of concrete and steel structures. The teamwork required for dam design and construction is critical not only because of the enormous costs of these projects but because the safety

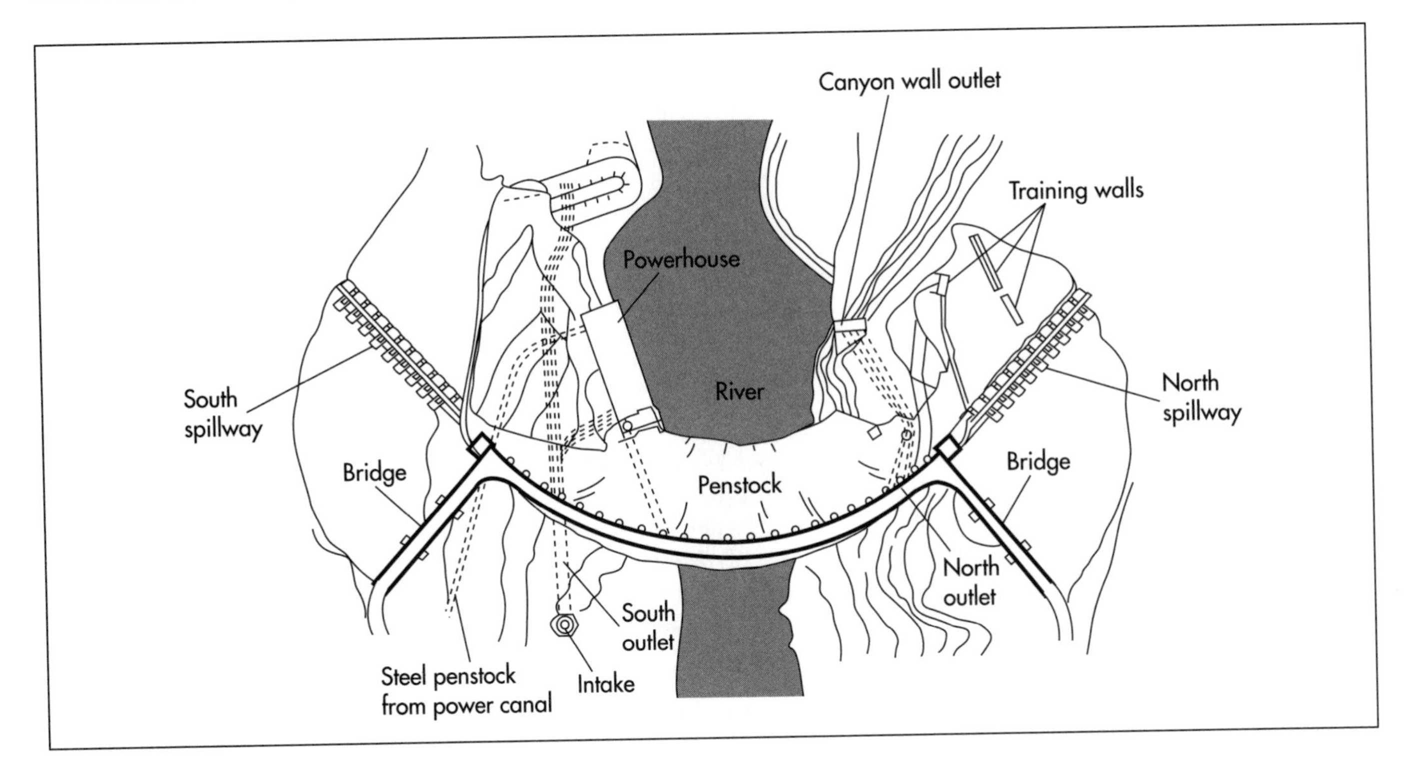


An example of a typical concrete arch gravity dam plan.

of persons and property downstream demands perfection.

## The Construction Process

1 Before construction can begin on any dam, the water in the streambed must be diverted or stopped from flowing through the site. As in the case of fill dams, a cofferdam (a temporary structure to impound the water)must be built or the water must be diverted into another channel or area downstream from the dam site. For large projects, this construction may be done several seasons before building of the dam begins. The flow of water is closed off at the very last moment.

2 The foundation area for any concrete dam must be immaculate before the first concrete for the dam is placed. As for fill dams, this is a detailed process of excavating, cleaning, and repairing the rock throughout the foundation "footprint" and on both abutments (the sides of the canyon that form the ends of the dam). Sites immediately downstream of the dam for any powerplant, stilling basin, or other structure must also be prepared.

At some sites, extensive work may be required. If the rock in the foundation or abutments is prone to fracturing because of the load imposed by the dam and its reservoir, earthquake activity, or the properties of the rock, it may be necessary to install extensive systems of rock bolts or anchor bolts that are grouted into the rock through potential fracture zones. On the abutments above the dam, systems of rock bolts and netting may be required to keep large rock fragments from falling onto the dam. Instruments to monitor groundwater levels, joint movement, potential seepage, slope movements, and seismic activity are installed beginning during the early stages of foundation preparation through completion of the dam.

A cutoff wall may be excavated deep into rock or holes may be drilled in the foundation for the installation of reinforcing steel, called rebars, that extend up into the dam and will be tied to the steel inside the first lifts of the dam. The idea is to build a reservoir that, like a bowl, is equally sound around its perimeter. The water is deepest and heaviest at the dam (when the reservoir is near capacity) so the dam and its foundation cannot be a weak point in that perimeter.

3 Forms made of wood or steel are constructed along the edges of each section of the dam. Rebar is placed inside the forms

and tied to any adjacent rebar that was previously installed. The concrete is then poured or pumped in. The height of each lift of concrete is typically only 5-10 ft (1.5 -3 m) and the length and width of each dam section to be poured as a unit is only about 50 ft (15 m). Construction continues in this way as the dam is raised section by section and lift by lift. Some major dams are built in sections called blocks with keys or interlocks that link adjacent blocks as well as structural steel connections.

The process is much like constructing a building except that the dam has far less internal space; surprisingly, however, major concrete dams have observation galleries at various levels so the condition of the inside of the dam can be observed for seepage and movement. Inlet and outlet tunnels or other structures also pass through concrete dams, making them very different from fill dams that have as few structures penetrating the mass of the dam as possible.

4 As soon as a significant portion of the dam is built, the process of filling the reservoir may begin. This is done in a highly controlled manner to evaluate the stresses on the dam and observe its early performance. A temporary emergency spillway is constructed if dam building takes more than one construction season; lengthy construction is usually done in phases called stages, but each stage is fully complete in itself and is an operational dam. The upstream cofferdam may be left in place as a temporary precaution, but it is not usually designed to hold more than minimal stream flows and rainfall and will be dismantled as soon as practical. Depending on design, some dams are not filled until construction is essentially complete.

5 The other structures that make the dam operational are added as soon as the elevation of the their location is reached as the dam rises. The final components are erosion protection on the upstream (water) side of the dam (and sometimes downstream at the bases of outlet structures), instruments along the crest (top) of the dam, and roads, sidewalks, streetlights, and retaining walls. A major dam like Hoover Dam has a full-fledged roadway along its crest; small dams will have maintenance roads that allow single-file access of vehicles only.

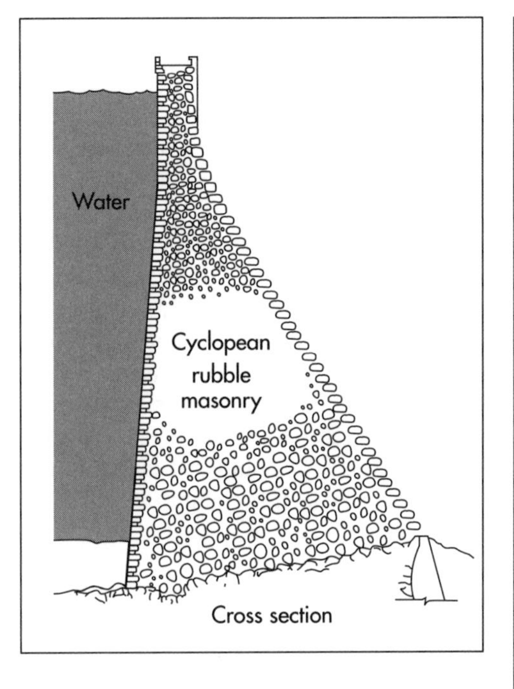


Cross section of a typical concrete arch gravity dam. The height is 280 ft (85 m). The thickness grows from 16 ft (4.9 m) at the top to 184 ft (56 m) at the base.

Away from the dam itself, the powerhouse, instrument buildings, and even homes for resident operators of the dam are also finished. Initial tests of all the facilities of the dam are performed.

6 The final details of constructions are wrapped up as the dam is put into service. The beginning of the dam's working life was also carefully scheduled as a design item, so that water is available in the reservoir as soon as the supply system is ready to pump and pipe it downstream, for example. A program of operations, routine maintenance, rehabilitation, safety checks, instrument monitoring, and detailed observation will continue and is mandated by law as long as the dam exists.

## Quality Control

There is no dam construction without intensive quality control. The process of building alone involves heavy equipment and dangerous conditions for construction workers as well as the public. The population living downstream of the dam has to be protected over the structure itself; the professionals who design and construct these projects are absolutely committed to safety, and they are monitored by local, state, and federal agencies like Divisions of Dam Safety, the U.S. Corps of Engineers, and the Department of Reclamation.

## Byproducts/Waste

There are no byproducts in dam design or construction although a number of other associated or support facilities may be needed to make the project work. Waste is also minimal because materials are too expensive for waste to be allowed. Also, locations are often remote, and the process of hauling waste away from the site and disposing it is prohibitive. Soil and rock that may be excavated from the foundation area, downstream sites, the abutments, or portions of the reservoir are usually used elsewhere on the project site. Quantities of materials cut away or placed as fill are carefully calculated to balance.

## The Future

The future of concrete dams is the subject of much debate. Each year, over 100,000 lives are lost in floods, and flood control is a major reason for building dams, as well as protecting estuaries against flooding tides and improving navigation. Lives are also benefited by dams because they provide water supplies for irrigating fields and for drinking water, and hydroelectric power is a non-polluting source of electricity. Reservoirs are also enjoyed for recreation, tourism, and fisheries.

However, dams are also damaging to the environment. They can change ecosystems, drown forests and wildlife (including endangered species), change water quality and sedimentation patterns, cause loss of agricultural lands and fertile soil, regulate river flows, spread disease (by creating large reservoirs that are home to disease-bearing insects), and perhaps even affect climate. There are also adverse social effects because human populations are displaced and not satisfactorily resettled.

For years before the start of construction in 1994 of the Three Gorges Dam in China, environmentalists the world over organized protests to try to stop this huge project. They have not succeeded, but controversy over this project is representative of the arguments all proposed dams will face in the future. The balance between meeting human needs for water, power, and flood control and protecting the environment from human eradication or encroachment must be carefully weighed.

## Where to Learn More

### Books


Bureau of Reclamation, U.S. Department of the Interior.*Design of Sam ll Dams*. Washington, DC: U.S. Government Printing Office, 1977.

Jansen, Robert B.*Dams and Public Safety*. Washington, DC: U.S. Dept. of the Interior, Water and Power Resources Service, 1980.

Krynine, Dimitri P. and William L. Judd.*Principles of Engineering Geology and Geotechnics*. New York: McGraw-Hill Book Co., Inc., 1957.

Smith, Norman. *A History of Dams*. Secaucus, New Jersey: The Citadel Press, 1972.


### Periodicals


Bequette, France. "Large Dams."*UNESCO Courier* (March 1997): 44.

Cotrim, John, et al. "Itaipu: South America's Power Play."*Civil Engineering* (December 1984): 40.

Fillon, Mike. "Taming the Yangtze: lauded and lambasted, China's Three Gorges Dam may be the biggest civil-engineering project ever."*Popular Mechanics* (July 1996): 52.

Moraes, Julival de, et al. "Itaipu: Part 1." [Itaipu Hydroelectric Project, Brazil/Panama] *Construction News Magazine* (March 1982): 18.

Moraes, Julival de, et al. "Itaipu: Part 2."*Construction News Magazine* (April 1982): 22.

"Power-poor nation taps jungle river for energy." [Concrete gravity arch dam, Paute River, Ecuador]*Engineering News Record* (December 13, 1979): 26.


*—Gillian S. Holmes*

# Cork

An incredibly versatile natural material, cork is harvested from living cork oak trees somewhat like wool is gathered from sheep. The trees are unharmed by the process, and they continue producing cork for an average of 150 years.

## Background

Cork is composed of dead cells that accumulate on the outer surface of the cork oak tree. Because of its honeycomb-like structure, cork consists largely of empty space; its density (weight per unit volume) is one-fourth that of water. Unlike a honeycomb, however, cork consists of irregularly shaped and spaced cells having an average of 14 sides. With 625 million of these empty cells per cubic inch (40 million per cubic centimeter), cork is like many layers of microscopic Bubble Wrap, making it an effective cushioning material. Its low density makes cork useful in products like life preservers and buoys. The large amount of dead-air space makes cork an effective insulation material for both temperature and noise. Furthermore, it is fire retardant; flames will only char the surface, and no toxic fumes are generated. Cutting the surface of cork turns many of the microscopic cells into tiny suction cups, creating an effective non-slip surface. In addition to being flexible, cork is highly resilient. After being crushed under a pressure of 14,000 lbs/in$^2$ (96,000 kPa), cork will regain 90% of its original size in 24 hours. Cork absorbs neither dust nor moisture, and it resists both rot and insects. Highly resistant to wear, it is used for polishing diamonds.

Among the many products made from cork are flooring materials (e.g., linoleum), shoe insoles, roofing panels, gaskets, safety helmet liners, bottle stoppers, dartboards, bulletin boards, and cores for golf balls and baseballs. Numerous artificial materials have been developed to substitute for cork in specific applications (e.g., a synthetic pea in a referee's whistle, foam insoles for shoes, or Styrofoam life preservers). However, no general substitute has been developed for cork that can be used in diverse applications.

## History

Cork bottle stoppers have been found in Egyptian tombs dating back thousands of years. Ancient Greeks used cork to make fishing net floats, sandals, and bottle stoppers. Two thousand years ago, Romans widely used cork in variety of ways, including life jackets for fishermen. For hundreds of years, Mediterranean cottages have been built with cork roofs and floors to keep out summer heat and winter cold—and to provide a soft walking surface.

Glass bottles were invented in the fifteenth century, but their use did not become widespread until the seventeenth century. The popularity of cork as a stopper led to deliberate cultivation of cork trees, which prior to about 1760 had simply been harvested wherever they happened to grow. The revolutionary crown cap—a metal lid lined with a disk of natural cork commonly known as a bottle cap—was invented in 1892.

A great deal of the cork harvest was wasted until around 1890, when a German company developed a process for adding a clay binder to cork particles and producing sheets of agglomerated (composite) cork for use as insu-

*Portugal's cork forests are the most productive. Accounting for 30% of the existing trees, they produce half of the world's harvested cork.*

*Bottle caps.*

Cork has been used since antiquity as a stopper for bottles because of its compressive abilities. During the Renaissance, cork stoppers were commonplace, and cork-oak trees were grown and processed in the Pyrenees Mountains especially for this purpose. Wine bottles were commonly sealed with oiled hemp. When Pierre Pérignon (1638-1715) invented champagne in 1688, he found that the gaseous pressure inside his bottles blew out the hemp stoppers. To solve the problem, he invented corks held in place by wire.

The modern metal bottle cap was developed by the prolific Maryland inventor William Painter, who patented his first stopper in 1885. By 1891, his definitive design, a cork-lined metal cap with a corrugated edge that is crimped around the bottle lip, appeared. Painter called his invention the "crown cap," founded the Crown Cork and Seal Company to market it, and became very wealthy from it.

The crown cap was the industry standard for nearly 80 years. In 1955, the crown cap's cork liner was replaced by plastic, and a high-speed machine to inspect crown seals was introduced in 1958. In the 1960s, the Coca-Cola company offered lift-top crown caps. The push-on, twist-off cap was first developed for baby food. Screw caps for carbonated beverages appeared in the 1960s and 1970s and are the standard today.

lation. The following year, an American named John Smith developed a technique for producing pure-cork agglomeration out of waste material by subjecting cork particles to heat and pressure without adding any binder. The next major development occurred in 1909 when Charles McManus invented a type of agglomerated cork that could be used to line crown caps. Since then, many other techniques have been developed to produce cork compounds with a variety of properties and uses.

## Raw Materials

The raw material for cork products is harvested from the cork oak tree (either the evergreen *Quercus suber* or the deciduous-*Quercus occidentalis*). The trees typically reach a height of 40-60 ft (12-18 m) and a trunk circumference of 6-10 ft (2-3 m). Virtually all of the world's commercial cork trees grow in the western Mediterranean region and the Iberian Peninsula. Portugal's cork forests are the most productive. Accounting for 30% of the existing trees, they produce half of the world's harvested cork.

A cork tree is ready for its first harvest when it is about 20 years old. The first harvest is of poor quality, and can only be used to make agglomerated cork products. Subsequent harvests occur at nine-year intervals, when the cork layer reaches a thickness of 1-2 in (2-5 cm). The harvest from a young tree yields about 35 lb (16 kg) of cork, while the yield for an older tree may be 500 lb (225 kg). Each tree has a productive life of about 150 years.

During the production of bottle stoppers, chemical baths are used to condition the corks. Among the more popular are a chlorinated lime bath followed by a neutralizing bath of oxalic acid, a hypochlorite bath neutralized by sodium oxalate, and a peroxide bath neutralized with citric acid.

Production of compound agglomerated cork involves adding a binder or adhesive agent to cork granules. Different binders are chosen, depending on the qualities desired in the ultimate product (e.g., flexibility, softness, resistance to wear). Among those frequently used are asphalt, rubber, gypsum, glue, and plastic.

## The Manufacturing Process

1 Using a specially designed hatchet, the harvester slices through the cork layer on the trunk of the tree, taking care not to cut deep enough to damage the living portion of the trunk. Horizontal cuts are made at the base of the trunk and just below the lowest branches. A few vertical cuts separate the

circumferential cork ring into sections of an appropriate size. Using the wedge-shaped handle of the hatchet, the harvester strips each panel of cork from the tree. On some large trees, cork is also stripped from the lower branches.

2 The cork planks are stacked outdoors and left to cure for a time ranging from a few weeks to six months. The fresh air, sun, and rain encourage chemical changes that improve the quality of the cork. By the end of the curing process, the planks have flattened out and lost about 20% of their original moisture content.

3 The planks are then treated with heat and water to remove dirt and water-soluble components like tannin, and to make the cork softer and more flexible. This process typically involves lowering stacks of cork planks into large copper vats filled with boiling water containing a fungicide. Heavy weights are placed on top of the cork to keep it submerged for 30-75 minutes.

4 When the planks are removed from the vat, a hoe-shaped knife is used to scrape off the poor-quality outer layer of cork, which amounts to about 2% of the volume of the plank but 20% of its weight. The planks are stacked in a dark cellar and allowed to dry and cure under controlled humidity for a few more weeks.

5 The cork planks are trimmed to a uniform, rectangular shape and are sorted by quality. The finest quality material will be used to make natural cork products like wine bottle stoppers. Poorer quality material will be ground and used to make composition or agglomerated cork.

### *Bottle corks*

6 Cork slabs of the desired thickness are placed in a steam chamber for 20 minutes to soften them. The slabs are then cut into strips whose width corresponds to the intended length of the bottle stoppers. The strips are fed through a machine that punches hollow metal tubes through them, removing cylinders of cork.

7 Although some beverage bottlers want cylindrical corks, others want tapered ones. To achieve this shape, the cylinders are arranged on a slanted conveyor that carries them past a rapidly rotating circular knife. As they pass the blade, the corks are also revolving on the conveyor, so they are trimmed to a taper.

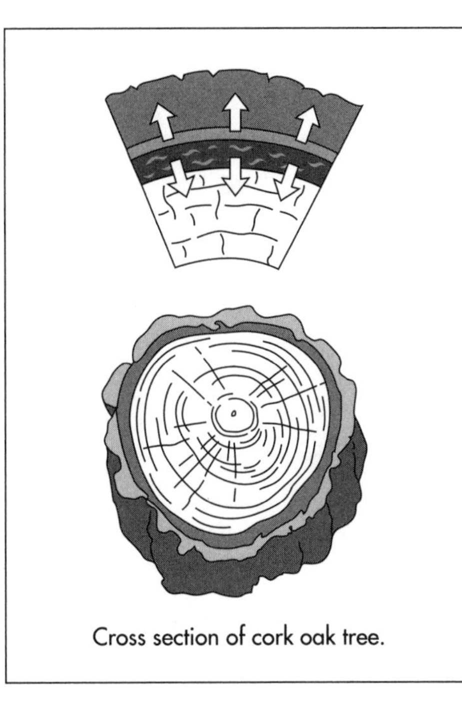

Cross section of cork oak tree.

8 Both cylindrical and tapered corks are washed, bleached, and sterilized in large vats. Rotating wooden paddles continually push the corks down into first a washing solution and then a neutralizing solution.

9 After being dried in a large centrifugal dryer, the corks may be marked with an identifying label (with ink or a hot-metal stamp). Some are also coated with a sealant such as paraffin or silicone. Then, they are packed in airtight bags in quantities of 1,000 or 1,500; the air is removed from the bags and replaced with sulfur dioxide ($SO_2$) to keep the corks sterile.

### *Agglomerated cork*

10 Waste cork is passed through a machine that breaks it into small pieces. The pieces are washed and dried, and then sent through two successive grinders to further reduce the particle size. After another washing and drying process, the particles are screened for uniform size.

11 Pure agglomerated cork is formed by packing cork particles into a mold and covering it securely. Superheated steam (600° F or 315° C) is passed through the

Cork is composed of dead cells that accumulate on the outer surface of the cork oak tree. Harvests occur at nine-year intervals, when the cork layer reaches a thickness of 1-2 in (2-5 cm). The harvest from a young tree yields about 35 lb (16 kg) of cork, while the yield for an older tree may be 500 lb (225 kg).

Cork intended to be used as bottle corks is first softened by steam and then cut into strips. Next, the strips are fed through a machine that punches hollow metal tubes through them, removing cylinders of cork.

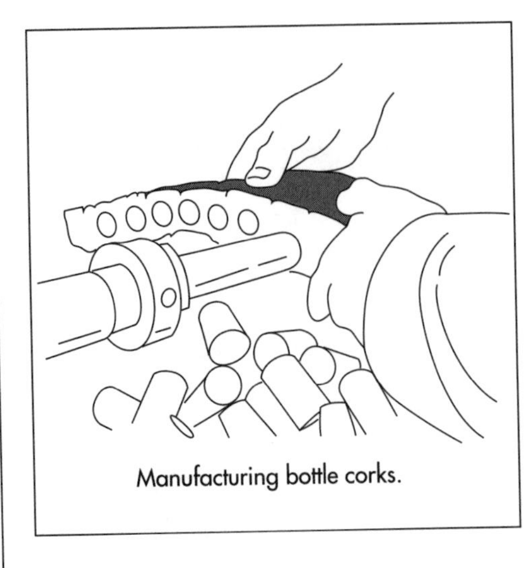

Manufacturing bottle corks.

mold. Alternatively, the mold is baked at 500° F (260° C) for four to six hours. Either process binds the cork particles into a solid block by activating their natural resins.

12 Compound agglomerated, or composition, cork is made by uniformly coating the cork granules with a thin layer of an additional adhesive agent. The coated granules are pressed into a mold and slowly heated (the temperature varies, depending on the adhesive used). When removed from the mold and cooled, the blocks are stacked to allow air circulation and are allowed to season.

13 The agglomerated cork is cut for its intended use. For example, sheets may be cut from rectangular blocks. Or if a tubular mold was used, the resulting cork rod may be sliced into discs. A large, cylindrical block might by revolved against a knife blade to shave it into a long, continuous sheet that is rewound into a roll.

## Byproducts/Waste

Cork waste generated during the manufacturing process is ground and used to make agglomerated cork products. Cork powder that is generated by the grinding process is collected and burned to help fuel the factory. Chemical components removed from cork during its processing can be recovered as useful byproducts and include tannin (used for curing leather), hard wax (used in products like paraffin, paint, and soap), resinous gum (helps vanish adhere to copper and aluminum), and phonic acid (used to make plastics and musk-scented toiletries).

## Where to Learn More

### Books

Cooke, Giles B. *Cork and the Cork Tree.* New York: Pergamon Press, 1961.

### Other

"Transformation Procedures for Natural Cork." Natural Cork Quality Council. http://corkqc.com/ctcor3.htm (February 1999).

Oliveira, Manuel, and Leonel Oliveira. "The Cork." http://www.portugal.org (February 1999).

"The Story of Cork." http://www.shofftackle.com/loadit.html (February 1999).

*—Loretta Hall*

# Cough Drop

A cough drop is medicinal tablet designed to deliver active ingredients which suppress or relieve the cough reflex. They are made just like hard candies; ingredients are mixed together, they are cooked, cooled, and packaged. First developed during the eighteenth century, cough drops have become a significant part of the $2 billion cough and cold market.

## Background

Anyone who has gotten sick knows the sensation of a cough. It is a natural reflex that helps protect the body from infections. It plays an important role in clearing the throat and other air passages of secretions and irritating particles. These particles include dust, food, liquids, and mucus. A cough occurs in three distinct steps. It typically begins with a deep breath which draws air into the lungs. The vocal cords spontaneously close thereby sealing the windpipe. Next, the air is compressed by the tightening of the expiratory muscles. The vocal cords are suddenly opened and the air trapped in the lungs is rapidly expelled along with any foreign debris in the windpipe.

Coughs associated with colds can be either productive or nonproductive. A productive cough helps clear the respiratory passages of the lung. A nonproductive cough is brought on by a minor irritation and has a limited benefit. It is the nonproductive cough that cough drops are designed to sooth or suppress.

The are two types of active ingredients in a nonprescription cough drop including expectorants and cough suppressants. An expectorant is a material that aids in the removal of phlegm from the respiratory tract. It works by blocking the sensory nerves that are involved in triggering a cough. While many expectorants are available, data about their functionality is not. Some clinicians even question whether expectorants are effective. Antitussives, which are cough suppressants, work in a variety of ways affecting either the lungs, muscles, or brain.

## History

Using syrups and herbal teas to control coughing has been known since antiquity. An ancient Hebrew text suggests the use of goat's milk for this reason. In the second century, Galen was perhaps the first to report an effective cough suppressant. Cough drops originally developed from candy. The first hard candies were produced during the fourth century. Since sugar was so expensive, these products were typically a luxury item available only to the rich. Over the years, sugar crops were planted in North America and throughout the world. Sugar refiners were established, and the price of sugar was reduced, making it available to everyone by the eighteenth century.

While the active ingredients in a cough drop were known for centuries, it was not until the nineteenth century that the cough drop was born. One of the first mass-produced cough drops was the Smith Brothers cough drop. According to the company, James Smith was operating a restaurant when a journeyman introduced him to a formula for a cough candy. He mixed up a batch in his kitchen and was able to quickly sell them. Demand for his product grew and he began advertising in 1852. He enlisted the aid of his two sons who helped mix batches and sell them on the streets of Poughkeepsie, New York.

*One of the first mass-produced cough drops was the Smith Brothers cough drop. According to the company, James Smith was operating a restaurant when a journeyman introduced him to a formula for a cough candy. He mixed up a batch in his kitchen and was able to quickly sell them. Demand for his product grew and he began advertising in 1852.*

They inherited the business in 1866 when James Smith died, and renamed the company Smith Brothers. During this time, they sold their cough drops in large glass bowls. To prevent imitators, they developed a unique package in 1872 that was filled at the factory. In 1922, menthol cough drops were introduced. Over the years, a variety of manufacturers have developed their own cough drop formulas. Each one has tried to improve the flavor and efficacy of their product.

## Raw Materials

Cough drops have two categories of ingredients. One type makes up most of the cough drop while the other is the active, or functional, ingredients. The major portion of cough drops is made up of ingredients found in typical hard candy recipes. The essential ingredients include sugar, corn syrup, acids, colors, and flavors. Sugar is a disaccharide compound called sucrose. It is obtained primarily from sugarcane or sugar beets by an extraction process. In a cough drop recipe, sugar crystals are usually used. Sugar is responsible for the physical structure of the cough drop along with its sweet taste and mouthfeel.

Corn syrup is a main component of cough drops. It is a mixture of sugars that is composed of polysaccharides, dextrose, and maltose. The main reason it is used is to control the crystallization of sugar. It also provides some sweetness and body to the cough drop. Additionally, it reduces the formation of dust from sugar during the blending stage.

To increase the visual appeal of the cough drop various dyes are added. In the United States, these dyes are strictly regulated by the government. Some that are allowed in food products include red dye #40, yellow dye #5, yellow dye #6, and blue dye #1. Natural colorants like caramel coloring are also used. Using only these colorants, the most popular cough drop colors, red and blue, can be produced.

To cover the taste of the active ingredients, various flavoring ingredients are put into cough drop recipes. Both artificial and natural flavors are used. Artificial flavors are mixtures of aromatic chemicals like methyl anthranilate and ethyl caporate. Natural flavors are derived from fruits, berries, and honey. Acids such as citric, lactic, tartaric, and malic acid are also included to modify the flavor.

Various active ingredients can be included in a cough drop recipe. As mentioned previously, these can be either expectorants or antitussives. Some common ingredients are volatile oils such as menthol or eucalyptus oil. Volatile oils, or essential oils, are obtained from parts of a plant through extraction or distillation processes. Menthol is typically isolated from the *Mentha arvensis* plant or distilled from peppermint oil. It may also be synthetically produced. Menthol has a cooling effect in the mouth that helps to relieve irritation. It is also thought to work as an expectorant. Eucalyptus oil is isolated from the eucalyptus plant. It is believed to have a medicinal effect functioning as an expectorant and a relief agent for minor mouth and throat irritations. Recently, companies have been including zinc in their cough drops. Certain evidence suggests that zinc may be beneficial in fighting symptoms of a cold. Vitamin C is another ingredient that has been included in some brands of cough drops. Other ingredients that may be found are herbals such as echinacea or ginko biloba. Peppermint oil, camphor, and sodium citrate have also been used.

## Design

Cough drops, or lozenges, are usually sold as small, hard candy pieces that slowly release their medicine as they melt in the mouth. Chemically speaking, they are a supersaturated solution of water molecules, sugar, and corn syrup. They can be either grained (opaque) or nongrained (clear). While all cough drops are designed to sooth and relieve coughing, some have added ingredients to help fight colds, freshen breath, or clear nasal congestion. Certain cough drops have reduced active ingredients and are created specifically for children. There are a wide variety of flavors, the most popular of which are cherry, honey, and menthol.

## The Manufacturing Process

The basic steps in producing a cough drop are mixing, cooking, cooling-working, forming, cooling, and packaging. Most man-

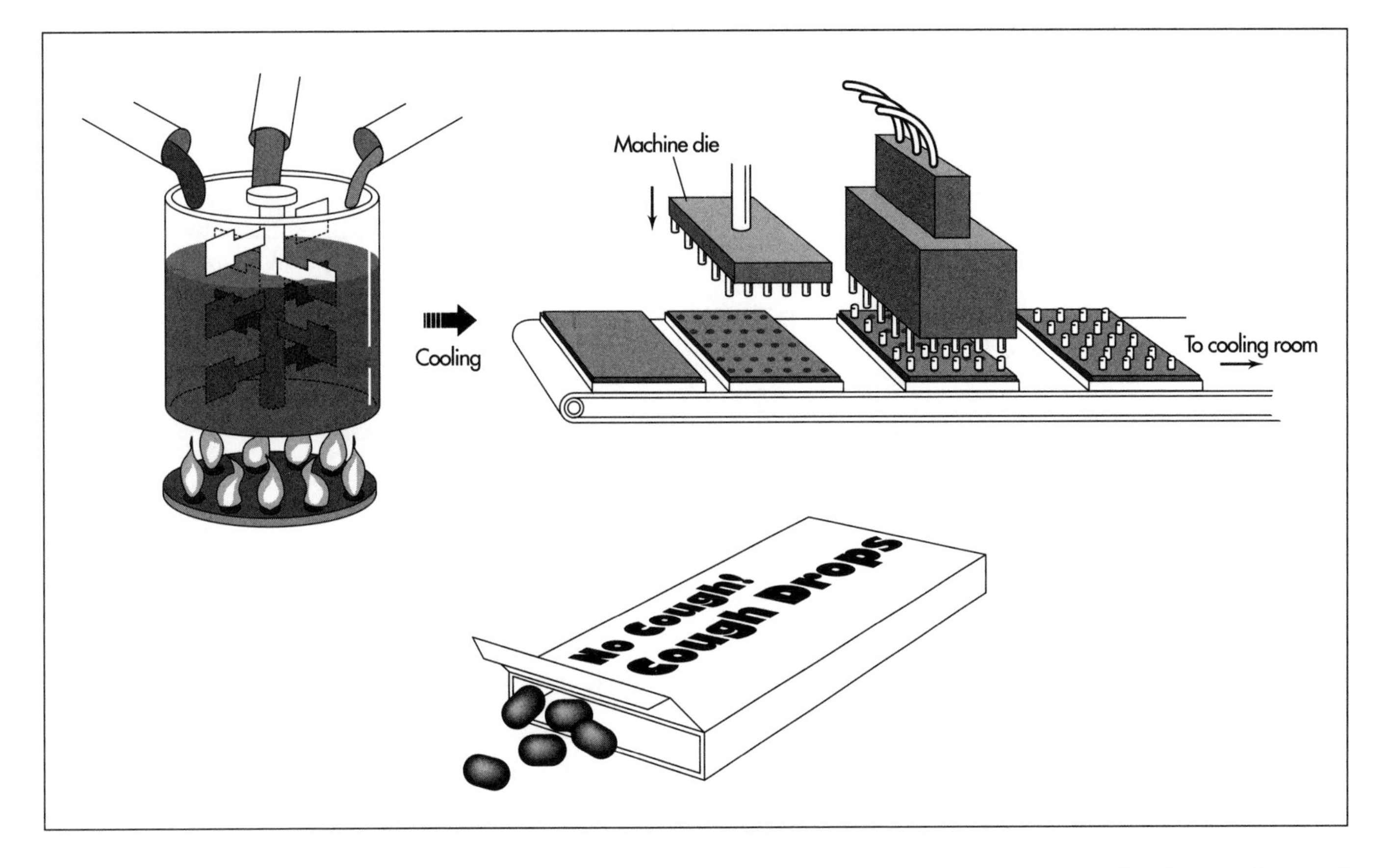


ufacturers have an automated production line connected by a conveyor system.

### Mixing

1 In this manufacturing step, the ingredients are combined in a large, stainless steel container by compounding personnel. Water is usually pumped directly in to the tank. The sugar, corn syrup, and certain other ingredients are then mixed until they are ready to be cooked.

### Cooking

2 To produce the cough drop, the moisture level of the mixture must be reduced. Some cough drops have a moisture content as low as 0.5%. There are three types of cookers that are employed. Batch cooking involves a direct-fire cooker. In this method, the heat is applied to the mixture from below and the moisture evaporates into the air. Quicker methods use either a semicontinuous cooker or a continuous cooker that operates under low pressure conditions. In the semicontinuous process, the cough drop mixture is first boiled under atmospheric pressure and then pumped into a second kettle that is a vacuum chamber. It is then rapidly cooked and drawn off for working. Continuous cookers use a scraped-surface heat-exchanger which can cook the cough drop within a few seconds.

### Cooling and working

3 After cooking, the cough drop mixture is conveyored to cooling slabs. It is cooled to facilitate incorporation of ingredients such as color, flavors, and active ingredients. The cooling slabs are composed of steel and have a built-in circulating water system. Candy plows are also involved. They move the mixture around so some of the hotter areas come in contact with the cooler surface of the slabs. As it cools, the mixture becomes plasticized and it looks like a workable mass similar to dough.

4 At this point, the rest of the cough drop ingredients are added. They are poured onto the batch and machines pull and twist the mass, working the ingredients throughout. When everything is adequately combined, the batch is split off and sent through a sizing machine. These devices roll the batch into a cylindrically shaped mass and reduce

Much like the manufacture of candy, the basic steps in producing a cough drop are mixing, cooking, cooling-working, forming, cooling, and packaging.

the diameter to a suitable size. From here, the mass is fed into the forming machines.

### Forming

5 Various forming machines are available. Typically, cough drops are tablet-shaped products. The cylindrical mass runs through this machine and is cut into smaller pieces. These pieces are then put into a die and stamped to produce the desired shape. The cough drops are then ejected from the die and moved to the finishing phases of production.

### Cooling and packaging

6 The formed cough drops are rapidly cooled to ensure that they maintain their shape. This is done on a conveyor belt that is equipped with rapidly blowing air jets. From the cooling area, the cough drops are put into packaging. Many manufacturers wrap each cough drop in a wax paper package to prevent them from absorbing moisture from the air. A number of these are then put into a larger bag for final sales. Other manufacturers do not individually wrap their cough drops, but store them in a bulk package. These are typically a wax-coated box that is sealed. To prevent them from sticking together, bulk packaged cough drops are often coated.

## Quality Control

As with all food and drug processing facilities, quality control begins by monitoring the characteristics of the incoming ingredients. These ingredients are taken to a quality control lab where they are tested to ensure they meet specifications. Tests include evaluation of the ingredient's physical properties such as appearance, color, odor, and flavor. Certain chemical properties of the ingredients may also be evaluated. Each manufacturer has their own tests that help certify that the incoming ingredients will produce a consistent cough drop. In addition to ingredient checks, the packaging is also inspected to ensure it meets the set specifications.

After production, the characteristics of the final product such as appearance, flavor, texture, and odor are also carefully monitored. The usual test methods involve comparing the final product to an established standard. For example, to make sure the flavor is correct, a random sample may be taken and compared to some set standard. Other qualities such as appearance, texture, and odor may be evaluated by sensory panels, a group of specialists that are trained to determine small differences. In addition to sensory tests, other instrumental measurements are taken.

## The Future

Cough drop recipes have changed little since they were first introduced. Most of the advancements have come in the design of the cookers and other processing equipment. It is expected that future improvements aimed at increasing the speed of production will continue to be found. Another area that will be expanded will be the addition of novel ingredients that may provide multiple benefits to the consumer. For example, some cough drop marketers have introduced vitamin C-containing products. These cough drops are intended not only to sooth a cough, but also relieve some of the symptoms of a cold.

## Where to Learn More

### Books

Alikonis, J. *Candy Technology*. Westport, CT: AVI Publishing Co., 1979.

Covington, T. *Handbook of Nonprescription Drugs*. Washington, DC: American Pharmaceutical Association, 1993.

Mathlouthi, M. and P. Reiser, eds. *Sucrose: Properties and Applications*. London: Blackie and Sons, Ltd., 1995.

### Periodicals

Friedman, M. "As Temperatures Drop, Cough Remedies Flourish."*Adweek* (February, 1989).

Slezak, M. "Warm Weather Cools Cough and Cold Sales."*Supermarket News* (March 6, 1995).

*—Perry Romanowski*

# Cranberries

## Background

The cranberry is a slender, trailing native North American shrub (*Vaccinium macrocarpon*) that grows in moist, sandy soil. The fruit berry is small, red, quite tart, and high in vitamin C. The berries are used either fresh or in processed foods such as juices, jams, and jellies. Vines are planted in sandy, peat-rich soil that is acidic and often flooded with water for more efficient harvesting. Cranberries grow wild in the northern climes of North America. The cranberry is one of only three fruits native to North America, the blueberry (*Vaccinium angustifolium* and *V. corymbosum*) and the Concord grape (*Vitaceae lubrusca*) being the other two. There are about 1,200 cranberry growers in North America and cultivation occurs only in Massachusetts, Wisconsin, New Jersey, Oregon, and Washington State. Today, at least 200 billion cranberries are harvested each year and the annual sales exceed $1.5 billion dollars. Of that, Ocean Spray Cranberry Growers Cooperative markets 90% of the annual yield in the United States.

### *History*

Because cranberries are an indigenous fruit, the Native Americans used them in a number of ways—fresh, ground, and mashed and baked with cornmeal. They mixed dried cranberries with dried meats and melted fat to make pemmican—a survival food for winter. Cranberry poultices, medicinal teas, and dyes were also used. European settlers thought the flower of the cranberry vine resembled the crane's head—hence the name cranberry—and used them extensively by 1700. Sailors used them to stave off scurvy because of their high vitamin C content.

Commercial cranberry production began in Cape Cod in 1815 and early commercial producers picked cranberries by hand. Cultivated throughout the nineteenth century in the East as well as the Midwest, individual growers often devised their own innovative machinery to meet harvesting needs. Nearly all attempted to address the problem of laborious picking. Individuals rigged up suction pickers that worked like vacuum cleaners, motorized pickers, homemade weed whackers, and hand scoops for pulling cranberries out of the water during wet harvesting. Standardized water reels and dry-pickers are now commercially available and greatly increase efficiency.

*At least 200 billion cranberries are harvested each year in the United States.*

## Raw Materials

The raw material needs are modest. The grower requires vines, which often flourish without abating for over 75 years (many are a century old). The land must be peat-rich, sandy, and acidic The beds must be situated near water that can be used to flood the cranberry marshes. No additives are mixed with fresh cranberries prior to packaging.

## Processing Cranberries

### *Preparing the cranberry beds*

1 Cranberrry cultivation begins with the preparation of individual cranberry beds. Cranberries are not planted in bogs or grown underwater. Cranberries are grown on dry land that is flooded at various points during cultivation generally for ease of harvesting (and sometimes to protect the crop from

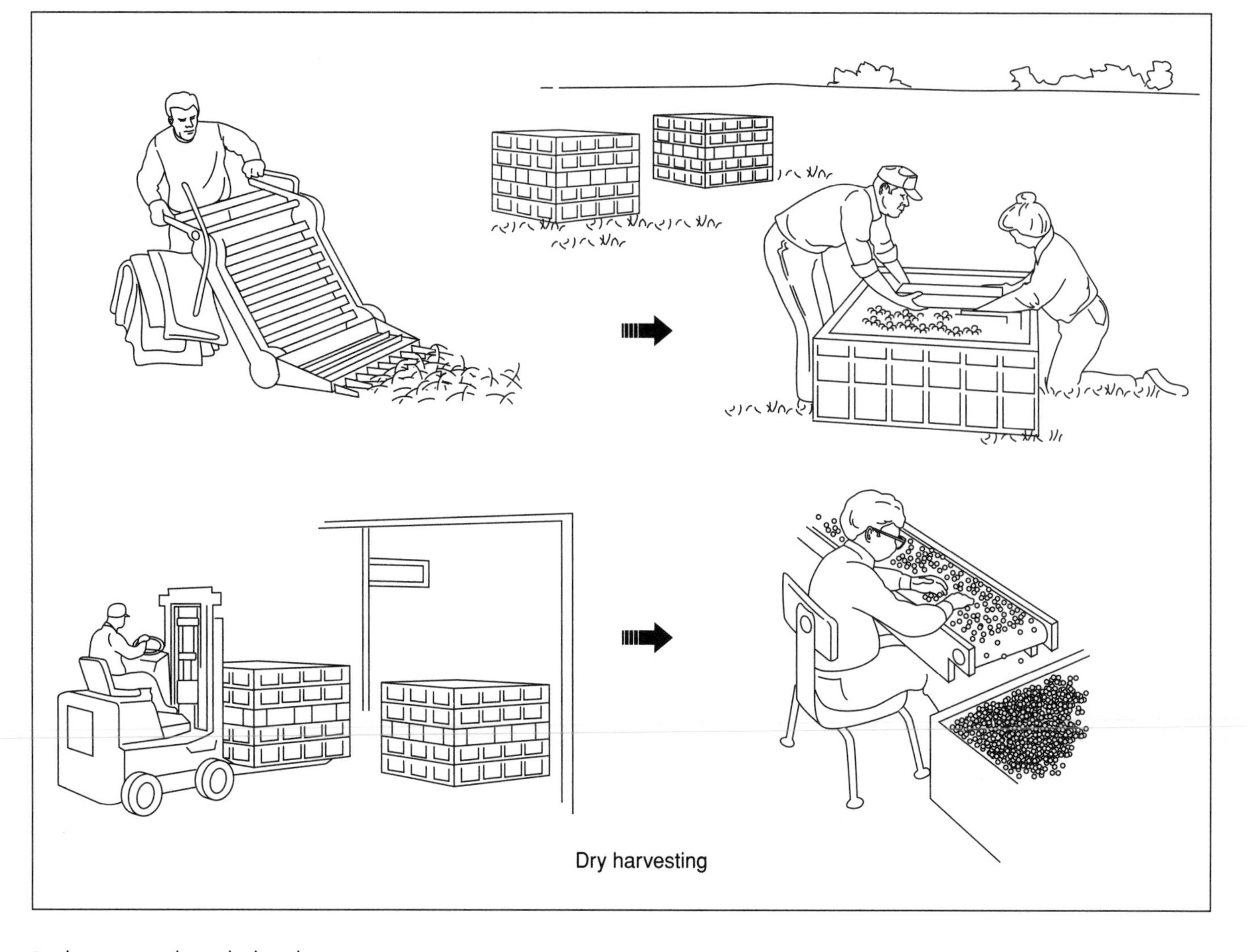

Dry harvesting

Dry harvesting is the method used to harvest cranberries that will be sold fresh. It involves collecting the ripe cranberries from the vines by using a mechanical picker that combs through the vines with its teeth. The machine deposits the berries in an attached burlap sack. Once full, a sack is emptied onto a metal screen that sifts debris from the berries. The berries are then crated and transported to the processing plant, where bruised and rotten berries are sorted out and discarded.

frost). One cranberry bed can take up few acres of land, a grower may have many beds in close proximity. The top layer of soil is sheared off and used to build earthwork dikes around the perimeter of each bed. Drainage ditches and canals are also constructed in order to manage and contain the water. The cranberry bed is leveled with laser-guided equipment.

### *Planting*

2 Cranberries are grown on vines. New plantings are developed from vine cuttings taken from well-established beds. In the spring cuttings are spread over the prepared soil bed and embedded into the dirt with a dull planting disk. By the end of summer, the vines have rooted. It takes three to five years for these cuttings to begin producing quantities of fruit for commercial processing, but once established, they will produce high-quality fruit for decades.

### *Caring for the crop*

3 The cranberry vine begins to bloom in early June. Cranberry vines blossom and flowers begin to open around mid-June. By the end of the month, all of the flowers are in full bloom. The vines must be pollinated by honeybees so they reach their full productivity. In order to ensure pollination, growers bring in about one to two beehives per acre. The result is a concentration of approximately nine million bees in a 100 acre conglomeration of beds. The honeybees continue to pollinate the flowers until early July, when blooming is complete.

4 Small berries appear in July and grow larger until they are harvested in October. During the growing season the vines, are irrigated and fertilized. Pest and weed populations are controlled in order to maximize fruit production.

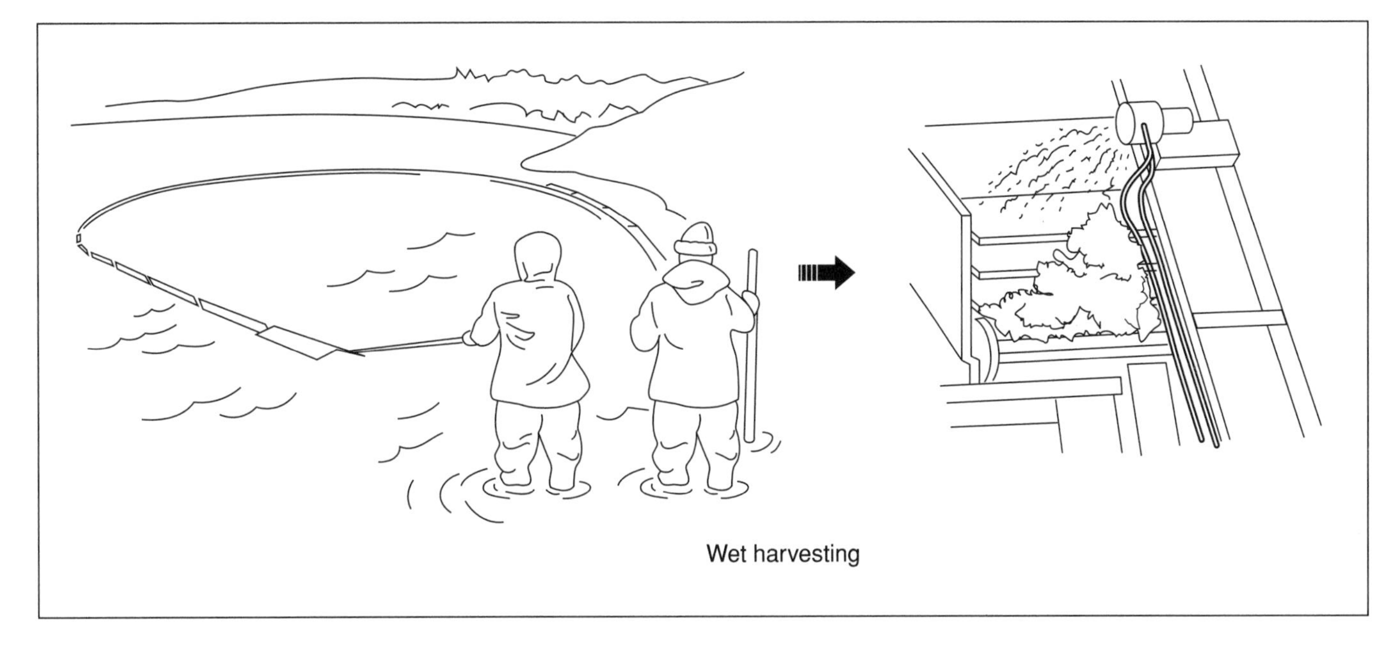

Wet harvesting

Wet harvesting methods are used when cranberries will be processed into jam, juice, and jellied sauce. The cranberry beds are flooded the night before harvesting. The next day, a machine called a water reel is driven through the water, its paddles churning up the water. The ripe cranberries are agitated just enough to separated from the vine and float to the surface of the water. The floating berries are corralled together with an inflatable boom and a large pipe siphons the berries into a metal box called a hopper, which separates the debris from the berries.

It is critical to protect the cranberry vines from frost. Temperature alarm systems are installed to alert the growers to low temperatures. When frost is impending, the irrigation system shoots sprays of water over the plants. Heat is released from the water before it freezes thus protecting the plants from damage.

### *Dry harvesting*

Dry harvesting is the best method for obtaining a firm, fresh bagged cranberry.

5 In dry harvesting, cranberry growers use a mechanical picker that resembles a reel lawn mower. Cranberry cultivators train the vines so that the fruit all lies in one direction, allowing the picker's metal teeth to comb the berries off the vine. Berries are placed onto a bucket conveyor that carries them to a burlap sack in the back of the machine. When the sack is full it is lifted off the conveyor and another is put in its place.

6 The sacks are emptied onto metal screens that are used to separate the berries from leaves and debris. Next, the berries are crated. When the crates are stacked three high, they are bundled together with a nylon belt. Many cranberry growers use helicopters to airlift crates to nearby flatbed trucks. Since harvesting occurs in the field, the use of helicopters eliminates the potential crushing of the cranberry vines.

7 Dry berries are taken to the plant to be sorted. Machines sort out unsuitable berries by bouncing them over a 1 in (2.54 cm) high "bounce board." Firm, round berries bounce over the board, while rotten or bruised berries remain trapped and are discarded. Berries that clear the bounce boards are carried away by conveyor belts and are packaged by machines that check weight and package accordingly.

### *Wet harvesting*

Wet harvesting is used when the final product will be juices, jams, and jellied sauces.

8 Wet harvesting is a quicker processing method than dry harvesting. Ninety-five percent of a crop can be wet harvested in 60% of the time it takes to dry harvest. However, since wet harvested berries are more perishable than dry harvested berries, they are only used for processed cranberry foods. Flooding occurs the night before harvesting is to begin. The beds are flooded with 18 in (45.72 cm) of water. The next day, a machine called a water reel (nicknamed egg beaters) is driven through the water, its paddles churning up the water. The ripe cranberries are agitated just enough to separated from the vine and float to the surface of the water. The floating berries are corralled together with an inflatable boom. A large pipe is placed just beneath the surface of the water in the center of the aggregation of gathered, floating cranberries. The

pipe sucks up the berries (along with impurities in the water such as leaves, water, twigs) into a metal box called a hopper.

9 The hopper separates the debris from the berries, pumps the water back into the bog, and pumps the berries into a trail truck. The loading of cleaned berries continues until thousands of pounds of berries are loaded into the truck and taken to the plant for processed cranberry foods.

### *After harvest*

10 From late December through March, the cranberry beds are flooded with water until they are completely frozen. The ice forms a protective layer around the plants, protecting them from dehydration and the weather. The plants remain dormant until the following spring.

11 Every four years, a layer of sand is added to the top of the frozen beds. As the ice begins to melt, the sand falls to the ground, creating a more established root system and promoting growth.

12 The beds are also periodically mowed to encourage shrub health and growth. Mowing occurs in spring and the plants do not produce fruit the year of a mowing.

## Quality Control

Quality control of agricultural products such as cranberries includes a wide variety of activities not always associated with what we might generally consider quality control. However, first and foremost, it is important to propagate new vines from old vines that have produced luscious, full, large berries. Second, the vines must be planted in soil that is truly sandy and full of rich peat to ensure necessary nutrients are provided. Then, the beds must be carefully constructed so that the earthworks and dikes adequately control the water flooding. Also, frost warning equipment which alerts that grower as temperatures drop dangerously low must be heeded and sprinkling equipment put into action or the entire crop is at risk.

As with other crops, pests are always a factor in quality control. The grower must be vigilant about checking for pests such as cranberry tipworm or cranberry fruitworm that can ruin a crop if left unchecked ( the fruitworm sucks out the meaty centers of berries, fills it full of its excrement and then encases the plants or vines in its silky web). Ultraviolet light traps can be used to attract the fruitworm and then spray the concentration of bugs with a USDA approved insecticide. In fact, insecticides are used for many different bugs but the industry tries to keep their use to a minimum. Fruit fungi and diseases are also issues for cultivators. The industry uses chemicals and carefully monitors weeds, over fertilization, and handling in order to reduce these diseases.

Finally, the machinery used in the processing of cranberries is constructed so as not to bruise or damage the berries. While these are fairly firm berries, they can be bruised if overhandled.

## Byproducts/Waste

While some cranberry growers do not use pesticides on their crops most do use chemicals to keep pests away from the berries. The United States Department of Agriculture (USDA) maintains strict controls on acceptable pesticides that may be used on fresh fruit and the growers must heed these guidelines. Washing and cleaning the fruit minimizes chemical residue remaining on the harvested fruit. Because the cranberry is harvested in or near water and some of the chemicals specifically formulated to fight diseases and berry blight are moderately toxic to fish, the cultivators must carefully monitor the effects of chemicals on the local fish population.

## The Future

Cranberry products are very popular and supply never has been equal to the demand. Since there are only 1,200 growers in North America and appropriate land available for cranberry crops is limited because of environmental factors such as wetland protection and limited water access, supply will not grow significantly. Presently, 90% of cranberry consumption occurs in North America. That demand will keep pricing competitive.

Cranberry juice has been an effective home treatment for urinary tract infections for some time and new studies suggest that cranberries are rich in cancer-fighting antioxidants.

## Where to Learn More


### Books

Jaspersohn, William. *Cranberries*. Boston: Houghton Mifflin, 1991.

### Periodicals

"Berry Good for You." *The University of California, Berkeley Wellness Letter* (February 1999): 2.

Mottau, Gary. "Water Berries." *Horticulture* (October 1984): 34.

### Other

Cranberries in New Jersey. http://www.burlco.lib.nu.us/.

Cranberry cultivation in Massachusetts. http://omega.cc.umb.edu/~conne/marsha/ccintro.html.

Maine Cranberry Growers. http://www.ne-maine.com/.

Ocean Spray. http://www.oceanspray.com/.

Oregon State University. http://osu.orst.edu/.

Wisconsin State Cranberry Growers Association. http://wiscran.org/.


*—Nancy EV Bryk*

# Crane

*As early as the first century, cranes were built that were powered by human beings or animals operating a treadmill or large wheel.*

## Background

A crane is a machine that is capable of raising and lowering heavy objects and moving them horizontally. Cranes are distinguished from hoists, which can lift objects but that cannot move them sideways. Cranes are also distinguished from conveyors, that lift and move bulk materials, such as grain and coal, in a continuous process. The word crane is taken from the fact that these machines have a shape similar to that of the tall, long-necked bird of the same name.

Human beings have used a wide variety of devices to lift heavy objects since ancient times. One of the earliest versions of the crane to be developed was the shaduf, first used to move water in Egypt about four thousand years ago. The shaduf consists of a long, pivoting beam balanced on a vertical support. A heavy weight is attached to one end of the beam and a bucket to the other. The user pulls the bucket down to the water supply, fills it, then allows the weight to pull the bucket up. The beam is then rotated to the desired position and the bucket is emptied. The shaduf is still used in rural areas of Egypt and India.

As early as the first century, cranes were built that were powered by human beings or animals operating a treadmill or large wheel. These early cranes consisted of a long wooden beam, known as a boom, connected to a rotating base. The wheel or treadmill powered a drum, around which a rope was wound. The rope was connected to a pulley at the top of the boom and to a hook that lifted the weight.

An important development in crane design occurred during the Middle Ages, when a horizontal arm known as a jib was added to the boom. The jib was attached to the boom in a way which allowed it to pivot, allowing for an increased range of motion. By the sixteenth century, cranes were built with two treadmills, one on each side of a rotating housing containing the boom.

Cranes continued to rely on human or animal power until the middle of the nineteenth century, when steam engines were developed. By the end of the nineteenth century, internal combustion engines and electric motors were used to power cranes. By this time, steel rather than wood was used to build most cranes.

During the first half of the twentieth century, European and American cranes developed in different ways. In Europe, where most cranes were used in cities with narrow streets, cranes tended to be built in the form of tall, slender towers, with the boom and the operator on top of the tower. Because quiet operation was important in crowded cities, these tower cranes were usually powered by electric motors when they became widely available.

In the United States, cranes were often used in locations far away from residential areas. Cranes tended to be built with the boom connected to a trolley, which could be moved easily from place to place. These mobile cranes tended to be powered by internal combustion engines. During the 1950s, the availability of stronger steels, combined with an increased demand for taller buildings, led to the development of cranes with very long booms attached to small trucks, or to crawlers with caterpillar treads. Mobile cranes and tower cranes of many different kinds are used extensively in construction sites around the world.

## Raw Materials

The most important substance used to manufacture cranes is steel. Steel is an alloy of iron and a small amount of carbon. For structures that do not require very high strength, a common form of steel known as carbon steel is used. By definition, carbon steel contains less than 2% of elements other than iron and carbon. Carbon steel exists in a wide variety of forms. The most important factor in determining the properties of carbon steel is the amount of carbon present, which ranges from less than 0.015% to more than 0.5%.

For structures that require great strength, particularly in cranes designed to lift very heavy objects, a variety of substances known as high-strength low-alloy (HSLA) steels are used. HSLA steels contain relatively low levels of carbon, typically about 0.05%. They also contain a small amount of one or more other elements that add strength. These elements include chromium, nickel, molybdenum, vanadium, titanium, and niobium. Besides being strong, HSLA steels are resistant to atmospheric corrosion and are better suited to welding than carbon steels.

Depending on the exact design of the crane, a wide variety of other materials may be used in manufacturing. Natural or synthetic rubber is used to make tires for mobile cranes. Certain structural components may be manufactured from various metals such as bronze and aluminum. Electrical components may include copper for wires and semiconducting elements such as silicon or germanium for electronic circuits. Other materials that may be used include ceramics and strong plastics.

## Design

Very few machines exist in as wide a variety of designs as cranes. Before the crane is constructed, the manufacturer must consider the site where it will be used and the weight it will need to lift. In addition, cranes are often modified to suit the needs of the user. For these reasons, it is not much of an exaggeration to say that no two cranes are exactly alike.

Cranes used for industrial purposes are generally designed to remain permanently in one location. These cranes often perform repetitive tasks that can be automated. An important type of industrial crane is the bridge crane. Traveling on tracks attached to two horizontal beams, known as a bridge, a trolley enables the movement of the bridge crane. Usually, the bridge itself can be moved along a pair of parallel rails, allowing the crane to reach a large, rectangular area. A bridge crane may also be designed so that one end of the bridge is supported by a central pivot while the other end moves on a circular rail, allowing a large, round area to be reached.

An overhead traveling crane is a kind of bridge crane in which the rails are located high above the ground. Usually supported from the ceiling of a building, an overhead traveling crane has the advantage of causing no obstruction in the work area.

Cranes used in construction often perform a variety of tasks and must be controlled by highly skilled operators. Construction cranes are divided into mobile cranes and tower cranes. Mobile cranes are mounted on trucks or crawlers in order to travel from place to place. An articulating crane is a mobile crane in which there is a joint between two sections of the boom, allowing it to move in a way similar to a knuckle in a human finger. Articulating cranes are generally used to lift objects located a relatively short distance away, but with a wide range of motion. A telescoping crane is a mobile crane in which two or more sections of the boom can extend and retract, changing the length of the boom. Telescoping cranes are less versatile than articulating cranes, but are usually able to lift heavier objects located a greater distance away.

Tower cranes are used in the construction of tall buildings. They are installed when construction begins and dismantled when the building is completed. An external tower crane is installed outside the building. As the building increases in height, the crane is raised by lifting the upper part of the crane and adding a new section of tower beneath it. An internal tower crane is installed within the

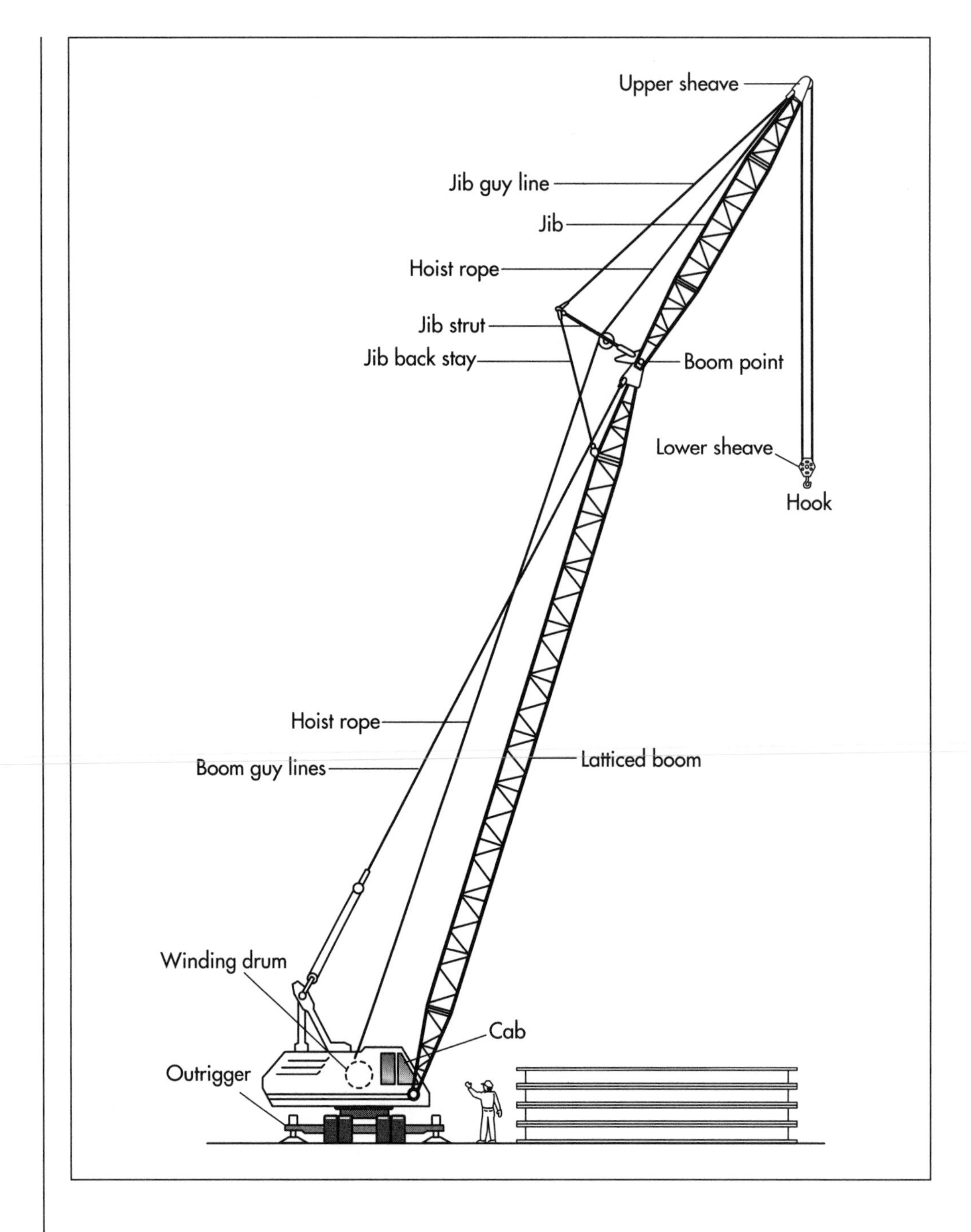


A mobile crane.

building. As the building increases in height, the crane is raised by lifting the base of the crane to a higher level within the building. .

## The Manufacturing Process

### *Making steel components*

1 Molten steel is made by melting iron ore and coke (a carbon-rich substance that results when coal is heated in the absence of air) in a furnace, then removing most of the carbon by blasting oxygen into the liquid. The molten steel is then poured into large, thick-walled iron molds, where it cools into ingots.

2 In order to form flat products such as plates and sheets, or long products such as bars and rods, ingots are shaped between large rollers under enormous pressure. Hollow tubes, such as those used to form the latticed booms of large cranes, may be made by bending sheets of steel and welding the long sides together. They may also be made by piercing steel rods with a rotating steel cone.

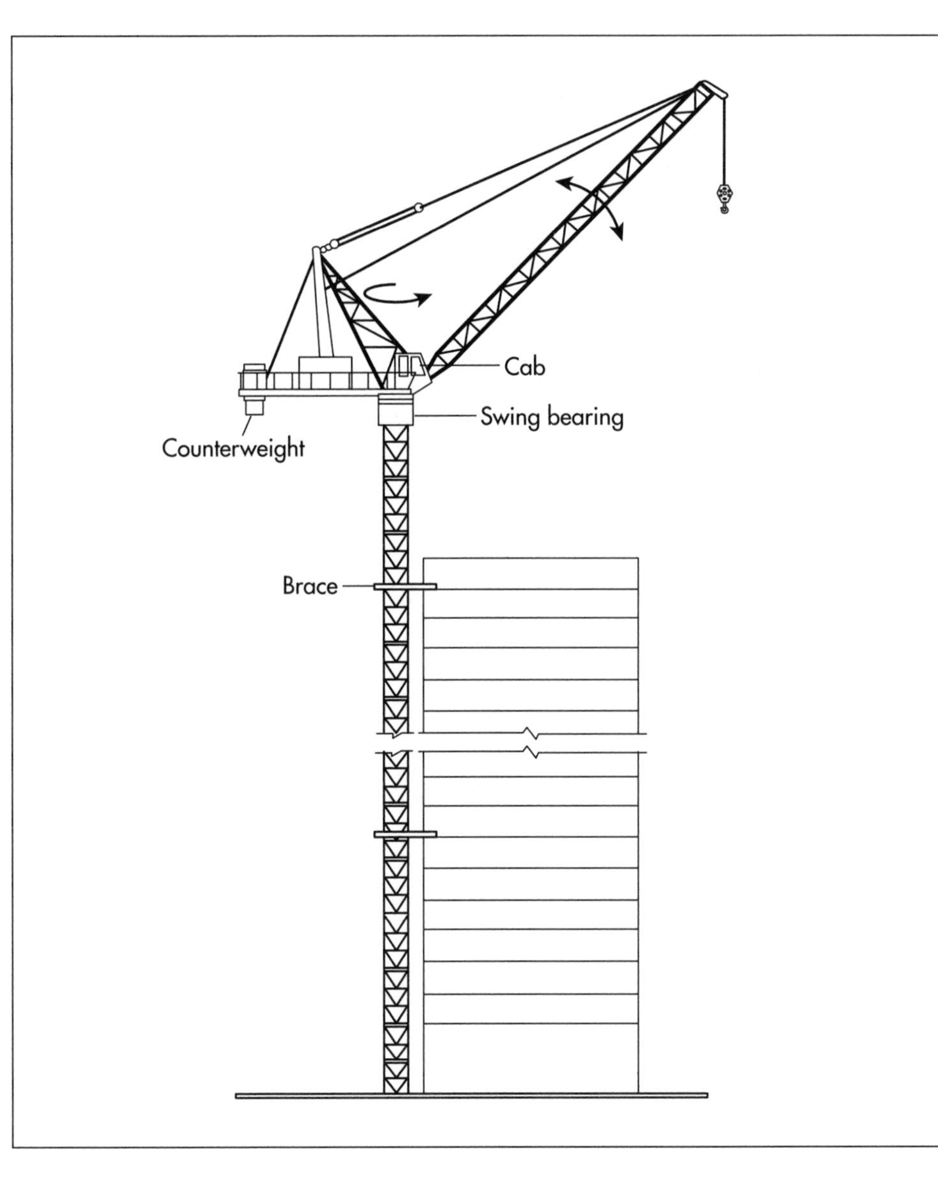


An external tower crane.

3 The cables used to lift weights are made from steel wires. To make wire, steel is first rolled into a long rod. The rod is then drawn through a series of dies which reduce its diameter to the desired size. Several wires are then twisted together to form cable.

4 Steel arrives at the crane manufacturer and is inspected. It is stored in a warehouse until it is needed. The many different components that will later be assembled into cranes are made using a variety of metalworking equipment. Lathes, drills, and other precision machines are used to shape the steel as required.

### *Assembling the crane*

5 A crane is put together from the necessary components. As the crane moves along the assembly line, the steel components are welded or bolted into place. The exact procedures followed during this process vary depending on the type of crane being assembled. For a mobile crane, the components are then assembled to a standardized truck or crawler of the appropriate type.

6 The assembled crane is tested and shipped. Depending on the size and type of crane, it may be broken down into subsections to be assembled on site. It may also be shipped whole on special large trucks.

## Quality Control

Safety is the most important factor to be considered during crane manufacturing. The steel used to make the crane is inspected to

An internal tower crane.

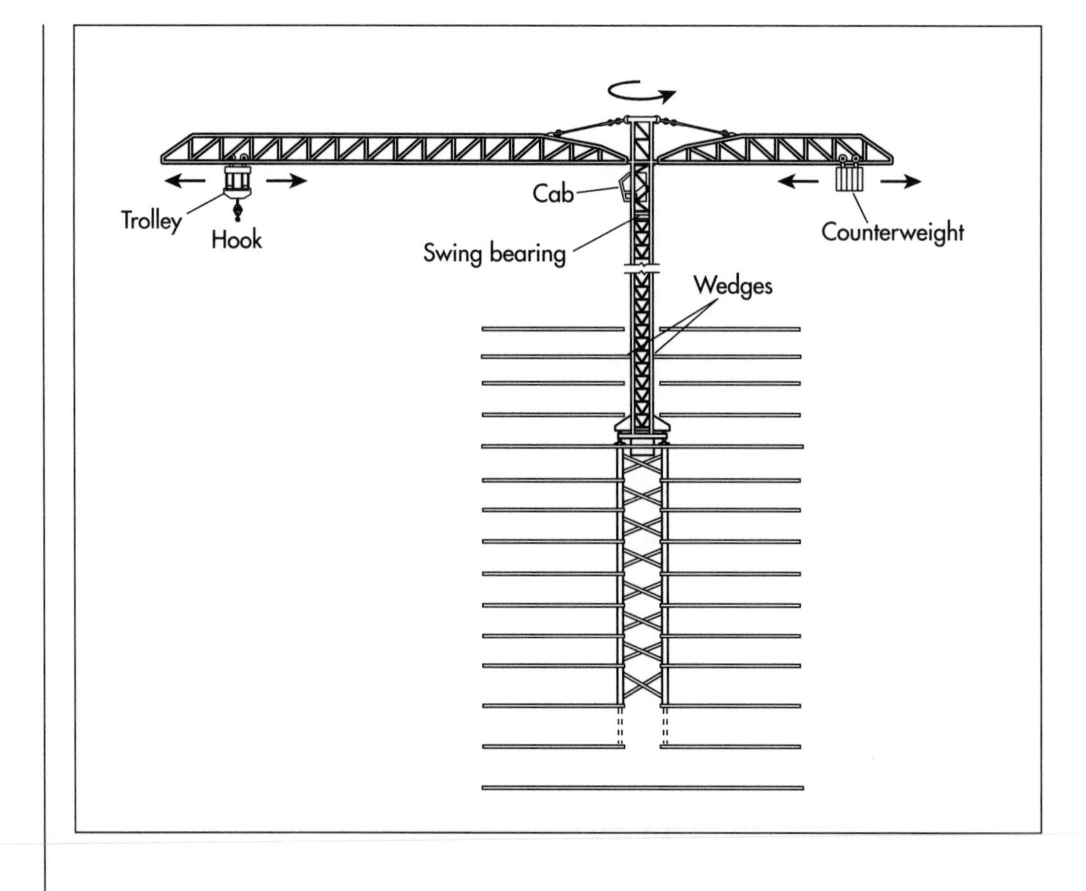


ensure that it has no structural flaws that would weaken the crane. Welds and bolts joints are inspected as well.

The United States government sets specific regulations through the Occupational Safety and Health Administration that limit the weight that a specific crane is allowed to lift. The Crane Manufacturers Association of America sets its own safety standards which exceed those required by the government. Special devices within the crane prevent the user from attempting to lift a weight heavier than that allowed.

A completed crane is first tested without a weight to ensure that all of its components operate properly. It is then tested with a weight to ensure that the crane is able to lift heavy objects without losing stability.

Safety ultimately depends on proper use of the crane. Crane operators must be specially trained, must pass specific tests, and must be examined for any visual or physical problems. The crane should be inspected each working shift, with a more thorough inspection of the motor and lifting apparatus on a monthly basis. Crane operators must be aware of changes in the environment in order to avoid accidents. For example, cranes should not be used during very windy conditions.

## The Future

Manufacturers of cranes are constantly seeking new ways to incorporate new technology into their products. Future cranes will have improved safety and versatility with computers and video screens that will allow operators to move heavy objects with increased accuracy.

Signs of the future can be seen in an unusual crane recently developed by James S. Albus, of the National Institute of Standards and Technology in Gaithersburg, Maryland. The Stewart Platform Independent Drive Environmental Robot (SPIDER) looks nothing like an ordinary crane. Instead, the SPIDER is shaped like an octahedron (a diamond-shaped solid consisting of eight triangles joined together in the form of two four-sided pyramids). Six pulleys support six cables from the top level of the SPIDER. The cables manipulate the lower level of the SPIDER, which is attached to tools or gripping devices. The six cables can be operated together or independently, allowing the lower level to be moved in all directions. The SPIDER can lift heavy objects to within 0.04 in (1 mm) of the desired location, and hold them within one-half of a degree of the desired angle. The SPIDER can lift up to six times its own weight.

## Where to Learn More


### Books

Jennings, Terry. *Cranes, Dump Trucks, Bulldozers, and Other Building Machines*. Kingfisher Books, 1993.

Shapiro, Howard I. *Cranes and Derricks*. McGraw-Hill, 1980.

### Periodicals

"The Crane: A Versatile Truck-Mounted Tool."*Public Works* (November 1988): 62-63.

Shapiro, Lawrence K. and Howard I. "Construction Cranes."*Scientific American* (March 1988): 72-79.

### Other

"Cranes." April 21, 1998. http://www.markocrane.com/crane.htm/ (June 29, 1999).


*—Rose Secrest*

# Crash Test Dummy

*Crash test dummies were developed in 1949 under contract to the U.S. Air Force for testing aircraft ejection seats that were mounted on rocket-propelled sleds on rails. "Sierra Sam" was the first dummy.*

## Background

Like a fashion mannequin, the dummy looks like a human, but its more-than-skin-deep beauties consist of high-tech instrumentation and a state-of-the-art physique. And like the ventriloquist's version, the crash test dummy can't speak except in a highly effective series of television commercials for seat belt safety.

Highway safety agencies around the world rely on the crash-worthiness test in which an automobile is rammed into a brick wall to observe damage and generate data about the car's performance. The occupants of the test vehicle are crash test dummies (properly called "Anthropomorphic Test Devices") that are fabricated to resemble and respond like human bodies and that are loaded with sophisticated instrumentation. The instruments record information on acceleration, speed, deceleration on impact, force of impact, and the various motions and deformations of each dummy's torso and limbs. These data are studied by safety engineers and related to behaviors of human occupants and their potential injuries.

The crash test method has major flaws despite its universal acceptance. The automobile (or other vehicle) is partially or fully damaged, depending on the test objective. The dummies can be retrofitted and reused many times, but their ability to respond with all the complexities of human bodies is always being perfected.

## History

Prior to the late 1940s, automobiles were tested using cadavers. Injuries could be observed on real bodies, but cadavers did not respond like breathing, flexible beings. Crash test dummies were developed in 1949 under contract to the U.S. Air Force for testing aircraft ejection seats that were mounted on rocket-propelled sleds on rails. "Sierra Sam" was the first dummy. The automotive industry later used the same type of dummy to develop lap seat belts and shoulder harnesses. Sierra Sam resembled the average adult male with a statistically correct weight and articulated limbs; however, Sam's spine and neck were rigid. The acceleration of his head as it followed the path of ejection could be measured, but this was far from sufficient for evaluating potential head injuries.

By 1952, Mark 1 was manufactured from a plaster cast of a live man, and this dummy marked a huge improvement in the state of the art. His skull housed sensors for measuring acceleration and the force of impact, and it was cast from two pieces of aluminum. Mark 1's spine consisted of a series of ball-and-socket joints with spacers to simulate the range of motion of a real backbone. The dummy also had a set of steel tubing ribs, vinyl skin, and foam flesh, but stiff limbs. Other dummies developed through 1956 were modified and instrumented to measure a selected range of motions. Dummies of sizes and weights other than average were made for the first time, and these post-Mark 1 dummies were used to test tractor safety, frogmen's suits for underwater escapes, and flight and safety aspects of space research for the National Aeronautics and Space Administration (NASA).

Space programs motivated the next advances in dummy design. To evaluate the effects of rocket thrust on astronauts strapped into the seats of space capsules, the

American and European space programs funded the invention of the Grumman-Alderson Research Dummy (GARD). The seated astronaut had to be perfectly aligned with the direction of rocket thrust, or both the man and seat would twist out of place. GARD was an important technological step because the processes of locating the center of gravity and evaluating the moment of inertia (the tendency to rotate with acceleration) were well instrumented in this durable dummy. GARD is still used to test ejection seats because the seats restrict motion ranges in this mannequin keeping them simple enough for engineers to measure rotations, acceleration history, and the stresses between the man and the seat.

Until 1966, the aviation industry, military aeronautics, and the space program led the development of test dummies. The automotive industry became the driving force in developing crash test dummies that year as automobile fatalities emphasized the need for improvements in the dummy's rib cage, spine, pelvis, and abdominal cavity to evaluate restraint systems. Mathematical models and experiments with separate ranges of the body showed that motions could be simulated with spring connections. The springs move on impact, and they also rebound so effects like whiplash can be studied. Tests of lap safety belts concentrated on injuries to the pelvis, but development of the shoulder harness required the dummy to have a breastbone (sternum), clavicles, and shoulder blades. The harnessed dummy was equipped with an instrumented visceral sac that imitated the motions of internal organs. Instrumentation was also installed in cavities in the thighs, chest, and head so that more complicated movements and force deflections could be studied.

By 1970, it was apparent that the adult-sized, male dummy did not accurately represent the smaller proportions of children and females. In fact some injuries to the smaller physiques were being caused by belts and harnesses that were proven safe by the average dummy. A larger male model and the first female dummy were produced in 1970 and were named Sierra Stan and Sierra Susie. That same year, they added to their family with Sierra Sammy (a six-year-old) and Sierra Toddler (a three-year-old). Neither child was correct in weight distribution, but at least they were represented and studied for the first time.

The problem of evaluating injuries to children was complicated by the fact that many children are injured in automobile accidents because they are standing or not sitting conventionally during those accidents. The range of motions and potential impacts are, therefore, far more variable for unrestrained children. Redesign of the child dummies followed immediately, and modifications to child dummies continued to be the most intensely pursued area of crash test design from the 1970s through the 1990s. In the 1970s, improved data gave the children more flexible skeletons and more supple limb joints than adults, made them adaptable to various positions in the car, and improved instrumentation so the broader range of variables could be measured or interpreted. Development and testing of child safety seats, booster seats, and airbags necessitated special attention toward infant, toddler, and youngster dummies.

Improvements in computer analysis were also revolutionizing crash tests by the mid 1970s. Computer methods allowed measurement of almost forty different parameters in the behavior of test dummies. Construction materials were similarly improved to make realistic, fully articulated dummies possible. To replicate the behavior of human bones, the dummy's bones were manufactured of fiberglass with greater breaking strength so the dummy could be used again. These many adaptations led to more and more specialized dummies, however, so models like the Supermorphic Dummy were made exclusively for car crash tests—they were too fragile for ejection testing.

Specializations in aircraft testing dummies included the Limb Restraint Evaluator (LRE) Anthropomorphic Manikin, which was created especially for testing restraint devices to prevent injuries due to flailing during ejection from military jets. Other specialized dummies tested experimental parachutes, helicopter crashes, and racing cars. For example, tests of crashes involving race cars (of the type raced at the Indianapolis 500-Mile Race or on the Championship Auto Racing Team [CART] circuit) showed that the heavy racing helmets worn by the drivers do protect their heads from impact damage, but the helmet's weight increases the risk of neck in-

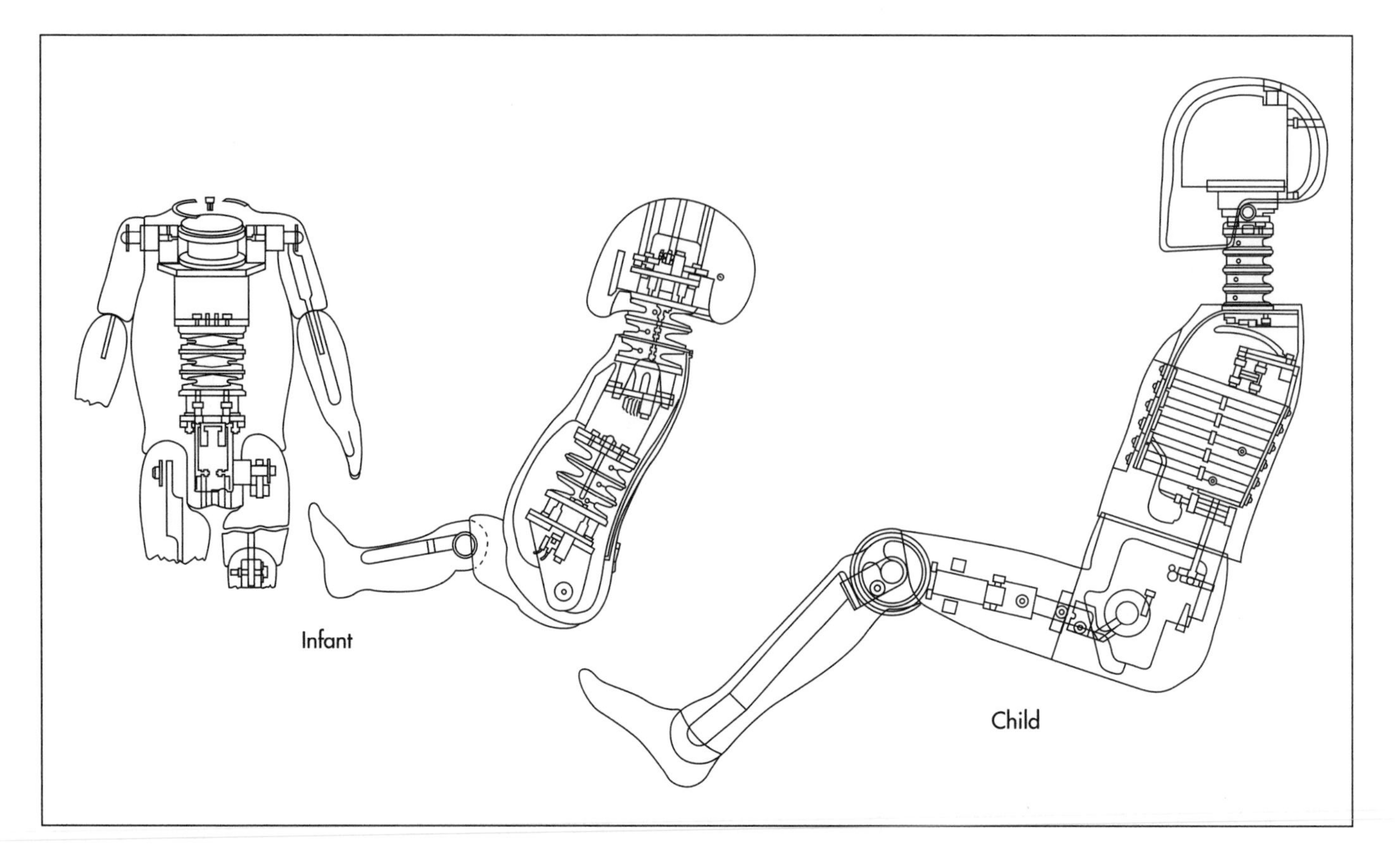


The first crash test dummies manufactured during the 1950s and 1960s did not address the differences between body types. Those first dummies were statistically correct adult males, and as such, they did little to represent the typical infant, child, and female body types. By 1970, the dummies' shortcomings were apparent and various body types were manufactured.

juries because it forces the neck to elongate. Impact of an Indy car into a concrete wall reduces the car's speed by 40 mi (64 km) per hour in 70 milliseconds, and the driver feels an impact of 60 times the force of gravity. The sensor "black box" in the dummy evaluates all of the forces at work 2,000 times per second while the crash is in progress.

Test dummy technology accelerated in development again with the design and testing of the airbag. The dummy family was diversified further to include still more body types. Fatalities from airbag deployment occurred most often among persons outside the statistical dimensions. These "small occupants out of position" or OOP occupants stimulated more evolutionary changes in the mannequins. Airbag deployment subjected the dummies to forces that exceeded their design capabilities. Significant upgrades to the dummy family had taken place by 1997 thanks to airbag testing and development of side impact airbags.

## Raw Materials

The body of the crash test dummy is made of metal parts consisting of aluminum, bronze, or steel (depending on design, purpose, and body parts) as well as metal plating materials. A wide variety of plastics are also incorporated in most designs, and the plastics include Delrin (a type of long-wearing acetate resin), urethane and polyurethane foam, and vinyl. Most dummies are clothed to simulate reality, and specific vendors supply clothing, paint, and adhesives for marking critical targets or measurement points on the dummies. In fact, not only are outside suppliers used for many dummy constituents, but the federal testing agencies that approve dummy designs dictate specific vendors that must be used for some materials.

Dummy manufacturers supply little or no instrumentation. Designs provide space for instrumentation to be secured, but instrumentation is so specific to actual testing conditions that the customers of the dummy builders install their own. Some small potentiometers are sometimes built into the manufacturer's lines, but more sophisticated instruments like accelerometers and load-measuring sensors are selected by the customer from their own instrument packages or specialized vendors.

## Design

The design of crash test dummies is an intricate and isolated process limited to government agencies, dummy manufacturers, and customers like automobile manufacturers. The dummy makers are not responsible for creating their own designs. Instead, they receive drawing packages, sets of construction specifications, and specifications for required testing from the National Traffic Safety Administration (NTSA). European agencies provide similar sets of documents to dummy manufacturers. Based on their experiences in building dummies, repairing or retrofitting them, and observing their post-accident traumas, manufacturers of dummies, as well as users, do have input into design changes. The NTSA bases its designs on a multitude of data including accident reports, location within a vehicle, variations in physique and physical development, autopsies, and simulations. Biomechanics experts analyze all the available data and create the specifications that are sent to the manufacturers.

## The Manufacturing Process

The manufacture of crash test dummies is highly proprietary because of the complexities of design and the small number of qualified manufacturers in the world. All design and construction meets an extremely high set of standards there are no "B-grade" crash test dummies.

1 When the drawing and specifications package for a new crash test dummy is received, the manufacturer begins by determining which parts and materials must be purchased from specific vendors, which must be acquired from a range of approved vendors, and which are to be manufactured in house. As examples, clothing for the dummies is furnished by a specific vendor; but, for the foam used to support the vinyl skin, the naming of specific vendors is avoided, and the manufacturer only has to meet certain equivalents.

2 Patterns and molds for metal pieces to be cut or cast are made from the design drawings, and necessary tooling is done to prepare to make the metal pieces. They are cast, cut, heat treated as many as three times, plated, and machined to trim, smooth, and refine the completed pieces. Finally, the metal pieces are assembled using fasteners that are also specified.

3 Plastics are similarly designed, tooled, and molded by injection molding or other methods suited to the part and material. Plastics are also machined and assembled or fitted to the metal parts.

4 After the metal and plastic body of the dummy has been assembled, the skin is fitted to the dummy. Some skins consist of vinyl that is molded on the dummy with foam injected under the vinyl to give a firmness and pliability to the skin that resembles the real thing. Other skins consist of pieces of vinyl backed with foam layers. These pieces can be slipped on and off like clothes. The skin that covers the head is all one piece that fits over the aluminum skull. Colors are also very important to dummy skin. Many customers want flesh tones, but skin with red or yellow colorations is sometimes preferred when the dummy will be used in testing that is filmed. The yellow and red tones make precise measurement of movements on film easier.

5 The dummy is designed with openings for instrumentation and cabling. These are in the larger parts of the body including the skull, chest, abdominal cavity, and the thighs. Clothing is sometimes fitted to the dummies in the factory or provided to the customer so the dummy can be dressed after the instrumentation has been inserted.

## Quality Control

Quality control is rigidly specified by the NTSA (or other agencies) before manufacture is even considered. Dummy assemblers are well aware that their "family" will be sacrificed for the safety of thousands of members of the public, so they feel an intense responsibility toward quality issues. When parts of the body are complete and when the entire body and head are assembled, the dummy goes through a rigorous set of tests in the calibration laboratory, where required tests are performed and measurements are made to confirm that the product conforms to every detail of the design drawings and specifications. Tests include a head drop test, thorax (chest) im-

The crash test dummy is made of metal parts consisting of aluminum, bronze, or steel as well as metal plating materials. A wide variety of plastics are also incorporated in most designs, and include Delrin (a type of long-wearing acetate resin), urethane and polyurethane foam, and vinyl. Most dummies are clothed to simulate reality, and specific vendors supply clothing, paint, and adhesives for marking critical targets or measurement points on the dummies. In fact, not only are outside suppliers used for many dummy constituents, but the federal testing agencies that approve dummy designs dictate specific vendors that must be used for some materials.

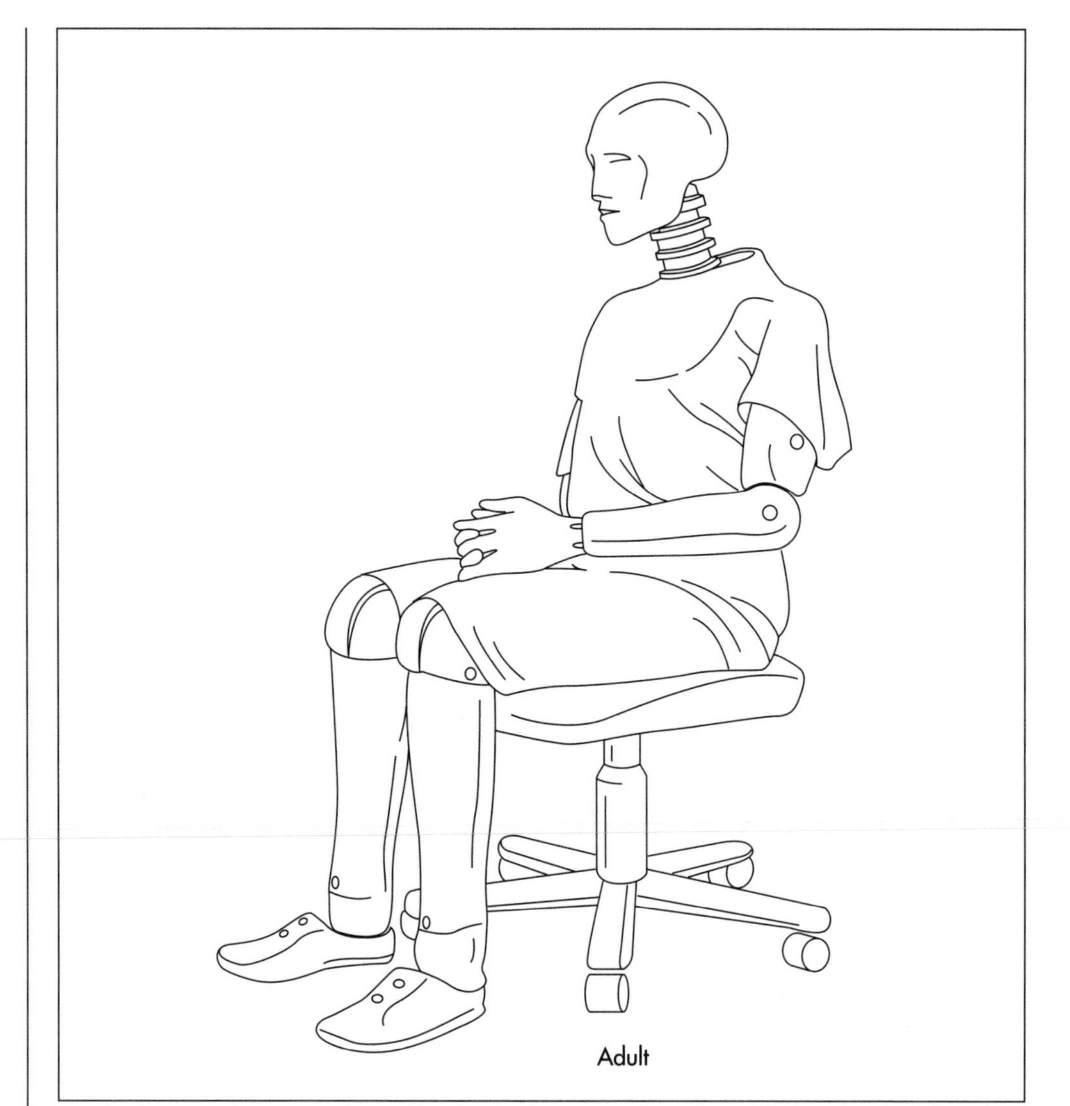


pact testing, and calibration of the skull, neck, knees, other major joints, and sometimes other parts of the limbs depending on the use of the dummy. Depending on the type of physique the dummy is intended to represent, the test results will vary. A child's head mass, for example, is dramatically smaller than an adult's, therefore the head drop test will be quite different. A set of certifications—a kind of birth certificate—stays with the dummy throughout its useful life to prove its calibration testing and for use in rebuilding, retrofitting, and recalibrating it.

## Byproducts/Waste

There are no byproducts from dummy manufacture. Waste is limited to metal and plastic that can be recycled.

## The Future

Computer technologies have made the crash test dummy adaptable to the extremes of human activity through design improvements, testing methods, miniaturization of instruments, and data analysis. Ultimately, however, the computer may prove fatal to the dummy family. Computer simulations are becoming so sophisticated that they can represent car crashes, damages to the vehicle itself, and injuries to the occupants. Simulations are attractive to auto manufacturers because they eliminate some, if not all, destructive testing, which costs approximately $750,000 per test. Vehicle occupants can be examined after simulated impact, and the details of injuries to the head and spine especially can be analyzed. Up to three months of design time can be removed from the design-production schedule for a new car model, and new designs may be safer for a wider range of different-sized drivers and passengers.

Dummy manufacturers, on the other hand, insist their families are here to stay. More modifications and complexities as well as technological applications have made crash test dummies reliable barometers for vehicle performance. Government agencies recognize this and are asking manufacturers to produce dummies and sensors that focus on very specific injuries. In 1995, this focus turned to lower legs, movement of the feet, and damage to Achilles tendons caused by frontal impacts. New sensors made the legs of the Hybrid 350 (certified in 1997) resemble those of a bionic man. In 1999, emphasis shifted to rear impact collisions and to developing state of the art vertebrae through cooperative research among agencies, universities, and manufacturers. Interest is also growing in skin simulants that will bruise and abrade during accidents; data from these developments will also aid doctors in improving treatments for skin injuries. Like its human counterparts, the crash test dummy is continuing to evolve for the purpose of saving our lives and limbs.

## Where to Learn More

### *Periodicals*

McCraw, Jim. "What happens when an Indy car crashes."*Popular Mechanics* (June 1995): 66.

"Virtual crash-test dummy."*Science News* (March 2, 1996): 138.

### *Other*

Applied Safety Technologies Corporation (ASTC).1999. http://www.astc.net. (June 29, 1999).

First Technology Safety Systems. June 14, 1999. http://www.ftss.com/ (June 29, 1999).

Insurance Institute for Highway Safety. http://www.hwy.safety.org/ (June 29, 1999).

National Highway Traffic Safety Administration. http://www.nhtsa.dot.gov/ (June 29, 1999).

Robert A. Denton, Inc. http://www.radenton.com/ (June 29, 1999).

TNO Crash Dummies BV. http://www.crashdummies.tno.nl/ (June 29, 1999).

*—Gillian S. Holmes*

# Cubic Zirconia

*In 1919, Marcel Tolkowsky, a third generation Antwerp-born diamond cutter and student of mechanical engineering, determined the proper proportions at which a diamond should be cut to obtain maximum fire and brilliancy. This ideal, known as the Brilliant Cut, is an objective measurable standard. Each brilliant cut diamond has 58 facets, cut at precise mathematically determined angles to reflect and refract maximum light rays.*

## Background

A gem or gemstone can be defined as a jewel or semiprecious stone cut and polished for personal adornment. Gemstones produced in the United States and other producing countries are of three types; natural, synthetic, and simulant. The natural gemstones are cut from minerals of crystalline form such as beryl, corundum, and quartz. (Diamond is a crystal of pure carbon.) Organic materials such as amber, coral, fossil, ivory, mother of pearl, natural and cultured freshwater pearls, and natural saltwater pearls are also considered natural gemstones.

Laboratory grown synthetic gemstones have essentially the same appearance, optical, physical, and chemical properties as the natural material that they represent. Synthetic gemstones produced in the United States include alexandrite, coral, diamond, emerald, garnet, quartz, ruby, sapphire, spinel, and turquoise. There are also synthetic stones that do not have a natural counterpart.

Simulants are laboratory grown gem materials that have an appearance similar to that of a natural gem material but have different optical, physical, and chemical properties. Cubic zirconia (CZ), a replacement for diamond, falls into this category and was first used for the production of jewelry stones in 1976. On the hardness scale for stones, the genuine diamond is a 10 compared to a hardness ranging from 8.5-9 for CZ. CZ has a refractive index (the ability to refract a ray of light into colors of red, orange, green, yellow, violet, and blue) of 2.15-2.18, compared to 2.42 for genuine diamond.

Predecessors to cubic zirconia as diamond imitations included strontium titanate (introduced in 1955) and yttrium aluminum garnet. However, strontium titanate was too soft for certain types of jewelry. Cubic zirconia became more popular since its appearance is very close to diamond as cut gems.

The gemstones simulants produced in the United States include coral, cubic zirconia, lapis lazuli, malachite, and turquoise. Additionally, certain colors of synthetic sapphire and spinel, used to represent other gemstones, would be classed as simulants. Colored and colorless varieties of CZ are the major types of simulants produced and have been on the market for over 30 years. As with genuine diamond, CZ is available in both higher and lower grades, ranging from several tens of dollars per carat to $100 per carat for the higher grades.

In the past decade, the use and consumer acceptance of synthetic and simulant gemstones have grown. Much of this growth is the direct result of the recognition of these gemstones for their own merits, not just as inexpensive substitutes for natural gemstones. Annual production of U.S. synthetic and simulant gemstones is currently valued at around $20 million, with production of natural gemstones at about two and a half times that.

## Raw Materials

Cubic zirconia is made from a mixture of high purity zirconium oxide powders stabilized with magnesium and calcium. The amount of each ingredient is carefully controlled, with certain additives sometimes being used to achieve a similar appearance to genuine diamonds.

## The Manufacturing Process

Synthetic and simulant gemstone producers use many different production methods, but they can be grouped into one of three types of processes: melt growth, solution growth, or extremely high-temperature, high-pressure growth. Solution techniques for making synthetic gems include flux methods for emerald, ruby, sapphire, spinel, and alexandrite. The other solution method is the hydrothermal method, often used for growing beryl (emerald, aquamarine, and morganite) and quartz. This method uses a large pressure vessel called an autoclave.

Other techniques involve solid- or liquid-state reactions and phase transformations for jade and lapis lazuli; vapor phase deposition for ruby and sapphire; ceramics for turquoise, lapis lazuli, and coral; and others for opal, or glass and plastics simulants or imitations. The Verneuil, Czochralski, and skull melting processes are the melt techniques most often used for gem materials.

French chemist Edmond Frémy produced the first commercial synthetic gemstones in 1877 by a melt growth method. These were small ruby crystals and were grown by fusing together a mixture containing aluminum oxide in a clay crucible, the process taking about eight days. These were termed reconstructed rubies. In 1885, larger synthetic rubies made their appearance using a flame fusion process and alumina powder. Later, sapphire, spinel, rutile, and strontium titanate were grown with this technique, also known as the Verneuil method.

The Czochralski pulled-growth method, developed around 1917 by a scientist of the same name, is used for ruby, sapphire, spinel, yttrium-aluminum-garnet (YAG), gadolinium-gallium-garnet (GGG), and alexandrite. In the Czochralski method, powdered ingredients are melted in a platinum, iridium, graphite, or ceramic crucible. A seed crystal is attached to one end of a rotating rod, the rod is lowered into the crucible until the seed just touches the melt, and then the rod is slowly withdrawn. The crystal grows as the seed pulls materials from the melt, and the material cools and solidifies. Yet, because of surface tension of the melt, the growing crystal stays in contact with the molten material and continues to grow until the melt is depleted.

Typically, the seed is pulled from the melt at a rate of 0.0394-3.94 in (1-100 mm) per hour. Crystals grown using this method can be very large, more than 1.97 in (50 mm) in diameter and 3.281 ft (1 m) in length, and of very high purity. Each year producers using this method grow millions of carats of crystals. The skull melt method is used for cubic zirconia and will be described in more detail below.

Certain gemstones pose unique problems when attempts are made to grow them. The problems arise because certain materials are either so reactive that they cannot be melted even in unreactive platinum and iridium crucibles or they melt at higher temperatures than the crucible materials can withstand. Therefore, another melting system must be used, called the skull melting system. Cubic zirconia, because of its high melting point, must be grown using this method.

### *Melting*

1 The "skull" is a hollow-walled copper cup. Water is circulated through the hollow walls to cool the inside wall of the skull. The cup is filled with powdered ingredients and heated by radio frequency induction until the powders melt. Because water cools the walls of the skull, the powdered materials next to the walls do not melt, and the molten material is contained within a shell of unmelted material. Therefore, the reactive or high-temperature melt is contained within itself.

2 When the heat source is removed and the system is allowed to cool, crystals form by nucleation and grow until the entire melt solidifies. Crystals grown using this system vary in size, depending on the number of nucleations. In growing cubic zirconia, a single skull yields about 2.205 lb (1 kg) of material per cycle.

### *Cutting*

The cut in any stone, whether natural or imitation, depends on the skill of the stone cutter. The cutter must evaluate a gem crystal carefully to determine how much of the crystal should be cut away to produce a stone or stones with good clarity. The cutter must also determine which stone shapes will make maximum use of the crystal. The cutter must make as much use of the crystal as he can, as diamond is too valuable to waste.

Proportion plays an important part in the cut of a diamond. An ideal stone is cut to mathematical specifications to allow a maximum amount of light to be reflected through the stone. This type of cut is known as the Brilliant cut. Variations from these set proportions can reduce the brilliance of the stone. When working with diamonds, a cutter might find it more cost wise to vary from these angles in order to remove a flaw or inclusion, yet still retain maximum carat weight. As counterfeit diamond crystals are more moderate in cost and almost flawless in clarity, larger stones can be cut to correct proportions.

3 There are several steps in cutting. First, the stone is marked to indicate cleavage planes and then cleaved to give initial shape to the stone. Since cubic zirconia readily splits in directions parallel to the octahedral crystal face, it can be cleaved using special tools impregnated with diamond powder. An alternative to cleaving is sawing, which is used to remove flawed areas (imitation stones won't have as many) using a small disk of phosphor-bronze impregnated with diamond on its edge. Precision programmable machines are now used to cut stones to a predetermined measurement (called calibration), so that every stone can be produced in the same size, shape, and depth.

4 The next step, called bruting or rounding up, consists of rounding the corners of the cleaved edges. This is accomplished using a machine equipped with a diamond tool called a sharp.

5 After bruting or rounding up, the stone is faceted using a cutting wheel made of cast iron. Polishing follows a secondary faceting operation.

6 The final step is a thorough cleaning by boiling in acid to remove all traces of oil, dirt, and diamond powder. After stones are inspected under a microscope, they are packed in a foam package.

## Quality Control

The quality of a real or imitation stone is determined by the four Cs: carat, color, clarity, and cut. It is the combination of the grades in all four that determine the final quality and hence value of both a genuine and an imitation diamond. The carat weight 0.0175 oz (0.5 g) of a genuine diamond is the weight of the stone in carat weight. Synthetic stones are always heavier in carat weight than genuine diamonds, as the material they are made of is more dense.

The color of a diamond can effect its value. Complete absence of color represents the high end of the scale, and pale, unevenly tinted stones comprise the lower end. Diamonds with an unusually high degree of color are known as fancies, and are graded by the evenness, rarity, and tone of the color. Both genuine and cubic zirconia diamonds are available in various colors ranging from palest yellows to brilliant reds.

The clarity is the clearness or transparency of a stone. In genuine diamonds, clarity is determined by nature, minute mineral traces, and small crystals of imperfections that can cloud a stone. The clearer the stone, the more valuable it is. The clarity of a counterfeit diamond can be controlled in the lab. However, bad melts can produce stones with small inclusions. Variations in the metal oxide mixtures can change the color of the stone. Uneven coloration is as undesirable in a counterfeit diamond as in a genuine diamond.

Of the four Cs, the cut is the most important in determining a diamond's brilliance. In 1919, Marcel Tolkowsky, a third generation Antwerp-born diamond cutter and student of mechanical engineering, determined the proper proportions at which a diamond should be cut to obtain maximum fire and brilliancy. This ideal, known as the Brilliant Cut, is an objective measurable standard. Each brilliant cut diamond has 58 facets, cut at precise mathematically determined angles to reflect and refract maximum light rays. This guarantees that a diamond has been cut to its best possible proportions for optimal beauty rather than simply to maintain maximum carat weight. In fact, diamond proportion and finish grades are defined in terms of the degree of departure from this standard.

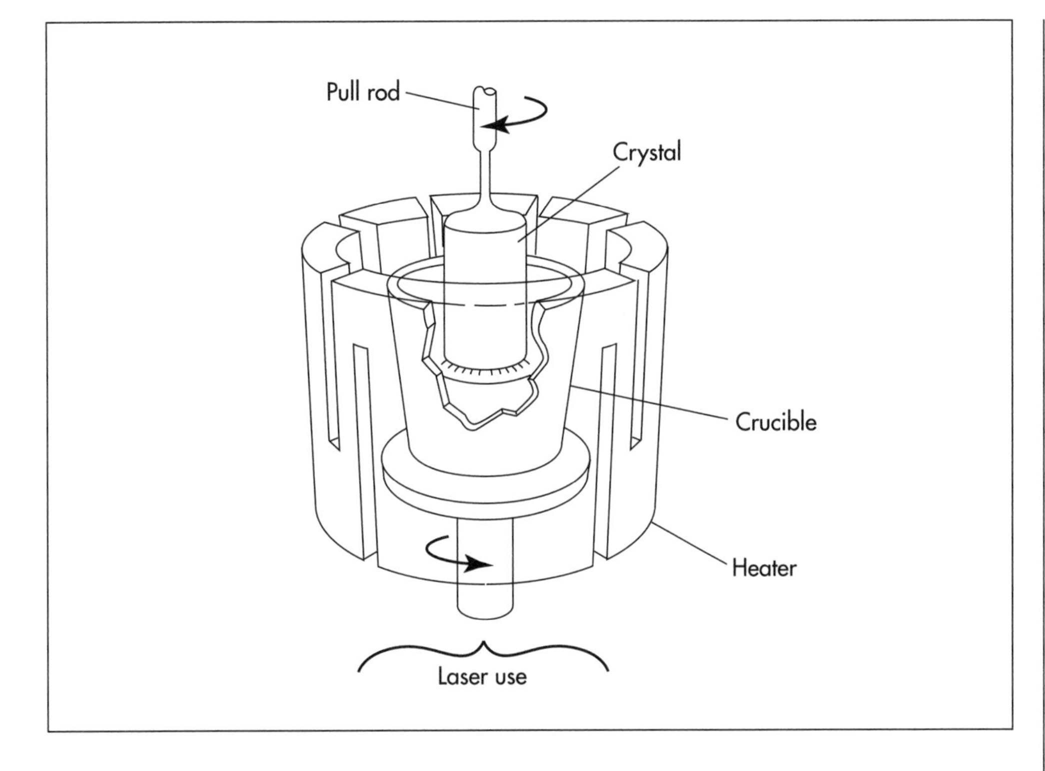


In the skull melt method to grow cubic zirconia, a hollow-walled copper cup is filled with powdered ingredients and heated by radio frequency induction until the powders melt. Circulating water within the hollow walls cool the inside wall of the skull. Because water cools the walls of the skull, the powdered materials next to the walls do not melt, and the molten material is contained within a shell of unmelted material. Therefore, the reactive or high-temperature melt is contained within itself. When the heat source is removed and the system is allowed to cool, crystals form by nucleation and grow until the entire melt solidifies. A single skull yields about 2.205 lb (1 kg) of cubic zirconia per cycle.

This standard is also applied to imitation diamonds made from cubic zirconia.

Although the 58 facet brilliant cut is known as the standard measurement of quality in the cutting industry, there are special instances when deviations are necessary. Too much light refraction in a small area reduces clarity, therefore extremely small stones may be cut with fewer facets. Larger stones may be cut with more facets for the opposite reason. The size of the stone may so increase the width of the facets that brilliance is lost. It is not unusual to find diamonds one carat and up cut this way. With the introduction of new cuts like the radiant, princess, and trillion, the demand for multifaceted stones has increased.

There are a variety of shapes that can be cut. The round delivers maximum brilliance, dispersion, fire, and is the most popular of all genuine and imitation diamond shapes. The oval has an oblong shape, a slightly elongated form. Facets around the top portion of the stone are very similar to the facet arrangement on the round stone. For this reason, a well-cut oval diamond gives off a sparkling appearance, catching and reflecting light from all directions. The time (eight to 10 hours for one carat) required to cut an oval is roughly twice the time needed for cutting a round stone because it takes more time to do the primary shaping.

Other testing methods are used to measure optical and physical properties. A binocular microscope is used to look for defects such as inclusions. The refractive index is measured using a refractometer. The specific gravity of a stone is determined by seeing whether it will sink, float up, or remain stationary in a liquid with a known specific gravity.

## The Future

According to industry experts, the market for cubic zirconia jewelry is expected to maintain a stable position as mass market retailers continue to offer it to their customers and price continues to decline. Cubic zirconia is also becoming more profitable as a substitute for genuine diamond in stud earrings, the diamond solitaire and the tennis bracelet. In other words, it is no longer being considered just a cheap imitation, especially with improvements in brightness and reflection. An overall growth rate of about 10% is thus predicted for the market. Though a new diamond simulant has recently been introduced called synthetic moissanite (a crystallized silicon carbide), the higher price and more difficult process of this material will limit it competing with cubic zirconia.

## Where to Learn More

### *Books*

Erem, Joel E. *Gems and Jewelry.* Geoscience Press Inc., 1992.

### *Periodicals*

Anonymous. "New Diamond Simulant to Debut."*Jewelers Circular Keystone* (December 1996): 32.

Janowski, Ben. "The Diamond Alternative."*Jewelers Circular Keystone* (June 1997): 146-148.

### *Other*

DeGrado, Inc. PO Box 1211. Mandeville, LA 70470-1211.

D. Swarovski & Co. http://www.swarovski.com.

Janos Consultants. (212) 288-1155.

U.S. Geological Survey. http://minerals.er.usgs.gov/minerals/pubs/commodity/gemstones/.


*—Laurel Sheppard*

# Dog Biscuit

Dog biscuits are a hard, dry, dog food product, typically composed of protein, carbohydrates, fat, and fiber. They are made in much the same way biscuits are made for human consumption. The raw materials are combined in a large container, the biscuits are shaped, cooked, cooled, and packaged. First developed accidentally during the 1800s, the dog biscuit snack market has grown to over $480 million in yearly sales.

*First developed accidentally during the 1800s, the dog biscuit snack market has grown to over $480 million in yearly sales.*

## Background

Dog biscuits were invented accidentally in a London butcher shop during the late 1800s. According to the story, the shop's owner was trying to expand his business by creating a new biscuit recipe for his customers. After baking a batch, he tasted them and thought they were terrible. He gave one to his dog, and the dog gobbled it right up. This gave him the idea of making biscuits especially for dogs. He made his biscuits in the shape of a bone and they began to sell rapidly. In 1908, his recipe was bought by an American businessman who introduced it to the United States. The F.H. Bennett Biscuit company was established, and they began selling the dog biscuit under the name Malatoid. In 1911, the recipe was granted a patent. The name was changed to Milkbone in 1915 to reflect the fact that cow's milk was one of the main ingredients.

The Milkbone dog biscuit brand was then acquired by Nabisco Biscuit Company and it dominated the dog biscuit market until the late 1960s. In fact, during most of this time, it was the only commercially available dog biscuit. Initially, it was marketed as a treat for dogs, but eventually the health aspects such as cleaner teeth and better breath were promoted. In the early 1970s, a number of manufacturers came out with competing products. This competition has remained, resulting in hundreds of different dog biscuit products.

## Design

It is estimated that over 50% of all dog owners regularly give their dogs treats. While treats are not essential to a dog's diet, they are typically given as a way to indulge or train their pets. The classic dog biscuit is a small, hard, bone-shaped product that is colored to reflect its flavor. Traditional flavors include beef, chicken, lamb, turkey, liver, cheese, and bacon. However, certain manufacturers have produced some interesting flavors such as oatmeal, raisin, spinach, peanut butter, and coconut. In addition to flavor variations, dog biscuits also are sold in different sizes; small biscuits for small dogs, large biscuits for large dogs. The shapes have also changed from the conventional bone shape. In an effort to differentiate their products, marketers have produced dog biscuits with shapes as varied as animals, people, bacon strips, and even fire hydrants. Since most treats are also sold as healthy food supplements, dog biscuits are typically fortified with vitamins and minerals essential in a dog's diet. Low fat varieties are also available for dogs that are overweight. The packaging for dog biscuits used to be strictly limited to cardboard boxes because this minimized breaking. However, advances in packaging technology have resulted in the use of foil packaging and plastic containers.

## Raw Materials

The primary ingredients in a dog biscuit recipe are carbohydrates, proteins, fats and oils, and fiber. These are combined with other ingredients that have a significant effect on the dog biscuit's final characteristics. The ingredients used for dog biscuits are specially tailored for dogs, and are chosen to be nutritious, easily digested, palatable, and economically feasible. While the materials have a high nutritional content, they are typically not as high quality as similar ingredients used in human food.

### *Carbohydrates*

Flour is one of the most abundant ingredients in most dog biscuit recipes and provides the bulk of the carbohydrates. It is obtained by grinding corn, wheat, or rice into a powder. This powder contains the three main parts of the seeds including the bran, germ, and endosperm. Of these, the endosperm is responsible for most of the important baking and shaping characteristics. Chemically, flour is composed of starch and protein. When mixed with water, it creates a mass called gluten that can be cut and formed into the dog biscuit form.

### *Proteins*

The proteins added to dog biscuits supply the animal with essential amino acids, improve taste, and in some cases supply energy. They are derived from a variety of sources including animal and plant. Typical plant protein ingredients found in a dog biscuit are such things as corn gluten meal and soybean meal. They are generally less expensive than animal proteins and are therefore used in greater quantities. A drawback to plant proteins is that they contain certain indigestible oligosaccharides that can cause undesirable flatulence. Animal proteins are added to improve the food characteristics of the biscuits. This includes ingredients such as poultry byproduct meal, meat and bone meal, and dried liver meal. Milk and egg are also used as a supply of protein.

### *Fats and oils*

Fats and oils are added to biscuit recipes to provide flavor and nutritional energy. They also have a significant impact on the finished texture. They are derived from plants, fish and animal sources. Much of the fats and oils used in pet food is obtained from the fast food industry.

### *Fibers*

Fiber represents another important type of nutritional ingredient found in dog biscuits. Normally, fiber is present in a variety of the other ingredients. Extra fiber, such as wheat bran or corn bran, is added to improve droppings and water absorption. This ingredient is more important in calorie-controlled recipes.

### *Other additives*

Other ingredients are added to dog biscuit recipes that do not have a direct influence on the animal's nutrition. Antioxidants are added to prevent the breakdown of vitamins and other nutrients present in the biscuit, and to maintain freshness. Examples of these materials include zinc oxide and manganous oxide. Flavor enhancers have become another important additive. They are used to make a biscuit more appealing to the pet. It has been found that dogs tend to prefer meat and fish flavors. Although dogs are essentially color blind, artificial color additives are added for the benefit of the pet's owner. Finally, preservatives such as BHA are added to prevent microbial growth.

## The Manufacturing Process

The speed of dog biscuit manufacture varies depending on the method used to produce them. The slowest methods involve manual mixing and oven baking. Other manufacturers use a highly automated process that can produce tons of product per hour. Both methods involve general steps include blending and conditioning of ingredients, shaping, baking, cooling, and packaging.

### *Ingredient handling*

1 In the automated method, an extrusion method is typically used. In the first step, the dry ingredients are mixed together. Water is then added to this blend to bring the moisture content up to about 30%. This creates a paste-like dough that is next pumped into a conditioner chamber and heated. Within this chamber, other ingredients such as animal byproducts or other moist ingredients are added. From this chamber, the batch is moved into an extrusion machine.

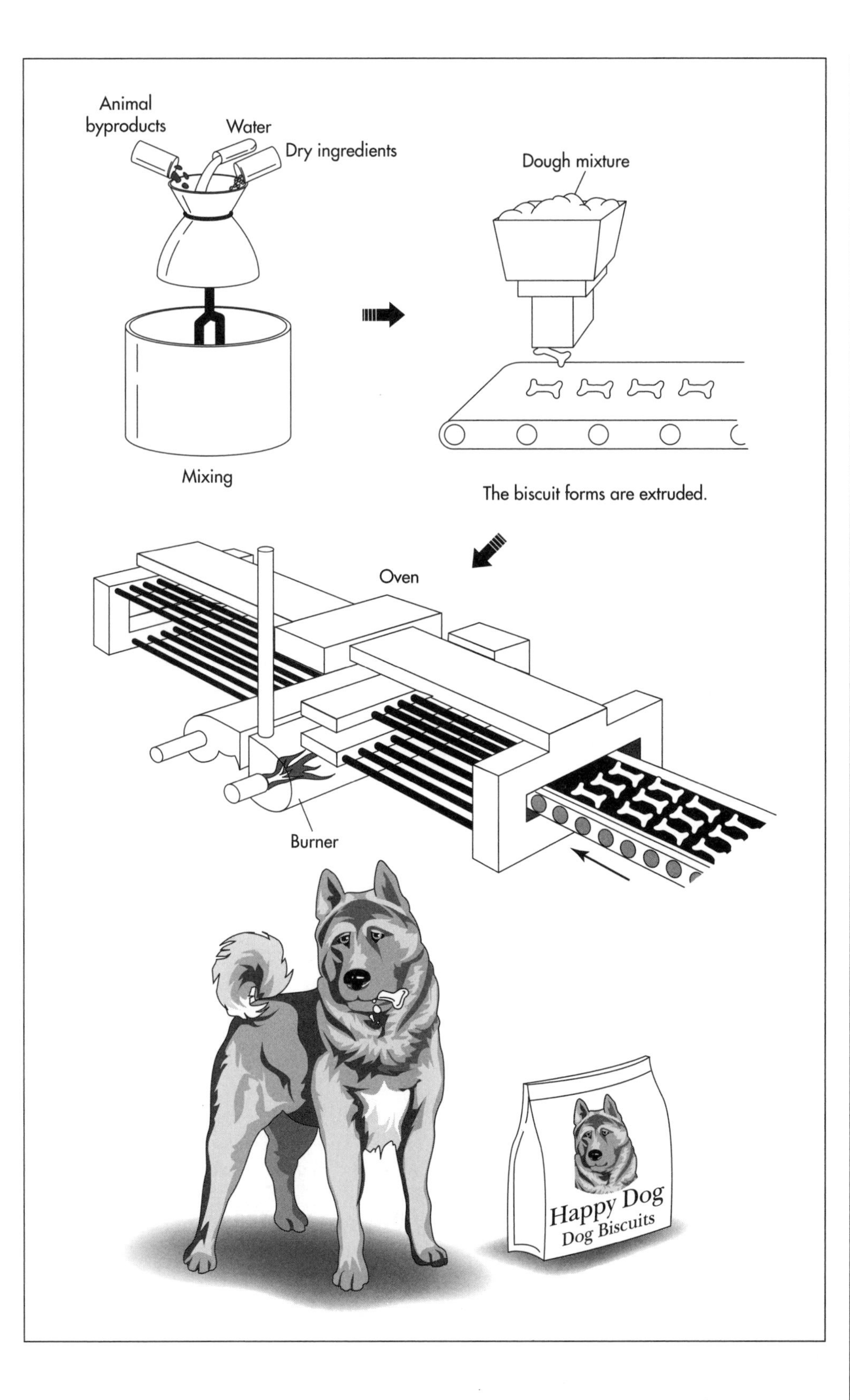


Commercial dog biscuit manufacturing operations begin by mixing the dry ingredients together in industrial-sized mixers. Water is added to increase the moisture content to 30%. Once all the additional ingredients are added and the batch is heated, the paste-like dough is transported to an extruder. The extrusion machine is a large, enclosed screw which has a tapered end. The screw is turned, pushing the dough through the extrusion chamber. As it is moved through, it heats up due to the friction generated. At the end of the extrusion chamber are the forming dies. These can have a variety of shapes such as bones, stars, circles, etc. As the batch is pumped out, automatically cut into individual biscuits, and laid on drying sheets. The biscuits are transferred to a metal conveyor belt and moved through an oven. Prior to packaging, the cooked biscuits may be coated with flavor or vitamins.

### *Shaping*

2 The extrusion machine is a large, enclosed screw that has a tapered end. The screw is turned, pushing the batch through the extrusion chamber. As it is moved through, it heats up due to the friction generated. At the end of the extrusion chamber are the forming dies. These can have a variety of shapes such as bones, stars, circles, etc. As the batch is pumped out, automatically cut into individual biscuits, and laid on

drying sheets. The sudden release of pressure causes the entrapped moisture to create an expansion of the biscuit much like exploding popcorn. From here, the biscuits are transferred to a metal conveyor belt and moved through an oven.

### Baking

3 The baking of the dog biscuits is done in a tunnel oven. The oven is anywhere from 100-300 ft (30.48-91.44 m) long. As the biscuits travel through the oven, the biscuit goes through a setting phase where it takes on the shape of the final product. Next, it is dried out to about 10% moisture content. The amount of time baking time is controlled by the speed of the moving conveyor belt.

### Finishing steps

4 The dog biscuits exit the ovens and travel through a series of conveyors to cool. They may be flipped to ensure that cooling is throughout. Depending on the recipe, additional coating ingredients may be added. These can include fats, vitamins, and other flavor enhancers. They are applied using depositor machines that spray coat the biscuits. If necessary, excess topping can be removed by forced air or shaker devices.

### Packaging

5 The final step in the manufacturing process is packaging. Because certain dog biscuits are fragile, the packaging is typically rigid. The biscuits are moved onto weighing bins. When the weight of the biscuits is heavy enough, they are dropped into boxes that are appropriately decorated to attract customers. These boxes are put into larger case boxes, stacked on pallets, and shipped to stores.

## Quality Control

At various points during manufacture, tests are performed on the raw materials and finished product to ensure that they conform to specifications. Before any ingredient is allowed to be used, the quality control lab evaluates its characteristics and properties. This involves studying things such as appearance, color, odor, and flavor. The particle size of the solids, pH, and viscosity may also be tested. During production, each batch of dog biscuit is also carefully monitored to make sure that the product being shipped to stores is of the same quality as the one developed in the food laboratory. The usual method for ensuring quality is to compare a random sample from the batch to an established standard.

## The Future

Since the 1800s, manufacturers have developed significant improvements in the production of dog biscuits. These improvements are likely to continue in the future resulting in faster and more efficient production. Improvements will also be made with new packaging and innovative merchandising techniques. The market for dog biscuits is expected to grow steadily, being considerably impacted by developments in the dog food market. Experts suggest that all natural, healthy treats are expected to have the greatest growth potential.

## Where to Learn More

### Books

*Kirk Othmer Encyclopedia of Chemical Technology.* John Wiley & Sons. New York: 1992.

### Periodicals

"Bone appetit! Kansas City's Three Dog Bakery pampers the palates of man's best friends."*People Weekly*, April 3, 1995.

Pavia, Audrey. "History of Dog Treats."*Pet-Product News*. July 1996.

### Other

United States Patent 5,501,868. 1996.

*—Perry Romanowski*

# Doorknob

## Background

There are 114 million existing doorways in the United States, with about two million new ones added every year. Doors equipped with suitable hardware are used to close off these openings and protect the interior of the building from the environment. Very early doors were merely hides or textiles. Wooden doors were also popular in ancient Egypt and Mesopotamia. Other materials used for doors include stone, metal, glass, and paper. Doors open by swinging, folding, sliding or rolling. Many swinging doors are installed with a lever or doorknob to open them with.

Door knobs have been used around the world for centuries, and were first manufactured in the United States in the mid-nineteenth century. Though spherical or ball-shaped door knobs are considered the hardest to turn, this shape is still the most common. Egg-shaped door knobs are the easiest for most people to use. Doorknobs have been made of many materials, including wood, ceramic, glass, plastic and different types of metal. Brass is one of the most popular materials because of its excellent resistance to rust.

The average doorknob is 2.25 in (5.715 cm) in diameter. The basic components are the knob rose, shank, spindle, and knob-top. The knob-top is the upper and larger part that is grasped by the hand. The shank is the projecting stem of a knob and contains a hole or socket to receive the spindle. The knob rose is a round plate or washer that forms a knob socket and is adapted for attachment to the surface of a door. The knob is attached to the spindle, a metal shaft that turns the latch of the lock.

## History

American doorknob designs and materials have changed throughout the years. In colonial times the first door hardware was made out of wood, and involved simple latches and strings. Round knobs first appeared around the time of independence. Decorative hardware, including knobs, emerged after the Centennial Exposition of 1876. Before this time, most door hardware was imported; 95% in 1838.

Glass knobs were rare until a faster and cheaper manufacturing method was developed based on pressing. Pressed glass knobs were popular from 1826-1850, followed by cut glass through 1910. Wooden knobs were introduced in the late 1800s and were phased out after 1910. China or ceramic knobs were mainly imported from France and England until the mid-1800s, when the first U.S. patent was granted for making knobs out of potters clay.

Before 1846, metal knobs were made from two pieces brazed together or three pieces soldered together. Cast metal knobs were introduced around 1846. In the late 1800s, composite metal knobs were introduced as a less expensive knob. The main body was made out of iron or steel, covered entirely or in part with a veneer of bronze or brass. During the last half of the nineteenth century, many patents were issued regarding the spindle methods of attaching metal knobs for lock use, as well as designs for ornamenting these knobs. In 1870, a compression casting method was introduced that accelerated ornamentation of hardware.

Many of the Victorian doorknobs were made of cast bronze with ornamental pat-

*There are 114 million existing doorways in the United States, with about two million new ones added every year.*

terns. During this period, a dozen major companies and many smaller firms produced hundreds of patterns of ornamented hardware, in addition to cast and wrought metal, glass, wood and pottery knobs. From 1830-1873, there were over 100 U.S. patents granted for knobs. Collectors have catalogued over 1,000 antique doorknob designs into 15 types based on shape, material, and design pattern. The best grade of knobs during this period were usually made from cast bronze or brass.

Around 1900, cast metal and glass knobs were introduced that incorporated ball bearings in the shanks of doorknobs. In operation, the knob shank rotated on sets of ball bearings fitted in the hardened steel cones. This reduced friction, assured closer adjustment, and eliminated endplay of the knobs. Other materials popular during the early 1900s included bronze and porcelain.

Most doorknobs come with some type of locking device. Machine processes for steel locks were first introduced in 1896. Today, the most common type of privacy lock is the spring lock, which uses a simple round, push button located in the center of the knob to control the bolt. It is easy to operate with a finger, closed fist, or elbow. Some locks come with both a spring lock and a dead bolt, which is operated by a key. Other locks have become more sophisticated, and use some sort of electronic device, such as a programmable computer chip that identifies users.

## Design

Door hardware selection is usually based on appearance, cost, and availability, rarely on function. In order to assure the most usable hardware, designers must carefully consider not only appearance but also the size, shape, and feel of each element of door hardware and how easy it is to use. Typical design features of a doorknob include: no sharp edges or ridges; a shape that is easily grasped or turned; a textured finish or non-slip coating on knob to improve grip; and a shaft long enough to fit hand behind knob.

If a new design is required, a two-dimensional model is usually made using computer-aided design software. A three dimensional prototype is then fabricated so that a mold or die can be made of the desired shape. If a metal casting process is used, a pattern in wood or clay is made from which to make the mold.

## Raw Materials

Most doorknobs are made of metal, with the most common type brass. The term brass refers to a group of alloys that contain a combination of varying amounts of copper and zinc. The material is usually received as a rod or billet of suitable diameter and is machine cut to the required length. The raw material must conform to standards developed by the American Society of Testing and Materials regarding physical, mechanical, chemical, thermal, and microstructural properties for each specific process.

## The Manufacturing Process

Though there are several processes used for metal doorknobs, including casting where a molten metal is poured into a mold, brass doorknobs are typically forged. Forging is a process in which heated metal is forced into shaped dies under very high pressure. Forging can produce products having superior strength, toughness, reliability, and quality (up to 250% stronger than castings). Forging can also be more efficient and economical.

### Forging

1 The billet must first be heated to 1,400° F (759.9° C) in a gas-fired furnace to soften the metal. The billet is then placed in a specially designed set of steel dies that are shaped to conform to the profile of the product being made. A press applies high pressure to force the heated billet into the die cavity, where the billet now takes the shape of the doorknob.

2 After the doorknob is ejected from the press, it is cooled, trimmed to size, tempered, and cleaned to remove heat scale produced during the process.

### Surface preparation

3 Next, the doorknob goes through a series of finishing steps. Separate coining, milling, drilling, and tapping processes produce a surface ready for the polishing opera-

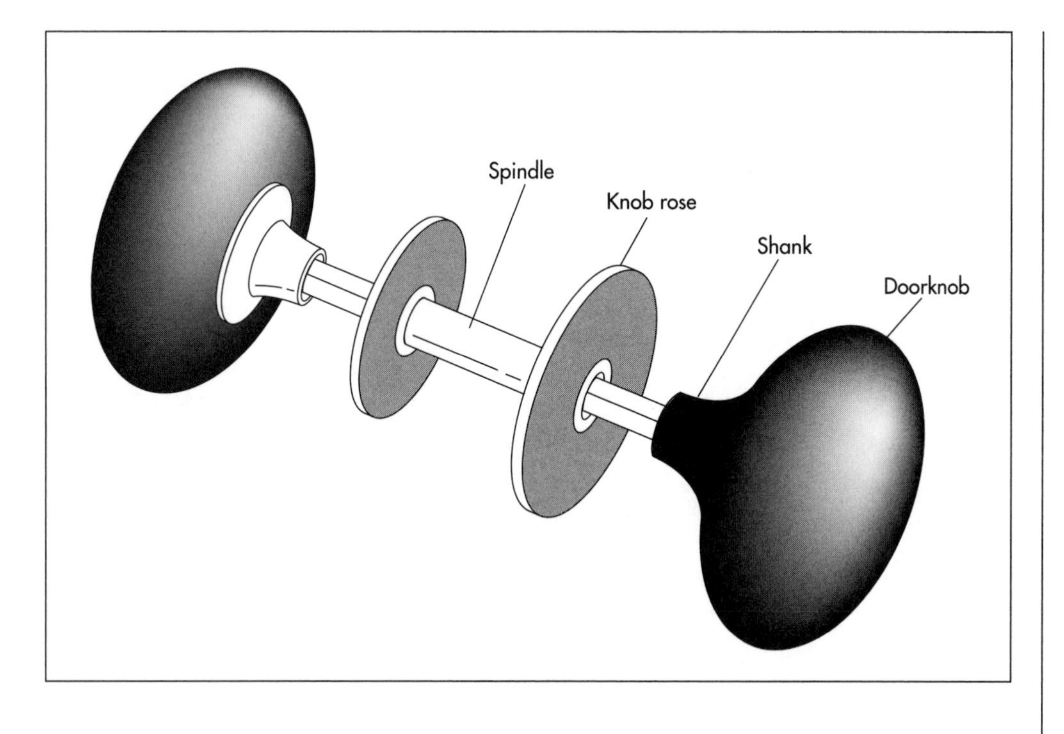


The average doorknob is 2.25 in (5.715 cm) in diameter. The basic components are the knob rose, shank, spindle, and knob-top. The knob-top is the upper and larger part that is grasped by the hand. The shank is the projecting stem of a knob and contains a hole or socket to receive the spindle. The knob rose is a round plate or washer that forms a knob socket and is adapted for attachment to the surface of a door. The knob is attached to the spindle, a metal shaft that turns the latch of the lock.

tion. One or more buffing steps are used to achieve a brilliant finish.

### *Coating*

4 For further protection, an organic or inorganic coating is applied using several different processes. Organic coatings include polyurethanes, acrylics, and epoxies. Because the solvents used in organic coatings can produce hazardous materials and quality problems, manufacturers are turning to inorganic coatings based on inert metals. These are applied using electroplating or physical vapor deposition (PVD).

5 PVD applies a coating produced by sputtering and thermal evaporators in an airtight chamber. The chamber is evacuated to high vacuum pressures (less than one millionth of an atmospheric pressure) by a series of pumps. A thin coating is deposited one molecule at a time. For successful PVD, the brass surface must first be extensively cleaned in a series of washing and agitating tanks, followed by electroplating with semiprecious materials.

## Quality Control

The raw material must be of suitable composition for the forging process, as established by the American Society for Testing and Materials. Various process parameters throughout the manufacturing process are monitored and controlled to ensure the final product meets quality standards. The finished doorknob is inspected for dimensions, surface finish, and other properties. Some of these properties may have to conform to certain building codes.

## Byproducts/Waste

Since forgings are designed to approximate final part shape, little waste is produced compared to other processes. The forging process also results in uniformity in composition, dimensions, and structure from piece to piece and lot to lot, which also minimizes rejects.

## The Future

Security and access control systems for doors will continue to become more sophisticated as the cost of electronics decreases. Though there will always be a demand for mechanical hardware, electrical hardware may have faster growth. New building codes may be required to accommodate this hardware.

The common doorknob will continue to play an important role in the building industry. Restoration and renovation of older buildings will continue to make antique doorknobs or their reproductions popular.

## Where to Learn More


### Books

Eastwood, Maud. *Antique Builders Hardware, Knobs & Accessories.* Woodinville, WA: Antique Doorknob Publishing Co., 1992.

Eastwood, Maud.*The Antique Door Knob.* Woodinville, WA: Antique Doorknob Publishing Co., 1976.

### Periodicals

*The Doorknob Collector* (January-February and March-April 1999).

Heppes, Jerry. "The Future of the Industry."*Doors and Hardware* (December 1998): 20-29.

### Other

The Antique Doorknob Collectors of America. PO Box 31, Chatham, New Jersey. (973) 635-6338. Fax: 973-635-6993.

Baldwin Hardware Corp. 841 Wyomissing Blvd., Reading, PA 19011. (800) 437-7448.

The Door and Hardware Institute. 14170 Newbrok Drive, Chantilly, VA 20151-2232. (703) 222-2010. Fax: 703-222-2410. Http://www.dhi.org/.


*—Laurel Sheppard*

# Doughnut

## Background

The doughnut is a fried ring or globule of sweet dough that is either yeast leavened or chemically leavened. The dough is mixed and shaped, dropped into hot oil and fried, and glazed. Jam-filled doughnuts are called bismarks. Batters vary and may be chocolate or lemon and include fruits such as blueberries, raisins, or nuts. Chemically-raised donuts are made with baking powder and are generally rather dense and cake-like. They are easily and quickly made. Yeast-raised doughnuts, which is leavened by the creation of carbon dioxide resulting from fermentation of yeast, are lighter in texture than chemically-raised doughnuts. They require several hours to produce.

These sweet treats are easily made at home using basic ingredients and require no special equipment. Doughnuts are baked and sold on premises at small, privately run bakeries, grocery stores, and in franchise operations that offer a standard product through the use of a pre-packed mix and carefully-controlled production. Large commercial bakeries make thousands of dozens of doughnuts each day, packaging them for distribution across vast regions.

Doughnuts are a beloved American snack. Children sing their praises in a song that begins "Oh I went downtown and walked around the block/I walked right into the doughnut shop..." Clark Gable taught Claudette Colbert how to dunk her doughnut in the classic 1934 movie "It Happened One Night." Many World War I and II veterans swear that doughnuts served in canteens got them through the roughest of times. Doughnut franchises have flourished in the United States since the 1930s. Despite their fat content (at least 3 g) and calorie content (a minimum of 200), Americans alone consume 10 billion doughnuts each year.

*Americans consume 10 billion doughnuts each year.*

## History

The doughnut supposedly came to us from the eighteenth century Dutch of New Amsterdam and were referred to as *olykoeks*, meaning oily cakes. In the nineteenth century, Elizabeth Gregory fried flavored dough with walnuts for her son Hanson Gregory, hence the name doughnut. By the late nineteenth century, the doughnut had a hole.

Doughnuts were a great favorite at lumbering camps of the Midwest and Northwest as they were easy to make and full of calories needed to provide quick energy for arduous logging jobs. "Doughboys" of World War I ate thousands of doughnuts served up by the Salvation Army on the French front. Soldiers reminisced that the doughnut was far more than a hot snack. The doughnut represented all the men were fighting for—the safety and comfort of mother, hearth, and home.

Soon after the doughboys returned, doughnut shops flourished. A Russian immigrant named Levitt invented a doughnut machine in 1920 that automatically pushed dough into shaped rings. By 1925, the invention earned him $25 million a year and it was a fixture in bakeries across the country. The machine-made doughnut was a hit of the 1934 World's Fair. Other machinery quickly developed for everything from mixing to frying. Franchises soon followed. By 1937, Krispy Kreme was founded on a "secret recipe" for yeast-raised doughnuts and Dunkin' Donuts (currently the franchise that sells the most doughnuts worldwide) was founded in Massachusetts. Presently, Krispy Kreme totals 147 stores in 26 states, while Dunkin' Donuts has 5,000 franchises in the United States and is present in 37 countries.

## Raw Materials

Ingredients vary depending on whether they are yeast or chemically leavened. Furthermore, homemade doughnuts generally include far few ingredients than mass- produced or those made from mixes. Chemically-raised doughnuts are made with ingredients such as flour, baking powder, salt, liquid, and varying amounts of eggs, milk, sugar, shortening and other flavorings. This type of doughnut uses baking powder in the batter to leaven the dough. Yeast-leavened doughnuts are made with ingredients that include flour, shortening, milk, sugar, salt, water, yeast, eggs or egg whites, and flavorings.

Doughnuts produced in sanitary baking conditions in grocery stores, bakeries, or franchises often come from pre-packaged mixes. These vary but can include: flour (wheat and soy flour), shortening, sugar, egg yolks, milk solids, yeast dough conditioners, gum, and artificial flavors. One franchise adds a yeast brew. Mixes require the bakeries to add fresh wet ingredients such as water, milk, and eggs in the mixing process. Doughnuts also require oil (usually vegetable oil) for frying. Glazes or frostings are often added after the product is fried and are made with flour, sugar, flavoring, and sometimes shortening.

## The Manufacturing Process

This process will describe the manufacture of doughnuts in a mechanized doughnut bakery that makes only yeast-raised doughnuts. Because yeast requires time for kneading, time to rest and additional time to rise or proof, it takes at least an hour to take dry pre-packaged mix to completed product.

### *Acquiring the ingredients*

1 Bakeries or franchises that do a brisk business (making hundreds of dozens in a day) acquire mixes in bags, often as large as 50 lb (22.7 kg). Chains have the ingredients shipped to them from company warehouses within the region and the mixes are stored on the premises and used as needed. The bakery must shop for large quantities of perishable fresh ingredients such as eggs and milk and keep them refrigerated.

### *Measuring the ingredients*

2 A batch is referred to by weight of dry ingredients put into the mixture. The weight of the batch varies with doughnut type and amount to be made. The pre-packaged mix is poured from a bag onto a scale and the precise amount measured.

### *Mixing and kneading*

3 The flour mixture is then poured into a large mixing bowl put onto an industrial mixer and the appropriate amount of wet ingredients are added depending on weight of the batch and type of doughnut in production. The wet yeast slurry (for leavening) is mixed separately and carefully added to the flour-water mixture at this time. The dough mixer then begins its work; a large dough hook first mixes and then simulates the human kneading process, pulling and stretching, as it homogenizes the ingredients and develops the dough by forming the gluten into elongated and interlace fibers that form the basic structure of the doughnut. The mix runs on an automatic timer and the entire mixture, including the softened yeast, is kneaded together for approximately 13 minutes.

### *Resting the yeast*

4 It is essential that yeast dough "rests" or simply sits for about 10 minutes after it is mechanically kneaded. As the yeast grows, it converts its food into carbon dioxide (this is called fermentation) and causes the yeast dough to rise. As the dough sits, it allows the gas to develop and the dough starts to rise, indicating the fermentation process of the yeast reacting to sugar in the mix is beginning. If this does not happen, the dough yields flat, tough doughnuts and the mix should be discarded. At the end of this period, a good-quality dough is spongy and soft.

### *Shaping the doughnuts*

5 The dough is then hoisted by hand and loaded into the hopper of a machine called an extruder—a machine that forms the individual doughnuts using a pressure-

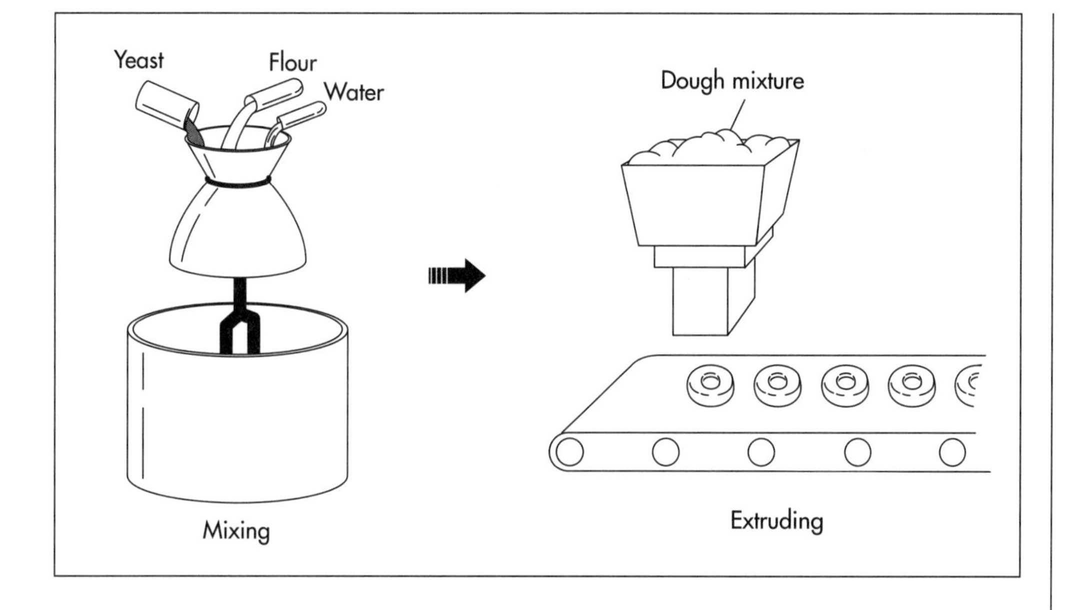


The premeasured flour mixture is mixed with the appropriate amount of wet ingredients The wet yeast slurry (for leavening) is mixed separately and carefully added to the flour-water mixture. Next, a large dough hook first mixes and then simulates the human kneading process, pulling and stretching, as it homogenizes the ingredients and develops the dough by forming the gluten into elongated and interlace fibers that form the basic structure of the doughnut. Once the yeast has had time to rise, the dough is loaded into a hopper that feeds the dough through an extruder. A cover is then placed on the machine and the machine is pressurized, forcing the dough into tubes that "plop" out a pre-determined amount of dough into the desired shape—rings for conventional doughnuts and circles for doughnuts that are to be filled with jam or creme.

cutter system. The batch of yeast dough is put into the top of the open machine. A cover is then placed on the machine and the machine is pressurized, forcing the dough into tubes that extrude a pre-determined amount of dough into the desired shape—rings for conventional doughnuts and circles for doughnuts that are to be filled with jam or creme. It takes about 15 minutes for the extruder to push out about 30 dozen doughnuts.

An automated doughnut stamper can also be used in conjunction with an extruder. In this case, the dough is extruded in a continuous, unshaped flow through a series of rollers that flatten the dough. Once flattened to 0.5 in (1.27 cm) thickness, the sheet of dough is stamped into doughnut shapes.

### Proofing

6 The extruder is attached directly to the proofing box (a warm, oven-like machine), which is a hot-air, temperature-controlled warm box set to approximately 125° F (51.6° C). Here, the thin doughnuts are slowly allowed to rise or proof as the yeast ferments under controlled conditions. Proofing renders the doughnuts light and airy. (Yeast doughs must be allowed to rise slowly and at just the right temperature. If the proofing box is too hot, the yeast bacteria will be killed and the doughnuts will not rise. If too cold, the yeast remains inactive and cannot ferment thus preventing leavening. A machine attached to the extruder pushes the rings or circles onto small shelves that move through the proof box for about 30 minutes. The shelves are chain-driven and move down, up, and over during this 30 minute period. After 30 minutes, they are quite puffy.

### Frying

7 Next, the raw doughnuts fall automatically, one row at a time, into the attached open fryer. It is important to drop just a certain amount of raw doughnuts into the grease at a time. If too many are placed in the fryer at one time, the oil temperature is drastically lowered, fry time is longer, and the doughnuts absorb too much oil. The frying oil is the most expensive ingredient in the production process, and if the doughnuts absorb too much oil, it reduces the profit margin on the batch. As the doughnuts move through the fryer, they are flipped over by a mechanism. After two minutes, the doughnuts have moved completely through the fryer and are forced into the mechanism that applies glaze.

### Glazing and drying

8 As the doughnuts leave the fryer, they move under a shower of glaze. Here, glaze is forced through holes from a bridge running several inches above the hot doughnuts. The glaze coats the top, sides, and part of the bottom of the doughnuts. The doughnuts are conveyored out of the production area to dry and cool.

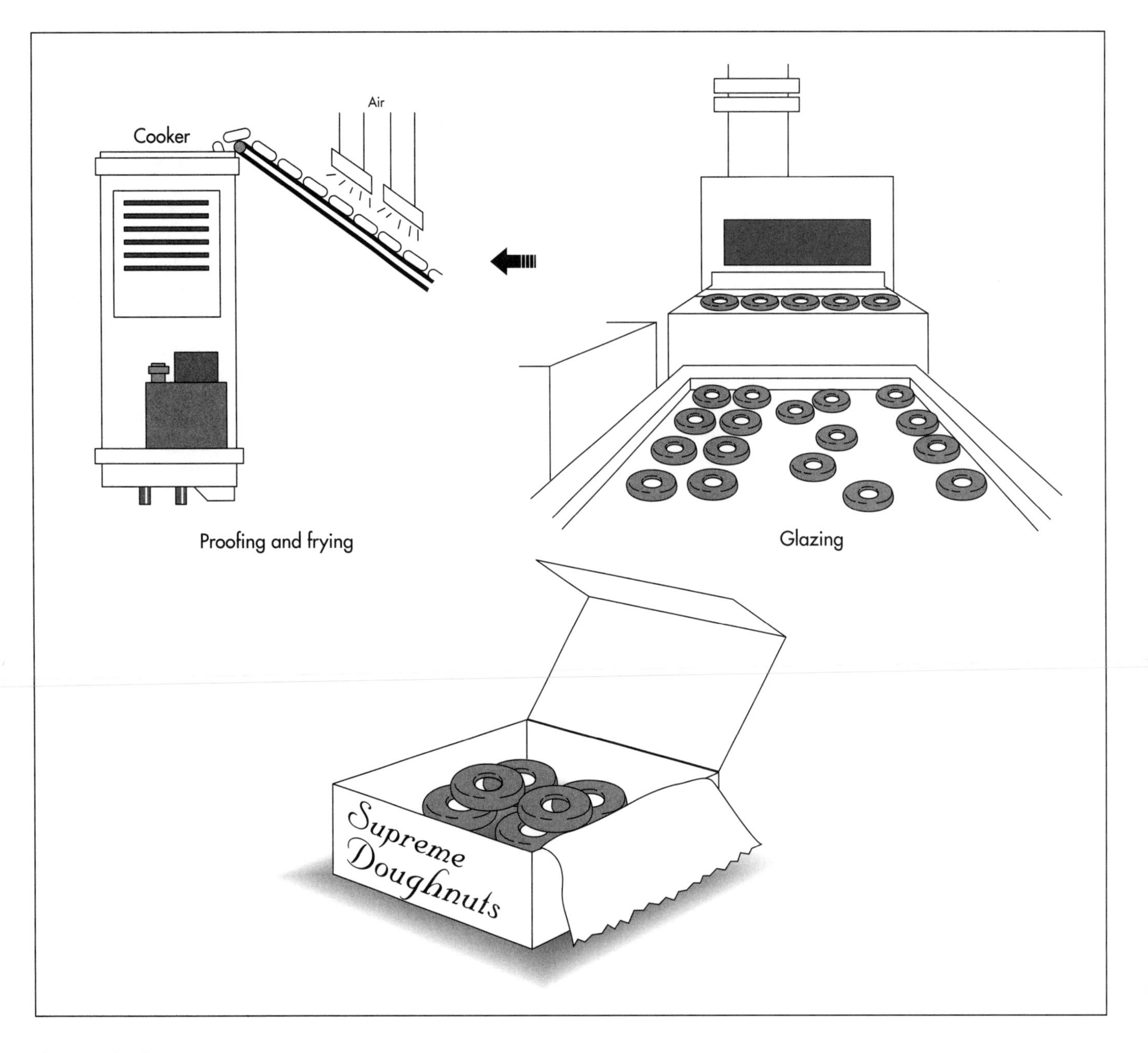


The raw doughnuts are conveyored to the proofing box, a warm, oven-like machine that slowly allows the doughnuts to rise or proof as the yeast ferments under controlled conditions. Proofing renders the doughnuts light and airy. After proofing, the raw doughnuts fall automatically, one row at a time, into the attached open fryer. It takes two minutes for a doughnut to move through the fryer. Next, the doughnuts move under a shower of glaze. The doughnuts are conveyored out of the production area to dry and cool.

### *Further finishing and sale*

9 Once conveyored to a finishing station, the doughnuts may be sprinkled with candies or nuts or are given a thicker frosting. The disk-like doughnuts (those with no hole) are forced onto a machine that injects two doughnuts at a time with the desired, pre-measured filling. The completed doughnuts are placed on trays for movement to the counter or packed into boxes for custom orders.

## Quality Control

Packaged dry mix is made to specifications and checked at the processing plant. Perishables must be purchased fresh and quickly used. The yeast brew must be precisely mixed and used within 12 hours. It is essential for employees to carefully monitor all intervals of time for kneading, resting, proofing, and frying.

Temperatures for proofing, baking, and frying machinery, liquid ingredients, and the production room are carefully monitored and maintained. Particularly important is adding the right temperature of water to the yeast brew and pre-packaged mix so the yeast is not inhibited or killed. The proofer must be precisely set at the right temperature—not too hot but warm enough to activate the yeast—or the yeast will be killed and the doughnuts will not rise. The fryer temperature is carefully determined so that the doughnuts will not absorb too much oil and be greasy. Employees must watch the ambient room temperature very carefully. If it is too hot in the room, it affects the rising of the yeast and may require re-calibration of the temperature of other machinery.

Finally, employees' senses tell them much about the quality of the dough. They can tell by the feel of the dough after it is mixed if the dough is spongy and the yeast is rising properly. Watching the doughnuts plump up in the proofer indicates the temperature is just right. They watch for the appropriate color of the frying doughnuts to ensure they're not overcooked. Occasionally, the manager may pull a doughnut off the drying conveyor and pull it apart to see if it is too greasy.

## Byproducts/Waste

Using the extruding device that simultaneously cuts the dough into individual doughnut shapes, alleviates much of the dough waste. The stamping mechanism leaves excess dough, but that dough can be re-mixed into the next batch.

## Where to Learn More


### Books

Fischer, Paul. "It's Time to Praise the Doughnuts."*Boston Magazine* (May 1991): 66.

Rombauer, Irma S. *The Joy of Cooking*. New York: The Bobbs-Merrill Co., Inc., 1953.

### Periodicals

Taylor, David A. "Ring King."*Smithsonian Magazine* (March 1998).

### Other

Dunkin' Donuts. http:///www.dunkindonuts.com/.

Krispy Kreme. http://krispykreme.com/.


*—Nancy EV Bryk*

# Eggs

*Annual consumption during the late 1990s averaged 245 eggs for each of the 265 million people in the United States.*

## Background

The unfertilized egg is considered an important and inexpensive food source, particularly high in protein, including 0.21 oz (6 g) of complete protein per two-ounce egg. However, it also includes 0.42 oz (12 g) fat, both saturated and unsaturated, which is nearly all located in the yolk. Therefore, if the yolk is separated from the albumen (white) of the egg and the white only used, the egg contains no fat and a fair amount of protein. The egg also contains significant amounts of iron, vitamins A and D, riboflavin, and thiamine. However, that nutritional value does vary somewhat depending on the diet of the laying hen. Annual consumption during the late 1990s averaged 245 eggs for each of the 265 million people in the United States.

While geese, squab, ducks, and turkeys supply edible eggs, the preponderance of eggs eaten come from domesticated chickens bred for laying eggs. The females (mature hens and younger pullets) are raised for meat and egg production and breeds have been developed to fulfill commercial needs. Each of the 235 million laying hens in the United States produces about 300 eggs a year. Farmers are careful to house and feed the chickens to maximize laying and ensure the hen has a relatively long and healthy life. Egg producers also have flocks of hens at different ages, ensuring they have a steady supply of eggs ready for market to provide year-round income.

Eggs are an important ingredient in many recipes. Because the protein in the egg white coagulates as it is heated, eggs are utilized in many recipes as a structural component. Stiffly-beaten egg whites expand as they are heated and are used as leavening in angel-food cakes, souffles, and meringues. In cake batters that utilize the entire egg, the egg acts as leavening as well as providing moisture and firm texture. Soups and sauces use eggs as a thickening agent. Ice creams often include eggs to prevent the formation of ice crystals which can ruin the product. But the plain egg is eaten by millions each day for its own unique flavor and nutritive value—they may be boiled, poached, fried, scrambled, or baked.

Fresh egg production is primary to the egg industry, however, a significant amount of egg production includes eggs purposely broken and used for powdered eggs, frozen eggs, or purchased by food producers for inclusion in food products. (In some fresh egg production plants, accidentally broken eggs are sold to bakeries or other food production plants.)

## History

The egg has been a protein source for centuries. Sometime in the second millennium B.C. Indian wild red jungle fowl, the ancestor of the modern chicken (*Gallus gallus*) was dispersed throughout Europe, China, and the Middle East. Chickens were brought to the New World on Columbus's second voyage in the late fifteenth century. These imported chickens laid eggs year-round and were considered most valuable for their egg production rather than for meat. Soon family farms raised chickens for the family's consumption of eggs and meat within the household—few families had laying flocks of any size. However, by 1800 chickens began being raised for meat and egg production in increasing numbers in the United States. Until World War II, egg production came from rather small flocks of less than

400 laying hens. After the war, automation and advances in breeding, feeding, and developing efficient henhouses gave rise to modern high-volume chicken farms. Today, a single egg producer may well have a flock of over 100,000 laying hens—and some have flocks of over one million.

## Raw Materials

The egg itself is the ingredient in egg production. Soaps are used in egg production facilities to clean the shell. Some processors coat the shell in a light film of oil.

However, the hen from which the egg drops can be considered an important part of the raw material. Certain breeds lay the majority of eggs in the United States. A particularly prolific laying hen is the Single Comb White Leghorn. This breed reaches maturity early and can begin laying at 19 weeks of age and continue to lay eggs for about a year. In addition, the Leghorn utilizes feed efficiently, is fairly small, can adapt to a variety of climates and is able to lay a relatively large amount of white eggs, the type most in demand. Plymouth Rock hens, Rhode Island Reds and New Hampshire hens produce brown eggs much favored in New England.

Feed for chickens is generally all-mash, consisting of sorghum grains, corn, and either cottonseed meal or soybean oil depending on availability. Farmers carefully mix the mash so that the chickens get just the right amounts of protein, fat, carbohydrates, vitamins, and minerals. This is essential in that the nutritional quality of the laid egg depends on the feed the chicken was given. All additives to chicken feed must be approved by the federal government (after research on toxicity to animals and humans is determined). Hormones are not given to chickens, but they occasionally require antibiotics. On average, a chicken requires about 4 lb (1.8 kg) of feed to produce a dozen eggs; a Leghorn hen eats about 0.25 lb (0.1134 kg) of chick mash a day.

## Processing Eggs

Some egg farmers have their own egg processing facility on premises. Others contract with egg processing firms in the locale who purchases eggs and processes them. Generally, an egg moves from the egg farm to being ready for public consumption in just a few days.

### *Laying the eggs*

1 Hens are kept in cages, usually in groups of three to five. After the egg is laid, the cage is devised so the egg rolls out for easy collection. Eggs are gathered twice a day, generally by automated machinery but are occasionally still gathered by hand. Eggs are gathered as soon as hens lay them because warmer temperatures encourage physical and chemical changes that affect freshness adversely. Thus, many egg farmers refrigerate the eggs immediately after gathering before they are packed for transport to the processor.

### *Packing the eggs on the farm*

2 The eggs are then packed on skids that are formed of layers of flats—eggs are packed on flats that include 2 .5 dozen eggs, with as many as six flats per layer (layers of six flats are separated by a board). A single skid holds about 30 cases or 900 dozen eggs. These eggs are then sent to the processing plant via truck.

### *Washing the eggs*

3 The skids are brought into the production room and the individual flats are put on the conveyor, one at a time. Individual eggs are grasped by small suction cups and placed onto another conveyor belt. Now, the eggs are moved into the grader where they are cleaned with a USDA approved cleanser. They are rotated as brushes and water jets move carefully across the eggs. A fan then dries the eggs.

### *Candling and grading*

4 The cleaned eggs are graded in a candling booth which is a dark cubical or room. A penetrating light is shined on the eggs in order to grade them. The egg processor is able to grade the egg during candling. The trained candler can see that an older egg has thinner albumen; thus, the yolk casts a sharp shadow and immediately indicates an older egg. Eggs are graded as "A" (sold for household use or at retail markets), grade "B" (used mostly for bakery operations), or grade "C" (sent to egg breakers who break the shell in order to convert it to other egg

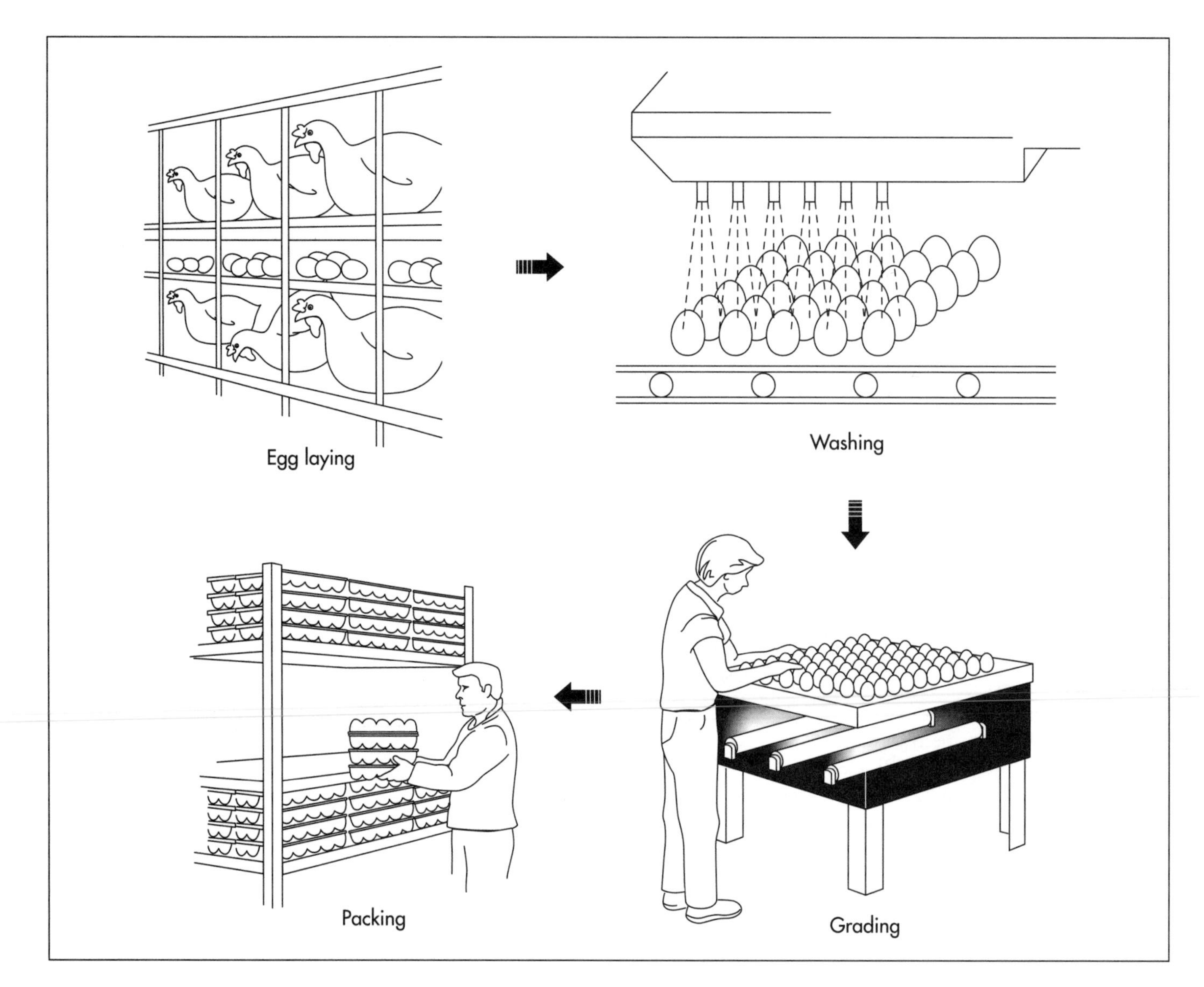


Commercial egg processing is a quick business that relies on speed to market in order to provide fresh, quality product. Hens are kept in cages that are devised so that when an egg is laid it rolls into a collection bin. The eggs are then packed on skids which are formed of layers of flats. Skids are individually placed on a conveyor belt. Each egg in the skid is grasped by a small suction cup and placed onto another conveyor belt. The eggs move into the grader where they are cleaned with a USDA approved cleanser. They are rotated as brushes and water jets move carefully across the eggs. A fan then dries the eggs. Once graded, the eggs are placed in cartons, packaged, and shipped to stores.

products); higher grade eggs have a thick, upstanding albumen, an oval yolk, and a clean, smooth, unbroken shell. Eggs with cracks that are not leaking are removed from the process at this point and are not packaged for household use or retail sale.

### Weighing and packing the eggs

5 The grading process actually includes the weighing of the eggs as well. At the grading station, a machine weighs each individual egg and remembers what each egg weighs. In the United States, eggs are sized (extra large, large, medium, small) on the basis of a minimum weight rather than by size of egg. Extra large eggs must weigh a minimum of 2.24 oz (64 g), large at least 1.96 oz (56 g), medium at least 1.72 oz (49 g), and small at least 1.47 oz (42 g). The packing machine then assembles cartons based on the weight of individual eggs; thus, the heaviest eggs are "found" and packed into the extra large cartons, the next heaviest tier are packed in to the large-size egg carton, etc. Packaging varies, but is generally either recycled cardboard or a colored polystyrene the egg producers purchase from a packaging manufacturer.

### Transportation to trucks

6 After packing, the cartons are placed on a conveyor and packed into flats by machine, put into trucks (generally refrigerated) and sent to be sold. One large Pennsylvania egg processing plant processes 45,000 dozen eggs per day.

## Quality Control

Quality control happens in all parts of the egg processing. First and foremost, the chicken farmer ensures that his hens are well fed with a diet specifically formulated to provide the best grade of egg. Shell strength, for example is determined by the presence of adequate amounts of vitamin D, calcium, and other minerals in the chick mash. Too little vitamin A can result in blood spots (not harmful to the consumer but render an egg undesirable and unusable to the consumer). Laying hens also require a good supply of clean fresh water. The henhouse is well insulated so that the farmer can control the temperature. Facilities are windowless so that light can be manipulated—egg production is spurred by maintaining a lighted environment 14-17 hours a day. The henhouse should be comfortable for the hens and well-ventilated. Birds are generally kept in cages because they are easier to clean and it is easier to collect eggs from cages; however, some hens are allowed to roam free.

Effective candling is essential to quality control as well. Candling reveals nearly everything that is need to know about the quality of the eggs—age, cracks, clarity (no blood spots). Furthermore, most egg processors can tell much about the quality of the egg just by looking at the shape and color of the shell.

Salmonella is a hazard of the egg industry. However, it is estimated that 90% of eggs are free from salmonella at the time they are laid. Salmonella bacteria occurs after laying. Proper washing and sanitizing of eggs with a government-approved soap eliminates most Salmonella and spoilage organisms that are deposited on the shell from the hen ovaries.

Egg farmers are also careful to refrigerate the eggs as soon as they are gathered just prior to packing. The egg processors, too, move eggs quickly through the processing to packaging to ensure the eggs are clean and fresh for the consumer.

Furthermore, government standards for the grade and size of eggs are strictly adhered to. Flocks are periodically monitored for proper feeding as well as acceptable facility standards. Extension services provide educational materials and new information about raising laying hens to farmers as needed.

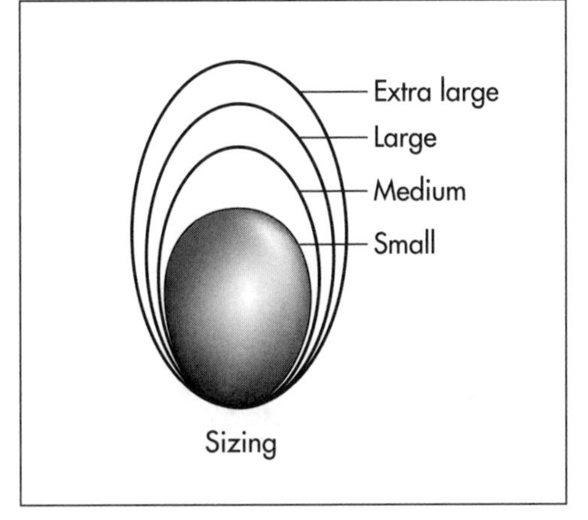

Sizing

## Byproducts/Waste

Eggs with cracks are removed from the egg processing line. Broken eggs are thrown into a bin and sold for utilization in dog food. Eggs that have cracks but are not leaking are taken away for pasteurization and transformation into liquid egg products (sold in cartons of plastic containers or perhaps even frozen). They may be sold to another processor who transforms them into powdered eggs, or they may be sold to local bakeries for use in goods there. Many egg processors are aware of the ecological problems created by the polystyrene cartons and egg processor encourage recycling of the product packaging.

## Where to Learn More

### Periodicals

Wexler, Mark. "Eggsquisite!" *World Magazine* (March 1989): 70.

### Other

American Egg Board. http://www.aeb.org/ (June 29, 1999).

Manitoba Egg Producers. http://www.mbegg.mb.ca/ (June 29, 1999).

*—Nancy EV Bryk*

In a darkened room, eggs are placed in a candling booth with penetrating lights shining up from below. A trained candler can see that an older egg has thinner albumen; thus, the yolk casts a sharp shadow and immediately indicates an older egg. Eggs are graded as "A" (sold for household use or at retail markets), grade "B" (used mostly for bakery operations), or grade "C" (sent to egg breakers who break the shell in order to convert it to other egg products); higher grade eggs have a thick, upstanding albumen, an oval yolk, and a clean, smooth, unbroken shell. Next, the eggs are individually weighed and assigned an appropriate size. In the United States, eggs are sized (extra large, large, medium, small) on the basis of a minimum weight rather than by size of egg. Extra large eggs must weigh a minimum of 2.24 oz (64 g), large at least 1.96 oz (56 g), medium at least 1.72 oz (49 g), and small at least 1.47 oz (42 g).

# Electric Automobile

*In an internal combustion-driven car, the engine, coolant system, and other specific powering devices total 25% of the weight of the car. In electric cars, the battery and electric propulsion system are typically 40% of the weight of the car.*

## Background

Unlike the gas-powered automobile, the electric automobile did not easily develop into a viable means of transportation. In the early twentieth century, the electric car was vigorously pursued by researchers; however the easily mass-produced gasoline-powered automobile squelched interest in the project. Research waned from 1920-1960 until environmental issues of pollution and diminishing natural resources reawakened the need of a more environmentally friendly means of transportation. Technologies that support a reliable battery and the weight of the needed number of batteries elevated the price of making an electric vehicle. On the plus side, automotive electronics have become so sophisticated and small that they are ideal for electric vehicle applications.

## History

The early development of the automobile focused on electric power rather than gasoline power. In 1837, Robert Davidson of Scotland appears to have been the builder of the first electric car, but it wasn't until the 1890s that electric cars were manufactured and sold in Europe and America. During the late 1890s, United States roads were populated by more electric automobiles than those with internal combustion engines.

One of the most successful builders of electric cars in the United States was William Morrison of Des Moines, Iowa, who began marketing his product in 1890. Other pioneers included S. R. and Edwin Bailey, a father-son team of carriage makers in Amesbury, Massachusetts, who fitted an electric motor and battery to one of their carriages in 1898. The combination was too heavy for the carriage to pull, but the Baileys persisted until 1908 when they produced a practical model that could travel about 50 mi (80 km) before the battery needed recharging.

Much of the story of the electric car is really the story of the development of the battery. The lead-acid battery was invented by H. Tudor in 1890, and Thomas Alva Edison developed the nickel-iron battery in 1910. Edison's version increased the production of electric cars and trucks, and the inventor himself was interested in the future of the electric car. He combined efforts with the Baileys when they fitted one of his new storage batteries to one of their vehicles, and they promoted it in a series of public demonstrations. The Bailey Company continued to produce electric cars until 1915, and it was among over 100 electric automobile companies that thrived early in the century in the United States alone. The Detroit Electric Vehicle Manufacturing Company was the last to survive, and it ceased operation in 1941.

Electric automobiles were popular because they were clean, quiet, and easy to operate; however, two developments improved the gasoline-powered vehicle so much so that competition was nonexistent. In 1912, Charles Kettering invented the electric starter that eliminated the need for a hand crank. At the same time, Henry Ford developed an assembly line process to manufacture his Model T car. The assembly was efficient and less costly than the manufacture of the electric vehicle. Thus, the price for a gas-driven vehicle decreased enough to make it feasible for every family to afford an automobile. Only electric trolleys, delivery vehicles that made frequent stops, and a few other electric-powered vehicles survived past the 1920s.

In the 1960s, interest in the electric car rose again due to the escalating cost and diminishing supply of oil and concern about pollution generated by internal combustion engines. The resurgence of the electric car in the last part of the twentieth century has, however, been fraught with technical problems, serious questions regarding cost and performance, and waxing and waning public interest. Believers advocate electric cars for low electrical energy consumption and cost, low maintenance requirements and costs, reliability, minimal emission of pollutants (and consequent benefit to the environment), ease of operation, and low noise output.

Some of the revived interest has been driven by regulations. California's legislature mandated that 2% of the new cars sold in the state be powered by zero-emissions engines by 1998. This requirement increases to 4% by 2003. Manufacturers invested in electric cars on the assumption that public interest would follow the regulation and support protection of air quality and the environment. General Motors (GM) introduced the Impact in January 1990. Impact had a top speed of 110 mph (176 kph) and could travel for 120 mi (193 km) at 55 mph (88 kph) before a recharging stop. Impact was experimental, but, later in 1990, GM began transforming the test car into a production model. Batteries were the weakness of this electric car because they needed to be replaced every two years, doubling the vehicle's cost compared to the operating expenses of a gasoline-powered model. Recharging stations are not widely available, and these complications of inconvenience and cost have deterred potential buyers. In 1999, Honda announced that it would discontinue production of its electric car, which was introduced to the market in May 1997, citing lack of public support due to these same deterrents.

## Components

Unlike primary batteries that have a limited lifetime of chemical reactions that produce energy, the secondary-type batteries found in electric vehicles are rechargeable storage cells. Batteries are situated in T-formation down the middle of the car with the top of the "T" at the rear to provide better weight distribution and safety. Batteries for electric cars have been made using nickel-iron, nickel-zinc, zinc-chloride, and lead-acid.

Weight of the electric car has also been a recurring design difficulty. In electric cars, the battery and electric propulsion system are typically 40% of the weight of the car, whereas in an internal combustion-driven car, the engine, coolant system, and other specific powering devices only amount to 25% of the weight of the car.

Other technologies in development may provide alternatives that are more acceptable to the public and low (if not zero) emissions. Use of the fuel cell in a hybrid automobile is the most promising development on the horizon, as of 1999. The hybrid automobile has two power plants, one electric and one internal combustion engine. They operate only under the most efficient conditions for each, with electric power for stop-and-start driving at low speeds and gasoline propulsion for highway speeds and distances. The electric motor conserves gasoline and reduces pollution, and the gas-powered portion makes inconvenient recharging stops less frequent.

Fuel cells have a chemical source of hydrogen that provides electrons for generating electricity. Ethanol, methanol, and gasoline are these chemical sources; if gasoline is used, fuel cells consume if more efficiently than the internal combustion engine. Fuel cell prototypes have been successfully tested, and the Japanese began manufacturing a hybrid vehicle in 1998. Another future hope for electric automobiles is the lithium-ion battery that has an energy density three times greater than that of a lead-acid battery. Three times the storage should lead to three times the range, but cost of production is still too high. Lithium batteries are now proving to be the most promising, but limited supplies of raw materials to make all of these varieties of batteries will hinder the likelihood that all vehicles can be converted to electrical power.

## Raw Materials

The electric car's skeleton is called a space frame and is made of aluminum to be both strong and lightweight. The wheels are also made of aluminum instead of steel, again as a weight-saving method. The aluminum parts are poured at a foundry using specially designed molds unique to the manufacturer. Seat frames and the heart of the steering wheel are made of magnesium, a lightweight metal. The body is made of an impact-resistant composite plastic that is recyclable.

Electric car batteries consist of plastic housings that contains metal anodes and cathodes and fluid called electrolyte. Currently, lead-acid batteries are still used most commonly, although other combinations of fluid and metals are available with nickel metal hydride (NiMH) batteries the next most likely power source on the electric car horizon. Electric car batteries hold their fluid in absorbent pads that won't leak if ruptured or punctured during an accident. The batteries are made by specialty suppliers. An electric car like the General Motors EV1 contains 26 batteries in a T-shaped unit.

The motor or traction system has metal and plastic parts that do not need lubricants. It also includes sophisticated electronics that regulate energy flow from the batteries and control its conversion to driving power. Electronics are also key components for the control panel housed in the console; the on-board computer system operates doors, windows, a tire-pressure monitoring system, air conditioning, starting the car, the CD player, and other facilities common to all cars.

Plastics, foam padding, vinyl, and fabrics form the dashboard cover, door liners, and seats. The tires are rubber, but, unlike standard tires, these are designed to inflate to higher pressures so the car rolls with less resistance to conserve energy. The electric car tires also contain sealant to seal any leaks automatically, also for electrical energy conservation. Self-sealing tires also eliminate the need for a spare tire, another weight- and material-saving feature.

The windshield is solar glass that keeps the interior from overheating in the sun and frost from forming in winter. Materials that provide thermal conservation reduce the energy drain that heating and air conditioning impose on the batteries.

## Design

Today's electric cars are described as "modern era production electric vehicles" to distinguish them from the series of false starts in trying to design an electric car based on existing production models of gasoline-powered cars and from "kit" cars or privately engineered electric cars that may be fun and functional but not production-worthy. From the 1960s-1980s, interest in the electric car was profound, but development was slow. The design roadblock of the high-energy demand from batteries could not be resolved by adapting designs. Finally, in the late 1980s, automotive engineers rethought the problem from the beginning and began designing an electric car from the ground up with heavy consideration to aerodynamics, weight, and other energy efficiencies.

The space frame, seat frames, wheels, and body were designed for high strength for safety and the lightest possible weight. This meant new configurations that provide support for the components and occupants with minimal mass and use of high-tech materials including aluminum, magnesium, and advanced composite plastics. Because there is no exhaust system, the underside is made aerodynamic with a full belly pan. All extra details had to be eliminated while leaving the comforts drivers find desirable and adding new considerations unique to electric automobiles. One eliminated detail was the spare tire. The detail of the rod-like radio antennae was removed; it causes wind resistance that robs energy and uses energy to power it up and down. An added consideration was the pedestrian warning system; tests of prototypes showed that electric cars run so quietly that pedestrians don't hear them approach. Driver-activated flashing lights and beeps warn pedestrians that the car is approaching and work automatically when the car is in reverse. Windshields of solar glass were also an important addition to regulate the interior temperature and minimize the need for air conditioning and heating.

Among the many other design and engineering features that must be considered in producing electric cars are the following:

- Batteries that store energy and power the electric motor are a science of their own in electric car design, and many options are being studied to find the most efficient batteries that are also safe and cost effective. An electric motor that converts electrical energy from the battery and transmits it to the drive train. Both direct-current (DC) and alternating current (AC) motors are used in these traction or propulsion systems for electric cars, but AC motors do not use brushes and require less maintenance.
- A controller that regulates energy flow from the battery to the motor allows for adjustable speed. Resistors that are used for this purpose in other electric devices are not practical for cars because they absorb too much of the energy themselves. Instead, silicon-controlled rectifiers (SCRs) are used. They allow full power to go from the battery to the motor but in pulses so the battery is not overworked and the motor is not underpowered.
- Any kind of brakes can be used on electric automobiles, but regenerative braking systems are also preferred in electric cars because they recapture some of the energy lost during braking and channel it back to the battery system.
- Two varieties of chargers are needed. A full-size charger for installation in a garage is needed to recharge the electric car overnight, but a portable recharger (called a convenience recharger) is standard equipment for the trunk so the batteries can be recharged in an emergency or away from home or a charging station. For safety, an inductive charger was created for electric cars with a paddle that is inserted in the front end of the car. It uses magnetic energy to recharge the batteries and limit the potential for electrocution.

## The Manufacturing Process

The manufacturing process required almost as much design consideration as the vehicle itself; and that design includes handcrafting and simplification as well as some high-tech approaches. The assemblers work in build-station teams to foster team spirit and mutual support, and parts are stored in modular units called creform racks of flexible plastic tubes and joints that are easy to fill and reshape for different parts. On the high- tech side, each station is equipped with one torque wrench with multiple heads; when the assembler locks on the appropriate size of head, computer controls for the machine select the correct torque setting for the fasteners that fit that head.

### Body shop

The body for the electric car is handcrafted at six work stations.

1 Parts of the aluminum space frame are put together in sections called subassemblies that are constructed of prefabricated pieces that are welded or glued together. The glue is an adhesive bonding material, and it provides a connection that is more durable and stiffer than welding. As the subassemblies for the undercarriage of the car are completed, they are bonded to each other until the entire underbody is finished.

2 The subassemblies for the upper part of the body are also bonded to make larger sections. The completed sections are similarly welded or glued until the body frame is finished. The body is added to the underbody. The adhesive used throughout staged assembly of the frame is then cured by conveying the body through a two-stage oven.

3 The roof is attached. Like other parts of the exterior, it has already been painted. The underbody and the rest of the frame are coated with protective sealants, and the finished body is moved to the general assembly area.

### General assembly

General assembly of the operating components and interior of the electric car is completed at eight other work stations.

1 At the first assembly station, the first set of the electric car's complex electronics are put in place. This includes the body wiring and seating of the Power Electronics Bay which holds the Propulsion Control Module, integrated drive unit, and a small radiator. The integrated drive unit consists of the alternating current induction motor and a two-stage gear reduction and differential. These units are all preassembled in their

The manufacturing process used to make an electric vehicle is as intricate as the vehicle design. It takes six work stations to create the body of an electric vehicle. Each station is equipped with one torque wrench with multiple heads; when the assembler locks on the appropriate size of head, computer controls for the machine select the correct torque setting for the fasteners that fit that head.

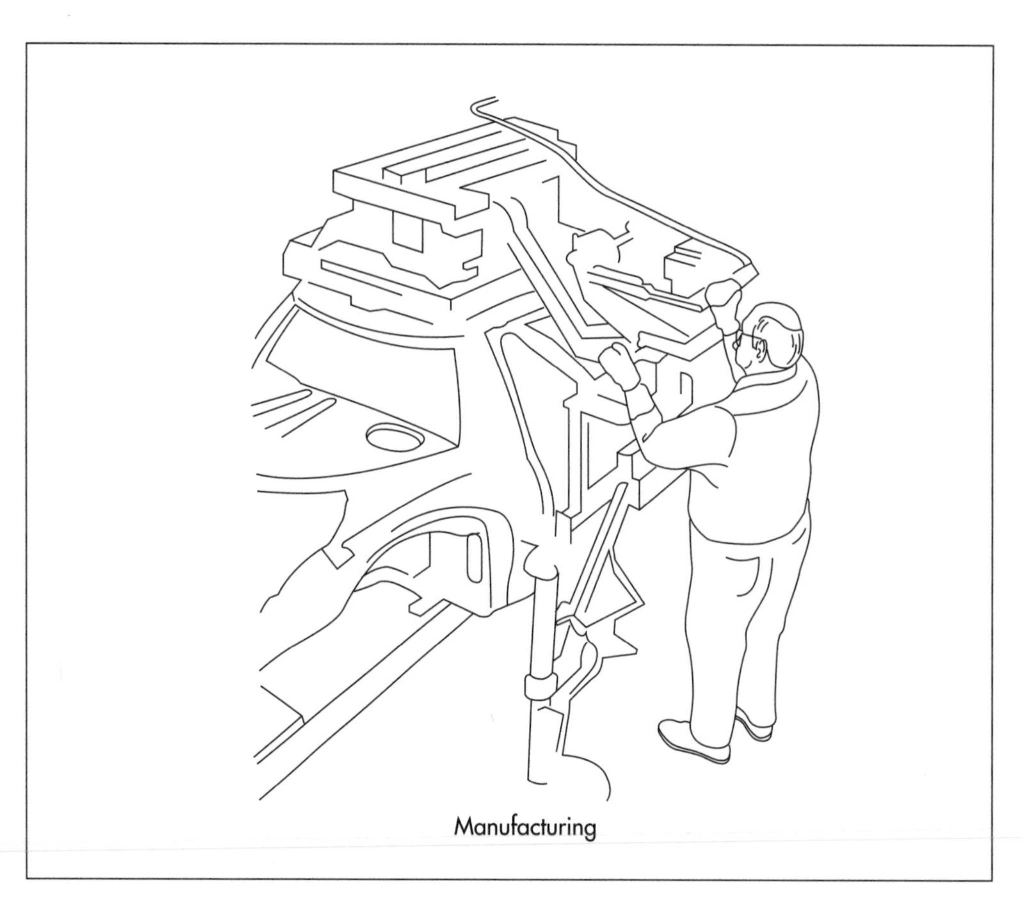

Manufacturing

own housings. The components of the control console are also installed.

2 The interior is outfitted. Flooring, seats, carpeting, and the console and dash are placed in the car. The process is simple because the instrument panel and console cover are made of molded, fiberglass reinforced urethane that has been coated with more urethane of finish quality and with a non-reflective surface. These two pieces are strong and don't need other supports, brackets, or mounting plates. Assembly is straight forward, and performance is superior because fewer pieces reduce possibilities for rattles and squeaks.

3 At the third work station, the air conditioning, heating, and circulation system is inserted, and the system is filled.

4 The battery pack is added. The T-shaped unit is seated by lifting the heavy pack using a special hoist up into the car. The pack is attached to the chassis, as are the axles complete with wheels and tires. With both batteries and the propulsion unit in place, the car no longer has to be moved from station to station on specially designed dollies. Instead, it is driven to the remaining work stations. The system is powered up and checked before it is driven to the next team.

5 The windshield is installed and other fluids are added and checked. The door systems (complete with vinyl interiors, arm rests, electronics, and windows) are also attached, and all the connections are completed and checked. The exterior panels are added. Similar to the roof and doors, they have been prepared and painted before being brought to the work station. The final trim is attached to complete the upper exterior.

6 At the final work station, the alignment is checked and adjusted, and the underbody panel is bolted into place. The process concludes with the last, comprehensive quality control check. Pressurized water is sprayed on the vehicle for eight minutes, and all the seals are checked for leaks. On a specialized test track, the car is checked for noises, squeaks, and rattles on a quality-based test drive. A lengthy and thorough visual inspection concludes the quality audit.

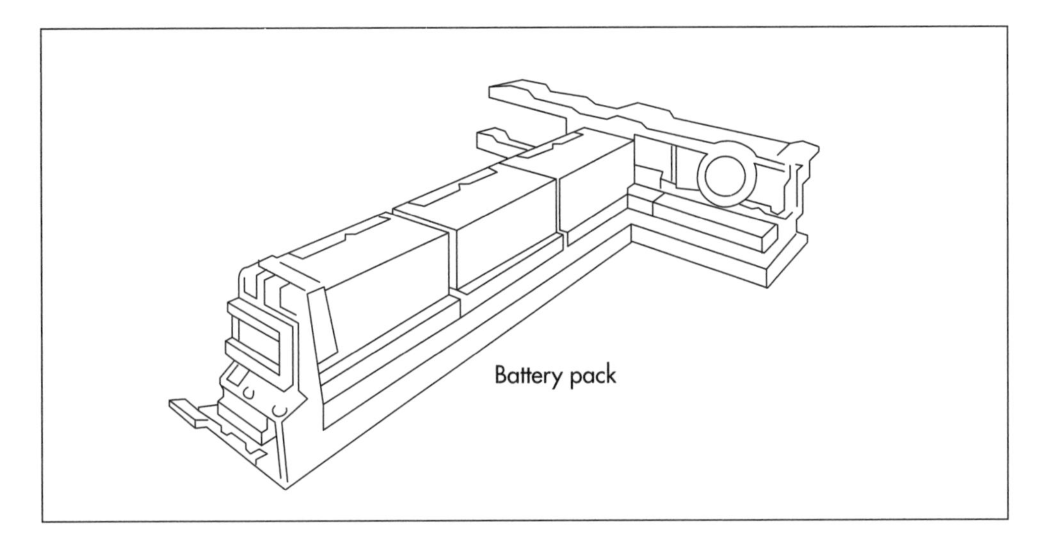


Unique to the electric vehicle is the battery pack. The battery pack is made up of rechargeable secondary-type batteries that act as storage cells. Batteries are situated in T-formation down the middle of the car with the top of the "T" at the rear to provide better weight distribution and safety.

## Quality Control

Industry has proven that work stations are a highly effective method of providing quality control throughout an assembly process. Each work station has two team members to support each other and provide internal checks on their part of the process. On a relatively small assembly line like this one for the electric car (75 assemblers in a General Motors plant), the workers all know each other, so there is also a larger team spirit that boosts pride and cooperation. Consequently, the only major quality control operation concludes the assembly process and consists of a comprehensive set of tests and inspections.

Unique to manufacture of the electric car, the operation of the car has been tested during the final assembly steps. The car has no exhaust system and emits no gases or pollutants, so, after the battery pack and propulsion unit have been installed, the car can be driven inside the plant. Proof that the product works several steps before it is finished is a reassuring quality check.

## Byproducts/Waste

There are no byproducts from the manufacture of electric cars. Waste within the assembly factory also is minimal to nonexistent because parts, components, and subassemblies were all made elsewhere. Trimmings and other waste are recaptured by these suppliers, and most are recyclable.

## The Future

Electric cars are critically important to the future of the automobile industry and to the environment; however, the form the electric car will ultimately take and its acceptance by the public are still uncertain. Consumption of decreasing oil supplies, concerns over air and noise pollution, and pollution caused (and energy consumed) by abandoned cars and the complications of recycling gasoline-powered cars are all driving forces that seem to be pushing toward the success of the electric car.

## Where to Learn More

### Books

Hackleman, Michael. *Electric Vehicles: Design and Build Your Own.* Mariposa, CA: Earthmind/Peace Press, 1980.

Shacket, Sheldon R. *The Complete Book of Electric Vehicles.* Northbrook, IL: Domus Books, 1979.

Whitener, Barbara. *The Electric Car Book.* Louisville, KY: Love Street Books, 1981.

### Periodicals

Associated Press. "Fuel-cell vehicles to take road test."*Daily Review* (Hayward, California), 21 April 1999, p. 3.

Associated Press. "Honda dumps electric cars."*Daily Review* (Hayward, California), 30 April 1999.

General Motors. *EVolution: The Official Publication of General Motors Advanced Technology Vehicles*, 1997.

Hornblower, Margot. "Is this clean machine for real?"*Time* (December 15, 1997): 62+.

## Other

Electric Vehicle Association of the Americas. http://www.evaa.org/.

General Motors (GM) Saturn EV1. http://www.gmev.com.

Honda EV Plus. http://www.honda1999.com/cars/ev.

*—Gillian S. Holmes*

# Envelope

## Background

An envelope is a flat, flexible container, made of paper or similar material, that has a single opening and a flap that can be sealed over the opening. The envelope is usually sealed by wetting an area of the flap. Some envelopes are sealed with a metal fastener. Others are sealed with a piece of string that wraps around flat, circular pieces of cardboard attached to the envelope. A recent development in envelopes is a thin strip of plastic, which is removed to reveal an area of the flap with an adhesive that does not need moistening.

Envelopes are almost always rectangular, but they exist in a wide range of sizes. The two main styles used are banker envelopes, which have the opening on the long side, and pocket envelopes, which have the opening on the short side. In the United States, standard sizes range from 3.5 x 6 in (89 x152 mm) to 10 x 13 in (254 x 330 mm). In Europe, sizes range from 3.2 x 4.5 in (81 x 114 mm) to 11 x 15.75 in (280 x 400 mm). Sizes are somewhat different in the United Kingdom, with the most common being 4.25 x 8.625 in (108 x 219 mm).

Some envelopes have one or more windows cut into the front to allow addresses written on sheets inside to be seen. These windows may be covered with a transparent material.

## History

The earliest ancestor of the envelope was used by the ancient Babylonians five or six thousand years ago. Messages were written on clay tablets, which were baked to harden them. The tablets were then covered with more clay and baked again. The inner tablet could only be revealed by breaking open the outer layer of clay, ensuring the security of the message.

True envelopes did not exist until much later, long after the invention of paper. The oldest form of paper was papyrus, first manufactured by the ancient Egyptians at least as early as 3000B.C. Papyrus was made from a fibrous material found within the woody stems of an aquatic, grassy plant (*Cyperus papyrus*). Long strips of this material were placed side by side, then covered with another layer of strips at right angles to the first. The sheet formed by the two layers was dampened, pressed, dried, flattened, then dried again. The resulting papyrus, if properly made, was pure white and free from spots and stains. An excellent writing material, papyrus was used extensively by the ancient Egyptians, Greeks, Romans, and Arabs. It continued to be used until paper made from other plant sources reached the rest of the world from China. Some papyrus was used in Europe as late as the twelfth century.

Early forms of Chinese paper, made from reeds and rice, date back as far as 1200B.C. A superior kind of paper, similar to modern paper, was first made about the year 105. Attributed to a court official named Ts'ai Lun, this improved paper was made from a mixture of materials, including mulberry and other woody fibers, hemp, rags, and fishing nets. Papermaking spread slowly from East to West, reaching Central Asia by 751 and Baghdad by 793. By the fourteenth century, there were several paper mills throughout Europe, particularly in Spain, Italy, France, and Germany. The development of the printing press in the 1450s greatly increased the demand for paper.

*By the late 1990s, nearly two hundred billion envelopes were made in the United States each year.*

The early history of the paper envelope is not known. Paper may have been used to wrap messages at a very early date in China. They did not appear in Europe until the seventeenth century, when they began to be used in Spain and France. Until that time, messages were simply folded and sealed. Even today, some stationery is designed to be folded and mailed without an envelope.

Cotton and linen rags were the main raw materials used to make paper until the early nineteenth century, when they were replaced by wood. At about the same time, papermaking by hand began to be replaced by papermaking machines. The emerging envelope industry was noted by Karl Marx in his book *Das Kapital* in 1867. Envelope manufacturers continued to increase the speed of production, from three thousand envelopes per hour at the time of Marx to more than fifty thousand per hour in the late twentieth century. By the late 1990s, nearly two hundred billion envelopes were made in the United States each year.

## Raw Materials

Most envelopes are made from paper. Some large, strong envelopes are made from synthetic materials, such as polyethylene. Polyethylene is a plastic made from ethylene, which is derived from petroleum.

Paper used for most envelopes is made from wood. Modern technology allows the wood to come from almost any kind of tree. Paper used to make very high quality envelopes, such as those used to enclose formal invitations, may be made partly or completely from cotton or linen. Some envelopes are made from manila, a fiber from the leaves of a plant found in the Philippines that produces a strong, yellowish paper. Most so-called manila envelopes, however, are made of paper derived from wood which only resembles true manila.

The glue applied to envelopes is of two basic types. The glue applied to the flap that is sealed by the consumer is usually a gum. A typical natural gum is gum arabic, derived from a substance produced by the acacia tree. Synthetic gums are often derived from dextrans, which are produced by the fermentation of sugar. The glue that holds the rest of the envelope must be stronger and more permanent. This glue is often derived from starches, which are obtained from corn, wheat, potatoes, rice, and other plants.

The fastener attached to some envelopes is made of aluminum or other metals. The string attached to other envelopes is made of cotton or other fibers. The material covering the windows in some envelopes is usually polystyrene. Polystyrene is a plastic made from styrene, a derivative of petroleum.

## The Manufacturing Process

### *Making wood pulp*

1 Mechanical methods can be used to transform wood into pulp, but this produces a relatively weak paper that is used for newspapers and similar products. Paper intended to be used for envelopes is made from pulp obtained by chemical means.

2 The most common chemical method used to make wood into pulp is known as kraft pulping. Chips of wood are placed in a large, sealed container known as a digester. The digester contains a strongly alkaline solution of sodium hydroxide and sodium sulfide. The mixture is heated to a temperature between 320-356°F (160-180°C) at a pressure of about 116 pounds per square inch (800 kilopascals) for about one-half to two hours.

3 Various methods exist to bleach the resulting pulp. Bleaching removes lignin, a substance found in wood pulp that gives paper a brown color. In general, bleaching involves mixing the pulp with a series of oxidizing chemicals that react with the lignin. After each mixture, the pulp is washed with an alkaline solution that removes the treated lignin.

4 In order to improve the brightness, opacity, and smoothness of the paper, fillers are added to the pulp. A typical filler is a clay known as kaolin. Other chemicals often added to pulp include various starches or gums to make the paper stronger. Rosin (a substance derived from pine trees) and alum (aluminum sulfate) are often added as sizers. Sizing makes the paper less absorbent,

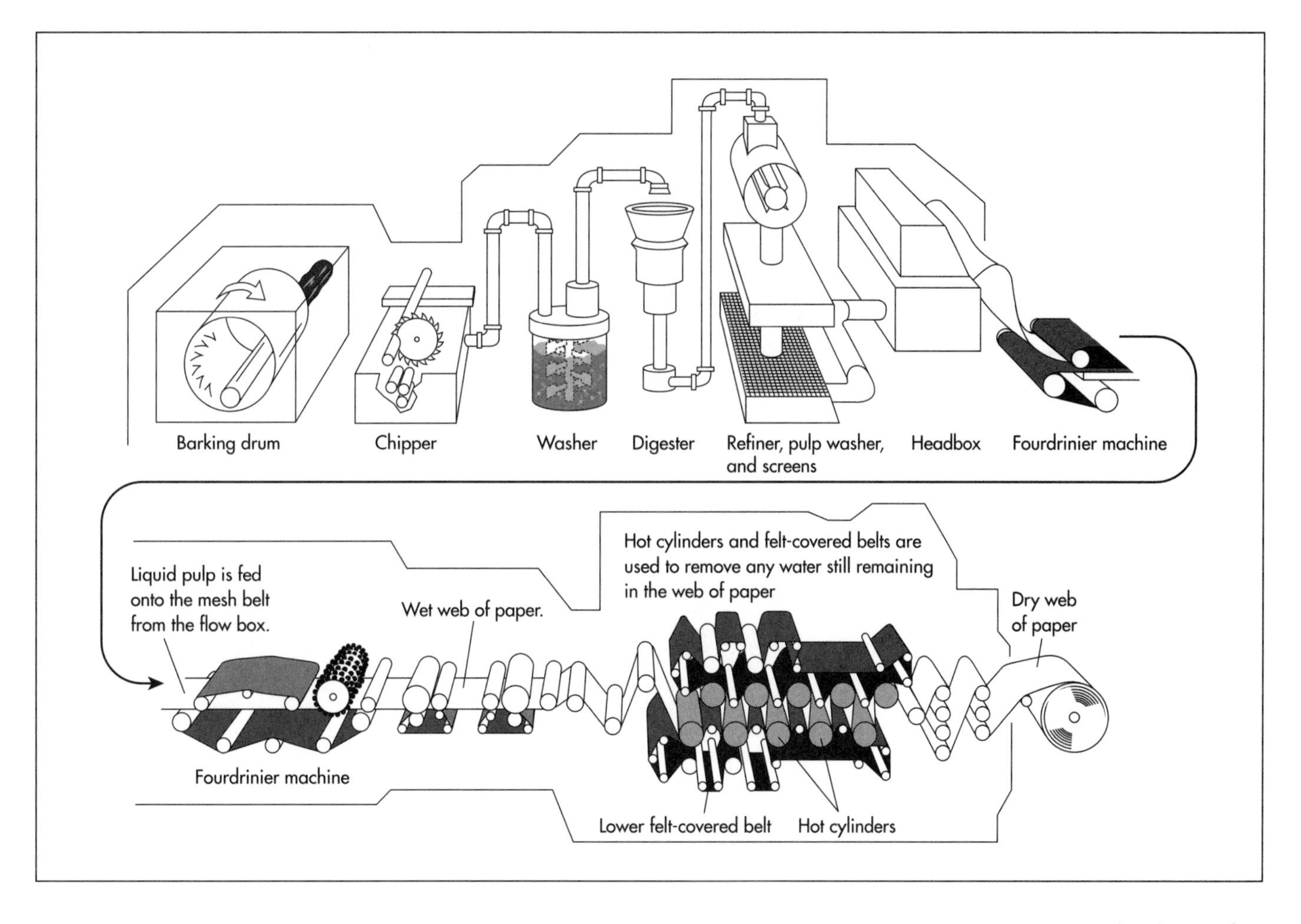


Most paper used in the manufacture of envelopes is derived from wood. The wood is mechanically or chemically processed to produce pulp that is then poured on mesh screening and squeezed through rollers to remove the moisture. The formed sheets of paper move through a series of heated cylinders that dry the paper further. The sheets are then wound on reels.

so that addresses written on the envelope in ink will not run and blur.

### Making paper

5 Pulp is added to water to form a very dilute slurry in order to make paper with an even density. The slurry is pumped onto a moving mesh screen. This screen is made up of very fine wires of metal or plastic. Water drains through the small openings in the mesh, forming a sheet of wet material from the slurry. Rapidly spinning rollers beneath the mesh create suction, a partial vacuum that removes more water from the mixture.

6 The sheet is moved on a belt made of felt containing wool and synthetic fibers. The felt absorbs water and prevents the sheet from being damaged as it moves between rollers, which squeeze out more water. The sheet then moves to a belt made of felt containing cotton and other fibers. This lighter felt allows water vapor to escape as the sheet is moved around a series of steam-heated rollers. As many as 40-70 rollers may be needed to dry the sheet.

7 The dried sheet moves between rollers known as calendars to make it smooth. It is then wound on a large reel. Variations in the papermaking process produce paper in a wide variety of basis weights. The basis weight of paper is the weight, in pounds, of a ream of 480 sheets cut to a size of 24 x 36 in (610 x 914 mm). Envelope paper usually has a basis weight between 16 and 40, with a basis weight of 24 being typical. Although many other kinds of paper are coated after being made, envelope paper is usually uncoated.

### Making envelopes

8 Rolls of paper, typically weighing 220 lb (100 kg), arrive at the envelope factory. The paper may need to be cut before it enters the automated machine that makes the envelopes, or it may be fed directly into the machine from the roll. If it is cut outside the machine, it is first cut by sharp

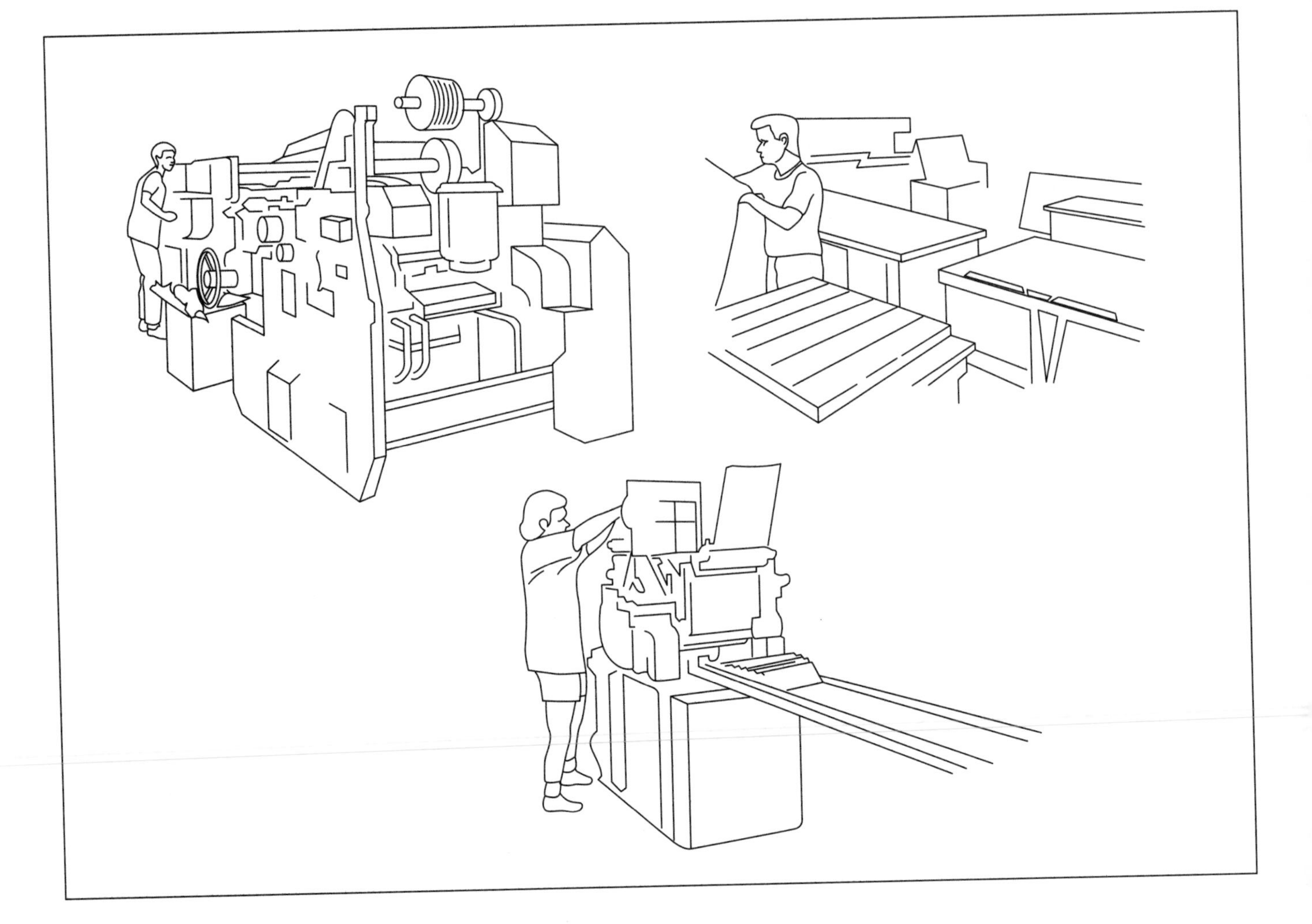

Rolls of paper, typically weighing 220 lb (100 kg), are either cut before they enter the automated machine that makes the envelopes, or fed directly into the machine from the roll. Once cut, the sheets are stacked and cut into blanks. A blank has the shape of an envelope with its flaps opened and laid flat. Blanks are generally shaped like diamonds and are cut from the sheets in such a way as to minimize waste. If the roll is fed directly into the machine, it cuts the paper into blanks very quickly with sharp blades. The machine also folds the blanks into envelopes at a very rapid pace. Strong glue is also applied to the places which will hold the envelope together. A weaker glue is applied to the flap that will be sealed by the consumer. The machine then folds the blank to form the envelope. The completed envelopes are filled in cardboard boxes and shipped to retailers.

blades into sheets of the proper size. The sheets are then stacked into large piles for further cutting. Strong blades then cut the pile of sheets into blanks. A blank has the shape of an envelope with its flaps opened and laid flat. Blanks are generally shaped like diamonds and are cut from the sheets in such a way as to minimize waste. If the roll is fed directly into the machine, it cuts the paper into blanks very quickly with sharp blades.

9 The machine performs all the operations needed to transform blanks into envelopes at a very rapid pace. Windows are cut if needed. If a transparent covering is needed for the windows, a strong glue is applied around them. The transparent material is then cut and glued in place. Strong glue is also applied to the places that will hold the envelope together. A weaker glue is applied to the flap that will be sealed by the consumer. The machine then folds the blank to form the envelope. Optional printing or fasteners are applied. The completed envelopes are filled in cardboard boxes and shipped to retailers.

## Quality Control

Modern envelope manufacturing is highly automated, and almost always results in a reliable product. Although constant testing is not necessary, certain factors are checked to ensure quality. Paper arriving at the factory is inspected to be sure that it has the correct weight. A very small number of sample envelopes are checked to ensure that they have the correct shape and size, and that adhesives have been applied in the correct places. Any printing that appears on the envelope must be in the correct position, of the correct color, and without printing errors. If any windows are cut in the envelope, they must have the correct dimensions and be in the correct position.

## The Future

Although major changes in envelope design are not expected, innovations are likely in the way paper is made. Manufacturers are constantly looking for ways to make paper that are more efficient, less costly, and result in less pollution. Genetic engineering may result in trees that grow faster and produce wood that is better adapted to producing pulp. A recent trend that is likely to continue is the increasing use of recycled paper as a raw material for making envelopes and other paper products.

## Where to Learn More


### Books

Biermann, Christopher J. *Essentials of Pulping and Papermaking*. New York: Academic Press, 1993.

Ferguson, Kelly, ed. *New Trends and Developments in Papermaking*. Miller Freeman, 1994.

### Periodicals

Kernan, Michael. "Pushing the Envelope."*Smithsonian* (October 1997): 30-31.

### Other

Ohio Envelope Manufacturing Company. http://www.ohioenvelope.com/ (September 30, 1998).


*—Rose Secrest*

# Eraser

*In 1770, the English scientist Joseph Priestley suggested that caoutchouc be named rubber, because of its ability to rub away pencil marks. In the United Kingdom, erasers are still known as rubbers.*

## Background

An eraser is a object that is used to remove marks from paper. Most erasers are designed to remove pencil marks. Other erasers are designed to be used on typewriter marks. Some special pens contain erasable ink that can be removed by erasers. While some erasers are sold separately in the form of wide, slender blocks, many more erasers are found permanently attached to pencils. Other erasers are made to temporarily attach to pencils. Some erasers are enclosed in wooden cases that resemble pencils. These erasers, designed to be sharpened like pencils, often have a brush attached. This is used to brush away small pieces of the eraser left behind after it removes a mark. This type of eraser is usually used to remove typewriter marks.

## History

The first erasers were pieces of bread. There was no better substance for removing pencil marks until rubber was available in the Old World. Rubber was known to the inhabitants of Central and South America long before Europeans came to the New World. As early as the eleventh century, it was used to coat clothing and to make balls. It was also used to make footwear and bottles by pouring the liquid form on earthen molds and allowing it to dry.

In 1735, the French scientist Charles de la Condamine described a substance known as *caoutchouc* and sent samples to Europe. Caoutchouc was derived from a fluid produced under the bark of a tree found in tropical areas of the New World. This milky liquid, known as latex, is still used to make natural rubber.

Caoutchouc was first suggested for use as an eraser in the Proceedings of the French Academy in 1752, probably by Jean de Magellan. In 1770, the English scientist Joseph Priestley suggested that caoutchouc be named rubber, because of its ability to rub away pencil marks. He also told readers of his book *Familiar Introduction to the Theory and Practice of Perspective* where to purchase "a cubical piece, of about half an inch, for three shillings." In the United Kingdom, erasers are still known as rubbers.

Until the late nineteenth century, pencils and erasers were always separate. In 1858, Hyman Lipman of Philadelphia patented a pencil with a groove in the tip, into which an eraser was glued. By the early 1860s, the Faber company made pencils with attached erasers. In 1862, Joseph Rechendorfer of New York City patented an improvement of Lipman's design and sued Faber. The United States Supreme Court determined that the idea of combining a pencil with an eraser could not be patented. The reason for this decision was the fact that combining the pencil and the eraser did not change the function of either. This decision opened the way for numerous companies to make pencils with erasers.

In 1867, a hollow eraser, into which a pencil could be inserted, was invented by J. B. Blair of Philadelphia. Earlier versions are also known to have existed. In 1872, the Eagle company made pencils with erasers inserted directly into the wooden case of the pencil. Other companies soon made similar pencils, which became known as penny pencils because they were inexpensive. The availability of pencils with attached erasers in schoolrooms was at first controversial. It was believed that the ability to correct errors easily would make students careless. Despite this concern, pencils with erasers were extremely popular. About 90% of modern American pencils are made with attached erasers. Pencils without erasers are somewhat more common in Europe.

## Raw Materials

The most important raw material in an eraser is rubber. The rubber may be natural or synthetic. Natural rubber is obtained from latex produced by the rubber tree (*Hevea brasilienesis*). Synthetic rubber exists in a wide variety of forms. The most common synthetic rubber is derived from the chemicals styrene and butadiene. Styrene is a liquid derived from ethylbenzene. Ethylbenzene is usually made from ethylene and benzene, both of which are derived from petroleum. Butadiene is a gas, derived either directly from petroleum or from substances known as butanes and butenes, which are derived from petroleum.

Other ingredients added to rubber include pigments that change the color of the eraser. White can be produced with zinc oxide and titanium oxide. Red can be produced by iron oxide. Many other colors can be produced with various organic dyes.

An important ingredient added to almost all rubber is sulfur. Sulfur allows rubber to be vulcanized. This process was invented by Charles Goodyear in 1839. It uses heat and sulfur to make rubber more durable and resistant to heat.

Various other ingredients may be added to rubber. These include vegetable oil, to make the rubber softer and easier to shape, and pumice, a natural mineral which makes the eraser more abrasive.

## The Manufacturing Process

### *Making natural rubber*

1 Rubber tree plantations are found only in tropical regions with high levels of annual precipitation. Malaysia is the leading producer of rubber trees. A rubber tree is tapped by cutting a thin strip of bark about 0.04 in (1 mm) deep off the tree as high up as the worker can easily reach. Later strips will be cut below the first one. Each strip reaches about halfway around the circumference of the tree and slants downward at an angle of about 30 degrees to allow the latex to drain into a container. If the latex is allowed to coagulate naturally, each cut will produce about 1 oz (28 g) of latex before the latex stops flowing after a few hours. A chemical may be applied to the bark to prevent the latex from coagulating, allowing it to flow for several days.

2 The collected latex passes through a sieve to remove foreign objects. Water is added to the latex and the mixture is pumped into large horizontal tanks containing aluminum partitions. Dilute acetic acid or formic acid is added to make rubber coagulate into slabs on the partitions. The slabs are sprayed with water while they pass through a series of rollers. Excess water is removed by another series of rollers. The slabs are packed in bales, usually weighing 225-250 lb (102-113 kg), in the shape of cubes about 2 ft (60 cm) on each side. The bales are coated with clay to prevent sticking, bound with metal straps, and shipped to manufacturers.

### *Making synthetic rubber*

3 Depending on what kind of synthetic rubber is being made, a wide variety of manufacturing processes may be used. The most common form of synthetic rubber, styrene-butadiene rubber, is usually made in an emulsion process.

4 Various chemicals are obtained from petroleum by fractional distillation. This process involves heating petroleum to about 600-700° F (315-370° C) and allowing the vapor to pass through a tall vertical tower. As the vapor rises through the tower, it cools. Chemicals with different boiling points change from gas to liquid at different points inside the tower and are collected. Chemicals with very high boiling points remain in the liquid state when the petroleum is heated and can be removed from the bottom of the tower. Chemicals with very low boiling points remain in the form of gases and can be removed from the top of the tower.

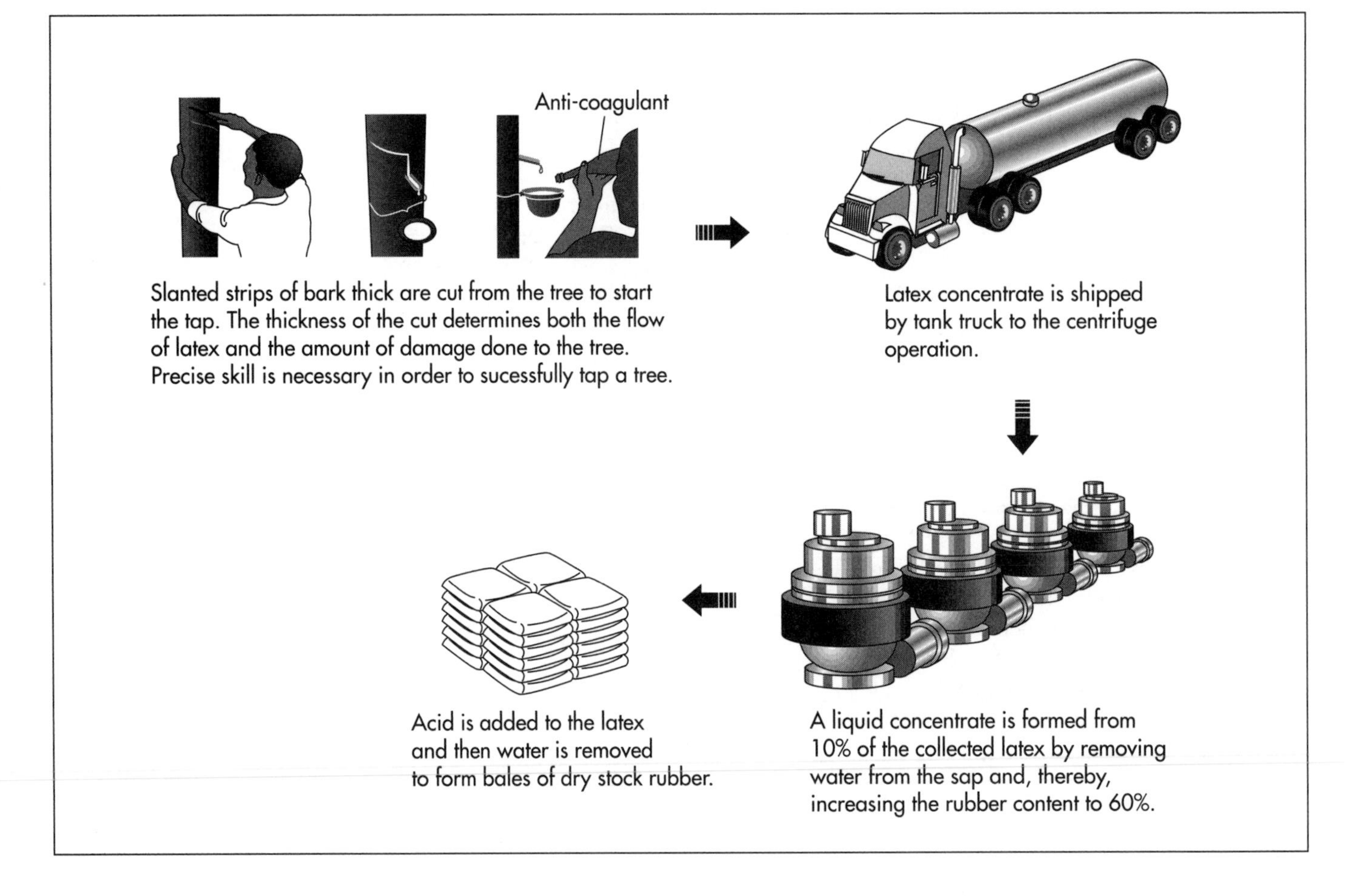


Erasers are made from either natural or synthetic rubber. Natural rubber is produced from latex that is collected from rubber trees. Water is removed from the latex, increasing the rubber content to 60%. Acid is added to the liquid concentrate in order to produce solid sheets of dry rubber.

5 Other chemicals are obtained by catalytic cracking. This process involves heating petroleum to about 850-900° F (454-510° C) under pressure in the presence of a catalyst. The catalyst causes chemical reactions to take place. The new mixture of chemicals are then separated by fractional distillation.

6 Styrene and butadiene are obtained by subjecting certain chemicals derived from petroleum to various chemical reactions. The styrene is a liquid under normal conditions, but the butadiene is a gas and must be stored under pressure to keep it in a liquid form.

7 The two liquids are pumped into a container and mixed with water, soap, and a catalyst. The catalyst causes the styrene and butadiene to react to form particles of synthetic rubber. The soap causes these particles to be dispersed in the water in a smooth emulsion. Constant agitation keeps the rubber particles from settling out.

8 Other chemicals added to the mixture include stabilizers, which prevent the rubber from breaking down, and modifiers, which change the properties of the rubber. Another chemical is added to stop the formation of rubber particles at an optimum point. Unconverted styrene and butadiene are removed and are reused. A coagulant is added to the emulsion, causing the synthetic rubber to be deposited. Styrene-butadiene rubber is generally shipped to manufacturers in bags made of polyethylene plastic, each holding 75 lb (34 kg) of rubber.

### *Making erasers*

9 Rubber arrives at the eraser factory and is mixed with pigments, vegetable oil, pumice, sulfur, and other ingredients that modify the properties of the final product. Synthetic rubber is easier to mix because it usually arrives as a powder or a liquid. Natural rubber usually arrives in bales and must be pulverized into powder or dissolved in a solvent before it can be mixed.

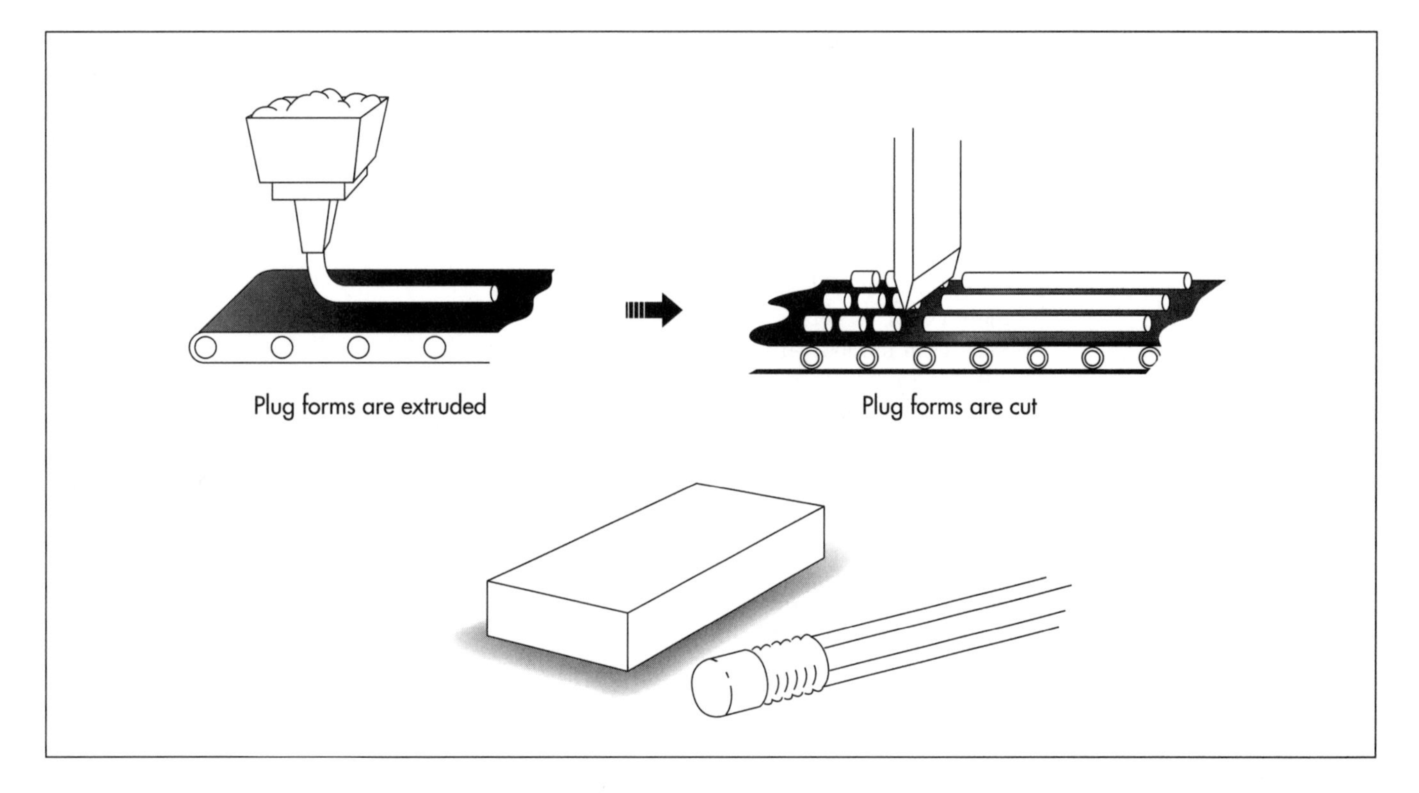


10 The mixture is heated, causing the sulfur to vulcanize it, making it more stable. To make plugs, which will be attached to pencils, an extrusion process is usually used. The mixture, in the form of a soft solid, is forced through a die to form a long cylinder. The cylinder is repeatedly cut as it emerges, forming plugs.

11 To make flats, which are not attached to pencils, an injection molding process is usually used. The mixture, in the form of a warm liquid, is forced into molds and allowed to cool into a solid. The flats are then removed from the molds.

12 Plugs are shipped directly to pencil manufacturers. They are attached to pencils by small, cylindrical, metal cases known as ferrules. Ferrules are made of plain aluminum for less expensive pencils, or painted brass for more expensive pencils. The ferrule is attached to the pencil with glue or with small metal prongs. The plug is inserted into the ferrule and clamped around it.

13 Flats may be marked with the name of the manufacturer or other markings. This may be done by stamping—pressing an inked stamp on the eraser. It may also be done by screen printing—moving an inked roller over a patterned sheet of silk or another material which covers the eraser. Three-dimensional markings can be made by embossing—cutting into the eraser with a sharp die. The completed flats are packed into cardboard boxes and shipped to retailers.

## Quality Control

The manufacturing of erasers is highly automated, with reliable products made in the millions each year. Experienced eraser manufacturers have refined the techniques used to the point where extensive inspection is not necessary.

The raw materials shipped to the manufacturer are supplied by companies that are known to provide substances with the proper characteristics. If a new substance is supplied, or if it comes from a new company, the eraser manufacturer may inspect it to be sure it meets all specifications.

Only a very small percentage of erasers need to be inspected to ensure that they have the proper physical properties. Flats must be the correct size to fit into boxes. Plugs must have the correct dimensions to fit into ferrules. The hardness of erasers is critical to how well they will work. Experienced inspectors can easily tell if an eraser is too hard or too soft.

Once the natural or synthetic rubber is mixed with pigments, vegetable oil, pumice, sulfur, and other additional ingredients, the mixture is heated and the erasers are formed. In order to make eraser plugs that will be attached to the ends of pencils, the rubber mixture is extruded and cut into plugs. To make flat, rectangular erasers, the mixture is injected into molds and then cooled.

## The Future

Erasers have remained mostly unchanged for many years. Improvements in eraser technology are likely to be made in the way rubber is produced. New chemical formulas are constantly being developed to produce synthetic rubber in ways that are more efficient, less costly, and which result in products with more useful properties. Genetic engineering may result in rubber trees that produce more latex, or trees that produce latex with physical properties that would make natural rubber production more efficient.

A hint of the future of eraser design is seen in the Ergoraser, a unique eraser from Levenger, a company specializing in very high quality writing supplies. The Ergoraser, developed after two years of research, is oval and curved, much like the shape of a spoon. The thumb fits inside the curve during use in a way which is designed to be comfortable and efficient. Although extremely expensive compared to ordinary erasers, the Ergoraser promises to play an important role in the future for those who demand the highest quality in simple objects.

## Where to Learn More

### Books

Petroski, Henry.*The Pencil.* Knopf, 1990.

### Other

"Eraser Certification." http://www.wima.org/consumer/c-obcertprograms/eraserprog.html/ (February 17, 1998).

*—Rose Secrest*

# Eyeglass Frame

## Background

American humorist Dorothy Parker (1893-1967) once wrote caustically that "Men seldom make passes at girls who wear glasses." Her comment tells as much about the eyeglass fashions available in her youth as about the customs of flirtation. Ms. Parker would be pleasantly surprised to visit any suburban shopping mall today and to see the wide variety of eyeglass frames now available. Frames have become hot fashion accessories just like jewelry or shoes, and the wearer can change them to match moods or to convey an image.

## History

The ancient Greeks made the first studies of vision and the workings of the eye. They also attempted to understand magnification and to use it to understand vision problems. Alhazen, an Arabian scientist who lived during the eleventh century, studied the refraction (bending) of light and the connection between optic nerves and the brain. It was the thirteenth century Polish scientist, Vitello, who first understood that the shapes of lenses could be used to control the focus of light rays.

In 1257, the English friar Roger Bacon explored so many aspects of science that he was imprisoned by the monks of his Franciscan order who were suspicious of his knowledge. While he was in prison, Friar Bacon sent Pope Clement IV some magnifying lenses for reading; despite Bacon's controversial standing, the monks who labored over detailed manuscripts and copy work quickly adopted the use of his spectacles. Bacon's work occurred at the same time as that of Salvino d'Armato of Florence, Italy, and several Chinese and German scientists. All can be thanked for their collective invention of spectacles.

The invention of devices to keep spectacles on the nose took several more centuries. And, despite Ms. Parker's rhyme, style and a variety of lens shapes and frames have been important since the beginning of eyeglass frames. The earliest eyeglasses were unframed lenses that were simply held by hand in front of the face. Alternatively, two lenses were mounted in a half frame that could be held with one hand. Spectacles also were attached to hats or tied around the head with bands made of leather or ribbon. Will Somers, a jester to the court of Henry VIII, sported a suit of armor with spectacles attached to the metal helmet with rivets. The painter El Greco portrayed Cardinal Niño de Guevara wearing glasses with cords that looped over his ears. The seventeenth century design called the forehead frame consisted of a metal band that encircled the head and had metal frames mounted to it.

The most common frames held two lenses on a frame that rode on the lower part of the nose. Lightweight materials were used to lessen the burden and pinching of these "nose frames." In the court of the Spanish King Philip V and Queen Marie-Louise (about 1701), all 500 of the queen's ladies-in-waiting wore tortoise-shell frames because of their light weight. This style saluted both fashion and superstition; the frames supposedly brought good luck because the tortoise is sacred in China. Attempts at stylistic designs were varied and clever. The bridge pieces that rest on the nose were decorated in endless ways. Lenses were mount-

*Three eye sizes are standard for the range of dimensions of corrective lenses. Each style is typically manufactured in four different colors, so a single style will result in 12 combinations of color and dimension.*

ed in fans, watch fobs, and on walking sticks. The status-conscious had their nose frames made of gold or other precious materials or employed artists to decorate the frames with coats of arms.

Other than nose glasses, lens wearers could choose the monocle (a single lens in a frame or holder), the lorgnette (a pair of lenses with a nose bridge and a single handle on one side), a quizzer or quizzing glass (a monocle that was mirrored so the wearer could see to the rear as well), the perspective glass (a single lens worn on a ribbon and used for distance vision), or scissors glasses that had two eyepieces mounted on a hinged handle that was held up in front of the nose. Finally, in 1728, Edward Scarlett of London developed temples for frames. These clamps gripped the temple area and held nose glasses more securely to the face. A loop at the end of each temple piece held ribbons that were tied around the head or wig. By the 1880s, temples were curved to extend and fit over the ears to hold spectacles in place.

In the Colonies, spectacles were imported and were very expensive until American glass-making skills improved enough to develop an eyeglass trade. Just as curved and fitted temples were developed and adopted all over the world, fashion reverted to a style called the oxford that consisted of nose glasses improved by a more elastic and wearable bridge. These glasses were also called pince-nez and had nose pads fitted to small springs on the flexible bridge. Presidents Teddy Roosevelt and Calvin Coolidge wore oxfords. During the 1900s, simple steel-framed glasses were the most common, although less expensive frames were available in a material called gutta-percha—a rubbery plastic-like substance. Tortoiseshell and horn-rimmed glasses became popular in the 1920s and 1930s, but many of these frames were actually made of celluloid, an early plastic that could be dyed and molded to resemble animal horn or tortoise shell. Steel-framed spectacles and sunglasses were issued to millions of servicemen during World War II.

The business of manufacturing eyeglass frames and lenses made its most dramatic leap in the twentieth century with the rise of plastics. Plastic lenses are lighter in weight and can be manufactured as bifocals, trifocals, and quadrifocals to correct a wider range of vision problems. Frames made of plastic are also less expensive. A broad range of styles and colors can be made in plastic and changed to suit wardrobes, fads, and moods. Sunglasses also became affordable, thanks to the plastics industry, but Hollywood was responsible for their popularity. Large, square-rimmed glasses like those worn by Clark Kent became popular among men in the 1950s, and the ladies favored "cat's-eye" glasses that angled up at the temples. Granny glasses with fine metal frames accompanied the flash of the "flower power" generation in the 1960s and may have been responsible for making antique eyeglasses popular collectibles. Although contact lenses were also developed during this century and have become very popular, the variety of available eyeglass frames has kept glasses fashionable.

## Raw Materials

Eyeglasses frames are typically made of either metal or a type of plastic called cellulose-acetate. Cellulose acetate is derived from cotton and is flexible and strong. It is produced in long narrow sheets that are slightly wider than eyeglass frames. The sheets are up to 3 ft (0.91 m) long and 0.33 in (0.84 cm) thick.

## Design

Eyeglasses manufacturers may retain their own staff of designers or use outside consultants to design frames. The consultants often include fashion designers, who create their own lines of eye wear that change along with trends in clothing design. The designers' names are important in selling eyeglasses and especially in interesting fashion-conscious buyers in multiple pairs of glasses or sunglasses. There are definitely trends or fashions in eyewear including light- or dark-colored frames, thick or delicate ones, and decorative shapes or ornament-bearing styles. Specialized frames for children and half-frames for reading glasses are also designed with an eye to style.

Designs also incorporate certain standards including bridge size and eye size. The bridge size allows for different thicknesses of the upper part of the nose where the nose

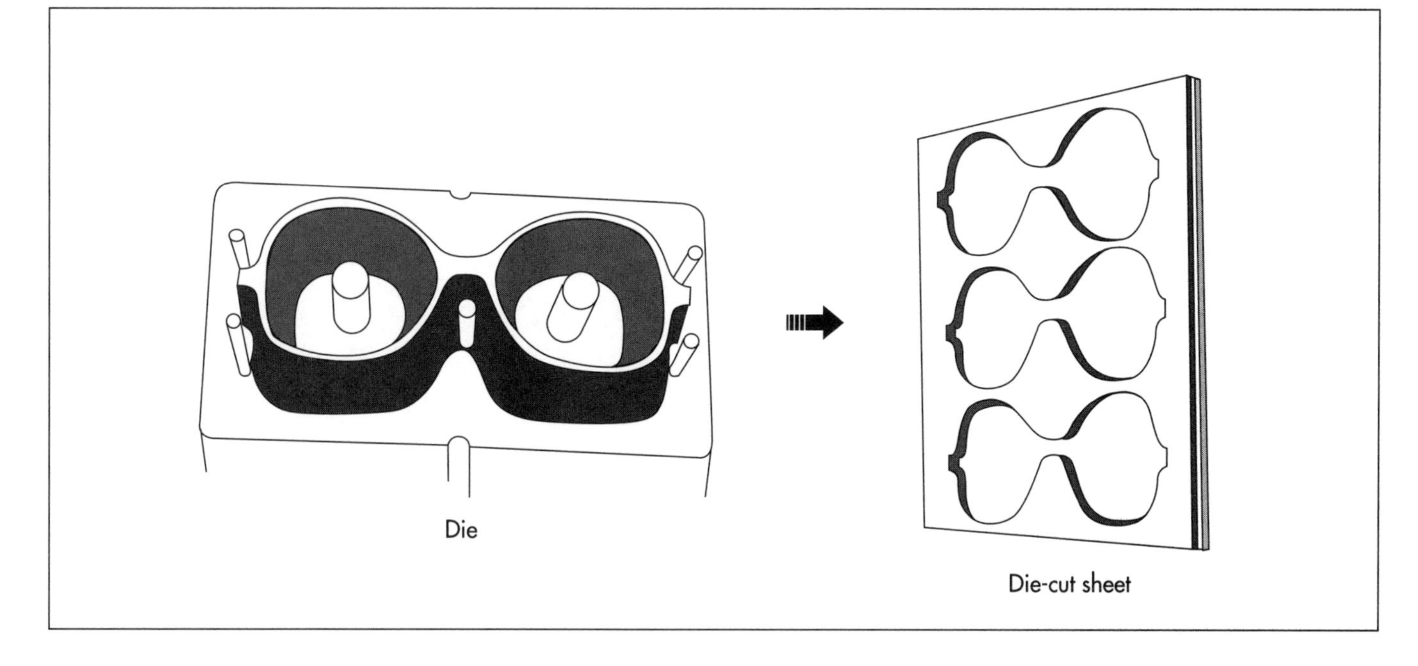


Blanks for plastic eyeglass frames are die cut from sheets of cellulose-acetate.

pads on the glass's rest. Three eye sizes are standard for the range of dimensions of corrective lenses. Each style is typically manufactured in four different colors, so a single style will result in 12 combinations of color and dimension. Frame designers and manufacturers typically produce a new style every few months and discontinue styles if they don't sell well.

## The Manufacturing Process

### *Die-cutting plastic frames*

1 After the design is decided, a die is made out of steel and is fitted in a blanking machine that punches blanks out of the sheets of cellulose-acetate. The edges of the steel rule die are sharp, and the dies have protruding rods that are used to remove the punched fronts where the lenses will be fitted. The acetate sheets are brought from a cool storage area to the blanking room where they are heated in small ovens to about 180° F (68° C) to soften the plastic. The soft sheets are fitted into a blanking machine, and, under several tons of pressure, the die cuts through the plastic to produce a blank. The machine is automated to lift the die and move it to the next portion of plastic. The blanks are produced quickly while the plastic is soft. The blanks are then removed from the sheet, and the lens portions are taken out of the frames. The lens blanks become scrap.

2 The blank frame fronts are finished in a series of operations. Grooves to hold the lenses are cut using a router. The frame is held tightly by a holding fixture made of aluminum and consisting of two pieces. The fixture is fastened around the frame and pressed against the router blade. The grooves are cut to 0.16 in (0.41cm) wide, which is an industry standard. If thick lenses are needed, they are ground down along the edges to fit the routered grooves.

3 The frames are then smoothed to remove rough edges by two different abrasive machines. One machine is specially shaped to smooth the edge of the frame that rests on the cheek and the second smoothes the area around the nose. The frame is then secured in a vise-like device while the nose pads or attachments that hold the nose pads are glued to the frames. Once the glue is allowed to cure for 24 hours, the glued areas are also smoothed.

### *Producing the temples*

4 The two side arms that curve around the ears are called temples. The temples are also punched out of sheets of acetate with blanking dies. The standard temple length ranges from 5-6 in (12.7-15.2 cm) and they are usually cut out of the same material or

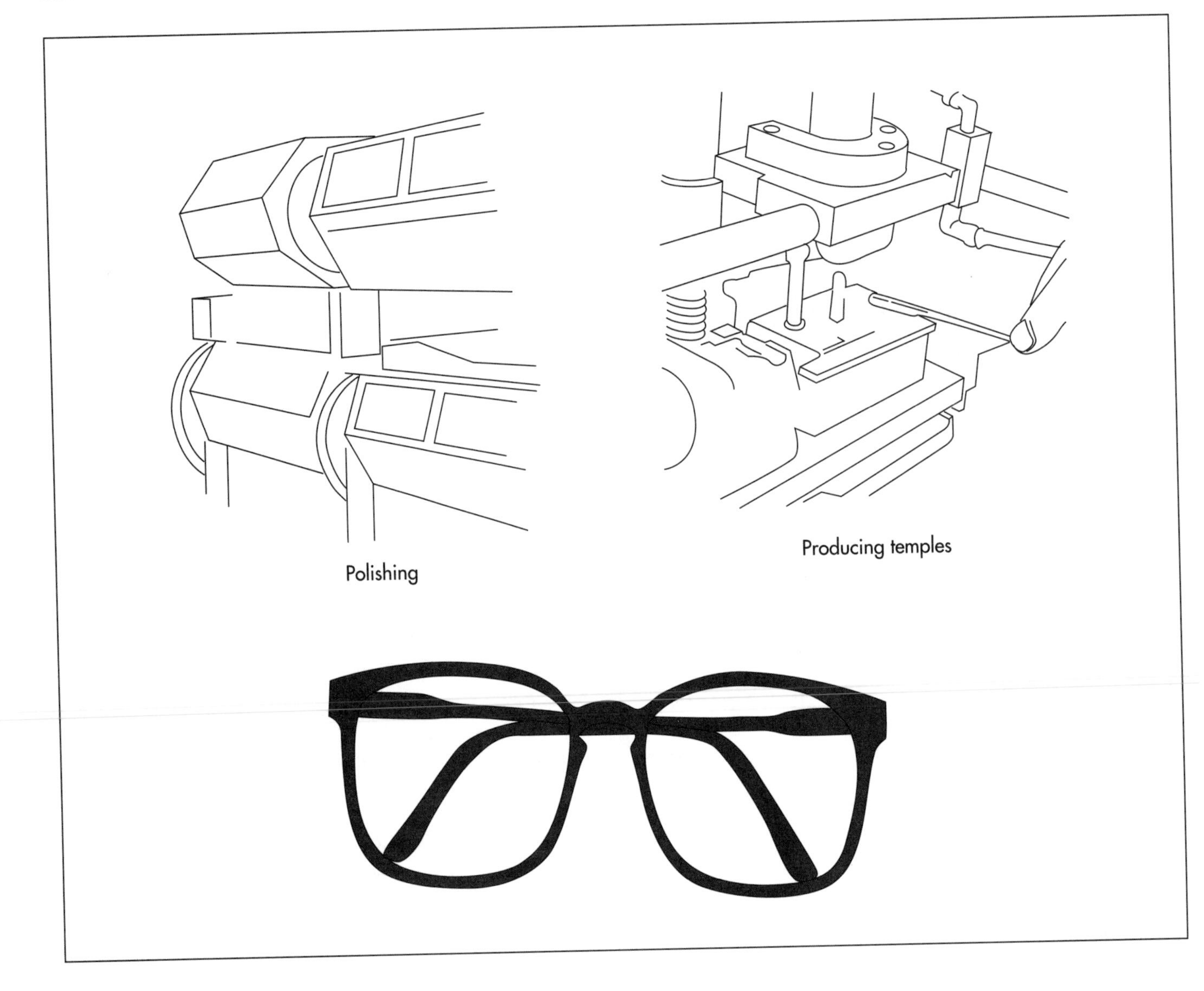


After the frames are smoothed, the temples are attached to the frames with a metal hinge. To make the temples, plastic temple strips are heated, and a narrow strip of steel called a core wire is also heated. When the right temperatures are reached, a core wire is inserted into the center of the softened temple.

complementary materials to match the fronts. The temples may be flat with angular edges or rounded in shape, depending on the frame style. The temples are heated, and a narrow strip of steel called a core wire is also heated. When the right temperatures are reached, a core wire is inserted into the center of the softened temple. Inexpensive glasses, like some non-prescription sunglasses, may be made without core wires, but they will also be less strong.

5 To attach temples to their frame, small slots are cut in the upper corners of the frame. A half of a metal hinge is put in each slot, and the frame and partial hinges are placed in a capitron machine. The capitron machine induces ultrasonic vibrations in the metal hinges and creates heat of friction. The friction causes the plastic of the frame to melt locally around the hinge to bond the hinge to the frame; this process is more secure than gluing or other types of bonding or mounting.

### *Finishing the fronts*

6 The fronts with hinges are then imprinted with the manufacturer's name or logo, the name of the style of the frame, and the size of the frame. Before the temples are attached to the frame, an angular fit is needed to make the frame front tilt inward toward the wearer's face from top to bottom. An automatic saw is used to cut the correct angles in the upper edges of the front. Caps are fitted over the hinges, while the front is polished. By this stage, the fronts are still flat with sharp edges except in the routered areas where the lenses fit. In the polishing

room, hundreds of fronts at a time are tumbled in a drum with pumice (soft stone that is ground to a powder and used as an abrasive) and small maple pegs that look like miniature kitchen matches with rounded edges and tips. The pumice adheres to the pegs, which grind against the fronts during a 24-hour-long process that smoothes the rough edges. Crushed coconut shells are sometimes used by manufacturers in the polishing process.

7 The smoothed fronts now have to be shaped to fit the curvature of the face. They are heated in an oven, mounted in a curved forming die (similar to the blanking die used to make the fronts), and placed in a press. Pressure is applied for about 30 seconds to produce a uniform curve in the front. The warm front is dipped in a cold water bath to harden it to the curved shape.

8 The shaped fronts are returned to the polishing room where they pass through a series of tumbling drums over a period of four days to add the finished sheen. Each drum contains pumice that is finer than the previous one; the final drum holds polishing wax. All of the polished fronts are inspected carefully for any scratches. They are placed individually in envelopes and filed by size, style, and color.

### *Finishing the temples*

9 Depending on the shape and style of the eyeglasses design and their temples, the temples are then ground and shaped during several operations. Grooves are cut into the ends of the temples, and the matching halves of the hinges are riveted into the temples. The ends are cut to match the angles of the finished fronts. Like the fronts, the temples are then finished during multiple operations, and pairs of polished temples are packed in envelopes by size, style, and color. Manufacturers store the envelopes containing fronts and temples until optometrists or optometric supply houses order them by size, style, and color. Sets of fronts and temples are then shipped.

## Quality Control

Eyeglasses frames must be manufactured with great attention to detail because they are critical in supporting lenses to improve vision, they must be comfortable for the wearer, and they are an accessory to professional dress and personal style. Although processes for making frames are performed by machines, operators are responsible for each step and are quality control checkers for their particular operations. The eyeglasses industry has become highly competitive because of the aspects of fad and fashion involved, but operators are well aware that their products provide vision care. Fronts and temples for eyeglasses can be rejected at any step in the process of manufacture.

## Byproducts/Waste

No byproducts result from the manufacture of eyeglass frames. Plastic waste is generated during blanking, with the bulk of the waste from the lens portion of the frame that is cut out. This waste is carefully collected and recycled.

## The Future

The past 50 years of eyeglass history have soundly established the future of frames. Despite the popularity of contact lenses and the advent of laser surgery to correct vision problems, many people will find eyeglasses necessary or desirable for their personal needs. Improved technology in the manufacture of plastic lenses and frames and in the comfort of fitted frames have made eyeglasses more enjoyable to wear. The fashion industry also actively supports eyeglass frames as an added avenue for expressive designs and a popular method of stating personal style.

## Where to Learn More

### *Books*

Corson, Richard. *Fashions in Eyeglasses.* Dufour Editions, Inc., 1967.

Goldstein, Margaret J. *Eyeglasses.* Minneapolis: Carolrhoda Books, Inc., 1997.

Gottlieb, Leonard. *Factory Made: How Things are Manufactured.* Boston: Houghton Mifflin Co., 1978.

Kelley, Alberta. *Lenses, Spectacles, Eyeglasses, and Contacts: The Story of Vision

*Aids*. New York: Elsevier/Nelson Books, 1978.

York, Alan. "Eyeglasses: Fads and Fashions in Spectacles." In *The Encyclopedia of Collectibles: Dogs to Fishing Tackle*. Andrea DiNoto, ed. Alexandria, VA: Time-Life Books, 1978.

### Periodicals

Morais, Richard C. "Luxottica's golden spectacles,"*Forbes* (May 20, 1996): 98.

### Other

The Antique Spectacles Home Page. 9 September 1998. http://web.ukonline.co.uk/christopher.ridings/ (March 11, 1999).

University of Waterloo Museum of Visual Science and Optometry. 6 December 1996. http://www.optometry.uwaterloo.ca/~museum/ (March 11, 1999).

*—Gillian S. Holmes*

# Fill Dam

## Background

Dams are among the oldest structures built by humans for collective use. A dam is a barrier that is constructed across a river or stream so the water can be held back or impounded to supply water for drinking or irrigation, to control flooding, and to generate power. The main kinds of dams are earth fill, rock fill, concrete gravity, concrete arch, and arch gravity. The last three types are all made of concrete, reinforced concrete, or masonry. (The term masonry can mean concrete, bricks, or blocks of excavated rock.) Fill dams include all dams made of earth materials (soil and rock) that are compacted together. One type of fill dam called a tailings dam is constructed of fine waste that results from processing rock during mining; at mine sites, this soil-like waste is compacted to form an embankment that holds water for the mining and milling processes or to retain the tailings themselves in water.

Of the main categories of dams listed above, all have been built since ancient times although many refinements were developed in the nineteenth and twentieth centuries with improved engineering technology. Dams that leak have failed to do their job, either because they simply can't hold water or because the water seeping through them eats materials away from the inside of the dam causing it to fail structurally. In modern times, most fill dams are also built with zones including a clay center or core, filter and drainage layers, coarser materials sandwiching the clay core, and rock on the upstream (water) face to prevent erosion. These zones can be seen clearly when a cross section is cut from the upstream to the downstream side of the dam. All fill dams depend on weight to remain stable.

Fill embankments are usually less expensive to construct than concrete dams. Soil or rock are present at the site, and construction techniques, though complex, are also less costly than for concrete construction. For these reasons of available materials, low cost, and stability with mass, fill dams are often built across broad water courses. They also are more flexible than concrete structures and can deform without necessarily failing if foundation materials under the dam compress with the weight of the dam and the water.

## History

Quite naturally, early dam builders began by using plentiful materials like sand, timber and brush, and gravel. Their construction method consisted of carrying the materials by the basketful and loosely dumping the fill, so many of these dams may have survived only a few years. Scientists have not been able to pinpoint dates for the earliest dam construction, but they do know dams were needed where food was grown and in areas prone to flooding.

Design of fill dams is based on experience; while failures are unfortunate and sometimes catastrophic, they are also the best teachers, and many engineering advances have been founded on careful study of earlier failures. The engineers of ancient India and Sri Lanka were the most successful pioneers of fill dam design and construction, and remains of earth dams can still be seen in both countries. In Sri Lanka, long embankments called tanks were built to store irrigation water. The Kalabalala Tank was 37 mi (60 km) long around its perimeter.

*The engineers of ancient India and Sri Lanka were the most successful pioneers of fill dam design and construction, and remains of earth dams can still be seen in both countries.*

The most famous earth fill dam recently constructed is the Aswān High Dam that was built across the Nile River in Egypt in 1970-1980. An earth fill dam was also the victim of a spectacular failure in June 1976 when the Teton Dam in Idaho eroded from within due to incorrect design of the zones inside the dam that allowed seepage, failure, and flooding of the valley downstream. Although earth dams tend to be short and broad, Nurek Dam in Tajikistan is 984 ft (300 m) high.

## Raw Materials

The materials used to construct fill dams include soil and rock. Soil is classified by particle size from the smallest, submicroscopic particles called clay; silt, which is also very fine; sand ranging from fine to coarse, where the fine grains are the smallest soil particles our eyes can see; and gravel. Coarser fragments called cobbles and boulders are also used in dam construction but usually as protective outer layers.

Specific soil types and size ranges are needed to construct the zones within the dam, and explorations of the dam foundation area, the reservoir where the water will be stored, and surrounding areas are performed not only for design of the dam but to locate construction materials. The costs of fill construction rise dramatically with the distance materials are hauled. Samples of potential construction materials are tested in a soil laboratory for grain size, moisture content, dry density (weight), plasticity, and permeability. Clay is not only very fine in size but has chemical characteristics that cause it to stick together. The combination of fine size and plastic behavior also causes the clay to be less permeable to water. If clay is available near the site, the dam can be built with an impermeable core or central zone that prevents water from passing through the dam; otherwise, the dam must be designed so water can seep slowly and safely through a different combination of materials in its zones.

Water is also a raw material. The various soil types have compaction characteristics that can be determined in the laboratory and used during construction. Soil can be compacted to its best functional density by adding moisture and weight and impact, called compactive effort. Large vibrating rollers press thin layers of soil into place after an optimal amount of water has been added. The water and weight bond the soil particles together and force smaller particles into spaces between larger particles so voids are eliminated or made as small as possible to restrict seepage.

Increasingly, fill dams also include geotextiles and geomembranes. Geotextiles are nonwoven fabrics that are strong and puncture-resistant. They can be placed between lifts as the dam is raised to strength weak materials. They are also used as filter fabrics to wrap coarser drain rock and limit the migration of fine soil into the drainage material. Geomembranes are made of high-density polyethylene (HDPE) plastic and are impermeable. They can be used to line the upstream face of a fill dam or even to line the entire reservoir.

## Feasibility and Preliminary Design

A specific need for a dam, whether it is water supply, storage of tailings or other materials, or flood control, stimulates the process of designing and building a fill dam. The need and the location are usually closely connected, so several sites may be considered. During feasibility studies, engineers identify these sites, make preliminary cost comparisons, decide on a probable design, and chose the best site for exploration. Feasibility certainly refers to the cost of building the dam, but it also includes the technical practicalities of site suitability, design, construction, and long-term maintenance and safety.

After a feasible site is chosen, a preliminary design of the dam is developed. The location of the dam is superimposed on a topographical map so the dimensions of the top of the dam relative to the tops of the adjacent hills and the proposed water level can be shown as well as the extent of the base of the dam in the stream channel. The proposed water level elevation shows the extent of the reservoir and determines—along with the shape of the basin—the quantity of water that the reservoir will hold. Quantities of water stored and materials used in constructing the dam help determine the value of the project and its costs. Sometimes multiple iterations of site selection, pre-design, and cost estimating are needed. Ideally, the foundation area under the dam will not require much excavation or grouting to prevent seepage, and the materials inside the reservoir area can be excavated and used to build the dam so that more reservoir storage is gained at the same time as soil or rock are excavated to construct the embankment.

When the optimal site is chosen on paper, an exploration program is developed and performed. During the exploration, test borings are drilled along the line of the axis of the dam across its proposed width, along or near the proposed upstream and downstream toes of the dam, at the site of the proposed spillway, and in the reservoir area. The borings are excavated deep into the foundation to evaluate its strength and permeability (potential for seepage) properties. As the borings are drilled through the overlying soil, it is also sampled and tested in the laboratory so it can be evaluated as potential dam construction material. Field tests of permeability are also performed at the site of the dam and in the reservoir area. If it is the source for construction materials, test pits are also dug in the reservoir area so that the volume of available soil (and related costs) can be estimated.

## Design

After the field exploration and laboratory testing are complete, the engineering team begins final design of the dam based on the preliminary assumptions, the findings in the field, and any changes in design or economics that are based on field findings. In designing a fill dam, engineers look at five critical considerations: the mass of the dam that will make it stable; design of a core and other interior zones to prevent seepage through the dam; design of a cut- off wall or other seepage prevention under the dam; erosion protection on the upstream face; and economics.

Fill dams are typically shaped like triangles with the apex or point at the top or crest of the dam and the broad base on the floor of the creek channel. The width of the base in cross section provides friction to prevent sliding, and the total mass of the dam makes it strong enough to resist the weight of water behind it. The foundation area is cleaned of soft, permeable, and compressible soil; and a cut-off wall is cut down to rock or firm soil. The cut-off wall can be constructed of steel sheet piling or concrete, but, for most fill dams built since about 1960, the cut-off wall is simply an extension of the clay core. Where foundation rock or soil contains voids or fractures, a series of holes may be drilled into the foundation, and concrete grout is injected in the holes to seal the fractures and help cut off seepage.

The zones of a fill dam may consist of a number of distinct layers from the center of the dam and moving upstream toward the water and a different set of layers from the center moving downstream. Materials for the zones are selected for strength properties and permeability characteristics, and the placement of one zone next to another is carefully governed by sets of calculations based on these properties. Filter and drainage zones are included so that any water succeeding in reaching the inside of the dam is channeled around the core and out through drainage layers at the base of the dam.

The upstream (water) face of the dam is sometimes protected with a concrete slab or an asphalt face. More commonly, cobble- and boulder-sized stones are placed on this face near the water surface; this facing is called riprap and prevents wave action at the water surface from eroding the dam construction materials. Other facilities for controlling the water level and any water movement through or over the dam, like an emergency spillway, are also designed specifically for the dam's location, uses, type and materials of construction, and water inflows into the reservoir.

The economics of dam construction are considered throughout the design process. Construction materials must be available at or near the site. Rock can be placed at steeper angles than soil, and it weighs more; so a dam built mostly of rock can be smaller in design section. Excavating and moving rock can be more expensive than soil, however, so the design engineers must consider cost factors. Other materials like asphalt, concrete, steel, and cement for grouting are also expensive. The proper balance of safety and economy must be determined by the engineers. Large earthmoving machines have made construction of zoned, fill dams more

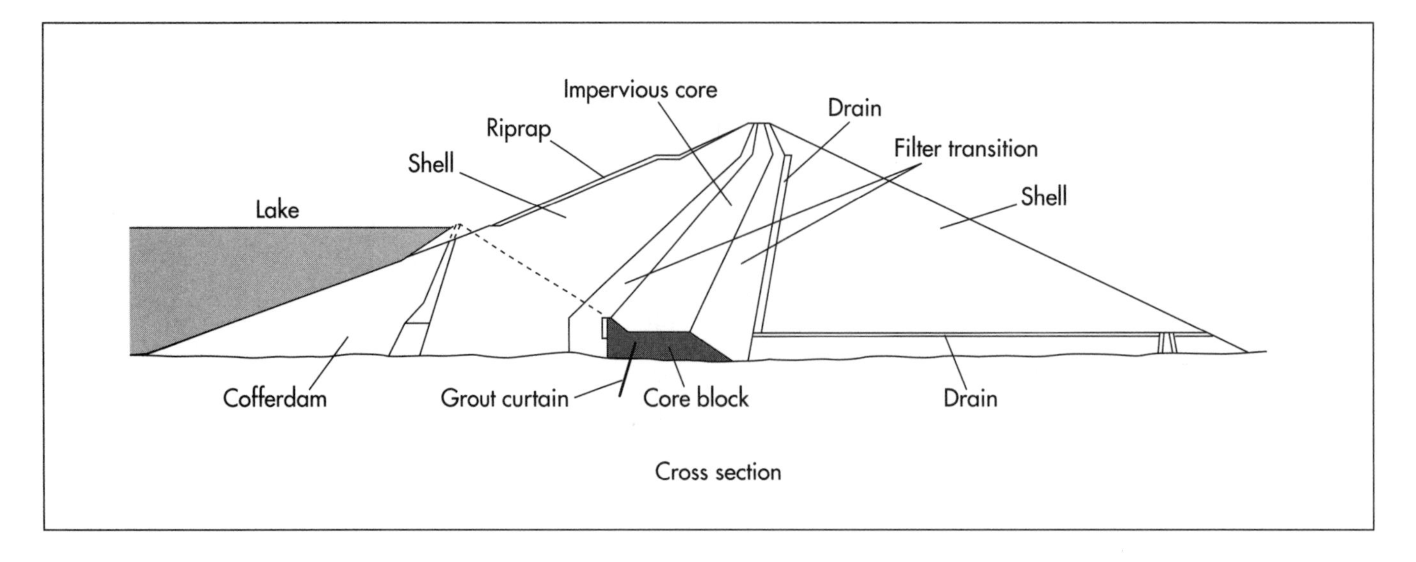


Cross section of a typical fill dam.

economical than construction of concrete dams at many sites.

## The Construction Process

1 Fill dams are constructed in the dry season when water levels in the river or stream are lower, rainfall on sources of fill material is less likely, and conditions are better for operating large construction equipment. Before construction actually begins, the site is surveyed to locate the dam alignment on the existing ground, the areas that will be excavated, and the borrow areas or sources for the soil or rock used in construction. Construction management facilities are set up; usually, the construction manager (a field engineer with years of similar experience) will work out of a trailer on site. Depending on the site, it may be necessary to install instruments to monitor the effects of dam construction on adjacent hillsides or other features and to measure groundwater levels throughout construction in the foundation and surroundings. And, of course, the flow of the stream that is being dammed through the site must be stopped. This can be done by a variety of methods including diverting the stream, perhaps to flow through a neighboring channel, or stopping it upstream with a temporary dam or cofferdam.

2 Before construction of the dam begins, the foundation area must be prepared. In rare cases, dams can be constructed directly on the existing materials in the channel floor; at most sites, these materials are compressible (and would cause the dam to settle irregularly) and permeable (allowing water to pass under the dam). The foundation area also includes the abutments, which are the hillsides forming the two ends of the dam. Soil and soft or highly fractured rock are excavated, sorted by type, and stockpiled for later use in dam construction. The surface of the foundation bedrock is cleaned to a surprising degree; it is broomed and hosed with water so that any voids or irregularities are visible and cleaned of soft soil. The foundation is carefully inspected before any construction work; additional exploratory drilling may be done if there are any questions about the foundation's condition. If the rock is fractured or contains voids or holes, these are sealed with cement grout that is injected through small diameter drill holes in a process called dental work.

3 The base of the dam must go down into the ground before it rises above it. A trench that is the full width of the dam (across the channel) is cut into firm rock. The trench is called a keyway or cutoff wall and may have several benches or notches into rock. It prevents the dam from sliding along a smooth foundation and also creates a longer path for any seepage to try to flow under the dam. The impervious clay that will make up the core of the dam is placed in the keyway and compacted and raised, layer by layer, until the top of the keyway or base of the majority of the foundation is reached.

4 The soil in the keyway and all the zones of the dam are raised to the same levels at the same time. Ramps may have to be cut

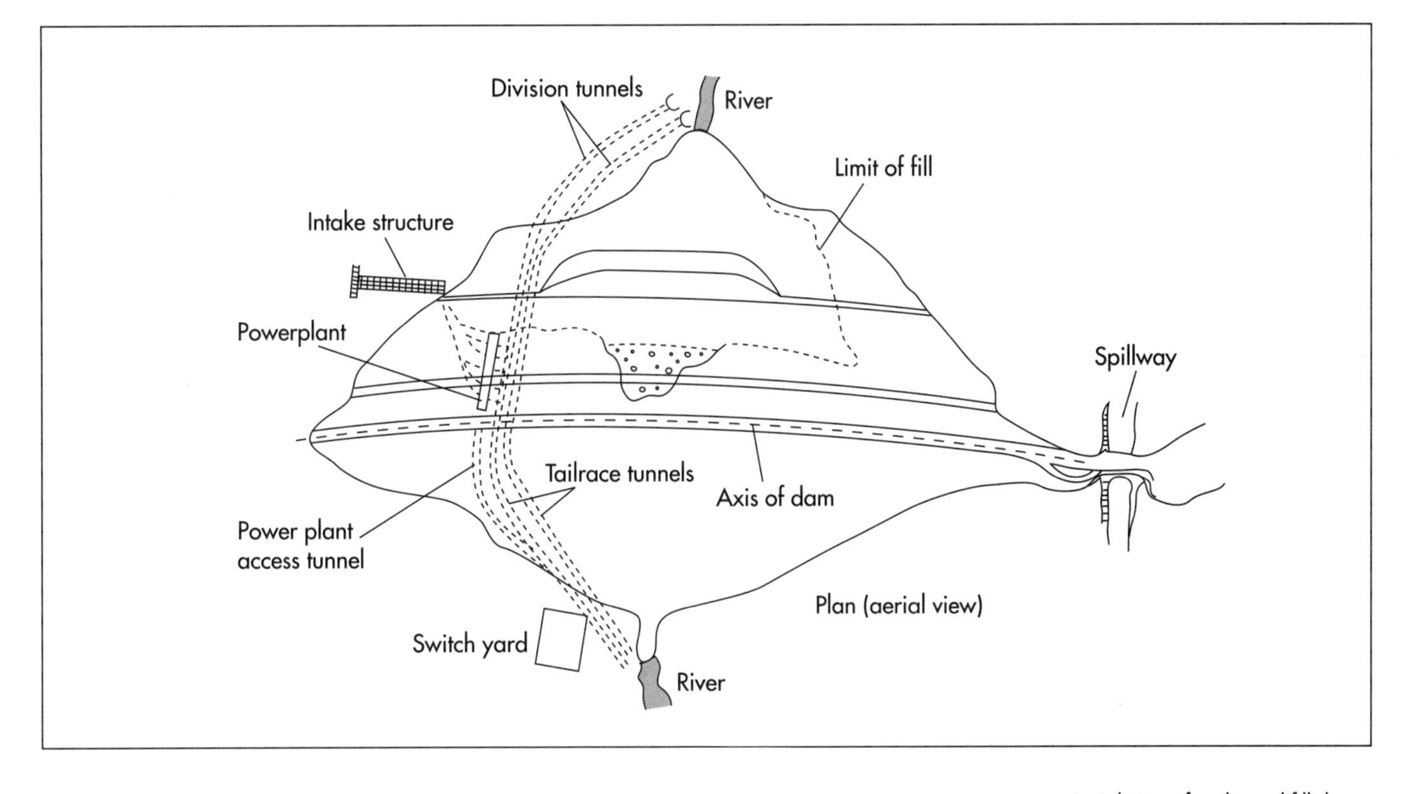


Aerial view of a planned fill dam.

into the keyway area for the construction equipment, and then they must be built up to the working surface of the rising top of the dam. Whenever possible, roads are cut in from the two sides (abutments) of the dam for the easiest access; eventually, an access road will be built on the crest of the dam and extending onto these abutments.

Large earthmovers haul the specific type of soil needed to raise the zone of the dam they are working on. The soil is spread in thin layers, usually 6-8 in (15.2-20.3 cm) thick, sprayed with water to the correct moisture content, and compacted with sheepsfoot rollers (compactive rollers with prongs resembling animal hooves mounted in rows around the roller that press and vibrate the soil firmly in place). If gravel is used in construction, a vibrating roller is used to vibrate the grains together so their angles intermesh and leave no openings.

Throughout the compaction process, inspectors approve the soil that is hauled on site and hauled to the particular zone of the dam. They reject material that is contaminated with grasses, roots, trash, or other debris; and they also reject soil that does not appear to be the proper grain size for that zone of the dam. For quality control, samples are collected and tested in the laboratory (for large dams, an on-site soil lab is installed in a construction trailer) for a variety of classification tests. Meanwhile, the inspector uses a nuclear density gauge to test the soil for density and moisture content when it has been placed and compacted. The nuclear density gauge uses a very tiny radioactive source to emit radioactive particles into the soil; the particles bounce back onto a detector plate and indicate the moisture and density of the soil in place. The process is not harmful to the environment or the operator (who wears a badge to monitor radioactive exposure) and provides data without having to excavate and sample. If the compaction requirements are not met, that layer of soil is excavated, placed again, and recompacted until its moisture and density are suitable.

Construction of the fill dam proceeds layer by layer and zone by zone until the height of each zone and, eventually, the crest of the dam are reached. If the entire dam cannot be built in one construction season, the dam is usually designed in phases or stages. Completing a construction stage (or the entire dam) is often a race against time, the weather, and the project budget.

5 Some earth dams have instruments installed in them at the same time as fill

placement is done, and the instruments are constructed to the surface in layers and zones, just like the fill. The condition of the dam is monitored throughout its lifetime, as required by federal, state, and local laws and by standards of engineering practice. Types of instruments vary depending on the location of the dam; almost all dams have settlement monuments that are surveyed to measure any settlement in the surface or zones of the dam, slope indicators to show if the sloping faces inside or on the surface of the dam are moving, and water-level indicators to monitor the water level in the dam's zones. Dams in seismically active areas may also be equipped with instruments to measure ground shaking.

6 Fill dams may have a variety of other facilities, depending on their, size, use, and location. An emergency spillway is required at all dams to allow for flood waters to flow over an escape route, rather than over the top of the dam. Other spillways for production of hydroelectric power may be designed and constructed at power-generating dams, and inlet and outlet tunnels are needed to release water for irrigation and drinking-water supplies at embankments built for those purposes. At fill dams, it is usually desirable to place these other facilities in excavations through the foundation or abutment rock; the process of compacting earth against structures that actually pass through the fill is tricky and allows for seepage paths.

7 Sometimes the reservoir area is also cleared when it is to be filled with water, particularly if lumber can be harvested. It is not necessary (and it is much too expensive) to clear it of all shrubs and grass. The process of filling the reservoir is relatively slow, so most wildlife will move as the water level rises; areas of concern include habitats for rare or endangered species, and drowning of these habitats has been a concern in the construction of a number of dams.

When the dam is complete, the water that was diverted from the stream channel is allowed to fill the reservoir. As the water rises, it is also rising in portions of the dam, and instruments within the dam are monitored carefully during the reservoir-filling period. Monitoring of the dam's performance, both by instruments and simple observation, is performed routinely; and safety plans are filed with local emergency services so that sudden changes in instrument readings or the appearance of the dam or its reservoir triggers actions to alert and evacuate persons living in the path of flood waters downstream. Repairs are also performed routinely.

## Quality Control

Quality engineering is essential in the construction of a fill dam because the materials used have lower strength properties than the steel and concrete required for concrete dams and because placement ultimately determine strength, potential for problems like seepage and settlement, and finally performance and safety. The geotechnical project engineer occupies the key role of making sure the design and earth materials match to make a safe product; but many other professionals including geologists, construction technicians, other engineers, and the representatives of overseeing agencies are fully committed to the same purpose.

## Byproducts/Waste

There are no byproducts in fill dam construction, although fill is sometimes generated for building access roads and other support structures. Waste is also minimal to nonexistent; excavation of excess soil and especially rock is very expensive as is hauling these materials so waste is engineered out of the design.

## The Future

Primarily due to environmental concerns, design and construction of any dam in the future will be a much-studied and controversial process. Fill dams, however, tend to be perceived as more environmentally friendly because they are made of earth materials and blend into the scenery better than monolithic concrete structures. Fill dams have proven useful and less expensive solutions to meeting human needs for water supply, and vast improvements in engineering technology have improved their safety record in the late twentieth century. Although many costs and agendas must be considered in building dams, fill dams have and will continue to prove themselves allies in the needs to provide drinking water, irrigation supply, and flood control.

## Where to Learn More

### *Books*

Bureau of Reclamation, U.S. Department of the Interior. *Design of Small Dams*. Washington, DC: U.S. Government Printing Office, 1977.

Jansen, Robert B. *Dams and Public Safety*. Washington, DC: U.S. Dept. of the Interior, Water and Power Resources Service, 1980.

Krynine, Dimitri P. and William L. Judd. *Principles of Engineering Geology and Geotechnics*. New York: McGraw-Hill Book Co. Inc., 1957.

Sherard, James, et al. *Earth and Earth-Rock Dams*. New York: John Wiley and Sons, Inc., 1963.

Smith, Norman. *A History of Dams*. Secaucus, New Jersey: The Citadel Press, 1972.

### *Periodicals*

Monastersky, Richard. "Dams on Demand." *Science News* (August 29, 1992):138.

*—Gillian S. Holmes*

# Fishing Fly

*Fishing flies are tied in over 5,000 patterns and sizes, and each has a specific name.*

## Background

A fishing fly is a hook that has been dressed with pieces of feathers, fur, thread, and other materials to resemble a literal fly or some other small insect or fish. Fishing flies are tied in over 5,000 patterns and sizes, and each has a specific name. The "Cosmo Gordon," for example, is a fishing fly that was made for salmon fishing in England around 1900 and is named for the legendary angler. The "Seth Green American Trout Fly" is named for a conservationist who, in the nineteenth century, restocked the rivers in New England with shad. The "Green Peacock" and the "Silver Grey" are named for their general appearance and coloring. Still others, like the mayfly, caddisfly, or stonefly resemble the insects for which they are named.

All varieties of sport fishing depend on the attraction of the fish to the lures. Fishing lures form a large category of artificial devices used to attract the fish's attention and lure it to the hook. Within the range of fishing lures, flies are the most beautiful. The three broad varieties of flies imitating the life cycles of flies and other insects; these are nymphs, wet flies, and dry flies. The nymph stage is the larval form of the insect that lives on stream bottoms. Wet flies are fished from below the surface and may look like dead or drowned bugs or insects that are either hatching themselves or laying their eggs. Dry flies look like insects floating on the water surface; they can resemble two stages in the fly's life cycle; either its first flying days after hatching from the nymph stage or the mature fly landing on the water after having mated.

Two other types of flies called streamers or bucktails don't imitate flying insects at all. Instead, they look like minnows, other baitfish, or leeches. Long feathers or hair from deers' tails (bucktails) suggest the shapes of these slender creatures.

Fly-makers, usually called tiers, advocate one of two schools. The traditionalists belong to the "match the hatch" school, and they try to make flies that look as much like the real thing as possible. These tiers observe the exact shade of this season's hatchlings along the trout stream and dye their flies to match. Other tiers simply believe the school of attraction. The flies they make may not look like any particular insect but are designed to be flashy or to move interestingly to stimulate the fish's interest.

## History

Fishing flies are known to be at least 1,700 years old. In the third centuryA.D., Claudius Aelianus describes the Hippouros fly made by Macedonian fly tiers of hooks, red wool, and wax colored feathers from a rooster's wattle. The next writings about fly fishing didn't appear until 1496 when a nun named Dame Juliana Berners wrote *The Treatyse of Fysshynge wyth an Angle*. She chronicled dozens of patterns of trout flies, suggesting that the years between Aelianus's observations and hers were filled with productive fly tying. Wet and dry flies and nymphs were all created and perfected in Europe and migrated to the New World with European colonists. However, streamers and bucktails are American developments.

Fishing flies have been an acknowledged art form for about 200 years. Late in the nineteenth century, fishing became an absolute craze in the British Isles, and dry flies were a part of this rage. Purists began to imitate nature's perfection, and some of the most beautiful and collectible specimens of artificial fly date from this era. Emphasis shifted around the 1920s when wet flies became more popular and were sometimes made with as many as 20 different kinds of feathers. After World War II, the availability of plastics and synthetics changed fly-tying interests again toward the attraction school of fly design, and enthusiasm for saltwater fishing also stimulated creation of new types of flies.

Designs of flies have diversified as sportsmen's interests in different fish have changed and even as the environment has caused shifts in the kinds of fish available. The different types of flies also require fishing at faster or slower speeds and copying the darting, diving, or twitching motions of the real-life insect or the dead drift of the dry fly; skill at manipulating the line and knowledge of the behaviors of both bait and prey are definitely parts of the art, history, and science of fly fishing.

## Raw Materials

There are four broad classes of materials used in fly-tying. Metal parts include hooks and lead wire to add weight; tiers purchase these from suppliers. Synthetic materials include plastics, poly-yarn with a silicon dressing, foam, and gold and silver mylar. Fabrics and threads are part of the tier's supply arsenal and are typically sewing thread, embroidery floss, and crochet thread; wool and burlap; and synthetics like chenille. Natural materials consist of many types of fur (rabbit, fox, seal, and mink), hair (elk, caribou, and deer), different types of fur or hair from various parts or the bodies of these animals and especially their under fur, and feathers (chicken, peacock, and marabou, among many), again from different parts of the bodies of these birds. Other incidental materials like beeswax, dyes, head cement (to coat the several wraps of thread that represent the head of the insect), and paint for eyes are used in tiny amounts.

## Design

At least 10,000 published fly patterns fill tiers' catalogs, and their names range from pure poetry to those that sound like members of the World Wrestling Federation. Flies that imitate nature can be miniature works of art that are tied, dyed, and painted in meticulous detail, like the "Lew Oatman's Brook Trout" fly that has tiny yellow and red spots of paint touched along its sides. Tiers in the realistic school rightly consider themselves artists. Tiers in the attraction school may go for simpler, more colorful materials including bits of plastic or metallic flash; they also tend to advocate flashier names for their creations. The "Mickey Finn" fly is devilishly detailed in its construction and ridiculously garish and yet this is among the most popular classic salmon flies.

Designs of flies use bits of feather and fur tied to the hook to look like the segmented bodies of insects. The basic segments of a dry fly are the long and stiff tail, the body, the hackle (a flayed section of feathers that looks like legs touching the water), and outstretched wings. A class of dry fly called the emerger fly rides on the water with its tail sunk below the surface; iridescent material is used to design the tail to resemble an egg sack. On a wet fly, the hackle is sparse, and the wing is folded back over the body. Nymphs, streamers, and bucktails have soft tail and body material, which is sometimes wrapped around lead to make them sink more quickly. They also have long, soft wings often made of marabou feathers that make the silhouettes like minnows.

## The Manufacturing Process

1 Essential tools of the fly tier are dextrous fingers, patience, excellent eyesight, creativity, and ability to visualize the spacing of the fly's segments as he or she ties from the tail (hook bend) toward the head (hook eye). The tools of the workbench consist of a special, y-shaped bobbin for holding a spool of thread, fine scissors, and a vise. A rotary vise is especially useful for turning the fly during the process.

2 The tier assembles a collection of materials specific to the type of fly, for example, a "Royal Coachman," which is an attractor dry fly with a red floss body, iridescent green collars made of peacock herl (part of the peacock feather below the eye that appears flat brown until light strikes tiny, curled fibers on the edges of the feath-

er's barbules that are bright green), white hair for the wings, and brown feathers for the pseudo-insect's hackle or legs. A right-handed tier will mount the hook (of the correct size and weight) in the vise with the eye to the right and the bent hook to the left. The y-shaped bobbin has a spool of thread clasped between the two branches of the "Y," and the sewing floss is threaded through the long leg of the "Y." It is held in tension in this devise but releases easily when pulled as the bobbin unit is either circled around the fly in a stationary vise or held while the vise itself rotates.

3 To create a secure working platform, the tier ties the thread near the eye of the hook and wraps thread around the hook all along its shank to the beginning of the bend in the hook and back toward the eye. To make the tail, four to six feather barbules (barbs that are the individual strands in a bird's feather) are tied close to the curved end of the hook with several wraps of thread that are drawn tight. Where the tail meets the main part of the body, a collar of peacock herl is tied on. Several wraps of thread fix this collar snugly to shank of the hook. The body is made with wraps of red floss. The red floss is extended toward the eye until about 60% of the length of the hook shank is covered. Another collar of peacock herl is wrapped around the hook to the right of the wrapped body.

4 The wings are created from a tuft of white hair from a calf's tail or other suitable material. The hairs must be even at the tips; the hairs are dropped in a hair stacker—a tube-within-a-tube that is tapped so all the hair tips are lined up. The tier pinches the squared hairs between the fingers and ties them to the hook with several thread wraps. The bundle of hair is split into two wings and figure-eight wraps of thread hold the two wings apart and angled up from the body.

5 A tannish brown chicken feather from a chicken that has been specially bred to produce neck feathers with hard, uniform barbules is selected for the hackle—the bug's legs. The quill end of the feather is stripped of soft feather barbules near the base. The center quill is wrapped around the fly to the right (eye end) of the wing attachment. The curving of the quill causes the barbules to flay out. A few wraps of thread fasten the quill tightly. When the fly is cast on the water later, the tips of the stiff feather barbules of the hackle will touch the water, support the fly by surface-water tension, and look from under the water surface to the unsuspecting fish like insect legs. The head is finished with three to five wraps of thread. Sometimes the head is coated with head cement to protect the threads.

6 The tier is ready for his next fly. If he or she chooses to make the leech-like "Woolie Bugger" as his next specimen, the tier will again assemble the needed materials. Unlike the "Royal Coachman," a dry fly that perches on the water surface, the "Woolie Bugger" is a bottom dwelling fly that looks like a succulent leech. To make the "Woolie Bugger" sink, the tier begins this fly by twisting two or three wraps of lead wire around the hook near the head end to make it heavier. A marabou feather is selected to imitate the leech body; when dry, the marabou feather is soft and fluffy, but when saturated in the stream, it forms a sleek leech. Alternatively, the tier may use chenille for the body and webby hackle material to make the "Woolie Bugger." The thread wrap is applied to the weighted hook, and construction of the fly proceeds similarly to the method described above.

## Quality Control

The tier's judgment, visual appreciation, and sense of touch are the controlling features in the quality of the handmade fly. The ultimate quality control assessment will be rendered by a trout, bass, or salmon who senses an invader in his territory or a tasty treat to line its belly.

## Byproducts/Waste

There are no byproducts from the hand tying of fishing flies. After materials are assembled, many varieties of flies can be made from similar supplies. Waste consists mostly of fine bits of thread, fur, and feathers.

## The Future

Fly fishing is one of the most popular sports in the United States, and the momentum shows no sign of slowing. In their search for the "perfect" fly, many fish-seekers turn to tying their own flies, especially as a winter

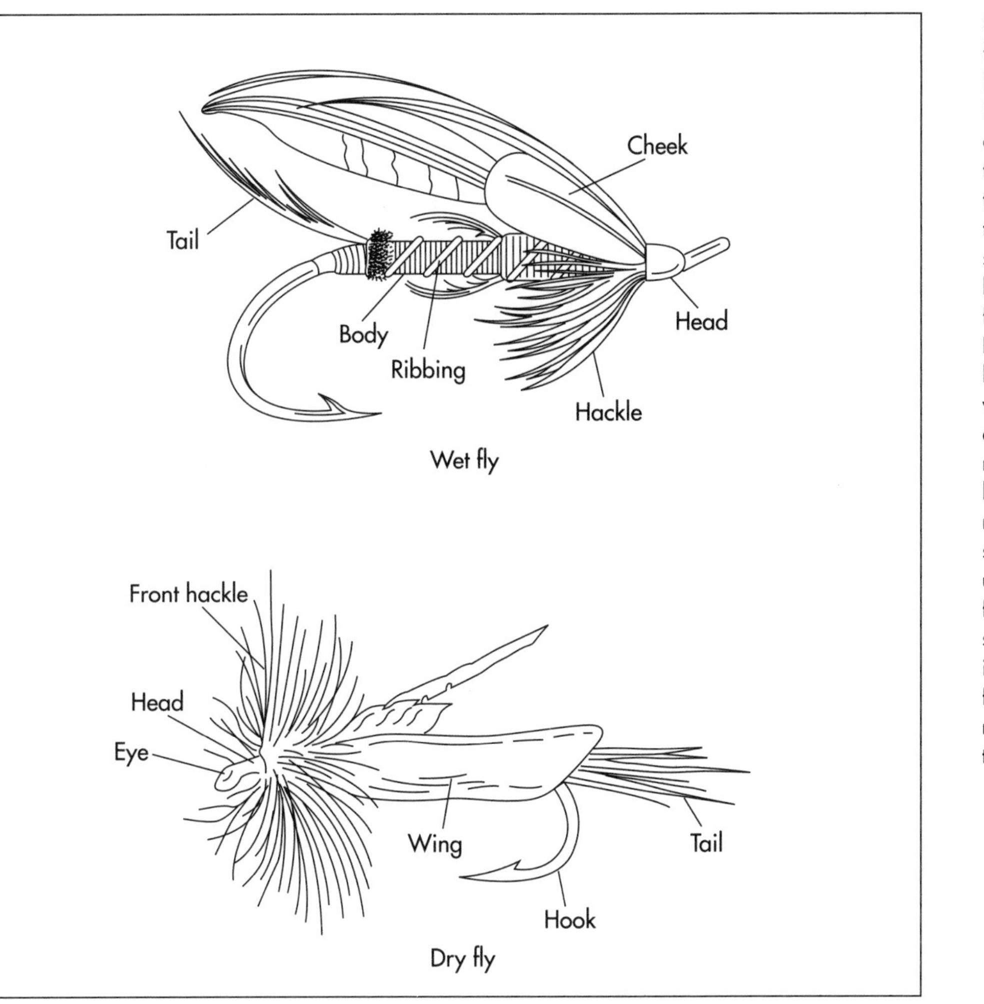


Fishing fly designs use bits of feather and fur tied to the hook to look like the segmented bodies of insects. The basic segments of a dry fly are the long and stiff tail, the body, the hackle (a flayed section of feathers that looks like legs touching the water), and outstretched wings. On a wet fly, the hackle is sparse, and the wing is folded back over the body. Nymphs, streamers, and bucktails have soft tail and body material, which is sometimes wrapped around lead to make them sink more quickly. They also have long, soft wings often made of marabou feathers that make the silhouettes like minnows. The tier uses a special, y-shaped bobbin for holding a spool of thread, fine scissors, and a vise. A rotary vise is especially useful for turning the fly during the process. Different materials are used depending on the type of fly.

preoccupation to wile away the days until the streams thaw and the fish run.

The future is also promising for fly tiers who dream of seeing their names in print; shelves of books on fly fishing and fly tying fill book stores and also help fishing hobbyists pass the dreary, fishing-free days. Historic interest in hand-tied flies has also become a mini-industry with museums and flies on Internet auction sites. Collectors of fishing flies value them by the person who tied the fly. The products of master tiers are artworks without signatures. Carrie Stevens of Maine (1882-1972) was acknowledged as a master tier, and she marked her flies with a tiny red band on the head of the fly. Other tiers' works must be identified by original paper backings or envelopes with the tier's name. Whether parts of a prized collection or residents in a tackle box, hand-tied fishing flies experience secure futures that would be the envy of their natural counterparts.

## Where to Learn More

### Books

Leonard, J. Edson. *The Essential Fly Tier.* Englewood Cliffs, NJ: Prentice-Hall, Inc., 1976.

Neff, Sid. "Fishing Tackle: Beautiful Tools of Sport." In *The Encyclopedia of Collectibles: Dogs to Fishing Tackle.* Alexandria, VA: Time-Life Books, 1978.

Shaw, Helen. *Fly-Tying.* New York: Nick Lyons Books, 1987.

Toth, Mike. *The Complete Idiot's Guide to Fishing Basics.* New York: Alpha Books, 1997.

Whitlock, Dave. *L. L. Bean Fly Fishing for Bass Handbook.* New York: Nick Lyons Books, 1988.

Zahner, Don, ed. *Fly Fisherman's Complete Guide to Fishing with the Fly Rod.* New York: Ziff-Davis Publishing Company, 1978.

### Periodicals

Raymond, Steve. "The best of flies are those that can catch fish and the imagination."*Sports Illustrated* (July 2, 1984): 76.

Wright, Leonard W. Jr. "The all-American fly: why has this classic lure stood the test of time? Because it works."*Field & Stream* (November 1995): 62+.

### Other

American Museum of Fly Fishing. 1998. http://www.amff.com/ (June 28, 1999).

Cabelas Inc.1999. http://www.cabelas.com/ (June 28, 1999).

The Virtual Flyshop.1994-1999. http://www.flyshop.com/ (June 28, 1999).

*—Gillian S. Holmes*

# Fishing Lure

## Background

The way to a fish's stomach is through his eyeballs, and fishing lures are objects that resemble any of the naturally occurring foods that fish might find attractive. The purpose of the lure is to use movement, color, and vibration to grab the fish's attention and cause him to bite the hook. Lures also seem to have the purpose of attracting the fishermen's attention. Sport fishing is now a huge business, with over $620 million spent on fishing lures in 1991.

Lures fall into several broad categories, each with specific characteristics that catch the fish's attention. All lures have to be kept moving when they are in the water to be effective in mimicking the actions of live bait.

Spoons are among the simplest of lure shapes. As their name suggests, they are rounded pieces of flat metal, just like the bowl of a dinner-table spoon. They can be colored or polished on both sides and use flashes of reflected light to resemble minnows. A single hook can be fastened inside the bowl of the spoon, or a three-pronged (treble) hook can be attached to one end through a small hole. A hole at the opposite end is used to attach the fishing line. The weight of the metal makes it easy to cast, troll, and retrieve.

Spinners are more complicated in appearance with several parts. A wire shaft forms the spine of the spinner, and it has loops or eyes at both ends, one for attaching the fishing line and the other for the hook. The body of the spinner is fixed along the metal spine. It can be made of a row of colored beads or collections of metal rings or cylinders that sparkle and glimmer. A skirt of hair made from a squirrel's tail may add interest. Near the top eye, a flat metal oval, similar to a spoon, is attached with a fine wire. This oval is called a spinner blade; in the water, it spins around the body to provide fish-luring movement.

Spinnerbaits are also known as hairpin lures because they have wire spines that are bent to form a v-shape, much like a spread hairpin. A spinner is wired to an eye at one end of the spinnerbait. At the other end, a hook is concealed in a skirt. The fishing line is tied to the bend forming the two arms of the lure. These lures are relatively large and have two points of action in the skirt and the spinner. Buzzbaits are close cousins of spinnerbaits and have small propellers on one arm instead of the spinner. These propellers attract the catch by vibration rather than flashing light.

Plugs encompass a wide variety of lures. Originally, a plug was a piece of wood or cork shaped like a minnow with hooks on the belly and tail. Now the term includes all wood or plastic objects that are shaped like minnows, other baitfish, and other prey ranging from crayfish and salamanders to small rodents. Plugs can be less than 1 in (2.54 cm) long and up to 8 in (20.32 cm) long. Poppers, also called topwater plugs, float on the water so they resemble frogs or surface-splashing baitfish. Plugs called floater-diver plugs may have several sets of treble hooks on their undersides as well as small metal or plastic scoops, termed lips, near the line end of the lure. This plug floats on the water surface until the fisherman begins to reel in the lure, then it dives below the water surface in imitation of the motion of a minnow. The lip causes the diving action. A crankbait also has a lip that

*Sport fishing is now a huge business, with over $620 million spent on fishing lures in 1991.*

causes it to dive when retrieved, but it has a wider body and dives deeper (up to 20 ft [6.1 m]) with a wobbling motion like the wriggle of a baitfish. Jerkbaits or stickbaits float but don't have any motion-causing devices like other plugs. The fisherman has to manipulate the rod tip to make these plugs life-like.

Jigs seem too simple to be true. They are weighted hooks with a lead head right behind the hook's eye. The heads and hooks come in many sizes, weights, and shapes with skirts mounted immediately behind the heads to camouflage the business end of the hook. For jigs (also called leadheads) to be effective, they need to dance on the stream bottom like minnows or crawdads, and the fisherman has to apply the right motion to the rod. This action, called jigging, gives them their name. Jigs are often fished with so-called "sweeteners" that are either live bait or plastic lures.

Plastic lures are imitations of worms, bait fish, bottom-dwellers, and even snakes and amphibians. They are molded of soft plastic, often in lurid colors, although blacks and blues seem to be preferred by fish. Again, the motion of the lure and its effectiveness depend on the dramatics of the angler.

## History

The caves and remains of habitations of ancient man have been found to hold fish hooks carved out of bone and molded out of bronze. The ancient Greeks and Romans both advocated fishing for sport, as well as for food, but Chinese and Egyptian archaeological digs have shown that fishing rods, hooks, and lines were known as early as 2,000B.C., or far earlier than the Greek and Roman civilizations. Bronze barbed hooks were used by the Egyptians; these hooks resulting from the alloy of tin and copper made hard, strong hooks that also could be worked until they were very thin and less visible to the fish. The Chinese spun fine fishing line from silk and used rice and small carp for bait.

Claudius Aelianus, a Roman who lived during the third centuryA.D., wrote of fly-fishing for trout and other kinds of sport fishing. He made lures of feathers, lead, bronze, and wild boar's bristles and used horsehair and twisted flax to make his fishing line. There is little documentation of advances in fishing tackle throughout the European medieval and Renaissance periods, but, in 1653, Izaak Walton wrote what is probably the most famous book ever to have been penned on fishing. His *The Compleat Angler* or *The Contemplative Man's Recreation* described all of the sport fisherman's necessities including fishing line, hooks, flies, and appropriate attitudes. He wrote about fishing for trout in streams in the English countryside, and his poetic style created an ideal that is associated with sport fishing today as well as describing the practical aspects of line and rod.

By the 1830s and 1840s in both England and America, the making of fishing tackle began to change from the monopoly of the individual craftsmen to commercial manufacturing ventures. From the early 1900s, the firm of Heddon and Pflueger in Michigan led the production of commercially made lures. These lures were often designed from proven lures that were simply pounded out of old kitchen spoons or whittled from pieces of wood. Rods and reels were handmade by jewelers and watchmakers from the early 1800s. This craft experienced many technical improvements as rod and reel production became commercial throughout the 1870s and 1880s. Advances in fishing line waited until after World War II when braided nylon followed by monofilament line improved the success rate in all types of fishing. Aided by the availability of more leisure time, fishing exploded as a hobby and sport.

## Raw Materials

Materials for the manufacture of fishing lures include metal, wood, cork, and plastic, depending on the type of lure. Most luremakers, whether commercial manufacturers or amateurs do not make the individual components themselves except for poured lead pieces and some molded plastic. Instead, specialty suppliers make the parts for manufacturers and hobbyists to assemble. Metal parts include hooks, wire, beads, blades, ball bearings, rings, loops, and spacers. Kinds of metal used are stainless steel or titanium, lead, and some brass. Pieces of cut, carved, and shaped wood and cork are used for the bodies of plugs. Plastic bodies are also popular, and many of the ornaments and attractions on lures are plastic, such as skirts and weed guards. Plastic lures themselves can be molded by the hobbyist with simple hinged metal molds and plastic that is melted and poured.

## Design

While designs of lures fall into some broad categories, as described above, they can also be highly individual.

## The Manufacturing Process

The manufacturing processes for two kinds of lures are detailed below. They are the "pulsator spinnerbait," a product of Nichols Lures Incorporated (one of few lure manufacturers remaining in the United States), and a simple jig. These are two of the most popular lures in use. The spinnerbait originated as a safety pin, and it can be fished on the surface, at depth, at speed, or on the bottoms of streams.

### *Making a spinnerbait*

1 Production of the spinnerbait begins with an "R"-bend wire or twist wire. It looks like a hairpin or bobbypin with a hook at the end of one leg and a dogleg bend in the other to form an overall shape that looks something like an "R" at the top. A twist wire is similar except that the top of the bobbypin is twisted around to form an eyelet shape. The wire is connected to the hook during construction of the lead head. The wire and a hook are linked at the hook eye and placed in a hinged mold. The mold has special channels to hold the wire that emerges from one end of the mold and the hook that has its bend protruding from the other side. The mold is sized to suit the wire and the hook. At the join between the wire and hook, the mold has a small cavity that will shape the head. The mold is closed and held by an insulated handle at melted lead is poured into a gated channel at the top of the mold. Mass-producers of lures have centrifugal molds in which a number of lures can be made at once, and the centrifugal action swirls the soft lead into all the nooks of the mold cavities. In the hand-molding process, molded heads are made individually.

2 When the lead has hardened and cooled, the head is finished by hand painting or spraying it, attaching prefabricated stick-on eyes, and finishing the heads with a clear coating that protects them. Usually, epoxy or polyurethane are applied.

3 At the end of the other leg of the lure, a flat metal piece called a blade or spinner is attached to the eye of wire leg. Several other connectors go between the blade and the wire, and the number and selection of pieces can vary considerably. For this example, a ball bearing, small shaft, and split ring are attached to the loop in the wire and crimped down. The blades also have their own personalities; the Colorado is round, the Indiana is tear-drop-shaped, and the willow looks like an elongated willow leaf. The action of the water against the blade causes it to spin in the water and flash in the light. Spinners are die-cut from stainless steel or titanium (for light weight), and sometimes they are painted to add visual interest for the fish.

Two of the most popular varieties of spinnerbaits are the double willow or the tandem. Both begin with a wire that has a longer blade end. The double willow has two willow blades. The first one is hung from a cleavis, a horseshoe-shaped device with two eyes that is threaded onto the wire. Beads, a shaft, and a split ring are connected to the wire end, and the second willow blade is added to the ring. The tandem spinnerbait is assembled similarly, except the front blade is a Colorado, and the second spinner is a willow. Again, the combination of beads and connectors can vary depending on the manufacturer's design.

4 The hook on the other leg of the wire is still exposed. To finish the spinnerbait, a skirt-keeper of plastic, rubber, or vinyl is slipped over the hook. The skirt-keeper looks like a piece of tubing except it may have a tiny hook-like extension or a small collar that will keep the skirt from sliding around on the hook. The skirt itself is also plastic, rubber, or vinyl that has die-cut slender bands or threads that wave in the water. Skirts come in slip-on varieties or can be

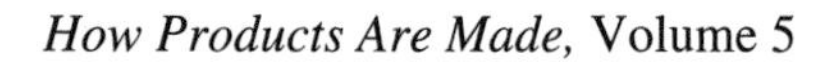

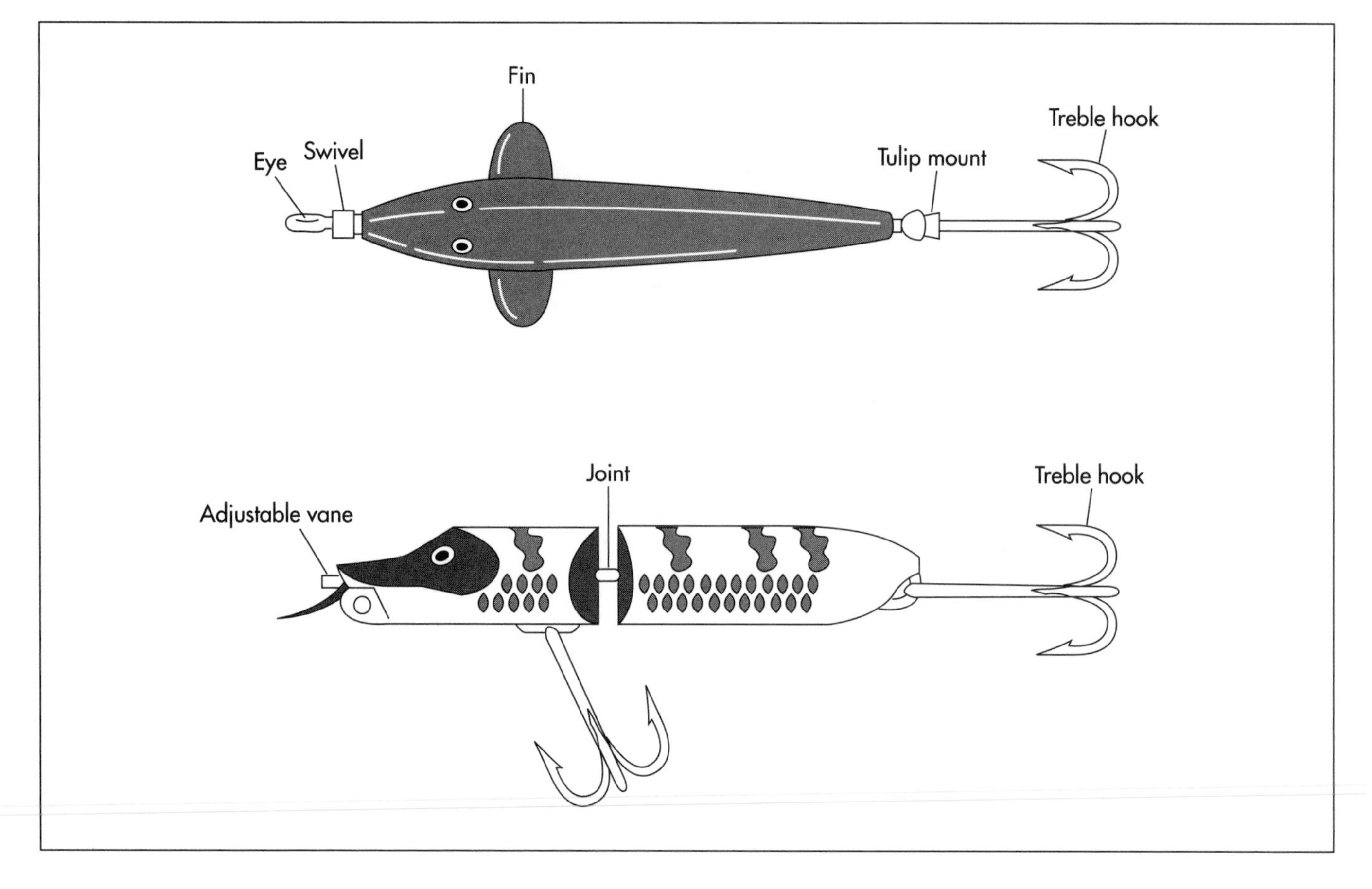


Materials used to manufacture fishing lures include metal, wood, cork, and plastic, depending on the type of lure. Most lure-makers, whether commercial manufacturers or amateurs, do not make the individual components themselves except for poured lead pieces and some molded plastic. Lures fall into several broad categories, each with specific characteristics that catch the fish's attention. They include spoons, spinners, spinnerbaits, plugs, jigs, and plastic lures. The spinnerbait and jig are the most popular designs.

tied on. They are also available in many colors, sizes, and varieties.

### *Making a jig*

5 The jig is also a lead head. A hook is placed in a lead mold; this mold is different because the eye of the hook will extend through the lead head. The lead is poured, cooled, and removed from the mold.

6 Like the spinnerbait head, the jig's head is painted and finished. A fiber weed guard is placed next to the head to prevent the jig from tangling in weeds or brush. A slip-on collar is fitted on the hook between the head and the bend in the hook, and a rattle is attached. Rattles, again, come in many sizes and forms, and their purpose is to imitate the sound of a crayfish or crawdad under the water.

7 A skirt-keeper and skirt are fastened on the hook. The jig itself is finished, but the jig can also be used as the hook end on a spinner rig to make a jig-spinner combination.

## Quality Control

Quality control is important to the manufacturer in producing a consistently high-quality product that has been proven in the field. Reputation is a significant selling point in this competitive business. Components are chosen with care.

## Byproducts/Waste

Byproducts do not result from lure manufacture, but the endless combination of components and possibilities keeps the creative lure-maker busy in producing a range of products. Small amounts of lead and metal waste are generated and are disposed; the wasted lead is not large enough in quantity to create a hazard. Similarly, exposure to melted lead or fumes during the head-molding process is not large enough to be measurable and does not present a safety hazard, either to the assembly-line worker or the home hobbyist.

## The Future

Lures have a sparkling future, thanks to the growth in popularity of the sport of fishing, which shows no signs of slowing down. The computer has ventured into lure design, and some of the latest innovations include scent-bearing lures, laser-honed hooks, and exterior finishes that are photographically imprinted and amazingly realistic. The future for lure manufacturers in the United States is not quite so bright. Although lures are mass-produced, the work still involves hand crafting and assembly. Manufacturers cannot pay laborers the mandated hourly rates and still profit, so much of the work has relocated to Central America, Southeast Asia, Taiwan, Haiti, and Mexico.

## Where to Learn More

### Books

Livingston, A. D. *Luremaking: The Art and Science of Spinnerbaits, Buzzbaits, Jigs, and Other Leadheads* Camden, ME: Ragged Mountain Press, 1994.

Mayes, Jim. *How to Make and Repair Your Own Fishing Tackle: An Illustrated Step-by-Step Guide for the Fisherman and Hobbyist.* New York: Dodd, Mead & Company, 1986.

Toth, Mike. *The Complete Idiot's Guide to Fishing Basics.* New York: Alpha Books, 1997.

### Periodicals

Almy, Gerald. "Golden oldies: 15 tacklebox classics."*Sports Afield* (February 1997): 106+.

### Other

Nichols Lures Inc. http://www.nicholslures.com/ (June 28, 1999).

Ray's Fishing Lures. April 15, 1999. http://www.raysfishing.com/ (June 28, 1999).

*—Gillian S. Holmes*

# Fishing Rod

*Thirty-one million fishing licenses were sold in the United States in 1992, compared to 20 million in 1959.*

## Background

A fishing rod is a device used in sport fishing that consists of a long pole with a line held in place alongside it with the use of guides. Usually the line is kept in storage on a reel that the angler spins to both take up and let out the line while casting. At the loose end of the line is a hook to hold the bait, either live worms and insects or artificial lures, as well as bobbers (or floats) and sinkers that keep the bait at the proper level in the water.

## History

Ever since humans began gathering foods, the living creatures in water have been considered possible sources of nutrition, and many ways were devised to catch a sufficient number of fish in a simple fashion. Nets and weirs, which are dams often woven from reeds and placed in streams, were both used to gather a great number of fish, which could then be sorted into edible and undesirable fish.

The very earliest rods were made of wood, bone, or stone and were called gorges. These gorges were only about 1 in (2.54 cm) long and were pointed at both ends. A bait and line were attached to these gorges, which soon were made of metal. Fishers would use gorges to fish from boats. Longer rods began to be used soon after, at first just a simple tree branch about a yard (0.9 m) long, and anglers could then fish from shore with ease. Pictures of angling show it as an accepted sport in ancient Egypt around 2000 B.C. A Chinese written account from around the fourth century B.C. describes a bamboo rod, with a silk line, a needle used as a hook, and rice as bait.

In ancient Greece and Rome, fishing with a rod was already a common activity. In Homer's time, or around 900B.C., and in Plato's time, or around 400B.C., fishing with a barbed hook, rod and line were known. The line was made from either horsehair or finely woven flax. The rod is believed to have been made from *Arundo donax*, a plant native to the Mediterranean area and the largest of European reeds. It grows up to 20-30 ft (6.1-9.1 m) long, and it is knotted. A plant known as ferula might have been the source for smaller rods, but cornel wood, a slender hardwood, or juniper might also have been used. The rod was jointed, either by tying the parts together with string, or tiny pieces of metal known today as ferrules could have been used. The bronze hook was not tied on; rather, the top of the hook extended over the string in a sheath to keep fish from biting through the line. Sometimes a piece of lead was attached to the hook to ensure depth, while fly fishing quickly developed when it was learned that some kinds of fish would eat flies off the surface of the water. The bait used in fly fishing was a piece of red wool with rooster feathers attached. The feathers presumably appeared as a waxy color to the fish, hence resembling fly's wings, and they helped keep the bait afloat. Anglers had no running tackle—they had to pull the line in using force.

Angling captured the imagination of Greek and Roman scholars. Plutarch wrote that a good rod must be slender and springy, because a heavy rod would cast a shadow and scare the fish. The line must not have knots or be coarse, and it should be white so as to match the color of the water. Dionysius wrote that an angler with two rods, four hooks apiece, and an assistant could catch more fish than a net, if good bait is used.

Fishing rods changed only slightly for more than a thousand years. In England in 1496, a nun named Dame Juliana Berners wrote *The Treatyse of Fysshynge wyth an Angle*. This book described artificial flies, some of which are still used today, and rods 18-22 ft (5.5-6.7 m) long with lines made of horsehair.

During the middle of the seventeenth century, the fishing rod was greatly improved by adding a wire loop or ring at the tip. This allowed the line to be let out and pulled in easily. Lines at least 26 yd (23.8 m) long were mentioned by 1667. Such lengthy lines led to the development of the reel. The first reels were wooden spools with a metal ring which fitted over the fisher's thumb. By 1770, fishing rods with guides along the length and reels were in common use.

Fishing rods were improved during this period by replacing heavy European woods with tough, elastic woods, such as lancewood and greenheart, both from the New World. Bamboo from the East was also used. By the late nineteenth century, hexagonal fishing rods were made by laminating six triangular strips of bamboo. At the same time, reels were greatly improved. Horsehair lines were replaced by silk coated with oxidized linseed oil.

During the twentieth century, fishing rods became shorter and lighter without losing strength. Bamboo was replaced by fiberglass or carbon fiber. Nylon became the dominant material for use in fishing lines after World War II, and plastic became used to make artificial flies. Fishing became an increasingly popular sport in the late twentieth century. Thirty-one million fishing licenses were sold in the United States in 1992, compared to 20 million in 1959.

## Raw Materials

Although some fishing rods are still made of bamboo, most modern rods are made of fiberglass or carbon fiber. The ferrules, which hold the portions of the rod together, are made of metal or fiberglass. The grips on fishing rods are usually made of cork, but are sometimes made of plastic, wood, or cloth. Reel seats are made of aluminum or other metals or plastic. Guides are made of chrome-plated brass or tungsten carbide and an alloy called nickel silver.

## The Manufacturing Process

Fishing rods consist of tubular sections, known as blanks, and various smaller components attached to them. Blanks are sometimes made of bamboo, but most blanks are made from strong, flexible fibers, such as fiberglass or carbon fiber.

### *Making bamboo blanks*

1 Bamboo is shipped to the fishing rod manufacturer in the form of hollow canes about 6-8 ft (1.8-2.4 m) long and about 1.25-2 in (3.2-5.1 cm) wide. The canes are split in half lengthwise with a heavy knife or a cutting machine. The interior of the split cane contains thin partitions spaced about 15-18 in (38.1-45.7 cm) apart. These natural partitions occur at the points where leaves emerge from the bamboo plant. They are removed with a chisel or cutting tool. The bamboo may then be heated briefly to harden it. Some makers of bamboo blanks soak the bamboo with a liquid plastic resin to strengthen it and to make it waterproof.

2 Precision cutting tools cut the bamboo into long strips. An accuracy of about one one-thousandth of an inch (0.03 mm) in the width of the strips is necessary to ensure that the strips will fit together properly. Six strips are usually used to make a hexagonal blank. Some manufacturers use five strips to make a pentagonal blank. The strips are glued together under heavy pressure. Strong thread is wrapped around the strips to maintain pressure at the points where they meet. The glue is allowed to dry, the string is removed, and the blank is lightly sanded to remove excess glue and to provide a smooth surface.

### *Making synthetic fiber blanks*

3 Fiberglass can be made from a variety of glasses. Sand (silicon dioxide) and limestone (calcium carbonate) are mixed with varying amounts of other ingredients, which may include sodium carbonate, potassium carbonate, aluminum hydroxide, aluminum oxide, magnesium oxide, or boric oxide. These ingredients are added to waste glass of the same type, known as cullet. The cullet acts as a flux, causing the other ingredients to melt together at a lower temperature than they would without it. The

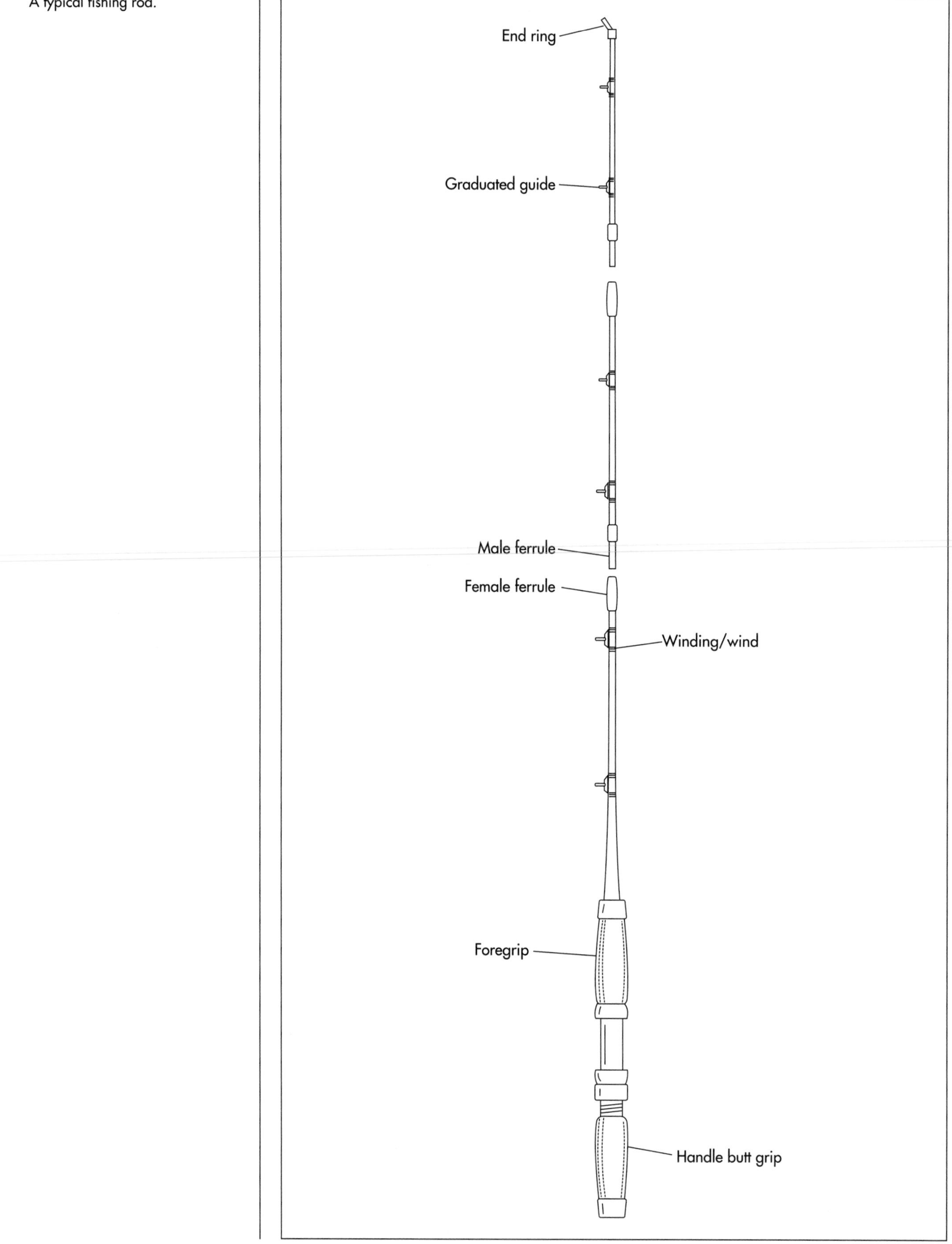


A typical fishing rod.

mixture is melted into a liquid in a furnace to form molten glass. The molten glass is then forced through a steel device containing numerous small holes known as spinnerets. The emerging glass cools into fibers, which are spun into yarn. The yarn is then woven into sheets.

4 Carbon fiber is derived from various synthetic fibers. These fibers are polymers, which consist of long chains of molecules which contain numerous carbon atoms. Carbon fiber can be made from rayon, a synthetic fiber derived from cellulose, a natural polymer found in plants. It can also be made from acrylic, a synthetic fiber derived from an artificial polymer of acrylonitrile molecules, which are obtained indirectly from petroleum. The synthetic fiber is heated, forcing out atoms other than carbon, resulting in long chains of carbon atoms. (If the fiber is heated too strongly, the carbon atoms will form sheets instead of chains, resulting in graphite.) The carbon fiber is spun into yarn, which is woven into sheets.

5 The fiberglass or carbon fiber sheet is dipped into a solution of liquid plastic resin, then squeezed between metal rollers to leave a controlled amount of resin in the sheet. The sheet is then heated to remove excess solvent and to partially harden the resin until it is slightly sticky. A metal template is laid on top of a stack of sheets. A sharp blade cuts around the template, producing several cut sheets of the same shape. The exact shape varies with the type of rod being made, but generally resembles a tapering rectangle.

6 One edge of the cut sheet is heated in order to attach it to a tapered steel rod known as a mandrel. The sheet is oriented so that the majority of the fibers line up along the length of the mandrel, with about one-tenth to one-sixth of the fibers at right angles to the rest. The mandrel is rolled between two heated metal rollers, known as platens, that apply pressure as layers of fiber are wrapped around the mandrel. A thin film of a synthetic polymer, such as cellophane or polyester, is wrapped around the layers of fiber.

7 The wrapped mandrel is heated in an oven to about 300-350° F (150-180° C) for about 30-60 minutes. The heat causes the polymer film to shrink, applying pressure to the fiber as the resin hardens. The mandrel is removed from the hardened fiber by using a pressurized ram to force it through a die. The polymer film is removed using a wire brush, a tumbler, high-pressure steam, splitting, or stripping. The blank is lightly sanded to remove excess resin and to provide a smooth surface. It is then coated with layers of various protective materials. The blank is buffed between each coating to give it a smooth finish.

### *Assembling the fishing rod*

8 Most fishing rods are made up of two or three blanks, allowing the rod to be disassembled for ease in storage and transportation. Usually the blanks are attached together with connectors known as ferrules. Ferrules are made from metal or fiberglass, and are attached to the ends of the blanks with strong cement.

9 Grips for the handles of fishing rods are made from natural cork, obtained from the outer bark of certain evergreen oak trees found in Mediterranean regions. They may also be made from synthetic foam rubber. The grip is attached to the end of the blank with epoxy glue.

10 Guides are small rings which are attached along the length of a fishing rod in order to control the line during casting. They also distribute the stress of the line evenly on the rod. The guides are made by cutting and bending wires of steel or chrome-plated brass. They made also be made from a combination of tungsten carbide and nickel silver. The guide is taped into position on the blank. Nylon thread is wrapped around the base of the guide to secure it in place. The wound thread is then coated with lacquer or varnish.

11 A reel seat is the part of a fishing rod to which a steel reel, containing nylon fishing line, can be attached. Reel seats are made from aluminum, chrome-plated brass, or plastic, then attached to blanks.

12 The fishing rod is packaged and shipped to retailers. The consumer assembles the fishing rod, attaches a reel, and threads the line from the reel through the guides.

Modern fishing rods are made using fiberglass or carbon fiber sheets. Coated with liquid plastic resin, the sheets are attached at one end of a steel rod called a mandrel. The mandrel is rolled between two heated metal rollers, known as platens, that apply pressure as layers of fiber are wrapped around the mandrel. The wrapped mandrel is heated, causing the resin to harden. Next, a pressurized ram removes the mandrel from the hardened fiber blank. The blank is lightly sanded to remove excess resin and to provide a smooth surface. It is then coated with layers of various protective materials. The blank is buffed between each coating to give it a smooth finish.

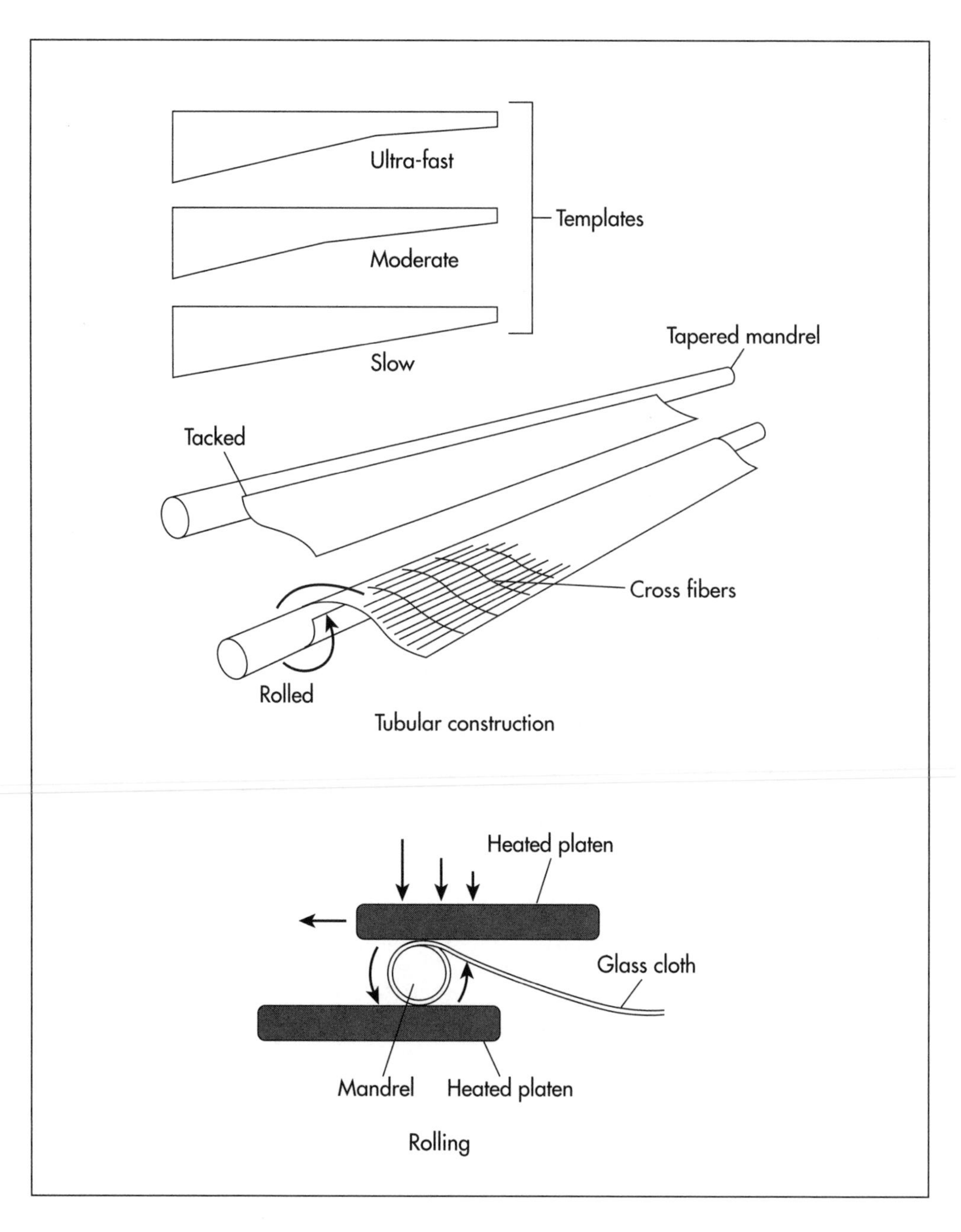


## Quality Control

Testing of a fishing rod begins soon after a new design is developed. A prototype of the new product is manufactured and used to catch fish in a variety of outdoor conditions. The design is altered as necessary, and the process is repeated until the new design meets its desired goals.

During the manufacture of the blank, the pressure applied to the fiber sheet as it is wrapped around the mandrel must be uniform, or the rod will be uneven. Protective finishings applied to the blank must be even and not too thick, or the rod will not function correctly.

During the assembly process, all the parts must fit together correctly. Ferrules must be lined up correctly and must have the proper dimensions to allow the fishing rod to be assembled easily without being too loose. Guides must be spaced the correct distance apart. If they are too far apart, the line will sag. If they are too close together, the line will not move smoothly.

## The Future

Manufacturers of fishing rods are constantly developing new products which enable consumers to cast lines farther and more accu-

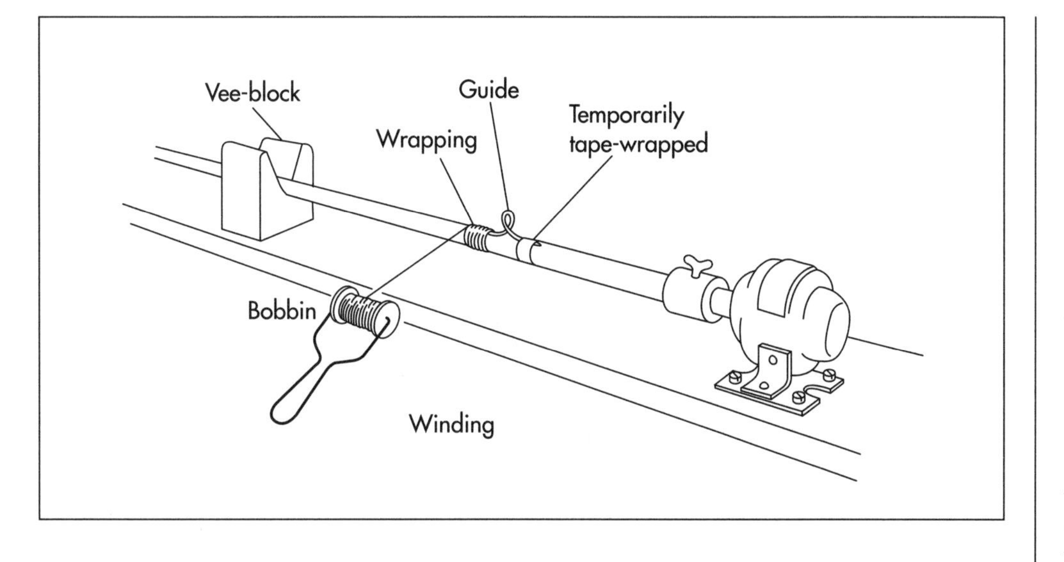


Most fishing rods are made up of two or three blanks, allowing the rod to be disassembled for ease in storage and transportation. Usually the blanks are attached together with connectors known as ferrules. Ferrules are made from metal or fiberglass, and are attached to the ends of the blanks with strong cement. Guides are small rings which are attached along the length of a fishing rod in order to control the line during casting. The guides are made by cutting and bending wires of steel or chrome-plated brass. Nylon thread is wrapped around the base of the guide to secure it in place. The wound thread is then coated with lacquer or varnish.

rately. The future is likely to see this trend continue. Fishing rods are also likely to become more and more specialized, with each rod designed to catch fish of a certain weight.

A recent development suggests that fishing rods may be very different in the near future. The Interline fishing rod, manufactured by Daiwa, has no guides. Instead, the line runs through the center of the rod and emerges from its tip. This revolutionary design avoids the common problems of broken guides and lines tangled in guides.

## Where to Learn More

### Periodicals

Gorant, Jim. "The Guideless Rod."*Popular Mechanics* (October 1997): 40-42.

Merwin, John, and Ken Schultz. "A Century of Piscatorial Progress."*Field and Stream* (October 1995): 38-41.

Pfeiffer, C. Boyd. "Fishing Rod Repair."*Outdoor Life* (June/July 1997): 128-129.

*—Rose Secrest*

# Flute

*Discovered in Slovenia during 1995, a fragment of a cave bear thigh bone containing two holes is believed by some scientists to be part of a flute used by Neanderthals more than 43,000 years ago.*

## Background

A flute is a musical instrument that produces sound when a stream of air is directed against the edge of a hole, causing the air within the body of the instrument to vibrate. Most flutes are tubular, but some are globular or other shapes. Some flutes are played by blowing air into a mouthpiece, which directs the air against the edge of a hole elsewhere in the flute. These instruments, known as whistle flutes, include the tubular recorder and the globular ocarina. Other flutes are played by blowing air directly against the edge of the hole.

Some flutes are held vertically and are played by blowing air against the edge of a hole in the end of the flute. These instruments include Japanese bamboo flutes and the panpipe. The panpipe, also known as the syrinx, consists of several vertical flutes of various sizes joined together.

Other flutes are held horizontally, and are played by blowing air against the edge of a hole in the side of the flute. These instruments, known as transverse flutes, include the modern flute used in orchestras.

## History

Flutes have existed since prehistoric times. A fragment of a cave bear thigh bone containing two holes, discovered in Slovenia in 1995, is believed by some scientists to be part of a flute used by Neanderthals more than 43,000 years ago. Flutes were used by the Sumerians and Egyptians thousands of years ago. Some ancient Egyptian flutes have survived, preserved in tombs by the arid desert climate. This Egyptian instrument was a vertical flute, about one yard (0.9 m) long and about 0.5 in (1.3 cm) wide, with between two to six finger holes. Modern versions of this flute are still used in the Middle East today.

The ancient Greeks used panpipes, probably indirectly influenced by more sophisticated Chinese versions. The transverse flute was used in Greece and Etruria by the second century B.C. and later appeared in India, China, and Japan. Flutes almost disappeared from Europe after the fall of the Roman Empire, until the Crusades brought Europeans into contact with the Arabs. Vertical flutes spread from the Middle East to Europe, and are still used in the Balkans and the Basque regions of Spain and France. Transverse flutes spread from the Byzantine Empire to Germany, then appeared in Spain and France by the fourteenth century.

During the Renaissance, transverse flutes consisted of wooden cylinders of various sizes, typically made of boxwood, with a cork stopper in one end and six finger holes. During the late seventeenth century, the Hotteterre family, noted French instrument makers, redesigned the transverse flute. Instead of a single cylinder, the flute consisted of a head joint, a body, and a foot joint. Modern flutes are still made in these three basic parts. The new flute also had a single key added, allowing more notes to be played. After 1720, the body was often divided into two parts of varying lengths, allowing the flute to be adjusted to play in various musical keys. By 1760, three more keys were added by London flutemakers, followed by two additional keys by 1780 and two more by 1800.

The transverse flute was completely redesigned in the middle of the nineteenth century by the German instrument maker Theobald Bohm. Bohm changed the position of the holes and increased their size. Because the new holes were impossible to reach with the fingers, new mechanisms were added to cover and uncover them as needed. The Bohm system is still used in modern transverse flutes.

## Raw Materials

Some modern flutes are made from wood that produces a different sound from metal flutes. These wooden flutes generally have metal keys and mechanisms.

Most flutes are made of metal. Less expensive flutes, intended for students, may be made from alloys of nickel and other metals. More expensive flutes may be plated with silver.

The pads attached to the surface of the keys in order to cover the holes are made of cork and felt. The springs that provide tension to hold the keys firmly against the holes may be made of steel, bronze, or gold. The pins and screws that hold the mechanism together are made of steel. The mouthpiece, containing the hole into which air is blown, may be made of the same metal as the rest of the flute, or it may be made of another metal, such as platinum.

## Design

Every flute is an individually crafted work of art. The flutemaker must consider the needs of the musician who will use the flute. Students need relatively inexpensive but reliable instruments. Professional musicians must have instruments of very high quality, often with special changes made in the keys to accommodate special needs.

The most individual portion of a flute is the head joint. Professional musicians often test several head joints before selecting one which produces the sound they prefer. Head joints are often manufactured to meet the special demands of individual musicians.

The material from which a flute is made greatly alters the sound which is produced. Wooden flutes produce a dark sound. Silver flutes produce a bright sound. The thickness of the metal used to make a flute also alters the sound it makes, as well as changing the weight of the instrument. All these factors may influence the design of a flute preferred by a particular musician.

A flute may be elaborately decorated. The physical appearance of a flute is an important consideration for professional musicians who perform in public. The most detailed designs are likely to found on the professional quality flutes. The process of forming these designs, known as chasing, requires the skill of an experienced artist, and makes the individual flute a truly unique instrument.

## The Manufacturing Process

### *Shaping the components*

A flute is made of hundreds of components, ranging from the relatively large body to tiny pins and screws. Although some of the small components are interchangeable and can be purchased from outside manufacturers, the vast majority of the components must be individually shaped for each flute.

1 Early flutes were made with hand-forged keys. The modern method is usually die casting. Molten metal is forced under pressure into steel dies. A group of connected keys may be made in one piece. Alternately, individual keys may be stamped out by a heavy stamping machine, and then trimmed.

### *Assembling the keys*

2 The components that will make up the keys are immersed in a flux solution, containing various chemicals that protect the metal and aid in the soldering process. The components are then soldered together into keys. They are also soldered to other components that will move the keys. The keys are then cleaned in a solution that removes grease and any excess flux. The clean keys are polished and inspected. Keys for inexpensive flutes may be placed in a tumbling machine, where friction and agitation of pellets in a revolving drum polish the metal. More expensive keys will often be buffed individually.

3 The keys are fitted with pads made from layers of cork and felt. Cork is cut to the proper size and attached to the surface of the

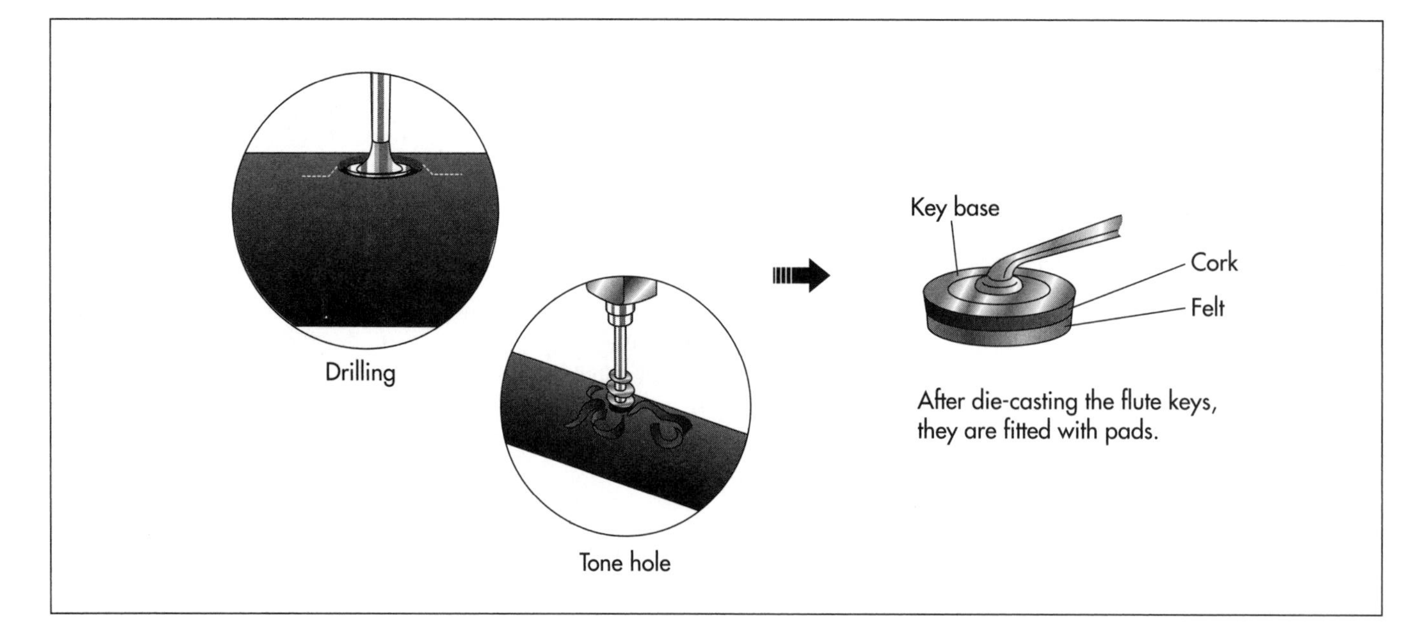

After die-casting the flute keys, they are fitted with pads.

Flutes are comprised of hundreds of components, ranging from the relatively large body to tiny pins and screws. The keys are die cast and fitted with pads made from layers of cork and felt. Tone holes are formed in the body of the flute by either pulling and rolling or by cutting and soldering. In the pulling and rolling method, the holes are drilled in the tube, and a machine pulls the metal from the edges of the hole and rolls it around the hole to form a raised ring. If the tone holes are to be cut and soldered, die cut metal rings are soldered to the drill holes.

key that will cover the tone hole. Felt is cut and attached to the cork, creating a protective covering.

### *Forming the tone holes*

Tone holes are formed in the body of the flute. They are formed by a process of pulling and rolling or by cutting and soldering. Either process requires great precision to ensure that the tone holes are located in the correct position and are of the correct size.

4 In the pulling and rolling method, the holes are drilled in the tube. Next, a special machine pulls the metal from the edges of the hole and rolls it around the hole to form a raised ring. The tone hole is then smoothed.

5 If the tone holes are to be cut and soldered, metal rings are die cut from sheet metal. The rings are then cleaned and polished. Holes are drilled in the tube and the rings are soldered around the holes to form the tone holes.

### *Mounting the keys*

6 Rods that support the keys are soldered to the body of the flute. Next, the keys are attached to the rods with pins and screws. Springs are attached to provide tension to hold the cork and felt pads firmly against the tone holes until they are lifted when the keys are operated by the musician.

### *Finishing*

7 The mouthpiece is shaped and soldered to the head joint. The head joint, body, and foot joint are fitted together and adjusted. The musician must be able to assemble and disassemble the flute easily, but the fittings should be tight. The flute is tested for sound quality. It is then disassembled, cleaned, polished, and packed into a special protective case.

## Quality Control

Constant inspection of each part of the flute during the manufacturing process is critical to ensure that the instrument will produce the proper sound. As the instrument moves from one position on the assembly line to the next, workmanship is reviewed.

The exact size, shape, and position of the keys and tone holes must be accurate to ensure that they will fit together correctly. The completed instrument is played by an experienced musician to ensure that it produces sound correctly. Because professional musicians often make special demands of flutes, flutemakers will often make small adjustments in flutes to satisfy them.

Much of the responsibility for maintaining the quality of a flute rests with the musician. Routine maintenance often prevents flaws from developing . Each time the flute is assembled, the connecting surfaces of the

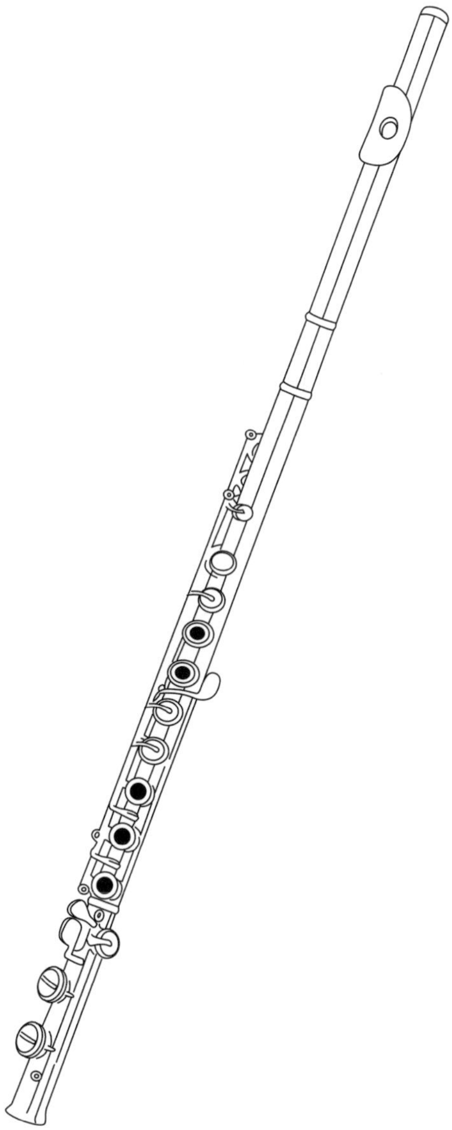

Most flutes are made of metal. Less expensive flutes, intended for students, may be made from alloys of nickel and other metals. More expensive flutes may be plated with silver. All flutes are individually assembled and play tested prior to sale.

Assembly

joints and body should be cleaned to prevent wear caused by dirt and corrosion. The interior of the flute should be swabbed each time it is played to remove moisture, which could cause the pads to swell so that they no longer fit the tone holes. Careful lubrication of the keys with a special lubricant is necessary about every three to six months in order to keep them working smoothly.

## The Future

Very few changes have been made in the basic design of the modern transverse flute since the middle of the nineteenth century. Flutemakers will continue to find ways to make small but critical changes in individual instruments to fit the needs of individual musicians.

Two seemingly opposite trends hint at the future of flutemaking. Many performers of music from the Renaissance, Baroque, and Classical periods prefer to use flutes that resemble the instruments used during those times. Such instruments are believed to be more suited to older music than modern flutes, which developed during the Romantic period. On the other hand, many performers of jazz, rock, and experimental music use electronic devices to alter the sounds of flutes in new ways. Despite these two trends, the instrument originally designed by Theobald Bohm is likely to dominate flutemaking for many years to come.

## Where to Learn More

### Books

Meylan, Raymond. *The Flute*. Amadeus Press, 1987.

### Periodicals

Wong, Kate. "Neanderthal Notes."*Scientific American* (September 1997): 28-29.

*—Rose Secrest*

# Foam Rubber

## Background

Foam rubber is found in a wide range of applications, from cushioning in automobile seats and furniture to insulation in walls and appliances to soles and heels in footwear. Foams are made by forming gas bubbles in a plastic mixture, with the use of a blowing agent. Foam manufacture is either a continuous process for making laminate or slabstock or a batch process for making various shapes by cutting or molding.

There are two basic types of foam. Flexible foams have an open cell structure and can be produced in both high and low densities. Applications include cushioning for furniture and automobiles, mattresses and pillows, automotive trim, and shoe soling. Rigid foams are highly cross-linked polymers with a closed cell structure that prevents gas movement. Their main application is as insulation for buildings, refrigerators and freezers and refrigerated transport vehicles.

## History

Originally, foam rubber was made from natural latex, a white sap produced from rubber trees. As early as 500B.C., Mayans and Aztecs used this latex for waterproofing purposes and also heated it to make toy balls. During the early 1900s, the first patent for synthetic rubber was granted and several decades later a process for foaming latex was invented. Another process was developed in 1937 for making foams from isocyanate-based materials. After World War II, styrene-butadiene rubber replaced the natural foam. Today, polyurethane is the most commonly used material for foam products. Foamed polyurethanes currently make up 90% by weight of the total market for polyurethanes.

Consumption of polyurethane in the United States during 1997 was estimated at around 4.8 billion lb (2.18 billion kg), up 13% over 1996 and representing about a third of global consumption. Canada consumed 460 million lb (209 million kg). The construction, transportation, furniture, and carpet industries are the largest users of polyurethane, with construction and transportation leading at 27% and 21%, respectively. Flexible foam is the largest end market, accounting for 44% of the total volume in the United States and 66% globally. Of the volume in the United States, slab materials accounted for 78% and molded products 22%. Rigid foam is the second-largest end product, accounting for 28% of the market in the United States and 25% globally.

*Consumption of polyurethane in the United States during 1997 was estimated at around 4.8 billion lb (2.18 billion kg), up 13% over 1996 and representing about a third of global consumption.*

## Design

The molecular structure, amount, and reaction temperature of each ingredient determine the characteristics and subsequent use of the foam. Therefore, each formulation must be designed with the proper ingredients to achieve the desired properties of the final material. For instance, a switch in blowing agent may require an increase in this additive to maintain thermal properties. Increasing the amount of blowing agent requires more water and a switch in surfactants to maintain optimum bubble sizes and formation rates during foaming. The density of the foam is determined by the amount of blowing. The stiffness and hardness of polyurethane can also be tailored by changing the level of flexible polyol in the chemical formulation. By mixing different combinations of the starting materials, the rates of the reactions and overall rate of cure during processing can be controlled.

## Raw Materials

Most foams consist of the following chemicals: 50% polyol, 40% polyisocyanates, and 10% water and other chemicals. Polyisocyanates and polyols are liquid polymers that, when combined with water, produce an exothermic (heat generating) reaction forming the polyurethane. The two polyisocyanates most commonly used are diphenylethane diisocyanate (MDI) and toluene diisocyanate (TDI). Both are derived from readily available petrochemicals and are manufactured by well-established chemical processes. Though MDI is chemically more complex than TDI, this complexity allows its composition to be tailored for each specific application. MDI is generally used in rigid foams, whereas TDI is typically used for flexible foam applications. Blends of MDI and TDI are also used.

Polyols are active hydrogen monomers based on polyesters, polyethers, or hydrocarbon materials that contain at least two active hydrogen atoms. The type of polyol used will determine whether the foam produced is flexible or rigid. Since most polyols immediately react with isocyanates when added together, it is easy to combine the polymerization and shaping processes into one step. During the polymerization proccess, the polyol and polyisocyanate molecules link and interconnect together to form a three dimensional material.

A wide range of additives are also used. Catalysts (tin and amines) speed up the reaction, allowing large volume production runs. Blowing agents that form gas bubbles in the polymerizing mixture, are required to produce foam. The amount of blowing can be tailored by adjusting the level of water. Flexible foams are typically made using the carbon dioxide formed during the reaction of water with isocyanate. Rigid foams use hydrochlorofluorocarbons (HCFCs), hydrofluorocarbons (HfCs), and pentanes as the blowing agents.

Surfactants are used for controlling the size of bubbles and include silicones, polyethers, and similar materials. Other additives that may be used include cross-linking agents, chain-extending agents, fillers, flame retardants and coloring materials, depending on the application.

## The Manufacturing Process

Polymerization rates for most industrial polyurethanes range from several seconds to about five minutes. Slower reacting formulations can be mixed and molded by hand, but require long cycle times. Faster systems provide shorter cycle times but must use machines for mixing. Polyurethane formulations are generally processed into a wide range of products by reactive molding, spraying, or open-pouring techniques.

### Material preparation

1 The liquid chemicals are delivered either by railroad tanker cars or tank trucks and pumped into large holding tanks. From there, the chemicals are pumped into smaller heated mixing tanks, and are kept separate if they react with each other. For continuous manufacture of foam such as slabstock, more than two monomer streams are typically used.

### Dispensing and mixing

2 Continuous dispensing (also called open pouring or free-rise) is used in the production of rigid and flexible low-density foams. A specific amount of each chemical, measured by metered pumps, is fed from the mixing tanks into a mixing head, where the blending of the chemicals take place. The reactive components are poured onto a moving surface or conveyor belt, where the foam rises and cures to form slabstock.

### Cutting and curing

3 As the foam moves toward the end of the conveyor belt, it is automatically cut by a horizontal bandsaw into smaller pieces, usually 12 ft (3.66 m) long sections. After cutting, the foam sections are cured at room temperature for 12 hours or more. They are not stacked since they are not firm enough to withstand any weight. After curing, a second automatic bandsaw cuts the sections into the desired thickness. Other shapes can also be cut.

### Other forming processes

4 A continuous lamination process is used to form rigid foam laminate insulation panels known as boardstock. For appliance insulation, liquid chemicals are injected be-

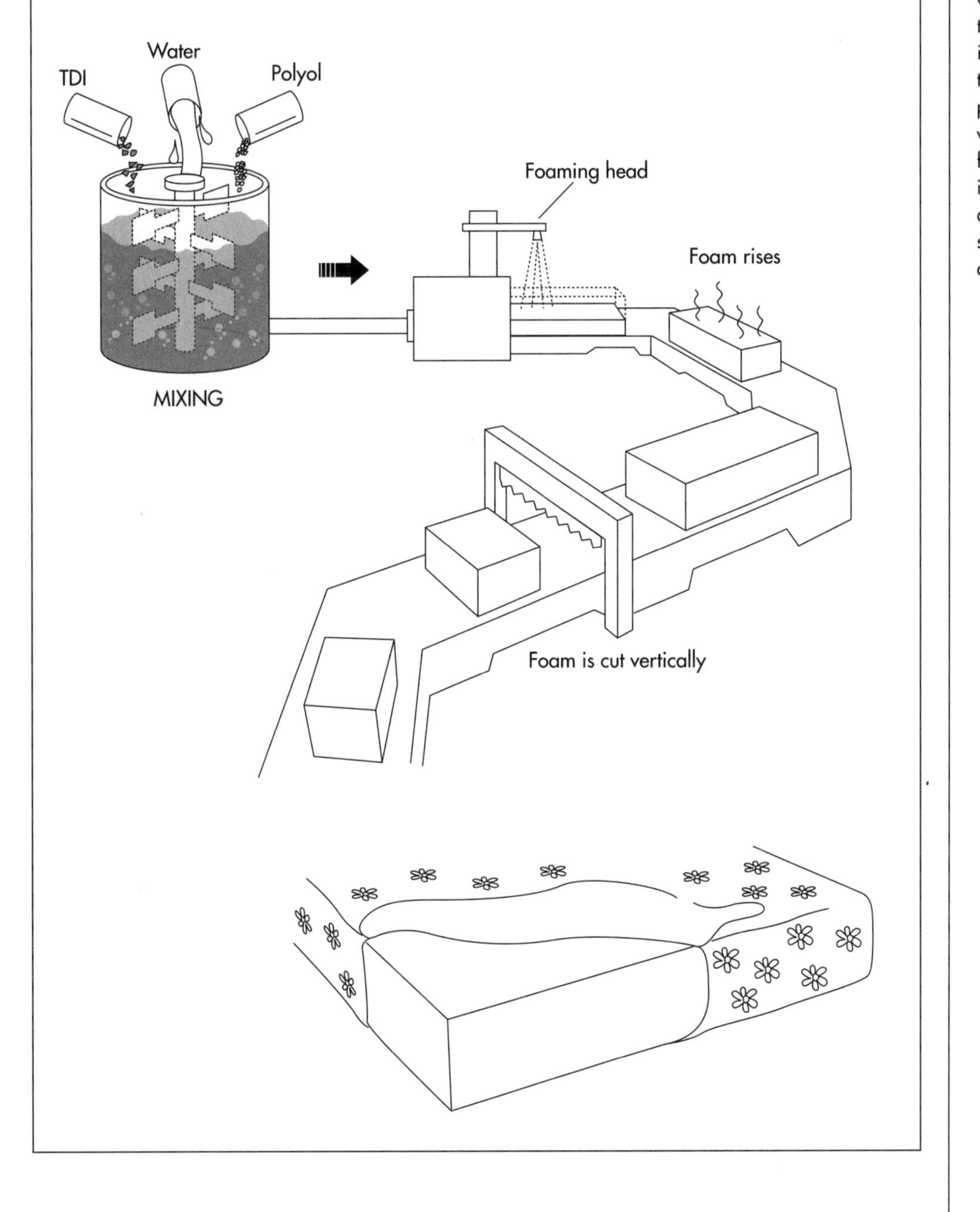


Chemicals are dispensed in a continuous fashion called open pouring or free-rise. Once blended, the reactive components are poured onto a conveyor belt, where the foam rises and cures to form slabstock. Next, the slabstock is conveyored through a series of automatic bandsaws that cut the slabstock to premeasured widths and thicknesses.

tween the inner and outer walls of the appliance cabinet, where they undergo the foaming process.

5 For flexible foam molding, dispensing machines are used to vary the output of chemicals or the component ratios during the pour. This allows the production of molded foams with dual hardness. Molded foam articles with solid surface skin are made from liquid chemicals in a single step, using carbon dioxide as the sole blowing agent. Automotive cushions are made by molding flexible foam behind a pre-shaped fabric cover. This process reduces the number of steps in the manufacture of car seats.

## Quality Control

In addition to monitoring the production process, the final product is inspected and tested for various physical and mechanical properties. An important property of foam rubber is called the indentation load deflection, which measures the spring tension or load-bearing quality of the material. The deflection is determined by the amount of weight needed to compress a circle of foam

50 sq in x 4 in (127 sq cm x 10.16 cm) thick by 25%. Most seat cushion foam has a rating of 35, which means it takes 35 lb (15.9 kg) of pressure to compress the circle of foam 1 in (2.54 cm).

## Byproducts/Waste

Since there is such a wide range of polyurethane chemistries, it is difficult to recycle foam materials using just one method. The majority of recycling involves reusing flexible slabstock foams for carpet backing. Over 100 million lb (45.4 million kg) of polyurethane carpet cushion is recycled every year into bonded carpet cushion. This involves shredding the scrap into flakes and bonding the flakes together to form sheets.

Other methods have recently been developed. One method involves pulverizing the foam into granules, dispersing these granules in a polyol blend, and molding them into the same parts as the original. Ground polyurethane can also be added to original systems as a filler in amounts of up to 10%. Another method called press bonding bonds granulated scrap using an isocyanate-based binder into large boards with densities ranging from 400-900 kg/$m^3$.

One manufacturer has patented a process that breaks down polyurethane into polyol using industrial waste or scrapped automotive parts; glycolysis processes are being used that generate polyol for use in automotive seating foam and cushions for furniture and bedding. In this process, polyurethane is pulverized into powder and then is heated in a reactor at 392° F (200° C) with glycol, which converts it into a raw material. During the reaction, chemical cleavage of the polymer chains (urethane) takes place.

## The Future

The foam rubber industry will continue to investigate methods for reducing and recycling scrap. One manufacturer is investigating a process for directly producing molded foam parts from shredded flexible polyurethane foam trimmings and production scrap.

The industry faces other environmental challenges that will require the development of new blowing agents. Hydrochlorofluorocarbons are scheduled to be phased out in the United States under the Montreal Protocol by the year 2003. Several international efforts may eliminate or even ban HCFCs before this date.

The market for flexible foam will continue to expand into nontraditional areas, such as shock absorption, acoustical applications, and toys. This will require improvements in material properties and processes, as well as the development of new chemicals to meet the demands of new applications for other types of foams. A new catalyst has already been developed for semi-rigid foam systems, which lowers densities and fill weights, as well as reduces demold times.

## Where To Learn More

### Books

*Modern Plastics Encyclopedia.* New York: McGraw-Hill, 1999.

### Periodicals

Myers, John. "PUR Recycling Advances on the Economic and Technical Front."*Modern Plastics* (October 1996): 2.

Pryweller, Joseph. "Recycler Seeks Revival for Polyurethane."*Automotive News* (October 27, 1997): 28.

Westervelt, Robert. "Survey Reveals Pop in Polyurethane Markets."*Chemical Week* (August 6, 1997): 33.

### Other

BASF Group. http://www.basf.com/ (February 4, 1999).

Foamex International. 1000 Columbia Ave. Linwood, PA 19061. (800) 776-3626. http://www.foamex.com/ (February 4, 1999).

Imperial Chemical Industries PLC. http://www.ici.com/ (February 4, 1999).

Polyurethane Foam Association. http://www.pfa.org/ (February 4, 1999).

*—Laurel Sheppard*

# Frisbee

## Background

Nearly 300 million frisbees have been sold since their introduction 40 years ago, for both organized sports and recreational play. According to Mattel, 90% of Americans have played with this flying toy at one time or another, translating to 15 million people enjoying the sport every year. Now, older versions of this toy have become collectors' items worth hundreds of dollars or more.

The frisbee's origins actually go back to a bakery called the Frisbie Pie Company of New Haven, Connecticut, established by William Russell Frisbie after the Civil War. The bakery stayed in operation until 1958, and during this period, the tossing of the company's pie tins, first by company drivers and later by Ivy League college students (some say it was cookie tin lids), led to frisbie becoming a well known term describing flying disc play in the Northeast.

Several years after World War II, Walter Frederick Morrison—the son of the inventor of the automobile sealed-beam headlight—and his partner Warren Franscioni, investigated perfecting the pie tin into a commercial product. First, they welded a steel ring inside the rim to improve the plate's stability, but without success. Then, they switched to plastic and the frisbee as we know it today was born.

The initial design, which incorporated six curved spoilers or vanes on the top, was vastly improved in 1951 and thus became the Pluto Platter, the first mass-produced flying disc. This design, which incorporates a slope on the outer third of the disc, has remained part of the basic design to this day. The Morrison Pluto Platter had the first true cupola (cabin in Morrison's terms) and resembled the concept of flying saucers (UFOs) depicted during this period complete with portholes. In 1954, Dartmouth University held the first frisbee tournament, involving a game called Guts.

The founders of Wham-O, a California toy company, became interested in this flying disc in 1955 and about a year later began production after acquiring the rights from Morrison. The name was changed to frisbee after the company heard about the pie tin game on the east coast called Frisbie-ing. (Wham-O first marketed the Pluto Platter in January of 1957, but didn't add the word frisbee until July 1957.) In 1959, the first professional model frisbee was produced.

It wasn't until the early 1960s when frisbees became the rage and soon organizations became established to promote sporting events, including the International Frisbee Association and the Olympic Frisbee Federation. The first game of Ultimate Frisbee, a sport of both distance and accuracy similar to football, was played in 1968 at a New Jersey high school. Now, it is played at nearly 600 colleges and in 32 countries. In 1969, the U.S. Army even invested $400,000 to see if flares placed on frisbees would stay aloft but without success.

During the 1970s, several organizations were formed to promote specific events, including disc golf, freestyle, and Guts. The Professional Disc Golf Association (PDGA) now has over 14,000 members in 20 countries playing on over 700 frisbee golf courses. Today, 40,000 athletes in 35 countries compete in Ultimate Frisbee. The formation of such associations led to world championships being held during the 1980s.

*The frisbee's origins actually go back to a bakery called the Frisbie Pie Company of New Haven, Connecticut, established by William Russell Frisbie after the Civil War. The bakery stayed in operation until 1958, and during this period, the tossing of the company's pie tins, first by company drivers and later by Ivy League college students (some say it was cookie tin lids), led to frisbie becoming a well known term describing flying disc play in the Northeast.*

Today, organized competitions in nine different events (including disc golf and freestyle) take place each year around the world, under the auspices of the World Flying Disc Federation (WFDF). Established in 1984, WFDF has member associations in 22 countries and provisional members in an additional 28 countries. During Operation Desert Shield in 1991, frisbee was used to boost the morale of the 20,000 U.S. soldiers on duty in Saudi Arabia.

## Design

Manufacturers of frisbees use computer aided design software to create a model. A prototype is then made to test the design. Sometimes, a wind tunnel and other sophisticated methods are used to test flying characteristics, depending on the type of frisbee. Manufacturing tolerances within a few thousandths of an inch are now incorporated into the design.

Designers are always looking for new ways to manipulate the physical properties that dictate flight characteristics by changing the design or shape in order to improve lift, drag, spin, angular momentum, torque and other forces that affect how an object flies. For instance, adding a small lip and concave edge to a disc greatly increases its stability in flight.

A major obstacle disk designers must overcome is this instability caused by gyroscopic precession, the tendency of spinning objects to roll right or left in flight, depending on the direction of their spin and where they get their aerodynamic lift. The closer the disk's center of gravity remains to its center of lift, the more stable and straight the flight.

As a spinning disk flies, its center of lift is near the front, or leading edge, of the disc and tends to pitch the disc upward. Because of the spin, much of the lifting force on a point near the disc's edge does not exert itself until about a quarter of a revolution later. Such gyroscopic precession pushes the disc up on the side, causing a sideways roll. This is why frisbees, which typically are thrown backhand to spin clockwise as viewed from the top, tend to roll left from the thrower's perspective in flight.

Once in the air, lift and angular momentum act on the frisbee, giving it a ballet-type performance. Lift is generated by the frisbee's shaped surfaces as it passes through the air. Maintaining a positive angle of attack, the air moving over the top of the frisbee flows faster than the air moving underneath it.

Under the Bernoulli Principle, there is then a lower air pressure on top of the frisbee than beneath it. The difference in pressure causes the frisbee to rise or lift. This is the same principle that allows planes to take off, fly, and land. Another significant factor acting upon the frisbee's lift is Newton's Third Law. It states that for every action there is an equal and opposite reaction. The frisbee forces air down (action) and the air forces the frisbee upward (reaction). The air is deflected downward by the frisbee's tilt, or angle of attack.

Spinning the frisbee when it is thrown, or giving it angular momentum, provides it with stability. Angular momentum is a property of any spinning mass. Throwing a frisbee without any spin allows it to tumble to the ground. The momentum of the spin also gives it orientational stability, allowing the frisbee to receive a steady lift from the air as it passes through it. The faster the frisbee spins, the greater its stability.

## Raw Materials

Frisbees have been made out of a thermoplastic material called polyethylene since the early 1950s. Polyethylene is the largest volume polymer consumed in the world. This material is derived from ethylene, a colorless, flammable gas. This gas is subjected to elevated temperatures and pressures in the presence of a catalyst, which converts the gas into a polymer. Other ingredients that may be added include colorants, lubricants, and chemicals to improve dimensional stability and crack resistance.

## The Manufacturing Process

To make a frisbee, a high-speed process called injection molding is used, which is based on the injection of a fluid plastic material into a closed mold, usually of the multi-cavity type. Once in the mold, the plastic is cooled to a shape reflecting the

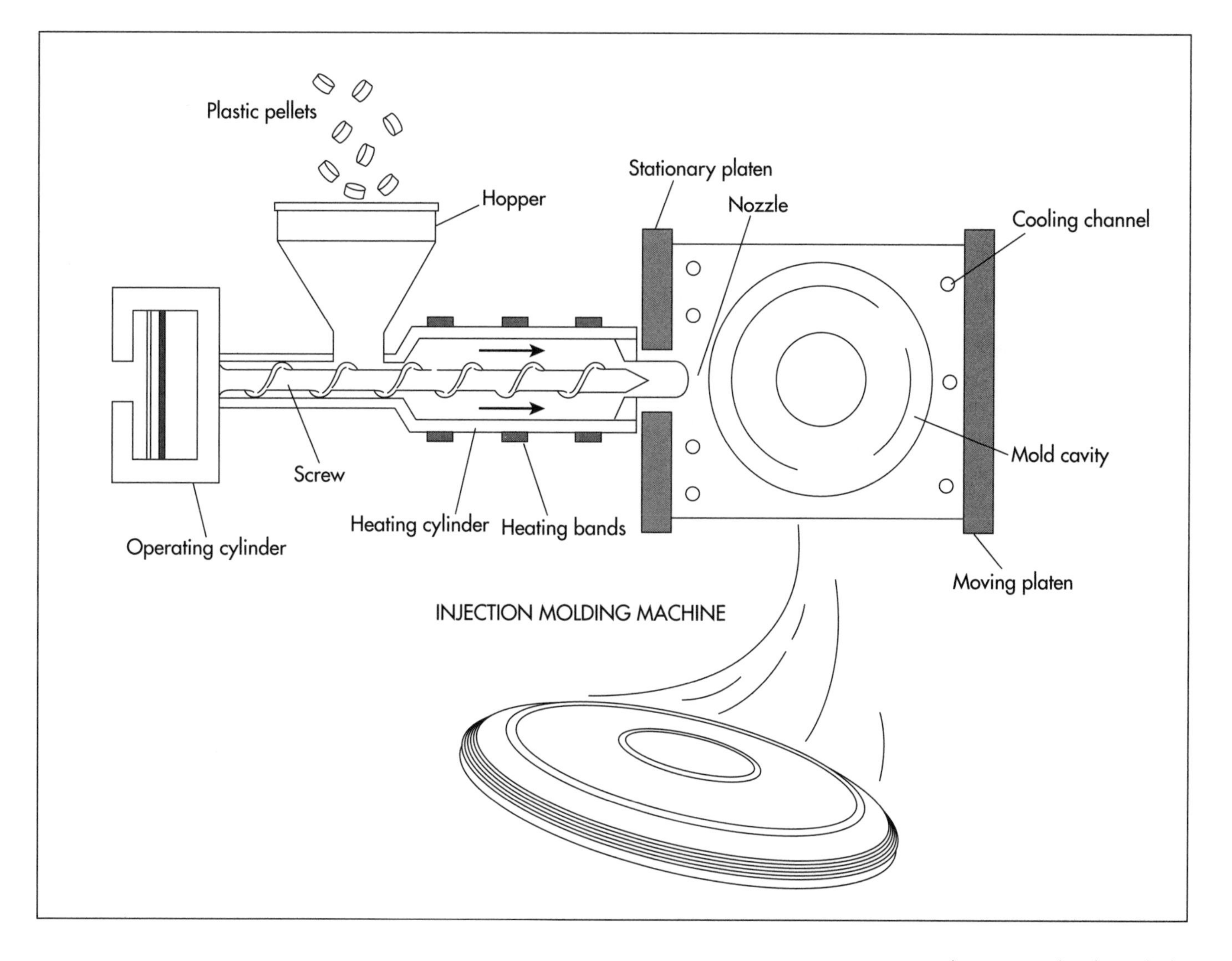


Frisbees are produced in a high-speed process called injection molding, which is based on the injection of a fluid plastic material into a closed mold, usually of the multi-cavity type. Once in the mold, the plastic is cooled to a shape reflecting the cavity.

cavity. Since complex shapes can be made using this process, minor trimming after removal from the mold is the only other finishing required. Because of these advantages, this process is used to manufacture a wide range of products, from various toys to automotive bumpers.

A molding machine equipped with a hopper, heated extruder barrel, reciprocating screw, nozzle, and mold clamp is typically used. This machine heats the plastic until it is able to flow readily under pressure, pressurizes this melt to inject it into a closed mold, holds the mold closed both during injection and solidification of the material, and opens the mold to allow removal of the solid part.

### *Raw material preparation*

1 A separate compounding operation is used to convert the form of a resin, while also introducing any additives, into one suitable for injection molding. For thermoplastics, this usually means forming the material into dry, free-flowing pellets by a combination of extrusion and drying steps, that usually take place at the plastic manufacturer. Once the compounding step is completed, the material is checked for moisture content and average molecular weight.

2 The plastic pellets are then shipped to the frisbee manufacturer in 50-100 lb (22.7- 45.4 kg) bags or in a 500-1,000 lb (227-454 kg) box. The plastic pellets are put into a large drum, to which pigments and weight-enhancing additives are added. A bonding agent such as oil is used to adhere the pigment to the pellets. The drum is vibrated to blend the materials together.

### *Feeding the material*

3 The unmelted pellets are placed into the hopper that feeds the material into the

barrel. The hopper may be equipped with a drying system to remove extra moisture and a magnet, to remove any iron contaminants. Sometimes, all or some of the compounding operation takes place during this step.

### Transporting and melting

4 During this step, the molding compound moves from the feed section into the heated extruder barrel that contains the reciprocal screw. Here, the material is gradually melted as it is conveyed down the barrel to the front. Tapered screws provide pumping, compression, decompression, and mixing, resulting in a pressure gradient that degasses the material with the help of vents in the screw. Thermoplastic materials require longer screws of compression ratios (open volume in the feed zone divided by the volume at the end of the screw) in the range of two to three or more. The proper compression ratio for an injection molding screw will be slightly greater than the bulk density of the as-received molding compound divided by the density of the melt.

5 At the end of the transport process, a volumetrically predetermined amount of compound accumulates somewhere in front of the screw as the screw slides out of the barrel. Called a shot, it is forced forward when the travel of the screw is reversed. To prevent the molding compound from flowing back down the screw during injection, a mechanical valve is placed at the tip of the screw.

### Injection

6 Once there is enough material to fill the mold, the screw rotation stops and the machine is ready for injection. The melted plastic is injected into the mold through the nozzle, under high pressure (typically, 10,000-30,000 psi) using a system of runners leading to the gate of the mold. Usually more than one gate is used to deliver the material into the mold, with each gate being fed by a channel or runner. For proper injection, the air within the mold must also be adequately vented.

### Cooling and removal

7 To speed up the solidification process, a cooling method is incorporated into the mold. This usually involves boring holes into the mold, through which a cooling fluid such as water can be circulated. After solidification, the mold clamp, which holds the halves of the mold closed against the injection pressure of the melt, opens the mold to allow retractable arms to remove the parts. During this step, the screw begins to rotate and melt new material for the next shot.

### Decorating

8 After the frisbee is removed from the mold, an imprint or decoration is applied with the name of the frisbee, manufacturer and other information or designs. There are three methods used, with hot stamping the most common. Other methods involve applying ink by using a silk screen or letter press machine. The letter press uses a different ink pad for each color.

### Packaging

9 After decorating, the frisbees are ready for packaging. One method involves placing each frisbee into a plastic bag to which a cardboard header is automatically attached with staples. Another method places the frisbee onto a cardboard backing, which is then shrink wrapped with a plastic film. Sometimes the frisbees are packaged in boxes.

## Quality Control

The weight is one of the most important properties and is automatically controlled during the injection molding process by the screw on the machine that pushes the proper amount of material into the mold. Another important property for golf frisbees is flexibility, which is tested by putting a weight on the edge of the disc and measuring the degree it bends. The PDGA only needs to test one sample of a new golf disc for flexibility.

## Byproducts/Waste

There is usually little waste produced during the manufacturing process. Sometimes after the frisbee is removed from the mold, it has excess material, called flash, around the edge. This is trimmed off, sent to a regrinding machine and then mixed with the virgin material. Other frisbees may be rejected if they are not the correct weight or have decorating defects. After the decoration is removed, by either cutting or by using a solvent depending on the application method, the rejects are also recycled in a similar way.

## The Future

The frisbee is expected to dominate the twenty-first century as one of the great sports and pastimes. Frisbee sports should continue to grow, as well as collector groups interested in preserving its history. Though other flying toys have come onto the market—such as boomerangs, cylinders, and rings—the flying disc will continue to provide more hours of entertainment to people around the globe than probably anything else ever invented.

Disc golf specifically is experiencing record growth. In each of the last three years the number of courses in the United States alone has increased by 10%. Such growth should continue since there are still many areas of the country that are not yet aware of disc golf and the benefits it offers.

Also, frisbee may some day be an Olympic sport. For a sport to be eligible for the Olympics, it must be played for two years in at least 50 countries. Ultimate Frisbee is already played in 35 countries and its popularity is growing, along with other frisbee sports.

## Where to Learn More

### Books

Horowitz, Judy and Billy Bloom. *Frisbee, More Than Just A Game of Catch.* Leisure Press, 1983.

Johnson, Dr. Stancil, ed. *Frisbee. A Practitioner's Manual and Definitive Treatise.* New York: Workman Publishing Company, 1975.

Malafronte, Victor. *The Complete Book of Frisbee: The History of the Sport & the First Official Price Guide.* American Trends Publishing Co., 1998.

Tips, Charles. *Frisbee by the Masters.* Celestial Arts, 1977.

### Periodicals

Leary, Warren E., Lift. "Drag, Spin and Torque: Sending Toys Aloft."*New York Times* (June 20, 1995).

Weismantel, Rick. "Part geometries, quality factors, and cycle times affect equipment choices."*Modern Plastics* (Mid-November 1994): D-74-D-79.

### Other

Innova-Champion Discs, Inc. 11090 Tacoma Drive, Rancho Cucamonga, CA 91730. (909) 481-6266. http://www.innovadiscs.com/.

Professional Disc Golf Association. http://www.pdga.com/.

Wham-O Inc. 3830 Del Amo Blvd., Suite 101, Torrance, CA. (415) 357-4200.

World Flying Disc Federation. Http://www.wfdf.org/ .

*—Laurel Sheppard*

# Frozen Vegetable

*The modern frozen food industry was made possible by Clarence Birdseye's invention of quick freezing in 1921.*

## Background

Frozen foods are ubiquitous in American supermarkets, and are increasingly a part of the food industry worldwide. Fruits and vegetables are usually frozen within hours of being picked, and when thawed, they are very close to fresh in taste and texture. The frozen meal is increasingly popular in time-starved American households. If the meal can be heated in a microwave, total time from freezer to table can be less than five minutes. Besides offering fresh taste and convenience, freezing is also a safe method of preservation, as most pathogens are inactivated at low temperatures.

The frozen food industry dates back to the early years of the twentieth century, when some foods were preserved by the so-called cold-pack method. Food handlers would wash and sort fruits or vegetables, then pack them in large containers holding from 30-400 lb (14-180 kg). The large containers were placed in a cold storage room for several days until the mass was frozen solid. Cold pack foods did not have the quality of modern frozen foods because of the time it took for the food to freeze. In slow freezing, the water in the food crystallizes, forming large needles of ice. These shards of ice destroy cells walls, and so when the food is thawed, it has deteriorated in taste and texture. The innovation that made the current frozen food industry possible was the invention of quick freezing by Clarence Birdseye.

Birdseye was born in Brooklyn, New York, in 1886 and studied biology at Amherst College before drifting to the Canadian Arctic to work as a fur trader and trapper. Living with his family in a remote Labrador cottage, Birdseye became fascinated with freezing food, and he experimented with many kinds of meats and vegetables. Birdseye noted that freshly caught fish that froze in seconds in the sub-zero arctic air tasted perfectly palatable when later thawed and cooked. He experimented with quick-freezing other foods, including fruits and vegetables, and soon became convinced that he had a viable commerical venture. Birdseye returned from Canada in 1917 and devoted himself to inventing a mechanical freezing device. He won his first patent in 1921, and established a frozen fish company in New York in 1923.

His first frozen food business failed to spark interest, and a second company he founded in Gloucester, Massachusetts also withered. Yet Birdseye continued to develop new freezing technology, decreasing the time it took to freeze foods. In 1929, the General Foods Corporation bought out Birdseye's enterprise, paying an enormous sum for his patents. General Foods made an intensive marketing push, installing freezers in grocery stores and developing freezer rail cars for long-distance distribution. Though American families still scoffed at frozen foods, the company began to make inroads with commercial food preparers such as hospitals and schools. American soldiers ate frozen foods during World War II, and after the war, the industry took off. Home freezers grew larger, and more and more items, from vegetables to pizzas to entire meals, became available in grocery stores.

Birdseye first froze fish and vegetables by immersing them in a circulating brine cooled to about -45° F (-42.8° C). Later he developed a so-called belt froster. This passed packages of food between two subzero metal surfaces, and so cooled top and bottom at the same time. This greatly decreased the time it took to freeze foods. Another innovation attributed to a General Foods scientist was the process of blanching vegetables before freezing. Blanching entailed immersing the vegetables in boiling water for a few minutes to halt the activity of certain enzymes. This preserved their flavor much more effectively. Current methods of freezing typically use the air blast method, where ultra-cooled air is blown on the food in a narrow tunnel, or by the indirect method, where the food is passed along metal plates cooled by a refrigerated liquid. Food can also be frozen cryogenically. In this method, the outer layers of the food are taken to far below their actual freezing point by passing quickly through a tunnel cooled by liquid nitrogen to as low as -80- -120° F (-62.2- -84.4° C). After the food exits the cryogenic tunnel, heat from the core of the food permeates to the outside, resulting in a final stable frozen state. Some products also use an immersion method. For foods with a viscous sauce or sticky surface, the surface might be immersed in an ultra-cooled liquid for only a few seconds, and then the food can be frozen by air-blasting.

Optimal freezing methods vary considerably with each food product. And not every food freezes well. Certain varieties of peas or strawberries for example have been found to freeze best. This might be because of their firm texture or specific sugar content. So farmers will grow these special varieties under contract with a frozen food company.

## Raw Materials

The raw materials for frozen foods include whatever is to be frozen, e.g. fish, chicken, green beans, pizza. In most cases, the food is specifically cultivated or adapted for freezing. In the case of frozen desserts such as cakes and pies or entrees such as meat loaf and gravy, the recipe must be tested and altered so that it freezes well. Large companies will order optimum ingredients according to standards they have established during their product testing. For example the noodles used in a frozen pasta entrée may be ordered in bulk from a distributor that makes them in a certain specified width or viscosity or flour content according to the precise need of the frozen food manufacturer. In this respect, ingredients in a frozen meal may differ from what a home cook would buy at the supermarket. But in general, frozen foods do not require a host of extra ingredients such as preservatives. Added ingredients are most frequently thickeners and stabilizers such as starch, xanthan gum, and carrageenan. These help retain the desired texture of the food after thawing. Recipes for foods destined for the freezer may also do better with the addition of a sauce or glaze, because this protects the food from dehydration when it is passed under the freezing air blast. Vegetables or fruits destined for freezing may also be picked at a different time than they would be if they were to be sold fresh, because they need to be at optimal tenderness.

The freezing equipment is typically made from stainless steel and other metals. The gas used for freezing is most commonly ammonia. Freon is used in some systems, though because it breaks down the ozone layer, ammonia is more environmentally sound. Cryogenic freezing uses liquid nitrogen.

## The Manufacturing Process

The actual process of freezing a food item varies somewhat depending on what is to be frozen. Peas are the most common frozen vegetable, having virtually replaced fresh peas in the American supermarket. The pea process is typical for many vegetables. A typical process for a frozen entrée follows.

### *Cultivating the peas*

1 Peas are grown principally in Washington and Oregon and in the northern Midwest, that is, Wisconsin and Minnesota. Food processors typically contract with farmers to grow their crops according to the specifications needed for freezing. The farmers sow a variety of pea that has been approved as a good freezer. Major varieties are Dark Skin Perfection and Thomas Laxton. The harvesting schedule needs to be agreed on by both the farmer and the producer. The producer may measure the tenderness of the peas, and will also evaluate how much volume the freezing plant can accommodate. Peas need to be frozen within hours of picking, and if a backlog develops

*Clarence Birdseye*

Born in Brooklyn, New York, Clarence "Bob" Birdseye attended Amherst College for two years before leaving in 1912 in order to indulge his spirit of adventure by fur-trading and trapping in Labrador, Canada. Birdseye returned to Labrador in 1916 with his new wife and infant. In order to preserve the few fresh vegetables that found their way to Labrador by ship, Birdseye began experimenting with the Eskimo method of quick-freezing foods. He stored fresh cabbages in a barrel with sea water which froze quickly in the subzero Arctic climate. Birdseye also experimented with quick-freezing fish and caribou meat. When thawed, these foods remained tender and fresh-flavored, unlike previous methods involving slow cold storage.

Birdseye returned to the United States in 1917 determined to develop commercial methods of rapid freezing, experimenting with an electric fan, cakes of ice, and salt brine. In 1923, he invested everything he had in Birdseye Seafoods, marketing frozen fish. In 1924, he and three partners founded General Seafoods in Gloucester, Massachusetts, which became the first company to use the technique of rapid dry freezing of foods in compact, packageable blocks.

The Postum Company bought Birdseye's business and 168 patents in 1929 for $22 million. The company renamed itself General Foods and marketed its frozen foods under the*Birds Eye* trademark.

After the sale, Clarence Birdseye devoted himself to more inventing, obtaining over 300 patents, including ones for an infrared heat lamp, a whale-fishing harpoon, a method of dehydrating foods, and a spotlight for store window displays.

at the freezing plant, some of the peas may deteriorate.

### Picking and washing

2 The peas may be picked by hand or automatically. Then, a machine called a viner removes them from their shells. If the processing plant is adjacent to the fields, the peas are carted there. If truck transport is necessary, they are cooled with ice water and then packed in ice for transport. At the plant, the peas are dumped into beds and sprayed with water to remove dust and dirt.

### Blanching

3 The cleaned peas are next passed into a vat of boiling water for a few minutes. This kills enzymes that effect the taste of the peas, but it does not cook them. After blanching, the peas are cooled with water and then passed to a specific gravity sorter.

### Sorting

4 The peas are next sorted to remove any old, starchy peas. They are immersed in water with a specified salt content. Tender peas float to the top of the brine tank, while peas with a high starch content sink to the bottom. The tender peas are then sprayed with clean water to remove the salt, and they pass to an inspection area.

### Inspection

5 In the inspection area, workers glance over the peas as they move along a belt. Nimble workers pick out any discolored or otherwise off peas, and also any rocks or other field detritus that may have made it this far.

### Packaging and freezing

6 Packaging may precede freezing, or the peas may be individually quick frozen and then boxed, depending on the processing plant. Freezing could be by any of the standard methods. If they are frozen before packing, the peas might pass through a blast tunnel where ultra-cooled air freezes them. Or they may be loaded on a belt that brings them into contact with metal plates cooled from below by chilled ammonia. If they are packaged before freezing, the sealed boxes may be loaded into trays. The trays are

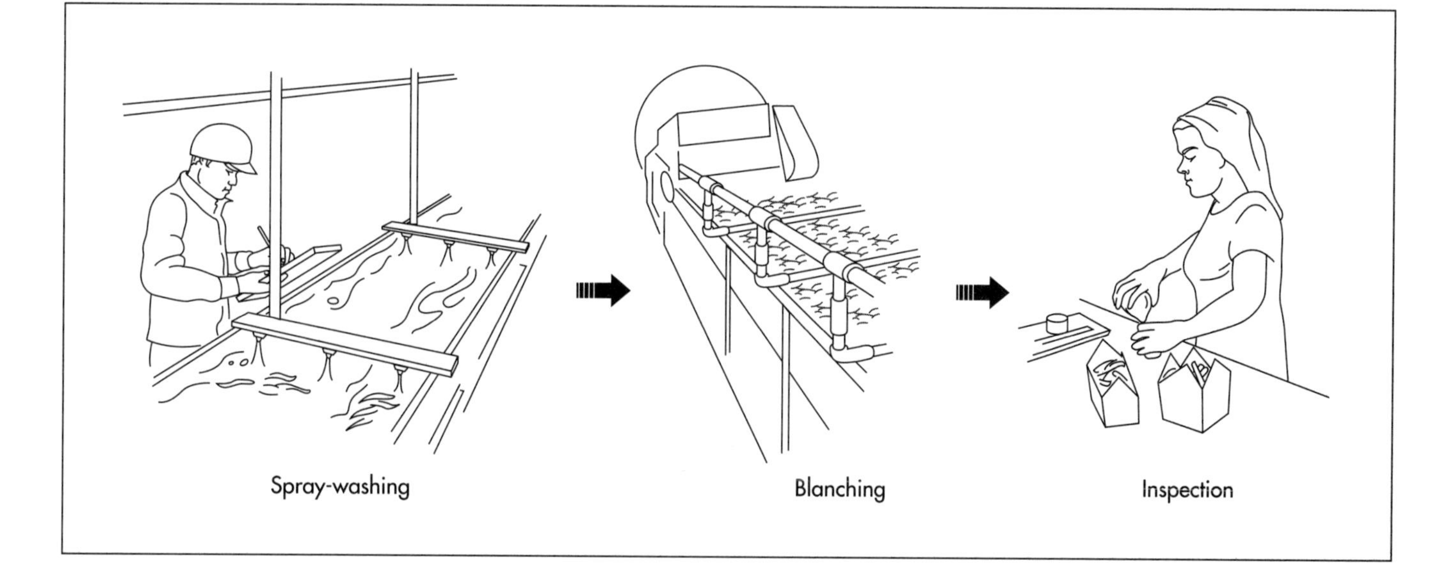
Spray-washing Blanching Inspection

stacked in a multi-plate freezer, which brings the pea packages into contact with chilled plates both above and below. Then workers load the frozen packages into shipping crates and move them to a cold storage room to await shipping.

### Testing the frozen entrée recipe

7 A large company that wishes to bring out a new frozen meal will first test the recipe extensively in a test kitchen. Ingredients from different distributors will be tried out to find which ingredients retain their qualities in the freezer, and which taste best. The company will garner consumer feedback by asking test customers to try the meal at home, heating it both in a conventional oven and in a microwave oven. The final recipe may be achieved after months of testing and evaluating.

### Pilot production

8 Before moving to full-scale production, a major manufacturer will devote some time to pilot production of the new frozen dish. Major frozen food producers may have a separate facility just for test runs, or the manufacturing plant may have one production line that it diverts. Here different cooking processes may be tried, for example to determine exactly how long to cook separate ingredients. All the bugs in the process should be worked out at this stage.

### From the oven to the freezer

9 When everything is running smoothly at the pilot production stage, the manufacturer begins to produce the frozen meal in quantity. The dish is cooked and assembled on a tray. Usually a hot meal does not need a cool-down period before moving to the freezer. It may be frozen by one of three standard methods. It can be "naked" air blasted, that is, sent through an air-blast freezer tunnel in its pre-packaged state. It may be cryogenically frozen in the same manner. Or it may be packaged, and then air-blasted.

### Packaging

10 In all but the very smallest frozen food operations, once the meal has passed through the freezer, all the packing stages are fully automatic. The frozen meal on its tray passes on a belt to mechanical equipment that bags it, puts it in a carton, and then stacks the cartons in a case. The cases are then put on pallets, and this stage too is often completely automatic. If workers are palletizing the cartons, they are dressed in cold-weather gear for protection. The pallets are stored in a warehouse cooled to between 0- -20° F (-17.8 - -28.9° C).

### Distribution

11 All further distribution of the frozen food should be carried out at 0° F (-17.8° C) or cooler. In other words, trucks or rail cars that carry the pallets should be kept

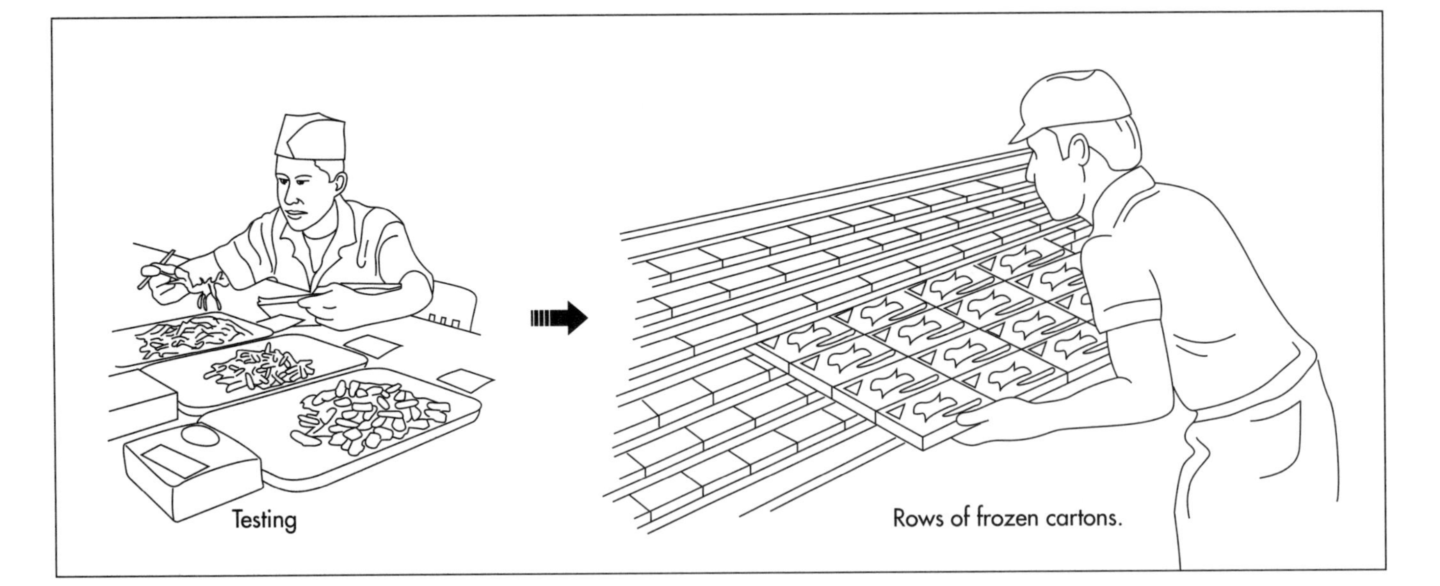


to this temperature, and so should warehouses, storerooms, and freezer cases where the cases are later stored.

## Quality Control

Frozen foods must be carefully inspected both before and after freezing to ensure quality. When vegetables arrive at the processing plant, they are given a quick overall inspection for general quality. The peas are inspected visually again as in step five, above, to make sure that only the appropriate quality peas go on to the packaging and freezing step. Laboratory workers also test the peas for bacteria and foreign matter, pulling random samples from the production line at various points. The packaged, frozen vegetables are also tested randomly by lab workers who cook and taste them. The freezing equipment is also cleaned at specified intervals, so that it is completely sterile. Manufacturers of freezing equipment work with food producers to develop machines that are easy to clean and maintain. The equipment manufacturers may also work with their customers to check and repair the machinery so that it works the way it is supposed to. For frozen meals, if any meat is used, the U.S. Department of Agriculture has oversight, and will send inspectors to make sure the manufacturer is maintaining its equipment properly and that the meat is kept at proper temperatures throughout the production process. However, if a frozen meal contains no meat, it is up to the manufacturer itself to maintain ideal conditions, and no government agency is directly responsible for quality control.

## The Future

In the late 1990s, the frozen food industry was expanding both within the U.S. market and abroad. The industry's biggest push was so-called home meal replacement, that is, whole frozen meals that took the place of cooking from scratch. More consumers were willing to trade the convenience of a frozen meal for the satisfaction of making their own dinner from fresh foods. This meant that the industry was challenged to come up with more elaborate frozen dishes, which required more testing and experimentation to pull off than the relatively simple frozen vegetables or waffles. Food scientists are still working out the chemistry and physics of frozen foods, studying for example the relationship between low-molecular weight sugars and high-molecular weight stabilizers in a recipe in order to better predict what foods will freeze well.

Cryongenic freezing is also a relatively new freezing method that may be gaining adherents. As not all foods benefit from being frozen this way, some equipment manufacturers are designing multiple-use machines that combine freezing methods.

## Where to Learn More

### *Periodicals*

Rice, Judy. "B-r-r-r-reakthroughs in Freezing Technology." *Prepared Foods* (November 1996).

Smith, Brian. "Cryogenic Food Freezing: A Guide to the Efficient Use of Food Freezing Tunnels." *Frozen Food Digest* (July 1994).

"Stabilizing the Big Freeze." *Prepared Foods* (January 1996).

*—Angela Woodward*

# Fruit Leather

*Mass-produced fruit leather is only about one third fruit puree and two thirds additives and sweeteners.*

## Background

Fruit leathers, sometimes referred to as fruit rolls or Fruit Roll-ups, are popular dried food snacks. They are formed when fruit is pureed (generally from a concentrate when mass-produced) cooked, dried, and rolled or cut out (for easy storage and packaging). The sticky solution is then spread on a non-stick surface on which it is dried. When dried, the fruit leather is firm to the touch (hence the name fruit leather) but malleable enough so that it can be rolled. The fruit leather can be easily cut in strips and shapes, according to the brand or variety of fruit leather. Fruit leather generally lasts quite a long time in this state and does not require refrigeration. The popularity of the fruit leather has increased significantly in the last 10 years because many view these snacks are more healthful than other confections because it is produced from fruit to which vitamins (particularly vitamin C) has been added.

Fruit leather is mass-manufactured by a number of different companies but can also be made rather easily at home. Recipes abound in cookbooks, including directions for making fruit leathers with grapes, raspberries, apples, and strawberries on the kitchen stove and using the oven or a food dehydrator to assist in the drying process.

## History

It is difficult to know when or who first developed fruit leathers. However, many believe that peoples of the Middle East were among the first to discover that fresh fruit could be utilized and preserved year round if pureed, cooked, and dried. It is likely that an early flavor for fruit leathers was apricot. Antiquarian cookbooks refer to fruit leathers as Persian or Middle Eastern, in fact. Armenian cookbooks refer to the treat as *bastegh* and give recipes for making them at home, discussing the "old ways" these fruit leathers were produced. The process recommends that the fruit treat be made in dry, sunny weather in that the cooked and pureed fruit be poured onto muslin sheets, hung outside to dry, sprayed with water on the reverse side so that it could be peeled off the muslin, and left to dry again outside in the sun. Recipes recommend that the fruit leather be brought in at nightfall but returned to sunshine the next day if not firm to the touch. The finished product was cut into desired shapes and placed into a glass jar for storage. Updated recipes have the cook pour the slurry onto waxy paper or plastic wrap and place it in the sun under cheesecloth, still others recommend the use of ovens or dehydrators for quick, reliable drying.

Recipes for fruit leathers are often found in organic and health-conscious recipe sources. These recipes call for using organically-grown fruit and eschew the inclusion of artificial ingredients, added vitamins or processed sugar, replacing it with honey (if any sweetener is used at all). In fact, mass-produced fruit leathers are only about one third fruit puree and two thirds additives and sweeteners.

Major food manufacturers have been making fruit leathers in this country for nearly two decades. These fruit leathers are available in a dizzying array of flavors (including watermelon) and have vitamins specifically added to them to make them more healthful. They are extremely popular with children and are designed with their interests in mind. These manufacturers have developed a range of bright colors for these fruit snacks (including hot colors and neon colors that are not natural fruit colors). Instead of cutting the leather into plain strips, some manufacturers cut cartoon figures onto the product to assist in the marketing and sales of the product. Packages feature prominent cartoon characters or movie characters, making the product more appealing to children.

## Raw Materials

Mass-manufactured fruit leathers contain three primary ingredients: fruit puree, a food additive called malto-dextrin and a sweetener of some sort. In some national brands, the fruit puree makes up only about one third of the product. As important as the fruit puree are the two other main ingredients, malto-dextrin and sweeteners. Malto-dextrin is a modified food starch that is added to a number of manufactured products. It is a white powder that mixes easily with any raw material and is cold-water soluble. When soluble, it turns transparent and is of low viscosity. The malto-dextrin additive is extremely important in fruit leathers in that it provides the soft texture and malleability required for the product. Sweeteners are the other significant additive to the product. Sweeteners generally include corn syrup or sugar, and in some products, include both. Many manufactured fruit leathers include a great many other additives but vary according to brand. These often include: partially hydrogenated cottonseed oil, glycerin or diglyceride, artificial and natural colors and flavors, pectin, gums, and added vitamins such as citric acid (vitamin C).

## The Manufacturing Process

Fruit leather is an extruded snack, meaning the fruit puree solution is forced through a metal die and cut into desired shape.

### *Cooking*

1 Fruit concentrate is generally purchased from a supplier. Once in house, the concentrate is placed in a vat and augmented with water. Corn syrup, malto-dextrin, and all of the various additives are added to the batch. Flavor enhancers and colors are added at this time as well. Only one fruit flavor is made at a time in a vat; thus, either many vats are working at one time or the vats are cleaned to receive a different-flavored puree. The fruit puree is cooked for about five minutes.

### *Extruding the puree*

2 The fruit slurry is then piped over to another large vat. The bottom of the vat is piped to an extruder. The liquid fruit solution is fed through a metal die that pushes the hot fruit puree out of the vat, flattens it, and extrudes it onto a thin, waxy paper. Some manufacturers offer different sizes or widths of the product. The shape of the extrusion die varies with the product being manufactured at that time.

### *Drying*

3 The fruit leather is still viscous at this time and must be and malleable enough to be able to be rolled in the packaging. It is conveyored through a drying tunnel that quickly solidifies the product. The tunnel does not use hot air to dry (as recommended in home recipes or traditional recipes) but cool air to drive off moisture and cool down the still-hot slurry. At the end of the tunnel, a mechanism rolls up the cooled fruit leather into a spool or roll (depending on the product). Some products are wrapped on the outside of the waxy paper with a strap to keep the spools closed (particularly useful for long, thin spools). In addition, some fruit leathers have patterns or images (often cartoon characters or even letters or numbers) cut into them with a die resembling a cookie cutter plate. This occurs before the leather is rolled.

### *Pouching*

4 At the end of the conveyor belt, the rolled leather is taken to a hopper. Individual spools are dropped into a pouching machine, in which the product is enclosed in a wrapper either of metallic foil or plain white paper (depending on the product). The pouch is automatically sealed as it goes through the machine.

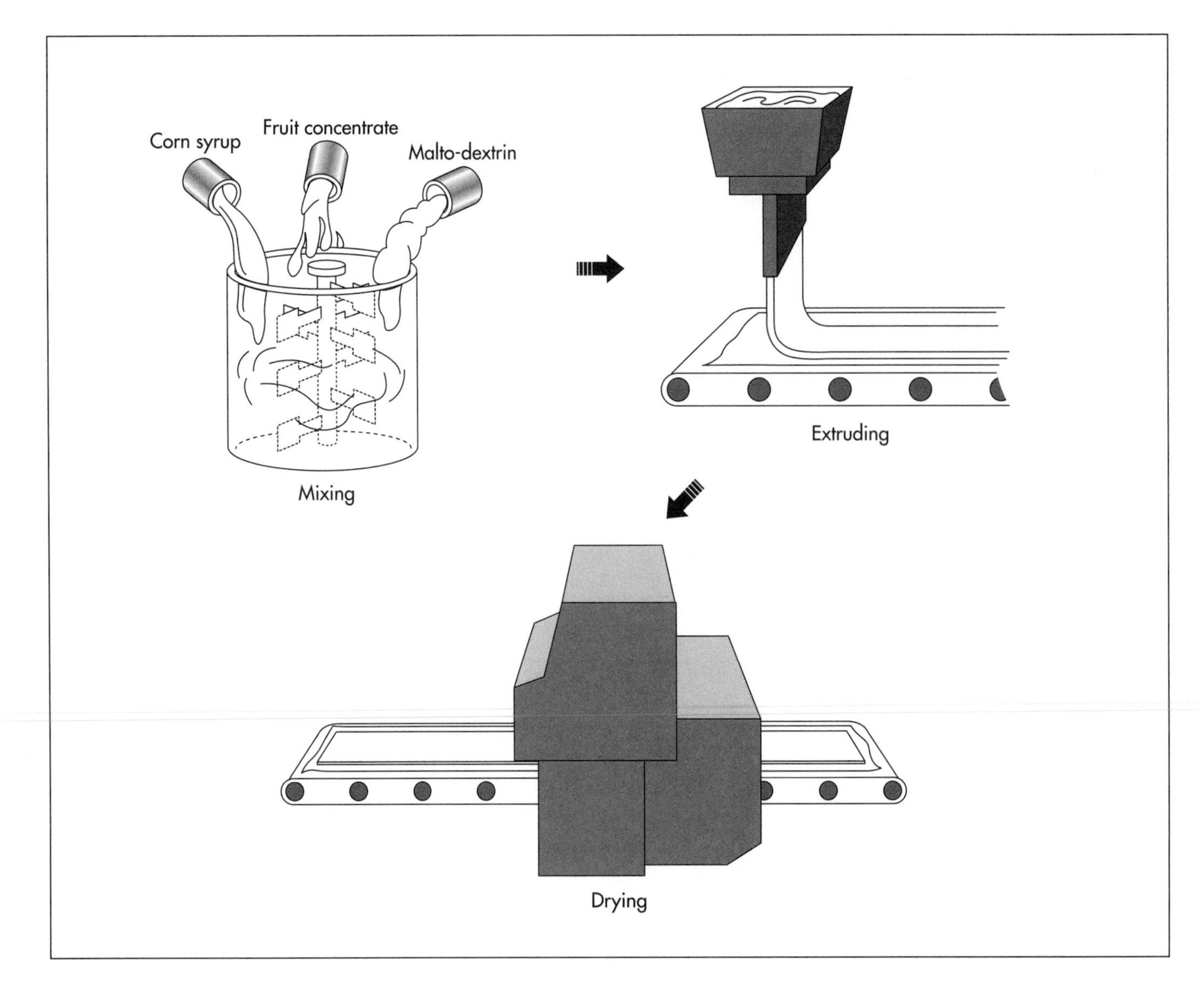


The manufacture of fruit leather involves mixing fruit concentrate with water, corn syrup, malto-dextrin, flavor enhancers, and coloring. Next, the fruit puree is cooked. Heating thickens the liquid into a slurry that is then extruded. The solution is fed to a metal die that pushes the hot fruit puree out of the vat, flattens it, and pushes it onto a thin, waxy paper. The fruit leather is conveyored through a drying tunnel that uses cool air to quickly solidify the product.

### *Packaging*

5 The pouches are then dropped into a larger machine that automatically groups and packages the product. This machine is programmed to fill cardboard packages by weight. The cardboard packaging is generally made out-of-house, and includes all important information such as nutritional information. Packaging is often very brightly colored and full of images of popular cartoon characters and is an essential part of product marketing.

## Quality Control

Quality control begins with the acquisition of high-quality fruit concentrate. Many purees are supplied by well-known fruit processors. Other quality control methods include careful calibration of all additives, particularly of those additives that affect hardening/malleability (malto-dextrin in particular). Also, cooking and drying temperatures are monitored closely to ensure moisture content. Scales are carefully calibrated so that each roll contains just the right amount of extruded product; similarly, the packaging machine is checked and re-checked so that each cardboard package includes the correct number of fruit leathers. Sample testing is performed periodically as well.

## Where to Learn More

### *Books*

Herbst, Sharon Tyler. *The Food Lover's Companion.* Barron's Educational Services, Inc.

### Periodicals

Atlas, Nava. "It's in the bag: nutritious school lunches and treats that score A+."*Vegetarian Times* (September 1998): 48.

Chandonnet, Ann. "Tlingit Food."*Skipping Stones* (November/December 1997): 23.

Creasy, Rosalind. "Tomato Leather."*Organic Gardening* (July/August 1989): 32.

Weddell, Leslie. "Home-made fruit leather is ideal for lunches, hiking trails."*Knight-Ridder/Tribune News Service* (August 30, 1993).

### Other

Unofficial Summary. *General Mills Canada, Inc. v. The Deputy Minister of National Revenue.* Appeal No. Ap-97-012.


*—Nancy EV Bryk*

# Galoshes

*The name for galoshes originated in the Middle Ages when many styles of boots from short to long were popular. The word came from Gaulish shoes or gallicae, which had leather uppers and soles carved of wood; when the Romans conquered the territory they called Gaul (France), they borrowed the Gaulish boot style.*

## Background

The name for galoshes originated in the Middle Ages when many styles of boots from short to long were popular. The word came from Gaulish shoes or *gallicae,* which had leather uppers and soles carved of wood; when the Romans conquered the territory they called Gaul (France), they borrowed the Gaulish boot style. Nobles wore red leather, which told observers of their aristocracy, and the wooden soles were often ornately carved.

Galoshes and boots are very closely linked in footwear history, and sometimes the words are used interchangeably. Properly, however, a galosh is an overshoe that slips over the wearer's indoor footwear but is made of waterproof material to protect the more delicate materials of the shoe as well as the wearer's foot from cold and damp. In the early days of boots, particularly those made for ladies, the boot was made of fabric and capped or "galoshed" with leather. Laced closures tightened the galoshed boot around the ankle just as fasteners on modern galoshes secure them.

## History

Cave paintings show that the first boots were worn as early as 13,000 B.C. The first attempts at galoshes may have originated more than 4,500 years ago when separate leggings were wrapped around the legs above moccasins to protect the wearer from cold, rain, and prickly plants. The moccasins were extended to form boots. However, most boots were made of pieces, and joins between the pieces allowed moisture and cold to creep in. The first boots were probably created in northern Asia, and, as the wearers migrated across the Bering Strait to the Americas, their creations traveled with them.

The Inuits of Alaska and North America imitated nature to create the perfect, weatherproof boot. They observed the polar bears dense, multilayered fur that kept the bear's skin from freezing. The Inuits used the complete paws and legs of these bears (with claws still attached) to create seamless boots. Similarly, the Ainu people of northern Japan used the complete leg skins of deer. Sealskin, caribou hide, and skins of other "waterproof" animals were used to make boots, but, again, these had to be cut in pieces and sewn. Multiple layers of wrappings inside the boots insulated the wearer from cold and seeping water but were awkward to wear and maintain. Many northern peoples stuffed their boots with grass for cushioning and insulation, but grass had to be cut and stored for this use during most of the year when it didn't grow. Curiously, some cultures made boots from fish skin but found they were useless in rainy weather.

In the Middle Ages in both Europe and Asia, people wore pedestal shoes outside to raise them above water or mud. The sole of the pedestal shoe was made of wood, and the upper was like a shoe and made of fabric or leather. The front and back ends of the pedestal were tapered so the walker could move forward by rocking the tall shoes. *Pattens* were similar overshoes with tall, shaped wooden bases and mules or slippers into which the wearer could slip her indoor shoes. Shoemaker Nicholas Lestage unknowingly borrowed a concept from the Alaskan Inuits when, in 1663, he made seamless, calfskin boots for King Louis XIV of France by taking the skin from a calf's leg and tanning and dressing it to form a seamless boot. He was forbidden by the king to make his boots for others or to reveal his secret, which was kept for 100 years.

Galoshes and many garments were also made by coating fabrics with waterproofing. Linseed oil was commonly used to coat fabric to make oilskins. Other oil or tar mixtures and nitrocellulose (mixed with castor oil and coloring) were used to impregnate or coat fabrics. The fabrics were then heated to oxidize the film left on the cloth and make it stable.

Plants, not animals, are the source of waterproof materials in nature. Historical legend has it that the Egyptians were the first to make galoshes from rubber. They made foot-shaped molds out of wood and poured liquid rubber over them. Interest in rubber generally languished for thousands of years. In 1823, Charles Macintosh discovered a way of waterproofing garments by using liquid rubber. He spread rubber that was mixed with a solvent on a marble slab. The solvent evaporated and left a thin sheet of rubber that was then stitched on a sheet of fabric that was then cut and sewn into a garment. Unfortunately, garments made this way did not wear well and were the victims of temperature changes as well as sunlight and grease.

It took one of America's greatest inventors to identify the rubber plant as the source of the "perfect" waterproof substance. Charles Goodyear is remembered more often for "footwear" for cars rather than people. Goodyear's dogged determination to find uses for the substance called "india rubber" occupied him for 20 years and exhausted his income. Goodyear was fascinated by properties of rubber including its elasticity, durability, lightweight, and the fact that it was waterproof.

The qualities of rubber, however, depend heavily on temperature; it becomes soft and tacky in the heat and turns hard and brittle in cold temperatures. The native tribes of the Amazon had protected their feet with rubber for generations by simply dipping their feet into liquid from the rubber tree and drying the custom-made galoshes over the fire, but the moderate temperatures during the rainy season in the Amazon jungle suited the properties of rubber. By contrast, when the first rubber coats were introduced in the United States in 1823, they were rigid and rattled like metal. Rubber-soled shoes also failed on their first introduction in 1832 because they stuck to floors in the heat and cracked in winter.

Goodyear persevered and patented the process of vulcanization in 1844. Vulcanization tempered the properties of rubber so that it was easily molded, durable, and tough. Among the thousands of products that Goodyear proposed to make from rubber were rubber shoes for children, waterproof boots, and rubber "chair" shoes for indoor wearers to eliminate static electricity, noise, and carpet wear. Rubberized elastic webbing was also made possible thanks to Goodyear's studies of rubber, and insets of such webbing were stitched in the sides of galoshes (circa 1890) to make them easy to pull on and off.

Thousands of rubber products appeared during the next 30 years as industrialization merged with vulcanization to make rubber products easy to cut, punch, and crimp by machine. Rubber-soled shoes answered the growing interest in sports and became the foundation for the huge sneaker industry. Rubber boots diversified to suit many forms and functions from galoshes to wellingtons, hip-boots, waders, and "body boots." Styles with rubber heels were also made.

An outdoorsman named Leon Leonwood Bean revolutionized the style of boots (and mail-order business) when, in 1911, he made a boot with waterproof leather uppers and rugged rubber-galoshes bottoms. The leather portion decreased the weight of traditional boots, but the rubber bottoms gave them the durability to withstand tough conditions. Bean's boots remain popular today and are available in a range of colors.

Natural rubber was largely replaced by synthetic rubber due to World War II; by the end of the war, 70% of all rubber was used in the making of tires, and the majority of the remaining 30% was used to manufacture footwear. Rubberized footwear has also been adapted to the workplace where steel-toed rubber galoshes and boots protect toes

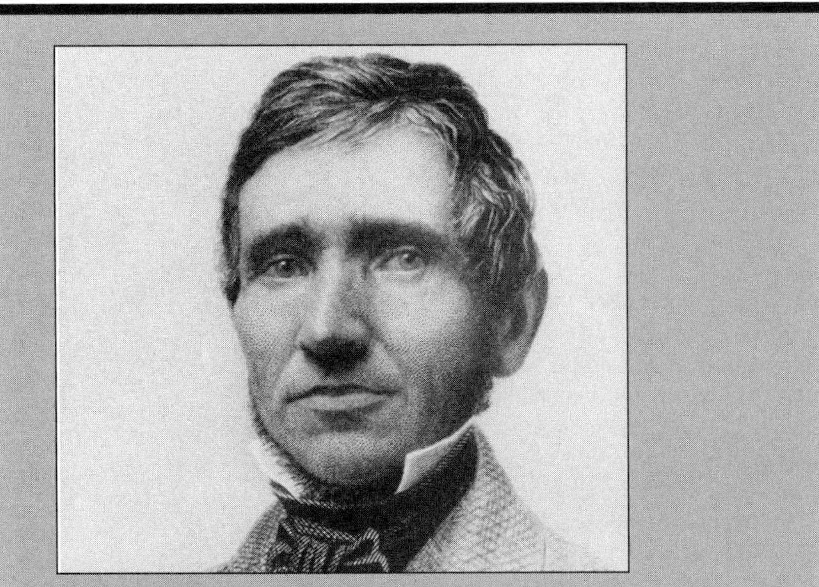

*Charles Goodyear*

Charles Goodyear was born in 1800 in New Haven, Connecticut, the son of a hardware manufacturer and inventor. In 1826, Goodyear and his bride, Clarissa, opened the first American hardware store as an outlet for the senior Goodyear's products. Both father and son went bankrupt in 1830.

In 1834, Goodyear bought a rubber life preserver from the Roxbury India Rubber Company in New York City and quickly invented an improved valve for the device. When Goodyear tried to sell his design to Roxbury, the manager told him it was the rubber itself that needed improving, not the valve. Consumers were fed up with the way rubber melted in hot weather and hardened in cold. Improving rubber's usability quickly became Goodyear's lifelong challenge

For the next five years, Goodyear dedicated himself to experimenting with rubber, in both his own and the debtors' prison kitchens. He had no idea of what to do, having no knowledge of chemistry; he had no money, and only the crudest equipment. The breakthrough finally came in 1839 when Goodyear accidentally discovered the *vulcanization* process—heating rubber-sulphur mixtures to yield a tough yet flexible product.

Goodyear struggled through five more years of destitution—at one point even selling his children's schoolbooks—before he was able to patent his process in 1844. Instead of profiting from his finally successful quest, Goodyear granted licenses for rubber manufacturing at ridiculously low prices, and he withdrew from manufacturing himself to contrive new uses for his product. Industrial pirates infringed on his patents, and he paid his attorney, Daniel Webster (1782-1852), more to secure his rights (successfully, in 1852) than he ever earned from his discovery. He was unable to patent his vulcanization process abroad; Thomas Hancock of England had already done so. In 1860, Goodyear died, leaving $200,000 in debts.

from heavy objects and the splash-resistant qualities of rubber protect workers from chemicals. Rubber also helps insulate against the cold, protect against abrasions, and guard against slippery surfaces.

Although galoshes have been considered practical, their fashion appeal has been limited until recently. Thanks to suppliers of outdoor wear like L.L. Bean and Land's End, new styles of outdoor shoes have become popular. The New England Overshoe Company brought the overshoe into the 1990s through their N.E.O.S. galoshes, which are sturdy, collapsible, practical, and fashionable.

## Raw Materials

Traditional galoshes are made of liquid rubber or sheet rubber. Some styles are lined with fabric and some have fasteners to tighten the relatively loose shape at the ankle.

New-style galoshes have rubber bottoms called outsoles. The uppers are made of microfiber fabrics or nylon that have been treated. Fasteners include elastic straps with plastic buckles to tighten the galoshes over casual or dress shoes and elastic fasteners at the tops of the boots to keep water out around the wearer's legs.

## Design

Until recently, design of galoshes has been limited to providing a waterproof covering to fit over shoes. Because galoshes are intended for limited use (as opposed to all day wear) and because they must fold to be carried in bags or briefcases, galoshes are usually thin or flimsy. Light treads are incorporated into the outsoles, and fasteners are needed to fit the loose design to the ankle. Some galoshes or rubber boots are made in bright colors, others are clear so the wearer's shoes show through, and still others are made in traditional shoe colors like black and brown.

N.E.O.S. and other new-style overshoes are lightweight and collapsible because of their microfiber uppers. The outsoles are harder rubber with deeper treads, made possible by the compensating lighter uppers. The ankle strap gives a stylish look, and colors are traditional.

## The Manufacturing Process

There are three different manufacturing processes for galoshes and rubber boots. Some are cut from sheets of rubber, others are made in a process called slush molding, and N.E.O.S. style overshoes are sewn from fabric with rubber outsoles attached.

### *Rubber boots made from sheets*

1 Larger forms of rubber boots are generally made from rubber that is poured in sheet form and uncured or semi-cured to give it workable properties. The nature of the rubber and the requirements for the curing process are part of a formula developed by the manufacturer.

2 The prepared rubber is rolled out and die cut in pieces. The pieces are fitted around aluminum lasts or forms made to suit the design of the boot and the foot and leg size. Sometimes these pieces are coated with talcum powder. The aluminum lasts complete with the fitted rubber are heated cured at a temperature of about 130°F (54°C) to complete the process of forming the rubber to fit the lasts and to meld the pieces together.

3 The talcum powder aids the process of removing the rubber boot from the last. The seams and other parts of the boot are trimmed, and any hardware is added.

### *Slush-molded boots and galoshes*

1 Shorter rubber boots and galoshes including very lightweight models like Totes galoshes are made by slush molding. For this design, a last is made, and an outer metal mold is also made that is perfectly sized to the last except that it is slightly larger.

2 The cavity between the last and the mold is filled with liquid that consists of polyurethane and other synthetics and a small percentage of rubber, and the mold and its contents are spun to spread the liquid uniformly throughout the cavity between the last and the mold.

3 The mold is removed, the formed boot is taken off the last, and the boot is trimmed and decorated, like its larger cousin.

### *Fabric and rubber galoshes*

1 Newer fashions in galoshes merge shoe and boot construction techniques. The outsoles are formed from hard, vulcanized rubber with deep treads. The uppers are laser cut from microfiber or nylon fabric based on computer designed shapes. These uppers may be lined with Polartec, thermal insulation, fleece, foam, or similar weatherproof and chill-proof insulating material that is cut to fit the galoshes. The outer fabric is treated to be weatherproof before it is cut.

2 The outsole is fitted to the last, and the upper is fitted and sewn around the upper part of the last. This fitting is fully computerized. Strong and temperature-tolerant adhesives are used to attach the outsole to the fabric upper. Depending on the manufacturer, waterproofing may be sprayed on the overshoe so that seams and joins have added protection against infiltrating water. Other makers use waterproof, hot-melt tape that is placed over the seams and melted into place to prevent moisture and cold from entering these joins. Manufacturers may also opt for leather uppers that have been tanned with waterproofing.

3 Heel cushions and steel shanks are other options in some models. Finally, the straps, buckles, or other hardware are added.

## Quality Control

Construction of rubber boots, galoshes, and rubber sections of other overshoes is carefully monitored by technicians trained in the processes of vulcanization, rubber curing, and slush molding. Cutting of rubber and fabric is sized and engineered by computer. Lasts for galoshes and rubber boots are usually only made in full sizes and a limited range of widths, and the variety of designs is more restricted than designs of shoes for example. These considerations help manufacturers keep costs down; overshoes are, for after all, a second set of footwear that are not necessarily chosen for appearance. Technicians also monitor trimming and attachment of fasteners or other accessories, whether these are done by hand or by machine.

Lightweight rubber boots and galoshes are manufactured using a process called slush molding. A last is made, and an outer metal mold is also made that is perfectly sized to the last except that it is slightly larger.The cavity between the last and the mold is filled with liquid that consists of polyurethane and other synthetics and a small percentage of rubber, and the mold and its contents are spun to spread the liquid uniformly throughout the cavity between the last and the mold. The mold is removed, the formed boot is taken off the last, and the boot is trimmed and decorated.

Slush molding

## Byproducts/Waste

Byproducts are not usually made by manufacturers of galoshes although they are always alert for marketing and fashion opportunities; for example, weather-proof clogs are made by some overshoe manufacturers to attract customers to a product that is easily put on and taken off and for gardeners and strollers who don't need the benefits of a full boot or the shoe protection of galoshes.

Rubber and fabric waste are minimized by computer-aided layouts and cutting. Waste that does result must be disposed.

## The Future

Galoshes have always been useful, if not prized for their beauty. Wily entrepreneurs are always on the alert for methods of modernizing reliable products and for recognizing ways of making them fashionable. The portability and lightweight of Totes, the outdoorsy appeal of products made by L.L. Bean and others, and the trendy combination of fabric and sturdy rubber in the N.E.O.S design are examples of how producers have made reliable "old" galoshes new again.

## Where to Learn More

### Books

Lawlor, Laurie. *Where Will This Shoe Take You? A Walk Through the History of Footwear.* New York: Walker and Company, 1996.

Moilliet, J. L., ed. *Waterproofing and Water-Repellency.* London: Elsevier Publishing Company, 1963.

O'Keefe, Linda. *Shoes: A Celebration of Pumps, Sandals, Slippers, & More.* New York: Workman Publishing, 1996.

Yue, Charlotte and David. *Shoes: Their History in Words and Pictures.* Boston: Houghton Mifflin Company, 1997.

### Periodicals

Canizares, George. "Galosh Revolution." *US Airways Attaché* (December 1998): 30.

### Other

Ben Meadows Company. http://www.benmeadows.com/.

MinAn Chemical Industrial Co., Ltd. http://www.minan.com.tw/.

New England Overshoe Company (N.E.O.S). http://www.overshoe.com/.

*—Gillian S. Holmes*

# Gelatin

Gelatin is a protein substance derived from collagen, a natural protein present in the tendons, ligaments, and tissues of mammals. It is produced by boiling the connective tissues, bones and skins of animals, usually cows and pigs. Gelatin's ability to form strong, transparent gels and flexible films that are easily digested, soluble in hot water, and capable of forming a positive binding action have made it a valuable commodity in food processing, pharmaceuticals, photography, and paper production.

As a foodstuff, gelatin is the basis for jellied desserts; used in the preservation of fruit and meat, and to make powdered milk, merinque, taffy, marshmallow, and fondant. It is also used to clarify beer and wine. Gelatin's industrial applications include medicine capsules, photographic plate coatings, and dying and tanning supplies.

## Background

Until the mid-nineteenth century, making gelatin was a laborious task. Calves' feet were loaded into a large kettle that was then placed over a fire. The feet were boiled for several hours after which the liquid was strained and the bones were discarded. After setting for 24 hours, a layer of fat would rise to the top. This was skimmed off and discarded. Sweeteners and or flavorings were added to the liquid and it was poured into molds and allowed again to set.

By the 1840s, however, some producers were grinding the set gelatin into a fine powder or cutting it into sheets. One of them was Charles B. Knox, a salesman from Johnston, New York, who hit on the idea of making gelatin more convenient after watching his wife Rose make it in their kitchen. Knox packaged dried sheets of gelatin and then hired salesmen to travel door-to-door to show women how to add liquid to the sheets and use it to make aspics, molds, and desserts. In 1896, Rose Knox published Dainty Desserts, a book of recipes using Knox gelatin.

The first patent for a gelatin dessert was issued in 1845 to industrialist and inventor Peter Cooper. Cooper had already made a name for himself as the inventor of the Tom Thumb steam engine. He had also made a fortune in the manufacture of glue, a process similar to that for making gelatin.

In 1897, Pearl B. Wait, a carpenter and cough medicine manufacturer, developed a fruit-flavored gelatin. His wife, May Davis Wait, named his product Jell-O. The new product was not immediately popular and Wait sold the rights to the process to Orator Francis Woodward, owner of the Genesee Food Company, for $450. Sales continued to limp along until 1902 when an aggressive advertising campaign in Ladies Home Journal magazine generated enormous interest. Sales jumped to $250,000.

The use of gelatin in food preparation increased six-fold in the 40-year period from 1936-1976. Today, 400 million packages of Jello-O are produced each year. Over a million packages are purchased or eaten each day.

In the field of photography, gelatin was introduced in the late 1870s as a substitute for wet collodion. It was used to coat dry photographic plates, marking the beginning of modern photographic methods. Gelatin's

*Presently, 400 million packages of Jello-O are produced each year. Over a million packages are purchased or eaten each day.*

Gelatin is a protein substance that is extracted from collagen, a natural protein present in skin, bones, and animal tissue. As a protein, it contains many amino acids that lend itself to a diverse amount of applications.

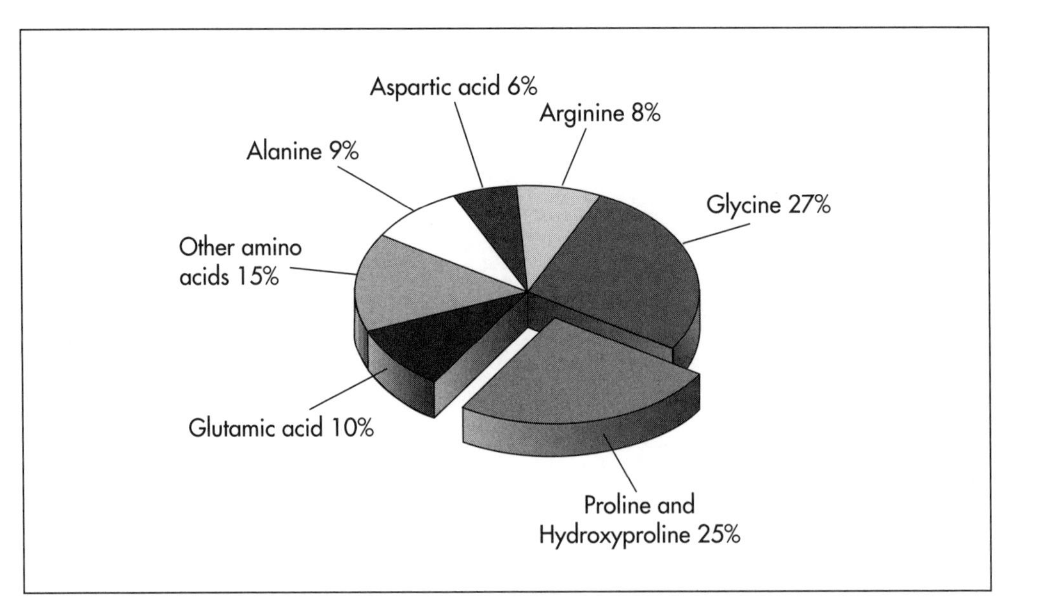


use in the manufacture of medicinal capsules occurred in the twentieth century.

## Raw Materials

Animal bones, skins, and tissue are obtained from slaughterhouses. Gelatin processing plants are usually located nearby so that these animal byproducts can be quickly processed.

Acids and alkalines such as caustic lime or sodium carbonate are used to extract minerals and bacteria from the animal parts. They are either produced in the food processing plant or purchased from outside vendors.

Sweeteners, flavorings, and colorings are added in the preparation of food gelatin. These can be in liquid or powdered forms and are purchased from outside vendors.

## The Manufacturing Process

### *Inspection and cutting*

1 When the animal parts arrive at the food processing plant, they are inspected for quality. Rotted parts are discarded. Then, the bones, tissues, and skins are loaded into chopping machines that cut the parts into small pieces of about 5in (12.7cm) in diameter.

### *Degreasing and roasting*

2 The animal parts are passed under high-pressure water sprays to wash away debris. They are then degreased by soaking them in hot water to reduce the fat content to about 2%. A conveyer belt moves the degreased bones and skins to an industrial dryer where they are roasted for approximately 30 minutes at about 200° F (100° C).

### *Acid and akaline treatment*

3 The animal parts are soaked in vats of lime or some other type of acid or akali for approximately five days. This process removes most of the minerals and bacteria and facilitates the release of collagen. The acid wash is typically a 4% hydrochloric acid with a pH of less than 1.5. The alkaline wash is a potassium or sodium carbonate with a pH above 7.

### *Boiling*

4 The pieces of bone, tissue, and skin are loaded into large aluminum extractors and boiled in distilled water. A tube running from the extractor allows workers to draw off the liquid that now contains gelatin. The liquid is sterilized by flash-heating it to about 375° F (140° C) for approximately four seconds.

### *Evaporating and grinding*

5 From the extractor, the liquid is piped through filters to separate out bits of

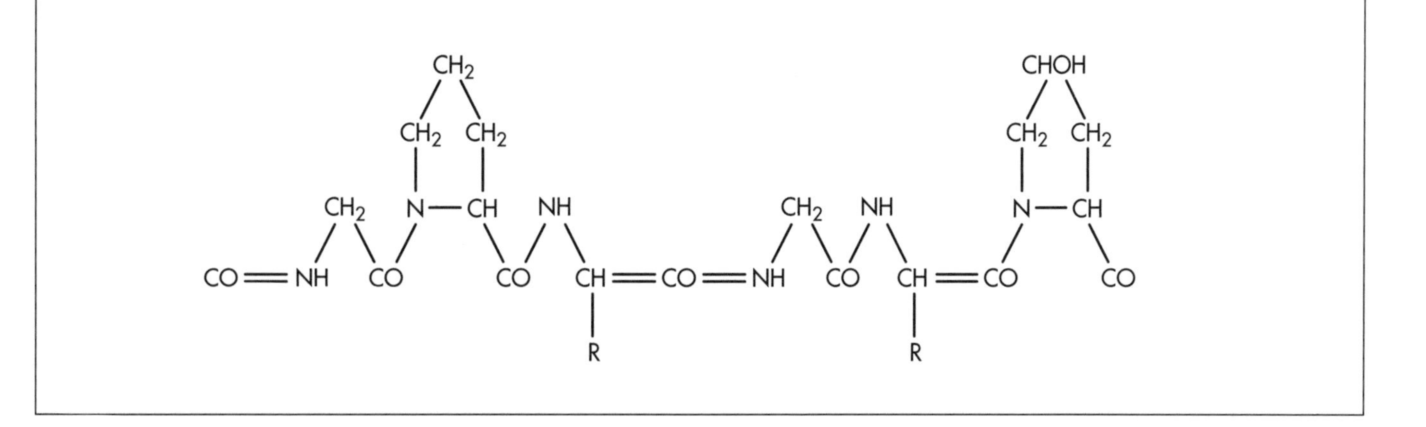


The chemical structure of gelatin is what makes gelatin water soluble; form digestible gels and films that are strong, flexible, and transparent; and form a positive binding action that is useful in food processing, pharmaceuticals, photography, and paper production.

bone, tissue or skin that are still attached. From the filters, the liquid is piped into evaporators, machines that separate the liquid from the solid gelatin. The liquid is piped out and discarded. The gelatin is passed through machines that press it into sheets. Depending on its final application, the gelatin sheets are passed through a grinder that reduces them to a fine powder.

### *Flavoring and coloring*

6 If the gelatin is to be used by the food industry, sweeteners, flavorings, and colorings may be added at this point. Pre-set amounts of these additives are thoroughly mixed into the powdered gelatin.

### *Packaging*

The packaging process is automated, with preset amounts of gelatin poured into overhead funnels through which the gelatin flows down into bags made of either polypropylene or multi-ply paper. The bags are then vacuumed sealed.

## Quality Control

Gelatin manufacturers must adhere to stringent national and international food processing requirements. These regulations include but are not limited to cleanliness of the plant, equipment and employees; and allowable percentages of additives, flavorings, and colorings.

Automated and computerized technologies allow the processors to preset and monitor ingredient amounts, time and temperature, acidity and alkalinity, and flow levels. Valves are installed along pipelines to allow for continuous sampling of the product.

Gelatin is processed to varying "bloom" values that measure the gel strength or firmness. The desired strength corresponds to the manner in which the gelatin will be used. The bloom value is technically measured and monitored throughout the production process.

### *The Future*

Since 1986 when the presence of bovine spongiform encephalopathy (BSE), also known as mad cow disease, was reported in Great Britain, there has been much concern about the processing of beef bones for the production of gelatin. In 1989, the United States Food and Drug Administration (FDA) banned the importation of cattle from the Department of Agriculture's list of of BSE-designated countries. However, a 1994 FDA ruling allowed the continued importation of bones and tissues for the production of pharmaceutical grade gelatin.

By 1997, however, the FDA held hearings to reconsider its decision. After interviewing gelatin processors, the agency found that while gelatin has not been implicated in the spread of BSE, officials are not convinced that the manufacturing processing is extracting all possible agents that are responsible for the disease. It was generally agreed that beef sources carry more of a risk than those from pork, that bones carry a higher risk than skins, and that alkaline processing is more effective than the acid-extraction method. These findings will certainly affect the gelatin-processing industry in the next century.

## Where to Learn More


### Books

Harvey Lang, Jenifer, ed. *Larousse Gastronomique.* New York: Crown Publishers, 1988, reprinted 1998.

### Periodicals

Marwick, Charles. "BSE sets agenda for imported gelatin." *Journal of American Medical Association* (June 4, 1997): 1,659.

### Other

Leiner Davis Gelatin. http://www.gelatin.com/ (June 29, 1999).

Kraft Foods. http://www.kraftfoods.com/ (June 29, 1999).

Sterling Gelatin. http://www.sterlinggelatin.com/ (June 29, 1999).


*—Mary McNulty*

# Glass Ornament

## Background

When Christmas trees were still a new custom, inventive methods of decorating trees with as much light as possible were developed. Actual lights, from candles to electric bulbs, were used, of course; but, to magnify the sparkle and fascinate the young, metallic tinsel and glass baubles became accessories for the well-dressed Christmas tree. Glass ornaments were a European invention and cottage industry from the rise in popularity of the Christmas tree in the mid-1800s to World War II. When European supplies were extinguished by the war, the manufacture of glass ornaments became a mass production industry in the United States. More recently, exclusive designers have created a demand for highly imaginative and colorfully decorated ornaments.

## History

Wreaths and garlands made of herbs and evergreens have been part of celebrations for thousands of years. Winter festivals in particular featured fragrant evergreens that symbolized everlasting life. From the Hebrews, Greeks, Romans, and Druids, early Christians borrowed the symbolism of holly, mistletoe, boxwood, rosemary, laurel, and the Christmas tree. In Europe, live trees were planted in tubs and brought indoors to inspire the residents. Indoor—and unornamented—Christmas trees were documented in Europe as early as 1521. Christmas trees were becoming widely popular in Europe by about 1755, and, also about this time, they were customarily decorated with candies, fruit, gilt nuts, dolls and other toys, and Christian symbols including figures of the Christ Child.

Queen Victoria's consort, Prince Albert, is generally credited with importing the Christmas tree custom from his native Germany in 1841. German immigrants, including Hessian soldiers, carried the heritage of the Christmas tree to America. Christmas markets in European towns and villages sold gold leaf for gilding fruits and nuts, dried fruits such as figs and prunes, strings of glass beads, tissue paper, all kinds of small toys from hobby horses to drums and soldiers, and tiny cakes and sparkling marzipan confections specifically for decorating the Christmas tree. Miniature candles were also made just for trees, and metal clips were made to hold the candles on the branches and collect the hot wax drips.

The annual Christmas markets demonstrated the growing demand for tree ornaments, and families sought heirlooms that could be passed to succeeding generations. In Victorian times, and in both Europe and the United States, Christmas lights made of glass began to replace the dangerous candles. These lights were tiny lamps made of pieces of colored glass and hung with loops of wire. Each lamp contained a wick and oil that floated on water. Some lights were hand-blown glass, and others made of patterned glass were popular because they reflected sparkling light.

Glass ornaments replaced mostly edible decorations on trees in about 1850-1860. A "cottage industry" blossomed in the Thuringian mountains of Central Germany, where peasant families manufactured hand-blown glass ornaments. Glass-making was a tradition in the town of Lauscha, for example, that turned its manufactures to ornaments in the 1860s. In a typical glass-making family, the father

*Glass ornaments were a European invention and cottage industry from the rise in popularity of the Christmas tree in the mid-1800s to World War II.*

and adult men blew glass tubing that was heated over a Bunsen burner into ornament shapes. The other family members applied a silver nitrate solution to the insides of the ornaments, so they would reflect light. Boards with rows of nails in them were hung from the cottage ceiling, and the coated ornaments (with stems of glass tubing still attached) were inverted over the nails and dried over night. Each ornament was then dipped in brightly colored lacquer and decorated with paint or fancy attachments like ribbon, spun glass, or feathers. The glass stem was cut, and a metal hanger was attached.

Balls and ovals certainly predominated, but, by making plaster or metal molds into which the glass was blown, many fanciful shapes were devised. Hunting horns, smokers' pipes, elaborate bells, delicate vases, and birds with tails made out of spun glass were especially cherished. The colors of these early ornaments imitated those of the colored sugars used to decorate confectionery that had ornamented trees of previous generations. Entrepreneurs, like Frank W. Woolworth, and shop owners who reached customers well beyond their storefronts via catalogs helped spread the sparkle of glass ornaments in the United States.

Germany was the exclusive producer of glass ornaments until 1925. German ornaments first came to America as prized parts of the heritage of immigrants. Later, the ornaments were imported. In 1925, Japan was the next country to produce significant quantities of ornaments; the cottage industry also suited Japanese families. Czechoslovakia and Poland, both countries with strong glass-making traditions, entered the marketplace in the late 1920s. By 1935, the United States imported over 250 million handmade ornaments but still had no industry of its own. In 1939, the commencement of World War II in Europe shut off supplies of glass ornaments among many other European imports. Corning Glass Works in New York entered the ornament business. Corning was skilled in the production of light bulbs, which used a ribbon machine to flow molten glass through an endless series of molds. The machine had been developed in 1926, and, by adapting it to the glass shapes needed for ornaments, Corning could produce over 2,000 ornament balls a minute; and about 100 million ornaments per production year could be generated by ribbon machines at the Corning Works. Today, ornaments are mass-produced by this same method or made by hand with blown glass and specially designed molds as they have been made for over 100 years.

## Raw Materials

Raw materials for ornaments are the same whether they are hand- or machine-made. Glass, in bulk form for ribbon machine manufacture, or in slender tubes in a range of gauges or sizes for blowing is the basic material. The inner reflective coating is called silvering solution; its chemical makeup varies among manufacturers and, in modern production, is proprietary. Lacquers, paints, frosting powders, glitter, and a wide variety of attachments made of other materials (fabric, ribbon, silk flowers, glass beads, etc.) are used to decorate the shapes. Metal catches are punched out of tin or aluminum by specialty manufacturers, as are metal hooks that are mounted inside the catches.

## Design

Mass-produced ornaments tend to be more standard in shape than the ornaments produced by exclusive designers. Balls and ovals are the most common mass-produced designs; but tree-toppers, tree shapes, bells, teardrops, stars, and icicles are among the common shapes that can be molded and finished on a large scale. For the artists who produce collectible designs, their ornament shapes, sizes, and colors are as limitless as their imaginations. World-renowned designer Larry Fraga of Dresden Dove declares that the only requirement for designing ornaments is "to be a kid at heart." His bubble gum colors and humorous designs have vaulted him into the upper echelon of ornament artists who are hired to produce exclusive ornaments for major department stores, to appear at signing events, and to hand-paint limited editions that command top dollar in an $800-million-per-year industry.

## The Manufacturing Process

### *Mass-produced ornaments*

1 In the factory, bulk quantities of glass are melted and flowed in a ribbon over a series of molds.

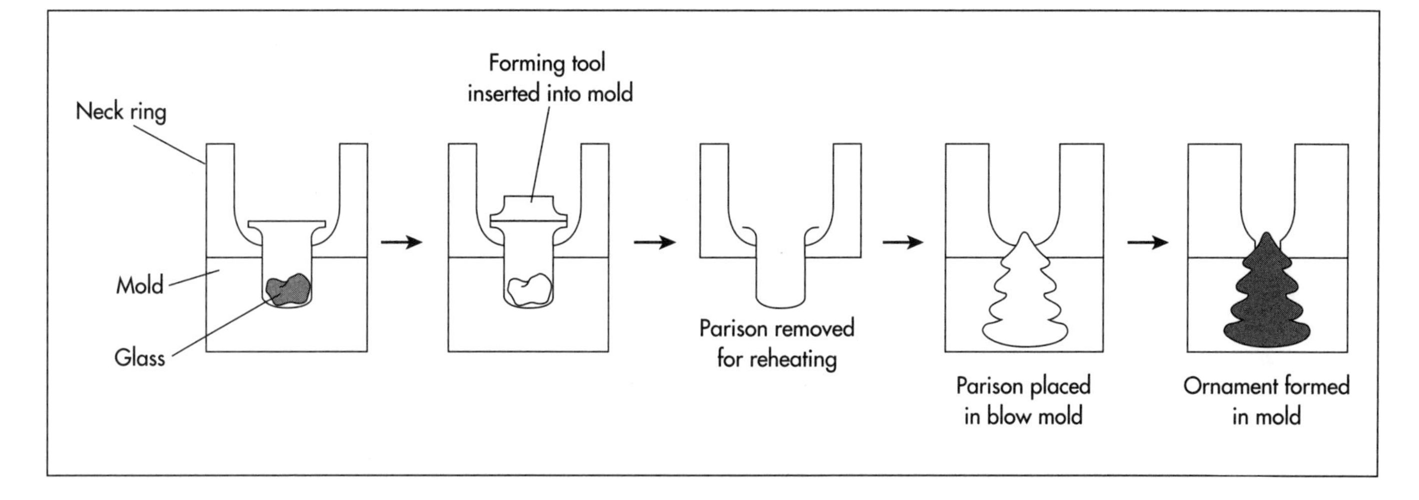


Diagram of a typical automated glassblowing process.

2 As each mold moves into position in front of the stream of glass, compressed air is blown into the mold to force the glass to uniformly take the shape of the mold. Clear glass is used, and sizes range from about 1.75-5 in (4.4-13cm) in diameter.

3 The ornaments move by conveyor to stations where they coated on the inside with silvering solution to provide the mirror-like reflective properties that will show through the exterior coatings.

4 Then, they are coated on the outside by dipping them into a white undercoat or base coat.

5 After the undercoat dries, the balls are transported by conveyor to the paint station where they are dipped in lacquer. Red and blue are the most common colors.

6 Decorations may be added by machine or by hand and may include painted designs, frosting, glitter, or glued-on decorations. Glass manufacturers can also produce spun glass or fiberglass to decorate the ornaments. Plain ornaments are also made for those who like to decorate their own ornaments at home.

7 Metal catches and hooks are prefabricated to the standard sizes of the ornament tops and are attached by machine after the ornaments are decorated; they are made of lightweight metal like aluminum or tin so they are not too heavy for the ornament.

8 The finished bulbs are then transferred to packing stations where specially designed packing materials are used to cushion and display the ornaments for sale.

### *Handmade ornaments*

1 The modern glass blower begins production of a handmade ornament with tubes of glass manufactured by suppliers. The craftsman can melt or cut the tubes into the desired quantity of glass needed for a specific ornament. By spinning the tube over a gas-powered torch, a portion of the glass is softened and kept at a relatively uniform temperature.

2 When the glass is ready to be molded, the operator depresses a foot pedal that opens the mold. The molds can be made of plaster, cast iron, graphite, or porcelain. They can have conventional or highly detailed shapes etched into the molds by laser beams. The soft glass is inserted in the mold as the blower puffs on the glassblowing pipe to expand the glass to fit the mold; the glassworker has three seconds to complete this process because, as soon as the glass touches the mold, it cools and forms. The finished object has all the detail of the mold and is called a hard casting. It also still has a length of tubing called a stem attached to the top, like a rigid puppet on a stick.

Mass-producers of ornaments claim that hand blown ornaments have an inherent disadvantage in that the thickness of the glass may not be uniform, making it subject to breakage. In fact, the skilled artists in Germany have so perfected the combination of glass and glassblowing skills that handmade ornaments may be more durable.

Bulk quantities of glass are melted and flowed in a ribbon over a series of molds.

Automated glassblowing process

3 In the next step, silvering solution is injected down the stem and swirled to coat the inside of the ornament; the silvering solution can be omitted to produce a translucent ornament that only takes the color of the outer paint and has less of a reflective quality. The silvered hard casting is dipped in white undercoat and allowed to dry.

4 Designer ornaments use a palette of colors and details to achieve their uniqueness. The paints used for ornaments are slow drying and tend to run together, so the ornaments must be painted in a hop-scotch fashion leaving adjacent areas untouched until the painted areas are dry. The artist then paints the alternating areas later. Decorations including glitter and ballo, a glitter-like substance that resembles fine sugar crystals, are applied after paints have dried.

5 An ordinary glass cutter is then used to cut the stem from the ornament, and the metal cap or catch is snapped in place on the remaining stub. Tags and special packaging to identify and protect the individual ornament are added before shipping.

## Quality Control

Quality control at the ornament factory is ensured by inspections and testing at various stages of manufacture. Several instruments monitor the operation of the ribbon glass machine, providing uniform temperature and viscosity (flow) of the molten glass. After the ornaments are molded, some are selected for testing in a compression machine to verify that the walls of the ornaments are uniform and less subject to breakage. Quality in the application of coatings and decorations is also checked by random inspection.

For handmade ornaments, quality control is in the hands of the designer who commissioned production of the ornaments. The glassblowers and crafts persons who handle the ornaments during the process are paid on a piece-part basis, so quantity tends to be more important to the workers than quality. If significant quantities of the ornaments are returned, however, the factory operator stands to lose his profit. Thus, it is in his best interest to correct errors in the handling of the ornaments.

## Byproducts/Waste

There are no true byproducts from the manufacture of glass Christmas ornaments, although the ornaments themselves may have originated as byproducts of other types of glass manufacture. Waste is minimal, although there is a considerable amount of breakage. Stems, glass tubing pieces, and broken ornaments can be recycled for use by other types of glass factories.

## The Future

The design and production of glass ornaments for Christmas trees has become a huge industry, thanks to vivid new colors, creative and clever designs, a revival of interest in the uniqueness of hand blown and decorated ornaments, and the explosion of the holiday industry in general. Not only are ornaments prized as family keepsakes and heirlooms, but there are many collectors who seek one-of-a-kind ornaments for year-round display.

Designs have gone beyond traditional Christmas themes and encompass other traditions and a considerable variety from endangered species, fictional characters, elaborate treetoppers, and rainbows to AIDS and other awareness ribbons. Other countries with strong glass making traditions and inventive handicrafts have entered the booming market with Czechoslovakia, Russia, and Italy making inroads in this field. Design houses often host receptions where collectors can meet the designers for ornament signings, and limited editions of designers' work guarantee their collectibility and value. With a blending of rich traditions as background and artistic geniuses leading the design front, glass ornaments are at the cutting edge of the holiday business.

## Where to Learn More

### Books

Foley, Daniel J. *The Christmas Tree.* Philadelphia: Chilton Company, 1960.

*The Merriment of Christmas: The Life Book of Christmas.* New York: Time Inc., 1963.

Rogers, Frances and Alice Beard. *5000 Years of Glass.* New York: Frederick A. Stokes Co., 1937.

### Periodicals

*Glitter.* Bi-monthly newsletter, Raleigh, North Carolina.

### Other

Old Europe Creations Ltd. http://www.polish-ornaments.com/.

Ornament Gallery. http://ornamentgallery.com/.

Rblooms Collector's Corner. http://www.rblooms.com/.

Sacramento Valley Ornament Collectors' http://www.calweb.com/~jtech/pages5.html.

*—Gillian S. Holmes*

# Glue

*The earliest evidence of use of glue can still be observed in the cave paintings made by our Neanderthal ancestors in Lascaux, France.*

## Background

It is estimated that about 40 lb (18.2 kg) per year of glue are used for every person in America, and it is easy to see how and why when one looks at the extent of uses. Furniture, plumbing, shoes, books, buildings, and automobiles all use glue in some part of their construction.

Glues are part of a larger family called adhesives. The two classes are distinguished by the fact that glue comes from organic compounds while adhesives are chemical-based. Adhering materials called epoxies, caulks, or sealants are also chemical compounds that have special additives to give them properties suitable for particular jobs or applications.

Glue came into being when ancient tribes discovered that the bones, hides, skin, sinew, and other connective tissues from animals could be processed to remove collagen, the protein in these tissues. The collagen was sticky and was useful for holding things together. Milk solids, known as casein, and blood albumin can also be used as a basis for glue. Dried serum from cows' blood yields albumin that coagulates (clumps together) when it is heated and becomes insoluble in water.

Fish glue was also made from the heads, bones, and skin of fish, but this glue tended to be too thin and less sticky. By experimenting, early man discovered that the air bladders of various fish produced a much more satisfactory glue that was white and tasteless. It eventually was named isinglass or ichthocol.

There are three classes of substance that are called glues and that do not contain chemicals, compounds, or high-tech additives; these are bone glue, hide or skin glue, and fish glue. Technically, other sticky substances are adhesives, gums, or cements, although consumers tend to use these terms interchangeably.

Plants have also been used to produce glues collectively called vegetable glues. These materials are dispersible or soluble in water and are usually made from the starches that compose many grains and vegetables. The natural gums include agar, from colloids in marine plants, algin that is derived from seaweed, and gum arabic, an extract of the acacia tree (also known as the gum tree). The substance called marine glue is used to caulk seams, but it consists of tar or pitch and is not truly a glue.

## History

The earliest evidence of use of glue can still be observed in the cave paintings made by our Neanderthal ancestors in Lascaux, France. These early artists wanted their work to last and mixed glue with the paint they used to help the colors resist the moisture of the cave walls. Egyptian artifacts unearthed in their tombs show many uses of glues; perhaps the most striking are the veneers and inlays in wood furniture, which was made using glue as early as 3,000 B.C. The Egyptians also used glue to produce papyrus. Greek and Roman artists used glues extensively; mosaic floors and tiled walls and baths are still intact after thousands of years.

Furniture-making relies heavily on glues. Although there are many techniques for fastening pieces together, glue is often used either permanently or to align pieces while other connections are put in place. All of the great cabinetmakers from the sixteenth through the nineteenth centuries used glue in furniture construction, including Chippendale, Hepplewhite, Duncan Phyfe, the Adams brothers, and Sheraton. The glues used by these cabinet makers were made from animal hides, hooves, and other parts that had been reduced to jelly, then dried. The jelly was ground into power or flakes. It was remixed with water and heated gently in a glue pot. This product was brown, brittle, hard, and not waterproof. Yet this glue was the only glue available until World War I. At that time, casein glues made of milk and nitrocellulose glues were first manufactured.

In the 1930s, advances in the chemical and plastics industries led to development of a wide range of materials called adhesives and plastic or synthetic resin glues. World War II led to a further flowering of this industry when neoprenes, epoxies, and acrylonitriles were invented. These were used by the military and were not available for commercial use until the late 1940s or 1950s. Since that time, highly specialized, waterproof adhesives have been developed for many industries and unique applications including construction of the Space Shuttle. Glues are still used in woodworking and the manufacture of abrasives like sandpaper. They are also used as a colloid in industrial processes; colloids are added to liquids to cause solid particles that are suspended in the liquid to separate out so they can be recovered, either to clean the liquid or process the solids.

## Raw Materials

Glue manufacturers obtain bones and tissues of animals from slaughterhouses, tanneries, and meat packing companies; it is no coincidence that the world's largest glue manufacturer is the dairy called Borden Company. The animal remains that are the raw materials for glue may include ears, tails, scraps of hide or skin, scrapings from the fleshy sides of hides, tendons, bones, and feet. Similarly, manufacturers of fish glue obtain bones, heads, scales, and skins of fish from canneries and other processing plants.

*Peter Cooper*

Best remembered as a philanthropist, Peter Cooper was a prolific inventive genius and a highly successful manufacturer. Cooper was born in New York City, the son a Revolutionary army soldier who was active in numerous enterprises and involved young Peter in all of them. Although Cooper had only one year of formal education, his early experiences with his father prepared him for success in his varied business career. Apprenticed to a coachmaker at the age of 17, Cooper did so well that his employer paid him a salary and offered to back him in his own enterprise. Instead, Cooper went into the cloth-shearing business, in which he prospered. He then bought the rights to a glue-making process, improved it with his own invention, began operating a glue factory, and secured a virtual monopoly of the American glue business.

In 1828 Cooper moved into iron manufacturing, building the Canton Iron Works in Baltimore, Maryland, intending to supply the Baltimore & Ohio Railroad. The railroad was on the verge of failure, however, because of the twisting and hilly route its tracks followed. Most engineers at that time held that locomotives couldn't run on such terrain. Cooper promptly built America's first steam locomotive, which was small but powerful. In 1830, this "Tom Thumb" pulled 40 passengers at a speed of 10 miles per hour and proved that railroads could run on track that curved.

Cooper's business enterprises grew rapidly after this success. His iron business expanded into mines, foundries, wire manufactories, and rolling mills. In 1854, Cooper's Trenton factory produced the first iron structural beams for use in erecting fireproof buildings. Cooper became a principal backer and unwavering supporter of Cyrus Field's (1819-1892) project for laying the Atlantic telegraph cable. As president of the North American Telegraph Company, Cooper owned and controlled half of the telegraph lines in the United States. As an inventor, Cooper designed an early washing machine and various engines for powering watercraft.

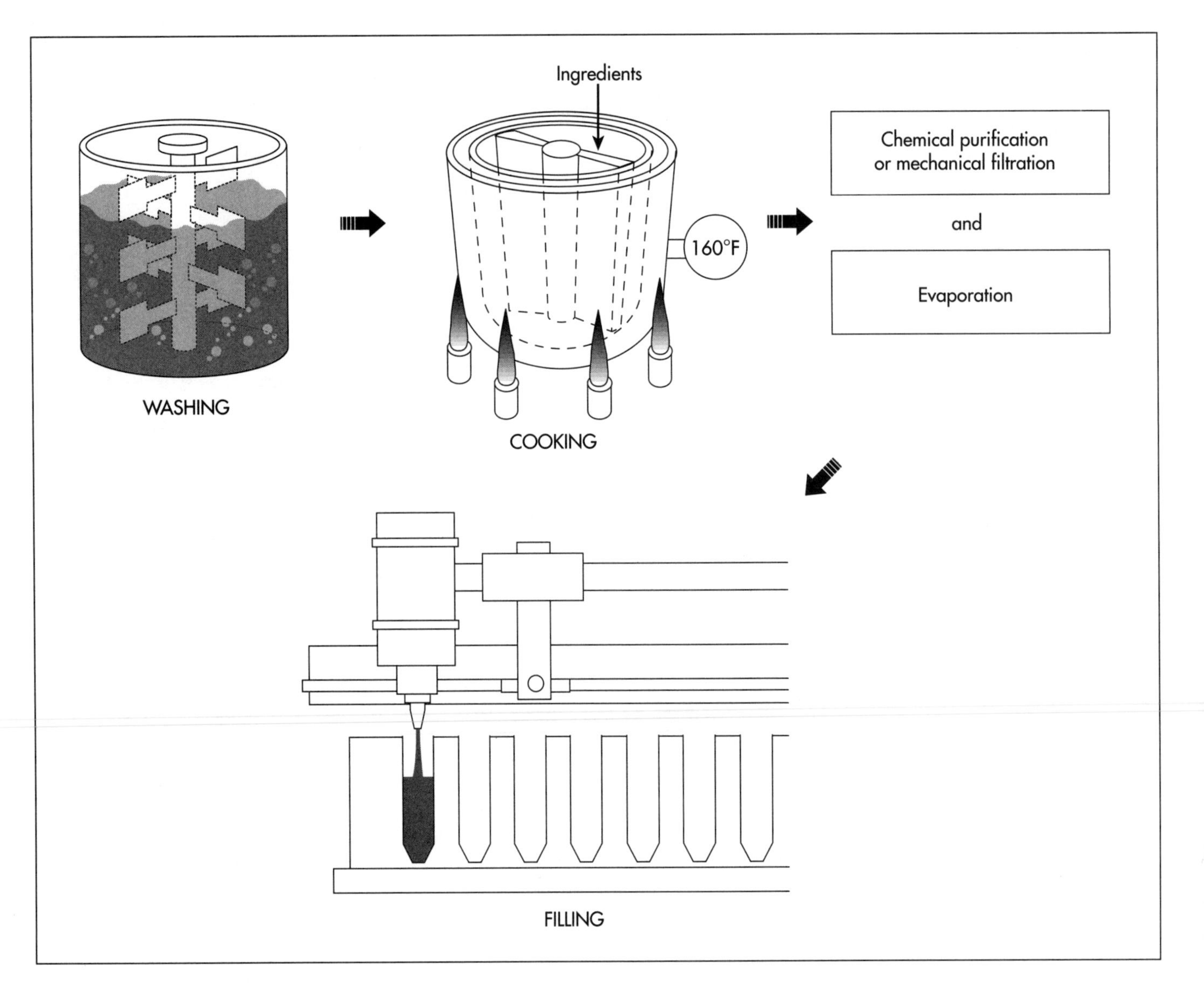


With only minor variations, the same basic processes are used to make bone glue, hide or skin glue, and fish glue. The hides and other scraps are washed so that dirt is removed, and they are soaked to soften them. This material is called stock, and it is cooked either by boiling it in open tanks or cooking it under pressure in autoclaves. The resulting liquid, called "glue liquor" is extracted and reheated again to thicken the glue. When cooled, this material looks like jelly and is solid. To remove the impurities and make the glue clear, chemicals like alum or acid followed by egg albumin may be added. These chemicals cause the impurities to precipitate, or fall out, of the glue. The glue is made more concentrated in vacu-

## The Manufacturing Process

### *Making hide or skin glue*

1 With only minor variations, the same basic processes are used to make bone glue, hide or skin glue, and fish glue. The hides and other scraps are washed so that dirt is removed, and they are soaked to soften them. This material is called stock, and it is passed through a series of water baths in which more and more lime is added to make the hides and skins swell and break them down. The swollen hides are rinsed in a large washing machine to remove the lime. The last traces of lime are eliminated by treating the stock with weak acids like acetic or hydrochloric acid. Finally, the stock is cooked either by boiling it in open tanks or cooking it under pressure in autoclaves.

2 Cooking at the correct temperature and for the right length of time breaks down the collagen and converts it into glue. If the temperature or timing is off, the quality of the glue will be ruined. Large steam coils in the open tanks heat the water and product to 160°F (70°C). Three or four treatments with clean water are performed at increasing temperatures (or pressures if a pressurized system is used). The resulting liquid, called "glue liquor" is extracted and reheated again to thicken the glue.

3 When cooled, this material looks like jelly and is solid; although it looks like the kind of gelatin used in food, it contains impurities. To remove the impurities and

make the glue clear, chemicals like alum or acid followed by egg albumin may be added. These chemicals cause the impurities to precipitate, or fall out, of the glue. Mechanical methods can also be used to clean the glue. These include passing the glue through a series of mechanical filters or through paper filters or ground bone called bone char.

4 Different additives are mixed with the glue liquor to make brown, clear, or white glue. Sulfurous acid, phosphoric acid, or alum are among these additives. Zinc oxide is added to produce white "school glue."

5 To this point, the glue is a weak, runny liquid. It is made more concentrated in vacuum evaporators and dried in one of several methods. The glue can be chilled into either sheets or blocks then suspended on nets to dry and become still more concentrated. The glue can also be dropped as beads or "pearls" into a non-water bearing liquor that further dries the concentrated beads. The pearls, blocks, or sheets are then mixed to the right consistency and pumped into bottles or jars for sale.

### *Making bone glue*

Manufacture of bone glue is somewhat more complicated. Bones are processed most often in pressure tanks, but additional processing is needed to remove the minerals. The bones are degreased with solvents, then hydrochloric acid in an 8% solution is applied to the bones. The acid removes calcium phosphate and other minerals and leaves collagen in the same shape as the piece of bone. The acid is removed from the collagen, and it is dried to produce commercial-grade ossein or bone protein (also termed acidulated bone) that is the basis for bone glues. After the ossein is created, it can then be processed in the open-tank method and the subsequent steps used to make glue from hides, as described above.

## Quality Control

All processes in the manufacture of glue are monitored carefully using instruments, computerized controls, and observation. Improper temperatures or pressures will ruin large quantities of stock that must then be wasted; manufacturers will not risk such errors. Safety and sanitation are also major concerns. Glue manufacturers tend to be located very close to supplies of hides and other raw materials to prevent disease, vermin, contamination, and major costs like transportation. Workers' safety is carefully monitored, as is the production of a pure glue.

## Byproducts/Waste

Glue itself is a byproduct of dairies, meat processing plants, and other facilities that generate the raw materials needed for glue production.

## The Future

Glues are essential to our future. More and more manufacturing processes are using various forms of glue (and including adhesives) to replace stitching, stapling, and more expensive (and less effective) forms of fastening. Experiments with medical glues suggest that one-third of all wounds may be "stitched" with glues in the next few years. Glues have proven to be so versatile that scientists are constantly watching for new applications that will make our lives simpler.

## Where to Learn More

### *Books*

Giles, Carl and Barbara. *Glue It!* Blue Ridge Summit, PA: TAB Books Inc., 1984.

Miller, Robert S. *Adhesives and Glues: how to choose and use them.* Columbus, OH: Franklin Chemical Industries, 1980.

### *Periodicals*

Allen, Laura. "Sticky stitches." *Science World* (October 6, 1997): 6.

"Sticky stuff." *Science Weekly* (January 10, 1996): 1.

### *Other*

Advanced Adhesive Technology, Inc. http://aatglue.com/.

American Chemical, Inc. http://www.glueit.com/.

The Gorilla Group. http://www.gorillaglue.com/.

*—Gillian S. Holmes*

# Golf Tee

*It is estimated that the U.S. golf markets use about two billion wood golf tees per year.*

A golf tee is a small device used to prop up a golf ball. It is typically used on the first shot of each new hole during a game of golf. Golf tees are made in a highly automated fashion. The exact method depends on the material; however, they are generally formed or cut, then finished and packaged. First patented in the late nineteenth century, the design of a golf tee has changed only slightly. It is estimated that the U.S. golf markets use about two billion wood golf tees per year. With the expected increase in golf popularity, this number is expected to rise significantly.

## History

While the basic design of the golf tee has changed little since the late nineteenth century, it has evolved significantly throughout the game's history. There are a variety of theories which suggest how and were golf began. The earliest evidence is a game similar to golf played by the ancient Chinese as early as 300 B.C. According to some, the modern game of golf finds its origins with a game played by during Roman times. This game called *Pagancia* involved hitting a leather ball stuffed with feathers. Others suggest that it was derived from the French game *chole* played in the 1300s or the English game *cambuca.* Both of these involved hitting a ball with a stick. The most direct ancestor to golf was a game called *kolfspel* played in Holland around 1295. This game required a player to hit a ball with a wooden club, or kolf, into a series of targets. To get a clear shot, players were allowed to elevate the ball on a *tuitje,* which was a small pile of sand. This was the first type of golf tee known. The modern game of golf developed in Scotland during the 1400s. By 1735, the first golf organization was formed. The first tournament was played in 1744. This tournament followed a set of 13 rules that became standard in 1754. The one exception was that a ball could not be placed on a tee to improve a watery lie.

In 1899, the first patent for a golf tee was issued to George Grant. Since this time a variety of innovations have been introduced. Some of these innovations relate to the tee's shape. For example, during the 1970s two patents were granted for golf tees that had unique designs which were supposed to help improve the flight of the ball. Other patents were issued for golf tees that stayed in the ground more consistently after the ball was hit. In the 1980s, an angled golf tee was introduced. The composite material of certain golf tees has also changed. For instance, a 1991 patent describes a golf tee made out of a biodegradable resin. Additional materials introduced recently include clay tees, tees made from corn, and tees composed of animal byproducts. Although more than 25 patents have been issued for improved golf tees in the last 20 years, the most popular tees still have the same basic shape as the ones first made a century ago.

## Raw Materials

Conventional golf tees are generally formed from wood or plastic. The primary wood used is cedar wood. One source is the red cedar, an evergreen conifer grown in the eastern United States. A variety of synthetic plastics have also been used to construct golf tees. Plastics are high molecular weight polymers that are formed through various chemical reactions. Most plastic golf tees are made using polypropylene or high density polyethylene (HDPE).

To make the primary raw materials easier to work with other materials are used. For example, fillers are added to change the plastic's properties. These fillers help control the flexibility of the plastic and make them more lightweight. They can even help make the tees less prone to breaking. For decorative purposes, colorants may be added to the plastic to modify the tee's appearance. Paint is used to coat wooden tees making them more attractive and weather resistant. Glue is also used for the production of wooden tees. Rubber has also been used to modify the characteristics of certain golf tees.

A variety of biodegradable tees have been recently introduced. One type of biodegradable tee is based on animal byproducts. By combining materials derived from skin, scales, bone, and soft tissue manufacturers have been able to produce an environmentally benign product that degrades automatically. Another type of biodegradable tee is based on a corn derivative. When this tee gets wet its molecular structure breaks down and it gets slowly washed away. Still other golf tees of this type are mixed with fertilizer and grass seed to improve their compatibility with the course.

## Design

A typical golf tee is a small, solid piece of wood or plastic with a wide, flattened head and a thin, pointed shaft. The head is grooved slightly to accommodate a golf ball. The pointed tip is designed to be inserted perpendicularly into the ground. The most common size for a golf tee is 2.125 in (5.4 cm) long. Tees are available in a wide range of colors including white, yellow, orange, light blue, pink, and green. Red, dark blue, black, silver, and natural are also available.

Another type of golf tee is an angled golf tee. This tee is designed to be inserted into the ground at an angle. The ball receipt surface is varied and can be more recessed or more perpendicular to the longitudinal axis. This golf tee can help reduce the effects of undesirable rotation that arises from contact with the head of the golf club. In this way, it can reduce hooking or slicing of the ball. Some tees are shaped differently to prevent disruption of the ball trajectory. Other tees have been invented to increase ball distance and improve accuracy. Some golf tees have adhesives applied to reduce rotational spin.

One problem with traditional golf tees is that after a player makes a shot, the tee takes flight with the ball and is lost. For this reason, golf tees that have a vented side wall are available. The venting is thought to minimize adverse effects of depressurization between the golf tee and the accelerating golf ball. The tee tends to remain stationary and the ball retains more of its momentum from the club. Another fix to this problem is found in golf tees that have barbs on their tips so they resist movement when the ball is hit.

For plastic tees, the most critical part of the manufacturing process is designing the mold. A mold is a cavity carved into steel. When molten plastic is introduced into the mold, it takes on the mold's shape when it cools. During manufacture, the mold cavity is highly polished because any flaw on the surface will be reproduced on the plastic. For making golf tees, a two piece mold is used. The pieces are joined together briefly to form the tee and then released. Special release agents are used to make the tee easier to remove. Steel molds are highly precise and can produce exact tees each time. When molds are designed, however, they are made slightly larger to compensate for the fact that the plastic shrinks as it cools

## The Manufacturing Process

The process of making a golf tee can be broken down into three basic segments. First the raw materials are prepared. Next, the golf tee shape is formed. Finally, the golf tee is finished and put into packaging.

### *Material preparation*

1 To begin the process of producing golf tees, the incoming raw materials must be prepared. For golf tees made of wood, this means that large logs are stripped of their bark and cut down to smaller wood blocks. For plastic tees, plastic pellets are mixed with colorants and other strengthening ingredients. The raw materials for biodegradable tees are mixed together in a large tank in the proper proportions. In all cases, after

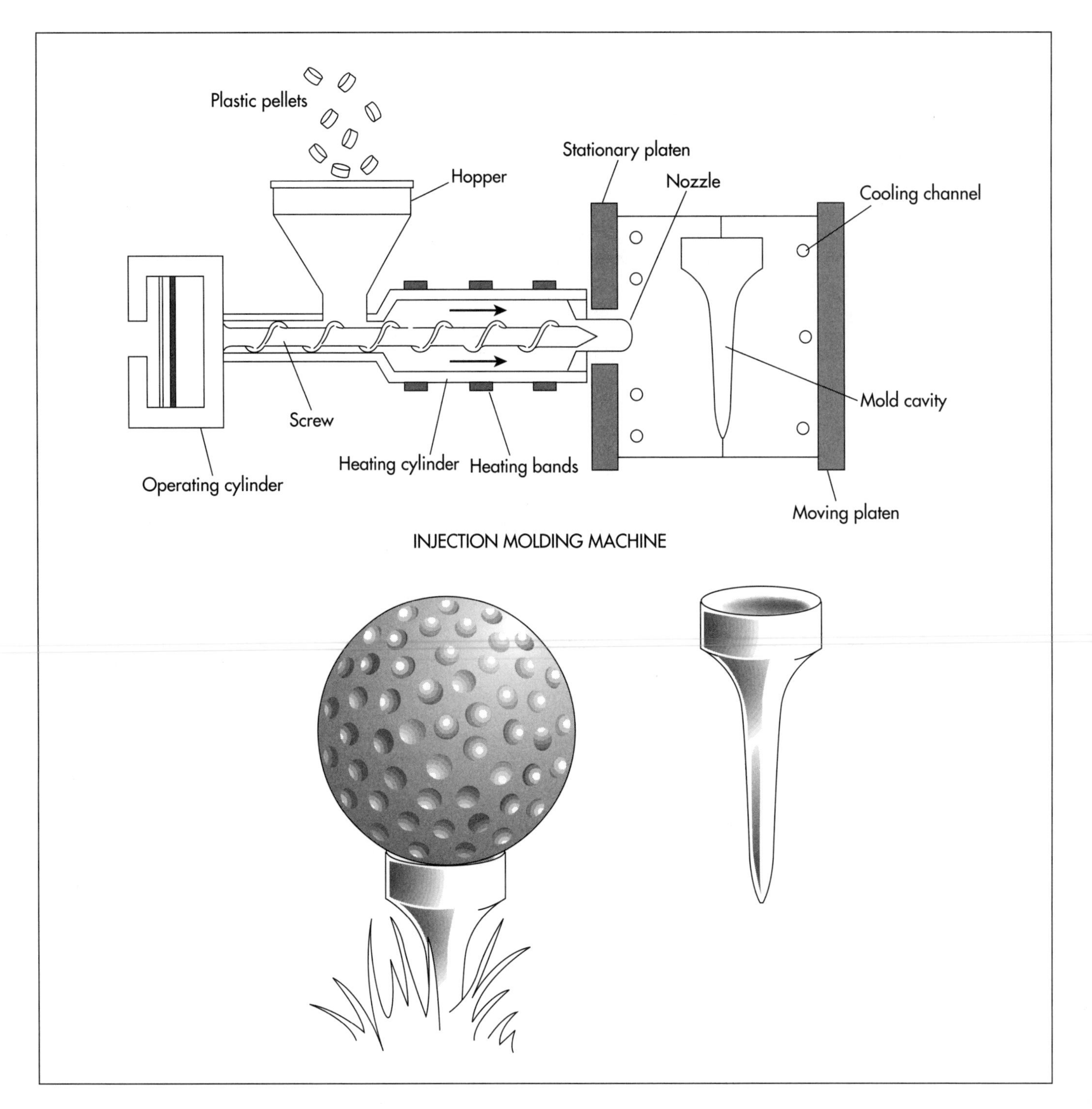


Plastic golf tees are manufactured using an injection molding machine. In this process, plastic pellets are loaded into a hopper and passed through a hydraulically controlled screw and melted. The liquid plastic is forced into a golf tee mold and held under pressure until it is set. As the plastic cools, it hardens into the shape of the mold. The mold opens and ejects the tee onto a conveyor belt.

this step the starting raw materials are sent to the forming stage.

### *Forming the tee*

2 In the forming section, the raw materials are shaped into tees. This can be done through injection molding or cutting. Injection molding is used for most non-wood tees. In this manufacturing step the plastic pellets are loaded into the hopper of an injection molding machine. They are passed through a hydraulically controlled screw and melted. The screw is then turned and the plastic is physically injected into a mold. Just prior to injection, the two halves of the mold are brought together. They form a cavity which has the same shape as a golf tee. The plastic is held in the mold under pressure and allowed to cool. While cooling, it hardens. The tee is pushed out onto a conveyor belt after the mold is

opened. The mold is then brought together and the process repeats. Extraneous plastic is trimmed from the molded tees.

3 Cutting is utilized for wood tees. The wooden blocks run on a conveyor under a machine which carves out tee slats. These are shapes that look like one half of the tee if cut longitudinally from the head to the tip. The tee slats are then glued together like a sandwich to form a complete tee.

### *Coating and packaging*

4 After the primary tee shape has been formed the tees are sent through other processing steps before being packaged. Wooden tees may be cut and shaped further to meet design characteristics. They are then sanded smooth. Plastic tees can be shaken and cleaned to remove residual plastic pieces.

5 The tees are then put through the decorating phase. Wood tees are coated with a weather resistant paint. They may then be passed through a lettering machine that paints on a company name or logo. Certain higher priced tees have the lettering engraved prior to painting. Plastic tees can have silk-screened lettering; however, this is more expensive and is typically not done.

6 The finished tees are then put into the final packaging. This can be a small plastic bag or a large box. To ensure that the correct number of tees are dropped into the package, they are first weighed. The bag is sealed and placed into larger cases. These cases are stacked on pallets and shipped to distributors.

## Quality Control

The quality of the golf tees are checked during each phase of manufacture. Since thousands of tees are made daily, a complete inspection of each piece is not possible. However, using statistical models as a guide, production line inspectors randomly pull samples of tees at fixed time intervals. These samples are checked to ensure they meet size, shape, and consistency specifications. By following this sampling method, the manufacturer can determine the overall quality of the entire production run. The primary test method is visual inspection. Quality control workers check for things such as deformed, damage or improperly labeled tees. Additionally, more rigorous measurements can also be performed. Measuring equipment is used to check the length, width and thickness of each part. Typically, devices such as a vernier caliper, a micrometer, or a microscope are used. Each of these differ in accuracy and application. If a tee is found to be out of the specification range, it is removed and set aside to be remelted or reformed into a new tee. The machinery is recalibrated and tested in order to eliminate the defect.

## The Future

With the continued growth in popularity of the sport of golf it is anticipated that golf tee manufacturers will try to improve on their product. Many of them have already introduced biodegradable products to help reduce the number of trees destroyed yearly and relieve the litter caused by the increased number of broken tees left on the golf course. One of the drawbacks to the currently available biodegradable products is that they break more easily than wooden or plastic tees. Manufacturers will no doubt address this problem in the near future.

## Where to Learn More

### *Books*

Purkey, Mike. "Good Form." *Golf Magazine* (April 1998): 170.

Zumerchick, J. editor. *Encyclopedia of Sports Science*. Simon & Schuster MacMillan, 1997.

### *Other*

United States Golf Association (USGA). 1996. http://www.usga.org/ (June 28, 1999).

*—Perry Romanowski*

# Green Tea

A total of about 100,000 tons of green tea is produced per year from 60,000 hectares of tea fields. Only green tea is produced in Japan.

## Background

In 1992, global production of all tea was almost 2.5 million tons. The majority of tea production occurs in the subtropical areas of Asia, including China, India, Sri Lanka, Japan, and Indonesia. More than 35 countries now produce tea, with India, China, and Sri Lanka the leaders. Black tea is the most produced, followed by oolong and jasmine tea. Besides the distinction between varieties of tea, the major difference between the type of teas is the processing method. Green tea leaves are picked and immediately sent to be dried or steamed to prevent fermentation, whereas black tea and other types are left to ferment after they are picked.

Green tea originated in China for medicinal purposes, and its first recorded use was 4,000 years ago. By the third century, it became a daily drink and cultivation and processing began. Today, China has hundreds of different types of green teas. Other producers of green tea include India, Indonesia, Korea, Nepal, Sri Lanka, Taiwan, and Vietnam.

Green tea was first introduced in Japan during the Nara period (710-794), when numerous Japanese Buddhist monks visited China and brought tea seeds back to Japan. The Japanese tea industry is said to have begun in 1191, when the monk Eisai planted tea seeds from China on temple land. He then encouraged the cultivation of tea in other areas of Japan by extolling the health benefits of tea drinking.

The making and serving tea as an art form (*sado*, the way of tea) was introduced in Japan during the eleventh century. The origins go back to China's Tang dynasty (618-907), when a ritual was performed in Buddhist temples. A brick of tea was ground to a powder, mixed in a kettle with hot water, and ladled into ceramic bowls.

One of the first Japanese uses of the tea ceremony in public was when Toyotomi Hideyoshi, then the most powerful warlord in Japan, held a tea party in his camp the evening before a large battle in order to calm his warriors and inspire morale. Hideyoshi's own sado teacher, Senno Rikyo, is also credited with elevating tea from a simple beverage to a highly respected method of self-realization. Today, there are tea schools in Japan to learn the proper methods of the tea ceremony or *chanoyu*. The Urasenke School is the most active and has the largest following.

The form of chanoyu that is practiced today was established in the second half of the sixteenth century by Rikyu. Chanoyu involves more than merely enjoying a cup of tea in a stylized manner. The ceremony developed under the influence of Zen Buddhism aims to purify the soul by becoming one with nature. The true spirit of the tea ceremony has been described by such terms as calmness, rusticity, and gracefulness. The rules of etiquette are carefully calculated to achieve the highest possible economy of movement.

For some 500 years after tea was introduced to Japan, it was used in its powdered form only. It was not until the mid-sixteenth century that the processing method for conventional green tea was invented. Prior to the Edo period (1600-1868), the consumption of tea was limited to the ruling class. Only after the beginning of the twentieth century, with the introduction of mass production techniques, did tea achieve widespread popularity among the general population.

Today, tea leaves for green tea are grown in the warmer southern regions of Japan, with about half produced in Shizuoka Prefecture. Uji, a district near the ancient city of Kyoto (and the district from which the finest Japanese tea comes from to this day) became the first tea-growing region in Japan. Later, tea plantations were planted in Shizuoka Prefecture and, finally to surrounding regions. A total of about 100,000 tons of green tea is produced per year from 60,000 hectares of tea fields. Only green tea is produced in Japan.

Though traditionally green tea was produced manually, the process has been fully mechanized in Japan. The various types of tea now produced differ according to cultivation practices and processing methods. Sencha is a tea with three quality levels: high, medium, and low. It is manufactured from the tender top two leaves and the shoots for the high and medium grades and from the third from the top leaf for the low grade.

Sencha, which comprises 80% of all green tea production, consists of tiny dark green needle-shaped pieces. Almost immediately after picking, the leaves are steamed for about 30 seconds to seal in the flavor, followed by drying, pressing, and rolling steps.

Gyokura is the highest grade of tea and is made from the most tender leaves that are grown under 90% shade using bamboo blinds. Matcha is made from similar leaves and is processed into a powder form for exclusive use in the tea ceremony. Bancha is a low-grade coarse tea made from older leaves picked after Sencha leaves are picked or picked in the summer. It is generally composed of lower grade tea leaves, which are divided into two kinds: large leaf, and small leaf.

Houjicha is a wedge-shaped tea made from Bancha that is roasted at 302° F (150° C) to prevent fermentation and produces a light golden color when made. Kamairicha comes from northern Kyushu and is first roasted at 392-572°F (200-300°C) followed by cooling at 212°F (100°C). Green tea is traditionally served without sugar, milk, or lemon since these would destroy the true flavor and aroma of the tea.

## Raw Materials

Green tea is made from the top two leaves and buds of a shrub, *Camellia sinensis*, of the family Theaceace and the order Theales. This order consists of 40 genera of trees or shrubs that have evergreen leaves, flowers with five sepal or leaf-like structures and petals. The genus *Camellia* consists of 80 species of East Asian evergreen shrubs and trees. Besides the leaves, other ingredients may be added to create special scents or flavors during the drying process, such as jasmine, flowers, or fruits.

The tea plant originates in an area between India and China. There are three main varieties of this plant—China, Assam, and Cambodia—and a number of hybrids in between. The China variety grows as high as 9 ft (2.7 m) and has an economic life of at least 100 years. The Assam variety is a tree that grows as high as 60 ft (18.3 m), with an economic life of 40 years dependent upon regular pruning and plucking. The 16 ft (4.9 m) high Cambodia variety is naturally crossed with other varieties.

## The Manufacturing Process

### *Cultivation and harvesting*

1 A suitable climate for cultivation has a minimum annual rainfall of 45-50 in (114.3-127 cm). Tea soils must be acid since tea plants will not grow in alkaline soils. A desirable pH value is 5.8-5.4 or less. Tea can be cultivated up to 7,218.2 ft (2,200 m) above sea level and can grow between the equator and the forty-fifth latitude. The plants are reproduced through tile-laying or through seeds from trees that have grown freely.

2 A crop of 1,500 lb (681 kg) of tea per acre requires up to two workers per acre to pluck the tea shoots by hand and maintain the field. The tea plant is generally plucked every five to 10 days, depending on where it grows. The length of time needed for the plucked shoot to redevelop a new shoot ready for plucking varies according to the plucking system and the climatic conditions. Intervals of between 70-90 days are common.

Harvesting an average tea planting of 1,500 lb (681 kg) of tea per acre requires up to two workers per acre to pluck the tea shoots by hand and maintain the field. The tea plant is generally plucked every five to 10 days, and the length of time needed for the plucked shoot to redevelop a new shoot ready for plucking varies according to the plucking system and the climatic conditions. Intervals of between 70-90 days are common. A bud and several leaves are picked from each plant. In Japan, the first crop is harvested in April and May, the second crop in June, the third crop in July and the final crop in September.

Harvesting

3 In Japan, the tea harvesting begins around the end of April, with the leaves picked by hand or machine. A bud and several leaves are picked from each plant. The first crop is harvested in April and May, the second crop in June, the third crop in July and the final crop in September. For gyokuro or matcha tea, the plants are shaded for two weeks after the first bud comes out in spring before picking. The leaves are then shipped to the factory for processing. Since not all can be processed at once, the leaves are stored in a large bin that is kept at the proper temperature by blowing cool air into the bottom.

### Drying

4 After the tea leaves are plucked, they must be dried to prevent fermentation, which stops any enzyme activity that causes oxidation. In China, green teas are often pan-fired in very large woks, over a flame or using an electric wok. The tea leaves must be stirred constantly for even drying. Withering is also used, which spreads the tea leaves on racks of bamboo or woven straw to dry in the sun or using warm air. Again, the leaves must be moved around to ensure uniform drying.

5 In Japan, steaming is normally used. Before the steaming process begins, the tea leaves are sorted and cleaned. The steaming time determines the type of tea that is produced. Sencha tea is normally steamed for 30-90 seconds. Another type of sencha called fukamushi is steamed for 90-150 seconds to produce a flaky light yellowish green tea. Steaming is conducted in a bamboo tray over water or by a revolving or belt-conveyor type machine. After mechanical steaming, the leaves go into a cooling machine that blows the water from the leaves.

### Shaping

6 In most countries, rolling or shaping green tea leaves is done by machinery. In China, high-end leaves are hand-rolled into various shapes, including curly, twisted, pointed, round, and more. Rolling the tea creates a distinctive look, as well as regulates the release of natural substances and flavor when it is steeped in the cup.

7 In Japan, a number of rolling and drying steps take place. A special machine is used to accomplish the first rolling and drying steps simultaneously and takes about 48 minutes. The tea leaves are dried to improve their strength so they can be pressed during the next drying process. Moisture from both the surface and from the inside of the tea leaves is removed using this machine.

8 This machine consists of a spindle with finger-shaped extensions that stir the leaves while heated air (at 93.2-96.8° F [34-36° C]) is blown into the machine. Though the rolling temperature is automatically controlled by the computer, it is still important

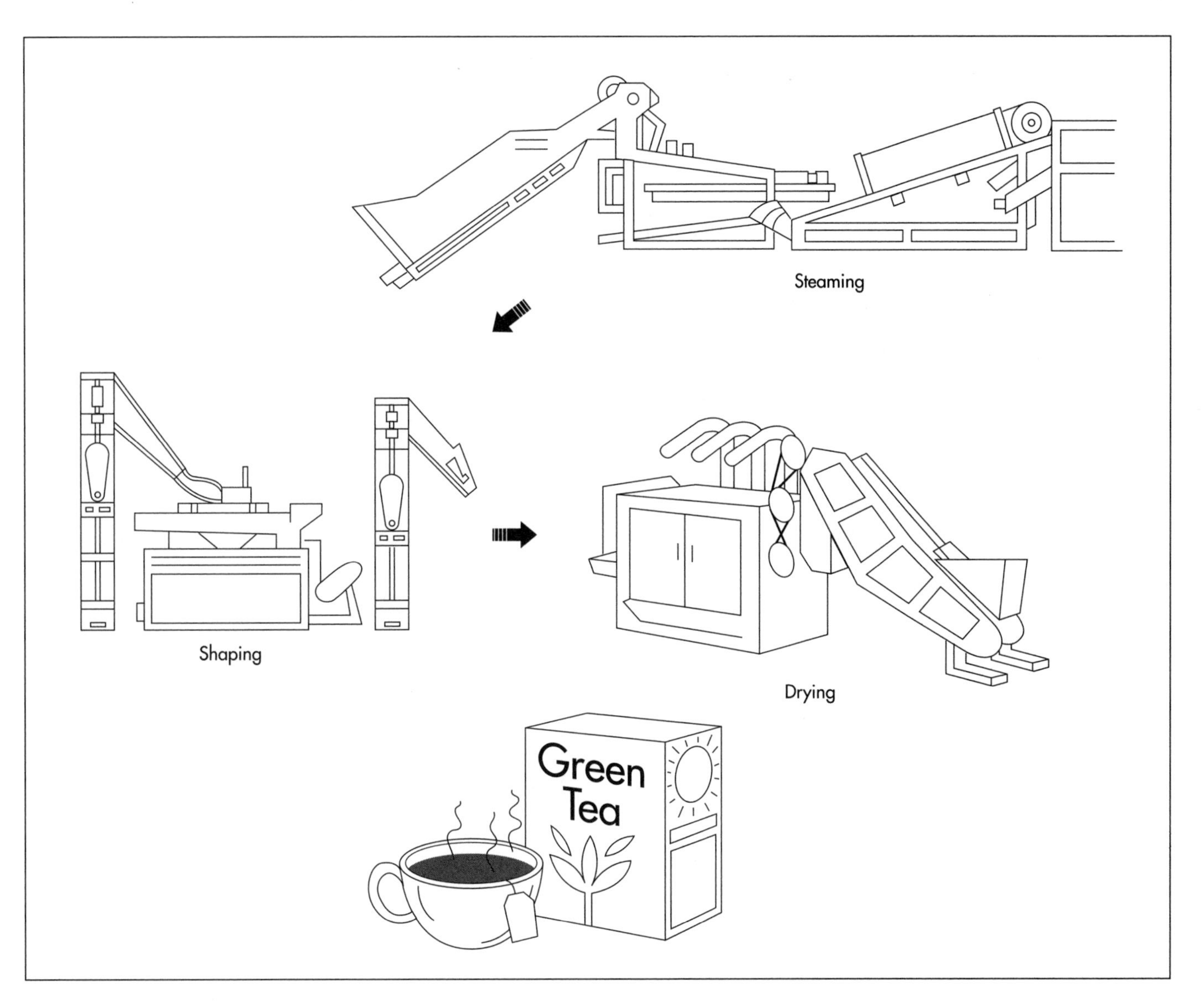


for the operator to touch the tea by hand to make sure it feels right.

9 Since the moisture level still varies for different parts of a leaf or from one leaf to another at the end of the first step, another rolling process takes place to uniformly distribute the remaining moisture in the leaves. This process rolls the leaves by pressing under a rotating disk to bring the moisture from the center of the leaves to the surface. The process is conducted at room temperature for 24 minutes.

10 Next, the leaves go to another rolling/drying machine, which uses a spinning pedal inside of a revolving drum to convert the leaves into a round shape. This process takes about 40 minutes. It is very important to take out the leaves at the same moisture level every time.

11 The tea leaves are removed from the previous machine, separated into small portions and placed in pots. They are gradually rolled into tiny round or needle shapes using a weight. This step takes 40 minutes and removes most of the moisture. The total process thus far takes about three hours compared to hand rolling and heating which can take up to 10 hours.

### *Final drying*

12 In Japan, green tea must be dried for about 30 minutes after the final rolling step for storage. The tea is spread on a caterpillar-type device and dried slowly to about 5% moisture content or less. At this stage

Once harvested, the tea leaves are dried in order to prevent fermentation. In China, pan firing is used. In Japan, the leaves are steam dried. The dried leaves are then shaped or rolled in order to regulate the release of natural substances and flavor when it is steeped in the cup, as well as create a uniformly rounded leaf. In Japan, the tea is dried once more to reduce its moisture content to about 5% before packaging.

the half-processed tea, called aracha, is shipped to tea merchants or wholesalers for final processing. Aracha is not uniform in size and still contains stems and dust.

### *Post-processing*

13 After the tea is shipped to the wholesalers in Japan, it undergoes several other steps to produce the final product. A special machine grades and cuts the tea by particle size, shape, and cleanliness, depending on the final qualities desired. The machine uses mechanical sieves or sifters fitted with meshes of appropriate size, as well as cutting devices to achieve a quality tea. Another drying step follows to produce the aromatic flavor, followed by blending per customer's specifications, packing and finally shipping to retail shops. In other countries, similar sorting, weighing, and packaging steps occur after the shaping process.

## Quality Control

The quality of green tea depends first on using good tea leaves. The natural quality of the leaf, including color and aroma, must then be preserved during the manufacturing process to produce a good green tea. In Japan, this involves controlling the temperature to 93.2-96.8° F (34-36° C) during rolling, drying, and storage. Since tea leaves can generate their own heat, cool air is blown into the bottom of the container to keep the leaves at the proper temperature during storage.

The Japanese government also subjects all exported tea to a strict inspection. Standard samples, which are established at the beginning of the tea season each year, are used to compare various properties of the finished product with the samples. Leaves, stems, moisture, content, flavor, taste, and color are all rigidly examined. There is also a stringent chemical analysis to determine tannin, caffeine, vitamin, and mineral contents. Tea is exported only after passing these tests.

## The Future

Though the health benefits of green tea have been known for centuries, recent research is providing concrete evidence of these benefits. Studies have shown that green tea can prevent cancer since it contains catechin, the major component of tea. A study in Japan showed that residents in areas devoted to green tea production in the central and western regions of Shizuoka Prefecture, who drink the tea daily, have a significantly lower death rate for all types of cancer compared to other regions.

These findings were supported by animal experiments that showed green tea reduced the growth of tumors. Other research has shown that green tea consumption may inhibit nitrosamine formation—known carcinogens or cancer-causing chemicals.

Green tea catechin has also been shown to limit the excessive rise in blood cholesterol in both animals and humans, as well as prevent high blood pressure. Other benefits of catechin include killing bacteria and influenza viruses, preventing halitosis, inhibiting increase of blood sugar, and fighting cariogenic bacteria. Green tea (especially matcha) also contains important vitamins (C, B complex, and E), fluoride (for preventing cavities), amino acids (for lowering blood pressure), and polysaccharides (lowers blood sugar). Green tea is a strong antioxidant as well and is even more powerful than vitamin E or vitamin C due to the presence of polyphenols, such as epigallocatechin gallate (EGCG).

Extracts of green tea may also make strains of drug-resistant bacteria that cause skin infections more sensitive to penicillin, British researchers report. The investigators also found that diluted tea extract acted synergistically with antibiotics, making them more potent against particular strains of this type of bacteria.

In addition to preventing or curing these more common diseases, preliminary research indicates the antiviral capability of green tea catechin may have some beneficial effect in fighting AIDS. Laboratory tests have verified that catechin can inhibit the activity of the AIDS virus. Instead of simply being known as a popular Japanese beverage, green tea may thus become an important "new" medicine of the twenty-first century for the entire world.

## Where to Learn More

### Books

Mitscher, Lester A. and Victoria Dolby. *The Green Tea Book: China's Fountain of Youth.* Avery Publishing Group, 1997.

Oguni, Dr. Itaro. *Green Tea and Human Health.* University of Shizuoka, Japan Tea Exporters' Association.

Okakura, Kakuzo. *The Book of Tea.* Dover Publications Inc., 1964.

Rosen, Diana. *The Book of Green Tea.* Storey Books Inc., 1998.

### Other

Japan Tea Exporter's Association. 17 Kitabancho, Shizuoka, Japan 420-0005. +81-54-271-3428. Fax: +81-54-271-2177.

Maruichi Green Tea Farm. http://www.maruichi-jp.com (February 2, 1999).

The Teaman's Tea Talk. http://www.teatalk.com (June 30, 1998).

The Teapot Salon. http://www.iris.or.jp/~hamadaen/ (1996).

*—Laurel Sheppard*

# Greeting Card

*Over 1,500 greeting card manufacturers sell an estimated seven billion cards each year. Each household receives an average of 80 cards annually.*

## Background

Greeting cards are pieces of paper or cardboard upon which photos, drawings, and a verse of cheer, greeting, celebration, condolence, etc. have been printed or engraved. Greeting cards are decorated with a variety of images and include messages to appeal to diverse audiences, sentiment, and occasion to be remembered. Greeting cards are easily made at home using pen and paper or software sold by greeting card and other companies. Recently, virtual cards that include images and verse can be sent to someone by way of the Internet and e-mail and may be printed out on paper by the receiver. Despite the electronic availability of these cards, the greeting card industry continues to sell cards in retail store in huge numbers. Over 1,500 greeting card manufacturers sell an estimated seven billion cards each year. Each household receives an average of 80 cards annually.

The market research associated with the development of a successful greeting card is just as important as attractive graphics or appropriate verse. Research has pushed large greeting card companies to expand traditional product lines and offer cards for pets, step-siblings, divorce, weight loss encouragement, company lay-offs, and more. Some smaller greeting card companies specialize in the production of cards that appeal only to one or two specific markets. Greeting card companies require a diverse talent pool in order to produce commercially-successful product, and these forms employ everyone from cartoonists to market researchers to pressman who print the cards.

## History

Some speculate that ancient Egyptians may have recorded greetings upon papyrus and sent them via messenger to the intended parties, and it seems plausible that the ancient Greeks recorded sentimental verse on scrolls, as well. By the late Middle Ages, letters and messages of love, including romantic verses sent near St. Valentine's Day, were exchanged throughout Europe. Personal messages of greeting and sentiment were individually crafted until the mid-nineteenth century. The first commercially-produced greeting card was a Christmas card invented in 1846 by British businessman Henry Cole who asked a printer to produce a printed Christmas greeting he could quickly send to friends. The idea caught on and mass-produced Christmas cards were popular by the 1860s. Louis Prang, an American printer who invented a multi-color printing process called chromolithography, fashioned beautifully colored cards by the 1870s. Cards for Easter, birthdays, baby arrivals, etc. soon followed. The larger American card companies were founded in the early years of the twentieth century and a number exist today and remain leaders in card sales. Innovations in card production have primarily revolved around developing efficient printing methods, diversifying the product offering by nurturing a large creative talent pool, and devising more effective point-of-sale displays so consumers can easily see the products in an attractive display.

## Raw materials

Greeting cards are made of card stock that may be of wood pulp or part "rag" (textile waste)—sturdy, fairly expensive paper. Increasingly, these card stocks are being made with recycled materials. Many, but not all, of the companies put a glossy aqueous coating consisting of water and a water-based acrylic coating on the stock after printing particularly when a photograph is featured. Inks vary as well. Many companies are moving toward the use of soy inks, containing water-based solvents and are more easily cleaned, recycled, or disposed of than oil-based solvent inks. Soy ink composition varies with the printing process; cards are most often printed using sheet-fed printing and the soy ink for that includes between 20%-30% soybean oil, resins, pigments, and waxes.

## The Manufacturing Process

The manufacture of greeting cards varies greatly depending on the size of the corporation. Successful greeting card companies put a great deal of importance on business research, marketing, and creative design because these help determine what cards will sell well.

### *Research and marketing*

1 Generally, before the artists and verse writers begin to put pen to paper, large companies support in-house researchers that learn all they can about potential buyers. These researchers find out all they can about consumers' wants, needs, and concerns that can be addressed in a greeting card not already in production. The researchers use statistical analysis, market research, and research on lifestyle changes. Once the Research Department has an idea for a new card line, they utilize focus groups, surveys, and controlled store tests to gauge the potential of the new product. For example, research may indicate that changes in the American family calls for cards that acknowledge step-siblings, or suggest that soaring numbers of cat owners will lead to a successful line of cards offering sympathy on the death of a pet.

### *Designing the card*

2 Smaller greeting card companies sometimes contract designers to provide sketches and ideas they feel will be good sellers and fit their niche markets. However, in larger companies, the Research and Marketing departments work closely with the Creative Department in order to collaboratively devise a new card. These larger companies employ an in- house creative staff that includes artists, graphic designers, photographers, writers, editors, and copywriters. This staff provides the illustrations and verse featured in the product. The Creative Department "marries" the sketch to the appropriate verse and creates a handmade card. Once the Marketers and Researchers are pleased with these mock-ups are examined and rated by consumer panels or focus groups. The prototypes deemed most marketable are then moved into technical production.

### *Graphic design and production preparation*

3 When the designs are approved to for production, graphic artists and technical production assistants are key to translating original artwork and scrawled words into a pleasing, coordinated product that can be mass produced. Thus, graphic designers might re-size artwork to make it fit a card, add color underneath or on top, combine images with appropriate typefaces for the verse inside, etc. The graphic designers must understand the capabilities of the printing machines and use only those numbers of colors that can be successfully and economically printed. Artwork, transparencies, etc. and verse are united in a mock-up that is approved for further development. When approved, all the specifications for the approved card—everything from the illustration on the front to the verse inside to the UPC code and price on the back—are scanned or input onto a computer disk and sent to the printer.

### *Producing the printing plates*

4 The printing process in controlled digitally. In the most modern printing facilities, plates are created directly by exposure to lasers. A computer disk has "recorded" the image to be reproduced. A plate is run through a machine in which a computer is used to direct lasers to burn an image onto the metal plate. Each color requires its own printing plate and the computer disk is programmed so that it outputs plate specifications for each individual color. It is most economical to print no more than four colors on a card; thus, in most cases a disk pro-

Large greeting card manufacturers have creative departments that design both the prose and artwork of a card. After the artists and writers produce their portion of the greeting card, each element is incorporated into a handmade mock up of the final product. Next, graphic designers translate the mock up into a card that can be mass produced. Oftentimes, artwork is resized, color is adjusted, and the fonts used for the typeface is changed. When the mock up receives final approval, it is digitized and sent to the printer.

duces four plates per card. The plates are now ready to print.

### Printing the cards

5 It is important to note that before an entire run of cards is processed, a couple of examples are run off and submitted for "proofing." The designers, marketers, graphic artists, press operators, etc. examine the card and check it carefully to ensure the imprint is of acceptable quality. Minor color corrections or ink adjustments occur before the print run can proceed. When the proof receives the sign-off, mass-printing begins.

Greeting cards, often printed in runs of 400,000 or more, are often printed using sheet fed offset printing that permits the printer to print between 4,000-18,000 sheets per hour. When ink and a fountain solution (water with chemical additives) are applied to the laser-burned metal plate in the right proportion, the image to be printed accepts ink but repels the fountain solution. The non-image (white or background area) attracts the fountain solution and repels the ink and is left unprinted. From the plate, the image is applied under pressure to a rubber canvas called a blanket. From the image is then transferred onto a sheet of paper. The sheets of paper to be printed, about 20 x 35 in (50.8 x 88.9 cm), are put on the press mechanically. The press grabs a single sheet of paper at a time, generally printing all of the black images and words first, then moving quickly to the next metal plate (which applies a different color of the design) without allowing time for drying the just-applied inks. While on the press, an aqueous coating (that provides shine) is applied to the just-printed card by another plate. The cards then air dry for approximately five to six days.

### Cutting apart and packaging

6 After the sheeted cards are completely dried, they are cut into individual cards from the larger sheet by a die cutter. The sheets are inserted into the die cutter and an apparatus that resembles a cookie cutter cuts them apart in one stroke. The cards are now a long strip that is yet unfolded.

7 The card is then sent to the folding machine where it is creased and folded automatically. Cards are often packed by hand and assembled with their envelopes in carton quantities of 700.

## Quality Control

The production process is carefully monitored. There are at least two submissions of proof copies to the product development teams—before the product is put into disk and then at first printing—to ensure the product is designed as was envisioned and can be printed to quality specifications. Pressmen check color, inks, and completed sheets throughout the printing process. The die-cutter is able to watch the process and makes sure the cutting is done correctly so that straight clean cuts are made. Finally, the

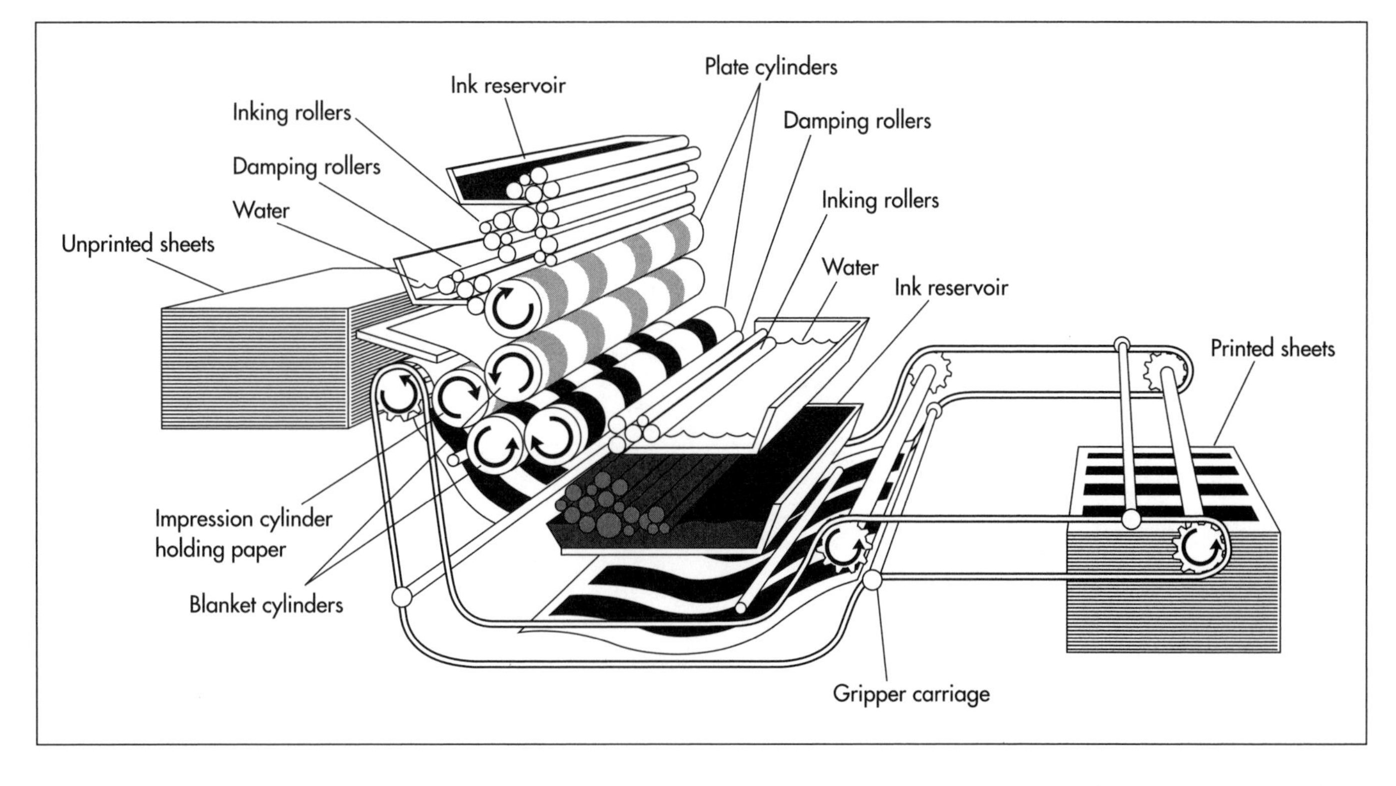


At the printer, greeting cards are printed on a sheet-fed offset printing machine. When ink and a fountain solution are applied to a metal plate that has had the greeting card pages laser burned onto it, the image to be printed accepts ink but repels the fountain solution. The non-image (white or background area) attracts the fountain solution, repels the ink, and is left unprinted. From the plate, the image is applied under pressure to a rubber canvas called a blanket and it is transferred to the paper. The sheets of paper are mechanically fed into the printer. The press grabs a single sheet of paper at a time, generally printing all of the black images and words first. A different metal plate is used for each color of the design. Inks are applied in sequence without allowing time for just-applied inks to dry. A top gloss coating is applied as a final step.

folding machine operator monitors the quality of the completed card. He or she is able to now see the card completed and ready for sale; this person has the prerogative to pull inferior cards (poorly printed, badly cut or folded) out of the line and jettison it.

## Byproducts/Waste

The use of soy ink has greatly decreased solvent disposal problems for the printer. Soy ink does not release a significant volume of volatile organic compounds (VOCs) into the air when it dries. VOCs are chemical compounds that evaporate and react in sunlight and form smog at lower atmospheric levels. Soy ink is not 100% biodegradable, but it is far more biodegradable than oil solvent-based inks. Furthermore, soy ink is more effectively removed from papers during recycling, resulting in less paper damage and a brighter paper. The use of soy ink helps printers meet the federal Environmental Protection Agency's clean-air standards. Waste papers are generally recycled by the printer as well.

## The Future

The future of greeting cards seems strong, with card sales for one company alone reaching nearly $4 billion last year. However, two very recent developments may affect the future of the greeting card industry tremendously—perhaps adversely. The first is the software available that allows consumers to create their own greeting cards (with the assistance of easy-to-use programs) using a personal computer and a color printer. Some card manufacturers created this software for sale; so, they see profits when the software is purchased, but it may reduce sales of store-bought cards. The second development is that anyone with Internet connections may send a free cybercard that can be taken off of a website and sent via e-mail. They may be obtained through any one of dozens of websites, require no postage stamp, and may be sent in an instant. In addition, many such cybercards feature sound, music, and animation. Even the large card companies offer such cards and they are often free-of-charge. It remains to be seen whether the new cybercard will affect over-the-counter sales of paper greeting cards.

## Where to Learn More

### *Periodicals*

Hirshey, Gerri. "They Feel For You: Hallmark writers." *The New York Times Magazine* (July 2, 1995): 20.

### Other

American Greetings. http://www.american-greetings.com/ (June 28, 1999).

Greeting Card Association. http://www.greetingcard.org/ (June 28, 1999).

Hallmark Card. http://www.hallmark.com/ (June 28, 1999).


*—Nancy EV Bryk*

# Hang Glider

## Background

A hang glider is an unpowered heavier-than-air flying device designed to carry a human passenger who is suspended beneath its sail. Unlike other gliders that resemble unpowered airplanes, hang gliders look like large kites. Other gliders are usually towed by a car a airplane or otherwise launched into the air from the ground. Hang gliders, on the other hand, are usually launched from a high point and allowed to drift down to a lower point.

## History

Human beings have attempted to fly using devices similar to hang gliders for at least one thousand years. Oliver of Malmesbury, an English monk, is said to have leapt from a tower with wings made of cloth in the year 1020. He supposedly glided for about 600 ft (180 m) before landing and breaking both legs. Similar brief flights are said to have been made in Constantinople in the eleventh century and in Italy in 1498. The Italian artist and scientist Leonardo da Vinci made detailed sketches of various flying machines, but these devices were never built.

The modern history of gliding begins with the English inventor Sir George Cayley. By 1799, Cayley had established the basic design for gliders that is still used today. In 1804, he flew his first successful model glider. In 1853, Cayley achieved the first successful human glider flight with a device that carried his coachman several hundred feet.

The next important pioneer in glider research was the German inventor Otto Lilienthal. In the 1890s, Lilienthal built 18 gliders, which he flew himself. He also kept detailed records of his work, influencing later inventors. After making more than two thousand successful flights, Lilienthal was killed in a crash in 1896.

Inspired by Lilienthal, the American engineer Octave Chanute and his assistants made about two thousand glider flights from sand dunes on the shores of Lake Michigan at the turn of the century. Chanute's work had an important influence on Orville and Wilbur Wright, who invented powered flight soon after. The rapid development of powered flight in the twentieth century led to a decreased interest in gliders until after World War II. At this time, light, smooth wings made of fiberglass were developed for gliders.

The most important innovation in the development of the hang glider was made by the American inventors Gertrude and Francis Rogallo. In 1948, the Rogallos applied for a patent for a flexible kite called a para-wing. Unlike other kites, the Rogallo design had no rigid supports. Instead, it remained limp until it was given firm but temporary shape by the wind in flight. The development of Mylar, an extremely light, strong plastic, improved the performance of the Rogallo kite.

In the late 1950s, the United States government took an interest in the Rogallo design for use in parachutes designed to return spacecraft to Earth. Experiments were also made in building large powered Rogallo kites for military transportation. Inspired by reports of these experiments, the American engineer Thomas Purcell build a 16 ft (4.9 m) wide Rogallo glider with an aluminum frame, wheels, a seat to hold a passenger, and basic control rods in 1961. This was the

*Seven fatalities from hang gliding were reported in 1995, compared to 40 in 1974.*

*Samuel Langley*

Samuel Langley was born in Roxbury, Massachusetts, in 1834. As a child, Langley became interested in studying the stars and, despite that fact that he never attended college, went on to become a professor of astronomy and physics. Langley made many contributions to the field of astronomy, one of the most significant being the invention of the bolometer—an instrument able to detect and measure electromagnetic radiation.

While serving as secretary of the Smithsonian Institute, Langley developed an interest in aeronautics and obtained a $50,000 grant from the United States War Department to study the possibility of manned flight. He began building large, steam-powered models of an aircraft he named *Aerodrome,* without taking the time to first test his theories on gliders. By 1891 he had begun building *Aerodrome* models which were to be catapulted off the roof of a houseboat. The first five models failed, but his 1896 model flew more than half a mile. Later that year one remained airborne for nearly two minutes.

Finally, on October 7, 1903, Langley was ready to fly his first full-scale *Aerodrome* from a houseboat in the Potomac River. With the press in attendance, the machine was launched and promptly fell into the river. Langley contended that the launching mechanism was at fault, but further attempts produced the same results, and his funding was soon depleted. Just a few months later, Orville and Wilbur Wright achieved the first powered flight at Kitty Hawk, North Carolina. Until his death in 1907, Langley maintained that if accidents had not depleted his funds, he would have achieved the fame accorded to the Wright brothers. A few years after Langley's death, experimenters did succeed in flying his *Aerodrome*after attaching a more powerful engine to it. Today, Langley Air Force Base in Virginia is named for this aviation pioneer.

first true hang glider. Early hang gliders were also built in the United States from bamboo by Barry Hill Palmer in 1961 and in Australia from aluminum by John Dickenson in 1963.

Although the United States government abandoned using the Rogallo design for spacecraft parachutes in 1967, hang gliders using the same design became popular in the 1970s. In 1971, the United States Hang Gliding Association was formed. While California is the favored spot for hang gliders of the West, Dunlap, Tennessee, claims to be the hang gliding capital of the eastern United States, thanks to its location high atop the Cumberland Plateau. Over the next several years, hang gliding became less of a dangerous fad and more of a serious sport. Seven fatalities from hang gliding were reported in 1995, compared to 40 in 1974.

## Raw Materials

A hang glider consists of a wing, a frame, cables, and items to hold these parts in place. The wing, also known as the sail, is made from a strong, light plastic. Usually a polyester cloth is used. Polyesters are polymers—they are large molecules made by linking many small molecules together. Polyesters are usually derived from ethylene glycol and terephthalic acid, or similar chemicals. The most common polyester for use in making hang gliders is polyethylene terephthalate, known by the trade name Dacron.

The frame of a hang glider, also known as the airframe, is made from an alloy of aluminum and other metals, such as magnesium, zinc, and copper. The cables and the parts that hold the hang glider together are made of stainless steel. Stainless steel is an alloy of iron, a small amount of carbon, and 12-18% chromium.

## The Manufacturing Process

### *Making polyester cloth*

1 Like many plastics, polyesters are made from chemicals derived from petroleum. In general, these chemicals are obtained by heating the petroleum with a catalyst, a process known as cracking. The resulting

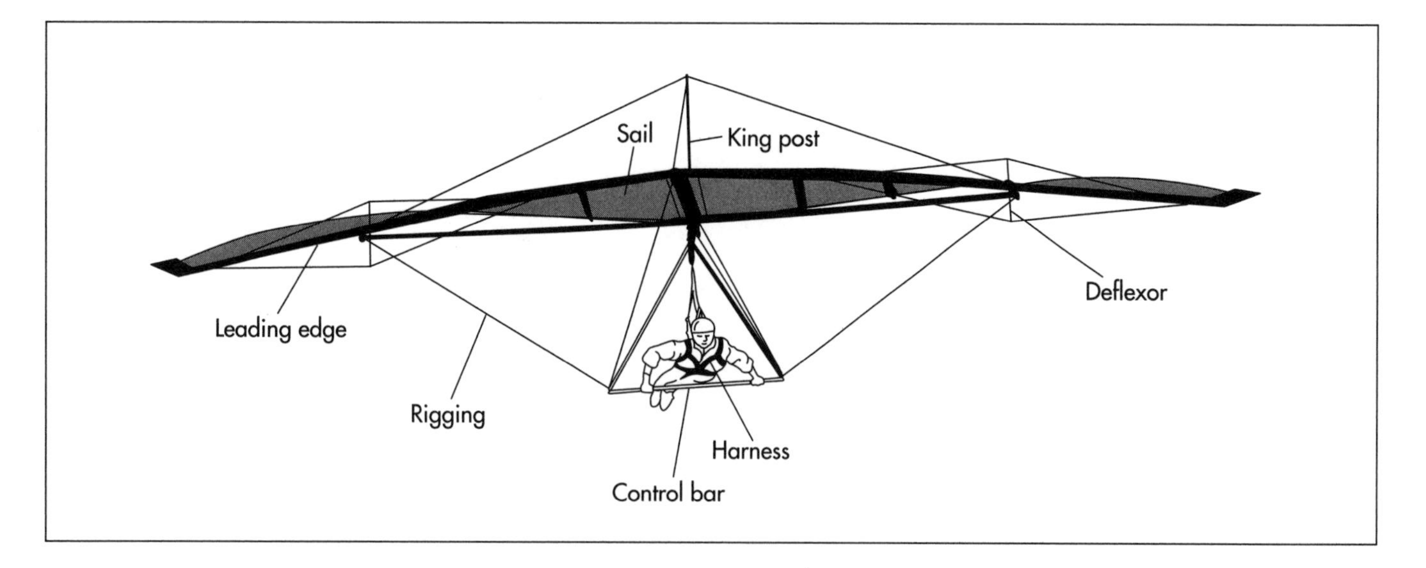


substances are then separated and subjected to various chemical reactions to obtain the desired chemicals.

2 Polyethylene terephthalate is made by combining ethylene glycol with terephthalic acid or with dimethyl terephthalate. It is then heated until it is a liquid. The molten polymer is sprayed through a device containing many small holes, known as spinnerets. As the liquid plastic emerges, it cools into long, then, solid filaments. The filaments are wound together into yarn. The yarn is then stretched to about five times its original length at an elevated temperature to increase its strength. The strengthened yarn is then woven into cloth and dyed with bright colors. This cloth is then put onto large rolls and shipped to the hang glider manufacturer.

### *Making the wing*

3 Rolls of polyester cloth are typically 54 in (137 cm) wide and 100-300 yd (91.4-274.3 m) long. Cloth is cut from the roll as needed and placed on an X-Y cutter. This device is a flat table, typically 40 ft (12.2 m) long and 5 ft (1.5 m) wide, with numerous small holes in the surface. As the cloth is placed on the cutter, vacuum pressure pulls it flat against the holes. A system of gears, belts, and rails carries a sharp blade across the table horizontally (the "X" direction) and vertically (the "Y" direction). A computer controls the movement of the blade so that complex shapes can be cut to within one one-thousandth of an inch (0.0025 cm).

4 The cut pieces of cloth, which may number in the hundreds, are marked so that they can be properly aligned with each other. They are then sewn together on industrial sewing machines. The process is repeated with the pieces of cloth of varying shapes and colors until the wing has been formed.

### *Making the frame*

5 Aluminum alloy tubing, typically 1.5 in (3.8 cm) in diameter and 10-20 ft (3-6 m) long, arrives at the hang glider manufacturer. The tubes are cut as needed by electric saws. Electric drills are then used to form holes where the frame will be held together.

6 Stainless steel cable arrives at the hang glider manufacturer in large spools that typically hold 5,000 ft (1,524 m) of the cable. The cable is cut as needed with large, sharp pliers.

7 The cable and tubing are assembled by hand to form the frame. Stainless steel nuts and bolts are used to hold the parts in place.

### *Assembling the hang glider*

8 The sail and frame are attached together. After being fully assembled, inspected, and tested, the hang glider is partly disassembled for ease of storage and transport. The disassembled hang glider is placed in a cylindrical container and shipped to the retailer or consumer.

A hang glider consists of a wing, a frame, cables, and items to hold these parts in place. The wing, also known as the sail, is made from a strong, light plastic. The frame of a hang glider, also known as the airframe, is made from an alloy of aluminum and other metals, such as magnesium, zinc, and copper.

The cable and tubing are assembled by hand to form the frame. Stainless steel nuts and bolts are used to hold the parts in place.

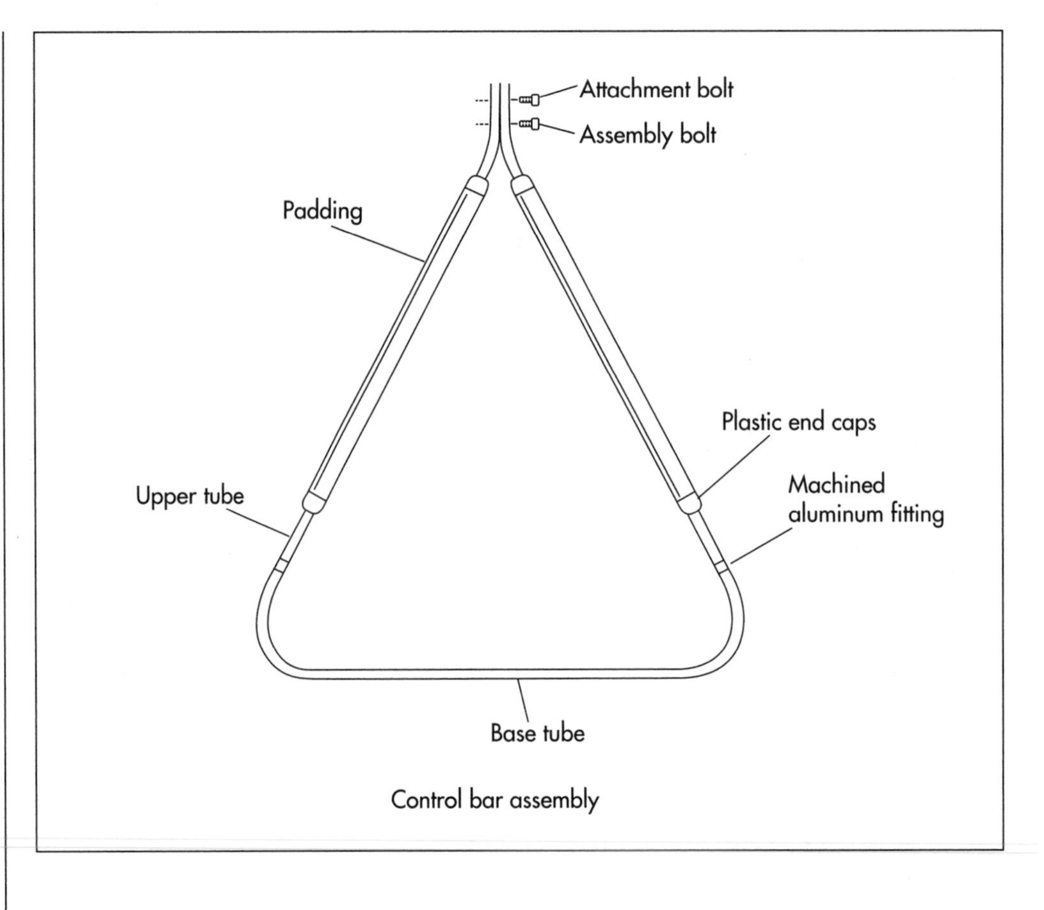

Control bar assembly

## Quality Control

Because hang gliding is a hazardous activity, ensuring that the equipment used is as safe as possible is critical. Manufacturers must meet the requirements for civil and/or military aviation equipment established by the federal government, under the supervision of the Federal Aviation Administration (FAA). Before manufacture begins, the hang glider maker inspects all raw materials. The aluminum tubing must be straight and free from dents. The stainless steel cable must be free of visible flaws. The polyester cloth must be properly woven and free of holes. Tensile test machines measure fabric strength and the amount of air that can pass through the fabric.

The most important part of the manufacturing process is making the wing. At each step in the procedure, the wing is given a full visual inspection to ensure that it is being sewn together properly, with no weak seams. It is carefully inspected on a lighted inspection table to make certain that the seams are correctly folded and sewn and that there are no flaws in the cloth. At the end of the manufacturing process, the hang glider is fully assembled for a final visual inspection.

Before being disassembled for shipping, every hang glider is given a full flight test by an experienced pilot. The hang glider must have the proper "feel" and respond to the pilot's movements correctly. It must be able to fly in a straight line at a steady speed, with no sudden unexpected changes in motion.

## The Future

Although fewer people are flying hang gliders now than in the 1970s, the technology has improved greatly. Today's hang gliders are able to fly more safely, for longer distances, for longer periods of time, and from greater heights. No doubt all hang gliding records will be broken in the near future.

## Where to Learn More

### *Books*

Whittall, N. *Complete Hang Gliding Guide.* Sterling Publishing Company, 1985.

The cables and the parts that hold the hang glider together are made of stainless steel. Stainless steel is an alloy of iron, a small amount of carbon, and 12-14% chromium.

Wolters, Richard A. *The World of Silent Flight*. McGraw-Hill, 1979.

### Periodicals

Zimmerman, Robert. "How to Fly Without a Plane." *Invention and Technology* (Spring 1998): 22-30.

### Other

"A Brief History of Hang Gliding, Paragliding, and Wills Wing." June 19, 1999. http://www.willswing.com/articles/history.htm/ (June 28, 1999).

*—Rose Secrest*

# High Heel

*Catherine de Medici (1519-1589) is credited with wearing the first true high heels and with taking the style to France in 1533 when she married the Duc d'Orléans, who was to become France's King Henry II.*

## Background

Shoe height has historically reflected nobility, authority, and wealth. France's King Louis XIV (1638-1715) was only 5 ft 3 in (1.6 m) tall until he donned specially-made high-heeled shoes with curved heels constructed of cork and covered with red-dyed leather, with the red color symbolizing nobility. On special occasions, his 5 in (12.7 cm) high heels were ornamented with hand-painted scenes of his military victories. Today, curved heels preserve his legacy and are known as Louis or French heels. Other heel-wearers used their footwear to boast of their wealth; the heels were so high that servants had to break them in, so to wear high heels also proved one could afford servants for this task.

Today, heels are blessed for the elegance they lend to the wearer's appearance and cursed for the damage they inflict on ankles, calves, and backs.

## History

The need to gain height above the ground may have originally been inspired by the weather and street conditions rather than money or vanity. During medieval times, special wooden soles called pattens were attached to the bottoms of fragile, expensive shoes made for wearing indoors so they could be kept out of the mud and damp when converted for outdoor use. Pattens were elevated in the heels and under the ball of the foot so the wearer could walk more easily by rocking forward on them; these shapes clearly foretold of high heels.

The entire shoe was elevated in the style called the *chopine* that originated in Turkey in about 1400. These shoes were effectively miniature stilts that were flat on the bottom and made of cork and covered with leather or fabric. The wearer slipped her feet into the tops that were open-backed slippers called mules or straps similar to sandals. Chopines were typically 7-8 in (18-20 cm) high, but, in the extreme, they were as much as 18 in (46 cm) tall. Chopines kept the wearer's skirts out of the mud, assuming the lady could walk at all. When the style became fashionable in sixteenth century Venice, chopine-shod ladies walked with a servant on either side of them so they wouldn't fall. The ladies loved the attention and the additional height, but chopines were so restrictive that women were also forced by their footwear to stay at home. The 18 in (46 cm) extreme was reached in France and England where the fashion spread from Italy.

Catherine de Medici (1519-1589) is credited with wearing the first true high heels and with taking the style to France in 1533 when she married the Duc d'Orléans, who was to become France's King Henry II. Italian designers created the high heel by modifying the chopine to eliminate its awkwardness while still raising the height of the wearer. A cork wedge was placed under the front of the shoe, with a high section under the heel. These high heels served vanity another way by making the feet appear smaller and the arch of the foot higher; both of these physical attributes were considered signs of noble birth. Catherine's Italian style was quickly adopted by the French court.

The French Revolution caused a revolution in footwear as well, and many shoe fashions vanished temporarily in the name of democracy. The red heels of the nobility disappeared completely, and showy buckles and rosettes were replaced by ribbon or cord ties. Flat shoes or very low-heeled shoes known as pumps replaced the arrogance of high heels, although high-heeled shoes and boots were restored to respectability by the mid-1800s.

Until high heels were invented, shoe soles for the left and right feet were identical and were called straights; shoes were formed on a single mold, called a last, for both feet. Shoes were bought not as a pair but as two single shoes of matching size and style. The arch shape of high heels, however, required different soles for the two feet, so, from 1818 onward, lasts were designed specifically for the left and right foot, and shoes were sold in pairs. Early pairs of shoes were termed crookeds, as opposed to the old-fashioned straights. The high-heel portions were originally made of wood or cork and were up to 6 in (15 cm) high. The French called them *chaussures à port* or bridge shoes, because of the open arch, or *chaussures à cric*, meaning clicking shoes for their sound. Usually, heels on men's shoes were larger in shape and heavier. The extreme heights of narrow heels were popular among gentlemen as well as ladies, and eighteenth century Englishmen who wore 6 in (15 cm) high heels usually walked with canes to be able to walk at all.

After World War II, the high heel regained its popularity primarily because of the growth in consumer spending and the variety and availability of designs produced. Stiletto heels, named for the narrow-bladed knives, soared into fashion in the 1950s. These 4 in (10 cm) spiked heels narrowed to pinpoints; they were made possible by seating a thin metal rod in the broader part of the wood or plastic heel that was attached to the shoe. A plastic tip was attached to the metal end, but these tips often fell off causing floors to be gouged and carpets to be ripped. Some office buildings provided overshoes for women to wear over their stiletto heels to prevent this damage. In the 1960s, stiletto heels were attached to 'wet-look' boots that enhanced the effects of miniskirts. Today's designers experiment with every material and type of ornamentation to create and embellish high heels. Heels have even been made of the lightweight aluminum used to manufacture airplane fuselage to give them strength in slender shapes.

## Raw Materials

Raw materials for the manufacture of high heels include plastic, leather, wood, fabric, animal hides, paper (for patterns and labels), and various cements and glues, depending on the component materials. Nails, screw nails, and tacks are used to hold fabric or leather in place and to attach heels to the shank of the shoe. Fabric and feathers, tree branches and sequins, faux pearls, and genuine diamonds have all been used to decorate high heels.

## Design

High-heel designers may be employed by the shoe manufacturer, or, more likely, are independent designers (sometimes connected to well-known fashion houses) who contract with the manufacturer to produce designs or lines of shoes bearing the designer's name. Designers work very closely with the master shoemaker who oversees the practicality of all designs for the shoe manufacturer. The designer may have an image or style to convey and a particular choice of materials, and the master shoemaker tells the designer what can be made or what production limitations are involved in the design. For example, the height of the heel may be restricted by the overall shape of the shoe or the number of stitches needed to make the shoe may affect its finished appearance (or may be impractical to manufacture). The designer and master shoemaker exchange ideas over the course of several months before they arrive at a satisfactory design.

One shoe is then made as a prototype; it is dissected into its various pieces and parts, and a pattern is made of paper. From the paper original, a master template is made of fiberboard and piped in copper so it will wear well. The fiberboard master represents the average size 8 shoe. From this master, a pantograph (a drawing tool that makes an exact copy of an outline but in a smaller or larger size) is used to outline masters on fiberboard for all other sizes in the range offered by the manufacturer in the new design. Metal dies are made to stamp out pieces in all these sizes, and the assembly line is set up to manufacture the new design. Alternatively, components of designs can be measured and scaled by computer, and the data are used to cut pieces with lasers controlled

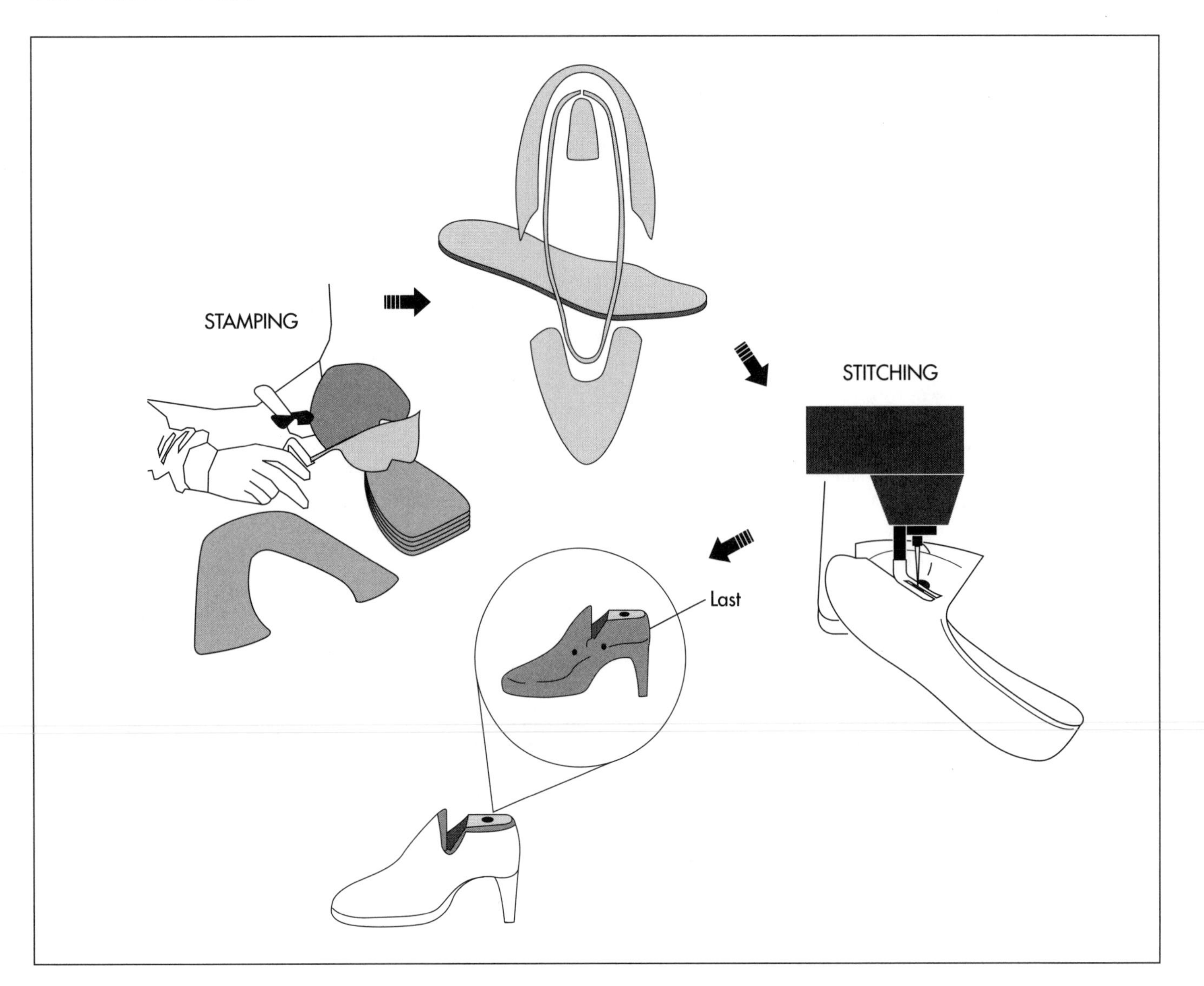


The first step in high heel manufacture involves die cutting the shoe parts. Next, the components are drawn into a machine equipped with a number of lasts—a shoe mold. The parts of the high heel are stitched or cemented together and then pressed. Lastly, the heel is either screwed, nailed, or cemented to the shoe.

by the computer. The completed designs and templates are copyrighted and registered by the designer and manufacturer.

## The Manufacturing Process

### *Handmade high heels*

1 Although most shoes today are mass-produced, handcrafted shoes are still made on a limited scale especially for performers or in designs that are heavily ornamented and expensive. The hand manufacture of shoes is essentially the same as the process dating back to ancient Rome. The length and width of both of the wearer's feet are measured. Lasts—standard models for feet of each size that are made for each design—are used by the shoemaker to shape the shoe pieces. Lasts need to be specific to the design of the shoe because the symmetry of the foot changes with the contour of the instep and distribution of weight and the parts of the foot within the shoe. Creation of a pair of lasts is based on 35 different measurements of the foot and estimates of movement of the foot within the shoe. Shoe designers often have thousands of pairs of lasts in their vaults.

2 The pieces for the shoe are cut based on the design or style of the shoe. The counters are the sections covering the back and sides of the shoe. The vamp covers the toes and top of the foot and is sewn onto the counters. This sewn upper is stretched and fitted over the last; the shoemaker uses stretching pliers to pull the parts of the shoe into place, and these are tacked to the last.

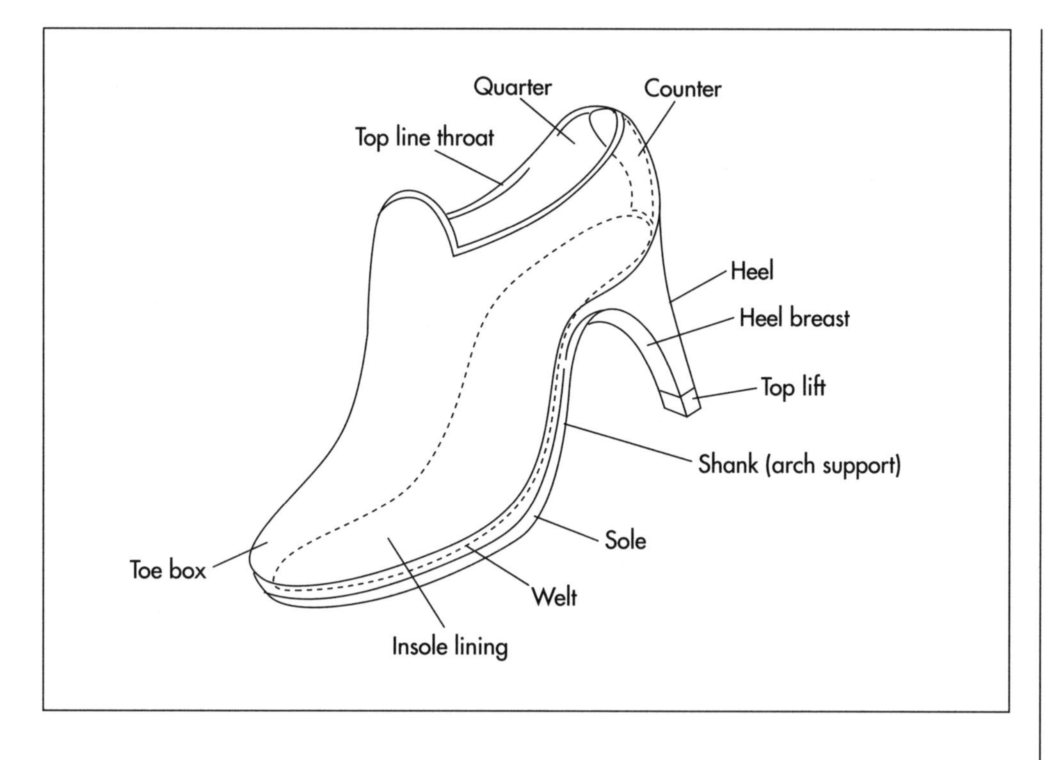


A typical high-heeled shoe.

Soaked leather uppers are left on the lasts for two weeks to dry thoroughly to shape before the soles and heels are attached. Counters (stiffeners) are added to the backs of the shoes.

3 Leather for the soles is soaked in water so that it is pliable. The sole is then cut, placed on a lapstone, and pounded with a mallet. As the name suggests, the lapstone is held flat in the shoemaker's lap so he can pound the sole into a smooth shape, cut a groove into the edge of the sole to indent the stitching, and mark holes to punch through the sole for stitching. The sole is glued to the bottom of the upper so it is properly placed for sewing. The upper and the sole are stitched together using a double-stitch method in which the shoemaker weaves two needles through the same hole but with the thread going in opposite directions.

4 Heels are attached to the sole by nails; depending on the style, the heels may be constructed of several layers. If it is covered with leather or cloth, the covering is glued or stitched onto the heel before it is attached to the shoe. The sole is trimmed and the tacks are removed so the shoe can be taken off the last. The outside of the shoe is stained or polished, and any fine linings are attached inside the shoe.

### *Machine-made high heels*

5 The design developed by the master shoemaker and designer can be mass-produced from the components committed to fiberboard masters or from computerized data. If metal dies are used to cut the pieces, leather (or other material) is fed into a 20-ton (18-metric-ton) press that has been equipped with one or more dies, and the pieces are punched out. The process is even more simplified when a computer-controlled laser beam is used to cut the pieces out.

6 When all the components are cut out, the material is drawn into a machine equipped with a number of lasts. The machine pulls the piece of material tightly over the last, and thermal cement is used to hold it together temporarily. Before synthetic adhesives were widely available, tacks were used exclusively. Now, only a few tacks may be needed along with the adhesives. High heels are made in three separate sections—the sole, the upper, and the heel. When the components have been assembled or stitched by computer-controlled machines into each of these three sections, the sections are conveyed to another machine for assembly. Strong cement is used to bond them together, and they may be put in a press for approximately 15 seconds to press the cemented pieces firmly.

7 The heel itself may be attached to the shank of the shoe using nails, screw nails, tacks, cement, staples, or sets of molded prongs (or some combination), depending on the style of the shoe, the height of the heel, the materials used in construction, and other factors including cost. Nine out of 10 heels are made of plastic and covered with material to match or compliment the shoe uppers. Plastic is used because it is lightweight and inexpensive.

8 After the sections are bonded to make the finished shoe, labels and stamped size notations are added to the insides of the shoes. The completed pair of heels is stuffed and wrapped with tissue paper and put in an appropriately labeled box. Cartons of boxes of the same style of shoe are packaged and prepared for shipment to the retailer.

## Quality Control

Although the mass-production of high heels is now done largely by machine, quality control personnel oversee all aspects of design and production. Materials are carefully inspected when they are received so that imperfect leathers, badly dyed fabrics, and other faulty items can be returned to suppliers immediately. Components that may be made by outside suppliers, such as wooden heels, are also inspected and accepted or rejected. During an initial production trial, the sections cut and assembled by computer are compared to design details and examined for flaws. When production is approved, quality control engineers also monitor all facets of production and spot-check components and completed sections and shoes. Testing laboratories are also used to evaluate the quality of materials before they are incorporated in the construction of shoes and for testing the durability of finished models in the prototype stage. Depending on the style, finished shoes may be polished or treated, and these steps are also carefully inspected. Boxed shoes may be opened and randomly checked before shipment.

## Byproducts/Waste

The use of computer-aided design and precise machines such as lasers has greatly limited the waste from shoe production. The waste that does result must be disposed and may contain synthetic adhesives and other materials that prevent recycling. Plastic waste can be remolded.

Byproducts usually do not result from the manufacture of high heels, but makers often market matching products like handbags. By fitting pieces from several types and sizes of product onto a single piece of leather, for example, waste is further reduced.

## The Future

Although high heels may be considered a modern fashion statement, their history proves that they have existed for centuries and will continue to do so. Wearers have followed King Louis XIV's example and have seen that heels can enhance height and flatter appearance. The heights and styles may change from season to season, but the high heel in some form is sure to be part of fashion's future.

## Where to Learn More

### Books

Lawlor, Laurie. *Where Will This Shoe Take You? A Walk Through the History of Footwear.* New York: Walker and Company, 1996.

Nichelason, Margery G. *Shoes.* Minneapolis: Carolrhoda Books Inc., 1996.

O'Keefe, Linda. *Shoes: A Celebration of Pumps, Sandals, Slippers & More.* New York: Workman Publishing, 1996.

Wilson, Eunice. *A History of Shoe Fashions.* New York: Theatre Arts Books, 1968.

Yue, Charlotte and David. *Shoes: Their History in Words and Pictures* Boston: Houghton Mifflin Company, 1997.

### Periodicals

Iverson, Annemarie. 'Manolo Blahnik.' *Harper's Bazaar* (July 1997): 110.

### Other

Action Shoes of India. http://www.action-shoes.com/process.html.

Nine West. http://www.ninewest.com.

*—Gillian S. Holmes*

# Holiday Lights

## Background

Festivals in a number of ancient civilizations were celebrated with lights; any and all of these may have been the inspiration for the lights we use to decorate Christmas trees and the exteriors of homes. The Druids in both France and England believed that oak trees were sacred, and they ornamented them with candles and fruit in honor of their gods of light and harvest. The ancient Roman festival of Saturnalia included trees decorated with candles and small gifts. The worship of trees as the homes of spirits and gods may have led to the Christmas tree tradition and that tradition has long been accompanied by the companion custom of decking the tree with brilliant lights evoking stars, jewels, sparkling ice, and holiday cheer.

## History

From the beginnings of Christianity to about 1500, trees were sometimes decorated outdoors, but they were not brought into homes. One legend has it that Martin Luther (1483-1546), the father of Protestantism, was walking through an evergreen forest on Christmas Eve. The beauty of the stars sparkling through the trees touched him, and he took a small tree home and put candles on its branches to recreate the effect for his family. The Christmas tree became a custom in Germany, and German-born Prince Albert took this tradition with him to England when he married Queen Victoria during the mid-nineteenth century. The first Christmas tree in Windsor Castle was decorated with candles, gingerbread, candies, and fruit.

Similarly, German settlers brought the Christmas tree to America where the first tree was displayed in Pennsylvania in 1851. Candles were attached to the boughs of the trees with increasingly extravagant candleholders, some with colored glass that made the lights appear colored. Of course, the practice of using candles was hazardous; many fire brigades were called to extinguish fires started by candles that had ignited the trees or the long hair or dresses of the ladies. Candles on trees were lit for several minutes only and sometimes only on Christmas Eve or Christmas Day; the custom of lighting trees for extended periods of time had to wait until the invention of the electric light bulb.

Candles were expensive in the mid-1800s, and tallow lights or nutshells with oil and floating wicks were also used. Various forms of brackets and hoops to hold candles, drip dishes hung under the candles, and twisted candles to channel melted wax were attempts to limit the fire hazard and the mess from dripping wax. In 1867, Charles Kirchhof of Newark, New Jersey, invented the counter-weighted candle holder that had a weight for balance suspended under the branch beneath the candle. The balance held the candle upright, and it was also brightly painted or decorated to add another spot of color to the tree. In 1879, the spring-clip candleholder was created by Frederick Arzt of New York. This clip was much lighter in weight than the counterbalance, and it was used until the 1920s when electric lights finally extinguished the use of candles on Christmas trees.

The first electric lights for Christmas debuted only three years after Thomas Alva Edison invented the lightbulb in 1879. Edward Johnson, a resident of New York and a colleague of Edison's, was the first to have an electrically lighted Christmas tree in his

*The first electric lights for Christmas debuted only three years after Thomas Alva Edison invented the lightbulb in 1879.*

home in 1882. The tiny bulbs were hand blown and the lights were hand-wired to make this event possible, but it opened an avenue for Edison's electric company that produced miniature, decorative bulbs for chandeliers and other uses from its earliest days. Electric lights appeared on the White House Christmas tree in 1895 when Grover Cleveland was President.

General Electric (GE) bought the rights to light-bulb production from Edison in 1890, but GE initially only made porcelain light bulbs. To light a tree, the family had to hire a "wireman" who cut lengths of rubber-coated wire, stripped the ends of the wires, fastened them to sockets with brass screws, fitted a larger socket to a power outlet or light fixture, and completed assembly of a string of lights. This was too expensive and impractical for the average family. In 1903, the Ever-Ready Company of New York recognized an opportunity and began manufacturing festoons of 28 lights. By 1907, Ever-Ready was making standard sets of eight series-wired lights; by connecting the sets or outfits, longer strings of lights could be made.

Ever-Ready did not have a patent on its series-wired strings of lights, and this basic wiring system was adapted by many other small companies. These sets were not always safe, and episodes of tree fires raised public alarm. In 1921, Underwriters' Laboratories (UL) established the first safety requirements for Christmas lights. A number of light manufacturers merged in 1927 to form the National Outfit Manufacturers Association (NOMA), which went on to dominate the Christmas light business, with GE and Westinghouse as the leading bulb makers. Also in 1927, GE introduced parallel wiring that permitted light bulbs to keep glowing when one on the string burned out.

Bulb shapes also evolved. In 1909, the Kremenetzky Electric Company of Vienna, Austria, began making miniature bulbs in the shapes of animals, birds, flowers, and fruit. Companies in the United States, Japan, and Germany also made figurative bulbs, but Kremenetzky consistently made the most beautiful glass that was hand-painted. World War I ended the influx of Austrian lights. GE made machine-blown shapes beginning in 1919, and the Japanese light-bulb industry, then in its infancy, began filling the void left by the Austrians. The Japanese techniques continued to improve and were quite sophisticated by 1930, but this trade ended with World War II.

NOMA started to make tiny lampshades with Disney figures on them to fit over standard miniature bulbs in 1936. The most spectacular miniature bulb success was the bubble light. Carl Otis invented it in the late 1930s, but World War II also interrupted this development. Bubble lights were finally introduced in 1945, peaked in popularity in the mid-1950s, and declined by the mid 1960s. So-called midget lights, midget twinkle lights, or miniature Italian lights began arriving from Europe in the 1970s and became the best sellers of all time in the Christmas tree light business.

## Raw Materials

Holiday lights are made of three sets of materials. The strings are composed of 22-gauge copper wire that is coated in green or white polyvinyl chloride (PVC) plastic. Specialized manufacturers supply the wire on spools that hold 10,000 ft (3,048 m) of wire. Two plugs begin and end each set of lights, and they are made of injection-molded plastic. The lights are held in lamp holders that are also injection-molded plastic and contain copper metal contacts.

The second set of materials goes into the making of light bulbs. The bulbs are made of blown glass, metal filaments, metal contact wires, and plastic bases. Bulbs are made in clear glass to produce white light, or they are painted to shine in assorted colors.

Finally, the finished sets of lights require packing materials. These include a molded plastic tray, a folded cardboard display box, and shipping cartons that hold multiple sets of boxed lights. The shipping cartons are made of corrugated cardboard. Each set is also packed with adhesive-backed safety labels and paper instruction and information sheets. All of the paper goods are made by outside suppliers and are produced from recyclable materials.

## Design

The basic design for holiday lights consists of a tried-and-true string of green plastic-covered wires with clear or colored light bulbs. Design aspects include the number of lights on the string (in multiples of 25 with 25, 50, 100, or 125 bulbs) and whether the string contains only clear bulbs, bulbs of a single color, or assorted colors of lights.

Green wires were made originally to blend in with the green branches of evergreens, either as indoor Christmas trees or outdoor shrubs. The tiny lights are used for many other holidays and for garden displays, so strings with white wires are made for other decorating uses. Plastic covers for the lights are also designed with Christmas and childhood themes as well as an extraordinary range for party decorating from aquarium fish to chili peppers.

The newest designs to take the decorating market by storm are nets of lights that can be spread over shrubs to save time in decorating, and icicle lights that look like long white icicles hanging from house eaves. Fiber-optic lights also became available in the 1990s; they are basic strings of wire and light bulbs, but each bulb is the source of light that passes through clusters of fiber-optic wire held in plastic covers that clip onto the bulb. Usually, they resemble flowers or other designs that take advantage of the cluster-like display of optic wires.

## The Manufacturing Process

### *Manufacturing the wire string*

1 A supplier of copper wire delivers large spools of wire to the holiday light manufacturer. The light maker may coat the wire in PVC plastic or purchase it already coated. The plastic resin used to form the wire coating contains additives that make the plastic resistant to ultraviolet (UV) rays in sunlight and to hot or cold temperatures; the strings can, therefore, be used in most climates year-round and can be left in the sun without breakdown of the plastic.

2 Each light set is made in series-parallel construction with twin wires. One wire is cut to the full length of the light string. At one end, the molded load fitting or wall plug is molded to the wire; the end connector is molded to the other end of the string. The end connector is made so another light set plug can be added to it, and multiple strings can be attached to each other. Each plug has a three-amp fuse in it; if too many sets are strung together, the fuse will blow before the sets overheat and cause a fire.

3 The second length of wire is assembled in 6 in (15.24 cm) long segments with the lamp holders as the links between each pair of segments. For example, a 100-light string will have 99 of the 6 in (15.24 cm) long parts. Manufacturers call the entire assembly of wire segments and lamp holders a membrane. The membrane is made by hand assembly. A worker takes one 6 in (15.24 cm) segment and places an end in a machine that strips the wire by trimming away the plastic from the wire end. A copper metal contact or flange is contact-welded to the wire end by the same machine.

4 The tiny lamp holders have been made of injection-molded plastic, again by the manufacturer or by a specialty supplier. The end of the wire with the metal flange is fitted into the lamp holder. It is pushed into place, and the flange snaps into the lamp holder. Another 6 in (15.24 cm) wire is hand-fitted into the other side of the lamp holder; this time, the stripped end of the wire without the copper flange is pushed snugly into the lamp holder. The flanged end of that wire goes into another lamp holder and so on.

5 When the 6 in (15.24 cm) segments have all been linked to lamp holders, the segmented wire is paired with a continuous wire. The two segments at the ends of one string are fitted into the load fitting (wall plug) and the end connector like the ends of the continuous wire. The double string is then taken to a twisting machine that twists the two wires together so they will be easy to string on a tree or other application without gapping apart. Light bulbs will be added to the string in the final assembly before packaging.

### *Manufacturing the light bulbs*

6 The light bulbs are made at a different factory where they are manufactured by the billions. Each tiny bulb uses only 600 milliamps of electricity, a very low electric current compared to other kinds of larger light bulbs. Glass modules specially

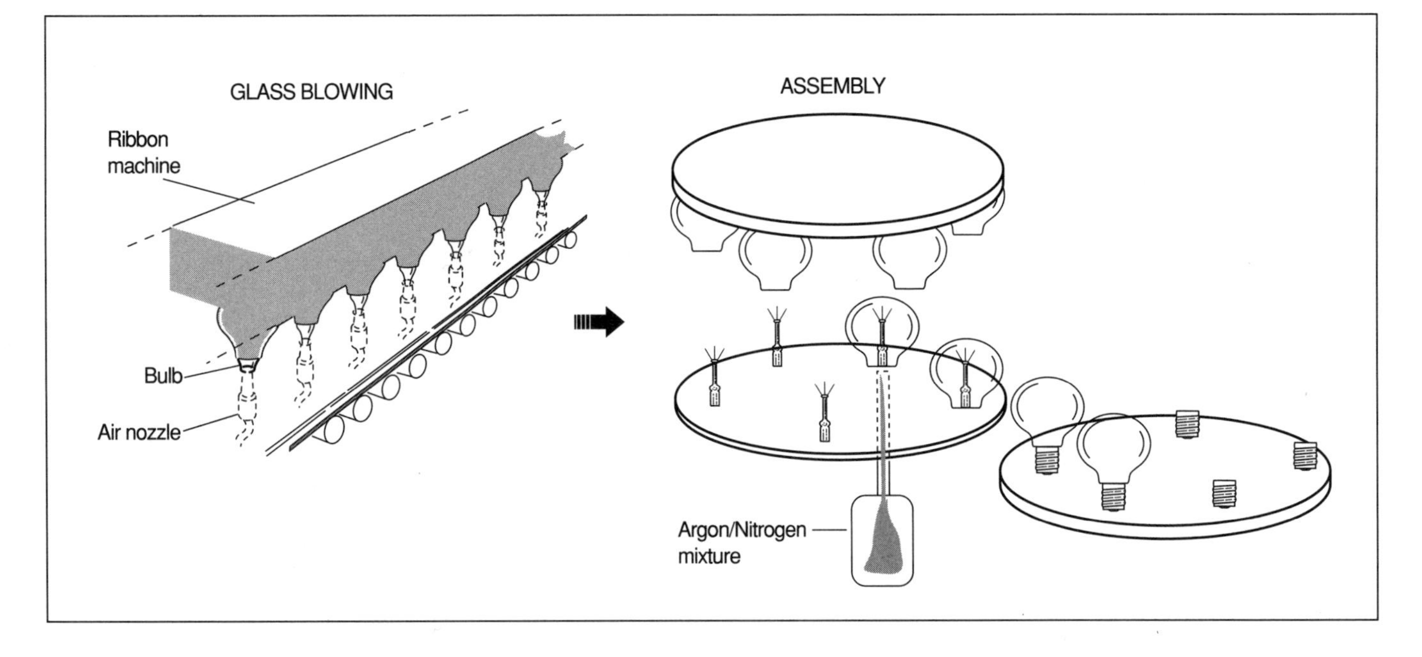


The light bulbs used in the manufacture of holiday lights are made at a circular table. The table rotates, and, as it turns, each glass module is processed into a completed bulb. First, the module is heated, then it is blown into the bulb shape. As it continues on its circular path, the filaments and elements are added to the inside of the bulb. The hot glass is pinched down to be sealed, but, just before it is sealed, a vacuum is applied to the bulb to remove the oxygen. This step is precisely done so the vacuum inside the bulb will be correct for the lighting filament to function.

processed to contain exact amounts of glass for the bulbs are vibrated into slots on a round table. The table rotates, and, as it turns, each glass module is processed into a completed bulb. Different operations are stationed around the table much like hour markers on a clock. First, the module is heated, then it is blown into the bulb shape.

7 As it continues on its circular path, the filaments and elements are added to the inside of the bulb. The filament is a metal subassembly that has been prefabricated and that will glow and produce light when electricity is applied to the bulb. One of the elements in the filament is a shunt at the base of the filament; if the filament breaks, the shunt drops into place and completes the electrical connection that previously was made by the working filament. The completed electrical connection keeps the string of lights operating when that single filament breaks and its bulb doesn't light.

8 The hot glass is pinched down to be sealed, but, just before it is sealed, a vacuum is applied to the bulb to remove the oxygen. This step is precisely done so the vacuum inside the bulb will be correct for the lighting filament to function. The end is sealed, and two wires are pressed into the hot glass. The finished, clear bulb is forced out of its slot on the table after it has completed the full rotation of the table, and another module of glass drops into that slot to begin the process again.

9 The clear glass bulbs are complete, but colored bulbs have to be painted. Magnets pick up bulbs to be painted by sticking to the wires protruding from the bases of the bulbs. The magnetic line feeds the bulbs through a paint tunnel; all the bulbs on one line are painted the same color. They are left to dry then deposited in a bin for further processing.

10 On the assembly line, the plastic bases are added to the bulbs. The bases have been formed by injection molding to fit the lamp holders for the particular manufacturer's range of lights. Assembly line workers feed the two wires sticking out of the bottom of the bulb into the plastic base and fold the two wires by hand against the sides of the bases where they will make contact with the metal flange in the lamp holder. The bins of completed bulbs in the bases are transferred to the factory where the assembled wire strings await.

### *Assembly and packaging*

11 The bulbs are inserted into the lamp holders by hand. The finished strings are folded into the correct pattern to fit the plastic insert trays, and they are fitted into the trays. The trays are conveyed to the final testing line, and every set is plugged in to confirm that it lights.

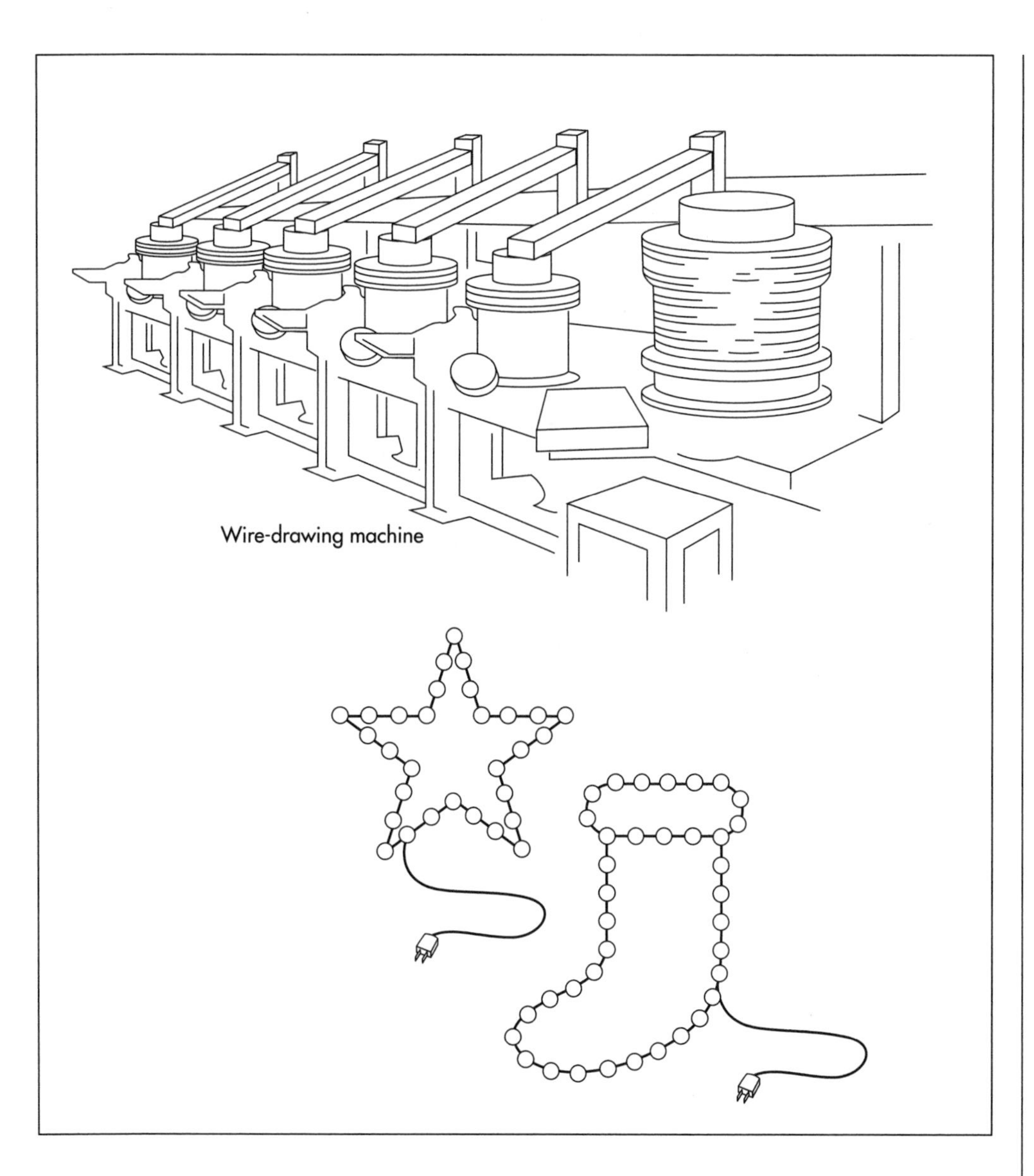


Copper wire is used in the manufacture of holiday lights. Spools of wire are coated with PVC plastic that is resistant to sunlight and to hot or cold temperatures. The series-parallel construction is made with twin wires. One wire is cut to the full length of the light string. At one end, the wall plug is molded to the wire; the end connector is molded to the other end of the string. The second length of wire is assembled with the lamp holders as the links between each pair of segments. The wires are connected with a copper metal contact or flange that is contact-welded (heated and melded) to the wire.

12 The tested sets in their trays are conveyed to the packaging department. Caution labels are wrapped around the plug end of each string, the tray is put in the cardboard display box, an instruction sheet with manufacturing and safety information is added to the box, and the box is sealed. The individual boxes are packed in shipping cartons with many other sets for overseas shipment to distribution centers.

## Quality Control

Top-quality UV additives, plastic resin, copper wire, and glass are chosen to make lights that will last for many years of use. Manufacture of the wire string is monitored continuously by the assembly line workers. The bulb-making process is inspected when the bulbs are forced out of the circular table where they were formed and fitted with filaments. Painted bulbs are inspected again after the painting process. The completed string is tested after all the bulbs have been inserted in their holders and the string has been packed in the insert tray. The test consists of actually lighting the string. Underwriters' Laboratories (UL) also establishes standards for the manufacture of holiday lights and tests them for quality and safety.

## Byproducts/Waste

Manufacturers of holiday lights make all kinds of sets including sets of clear and colored bulbs, sets with a variety of numbers of bulbs, fish-net style strings, icicle lights, sets with bulb covers in an amazing array of styles, and other holiday lighting products.

Waste consists of broken glass bulbs, plastic stripping from the wires, broken plastic packing trays, broken plastic lamp holders and trimmings from them, and bits of copper wire. All amounts are small, and all materials can be recycled.

## The Future

Manufacturers of holiday-related products know that trends in these products typically last three to five years. Tinsel and glass ornaments were out of style until the mid-1990s when they experienced a renaissance. Stringed lights have also gone through their own rebirth and are very popular for use in gardens and houses as year-round accessories, not just holiday decorations. Ongoing research and development is attempting to produce brighter lights. High-tech titanium and tungsten filaments are in the development stage, as of 1999, to make miniature lights that are twice as bright as anything previously produced. Like many other products, trends and popular demand will phase out existing products in popularity and introduce new ones to brighten parties, decor, and holidays far into the future.

## Where to Learn More

### Books

Rogers, Barbara Radcliffe. *The Whole Christmas Catalogue*. Los Angeles: Price Stern Sloan, 1988.

Snyder, Phillip V. *The Christmas Tree Book.* New York: Penguin Books, 1977.

### Other

Brite Star Company, Inc. http://www.britestar.com/ .

*—Gillian S. Holmes*

# Honey

Honey is a sweet syrupy substance produced by honeybees from the nectar of flowers and used by humans as a sweetener and a spread. Honey is comprised of 17-20% water, 76-80% glucose, and fructose, pollen, wax, and mineral salts. Its composition and color is dependent upon the type of flower that supplies the nectar. For example, alfalfa and clover produce a white honey, heather a reddish-brown, lavender an amber hue, and acacia and sainfoin a straw color.

## Background

Honey, golden and sweet, has always been held in high regard. The Bible refers to heaven as the "Land of Milk and Honey." In ancient times, honey was considered the food of the gods and the symbol of wealth and happiness. It was used as a form of sustenance and offered in sacrifice. In the Middle Ages, honey was the basis for the production of mead, an alcoholic beverage. Because of its antiseptic qualities, physicians found it a perfect covering for wounds before the advent of bandages. Even Napoleon was enchanted by it, choosing the honeybee for his personal crest.

Beekeeping is one of oldest forms of animal husbandry. Early beekeepers encouraged the establishment of bee colonies in cylinders of bark, reed, straw, and mud. However, when the honeycomb was removed from the cylinders, the colony was destroyed.

Honeybees were brought to North America in the mid-1600s. Although there were bees on the continent, they were not honeybees. Early settlers took note of the bees' penchant for hollow logs. They developed a "bee gum," by placing sticks crosswise over the opening of the logs to support the honeycombs. This not only allowed for the comb to be removed from one end, but also kept the comb intact so that the colony could use it again.

In Europe, beekeepers working toward a similar goal, developed a device called a skep. It was essentially a basket placed upside-down over the beehive. The full honeycombs were removed from underneath. A further innovation called for cutting a hole in the top of the hive and placing a straw or wooden box over the hole. The box would eventually fill with honey as well. It could then be removed without harming the comb.

In the mid-nineteenth century, an American named Moses Quimby improved upon the beekeeping system by layering a number of boxes over the main chamber. But it was the Reverend Langstroth who was responsible for creating the basis for the method that is currently used. Langstroth's moveable frame hive allowed for easy extraction and reinsertion of the combs. It consisted of a base, a hive body fitted with frames that contained the brood chamber, one or more removable sections (called supers) that were also fitted with frames for honey storage. The entire system is protected with waterproof covers.

Another popular type of hive is the leaf hive. This is a wooden box divided by means of a metal grid into an upper (honey) chamber and a lower (brood) chamber. Just above the floor and above the grid are racks of horizontal metal bars. Frames that hold the hanging honeycombs slide onto the racks.

## Raw Materials

An average bee colony produces 60-100 lb (27.2-45.4 kg) of honey each year. Colonies

*An average bee colony produces 60-100 lb (27.2-45.4 kg) of honey each year. Colonies are divided by a three-tier organization of labor: 50,000-70,000 workers, one queen, and 2,000 drones.*

are divided by a three-tier organization of labor: 50,000-70,000 workers, one queen, and 2,000 drones. Worker bees only live for three to six weeks, each one collecting about one teaspoon of nectar. One pound (0.454 kg)of honey requires 4 lb (1.8 kg) of nectar, which requires two million flowers to collect.

When the worker bees are about 20 days old, they leave the hive to collect nectar, the sweet secretion produced by the glands of flowers. The bee penetrates the flower's petals and sucks the nectar out with its tongue and deposits the nectar into its honey sac or abdomen. As the nectar journeys through the bee's body, water is drawn out and into the bee's intestines. The bee's glandular system emits enzymes that enrich the nectar.

Pollen attaches to the bee's legs and hairs during the process. Some of it falls off into subsequent flowers; some mixes with the nectar.

When the worker bee cannot hold anymore nectar, she returns to the hive. The processed nectar, now on its way to becoming honey, is deposited into empty honeycomb cells. Other worker bees ingest the honey, adding more enzymes and further ripening the honey. When the honey is fully ripened, it is deposited into a honeycomb cell one last time and capped with a thin layer of beeswax.

## The Manufacturing Process

### *Full honeycombs removed from hive*

1 To remove the honeycombs, the beekeeper dons a veiled helmet and protective gloves. There are several methods for removing the combs. The beekeeper may simply sweep the bees off the combs and guide them back into the hive. Alternately, the beekeeper injects a puff of smoke into the hive. The bees, sensing the presence of fire, gorge themselves on honey in an attempt to take as much as they can with them before fleeing. Somewhat tranquilized by engorgement, the bees are less likely to sting when the hive is opened. A third method employs a separator board to close the honey chamber off from the brood chamber. When the bees in the honey chamber discover that they have been separated from their queen, they move through a hatch that allows them to enter the brood chamber, but not reenter the honey chamber. The separator board is inserted approximately two to three hours before the honeycomb is to be removed.

The majority of the cells in the comb should be capped. The beekeeper tests the comb by shaking it. If honey spurts out, the comb is reinserted into the honey chamber for several more days. Approximately one-third of the honey is left in the hive to feed the colony.

### *Uncapping the honeycombs*

2 Honeycombs that are at least two-thirds capped are placed into a transport box and taken to a room that is completely free of bees. Using a long-handled uncapping fork, the beekeeper scrapes the caps from both sides of the honeycomb onto a capping tray.

### *Extracting the honey from the combs*

3 The honeycombs are inserted into an extractor, a large drum that employs centrifugal force to draw out the honey. Because the full combs can weigh as much as 5 lb (2.27 kg), the extractor is started at a slow speed to prevent the combs from breaking.

As the extractor spins, the honey is pulled out and up against the walls. It drips down to the cone-shaped bottom and out of the extractor through a spigot. Positioned under the spigot is a honey bucket topped by two sieves, one coarse and one fine, to hold back wax particles and other debris. The honey is poured into drums and taken to the commercial distributor.

### *Processing and bottling*

4 At the commercial distributor, the honey is poured into tanks and heated to 120°F (48.9°C) to melt out the crystals. Then it is held at that temperature for 24 hours. Any extraneous bee parts or pollen rise to the top and are skimmed off.

5 The majority of the honey is then flash-heated to 165°F (73.8°C), filtered through paper, then flash cooled back down to 120°F (48.9°C). This procedure is done very quickly, in approximately seven sec-

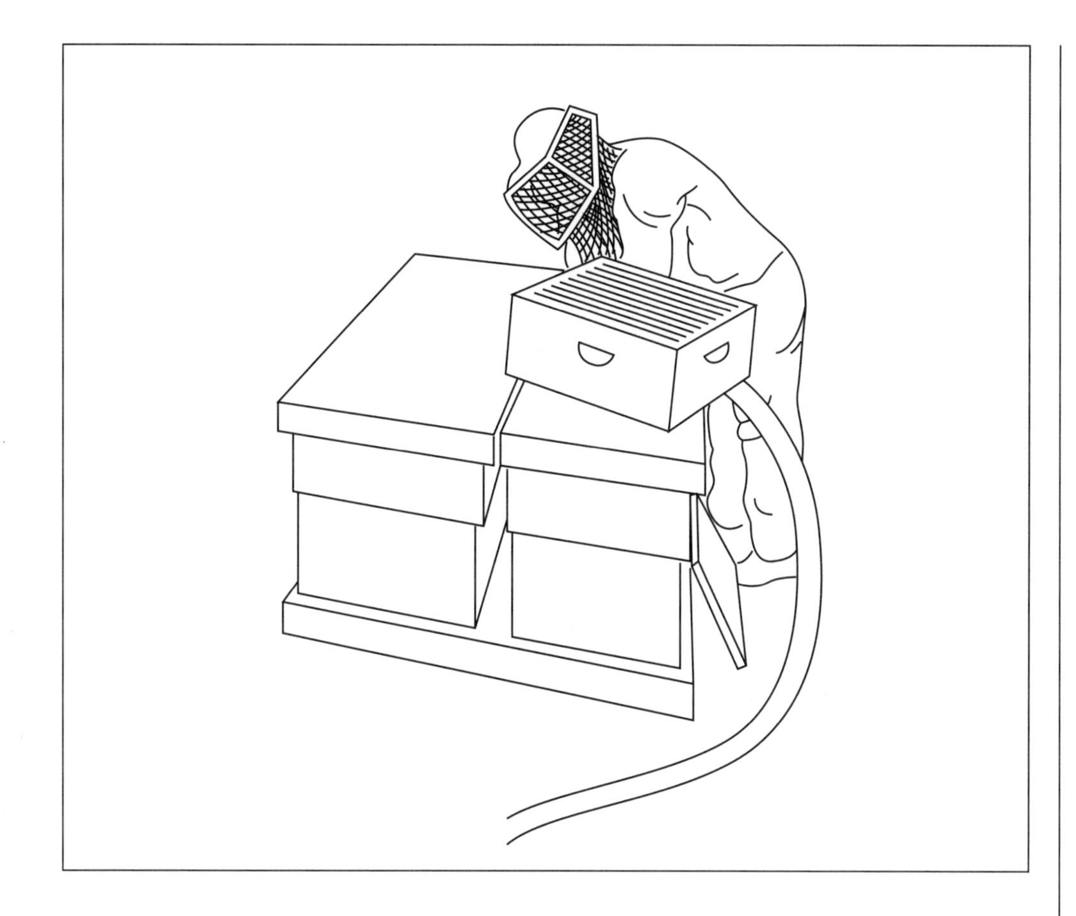

There are several methods for removing honey combs. The beekeeper can either sweep the bees off the combs and guide them back into the hive or inject a puff of smoke into the hive. When the bees sense the presence of fire, they gorge on honey in an attempt to take as much as they can with them before fleeing. Somewhat tranquilized by engorgement, the bees are less likely to sting when the hive is opened. Alternately, a separator board can be placed between the honey chamber and the brood chamber. When the bees in the honey chamber discover that they have been separated from their queen, they move through a hatch that allows them to enter the brood chamber, but not reenter the honey chamber.

onds. Although these heating procedures remove some of the honey's healthful properties, consumers prefer the lighter, bright- colored honey that results.

A small percentage, perhaps 5%, is left unfiltered. It is merely strained. The honey is darker and cloudier, but there is some market for this unprocessed honey.

6 The honey is then pumped into jars or cans for shipment to retail and industrial customers.

## Quality Control

The maximum USDA moisture content requirement for honey is 18.6%. Some distributors will set their own requirements at a percent or more lower. To accomplish this, they often blend the honey received from various beekeepers to produce honey that is consistent in moisture content, color, and flavor.

Beekeepers must provide proper maintenance for their hives throughout the year in order to assure the quality and quantity of honey. (pest prevention, health of the hive, etc.) They must also prevent overcrowding, which would lead to swarming and the development of new colonies. As a result, bees would spend more time hatching and caring for new workers than making honey.

### *Byproducts/Waste*

Four major byproducts of the honey-making process: beeswax, pollen, royal jelly, and propolis. Beeswax is produced in the bee's body as the nectar is transforming into honey. The bee expels the wax through glands in its abdomen. The colony uses the wax to cap the filled honeycomb cells. It is scrapped off the honeycomb by the beekeeper and can be sold to commercial manufacturers for use in the production of drugs, cosmetics, furniture polish, art materials, and candles.

Pollen sticks on the worker bee's legs as she collects flower nectar. Because pollen contains large amounts of vitamin $B_{12}$ and vitamin E, and has a higher percentage of protein than beef, it is considered highly nutritious and is used to the dietary supplement. To collect it, the beekeeper will force the bees through a pollen trap—an opening

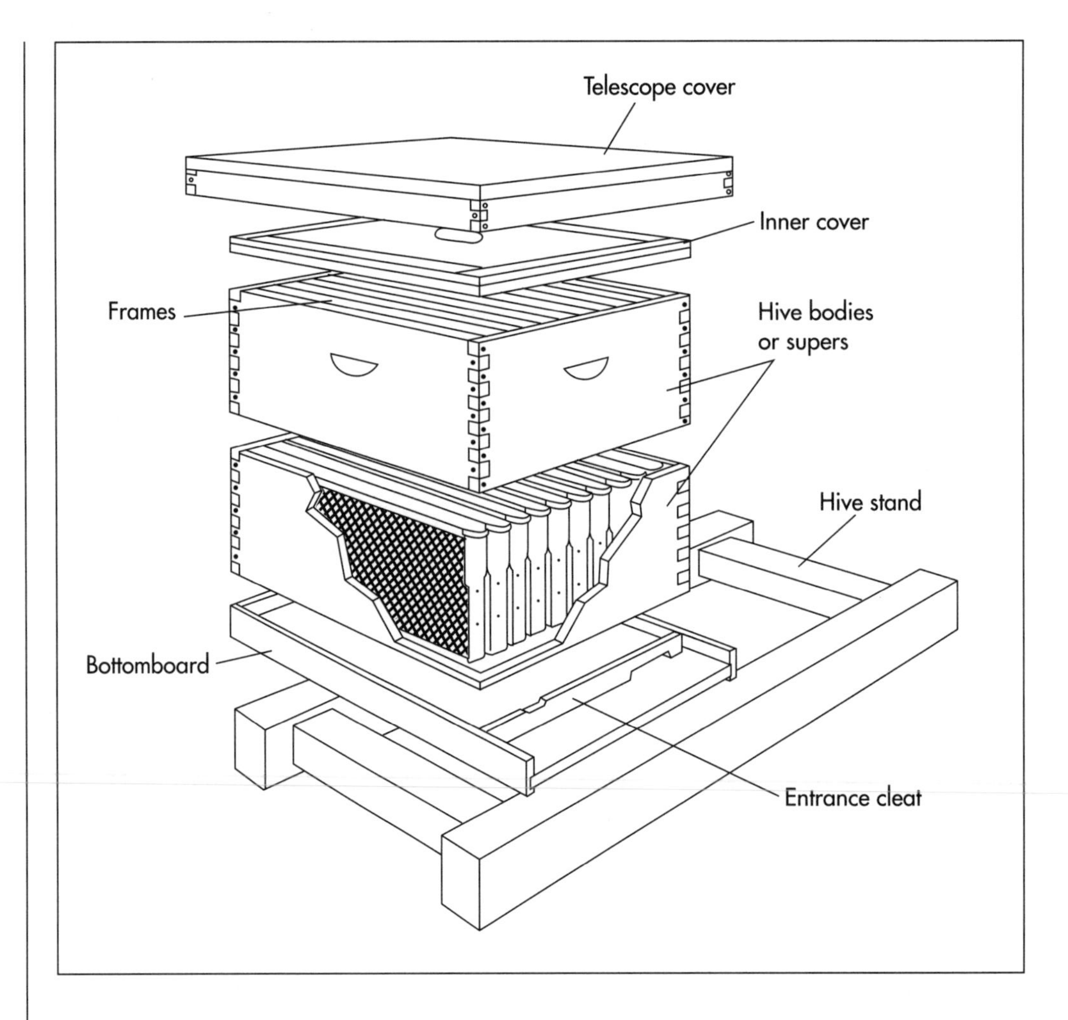


A typical hive used in beekeeping.

screened with five-mesh hardware cloth or a 0.1875-in (0.476-cm) diameter perforated metal plate. The single- or double-screened opening allows the pollen to drop from the bees' legs as they fly through. The pollen drops into a container and is immediately dried and stored.

Royal jelly is a creamy liquid produced and secreted by the nurse bees to feed the queen. Nutrient rich with proteins, amino acids, fatty acids, sugars, vitamins, and minerals, it is valued as a skin product and as a dietary supplement. Proponents believe it prolongs youthfulness by improving the skin, increases energy, andhelps to reduce anxiety, sleeplessness, and memory loss.

Propolis is plant resincollected by the bees from the buds of plants and then mixed with enzymes, wax and pollen. Bees use it as a disinfectant, to cover cracks in the hive, and to decrease the hive opening during the winter months. Commercially it is used as a disinfectant, to treat corns, receding gums, and upper respiratory disease, and to varnish violins.

## The Future

In the latter part of the twentieth century, the honeybee industry has been hard hit by two factors: parasitic mites and Africanized bees. Mites, primary the tracheal and varroa types, have destroyed thousands of bee colonies. The destruction of bee colonies not only affects honey sales, but the fruit and vegetable farmers who depend on bees to pollinate their crops. It is estimated that the value of bee pollination reaches $10 billion annually. At the close of the century, researchers were exploring ways to prevent the mite infestation without chemical intervention.

Africanized bees were first detected in North America in the early 1990s. Their presence has been detected in Texas, southern California, New Mexico, and Arizona, but further migration has not been detected. As a subspecies of honeybee, Africanized bees can only be distinguished from the European honeybee by its more aggressive stinging behavior. Hence, they have earned the appellation "killer bees." Africanized honeybees can mated with the European honeybee, creating a hybrid with the more aggressive stinging behavior. By the early 1990s, almost 100% of honeybees in Mexico carried the aggressive gene. In tropical climates, the aggressiveness gene is a dominant trait. Scientists have isolated five genes linked to the aggressiveness, one of which triggers stinging behavior. The goal is to use such findings to limit the spread of the Africanized trait throughout the Western Hemisphere and the U.S. honeybee population.

Although it has long been known that the antioxidants in honey prevent the growth of bacteria, the use of honey to retard food spoilage has not garnered widespread support. In the late 1990s, proponents began to gather statistical evidence to support their case.

## Where to Learn More

### Books

Bonney, Richard E. *Hive Management.* Pownal, VT: Garden Way Publishing, 1990.

Diemer, Irmgard. *Bees and Beekeeping.* London: Merehurst Press, 1988.

Melzer, Werner. *Beekeeping: A Complete Owner's Guide.* Hauppage, NY: Barron's Educational Services, Inc. 1986.

### Other

Cyberbee. http://www.cyberbee.net/ (January 16, 1999).

International Bee Research Association. 10 North Road, Cardiff CFI 3DY, UK. (+44)1222 372409. ibra@cardiff.ac.uk.

Sioux Honey Association. Sioux City, IA. (712)259-0638.

*—Mary McNulty*

# Hourglass

*Some hourglasses had dials with pointers, so with each turning of the glass, the number of turns could be shown with the pointer to mark the cumulative passage of time.*

## Background

Before the invention of mechanical clocks, timepieces used the sun's motion or simple measurement devices to track time. The sundial may be the best known ancient keeper of time, and it is still manufactured as a popular garden accessory—but for its visual interest, not for practical time measurement. Stonehenge, the giant monument built of upright stones on the Salisbury Plain of Wiltshire, England, may have been used as a sundial and for other time and calendar purposes. Sundials have obvious disadvantages; they can't be used indoors, at night, or on cloudy days.

Other simple measurement devices were used to mark the duration of time. Four basic types could be used indoors and regardless of the weather or time of day. The candle clock is a candle with lines drawn around it to mark units of time, usually hours. By observing how much of the length of a candle burned in one hour, a candle made of the same material was marked with lines showing one-hour intervals. An eight-hour candle showed that four hours had passed when it had burned down beyond four marks. The clock candle had the disadvantages that any changes in the wick or wax would alter burning properties, and it was highly subject to drafts. The Chinese also used a kind of candle clock with threads used to mark the time intervals. As the candle burned, the threads with metal balls on their ends fell so those in the room could hear the passage of the hours as the balls pinged on the tray holding the candle.

The oil lamp clock that was used through the eighteenth century was a variation and improvement on the candle clock. The oil lamp clock had divisions marked on a metal mount that encircled the glass reservoir containing the oil. As the level of oil fell in the reservoir, the passage of time was read from the markings on the mount. Like the candle clock, the oil lamp clock also provided light, but it was less prone to inaccuracies in materials or those caused by drafty rooms.

Water clocks were also used to mark the passage of time by allowing water to drip from one container into another. The marks of the sun's motion were made on the first container, and, as water dripped out of it and into another basin, the drop in water level showed the passage of the hours. The second container was not always used to collect and recycle the water; some water clocks simply allowed the water to drip on the ground. When the eight-hour water clock was empty, eight hours had passed. The water clock is also known as the clepsydra.

## History

Hourglasses (also called sand glasses and sand clocks) may have been used by the ancient Greeks and Romans, but history can only document the fact that both cultures had the technology to make the glass. The first claims to sand glasses are credited to the Greeks in the third century B.C. History also suggests sand clocks were used in the Senate of ancient Rome to time speeches, and the hourglasses got smaller and smaller, possibly as an indication of the quality of the political speeches.

The hourglass first appeared in Europe in the eighth century, and may have been made by Luitprand, a monk at the cathedral in Chartres, France. By the early fourteenth century, the sand glass was used commonly

*John Harrison*

John Harrison and his brother James were introduced to clock repair by their father, Henry. At the time, clock making, or horology, was undergoing a developmental revolution. Mechanical clocks had existed since the fourteenth century, but had remained rather primitive in their operation until Christiaan Huygens invented the weight-and-pendulum clock in 1656. One limitation was that they were totally dependent upon the earth's gravity for their operation. This meant that they could not keep accurate time at sea, and could not be adapted for portability. Even moving them across a room would require adjustment.

The Harrison brothers set to work on developing a marine chronometer in 1728. The motivating factor was money. In 1714, the English Admiralty set up an award of £20,000 for anyone who could provide mariners with a reliable clock that, when used with celestial sightings, could keep them informed of their longitude at sea. Mariners had to rely heavily on dead reckoning to find their way, often leading to tragic results.

The Harrison strategy was to design an instrument that was not only internally accurate but also externally stable. The Harrisons made several models of marine chronometers. The fourth model proved to be the most successful. On a nine-week voyage from England to Jamaica in 1761, the device had only a five-second error.

The Board of Longitude, apparently miffed that a common artisan had achieved the coveted goal, reluctantly gave up only half of the prize. John, minus his brother, refused to accept only half of the reward and persisted until the other half was relinquished .

The Board subjected his invention to undue scrutiny and required him to design a fifth model. This time, Harrison outdid himself by designing a compact timepiece that resembled a modern day pocket watch. It was far more convenient than the previous models, which were heavy and bulky. The Board still refused to capitulate. Finally, only a personal appeal to King George III and the King's intervention could set things right, and Harrison received the full reward in 1773 at age seventy-nine. Harrison lived only three more years.

in Italy. It appears to have been widely used throughout Western Europe from that time through 1500. The hourglass or sand clock follows exactly the same principle as the clepsydra. Two globes (also called phials or ampules) of glass are connected by a narrow throat so that sand (with relatively uniform grain size) flows from the upper globe to the lower. Hourglasses were made in different sizes based on pre-tested measurements of sand flow in different sizes of globes. A housing or frame that enclosed the globes could be fitted to the two globes to form a top and bottom for the hourglass and was used to invert the hourglass and start the flow of sand again. Some hourglasses or sets of hourglasses were set in a pivoted mount so they could be turned easily.

The earliest writings referring to sand glasses are from 1345 when Thomas de Stetsham, a clerk on a ship called *La George* in the service of King Edward III ( 1312-1377) of England, ordered 16 hourglasses. In 1380, following the death of King Charles V (1337-1380) of France, an inventory of his possessions included a "large sea clock . . . in a large wooden brass-bound case."

These two early associations of sand clocks with the sea show how navigation had become a time-dependent science. Compasses and charts, developed in the eleventh and twelfth centuries, helped navigators determine bearings and direction, but time measurement was essential to estimating distance traveled. The sand glass may have been invented—or perfected—for use at sea where equal units of time were measured to estimate distance; by contrast, on land, unequal time measurements were more important because activities depended on the length of day.

The great advances in maritime science occurred in the twelfth century with the development of the magnetic compass in Amalfi, Italy. Other Italian port cities like Genoa and Venice contributed to the astronomical advances in navigation, and, by coincidence, Venice was the world's greatest glass-blowing center. Furthermore, the fine marble dust from the quarries at Carrara was perfect for use as sand in navigational sand clocks. As well as measuring time as distance at sea, hourglasses were used by the navies of several nations to "keep the watch" or measure the time the crew worked. The ship's boy was in charge of turning the hourglass; to get off work early, he would "swallow the sand" or turn the glass before it was empty.

The most extraordinary hourglasses were made as gifts for royalty. Charlemagne (742-814) of France possessed a 12-hour hourglass. In the sixteenth century, Holbein (1497-1543) the artist made spectacular hourglasses for Henry VIII (1491-1547) of England. Other sand glasses contained multiple instruments. For example, a sand glass made in Italy in the seventeenth century contained four glasses. One had one-quarter hour of sand; the second, a half-hour of sand; the third, three-quarters of an hour of sand; and the fourth contained the full hour's measure of sand. Some glasses also had dials with pointers, so, with each turning of the glass, the number of turns could be shown with the pointer to mark the cumulative passage of time.

The upper and lower globes of each glass were blown separately with open apertures or throats. To join them so that sand could flow from the upper globe to the lower, the two halves of the glass were bound together with cord that was then coated with wax. The two-coned glass phial could not be blown as one piece until about 1800.

In about 1500, the first clocks began to appear with the invention of the coiled spring or mainspring. Some weight-powered clocks had been made before 1500, but their size limited their practicality. As the mainspring was improved, smaller, tabletop clocks were manufactured and the first watches were made. Mainspring-driven clocks made curiosities out of clepsydras and sand glasses, but, interestingly, the most beautiful hourglasses were made after 1500 as decorative pieces. These are the hourglasses that are displayed in museums.

By the 1400s, many private homes had sand clocks for household and kitchen use. Sermonglasses were used in churches to track the length of the minister's sermon. Hourglasses were also routinely used in the lecture halls of Oxford University, craftsmen's shops (to regulate working hours), and in England's House of Commons where bells to signal voting and lengths of speeches were timed based on sand clocks. During the height of the sand glass, doctors, apothecaries, and other medical practitioners carried miniature or pocket sand glasses with durations of one-half or one minute to use when timing pulses; the practice of carrying these continued until the nineteenth century. Today, miniature versions containing three minutes worth of sand are sold as egg timers and as travel souvenirs. Larger sand clocks are still made today of ornamental materials and in interesting styles for use as decoration. All of these measuring devices (clock candles, water clocks, and sand clocks) have the disadvantage that they must be watched carefully.

## Raw Materials

Glass for hourglasses is the same material as that used for other blown glass. It is manufactured in tubes of varying lengths by specialized suppliers for firing and shaping by machine or by mouth-blowing. Pre-formed light-bulb blanks can also be transformed into hourglasses by joining them together at the bases of the bulbs. Similarly, jars can be hooked together at their necks to make hourglasses; these can range in appearance from rustic to modern depending on the "character" of the jars.

The frames or housings for hourglasses are open to the designer's whims. Raw materials most often consist of pieces of fine wood that can be crafted or carved to suit a particular style, decor, design, or theme. Bamboo, resin, and various metals like brass, bronze, and pewter are also beautiful framing materials. Specialized hourglasses are made in such small numbers that raw materials are purchased from outside sources for limited

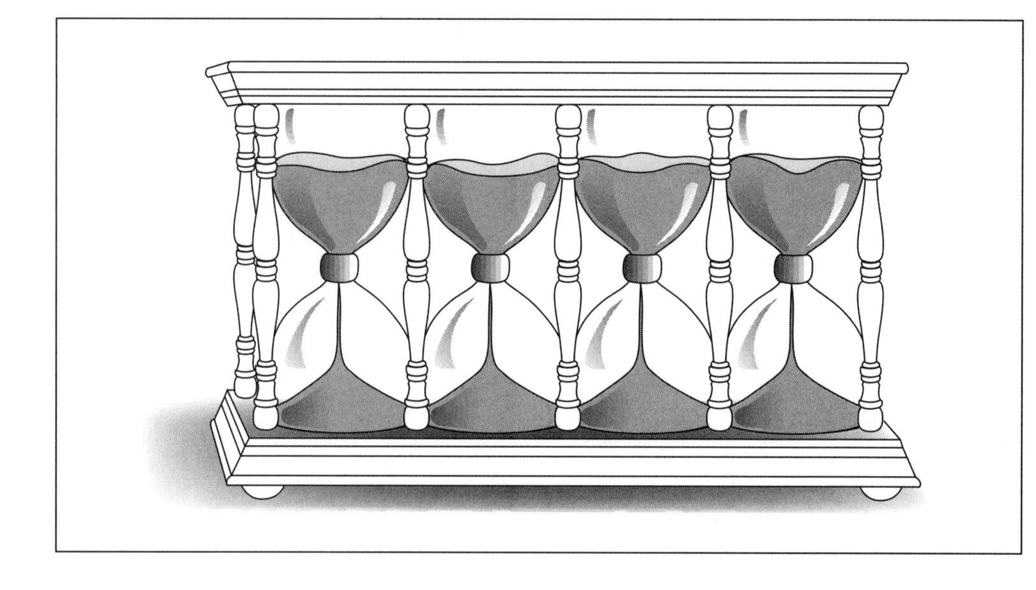

The hourglass was widely used as a timekeeping device up until the sixteenth century in Western Europe. Its design was simple. Two globes (also called phials or ampules) of glass were connected by a narrow throat so that sand (with relatively uniform grain size) flowed from the upper globe to the lower. Hourglasses were made in different sizes based on pretested measurements of sand flow in different sizes of globes. A frame that housed four hourglasses was made in Italy during the seventeenth century. Each hourglass contained different amounts of sand. One had one-quarter hour of sand; the second, a half-hour of sand; the third, three-quarters of an hour of sand; and the fourth contained the full hour's measure of sand.

issue. Sometimes customers provide their own materials to hourglass makers. Egg-timer hourglasses are also framed in wood or plastic. For these small examples, manufacturers purchase plastic chips from suppliers and produce frames in their factories by injection or extrusion molding.

Sand is the most complex of the components of hourglasses. Not all types of sand can be used because the grains may be too angular and may not flow properly through the neck of the hourglass. White quartz sand, the sand found on sparkling white beaches, is attractive but not the best for hourglass-making because it is too angular and does not flow smoothly. Marble dust, other rock dust and rock flour—powder from glass cutting—and round sand grains, like those of river sand, are best for sand clocks. During the Middle Ages, books for homemakers included recipes not only for cooking but for making glue, ink, soap, and also sand for hourglasses. Perhaps the best sand isn't sand at all; ballotini, tiny glass beads or shot (like miniature marbles about 40-160 microns [0.00016-0.0063-in or 0.0004-0.0016-cm] in diameter) are used in hourglasses because their round edges flow smoothly through the glass. In addition, ballotini can be made in different colors so the sand in the hourglass can be chosen to match room decor or some other color preference.

## Design

Design and conceptualization are usually the most complicated part of hourglass making. The hourglass maker must be craftsman, artist, and public relations expert in advising clients of the practicalities in hourglass design and construction. Businesses are commissioning hourglasses as gifts related to the year 2000, but they also want to reflect the character of their business or incorporate materials associated with their products. After the design is finalized, actual construction of the hourglass is relatively straightforward.

Sand-clock designs can also vary considerably in size. The smallest known hourglasses are the size of a cufflink, and the largest are up to 3 ft (1 m) tall. The glasses can have various shapes from round to oblong and can be engraved. Multiple (more than two) glass gloves can be linked together, and several hourglasses can be mounted in the same frame and turned on a turnstile.

According to one craftsman, hourglass design has no limits. He develops his own designs, makes hourglasses based on designs or requests provided by his customers, or creates designs to suit a particular market interest. He has sculpted pairs of his own hands as frames to hold the glasses, used unusual materials like bamboo or marble to craft the frames, and called on other hourglasses, such as the gargoyle-framed hourglass in the movie version of *The Wizard of Oz*, for his inspirations. He has made hourglasses containing coal sand, mining slag, sand from the Great Wall of China, and polyethylene resin sand. Shopping networks on television are currently selling hourglasses that are futuristic in design in keeping with interest in the new millennium.

## The Manufacturing Process

1 After the design and materials are selected, the body of the hourglass is blown on a glass lathe to a size appropriate for the size (time interval) of the hourglass.

2 The frame is made; depending on its design, it may be a single piece or multiple pieces including a bottom, top, and three or four posts. This manufacture depends on the material. If the frame is made of resin, molds may be constructed, the resin is poured in and allowed to cure, the pieces are sanded or otherwise smoothed and polished, and they are fitted together. Frame pieces may be fitted to interlock; or they may be glued, bonded, or welded, again depending on the materials involved.

3 One of the most common misconceptions about hourglasses is that there is a formula for the quantity of sand contained in the glass. The sand quantity in a given hourglass design or shape is not based on science or a measurement formula. The types of grains, the curves of the glass, and the shape and size of the opening impose too many variables on the rate of flow of the sand through the glass, so the amount of sand can not be mathematically calculated. Before the top of the frame is sealed, sand is added and allowed to flow through the glass for its prescribed time interval. At the end of that time period, sand remaining in the top of the glass is poured off and the glass is sealed.

## Quality Control

Quality control is inherent in the manufacture of hourglasses because the designer or manufacturer does all aspects of the work. The customer is also involved in conceptualizing the design and choosing materials and colors. The end result is that customers receive handmade products that suit their requirements and evoke historic and artistic associations; hourglasses are aesthetically pleasing ornaments, rather than accurate timepieces.

## Byproducts/Waste

Minor quantities of waste result from hourglass construction, depending on the types of materials used. Wood that is carved to make the hourglass frame will create some waste, for example. Glass that is too thin or flawed can be melted and blown again. Excess quantities of sand can be saved for future use.

## The Future

The hourglass would seem to have no future. In fact, the beautiful shape of the glass itself and its custom-made frame and colored sand can be selected to suit decor, atmosphere, or occasion. While future production may be limited, the hourglass as an object with ancient associations as well as built-in elegance will always appeal to collectors and those who appreciate the mysteries of art and time.

## Where to Learn More

### Books

Branley, Franklyn M. *Keeping Time: From the Beginning and into the Twenty-first Century.* Boston: Houghton Mifflin Company, 1993.

Cowan, Harrison J. *Time and Its Measurement: From the Stone Age to the Nuclear Age.* New York: The World Publishing Company, 1958.

Guye, Samuel and Henri Michel. *Time & Space: Measuring Instruments from the Fifteenth to the Nineteenth Century.* New York: Praeger Publishers, 1970.

Smith, Alan. *Clocks and Watches: American, European and Japanese Timepieces.* New York: Crescent Books, 1975.

### Periodicals

Morris, Scot. "The floating hourglass." *Omni* (September 1992): 86.

Peterson, Ivars. "Trickling sand: how an hourglass ticks." *Science News* (September 11, 1993): 167.

### Other

The Hourglass Connection. http://www.hourglass.com/ (June 29, 1999).

*—Gillian S. Holmes*

# Incense Stick

When the Three Wise Men brought their most precious gifts to Bethlehem, two of them—frankincense and myrrh were resins used to make incense. The third gift was gold, but it was the least valuable of these substances at that time. This relationship shows the importance that incense once held in our world. In modern times and in the Western world, the vast fragrance industry is dominated by perfumes, and incense manufacture and use is comparatively small. But, in fact, incense is the parent of perfumes. Incense is also the ancestor of many other products related to good smells. The importance of clean-smelling personal hygiene, bathrooms with sweetened air, laundry evoking the great outdoors, and romance-inspiring aromas originated with the powerful effects of incense in religious and public ceremonies.

## Background

Incense comes from tree resins, as well as some flowers, seeds, roots, and barks that are aromatic. The ancient religions associated their gods with the natural environment, and fragrant plant materials were believed to drive away demons and encourage the gods to appear on earth; they also had the practical aspect of banishing disagreeable odors. Fashion is intensely linked to scents, and designers include signature scents to evoke the spirit of their clothes. That perfume originated from incense shows in the word itself; *per* and *fumum* mean through and smoke in Latin.

There are two broad types of incense. Western incense is still used in churches today and comes almost exclusively from the gum resins in tree bark. The sticky gum on the family Christmas tree is just such a resin, and its wonderful scent evokes the holidays. The gum protects the tree or shrub by sealing cuts in the bark and preventing infection. In dry climates, this resin hardens quickly. It can be easily harvested by cutting it from the tree with a knife. These pieces of resin, called grains, are easy to carry and release their fragrance when they are sprinkled on burning coal.

Eastern incense is processed from other plants. Sandalwood, patchouli, agarwood, and vetiver are harvested and ground using a large mortar and pestle. Water is added to make a paste, a little saltpeter (potassium nitrate) is mixed in to help the material burn uniformly, and the mix is processed in some form to be sold for burning. In India, this form is the *agarbatti* or incense stick, which consists of the incense mix spread on a stick of bamboo. The Chinese prefer the process of extruding the incense mix through a kind of sieve to form straight or curled strands, like small noodles, that can then be dried and burned. Extruded pieces left to dry as straight sticks of incense are called joss sticks. Incense paste is also shaped into characters from the Chinese alphabet or into maze-like shapes that are formed in molds and burn in patterns believed to bring good fortune. For all incense, burning releases the essential oils locked in the dried resin.

## History

Incense has played an important role in many of the world's great religions. The Somali coast and the coasts of the Arabian Peninsula produced resin-bearing trees and shrubs including frankincense, myrrh, and the famous cedars of Lebanon. The cedar

*Western incense is still used in churches today and comes almost exclusively from the gum resins in tree bark. Eastern incense is processed from other plants.*

wood was transported all over the Tigris and Euphrates valleys, and the name Lebanon originated from a local word for incense.

The ancient Egyptians staged elaborate expeditions over upper Africa to import the resins for daily worship before the sun god Amon-Ra and for the rites that accompanied burials. The smoke from the incense was thought to lift dead souls toward heaven. The Egyptians also made cosmetics and perfumes of incense mixed with oils or unguents and blended spices and herbs.

The Babylonians employed incense during prayers and rituals to try to manifest the gods; their favorites were resins from cypress, fir, and pine trees. They also relied on incense during exorcisms and for healing. They brought incense into Israel before the Babylonian Exile (586-538 B.C.), and incense became a part of ancient Jewish worship both before the Exile and after. True frankincense and myrrh from Arabia were widely used in the temples in Jerusalem during the times of Christ's teachings, although incense has fallen out of use in modern Jewish practice.

Both the ancient Greeks and Romans used incense to drive away demons and to gratify the gods. The early Greeks practiced many rites of sacrifice and eventually began substituting the burning of incense for live sacrifices. As a result of his conquests, Alexander the Great (356-323 B.C.) brought back many Persian plants, and the use of incense in civic ceremonies became commonplace in Greek life. Woods and resins were replaced by imported incense as the Roman Empire expanded. The Romans encountered fine myrrh in Arabia, and the conquerors carried it as incense with them across Europe.

By the fourth century A.D., the early Christians had incorporated incense burning into their practices, particularly the Eucharist when the ascending smoke was thought to carry prayers to heaven. Both the Western Catholic Church and the Eastern Orthodox Church used incense in services and processions, but incense has always been more intensely applied in the Eastern services. The rite of swinging the censer was and is used in many religions; the censer (also called a *thurible* in the West and a *k dan* in Japan) is suspended on chains and carried by hand. The Reformation ended the presence of incense in Protestant church practices, although its use returned to the Church of England after the Oxford Movement in the nineteenth century.

Incense was always employed more extensively in eastern religions. The Hindu, Buddhist, Taoist, and Shinto religions all burn incense in festivals, processions, and many daily rituals in which it is thought to honor ancestors. Incense burners, which are containers made of metal or pottery in which incense is burned directly or placed on hot coals, were first used in China as early as 2,000 B.C. and became an art form during China's Han dynasty (206 B.C.-220 A.D.). These vessels had pierced lids to allow the smoke and scent to escape, and the designs from this period through the Ming dynasty (1368-1644) became increasingly ornate with smoke-breathing dragons and other imaginative creations. The Chinese also applied incense to a wide variety of uses including perfuming clothes, fumigating books to destroy bookworms, and scenting inks and papers. Even the fan (an import into China from Japan) was constructed with sandalwood forming the ribs so the motion of the fan would spread the fragrance of the wood.

In Japan, incense culture included special racks to hold kimonos so the smoke from burning incense could infiltrate the folds of these garments. Head rests were also steeped in incense fumes to indirectly perfume the hair. Clocks were made of incense sticks; different scents from the sticks told those tracking the time of the changing hours.

## Raw Materials

Stick incense is made with "punk sticks" and fragrance oils. All the components are natural materials. The sticks themselves are imported from China and are made of bamboo. The upper portion of each stick is coated with paste made of sawdust from machilus wood, a kind of hardwood. The sawdust is highly absorbent and retains fragrance well. Charcoal is also used to make the absorbent punk, and it is favored in incense sticks made in India.

The fragrant oils are made of oil from naturally aromatic plants or from other perfumes or fragrances that are mixed in an oil base. Small quantities of paint are used to color-

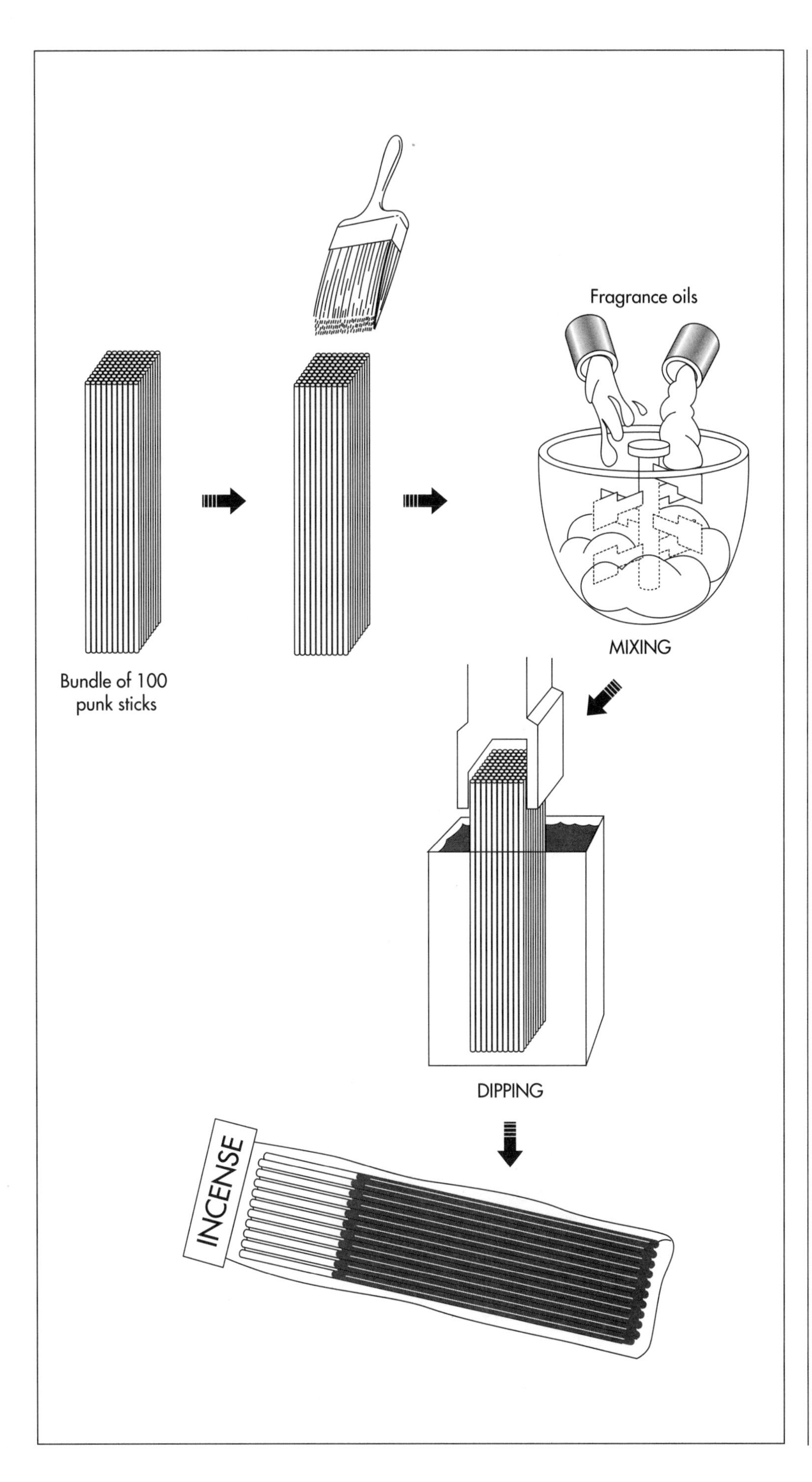


Stick incense is made with punk sticks imported from China and fragrance oils. Bundles of one hundred punk sticks are painted with a color unique to the intended fragrance. Next, fragrance oils are mixed, and the punk-covered ends of the bundles are dipped in the fragrance. Once dry, the bundles are each wrapped in wax paper and sealed in plastic bags. The bags are placed in bins. As orders are received for the incense sticks, they are individually packaged, packed in boxes made of recycled cardboard, and shipped for sale.

code the ends of the incense sticks depending on their fragrance.

## Design

The design of incense is based almost solely on fragrance. Incense makers carefully monitor trends in fragrances by obtaining samples from fragrance houses, discussing fashions and interests with their customers, and even noting those fragrances that are used in detergents, fabric softeners, and room air fresheners. When a fragrance seems like a possibility for use in incense, the manufacturer makes test batches of oils and incense sticks and gives employees and customers samples to burn at home. Positive feedback helps them select new incense fragrances.

## The Manufacturing Process

1 At the incense factory, bundles of punk sticks arrive from a specialized supplier in China. Each bundle consists of 100 sticks. The ends of the sticks are cleaned by pounding the end of each bundle in front of a vacuum cleaner that sucks up the dust. The bundles are selected for a particular fragrance, and the even ends of the sticks, still tightly bundled, are painted with a color unique to that fragrance. The number of bundles designated for a particular fragrance is based on the popularity of the fragrance. For example, the factory may make 1,200 bundles (12,000 sticks of incense) with vanilla fragrance that is very popular; and it may make only 300 bundles (3,000 sticks of incense) in the green apple fragrance, which doesn't sell as well. After the ends are painted, the bundles are left overnight for the paint to dry.

2 The next day, fragrance oils are mixed, and the punk-covered ends of the bundles are dipped in the fragrance. They are again left on shelves to dry overnight. A typical incense maker may stock hundreds of fragrances, some of which contain hundreds of elements to make their perfume. Many Indian scents are complex combinations of ingredients.

3 The dried bundles are each wrapped in wax paper and sealed in 12 x 3 in (30.5 x 7.6 cm) ziplocking plastic bags. The bags are placed in bins. As orders are received for the incense sticks, they are individually packaged, packed in boxes made of recycled cardboard, and shipped for sale.

## Byproducts/Waste

No byproducts are made, although incense manufacturers may make a wide range of fragrances in stick, cone, or powder form. Dust is the primary waste material and is contained by vacuuming and excellent ventilation. All paper goods are recyclable.

There are no safety hazards to employees, but there is a considerable risk to those with allergies. Potential employees are warned that the natural components of the sticks and the fragrances may cause allergic reactions, and some find they are unable to work in the factory because of this consideration.

## The Future

Customs in incense manufacture have changed little over the centuries except in the range of fragrances offered. In ancient times, only naturally fragrant resins or woods like sandalwood and patchouli were used for incense. Modern fragrance production allows virtually any scent to be duplicated, and fragrances are available now that couldn't be offered before. Examples include green tea, candy cane, blueberry, pumpkin pie, and gingerbread incense.

The custom of use of incense is also likely to change in the future and in Western culture. In India, two or three sticks of incense may be burned every day in a typical home, while, in the United States, users of incense may only burn one stick a week. Incensemakers hope the variety, effectiveness, and low cost of incense sticks will make them more popular than air fresheners and room deodorizers made with artificial perfumes. Also, the popularity of meditation and aromatherapy have spurred incense sales among clients who want their rare moments of quiet and relaxation to be healing and beautifully scented.

## Where to Learn More

### *Periodicals*

Casson, Lionel. "Points of origin." *Smithsonian* (December 1986): 148+.

Karban, Roger. "Use of incense has unsavory past." *National Catholic Reporter* (December 8, 1995): 2.

Morris, Edwin T. *Fragrance: The Story of Perfume from Cleopatra to Chanel.* New York: Charles Scribner's Sons, 1984.

### Other

Wild Berry Incense, Inc. 1996. http://www.wild-berry.com/ (June 29, 1999).

*—Gillian S. Holmes*

# Jam and Jelly

*In the United States, food processing regulations require than jams and jellies are made with 45 parts fruit or juice to 55 parts sugar.*

Jams and jellies are spreads typically made from fruit, sugar, and pectin. Jelly is made with the juice of the fruit; jam uses the meat of the fruit as well. Some vegetable jellies are also produced.

## Background

It is difficult to pinpoint when people first made a fruit spread. Ancient civilizations were known to set a variety of foods in the sun to dry in order to preserve them for later use. One of the first recorded mentions of jam making dates to the Crusades whose soldiers brought the process back from their journeys in the Middle East.

Preserving foods was a home-based operation until the nineteenth century. Even today, millions of people make fruit preserves in their own kitchens. Whether in the home kitchen or in a modern food processing plants, the procedure is essentially the same. Fruits are chopped and cooked with sugar and pectin until a gel is formed. The jam or jelly is then packed into sterilized jars.

Spoilage prevention is a major concern for both the home and the commercial jam producer. An important innovation in food preservation occurred in 1810. Nicolas Appert, a French confectioner, determined that by filling jars to the brim with food so that all air is expressed out and then placing the jars in boiling water would prevent spoilage.

In the early 1800s in the United States, the country was experiencing a surge westward. Of the many legendary characters to emerge during this period was John Chapman, better known as Johnny Appleseed. A nurseryman from western Pennsylvania, Chapman walked through the Midwest planting apple orchards. His purpose was to provide crops for the coming pioneers.

One of those pioneers was Jerome Smucker of Ohio who used Chapman's apples to open a cider mill in 1897. Within a few years, he was also making apple butter. Smucker blended the apple butter in a copper kettle over a wood stove. He and his wife ladled the apple butter into stoneware crocks. She then sold it to other housewives near their home in Wayne County, Ohio.

Fifty years earlier in Concord, Massachusetts, Ephraim Wales Bull finally achieved his goal of cultivating the perfect grape. His rich-tasting Concord grape became enormously popular. In 1869, Dr. Thomas Branwell Welch used the Concord grape to launch his grape juice company. When, in 1918, Welch's company made its first jam product, Grapelade, the United States Army bought the entire inventory. The company's trademark Concord grape jelly debuted in 1923.

After World War II, food scientists developed the process of aseptic canning: heating the food and the jar or can separately. For sensitive foods such as fruits, this allowed for high-temperature flash cooking that preserved taste and nutritional value.

When sugar prices soared in the early 1970s, high fructose corn syrup (HFCS) became a popular substitute. Several major food processing companies, including Archer Daniels Midland, Amstar CPC International, Cargill, H.J. Heinz, and Anheuser Busch opened HCFS plants.

## Raw Materials

Jams and jellies are made from a variety of fruits, either singly or in combination. Most of the fruits are harvested in the fall. The level of ripeness varies. Pears, peaches, apricots, strawberries, and raspberries gel best if picked slightly underripe. Plums and cherries are best if picked when just ripe. The fruit is purchased from farmers. Most jam and jelly producers develop close relationships with their growers in order to ensure quality. The production plants are built close to the fruit farms so that the time elapsed between harvest and preparation is between 12-24 hours.

Sugar or high fructose corn syrup, or a combination of the two are added to the fruit to sweeten it. Cane sugar chips are the ideal type of sugar used for preserving fruit. Granulated and beet sugar tend to crystallize. Sugar is purchased from an outside supplier. High fructose corn syrup is processed by fermenting cornstarch. It is purchased from an outside supplier

The element that allows fruit to gel, pectin is present in varying degrees in all fruit. Apples, blackberries, cherries, citrus fruits, grapes, quinces, and cranberries have the best natural gelling properties. Strawberries and apricots are low in pectin. Jams made from such fruits are either blended with fruits high in pectin, or extra sugar is added to the mixture. Sometimes pectin is extracted industrially from dried apples.

Citric acid is added to obtain the correct balance needed to produce the jam or jelly. Lime and lemon juice are high in citric acid, therefore they are the most prevalent source used. Citric acid can also be obtained by the fermentation of sugars. It is purchased from outside suppliers.

Other flavorings, such as vanilla, cinnamon, mint, alcoholic beverages such as rum or Kirsch, can be added to the jam or jelly. These flavorings are purchased from outside suppliers.

## The Manufacturing Process

The ingredients must be added in carefully measured amounts. Ideally, they should be combined in the following manner: 1% pectin, 65% sugar, and an acid concentration of pH 3.1. Too much pectin will make the spread too hard, too much sugar will make it too sticky.

### *Inspection*

1 When the fruit arrives at the plant, it is inspected for quality, using color, ripeness, and taste as guides. Fruit that passes inspection is loaded into a funnel-shaped hopper that carries the fruit into pipes for cleaning and crushing.

### *Cleaning, crushing, and chopping*

2 As the fruit travels through the pipes, a gentle water spray clears away surface dirt. Depending on whether the finished product is to be jam or jelly, paddles push the fruit and or just its juice through small holes, leaving stems and any other excess debris behind. Some fruits, such as citrus and apples may be manually peeled, cored, sliced and diced. Cherries may be soaked and then pitted before being crushed.

### *Pasteurizing the fruit*

3 The fruit and/or juice continues through another set of pipes to cooking vats. Here, it is heated to just below the boiling point (212° F [100° C]) and then immediately chilled to just below freezing (32° F [0° C]). This process, pasteurization, prevents spoilage. For jelly, the pulp is forced through another set of small openings that holds back seeds and skin. It will often then be passed through a dejuicer or filter. The juice or fruit is transferred to large refrigerated tanks and then pumped to cooking kettles as needed.

### *Cooking the jam and jelly*

5 Premeasured amounts of fruit and/or juice, sugar, and pectin are blended in industrial cooking kettles. The mixtures are usually cooked and cooled three times. If additional flavorings are to be included, they are added at this point. When the mixture reaches the predetermined thickness and sweetness, it is pumped to filling machines.

### *Filling the jars*

6 Presterilized jars move along a conveyer belt as spouts positioned above pour premeasured amounts of jam or jelly into them.

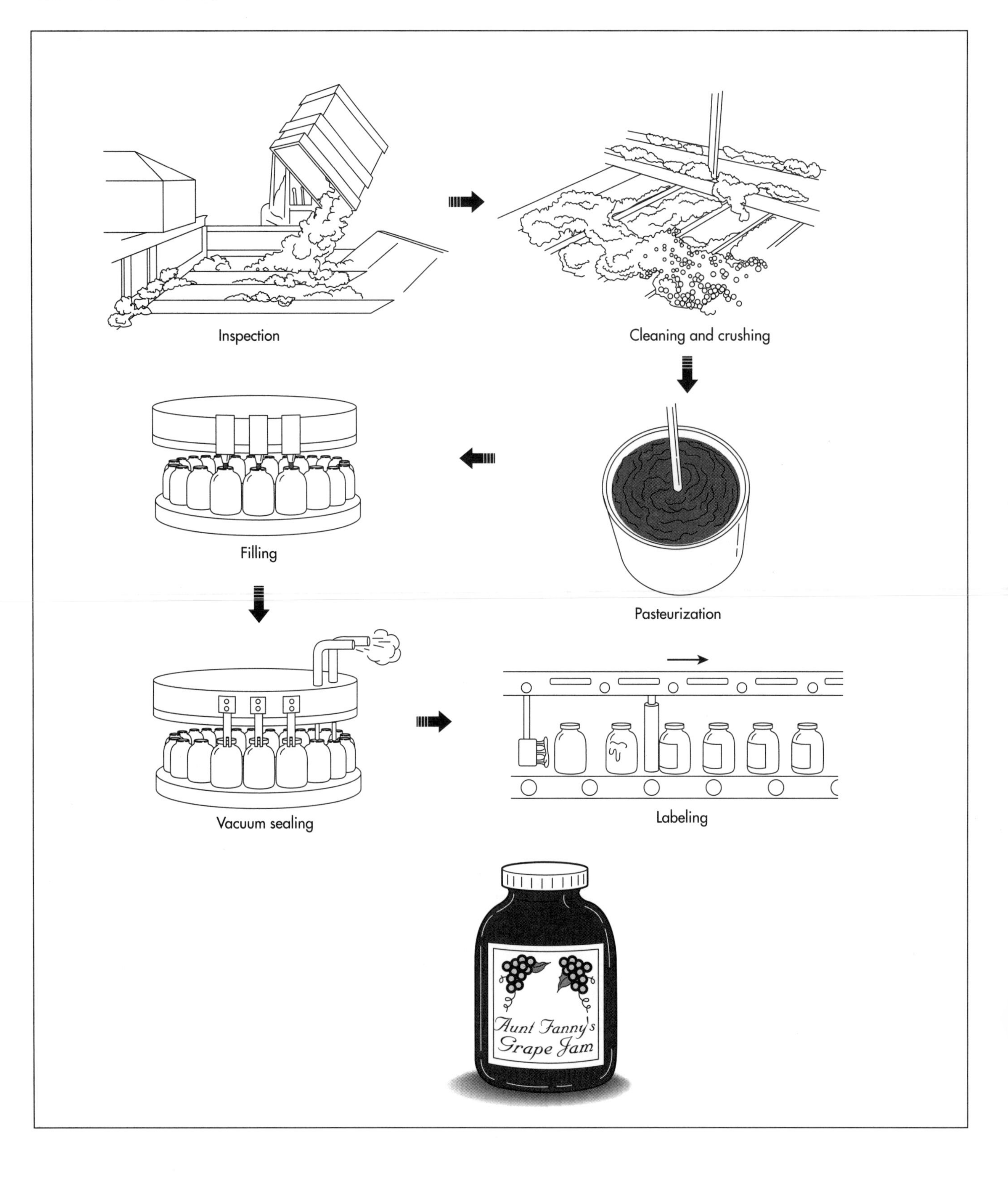


Metal caps are then vacuumed sealed on top. The process of filling the jars and vacuum packing them forces all of the air out of the jars further insuring the sterility of the product.

### *Labeling and packaging*

7 The sealed jars are conveyed to a machine that affix preprinted labels. According to law, these labels must list truthful

and specific information about the contents. The jars are then packed into cartons for shipment. Depending on the size of the producer's operation, labeling and packaging is either achieved mechanically or manually.

## Quality Control

In the United States, food processing regulations require than jams and jellies are made with 45 parts fruit or juice to 55 parts sugar. The federal Food and Drug Administration (FDA) mandates that all heat-processed canned foods must be free of live microorganisms. Therefore, processing plants keep detailed lists of cooking times and temperatures, which are checked periodically by the FDA.

Requirements also exist for the cleanliness of the workplace and workers. Producers install numerous quality control checks at all points in the preparation process, testing for taste, color and consistency.

## The Future

Because it is a relatively simple process, the production of jams and jellies is not expected to change dramatically. What is apparent is that new flavors will be introduced. Certain vegetable jellies such as pepper and tomato have been marketed successfully. Other, more exotic types including garlic jelly are also appearing on grocery shelves.

## Where to Learn More

### Books

Coyle, L. Patrick. *The World Encyclopedia of Food.* New York: Facts on File, 1982.

Lang, Jenifer Harvey, ed. *Larousse Gastronomique.* New York: Crown, reprinted 1998.

Trager, James. *The Food Chronology.* New York: Henry Holt, 1995.

### Periodicals

Anusasananan, Linda Lau. "Why?" *Sunset* (June 1996): 142.

Kawatski, Deanna. "Canning: a modest miracle." *Mother Earth News* (August-September 1996): 52.

### Other

J.M. Smucker Co. 1999. http://www.smucker.com/ (June 28, 1999).

Welch's Co. http://www.welchs.com/ (June 28, 1999).

*—Mary McNulty*

Opposite:
When the fruit arrives at the plant, it is inspected for quality, using color, ripeness, and taste as guides. Fruit that passes inspection is cleaned, crushed, and pasteurized. Next, the premeasured mixture is cooked with added sugar and pectin until it reaches the appropriate thickness and taste. Then it is vacuum-packed in jars and labeled.

# Jukebox

*In its height of popularity in the mid-1950s, approximately 750,000 jukeboxes were in use across the United States.*

## Background

A jukebox is a coin-operated machine that plays music from a record or compact disc (CD) once a selection is made. Originally called nickelodeons, the term jukebox did not appear until the late 1930s and its origins are in dispute. Some believe it is derived from the African word jook, meaning to dance. Others link it to the juke joints—roadside bars located in the South and frequented by African Americans—that were popular at that time.

In its height of popularity in the mid-1950s, approximately 750,000 jukeboxes were in use across the United States. That number dipped during the 1970s and 1980s, but with the advent of CD technology and a growing antiques market, the number of jukeboxes presently in use is a solid 250,000.

## History

In 1877, Thomas Edison invented the phonograph, a coin-operated music machine that played music from a wax cylinder. On November 23, 1889, Louis Glass installed a coin-operated phonograph in his Palais Royale Saloon located in San Francisco. It was called "nickel-in-a-slot" because that was the amount of money needed to make a selection. Later, the term was shortened to nickelodeon. In 1906, John Gabel invented the "Automatic Entertainer," a music machine that replaced the wax cylinder with 78-rpm disc recordings and offered several selections of records that could be played. Gabel's Automatic Entertainer dominated the market until the mid-1920s.

The jukebox remained something of a novelty arcade item until the invention of the electric amplifier. Without amplification, it was impossible for a large group of listeners to enjoy the music played by the jukebox. When Automated Musical Instruments Inc. (AMI) developed an amplifier in 1927, the popularity of the jukebox surged. It was especially popular in the illegal speakeasies of the Prohibition Era because it provided a cheap form of entertainment. AMI sold 50,000 of its amplified machines in one year, bringing to life the age of the jukebox.

During the Depression, record sales plummeted from $75 million in 1929 to $5 million in 1933. The growing popularity of the jukebox and the purchases by store owners that went along with it resurrected the waning music business, and by 1938, the industry had resurfaced at $25 million in sales. By 1940, there were 400,000 jukeboxes in use in the United States.

Three names were made during the 1940s and they remain synonymous with the jukebox industry. Seeburg, Rock-Ola, and Wurlitzer all manufactured jukeboxes at this time. Each company began by creating jukeboxes in the likeness of the radio, but in the 1940s, jukebox design came into its own with the help of a few great designers employed by the companies. Perhaps the best known is Paul Fuller, the designer behind the Wurlitzer models that pushed Wurlitzer to the top of the industry in the late-1940s and 1950s. With the use of rotating lights, art deco styled cabinets, and bubble tubes, Wurlitzer models were works of art. The most popular design was the Wurlitzer 1015 that was introduced in 1946 and became the biggest selling jukebox in history. In its original run, it sold a total of 56,246 boxes. In 1948, Seeburg offered its own innovation to the jukebox industry with the introduction of its Select-O-Matic 100, the first jukebox to include 100 selections. This technology allowed popular music to be played in the same venue as regional country, folk, jazz, and blues music—a variety that changed the music industry and its development completely. By 1956, jukeboxes with 200 selections were being manufactured.

Just as the proliferation of fast-food restaurants such as McDonald's and chain restaurants such as Houlihan's spelled doom for mom-and-pop establishments, the taped music played in the new gathering places signaled the end of the jukebox's glory. The introduction of cassette tapes and the declining production of 45-rpm records also added to the decreased popularity of jukeboxes. By the mid-1970s, the number of jukeboxes had fallen to 225,000.

The jukebox industry waned through the 1980s until a growing antiques market and new technology revived the industry. Refurbished classic models are collectables, and a Wurlitzer 1015 that first sold for $750 is now approximately $12,000. CD technology has breathed new life into the primary market, creating new models that house 100 CDs totaling 1,000 song selections. Since the late 1980s, the number of jukeboxes has creeped back up to 250,000. A remake of the Wurlitzer 1015 is even being manufactured by Wurlitzer of Germany. Rock-Ola machines are also still produced as well.

## Raw Materials

Each jukebox is comprised of 700-800 different components, including wood cabinetry; injection-molded plastic pieces; electronic stereo equipment such as amplifiers, woofers and tweeters, turntable or disk player; lighting; mirrors; records or compact disks; and the selection mechanism. In some cases, the bulk of the components are purchased from outside suppliers. Other manufacturers create everything in-house except the records or compact disks.

The cabinets are constructed of multiple layers of wood, usually Italian poplar, Finland ply, walnut, olive ash, alder, maple, and Corinthian burl elm. Metal parts such as the grills, trim, and money changers are cast from metal dies. The grills and trim are plated with copper, nickel, and polished chrome.

## Design

Some jukeboxes are replicas of original designs, others are original. In either case, a designer creates a prototype, or sample, of the jukebox to be manufactured using CAD/CAM software. The company produces a half dozen or so to place in restaurants or taverns. The prototypes are test-marketed for several months before they are approved for mass production.

## The Manufacturing Process

Although many of the components are machine-crafted, each jukebox is hand-assembled. Therefore, the assembly line moves very slowly. The factory can produce approximately 10 jukeboxes per day.

1 Plywood or solid wood sheets are cured and molded into the basic jukebox shape. Slots are cut into the sheets for the side metal trim. The cabinets are varnished and stained. After the varnish and stain has dried, the side metal trim is riveted onto the cabinet.

2 All sheet metal parts are laser cut.

3 The florescent lighting fixtures and bubble tubes are installed in the cabinets. A polarized light system allows the colors to change.

4 The electronic components are screwed to the inside of the door. The door is then attached to the cabinetry. The record/compact disk storage and changer mechanisms are installed inside the cabinetry.

5 After the jukebox is completed, it is sent to the testing room for 24-48 hours. Inspectors check all of the components: lighting, sound, selection, money changing, etc., to insure that they are working properly.

## Quality Control

Quality control is key in the production of a jukebox. After each step in the production process, qualified personnel check the integrity of the work done. Final inspection of

A jukebox is made up of 700-800 different components, including wood cabinetry; injection-molded plastic pieces; electronic stereo equipment such as amplifiers, woofers and tweeters, turntable or disk player; lighting; mirrors; records or compact disks; and the selection mechanism. In some cases, the bulk of the components are purchased from outside suppliers. Other manufacturers create everything in-house except the records or compact disks. Although many of the components are machine-crafted, each jukebox is hand-assembled at workstations.

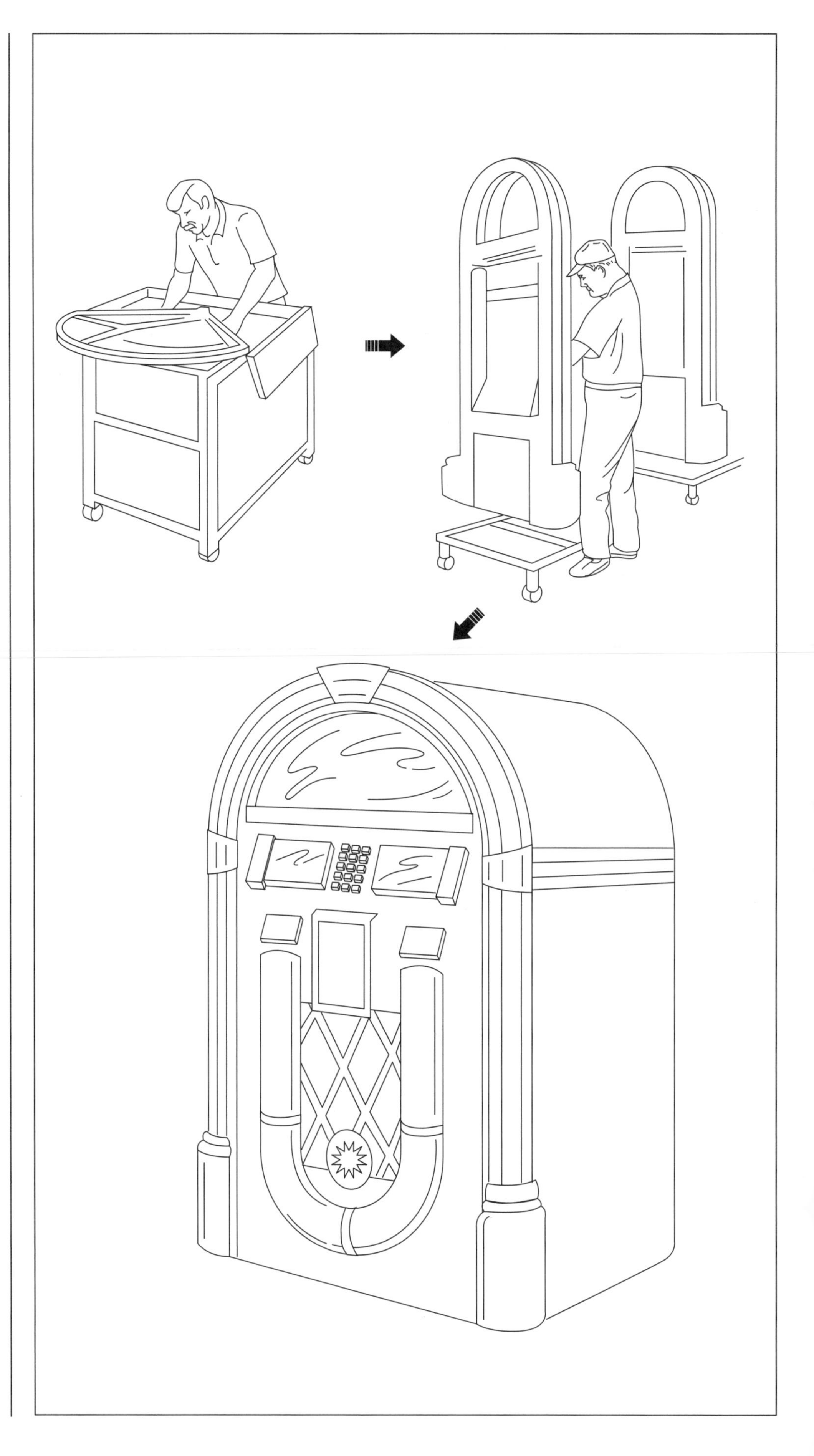

the jukebox is made on all its components prior to shipment.

## The Future

Since the late 1980s, the demand for new jukeboxes has remained steady. Collecting refurbished jukeboxes has also gained in popularity and is now a multimillion dollar secondary market. Aficionados created their own organization—American Historic Jukebox Society— and meet regularly at vintage jukebox shows around the United States. No doubt the jukebox is as American as apple pie, and while there is nostalgia for a time already past, there will also be a market for new and vintage jukeboxes.

## Where to Learn More

### Books

Bunch, William. *Jukebox America.* New York: St. Martin's Press, 1994.

### Periodicals

Barol, Bill. "The Wurlitzer 1015." *American Heritage* (September/October 1989):28.

Boehlert, Eric. "Put Another Nickel In." *Billboard* (November 1, 1994): 92.

Botts, Rick. "The Jukebox." *Popular Machanics.* (June 1995): 74.

"Classic Jukebox Goes Mod." *Design News* (March 9, 1987): 30.

Gustaitis, Joseph. "The Jukebox: America's Music Machine." *American History Illustrated* (November/December 1989): 44.

Russell, Deborah. "Juke Biz Finds New Life Via New Technology, Markets." *Billboard* (October 19, 1991): 10.

Webb, Marchus. "Classical Music: Antique Apparatus hits it big with reproductions of class jukeboxes." *RePlay Magazine* (January 1989): 113.

### Other

Amusement & Music Operators Association (AMOA). 401 N. Michigan Avenue, Chicago, IL 60611-4267. (312) 644-6610. Fax: (312) 321-6869.

Rock-OlaManufacturing Corporation. 2335 208th St., Torrance, CA 90501. (310) 328-1306. Fax: (310) 328-3736. http://www.rock-ola.com/ .

# License Plate

*License plate sizes were not standardized until 1957 when the dimensions of 6 in by 12 in (15.24 cm by 30.48 cm) were selected.*

## Background

Metal plates attached to motor vehicles are commonly called license plates, but this is a misnomer. The driver of the vehicle must be licensed, and the vehicle is registered; therefore, these plates are really registration plates. In some states, especially in the South, the plates are also called license tags. The information printed on the plate, either in the metal itself or on one or two attached stickers tells authorities about the registration of the vehicle and about the owner.

The American states and Canadian provinces have separate departments or administrations for motor vehicles and their own systems for coloring, issuing, and numbering the plates. For law enforcement, distinctions are made in color combinations, captions or words embossed on the plates, and combinations of alpha and numeric characters that provide identification to those trained in decoding them. Various classes of vehicles, weight and use restrictions, validity of registration, and information about the owner and the vehicle are communicated on the license plate.

Currently, all the states and provinces issue license plates that are permanent or semi-permanent; they are intended to last throughout the ownership of the vehicle or for some period longer than one year. To revalidate the plate when registration fees are paid annually, decals are issued for the vehicle operator to stick on the plate. Out of the 51 U.S. jurisdictions (the 50 states and Puerto Rico), 31 require two plates for each vehicle, one for the front and one for the back. The remaining jurisdictions only require a rear plate. The trend, however, may move to one plate only because of the cost of manufacture.

License plates on passenger cars may include information about the county of the owner's residence, the owner's occupation, expiration codes related to vehicle registration, government department codes, and special codes for officials and certain groups like disabled drivers. Commercial vehicles like buses and trucks usually have plates of a different color and with different codes than those for passenger vehicles; their license plates are also changed more frequently.

The reasons for these differences are taxes. Owners of commercial vehicles pay several different taxes, over and above the registration fees, depending on miles driven, fuel use, and vehicle weight. Taxes are computed based on formulas for these factors. Interstate traffic is also complicated because the taxes from vehicle registration and taxes are essential for highway maintenance and new construction. States apportion fees among themselves based on the vehicle's portion of highway use in each state; the displays of license plates and decals on some vehicles hint at these complexities.

## History

Before license plates came vehicle registration bureaus. States realized as early as 1892 that some form of regulation was becoming necessary with the increase in the number of motorized vehicles. Automobiles, horses, and pedestrians were causing roads to deteriorate rapidly, and regulations—and funds—were needed to correct the problem. Public safety was also a grave issue, and law enforcement officials needed a device to help them keep records regarding vehicle owners and their actions. License plates came into existence in 1903 when it became apparent that motor vehicles were sure to replace horse-drawn carriages and that a system of registering and taxing them and their drivers was needed. Massachusetts was one of the first states to issue licenses for drivers and registration plates for vehicles.

Funnily, many license plate terms refer back to the early history of the automobile or even to the days of horses and carriages. Vintage vehicles bear Horseless Carriage license plates in many states, and historical vehicle plates are issued to owners whose vehicles were manufactured after 1922 (varies by state) and are at least 25 years old. In Alabama, some trucks are licensed to operate in an area with a 15-mile radius. This is an outdated form of measurement based on the distance a mule can travel in one day, so plates on these trucks are termed mule tags. In the early dates of motorcycle registration, small plates were mounted on the motorcycles and the drivers were required to carry brass watch fobs bearing the registration information.

Vehicle operators were concerned that front-mounted plates would block the flow of air to the radiator, so some states made slotted plates to reduce this hazard; this practice was discontinued in about 1918. The first plates were made of sheet iron, but tin became the standard by about 1920. The State of Arizona made its tags from sheet copper in 1932-1934. Porcelain plates were also quite common in the early days of vehicle plates, and Delaware was the last state to make porcelain plates in 1942. During World War II, soybean-based fiberboard was used for license plates because of the need to devote all available metal to the war effort; goats were particularly pleased with this development because the license plates made tasty snacks.

Also in about 1920, the states began using the labor of inmates in their correctional institutions to manufacture registration plates to provide useful work for the prisoners and also to keep plate production costs down. The states began to require that automobile builders provide lights for illuminating license plates in about 1923. The first reflectorized plate was issued in the State of Georgia in 1941, and Georgia was also the pioneer in the use of decals to update registration information, rather than issuing new plates every year. Plate sizes were not standardized until 1957 when the dimensions of 6 x 12 in (15.24 x 30.48 cm) were selected.

From the early days of license plates to about 1965, many states also showed police troop codes, county designations, or congressional districts on their vehicle plates. By 1991, only 10 states continued any of these practices, although county name decals are affixed to the plates from some other states. Computer technology is largely responsible for this change because fewer codes are needed for quick identification of vehicles.

## Raw Materials

The raw materials used to make license plates include sheets of aluminum, preprinted and colored reflective and adhesive sheeting, and paint. The aluminum blanks are usually precut to size by metal manufacturers and supplied in this form to correctional institutions or other plate makers.

Decals for annual registration renewals, county designations, or other uses are made by specialty printers on reflective sheeting much like that used to cover the license plates themselves.

## Design

Design of license plates is limited to a standard size and thickness that will fit license-plate mounts on most vehicles. Other properties like colors and reflective coatings are continuously improved for visibility, primarily to aid in law enforcement.

Beyond these restrictions, license plate designs can be quite creative. Personalized plates, also called vanity, custom, or prestige plates, have become popular in the last three decades when states recognized the cash benefits of appealing to the public's individuality. Canada calls these personalized number plates (PNPs). In California, the revenue from vanity plates is dedicated to environmental projects, so such plates are called environmental license plates (ELPs).

Other special plates are made to survivors of the attack on Pearl Harbor and recipients of

Modern license plate manufacture involves applying a preprinted, reflective sheet to the unstamped metal blank. The sheet is affixed to the blank, the sheet-covered blank is stamped to create the raised characters, and the characters are colored with ink. The plates are also given a clear, protective coating.

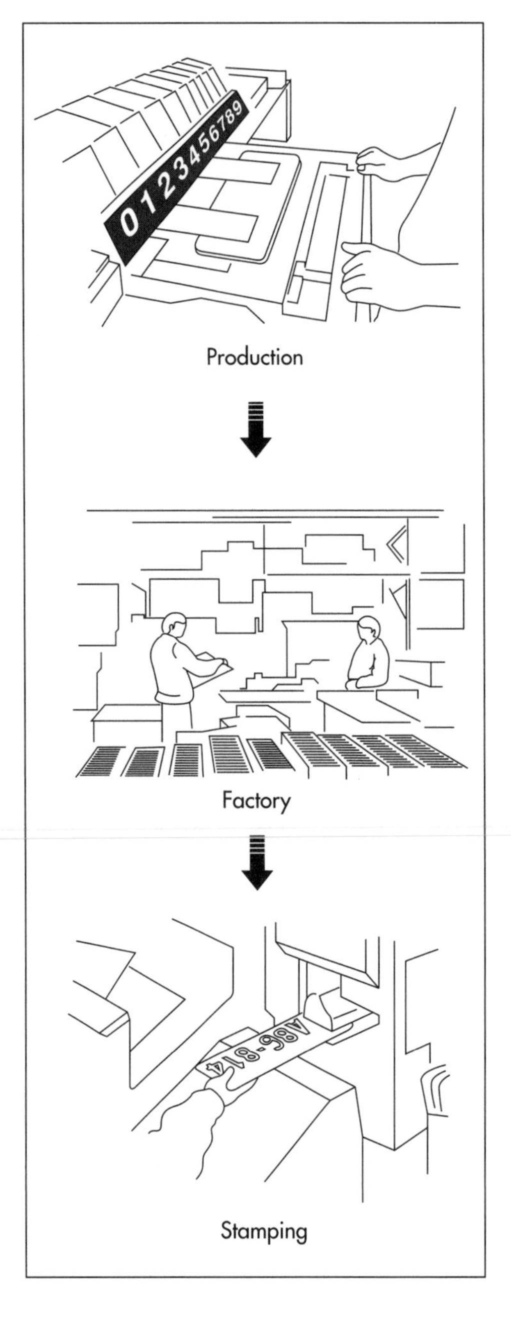


national honors like the Purple Heart and the Congressional Medal of Honor. Every state has its own list of special plates that changes—and tends to increase—every year. States often use license plates to promote tourism, and considerable thought goes into plate design so the plate mirrors the state's desired image.

Special category plates are another variety of license plate that is becoming increasingly popular. These include plates recognizing graduates of colleges and universities and veterans of military service and plates dedicated to a number of causes such as the Astronauts' Memorial commemorating the Challenger Space Shuttle disaster; Florida uses sales of this plate to raise funds for the memorial. Florida has also used other special category plates to raise funds to protect the manatee and the Florida panther and to commemorate the twentieth anniversary of Disney World and the five hundredth anniversary of Columbus's discovery of the New World; in fact, Florida has produced over 50 specialty tags. Again, computers have made these plates possible because it is no longer essential for a plate from California, for example, to be blue and gold for law enforcement professionals to recognize the home state of the vehicle.

## The Manufacturing Process

### Old method

Most license plates are still made in correctional institutions, although some states contract out plate manufacture to private companies.

1 Metal blanks are inserted into presses and stamped with rounded edging and a series of characters, usually including both numbers and letters.

2 The whole plate is painted in its main color, and the raised characters (and sometimes decorations) are painted in a contrasting color with an automated roller that is set to the correct height to only paint the raised elements.

3 The two-color plate is then treated with reflective coating. This is a paint-like substance made of extremely fine glass beads that refract (bend) light and bounce it back at many times the brightness of the paint alone. Plates made by this method are usually considered partially reflective because the application of the reflective coating over the dual levels of the plate is not uniform.

### New method

1 Modern license plate manufacture applies a preprinted sheet to the unstamped metal blank. This preprinted sheeting has the main color of the plate as well as multicolored decorations. It also has reflective material in the coating on the sheet. The sheet is affixed

to the blank, the sheet-covered blank is stamped to create the raised characters, and the characters are colored with ink.

2 The ink allows the reflective material to shine through completely and uniformly, so these plates are described as fully reflectorized. The ink is less durable than the paint, however, so these plates are also given a clear, protective coating. The reflective sheeting makes license plates easier for law enforcement officials to read by the light of headlights or flashlights, and it is more effective as a safety device on the highway. It does make license plates more expensive to manufacture.

## Quality Control

State motor vehicle departments establish the rules for license plate manufacture under the direction of the State legislature. The Interstate Commerce Commission also has requirements for plate manufacture particularly with respect to interstate traffic, licensing, and taxes. Number designations, the numbers and letters appearing on vanity plates, and special interest plates are all approved well in advance of production by the state motor vehicle authority, which also oversees manufacture in correctional institutions and private companies. Quality control during actual manufacture is by observation; the stamped metal plate is inspected, as is the coated and painted plate before packaging and distribution.

## Byproducts/Waste

No byproducts are made from license plate manufacture, although a considerable industry has arisen for making collectible plates for sports teams, those bearing first names and nicknames, and humorous plates with puns and jokes in the number and letter combinations.

Little waste results from plate making. The metal blanks are sized to limit metal waste. Minor amounts of trimmings from the sheets of reflective coating material are simply disposed.

## The Future

License plates are here to stay at least for the foreseeable future. This tried-and-true method of tracking vehicle ownership and operation is useful for law enforcement, a variety of taxes, registration, and other data. The future may see devices like bar codes or scanner codes incorporated into license plates or directly into vehicles. When the code is read by a scanner (a handheld model for police officers and fixed models for other applications), the complete history of the vehicle and information about its operation and ownership would be available by computer almost instantly. The bar codes might also include other fees like bridge tolls that would be billed to the vehicle owner.

License plates as we know them are also experiencing second lives, thanks to collectors. The Automobile License Plate Collectors Association (ALPCA) is one of several organizations for hobbyists who collect historical plates and those from different locations. ALPCA awards one state per year with a coveted "Best Plate Award" based on a competition among the discerning members of ALPCA. To encourage collectors and tourism, many states and provinces also sell sample plates. Their variety, colorful designs, and historic associations make license plates attractive both on the road and in the enthusiast's collection.

## Where to Learn More


### Books

Murray, Thomson C. *License Plate Book.* Jericho, New York: Interstate Directory Publishing Company Inc., 1992.

### Periodicals

Tooley, Jo Ann. "GR8 PL8S." *U.S. News & World Report* (August 12, 1991): 9.

### Other

American License Plate Collectors' Association. http://www.alpca.org/

State of Massachusetts Registry of Motor Vehicles. http://www.state.ma.us/rmv/


*—Gillian S. Holmes*

# Lock

*Linus Yale Jr., an American locksmith born in 1821, made a significant improvement in lock design in 1861 with his invention of the modern pin-tumbler lock.*

## Background

Locks have been used to fasten doors against thieves since earliest times. The Old Testament contains several references to locks, and the first archaeological evidence of locks are about 4,000 years old. These are Egyptian locks depicted in the pyramids. These earliest locks were of a type known as pin tumbler, and they are actually not very different from common door locks in use today. The Egyptian lock consisted of a heavy wooden housing mounted to the door. A wooden bolt passed through the lock and was held in place by iron pins which dropped into slots and held it firm. The key was a straight piece of wood with pegs projecting up from its end. When the key was inserted and pushed upwards, the pegs on the key lifted the pins in the lock, and the bolt was freed.

The Greeks developed a simple door lock by about 700 B.C. This used a latchstring to pull a bolt through brackets in the door. By pulling the string, the homeowner could lock the door from the outside. Then, the string was stuffed back through the keyhole. The key itself was a sickle-shaped piece of metal from two to three feet long. The key could be fitted into the hole in the lock to pull back the bolt from the outside. The major drawback to this lock was that anyone with a curved stick or their own key could open it. And, the large metal key was cumbersome.

Romans adopted the Greek lock system, but solved the problem of the heavy key by chaining it to a slave, and then chaining the slave to the doorpost. Eventually, Romans developed a new kind of lock, called the warded lock. In the warded lock, notches and grooves called wards were cut into the keyhole, and the key was cut with corresponding notches and grooves. Only the proper key could fit into the keyhole, and then its tip engaged the bolt and withdrew it. The warded lock was much smaller than its predecessor, and keys were small enough that no slave was needed to take care of it. But because the classic Roman toga had no pockets, the key still wasn't easy to carry, and so it was usually attached to a finger ring. Warded locks were widespread in Europe by the thirteenth century and remained in use well into the eighteenth century. They persisted in spite of the fact that they were easy to pick, and were barely an obstacle to determined thieves.

The Romans also used padlocks, in which a key turned a bolt releasing a spring on a shackle. These were used for locking trunks. Similar locks were invented in China, India and Russia during the same era. The Chinese also invented the combination lock. It had moveable rings inscribed with numbers or letters, and its hasp was released only when the rings were aligned in the proper sequence of symbols. Combination locks found their way to Europe, and were used in the Middle Ages especially on couriers' dispatch boxes.

European locksmiths in the Middle Ages made beautiful, intricate locks which took appallingly long hours of work to build and offered little real security. Locksmiths apprenticed for 10 years to reach the journeyman level. To reach the rank of master, the locksmith had to complete a masterpiece lock for approval by his guild. These masterpieces took thousands of hours to complete, and the results were generally much more decorative than functional. Locks that offered improved protection against theft were not developed until the late eighteenth century, when an English locksmith, Robert Barron, patented what was known as the double-action lever-tumbler lock in 1788. Barron's lock had two interior levers held by a spring. These levers, or tumblers, had notches that hooked over the bolt and held it shut. The key also had notches on it corresponding to the notches on the levers. When the right key was inserted, it would lift both tumblers, and the bolt could be drawn. Other inventors added many more tumblers to this design, and it proved much more difficult to pick than the earlier warded locks.

Linus Yale Jr., an American locksmith born in 1821, made a significant improvement in lock design in 1861 with his invention of the modern pin-tumbler lock. The design principal was similar to the Egyptian lock. This lock has a rotating cylinder which is held fast in the bolt by a series of five spring-driven pins of different heights. The key has five notches on it that correspond to the heights of the five pins. When the correct key is inserted, the pins line up level, and the cylinder can be turned to disengage the bolt. If the wrong key is inserted, the pins catch. Picking a Yale lock proved extremely difficult, and the parts for the lock could be inexpensively mass-produced by machine. Within several years of its invention, the Yale lock became the standard, replacing virtually all earlier lock technology.

Even more sophisticated locks were developed in the twentieth century, including timer locks used in bank vaults, push button locks, and electronic locks that operate with a credit card like key. The manufacturing process that follows is for a standard pin-tumbler lock. This is the kind of lock that may be found on any front door or file cabinet drawer.

## Raw Materials

Standard five-tumbler key locks are made of various strong metals. The internal mechanisms of locks are generally made of brass or die-cast zinc. The cam, which is the tongue that protrudes from the lock to secure it, is usually made of steel or stainless steel. The outer casing of a lock may be made of brass, chrome, steel, nickel or any other durable metal or alloy.

## The Manufacturing Process

### *Design*

1 Locks come in grades, from low-security to high-security. A low-security lock is generally made from cheaper materials, and its parts can be mass-produced. A company that manufactures low-security locks may have two or three available models, and keep in stock the parts needed to customize them. Beyond low-security, the lock manufacturer is generally what is called an original equipment manufacturer, meaning that they make the parts for their locks as well as the final products. This kind of manufacturer may keep only the most basic and common parts in stock, and most of its orders require custom design.

The process begins with the manufacturer assessing the customer's specifications. The customer orders a lock to fit a certain size door for example, and asks that the locks can be opened with a master key. The lock manufacturer then comes up with the best design for that customer's needs. In some cases, a customer may have purchased locks in the past from one company, and now wants more identical locks from a different manufacturer, who promises to make them more economically. Then, the lock manufacturer examines the customer's original locks and goes through what is known as a reverse engineering process. The manufacturer's design team figures out from the existing lock how to make their product match it. In many cases, the customer's first lock company has patented aspects of its lock construction. The second manufacturer may not duplicate it without infringing the other company's patents. So, the designers "design around" the first company's product, producing a lock that will match the customer's originals and serve the same purpose, but using different mechanisms. Medium and high security locks in most cases go through this design stage, making the production of locks a time-consuming process. A reputable manufacturer making anything but low security locks may take from eight to 12 weeks to produce locks for an order, from the time the specifications are given to when the locks are packed and shipped.

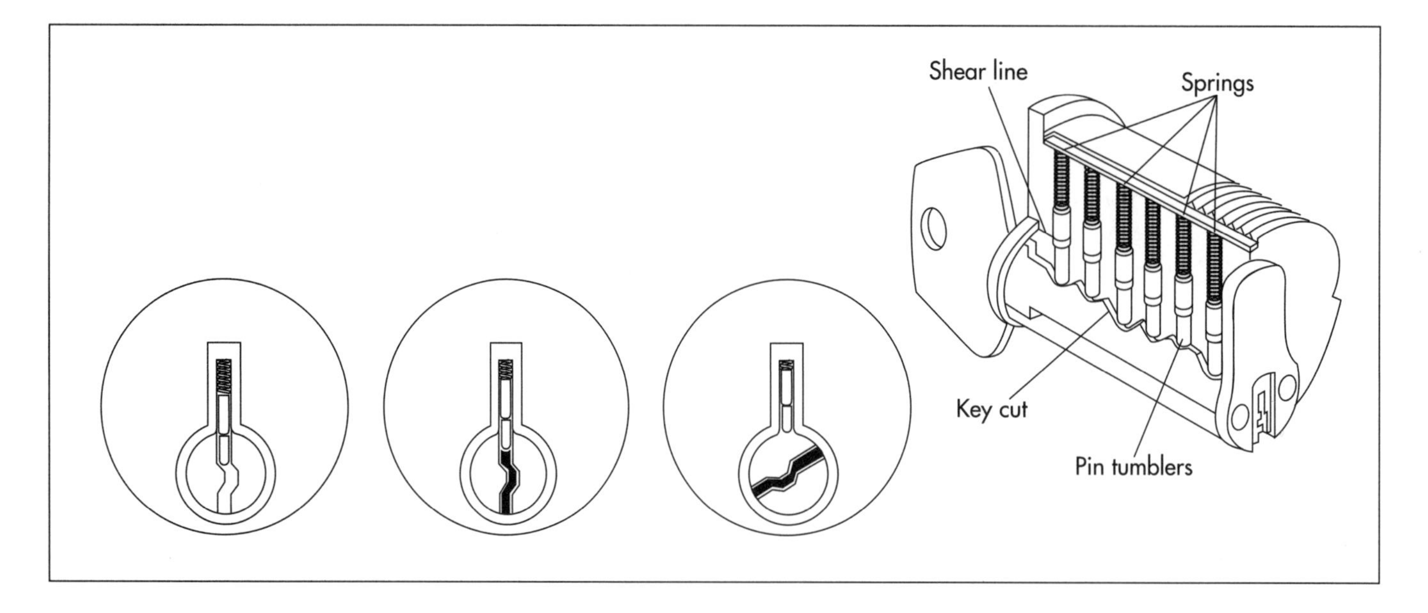


A cutaway of a standard Yale lock. This lock has a rotating cylinder that is held fast in the bolt by a series of five spring-driven pins of different heights. The key has five notches on it that correspond to the heights of the five pins. When the correct key is inserted, the pins line up level, and the cylinder can be turned to disengage the bolt. If the wrong key is inserted, the pins catch.

### The key

2 For the standard five-tumbler key lock, the key is made first. The lock manufacturer buys key blanks and cuts the ridges, or combinations, in each key. Each key has five bumps on it that are cut to different levels. These levels are designated by numbers. A low cut is one, next up is two, then three. In many cases, there are only four levels, though some manufacturers may use as many as seven. A five-tumbler key lock with four levels in the key yields four to the fifth power, or 1024, different possible combinations of ridges in the key. The five ridges are listed by the height of each level, yielding what is called the combination for the key. A key with the combination 12341 is cut with the first ridge at level one, the second at two, the next at three, and so on. The lock manufacturer chooses the combinations from a random list and cuts each key differently.

### Internal mechanisms

3 The internal mechanisms are made next. These have been designed to fit this particular lock order, and the machinery that makes them may have to be re-tooled or reset. Because the tiny interior parts, specifically the pins, must be manufactured to exceedingly fine tolerances, the machinists may make a trial run before starting a big job. Then the machines may be re-set if necessary. The machining of small brass parts takes many steps. They may be cast, then grooved, ridged, jeweled, and polished. Precision tools handle these jobs, cutting the metal to within tolerances of plus or minus 0.001 of an inch.

### Other parts

4 The manufacturer also makes the other parts of the lock. The cylinder, or plug, that the key fits into, guard plates, washers, the bolt or cam, and the casing, are all made according to design specifications, by die-casting and then further machining. The number of parts varies with the design of the lock, but even a small and relatively simple lock may have thirty separate parts, and some of these parts require multiple toolings. The process of making the lock components can take several weeks.

### Assembly

5 When all the parts are ready, the locks are assembled by hand. Lock workers sit at well-lit tables with a kit of the pieces of the lock in a bin, and the key on a stand in front of them. An experienced worker can tell the combination of the key just by looking at it. The worker first fills the plug, or cylinder, of the lock with the pins that correspond to the combination of the key. The worker inserts a tiny spring and then the lock pin, using a small tool called an assembly pick to hold the small parts. The assembly pick has a small screwdriver on one end and a point on the other, and the worker uses it to prod the delicate parts in where they belong. Once the

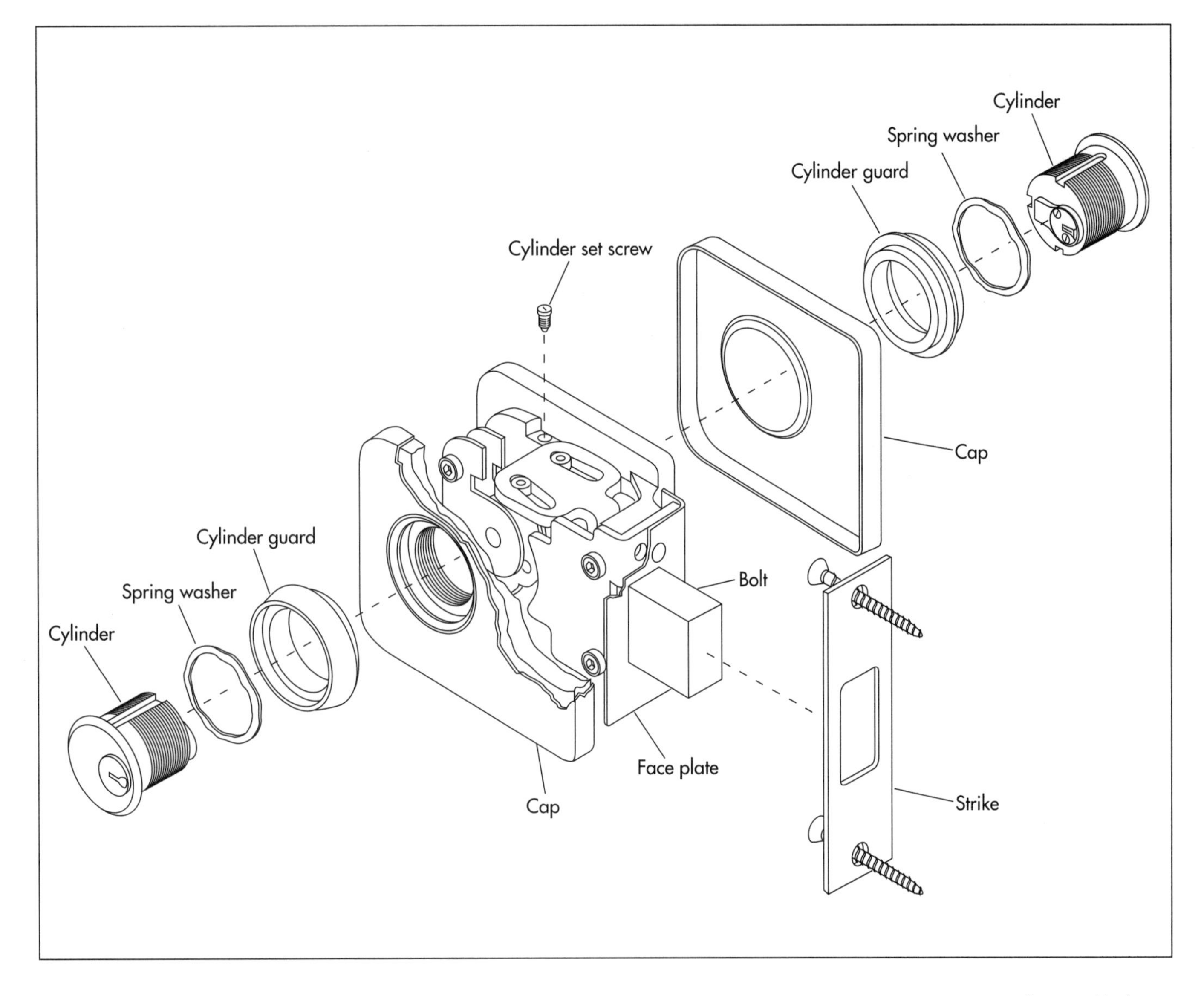


Cross section of a typical lock.

plug is filled according to the key combination, the worker snaps or screws together the other parts around the lock. Though it is skilled work, it takes no special training, and these workers are not locksmiths.

### *Final steps*

6 Once a lock is fully assembled, the worker checks it with the key to make sure it works. It may pass to a quality control station at this point, and then be dusted or polished. Workers package the completed locks and box them for shipment.

## Quality Control

The most important aspect of quality control in lock manufacturing is ensuring that the tiny machined parts are the exact sizes specified. For a new custom order, the machinists usually produce trial samples of the parts, and each one of these may be inspected and measured manually, using precise gauges. If all seems to be going well, the machinists will run the rest of the order, and then perhaps one of every 500 or 1,000 parts is checked. After the worker assembles the lock, he or she tests it with the key to make sure it works properly. A quality control specialist may also spot check the locks at this stage.

## The Future

Many entities such as universities and large corporate headquarters that use large numbers of locks are converting to electronic pass-key systems. These use a magnetic swipe card to open a door. The cards can have a bar code on them, and computers can be used to store information on who goes in through each door, raising privacy issues for some concerned people. Other high-tech locks open with voice activation or palm or fingerprint recognition. Such locks offer relatively high security, but are generally too expensive and elaborate for the ordinary citizen's home. However, the trend towards these kinds of electronic and computer-controlled locks is growing in the late 1990s, and they will undoubtedly be more prevalent in the future.

## Where to Learn More

### Books

Roper, C.A. *The Complete Book of Locks and Locksmithing.* Blue Ridge Summit, Pennsylvania: Tab Books, 1991.

Tchudi, Stephen. *Lock and Key.* New York: Scribner's, 1993.

### Periodicals

Belsie, Laurent. "Slide Toward Surveillance Society: New Technology Allows Government and Corporations to Cut Fraud and Boost Security, but Privacy Concerns Mount." *Christian Science Monitor.* (March 4, 1999).

Leigh, Bobbie. "An Alarming Trend: Bulletproof Living." *Wall Street Journal.* (November 29, 1996).

*—Angela Woodward*

# LP Record

## Background

Sound has always fascinated human listeners, but, until late in the 1800s, it eluded capture. This fact seems peculiar to us today because, with compact discs, cassette tapes, highly portable players, automobiles with lush sound systems, hundreds of radio stations on the dial, television stations devoted to music, and a myriad of other broadcast sounds, we are surrounded by sound.

Among the solid forms that music and other recordings have taken in their brief history, the long-playing phonograph record may be the most romantic and among the most cherished. Phonograph records are no longer manufactured except by private parties with the equipment and the interest, and most sound systems are not equipped with turntables. Long-playing records, known as LPs, are coveted by collectors, however, and there is a large secondary market in used records among aficionados of particular types of music like jazz or opera or performers like Frank Sinatra or the Beatles.

## History

The long-playing record was a direct descendant of the first record made and played on November 20, 1877, by Thomas Edison. Edison's bounty of inventions came from a thorough understanding of science. Edison knew that sound consists of a vibrating wave of air molecules that enters our ears, strikes the eardrum and sets up vibrations in the tiny bones of the inner ear, and passes along nerve endings to the brain. The brain decodes these vibrations as sounds. The number of vibrations per second is the frequency of the sound, and those vibrating waves have amplitude or size that we interpret as loudness or softness. Any and all sounds have these properties so, to record a bird's song, the symphony of vibrations produced by the instruments in an orchestra, or the voice of the lead singer in a rock band, the same techniques are used.

Edison's victrola recorded the sound and played it back. He used a metal cylinder with open ends that was wrapped with a sheet of tinfoil. By speaking into a "sounding disc" that vibrated and was attached to a stylus or needle, the vibrations Edison created by speaking were etched by the stylus onto the tinfoil. The etching looked like small hills and valleys that spiraled around the cylinder. To play back his recording, Edison moved the needle back to the start of the record of the vibrations and revolved the cylinder at the same speed as it had moved during recording. The vibrations came back out of the sounding disc and were amplified by the cup, or primitive microphone, into which Edison had spoken.

Following significant improvements to his phonograph, the first records were made of wax cylinders. Jules Levy, a coronet player, is credited as being the first recording artist. He played "Yankee Doodle" on his coronet, and the wax cylinder of his rendition could be played at home on the Edison Parlor Speaking Phonograph (the first home-use phonograph), which sold for $10 in 1878.

In about 1887, Valdemar Poulsen, a Danish scientist, used the same principles to record sound on a magnetic tape. At the turn of the century, the infant recording industry made cylinders of various materials with permanent recordings on them, but World War II pushed the magnetic tape into broad acceptance as the medium for recording sound

*Jules Levy, a coronet player, is credited as being the first recording artist. He played "Yankee Doodle" on his coronet, and the wax cylinder of his rendition could be played at home on the Edison Parlor Speaking Phonograph (the first home-use phonograph), which sold for $10 in 1878.*

*Thomas Alva Edison*

The American inventor Thomas Alva Edison is credited with inventing the phonograph, which was reportedly his favorite creation. Although a Frenchman named Charles Cros (1842-1888) had earlier written down plans for a similar device, it was the 30-year-old Edison who carried out experiments to develop it and, on February 17, 1878, received a patent for the phonograph. In late 1877, Edison had been working in his Menlo Park, New Jersey, research laboratory on improvements to the telephone (which had recently been invented by Alexander Graham Bell). Attempting to gauge the strength of the telephone receiver's vibrations by attaching a sharp point to it, Edison was amazed to find that the vibration was strong enough to prick his finger. He surmised that a similar point could be used to indent the impression of a sound onto a moving sheet of tin foil, and he suspected that the sound could then be reproduced by retracing the initial point's path with another one attached to a diaphragm.

Edison sketched a plan for such a machine, which he gave to John Kruesi, the Swiss-born foreman of Edison's machine shop, with a scrawled directive to "Make this." The device Kruesi built consisted of a brass cylinder inscribed with spiral grooves and wrapped with a sheet of tin foil; when turned by a hand crank , the cylinder simultaneously rotated and moved lengthwise. At each side was situated a diaphragm equipped with a stylus (needle). A receiver would carry sound waves to one needle, which would be applied to the tin foil as the crank was turned and would follow the cylinder's grooves. Then the cylinder would be reset to the beginning and the other point—attached to the device's amplifier—would turn into sound the vibrations etched into the tin foil. On December 6, 1877, Edison tested his device by reciting the nursery rhyme "Mary Had a Little Lamb." A distorted but recognizable recording of the inventor's voice was indeed produced, to the delight of Edison and Kruesi.

News of the ingenious talking machine spread rapidly, interesting not only the National Academy of Sciences and the Smithsonian Institution but President Rutherford B. Hayes, who is said to have sat up until 3 A.M. listening to the device.

and then transferring it to records. Leading recording companies like RCA Victor found that magnetic tape produced greater fidelity, or faithful reproduction of sound, than other methods. Also, tape can easily be cut and edited to shorten, lengthen, or remove performance errors from recordings.

Until just after World War II, records were available in only one playing speed and turned on their turntables at a rate of 78 revolutions per minute (rpm). In 1948, Peter Carl Goldmark (1906-1977), an American physicist who had been born in Hungary, invented a record that revolved at less than half that speed, at 33.33 times per minute. Improvements in production also allowed the track (the groove for the needle) to be narrowed, and these two developments allowed six times as much music to be recorded on a single record. Large-scale record production was ready for the age of Elvis and rock and roll, and entire symphonies could be reproduced on a single long-playing album instead of a set of 78s.

## Raw Materials

The raw materials for record manufacture were subdivided into those needed to make the master disc, those for actual pressing of the records, and the paper goods needed for labels, sleeves, and jackets. The master disc was made of black lacquer, so it could be etched with grooves to carry the sound. Silver was used to coat the finished disc, and chromium-plated nickel discs were used to press the "vinyl" records.

Records were most commonly made of black plastic, although some were produced in other colors. Recording companies developed the designs for their own labels, sleeves, and album jackets; however, manufacture of these was usually subcontracted to paper suppliers and printers.

## Design

Records evolved into three sizes and three forms of sound reproduction. Originally, records were played at a speed of 78 revolutions per minute (rpm) and were called 78s. The 78s were largely replaced by long-playing records, also called LPs and 33s because they revolve when played at 33.33 revolutions per minute. Records with a single song on each side were known as singles and also called 45s because their playing speed was 45 revolutions per minute.

In their early years, these records were monaural with sound that usually only came from one needle or speaker and seemed to have only one dimension or source direction. As technology improved, sound was recorded in stereo or quadrophonic sound that was also typically projected from two or four speakers and was more realistic because it captured sound as we hear it with two ears.

Standardized record players prevented much variation in physical design of the record. Creativity, instead, came from the recording studio but also from the artists, writers, and researchers who developed the artwork and text on the album jackets. Today's collectors are often as interested in the rare photos and drawings and historical narratives on the record jackets as they are in the music inside.

## The Manufacturing Process

### *Recording the sound*

1 In the recording studio, microphones are located in several different places depending on the acoustics (sound-bouncing properties) of the room and the music being recorded. There are different types of microphones: a specialized microphone for a vocal soloist and several others for instrumental backup, for example, are used. The microphones hear the sounds and translate them into bursts of electrical current that are fed to the recording head on a magnetic tape recorder. The head is made of layers of metal that formed an electromagnet, and the magnet transmitted the current as a pattern of sound waves to the magnetically sensitive tape. The flow of the current or magnetism varies with the intensity of energy picked up by the microphone as sound.

2 The magnetic tape consists of a long ribbon of 2 in (5 cm) wide plastic that is coated on one side with iron oxide. As the tape winds its way through the machine and across the face of the electromagnet, the iron oxide responds to the changes in current or magnetic flow so a permanent picture of the sound was formed on the tape by the rearranged particles. The pattern can be seen with a microscope but not with the naked eye. It is, however, permanent and very precise.

3 During a recording session, sound engineers monitor the work in progress to make sure that every note is captured on tape. The 2 in (5 cm) wide tape is divided into 16 separate tracks, each of which records particular instruments, voices, orchestra sections, or sound from different microphones. During recording, the sound engineer also manipulates the master control board to add special effects or modify the sound he hears from one instrument or section. The master control board also shows the recording levels on each track so these could be made softer or louder. The sound engineers then "do the mix" when the recording is finished to adjust the balance of the various instruments or singers. They may emphasize a particular instrument during one song, for example, and minimize it during another.

4 Sometimes the sound from a particular instrument or voice is not right for the finished recording, and the artist is called back to the studio to rerecord. This process is called overdubbing and adds another part to a separate track on the tape or to a multitrack master. If the tape is overdubbed, it may also have to be remixed. Sometimes, the collection of artists recording the music can not meet in the recording studio at the same time; in that case, the sound engineers record the rhythm tracks first, then the singers and the strings. This multiple process is called sweetening. The record

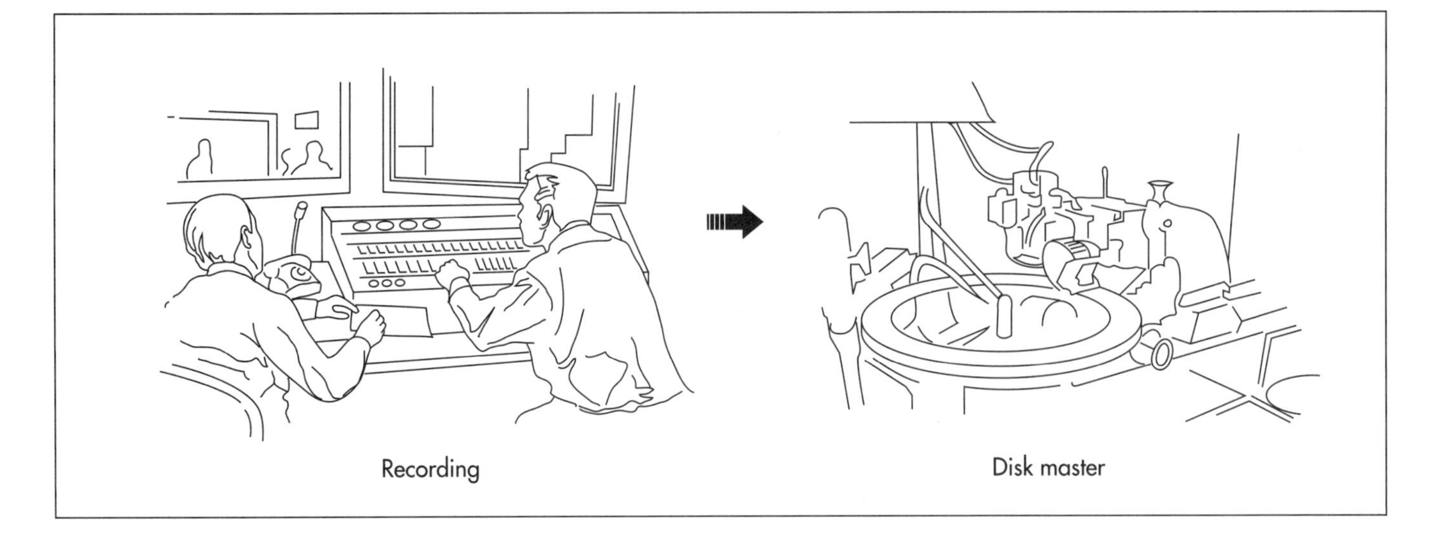


Recordings are made in a sound studio, where engineered monitor and manipulate the sound using highly technical sound recording equipment. Once the sound is recorded on magnetic tape, a master disc is made of aluminum coated with soft black lacquer and called a lacquer. The master is used to create mass quantities of LPs.

producer and the sound and mixing engineers work together on the final mix. The tape is then edited to produce the collection of sounds heard on the final recording. The finished tape, called the master tape, is used to make a master disc.

## Making the master disc

5 The master disc is made of aluminum coated with soft black lacquer and called a lacquer. The responsibility for making it rests with the mastering engineer. The mastering engineer fits all the sound for one side of the record in the specified width of the playing space. For instance, the sound for a 45-rpm record is allowed to occupy a 1.1875 in (3 cm) wide space for grooves on the record, regardless of whether the song was three to five minutes long. The mastering engineer experiments with the spacing of the grooves. The work of the mastering engineer is critical because the master disc he produces is used as the model for pressing thousands of records. Loud music requires large, fat grooves, while softer music takes narrow grooves.

6 The mastering engineer controls the space taken by the record grooves most easily by manipulating the volume; however, if more than one song appears on one side of the album, it is also important to keep the volume relatively constant. For the best sound quality, mastering engineers try to use the loudest possible volume. They also use microscopes to inspect the grooves, and they are very adept at recognizing sounds by their grooves.

7 Mastering engineers use a special grooving machine called a Variable Pitch Cutting Lathe that is equipped with an electronic cutting stylus to etch the grooves in a hard plastic disc. The master disc looks much like a record, but it is larger. A 7 in (17.8 cm) diameter, 45-rpm record is cut onto a 10 in (25.4 cm) diameter blank. A 12 in (30.5 cm) diameter, 33.33 LP is cut onto a 14 in (35.6 cm) diameter blank. The grooves are just like the patterns of iron oxide particles on the magnetic tape in that they imprison the sound vibrations in plastic. As the lacquer is cut, the stylus is heated to help it cut more smoothly. The cutting lathe also has a small vacuum-producing tube mounted next to the stylus. It vacuums up the continuous thread of black lacquer as the grooves are cut. This spiral of waste lacquer is called the chip.

8 The mastering engineer scribes (marks) the cut disc on the outer edge with identification information including the name of the song or album, the master number that also appears on the master tape, and the type of sound recording, which is monaural, stereo, or quadraphonic sound. At this point in the manufacture, the record producer and the artists may listen to a "reference acetate" or the master disc before it is completed in a final set of steps. After the master disc is cut and approved for production, the disc is plated with a very thin coat of silver. It is then called the metal master and is the basis for all the records manufactured.

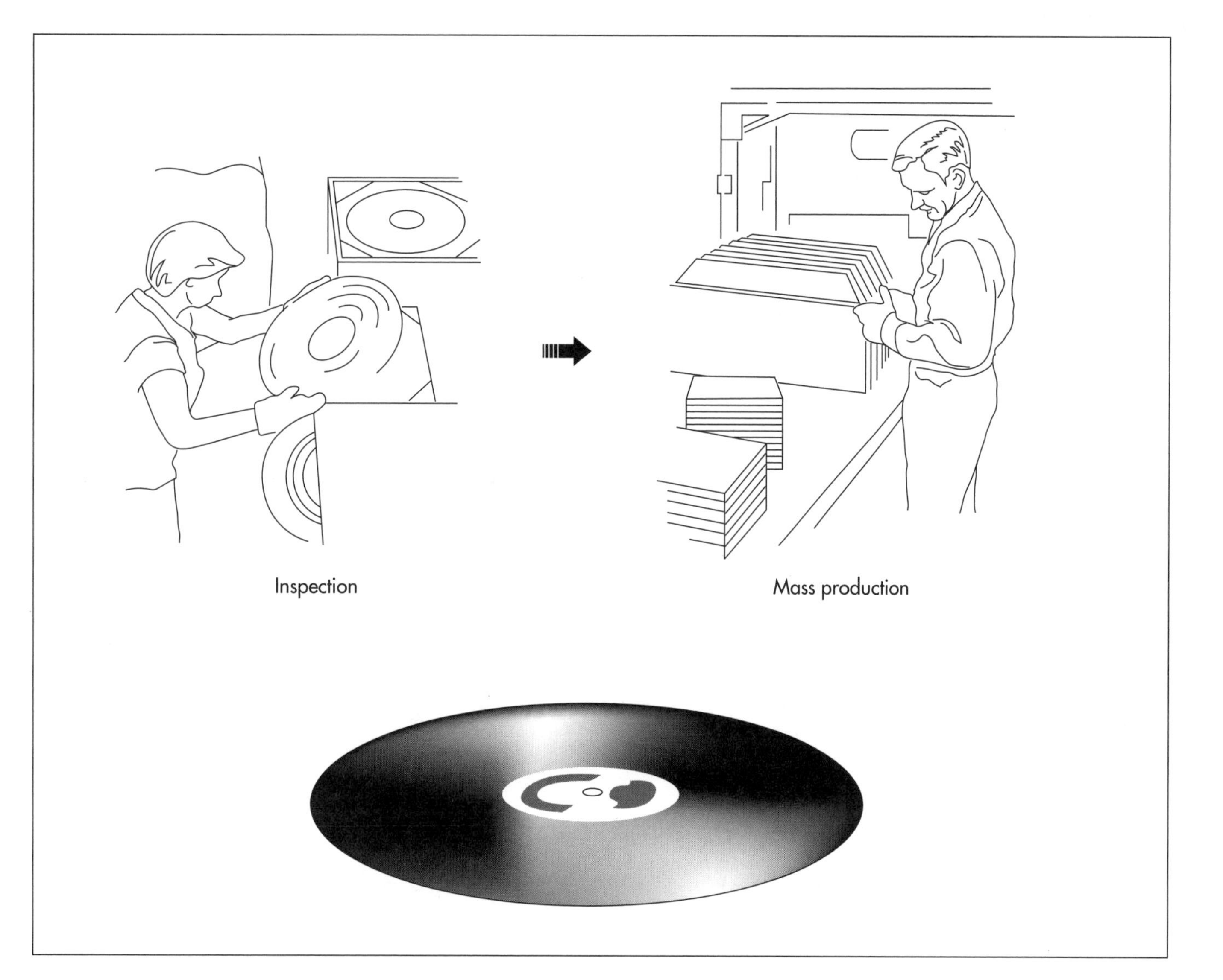


9 At the plating plant, a metal mold is formed from the metal master, and liquid nickel is poured into the mold to produce a nickel stamping record from each side of the metal master. These stampers are also electroplated with chromium that is less than one hundred thousandth of an inch ($2.5^{-5}$ cm) thick. The chromium coat protects the stampers from scratching.

### *Producing LPs*

10 LPs are produced in factories called pressing plants that usually are located some distance from the recording studio, the birthplace of the master disc, and the plating plant where the stampers are made. A pressing plant is capable of producing up to 185,000 records per day. The plastic or vinyl for the records is produced by melting plastic powder in a heated mixer. The plastic is melted and mixed until it has the consistency of jelly. It is then fed through a roller press that produces long, thin sheets within strict tolerances for the thickness and brittleness of the plastic. When the sheets are cooled, they are cut into squares called biscuits. An automatic press is fitted with the nickel stampers—one for each of the two sides of the record. The biscuits are reheated to soften them slightly, and they are fed into the press. The operator makes sure the biscuit is seated properly and activates the press. The grooves and the sound pattern are pressed into the soft plastic. This same process is used for both long-playing records and singles.

11 Still square shaped, the stamped biscuits are conveyed to another machine where

After the master disc is thoroughly inspected, a metal mold is formed from the metal master, and liquid nickel is poured into the mold to produce a nickel stamping record from each side of the metal master. These stampers are also electroplated with chromium in order to prevent scratching. The stampers are used to produce LPs from vinyl squares called biscuits.

the labels are pasted on, and the square corners are rounded. The edge of the disc is smoothed, and the center hole is drilled through the labels and the finished disc.

12 In an alternate version of the same process, the automatic press is fitted with the stampers (the two sides of the album), the round record labels, and a coil of black vinyl plastic. The press is heated to 300°F (149°C), causing the plastic coil to melt and spread between the stampers and into the grooves in a process similar to injection molding. This same machine forms the hole through the center of the record. A flash cutter is used to trim and finish the edge of the LP.

13 In the finishing department, each record is carefully inspected before packing. The newly pressed record moves to a packaging station where it is inserted in a paper or cellophane envelope or sleeve, slipped into the printed record jacket or album cover, and then shrink-wrapped with plastic. Packing boxes filled with the packaged record albums are shipped to distributors.

## Quality Control

Historically, sound engineers in the studio carefully monitored all aspects of recording to make sure the most desirable sound quality was recorded. The mastering engineer's job was to transfer that quality to a reproducable master disc within the technical constraints of the size of the record and its grooves. After a test pressing was made, the record producer (and sometimes the artists) had the opportunity for an important quality control check in reviewing and approving the test pressing.

In the record factory, operators checked the biscuits and the motions of the press and provided ands-on monitoring of the pressing of records. The finishing department also inspected the final product for scratches, bumps, and other irregularities and cleaned each LP before it was packaged. After the records were sealed in their jackets and boxed in bulk, an independent group of testers chose packaged records randomly and removed them from their packaging. These testers checked the packaging itself, played the records, and inspected them for any flaws.

## Byproducts/Waste

Flawed records were melted and pressed again, as were the square corners that were removed from the biscuits to make them into round LPs. The chip of waste lacquer from the making of the master disc was recycled, and any nickel or chromium from the metal processing portions of master disc production was carefully controlled and recycled.

## The Future

The manufacture of long-playing records is a thing of the past. Compact discs stepped to the forefront of recordings in the 1980s because they are not worn by playing, they are more convenient in size, and their sound reproduction quality is better. All sizes of vinyls, however, have many fans among collectors. Some recordings simply have not been remade in compact disc form and are only available on LPs. More often, collectors treasure the collectible character of these records for their sounds, the kinds of music they preserve, and the artwork and information on record jackets.

## Where to Learn More

### Books

Edmunds, Alice. *Who Puts the Grooves in the Record?* New York: Random House, 1976.

Miller, Fred. *Studio Recording for Musicians.* New York: Amsco Publications, 1981.

Wullfson, Don L. *The Kid Who Invented the Popsicle: And Other Surprising Stories About Inventions.* New York: Cobblehill Books, 1978.

### Periodicals

Althouse, Paul "Audio: whither LP?" *American Record Guide* (May-June 1994): 236.

Egan, Jack. "Where's the value in vinyl?" *U.S. News & World Report* (December 13, 1993): 106.

McKee, David. "The flip side." *Opera News* (October 1997): 70.

Scull, Jonathan. "All Sales are Vinyl." *Atlantic Monthly* (December 1997): 106-112.

### *Other*

"A Science Odyssey: Everyday Objects." http://pbs.org/wgbh/aso/tryit/tech/indext.html/ .

"Newton's Apple: Which sounds better: an LP or CD?" http://wwwO.pbs.org/ktca/newtons/11/cdlp.html/ .

"This Week in Music History." http://cgi.canoe.ca/MusicHistoryJune/june21.html/ .

*—Gillian S. Holmes*

# Lyocell

*Commercial production of a cellulosic fiber was first carried out by the French chemist Count Hilaire de Chardonnet. He exhibited his so-called artificial silk in Paris in 1889. Chardonnet built a plant in Besançon, France in 1891, and had great success bringing his new fabric, now named rayon, to the forefront of the fashion industry.*

Lyocell is a manmade fiber derived from cellulose, better known in the United States under the brand name Tencel. Though it is related to rayon, another cellulosic fabric, lyocell is created by a solvent spinning technique, and the cellulose undergoes no significant chemical change. It is an extremely strong fabric with industrial uses such as in automotive filters, ropes, abrasive materials, bandages and protective suiting material. It is primarily found in the garment industry, particularly in women's clothing.

## Background

Fabrics derived from cellulose date back to the middle of the nineteenth century, though no one commercially produced one until 1889. A Swiss chemist, George Audemars, was granted an English patent in 1855 for an artificial silk he derived from mulberry bark. Audemars attempted to reproduce the method the silkworm uses for making silk by dissolving the fibrous inner bark of mulberry trees to separate out the cellulose. To form threads, he dipped needles into the cellulose solution and drew them out. Another chemist, Englishman Joseph W. Swan, modified Audemars' technique by forcing the cellulose solution through fine holes. His main interest was in producing filaments for electric lamps, but Swan realized that it was possible to manufacture a cellulose textile using his extrusion method. He exhibited cellulose fabric in London in 1885, but he failed to kindle any interest, and the project died out. Commercial production of a cellulosic fiber was first carried out by the French chemist Count Hilaire de Chardonnet. He exhibited his so-called artificial silk in Paris in 1889, and contrary to Swan's experience, people were thrilled by his new fabric. Chardonnet built a plant in Besançon, France in 1891, and had great success bringing his new fabric, now named rayon, to the forefront of the fashion industry. In the United States, rayon production began in 1910 under the auspices of the American Viscose Company. This company was an affiliate of Samuel Courtaulds and Co., Ltd., the forebear of the primary developer of lyocell, Courtaulds PLC of the United Kingdom (now known as Accordis Fibers).

Chemists and manufacturers were intensely interested in manmade fibers in the twentieth century. Acetate, another cellulosic fiber, was first commercially produced in the United States in 1924. Chemists at the Du Pont company developed nylon in the 1930s, and it came into wide commercial use beginning in 1939. Acrylic and polyester were two other significant man-made fibers. These debuted in the 1950s. Nylon, acrylic and polyester differ from cellulosic fibers because they are derived from chemicals, and thus are totally manmade. Lyocell, rayon and acetate are based on the cellulose in wood pulp, and so these are often designated "natural" fibers, though the fibers would not occur except for a manmade process.

The manufacturing and processing of artificial fibers underwent much refinement throughout the twentieth century. The first rayon produced by Count Chardonnet, for example, proved to be highly flammable, and the rayon on the market today is quite different. In the late 1970s and early 1980s, researchers, principally at the leading rayon

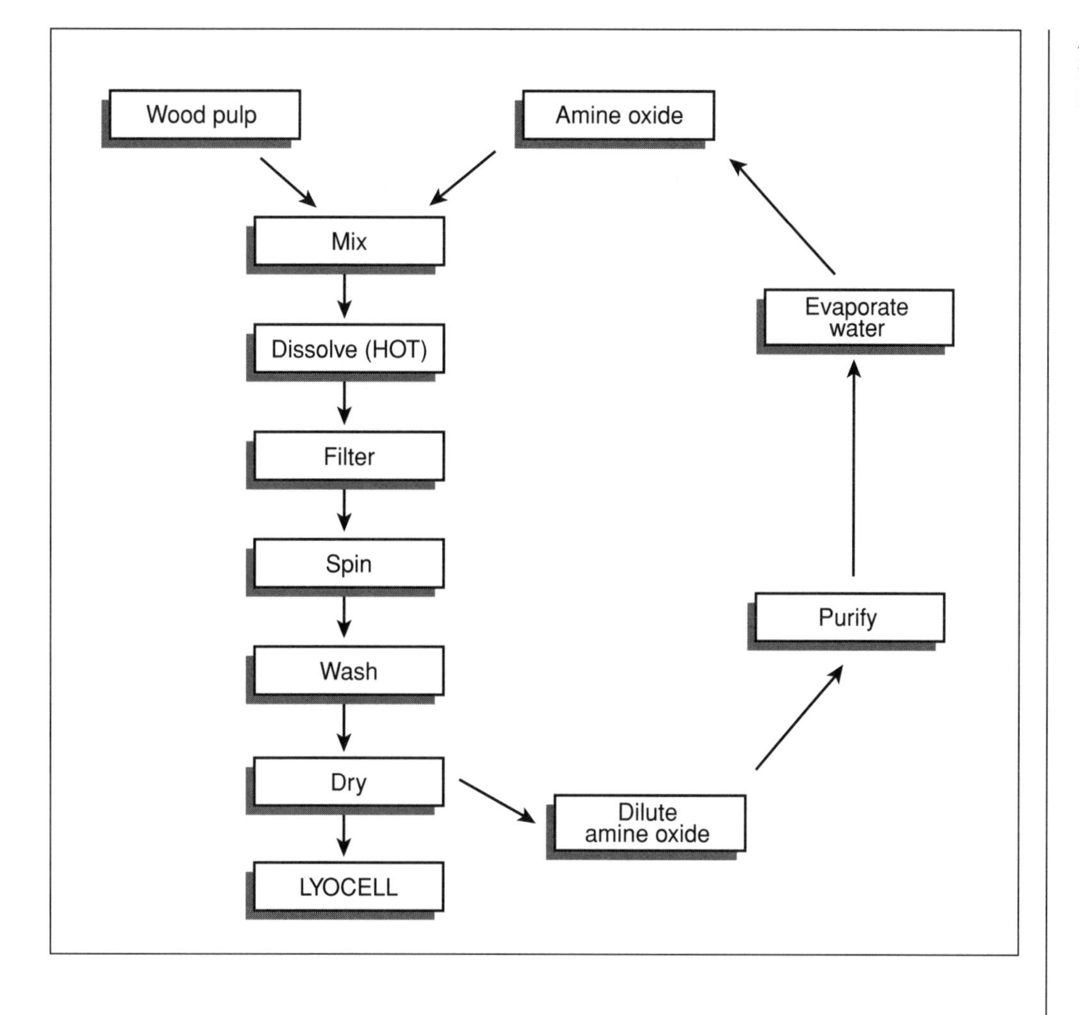


A diagram depicting the manufacturing processes used to make lyocell fiber.

producer Courtaulds Fibres and at an American firm American Enka, began investigating a new method of producing a cellulosic fiber through a solvent spinning technique. In this method, the cellulose is softened and then spun into fiber. Conventional rayon, by contrast, is called a "regenerated cellulose fiber," and it involves many more steps. The cellulose is first chemically converted into xanthate, then dissolved in caustic soda, then regenerated into cellulose as it is spun. The solvent spinning technique is both simpler and more environmentally sound, since it uses a non-toxic solvent chemical that is recycled in the manufacturing process. The solvent-spun cellulosic fiber lyocell was first produced commercially in the United States in 1992 by Courtaulds. The company used the brand name Tencel. The fabric was given the generic name lyocell in the United States in 1996. By this time, there was one other major lyocell producer in the world, the Austrian company Lenzing AG, which had acquired patents and research on lyocell from American Enka when it bought that company's rayon operation in 1992.

## Raw Materials

The main ingredient of lyocell is cellulose, a natural polymer found in the cells of all plants. It forms the basis for other plant-derived fibers such as cotton, hemp, and linen. The cellulose for lyocell manufacturing is derived from the pulp of hardwood trees. The pulp is typically from a mix of trees chosen for their cellulosic properties such as the color and amount of contaminants. Some common tree species used are oak and birch. The trees are grown on managed tree farms, generally on land that is not suitable for other agricultural uses. The solvent used in the manufacturing process is an amine oxide. Water is another key ingredient in producing lyocell fiber. A finishing agent is also used, and this varies, but is generally a lubricant such as soap or silicone. Lyocell fabrics are generally dyed

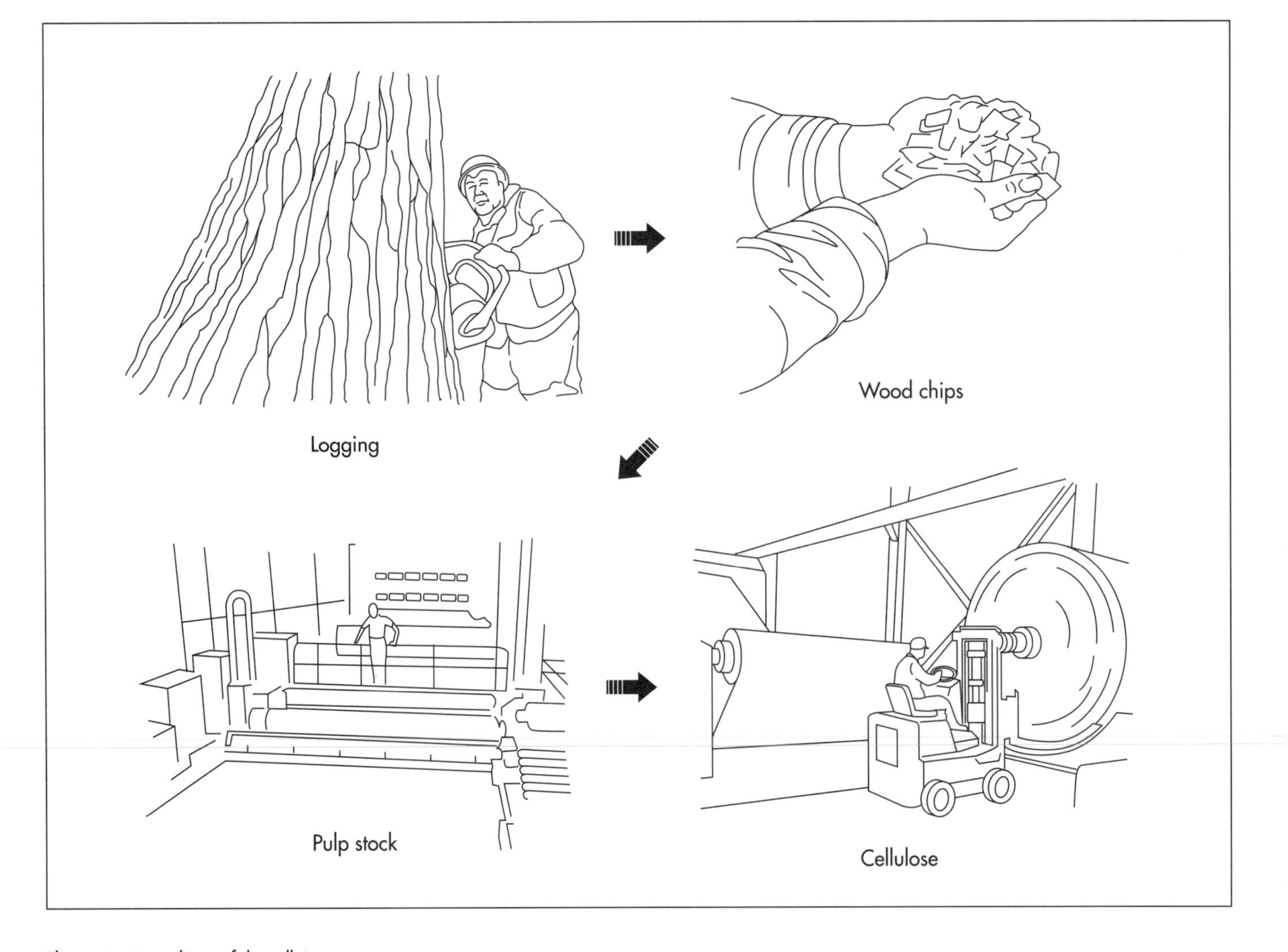


The main ingredient of lyocell is cellulose, a natural polymer found in the cells of all plants. The cellulose for lyocell manufacturing is derived from the pulp of hardwood trees. During the manufacturing process, the hardwood is broken down into chips and then fed into a vat of chemical digesters, which soften them into a wet pulp. The pulp is washed, bleached, and dried in a huge sheet that is rolled onto a giant spool.

with any dyes that are also compatible with cotton and rayon.

## The Manufacturing Process

### *Preparing the wood pulp*

1 The hardwood trees grown for lyocell production are harvested by loggers and trucked to the mill. At the mill, the trees are cut to 20 ft (6.1m) lengths and debarked by high-pressure jets of water. Next, the logs are fed into a chipper, a machine that chops them into squares little bigger than postage stamps. Mill workers load the chips into a vat of chemical digesters that soften them into a wet pulp. This pulp is washed with water, and may be bleached. Then, it is dried in a huge sheet, and mill workers roll it onto spools. The sheet of cellulose has the consistency of thick posterboard paper. The roll of cellulose is enormous, weighing some 500 lb (227 kg).

### *Dissolving the cellulose*

2 At the lyocell mill, workers unroll several spools of cellulose and break them into one inch squares. The workers then load these squares into a heated, pressurized vessel filled with amine oxide.

### *Filtering*

3 After a short time soaking in the solvent, the cellulose dissolves into a clear solution. It is pumped out through a filter, to insure that all the chips are dissolved.

### *Spinning*

4 Next, the solution is pumped through spinnerets. These are devices used with a variety of manmade fibers. Something like a showerhead, the spinneret is pierced with

small holes, and when the cellulose is forced through it, long strands of fiber come out. The fibers are then immersed in another solution of amine oxide, diluted this time. This sets the fiber strands. Then, they are washed with de-mineralized water.

### Drying and finishing

5 The lyocell fiber next passes to a drying area, where the water is evaporated from it. The strands at this point pass to a finishing area, where a lubricant is applied. This may be a soap or silicone or other agent, depending on the future use of the fiber. This step is basically a detangler, making the future steps of carding and spinning into yarn easier.

### Final steps

6 The dried, finished fibers are at this stage in a form called tow. Tow is a large untwisted bundle of continuous length filaments. The bundles of tow are taken to a crimper, a machine which compresses the fiber, giving it texture and bulk. The crimped fiber is carded by mechanical carders, which perform an action like combing, to separate and order the strands. The carded strands are cut and baled for shipment to a fabric mill. The entire manufacturing process, from unrolling the raw cellulose to baling the fiber, takes only about two hours. After this, the lyocell may be processed in a wide assortment of ways. It may be spun with another fiber, such as cotton or wool. The yarn can be woven or knit like any other fabric, and given a variety of finishes, from soft and suede-like to silky.

### Recovery of the solvent

7 The amine oxide used to dissolve the cellulose and set the fiber after spinning is recovered and re-used in the manufacturing process. The dilute solution is evaporated, removing the water, and the amine oxide is routed for re-use in the pressurized vessel in step 2. Ninety-nine percent of the amine oxide is recoverable in the typical lyocell manufacturing process.

## Quality Control

Lyocell is only produced at a few plants in the world. These are specially designed, state-of-the-art mills, and quality control is carried out by sophisticated computer monitoring systems. The computers continuously check a variety of key factors, such as the tenacity of the fiber, its color, the denier (a measurement of the fiber's diameter), elongation, moisture level, and level of the finish application. Computers also monitor for "trash" that results when one hole in the spinneret becomes blocked, and the filament comes out splintered or undrawn.

## Byproducts/Waste

The manufacture of lyocell produces no harmful byproducts and is significantly less toxic and wasteful than the manufacture of other cellulosic fibers. Its principal ingredient, cellulose, is easily obtained from managed tree farms, and the industry has not been accused of poor forestry habits. The amine oxide solvent is nontoxic, and because it is almost completely recycled during manufacturing, it is not released into the environment. Lyocell fabrics are also naturally biodegradeable. Manufacturing lyocell is also environmentally sound because less water and energy is used than in the manufacture of other manmade fibers.

## Where to Learn More

### Books

Kadolph, S.J. and Langford, A.L. *Textiles*. Upper Saddle River, NJ: Prentice-Hall, 1998.

### Periodicals

Maycumber, S. Gray. "Lenzing Developing New Lyocell." *Daily News Record* (October 14, 1993).

McCurry, John W. "Lyocell Disciples See Lofty Future." *Textile World* (March 1998).

Pfaff, Kimberly. "Courtaulds Tencel Plant Opens; Great Potential Seen for New Fiber Said to Feel Like Cotton, Wear Like Polyester." *HFD* (December 21, 1992).

*—Angela Woodward*

# Macadamia Nut

*Mauna Loa Macadamia Nut Corporation is the largest manufacturer of macadamia nuts in the world. The firm's plantation was founded in 1948, the trees began to bear fruit in 1954, and the first commercial crop was harvested and processed in 1956. Over 10,000 acres of rich, volcanic soil host Mauna Loa's orchards.*

## Background

In the world of nuts and berries, macadamia nuts are almost as precious as gold. These delicious, exotic nuts with their rich flavor and oil are considered delicacies and are served as dessert nuts. They are popular gifts at holiday times, both alone and when covered with chocolate. They are prized as souvenirs from Hawaii, and, thanks to Mrs. Fields' cookies, macadamia and chocolate chip cookies have brightened many afternoons at the local mall.

Macadamia nuts are associated in the minds of most Americans with Hawaii. Macadamias are a commercial crop in Hawaii, but they originated in northeastern Australia in the rain forests along the coast. The tree is from the family Protaceae and is one of about 10 species of which two grow best as commercially productive plants. The *Macadamia integrefolia* produces nuts with smooth shells, and the *Macadamia tetraphylla* has rough-shelled nuts.

Macadamia trees produce throughout their lives but they are slow growing. The demand for the rich nuts has outstripped growers' ability to produce them. Consequently, growers in many other countries including New Zealand, Zimbabwe, Malawi, South Africa, Kenya, Israel, Guatemala, Brazil, Mexico, and Costa Rica have begun planting large orchards. In the United States, California and Florida boast macadamia crops, along with Hawaii.

## History

The macadamia nut was discovered by British colonists in Queensland, Australia, in 1857. Walter Hill, who was the Director of the Botany Garden in Brisbane, found one of the nuts, cracked it open using a vise, and planted the seed. This "first" macadamia nut tree is still growing and producing nuts, although typically the trees only produce for about 70 years.

Hill had been traveling on a botanical expedition with Baron Ferdinand von Mueller, who is considered the father of Australian botany and who held the post of Royal Botanist in Melbourne at the time. He is credited with naming the tree after Scotsman John Macadam, a friend, physician, and member of the Philosophical Institute of Victoria. Mr. Macadam never tasted the nut that bears his name after a shipboard injury caused his premature death en route to New Zealand.

Of course, the trees had long been known to the native Australian aborigines who called the macadamia trees *kindal kindal* and who feasted on the nuts in winter. The colonists took the tree to their hearts and began to learn to propagate it. The first known macadamia orchard consisted of 250 trees planted in 1890 on the Frederickson Estate in New South Wales, Australia. The tree was heavily cultivated and hybrids were grown from seeds and, more often, by grafting. Australia remains one of the world's major producers.

The macadamia migrated to Hawaii courtesy of William Herbert Purvis who gathered macadamia nuts near Mount Bauple in Queensland, Australia, and brought them to Hawaii's Big Island in 1882. He nurtured the imported nuts and planted them as seedlings in Kukuihaele, Hawaii. One of Purvis's original seedlings is also still growing and producing nuts.

Today, Mauna Loa Macadamia Nut Corporation is the largest manufacturer of macadamia nuts in the world. The firm's plantation was founded in 1948, the trees began to bear fruit in 1954, and the first commercial crop was harvested and processed in 1956. Over 10,000 acres of rich, volcanic soil host Mauna Loa's orchards. The single largest planting is not in Hawaii but on 3,700 acres in South Africa. The hybrid grown in New Zealand produces the most expensive macadamia nuts in the world; the *Beaumont* sp. does not drop its macadamias, so they are expensive to harvest and are also among the finest in flavor.

## Raw Materials

The raw materials needed for commercial production of macadamia nuts are the nuts themselves, as well as salt and oil.

## Design

Macadamia nuts are sold in jars and cans for home consumption. These nuts have been roasted in oil in the factory and salted. Unsalted nuts are also packed for commercial use, primarily by bakeries. The flavor of the nuts is well suited for use in many kinds of desserts. Design can become elaborate if the nuts are used in candies, cookies, and other products. A wide variety of desserts, souvenirs of the islands, and other treats containing macadamias are available on the market; some are byproducts of the macadamia nut processors, but, more often, other firms use the nuts as ingredients in their own product lines.

## The Manufacturing Process

### *Cultivation*

1 Macadamia trees require rich soil, about 50 in (130 cm) of rain per year, and temperatures that are not only frost-free but that vary within a limited range. The soil must also drain well, so not all tropical zones are suited to macadamias as a crop. The trees are evergreen and everbearing; they have leathery leaves much like holly that are shiny and 7-12 in (20-30 cm) long. The trees themselves grow to as much as 60 ft (18.3 m) high. They produce clusters of flowers that are white or pink and fragrant; about 300-600 flowers appear in sprays. Each flower spray produces up to 20 nuts, which have green, fibrous husks and hard, outer shells called pericarps. The pericarps split as the nuts ripen on the trees. Each nut (including the kernel in its shell) is approximately 1 in (2.5 cm) in diameter.

2 The flowering of the trees occurs over a four to six month period. Consequently, the nuts mature at different times over the course of the year. They are also biennial, so alternate years produce light then heavy crops from a single tree. The trees require pollination during flowering, so beehives are usually imported into the orchards. Weeds grow heavily among the trees, and insects proliferate in the tropical climate; mowing and bug control are required. The trees are also fertilized with husks from their own nuts, chicken waste, and carefully selected and controlled chemicals. When the nuts are ready to be picked, the trees are pruned first so the nut clusters are easier to reach.

3 Ripened and unripened nuts look identical, so producers either wait until the nuts fall to the ground or they harvest infrequently, approximately five or six times a year. The nuts are harvested through a complicated process that includes gathering by hand, shaking the trees and picking the fallen nuts, or picking them with a mechanical picker. Blowers are sometimes used to blow the nuts and fallen leaves into windrows so they can be collected by machine.

### *In the factory*

4 The harvested macadamias are fed into large hoppers and then into a dehusker made of double rollers that strip the husk away. The husked nut has a moisture content of about 25%, and it has to be dried and cured to reduce its moisture content to about 1.5%. Drying is done in a greenhouse, and curing is accomplished by heating the nuts to 104-122° F (40-50° C). Some processors store the nuts in netted bags or onion bags during drying so the heated air can move freely through the nuts.

The lower moisture helps later in processing in separating the kernel from the shell. Macadamia nuts have the hardest shells to crack (although they are followed closely by the Brazil nut). The process of cracking the

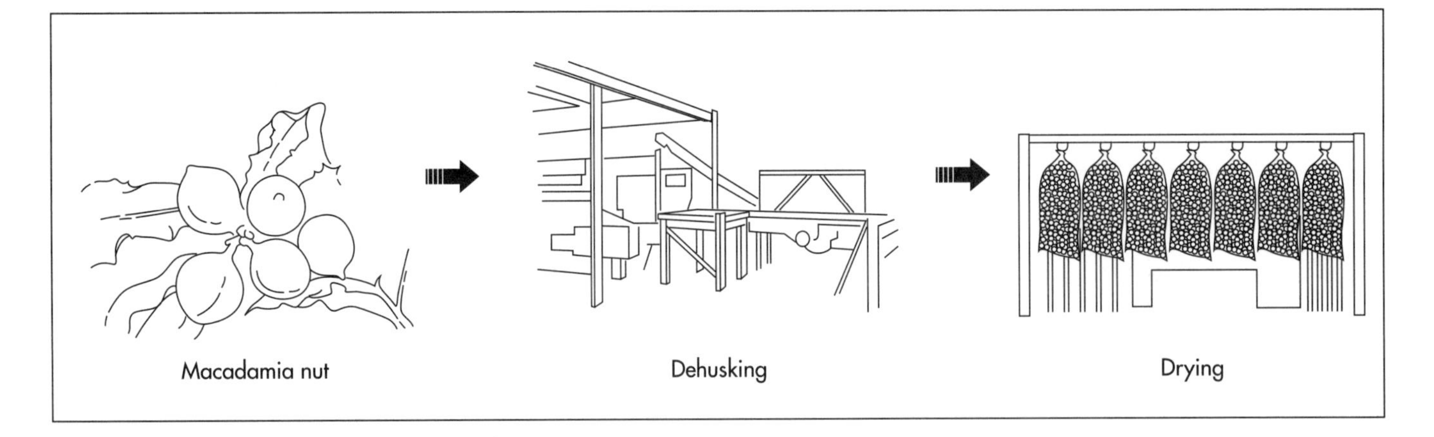


Macadamia nuts are harvested by hand, shaking the trees and picking the fallen nuts, or picking them with a mechanical picker. The harvested nuts are fed into large hoppers and then into a dehusker made of double rollers that strip the outer husk away. The husked nut has a moisture content of about 25%, and it has to be dried and cured to reduce its moisture content to about 1.5%. Drying is done in a greenhouse, and curing is accomplished by heating the nuts to 104-122° F (40-50° C). Some processors store the nuts in netted bags or onion bags during drying so the heated air can move freely through the nuts.

nuts also makes them rare and expensive. The hulls are too hard and smooth to be cracked by standard nutcrackers, rocks, or hammers. (In Hawaii, those who can collect the nuts from trees in their own neighborhoods often resort to driving cars over them.) In the factory, harvesting is complicated by the ever-bearing nature of the trees. Their nuts mature throughout the year (i.e., not seasonally), so the process of gathering and hulling the nuts is continuous and expensive.

5 To shell the nuts, they are passed through steel rollers that counter-rotate (rotate in opposing directions) and are carefully spaced and engineered to conform to the size of the macadamias. These rollers exert a pressure of 300 lb per sq in (21 kg per sq cm) on the shells and cause them to crack without damaging the kernels inside. An alternative process uses a cracking machine equipped with rotating knives that pin the nuts against a wedge shape and crack them. The kernels are passed through a series of blowers and trommels or gravity separators with holes that remove dust, dirt, any remaining bits of husk, or nuts that are substandard. Uncracked nuts are collected after they pass through the trommel and are recycled through the crackers. The air blowers separate the kernels from pieces of shell.

6 Optical devices inspect the nuts and sort them by color. Quality control inspectors also observe the passing flow of kernels and sort them by hand. Light-colored nuts are classified as Grade I or fancy nuts, and the darker-colored nuts or those that are not within a standard size range are sorted as Grade II. Grade I nuts are used for retail sales, while the Grade II nut is processed for commercial use where size and color aren't as important. The cracking and sorting machinery is cleaned and disinfected at the end of each day of operation.

7 Processing of the kernels includes grading and sorting of whole kernels as well as chips and halves. The kernels and pieces are sold in raw form, roasted and salted, or coated in a variety of products. Conveyors carry the kernels to different stations. Raw kernels are packed directly in cans or boxes. Kernels to be roasted are separated into small batches of about 1 lb (2 kg) of nuts, coated in coconut oil, and cooked for about three minutes. Salt is mechanically sprinkled on the nuts, and the excess salt and oil are blotted off so the nuts will remain crisp and flavorful in their packages. Some processors use a dry roasting process rather than introducing additional oil.

8 Macadamia nuts that are to be made into candy are sometimes processed on site by the manufacturer or done off site by contractors. Chocolate coatings of milk and dark chocolate, various forms of brittle, and honey sesame coatings are among the most popular.

## Quality Control

The cultivation and processing of macadamia nuts are persnickety by definition. Their unusual growing characteristics and their devilishly hard shells need careful attention. Quality control is essential in the orchard to produce nut clusters and collect them in a timely manner (and as cost effective as possible). During processing, machines are essential to remove both the husks and the hard shells, but observation is

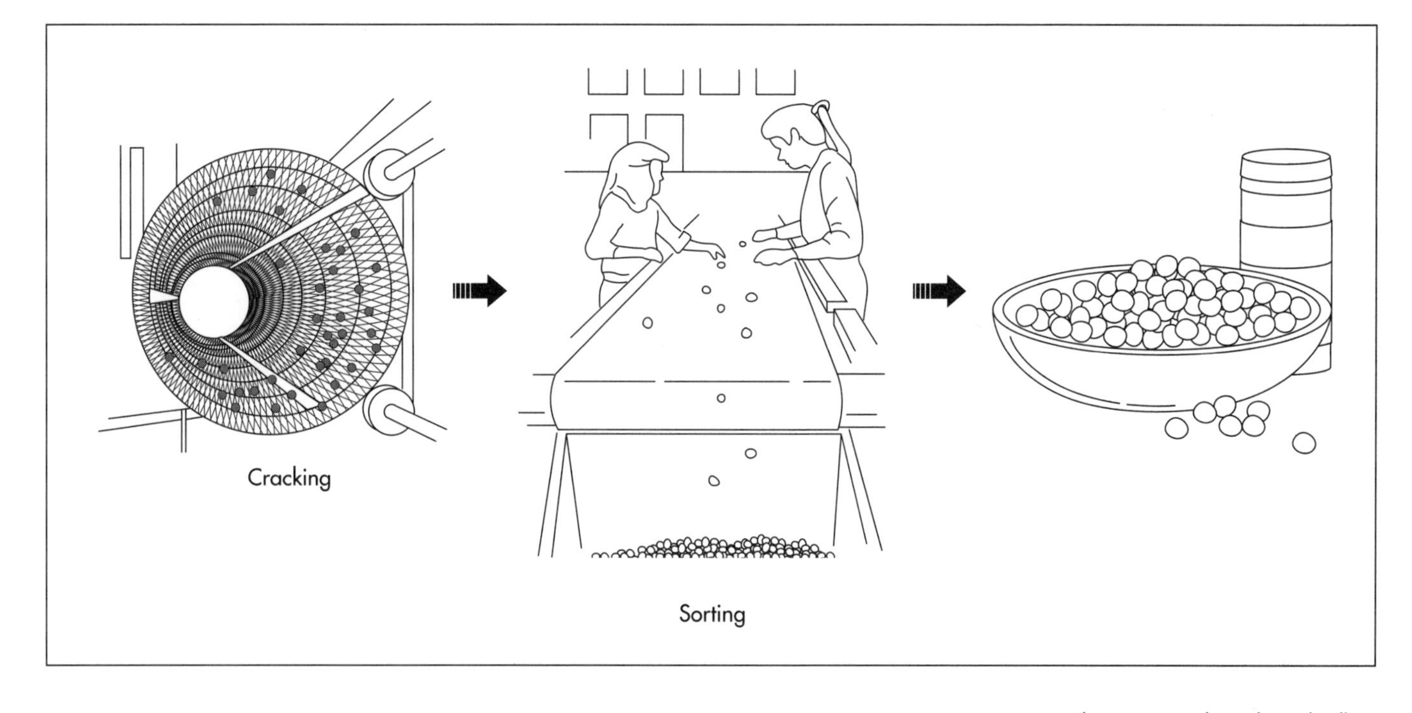


The nuts pass through steel rollers that counter-rotate and exert a pressure of 300 lb per sq in (21 kg per sq cm) on the shells, causing them to crack without damaging the kernels inside. The kernels passed through a series of blowers and trommels or gravity separators with holes that remove dust, dirt, any remaining bits of husk, or nuts that are substandard. Uncracked nuts are collected and recycled through the crackers. Optical devices inspect the nuts and sort them by color. Quality control inspectors also observe the passing flow of kernels and sort them by hand.

provided throughout by quality inspectors and by computer-controlled optical devices.

## Byproducts/Waste

Kernels are also ground and processed to produce macadamia nut oil as a byproduct. Bad kernels are not wasted and are often used as animal feed. The shells and other waste comprise almost 70% by weight of the macadamia nuts, and they also are collected for other uses. The countries and regions that produce macadamias also usually grow coffee, and shells can be burned as a wood substitute in coffee roasting. The husks are ground to produce organic waste for gardening, for mulch in the nut tree orchards, and for chicken litter that, after use, returns to the orchard for use as fertilizer.

## The Future

Macadamia nuts are an excellent source of iron, calcium, vitamin B, and phosphorus. Although they contain 73-80% fat, the fat is monosaturated or "good" and as acceptable as olive oil in many diets. Although macadamias have many healthful properties, their unusually rich flavor, crunch, and co-mingling with chocolate in a bounty of forms make them favorites among gourmets and snackers alike. Good taste is always in style, so the future of the macadamia nut is promising indeed.

## Where to Learn More


### Books

Thomson, Paul H. "Macadamia." In *Nut Tree Culture in North America.* Richard A Jaynes, ed. Hamden, Connecticut: The Northern Nut Growers Association, Inc. 1979, pp. 188-202.

### Periodicals

Sokolov, Raymond "Tough Nuts: Despite Some Superficial Similarities, Brazil Nuts And Macadamia Nuts Present A Study In Contrasts." *Natural History* (March 1985): 78.

### Other

California Macadamia Nut Society. http://users.aol.com/TecterJS/themac.html/.

Hamakua Macadamia Nut Company. http://206.127.252.21/hawnnut/.

Macadamia Miravalles S.A., Costa Rica. http://macadamia.co.cr/.

MacNut Farms & Café. http://macnut.co.nz/.

Mauna Loa Macadamia Nut Corporation. http://www.maunaloa.com/.


*—Gillian S. Holmes*

# Molasses

*With 425,000 acres devoted to sugar cane, the Florida Everglades is a major producer that annually yields 90 million gallons of black strap molasses.*

Molasses, from the Latin word *melaceres*, meaning honey-like, is a thick dark syrup that is a byproduct of sugar refining. It results when sugar is crystallized out of sugar cane or sugar beet juice. Molasses is sold both for human consumption, to be used in baking, and in the brewing of ale and distillation of rum, and as an ingredient in animal feed.

## History

The pressing of cane to produce cane juice and then boiling the juice until it crystallized was developed in India as early as 500 B.C. However, it was slow to move to the rest of the world. In the Middle Ages, Arab invaders brought the process to Spain. A century or so later, Christopher Columbus brought sugar cane to the West Indies. Another two hundred years later, cuttings were planted in New Orleans.

Molasses figured prominently in the infamous slave trade triangles of the late seventeenth century. English rum was sold to African slave traders who brought slaves to the West Indies and then brought West Indian molasses back to England.

Using sugar beets to produce sugar was not developed until the mid-1700s when a German chemist Andreas Marggraf discovered the presence of sugar in the vegetable. By 1793, another German chemist, Franz Karl Achard, perfected the process for extracting the sugar from the beets.

The first beet sugar factory opened in Prussian province of Silesia in 1802. During the Napoleonic Wars, the British blockaded France, cutting off French access to sugar imports from the West Indies. Napoleon then issued land grants and large sums of money to encourage the establishment of a beet sugar industry. One man who took Napoleon up on his offer was a French banker named Benjamin Delessert. Delessert set up several beet sugar factories at Passy and within two years produced four million kilos of sugar. For his efforts, Napoleon awarded Delessert with the medal of the Legion of Honor. By the end of 1813, 334 French sugar beet plantations were producing 35,000 tons of sugar.

In contrast, the beet sugar industry struggled in the United States until the end of the nineteenth century when a California factory finally turned a profit. At the turn of the century, the country had 30 beet sugar processing plants.

Molasses figured prominently in two peculiar events in United States history. The first was the Molasses Act of 1733, which imposed duties on all sugar and molasses brought into North American colonies from non-British possessions. The second was the Great Boston Molasses Flood of January 1919 when a molasses storage tank owned by the Purity Distilling Company burst, sending a two-story-high wave of molasses through the streets of the North End of Boston.

Before the advent of harvesting machinery, laborers performed the back-breaking work of cutting and stripping the sugarcane by hand. Mule-driven mills pressed the sugar cane to release the syrup, which was then cooked in large kettles over a fire until thickened. Workers, usually the farmer's

*When a giant, five-story steel tank suddenly fractured, millions of gallons of molasses were released, engulfing people, animals, and property.*

Boston's inner harbor was memorialized on January 15, 1919, by one of the most bizarre structural failures ever to occur anywhere. On that mild winter day, without warning, a massive tidal wave of nearly 12,000 tons of thick, brown, sugary molasses gushed from a fractured steel tank, leaving twenty-one dead, over one hundred and fifty injured, and many buildings crushed under.

The huge 50 ft (15.2 m) high, 90 ft (27.4 m) diameter steel tank had been used by the U.S. Industrial Alcohol Company for storing up to 15,000 tons of molasses. The tank was ordered from Hammond Iron Works in 1915 by the Purity Distilling Company on authorization of U.S. Industrial Alcohol. The treasurer of Purity ordered it without consulting an engineer. The only requirement used in making the order was that the tank have a factor of safety of three for the storage of molasses weighing 12 lb per gal (5.4 kg per l) (50% heavier than water).

All the steel sheets used in construction of the tank actually proved less thick than shown on the drawings used to obtain the building permit. For instance, the bottom ring—the most stressed part of the structure—was supposed to be 0.687 in (1.74 cm); in actuality it was only 0.667 in (1.7 cm). The steel thicknesses for the other six rings were likewise 5-10 percent less than shown on the permit plans. The bottom ring (ring one) had a 21 in (53.3 cm) diameter manhole opening cut out of it.

The tank was completed early in 1916 and tested only by running six inches of water into it. On a dozen occasions during the tank's three years of service it contained a maximum of around 1.9 million gallons (for periods up to twenty-five days). The content that dark day in January was some 2.3 million gal (8.7 million l)—near capacity. It had been in the tank for four days. Months later, at the legal proceedings, several recalled that the seams of the tank were leaking molasses, but no one seemed to be concerned

It was 12:40 P.M. when eyewitnesses said they heard sounds like machine-gun fire and then saw a torrent of molasses, two stories high, exploding out over adjoining property. Six children were immediately swallowed up.

The tank had fractured and burst open. A 2.5-ton section of the lower part of the tank was thrust out onto a playground 182 ft (55.5 m) away. Another section of the structure wrapped itself around and completely sheared off one column of an adjacent elevated railway. Traveling at about 35 mi (56.3 km) per hour, the molasses swept over and through everything in its path. At its most destructive moments, the gooey wave was 15 ft (4.6 m) high and 160 ft (49 m) wide.

People and animals lost their lives either by drowning or by being struck by wreckage. Occupied houses were demolished, while the cellars of others were filled with molasses. Rescue crews found one man and his wagon embedded in a mountain of molasses, man and horse both frozen in motion.

By midafternoon the flood had settled. It covered more than a two-block area and the general vicinity looked like it had been hit by a cyclone. Buildings were destroyed or bulldozed off their foundations, rails from the elevated railway were dangling in the air, and the tank itself lay on the ground, a heap of crumbled junk metal.

Police and firemen used huge hydraulic siphons around the clock to pump molasses out of flooded cellars. It was nearly a week before all the bodies were recovered and months before signs of the disaster disappeared.

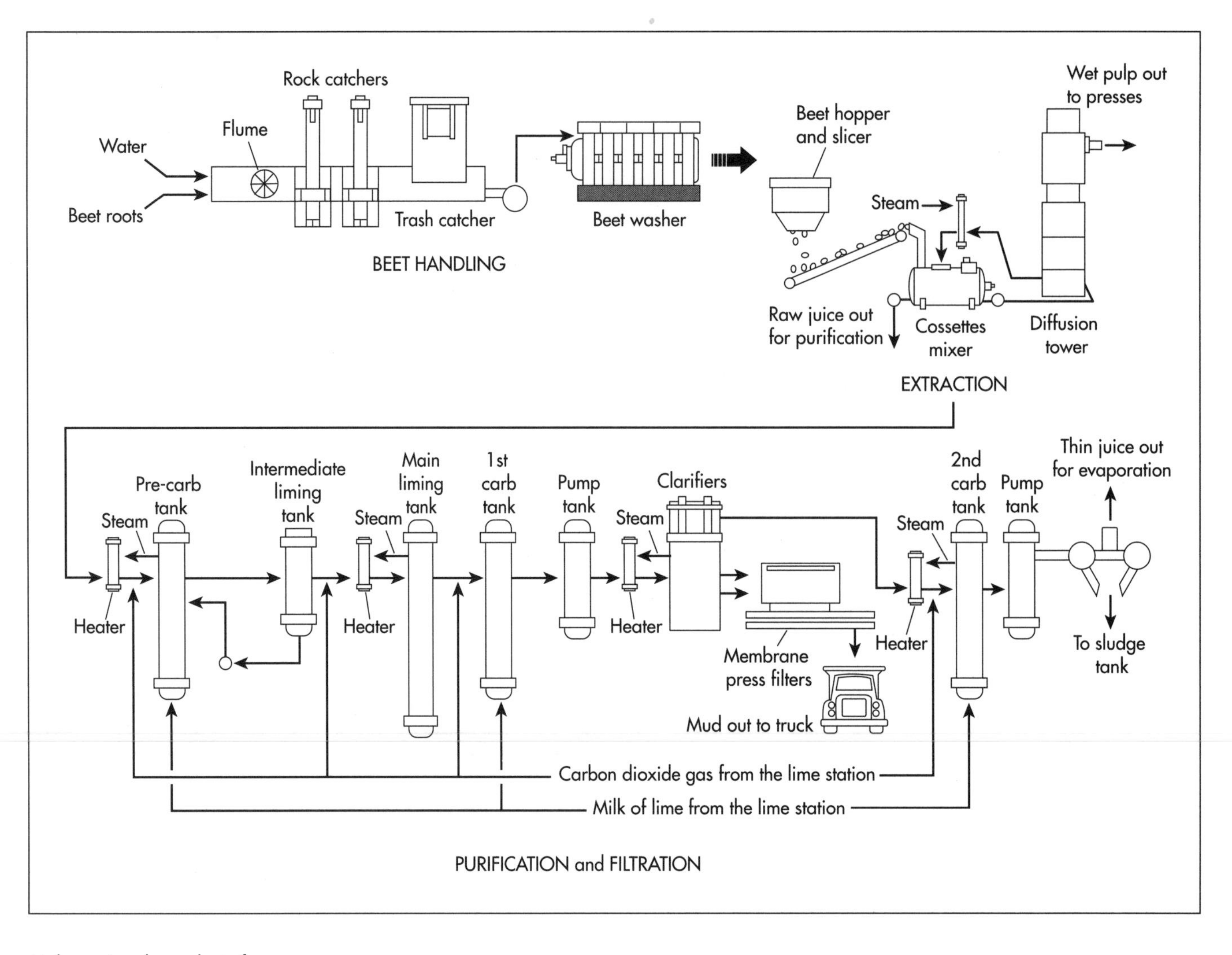


Molasses is a byproduct of sugar cane or sugar beet processing. In the later, beet roots are loaded into a flume, where they are separated from debris. Once washed, they are sliced and loaded into cylindrical diffusers that wash the beet juice out with the aid of hot water. The extracted juice is clarified by adding milk of lime and carbon dioxide, then it is heated and mixed with lime. The juice is filtered, producing a mud-like substance called carb juice. Next, the carb juice is heated and clarified, causing the mud to settle and the clear juice to rise. Once again the mud is filtered out, leaving a pale yellow liquid called thin juice. The juice is pumped into an evaporator which extracts the water until a syrup remains . The syrup is concentrated through several stages of vacuum boiling.

wife and children, poured the syrup into cans and covered them. The cans were loaded onto a platform and cooled by an overhead fan.

## Raw Materials

Sugar cane or sugar beets are the primary ingredient for the sugar process of which molasses is a byproduct. Sugar cane (*Saccharum officenarum*) is a tall thick perennial that thrives in tropical and subtropical regions. It can grow to heights range from 10-26 ft (3.05-7.9 m), and measuring 1-2 in (2.54-5.08 cm) in diameter. Colors range from white to yellow to green to purple. The Everglades of south Florida are a major producer of sugar cane with 425,000 acres grown annually that yields 90 million gallons of black strap molasses. At harvest time, the stalks are stripped of their leaves and trimmed.

Sugar beets (*Beta vulgaris*) can tolerate more temperate or colder climates than sugar cane. Therefore, the choices of growing areas is greater. At the end of the twentieth century, the leading sugar beet producers were Russia, France, the United States, and West Germany. The sugar is contained in the vegetable's root, approximately 15 teaspoons per beet root. At harvest, the tops are removed and used for cattle feed.

Milk of lime is used in the clarification process. Essentially burnt lime, it is produced in the factory by heating lime rock in a kiln. The lime rock is then mixed with sweet water—a byproduct of a previous clarification process.

Carbon dioxide is released in the lime milk process. It is purified in tanks and also used in to clarify the sugar juice.

## The Manufacturing Process

Whether the base is sugar cane or beets, the sugar extraction and refining process of which molasses is a byproduct is a circular path of washing and heating the cane and beets with hot water.

### *Washing and cutting*

1 The sugar cane stalks are loaded onto conveyer belts and subjected to hot water sprays to remove dirt and other field debris. Then, they are passed under rotating knife blades that cut the stalk into short pieces or shreds.

Beet roots are loaded into a tunnel-like machine called a flume, in which leaves, weeds, and rocks are separated out. A pump pushes the beets into a washer fitted with a large shaft that moves the beets through the water to remove any remaining dirt. The beets move through a slicer that cuts the beet roots into thin strips called cossettes.

### *Extracting the sugar juice*

2 In the sugar cane processing plant, extraction can be accomplished in one of two ways: diffusion or milling. By the diffusion method, the cut stalks are dissolved in hot water or lime juice. In the milling process, the stalks are passed under several successive heavy rollers, which squeeze the juice out of the cane pulps. Water is sprayed throughout the process to facilitate the dissolving of the juice.

In the sugar beet factory, the sliced beet roots, or cossettes, are loaded into cylindrical diffusers that wash the beet juice out with the aid of hot water. The discarded beet juice is used to pre-scald cossettes in the mixer so that they absorb even more of the sugar.

### *Clarifying the juice*

3 The extracted juice is clarified by adding milk of lime and carbon dioxide. The juice is piped into a decanter, heated and mixed with lime. The juice passes through carbon filters, producing a mud-like substance. Called carb juice, this mud is pumped through a heater and then to a clarifying machine. Here the mud settles to the bottom and the clear juice is piped to yet another heater and treated again with carbon dioxide. Once again the mud is filtered out, leaving a pale yellow liquid called thin juice.

### *Evaporating and concentrating the syrup*

4 The juice is pumped into an evaporator that boils the juice until the water dissipates and the syrup remains . The syrup is concentrated through several stages of vacuum boiling, a low temperature boil to avoid scorching the syrup. Eventually, the sugar crystallizes out of the syrup, creating a substance called massecuite. The massecuite is poured into a centrifuge to further separate the raw sugar crystals from the syrup. In the centrifuge, the sugar crystals fall away from the syrup that is being spun at a significant force. This remaining syrup is molasses, and it is forced out through holes in the centrifuge.

### *Storage and bottling*

5 The molasses is piped to large storage tanks. It is then pumped, as needed, to the bottling machine where pre-measured amounts of molasses are poured into bottles moving along a conveyer belt.

## Byproducts/Waste

In addition to molasses, which is itself a byproduct of the processing of sugar cane, there are several others materials that are used for other purposes. After pressing the juice out of the cane stalks, the dry stalk residue, called bagasse, are used as fuel in the plant. Beet pulp is used in the processing of pet foods. Cane wax, which is extracted from the dry residue, is used in the manufacture of cosmetics, polish, and paper coatings.

## Where to Learn More

### *Periodicals*

"Sugarcane and the Everglades." *Journal of the American Society of Sugar Cane Technologists* (1997): 9-12.

### Other

Florida Crystals. http://www.floridacrystals.com/.

Monitor Sugar Company. http://www.monitorsugar.com/htmtext.

Steen's. http://www.steensyrup.com.

*—Mary McNulty*

# Mousetrap

Scientists describe the mousetrap as a device that is "irreducibly complex." The mousetrap cannot be made more simply and still function, and, at the same time, it is so simple and does its job so well that it gives the illusion of being a profound achievement. "To build a better mousetrap" means to achieve an ideal, to reach a pinnacle of achievement, or to create the best possible device in an imperfect world.

*Since 1838 when the U.S. Patent Office was opened, 4,400 patents have been granted for mousetraps, although less than 25 such inventions have made their creators any profits.*

## Background

Traps as simple as pits dug in the ground have been used since humans began to hunt for food or to kill predators and vermin. The needs to catch prey and to protect the earliest settlements first motivated our ancestors to devise means of catching small and large animals for food and clothing. As early man learned to grow a wider variety of food and domesticate animals, trapping changed. Animals that raided farmyards, fields, and grain supplies had to be stopped, and the trapping of wild animals as a part of hunting became a sport, rather than a necessity.

More recently, animals have been trapped in the wild for medical research, although the most commonly used research animals—the laboratory mouse and rat—have been bred for this use. And city dwelling, heated houses, and stored food supplies have invited the "country cousins" of these laboratory animals to share our homes. When they feed, live, and breed where they are not wanted, traps, poisons, and the family cat have been used to keep their numbers down or eliminate them entirely.

The traditional snap mousetrap is a primitive device made of a combination of simple machines. It works—sometimes. However, it has enough flaws in design, operation, the process of baiting the trap, the resulting killing, and the calculated ability of the mouse to outsmart it that the objective of "building a better mousetrap" has become the inventor's watchword. Our culture has also evolved in its level of humanitarian concern and the desire to live trap even the most pesky animals so they can be released in the wild. The second major class of mousetraps that has been invented is the cage trap.

## History

In 1895, John Mast of Lancaster, Pennsylvania, invented the snap-trap. Many other inventors had produced devices for killing mice before that date, but Mast's design was brilliant in its simplicity—and it had the advantage of not catching in ladies' long skirts of the day. Mast's simple trap enticed the mouse with a bit of bait held in a bait pedal and dispatched it with a striker that struck within three milliseconds of the mouse's fatal nibble.

Mast made a small fortune from his invention and sold his company in 1907 to Oneida Community Ltd. Known for its production of silverware, Oneida was also a manufacturer of steel traps that used its profits to supplement local farmers in what is now called Oneida, New York. Woodstream Corp. in Lititz, Pennsylvania, purchased the mousetrap business from Oneida, and, today, Woodstream manufactures up to 10 million mousetraps per year that are close cousins to Mast's original invention. Part of the beauty of the snap trap is its low price. In 1900, it retailed for five cents and, in 1962, it was still available for seven cents.

Many have followed in Mast's footsteps and have attempted to improve on his idea. Since 1838 when the U.S. Patent Office was opened, 4,400 patents have been granted for mousetraps, although less than 25 such inventions have made their creators any profits. The Patent Office has 39 official subclasses for mousetraps that read like an index of the Chamber of Horrors and include choking, squeezing, impaling, non-return entrance, swinging killer bar, explosive, and constricting noose devices. Electric mousetraps that dispatch victims with a shock, and various shapes and sizes of plastic and metal traps that conceal the mouse remains have been created but have not achieved commercial success.

In 1924, a janitor in an Iowa school witnessed an infestation of mice in the school. He turned his attentions in the evenings to building a better mousetrap in his garage. The resulting invention, called the "Catch-All Multiple Catch Mousetrap" would trap a mouse without killing it, allow its release, and could be reset for the next capture. A. E. "Brick" Kness went on to found Kness Manufacturing in Albia, Iowa. Today, Kness Manufacturing builds a standard snap-trap that has a plastic base instead of a wood one and the metal and plastic "Catch-All." The company is one of approximately three in the world in this industry and markets to 14 countries.

## Raw Materials

The raw materials for the snap trap include pine for the base, a metal killer bar or striker that is driven by a 15-gauge coil spring, a thin metal rod called a trigger rod, and a bait pedal that is another small square of metal. Staples hold all the pieces together.

For the cage trap, sheet steel is used to form the basic box, and plastic is used to form the internal workings. Both materials are supplied by outside producers and formed in the mousetrap factory.

## Design

The design of the cage trap is also relatively simple. The cage typically has six sides, and one of these is the rodent's entrance—but not an exit. Some cages have two inescapable entrances. Most are equipped with handles and with sliding release doors so the captured rodent can be released in the wild. The design lies in understanding the habits of the animal being trapped. A large but slow animal requires a cage made out of stronger materials, but the door mechanism can close securely but simply. For the wily house mouse, an inexpensive, portable, lightweight trap with swift-action doors is needed to capture it before it escapes. Ease of baiting the trap, its reusability (called repeatability), the number of animals it can capture and hold, types of construction materials, and other factors are design considerations.

The cage also contains a trip pan where the bait is placed. The trigger rod is attached to the trip pan so, when the bait is taken, the exit slams shut. The tension required to activate the trip pan and its trigger rod are set in the factory and are designed for the weight of the animal. Cage traps for mice are made of steel to limit opportunities for the animals to try to chew through wire.

## The Snap Trap

The base of the snap trap is made of wood or plastic, depending on the manufacturer. Pine is used because it is solid and relatively inexpensive. Plastic can also be injection-molded with sites to hold the metal parts on the base, termed the mouseboard. The bait pedal is a small piece of die-cut metal with bait mount and a nub on it that allows it to pivot very slightly within the confines of its staples. The pivoting action releases the trigger rod and striker no matter how slightly the mouse moves the bait.

Some manufacturers use hand labor to assemble snap traps, however at least one maker has a fully automated process. In the automated process, copper-coated strands of steel are extruded and shaped into the trigger rod and striker. All metal parts are attached to the mouseboard with metal staples that extend through the mouseboard and crimped into the board so the pieces do not easily pull apart.

## The Manufacturing Process

The manufacturing process of the catch-all metal trap is described below.

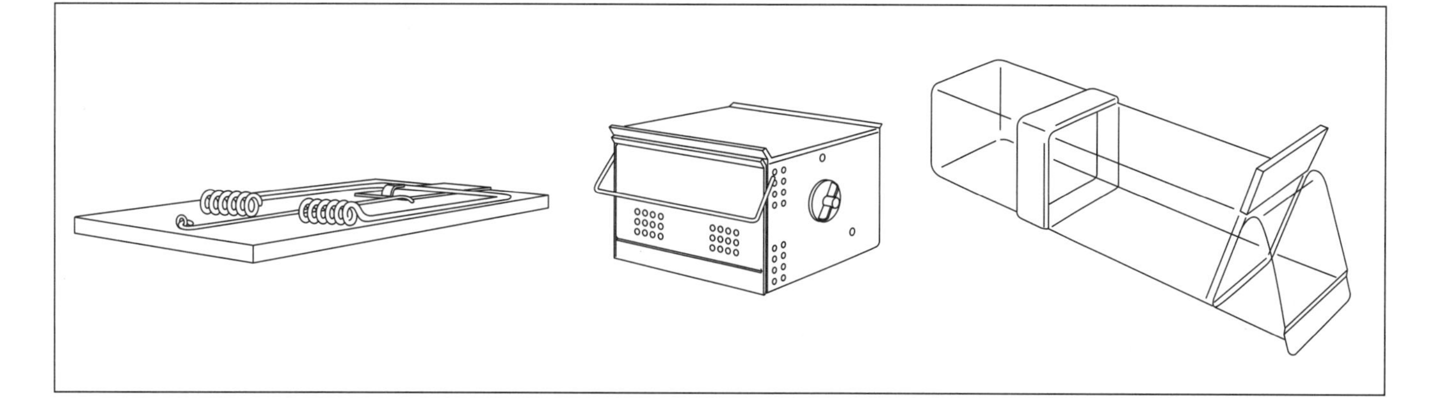

A traditional mousetrap and two types of cage traps.

1 Sheets of galvanized steel are delivered to the factory for die cutting. The steel is termed "G90 Lock-Forming Quality" steel that is chosen for its strength characteristics and durability. The sheets of steel are precut into six different widths. Outside suppliers also deliver pellets of high-impact polystyrene for use in making the internal parts of the cage.

2 The body, ends, lid, and paddle are all made of galvanized steel. The various widths of steel are run through dies specific to each width and trap part. These dies further cut, punch, notch, and form the pieces of the cage. The dies exert pressures ranging from 5-40 tons (4.5-36 metric tons), depending on their purposes. All of the metal piece parts go through the die machine.

3 Meanwhile, the plastic pellets are melted in an injection molding machine that forms the pieces of the trip assembly. All of the metal and plastic pieces are transported by conveyor belt to the assembly line.

4 The parts of the cage trap are assembled by hand. Operators at eight different stations snap the pieces of the body together first. The trip assembly is also snapped together and inserted in the body. The trip assembly operates because of a wind-up mechanism that also allows it to be rewound (that is, the tension is reset) and reused. At the last sequence of assembly, the trip is wound up and tested. One hundred percent of the cage traps are tested before they are placed back on the conveyor and transported to packing and shipping. The cage traps are wrapped singly and then packed in quantities for shipment and sale.

## Quality Control

Quality control for the cage traps is intensive. Because the traps can be used many times, they must be constructed to last for a long time and to maintain their appearance. Each piece part is checked by hand before assembly, examined for cosmetics, and measured for exact dimensions. The line operators are responsible for rejecting pieces or partially assembled cages if they do not snap together properly.

## Byproducts/Waste

There are no byproducts from manufacture of either type of trap, but most makers produce other varieties of traps for larger animals. Wood waste is disposed. Both metal and plastic waste are segregated and returned to their suppliers where they are remelted and recycled.

## The Future

The future of the two most successful forms of the mousetrap—the snap trap and the cage trap—will be secure as long as mice are with us. Both the snap and cage traps have seen minor evolutionary changes, but, essentially, they are near-perfect inventions. Agriculture is becoming more limited in area but generating higher density production, so protection of harvested and stored crops from vermin is essential. In urban settings, more crowded living conditions result in increases in rodent populations, making traps a necessity. People are also more interested in saving the lives of the creatures they capture, so humane traps are increasingly popular in the United States and other countries, thus broadening the marketplace for

American manufacturers. Competition in building the better mousetrap is limited except, perhaps, for that best mousetrap of all—the cat.

## Where to Learn More

### Books

Meyer, Steve. *Being Kind to Animal Pests: A no-nonsense guide to humane animal control with cage traps*, 1991.

### Periodicals

Fenn, Donna. "A better mousetrap." *Inc.* (March 1985): 69.

Hope, Jack. "A better mousetrap." *American Heritage* (October 1996): 90.

### Other

Kness Manufacturing Co., Inc. http://www.kness.com/.

"A Mousetrap Is Irreducibly Complex." http://www.ou.edu/engineering/cems/OKChE/jhh/Evol/sld038.htm.

Seabright Laboratories. http://www.seabrightlabs.com/mouse.html.

McGuire, Odell. "The Slidewhistle Mousetrap." December 21, 1995. http://www.wlu.edu/~omcguire/mousex.html.

*—Gillian S. Holmes*

# Mustard

A piquant condiment made from the seeds of the mustard plant. When the seeds are crushed, two elements, myronate and myrasin, are released, creating a fiery tasting essence. It is either left in a powdered form to which the consumer adds water; or it is mixed with water, wine, vinegar, or a combination of these ingredients, in a food processing plant.

## Background

Mustard seeds have been used for culinary purposes since prehistoric times. You will find mention of them in the Bible. The plants were cultivated in Palestine and then made their way to Egypt where they have been found in the pyramids.

The seeds were chewed during meals, quite possibly to disguise the rank flavor of spoiled meat. The Romans were known to crush the seeds and mix them with verjuice (unripened grape juice). Greek and Roman cooks used the seeds in a flour form, or mixed into a fish brine to flavor both fish and meat.

By the fourth century, mustard was being used in Gaul and Burgundy. Pope John XXII was so enamored of its flavor that he created a new office, grand *moutardier du pape* (great mustardmaker to the pope), and installed his nephew as the first moutardier.

In 1390, the French government issued regulations for the manufacture of mustard, decreeing that it contain nothing more than "good seed and suitable vinegar." Two hundred years later, corporations of vinegar and mustard manufacturers were founded at Orleans and Dijon.

Mustard popularity increased in the eighteenth century, thanks to two innovators. An Englishwoman named Clements developed a recipe that combined mustard powder with water. She traveled the countryside selling her product, keeping its ingredients a secret. King George I is said to have been a frequent customer. In Dijon, France, a mustard manufacturer named Niageon created a recipe for a strong mustard that combined black and brown seeds with verjuice.

In 1777, one of the most famous names in mustard was created when Maurice Grey, who had invented a machine to crush, grind and sieve seeds, joined forces with Auguste Poupon. The resulting Grey-Poupon Dijon mustard is made from brown or black mustard seeds that have been mixed with white wine.

In 1804, a British flour miller named Jeremiah Colman expanded his business to include the milling of mustard seeds. His process for producing his dry mustard is virtually unchanged since that time, with the only alteration being the use of brown seeds instead of black ones. Brown and white seeds are ground separately and then sifted through silk to filter out the seed hulls and bran. The two mustards are then blended and poured into tins.

By the turn of the century, an American named Francis French was also finding success making mustard. French's version was milder, made solely with white seeds, colored bright yellow with tumeric and made tart with vinegar.

The process by which mustard is made has not changed substantially over the years. The seeds are cleaned, crushed, sieved, and

*Eighty-five percent of the world's mustard seeds are grown in Canada, Montana, and North Dakota.*

sifted. A variety of liquids such as wine and vinegar are added to make prepared mustards. Just like Mrs. Clements of Great Britain, however, manufacturers are still secretive about the precise measurements of each ingredient.

Today, most of the work is done by sophisticated machinery. In the earliest times, the seeds were crushed and grinded by hand. Then, steam-powered stampers were used. Now, the seeds are loaded into roller mills that can flatten and hull them simultaneously.

## Raw Materials

Brown (*Brassica juncea*) and white (*Sinapis alba*) mustard seeds are used to make mustard. They are sown in March and April, the plants usually flower in June, harvesting takes place in September. It is important to harvest before the pods are fully ripe because they will split and spill the seeds out. An 8 oz (226.8 g) jar of mustard requires approximately 1,000 seeds.

Before the invention of modern farming procedures, much of the work was done by hand. Quality was difficult to assure. Today, plant breeding allows the farmer to produce a consistently high quality seed. Combines have eliminated the back-breaking work of hand-cutting the plants with sickles.

Eighty-five percent of the world's mustard seeds are grown in Canada, Montana, and North Dakota. Most mustard producers purchase seeds from a cooperative. The seeds are stored by the tens of thousands in silos until they are ready to be used. Samples are taken from each shipment and tested for quality.

Vinegar, water and/or white wine are purchased from an outside supplier and added to the milled mustard seed to make a paste. A variety of spices including tumeric, garlic, paprika, and salt are added to the mustard paste for flavoring and color. These are purchased from outside supplier. Other ingredients may be added to the mustard paste to create flavored varieties. These ingredients are purchased from an outside supplier and range from lemon to honey to horseradish.

## The Manufacturing Process

### *Seeds are examined, cleaned, dried, and stored*

1 When the seeds arrive from the harvester, they are visually examined for quality. They are then loaded onto conveyer belts and passed under water sprayers to remove dirt and other debris. After the seeds dry, they are stored in silos until ready to use.

### *Seeds are soaked*

2 Some companies soak the mustard seeds in wine and vinegar for lengths of time ranging from a few hours to several days. This softens the seeds, making the hulls easier to remove.

### *Seeds are crushed and ground*

3 The seeds are loaded into roller mills, where large wheels crush and grind them into a flour. Some companies subject the seeds to numerous rounds of crushing and grinding in order to obtain a desired degree of fineness.

### *Hulls and bran are sifted out*

4 The crushed seeds are passed through sieves, so that the hulls and bran fall to a tray underneath. Heartier varieties of mustard may include the hulls.

### *Liquids added to the seed flour*

5 The seed flour is loaded into large mixing vats and specific proportions of white wine, vinegar and/or water are added. The mixture is blended until a paste is created.

### *Seasonings and/or flavorings are added*

6 Pre-measured amounts of seasoning and/or flavorings are added to the paste and blended thoroughly.

### *Mustard paste is heated and cooled*

7 The mustard mixture is then heated to a pre-determined temperature and allowed to simmer for a pre-determined time. It is then cooled to room temperature. Some va-

Harvesting

Storage silos

Modern roller mills

Mustard is made from brown (*Brassica juncea*) and white (*Sinapis alba*) mustard seeds. They are sown in March and April, the plants usually flower in June, harvesting takes place in September. The manufacturing process has changed little. The seeds are cleaned, crushed, sieved, and sifted. A variety of liquids such as wine and vinegar are added to make prepared mustards, however, the quantity and variation of added ingredients are generally considered proprietary secrets.

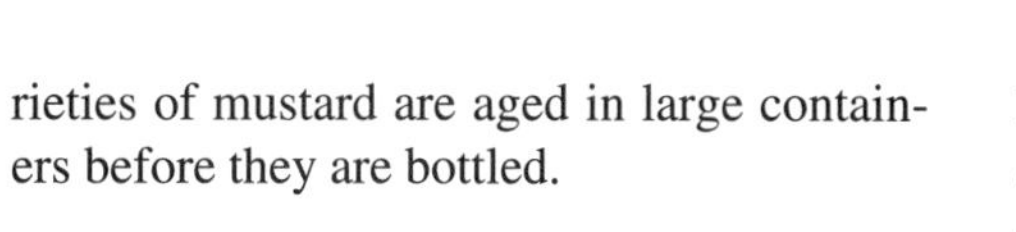

rieties of mustard are aged in large containers before they are bottled.

### *The mustard is bottled and packed for shipment*

8 Pre-measured amounts of mustard are poured into glass jars or plastic bottles that are moving along a conveyer belt. Lids are vacuum-sealed onto the tops of the containers. The containers are then loaded into cartons for shipment.

## Quality Control

All manufacturers check the mustard at each point in the process. Government food processing regulations set parameters for cleanliness in the plant. These regulations include all utensils and machinery, floors, and workers' garments

## The Future

In the United States, mustard is used more than any other spice except pepper. Mustard is also popular in Europe and Asia. By the late twentieth century, mustard cookery became a favorite of both professional and amateur chefs. Recipes were developed to use mustard as a marinade for meats and fish. Mustard sauces were developed for a wide variety of dishes. The number and types of flavors seem to be restricted only by the imagination.

## Where to Learn More

### Books

Jordan, Michele Anna. *The Good Cook's Book of Mustard.* Reading, Massachusetts: Addison-Wesley, 1994.

Lang, Jenifer Harvey, ed. *Larousse Gastronomique.* New York: Crown Publishers, 1988, reprinted 1998.

Roberts-Dominguez, Jan. *The Mustard Book.* New York: Macmillan, 1993.

### Other

Mount Horeb Mustard Museum. 109 E. Main Street, Mount Horeb, WI 53572. (608) 437-3986. http://www.mustardweb.com/index.html/ (June 29, 1999).

Nabisco. http://www.nabisco.com/museum/gpoupon.html/ (June 29, 1999).

Unofficial Colman's mustard site. http://il-hawaii.net/~danrubio/mustard/history/ (June 29, 1999).

*—Mary McNulty*

# Nuclear Submarine

## Background

A nuclear submarine is a ship powered by atomic energy that travels primarily underwater, but also on the surface of the ocean. Previously, conventional submarines used diesel engines that required air for moving on the surface of the water, and battery-powered electric motors for moving beneath it. The limited lifetime of electric batteries meant that even the most advanced conventional submarine could only remained submerged for a few days at slow speed, and only a few hours at top speed. On the other hand, nuclear submarines can remain underwater for several months. This ability, combined with advanced weapons technology, makes nuclear submarines one of the most useful warships ever built.

## History

The first serious proposal for a ship designed to travel underwater was made by the English mathematician William Bourne in 1578. Bourne suggested using two hulls, one of wood and one of leather, but this device was never actually built. The first working submarine was built by the Dutch inventor Cornelis Drebbel in 1620. Using a design similar to the one proposed by Bourne, this device was propelled beneath the surface of the Thames River by eight wooden oars.

During the early eighteenth century, several small submarines were built using similar designs. In 1747, an unknown inventor suggested attaching goatskin bags to a submarine. Filling the bags with water would lower the submarine, and ejecting water from the bags would raise it. The same basic concept is used in modern ballast tanks.

The submarine was first used in warfare during the American Revolution. The *Turtle*, designed by Yale student David Bushnell, attempted to attach an explosive to a British warship, but failed to penetrate the copper sheathing on the ship's hull. In 1801, the American inventor Robert Fulton built the *Nautilus*, a submarine constructed of copper sheets over iron ribs. The *Nautilus*, which could carry a crew of four, succeeded in sinking ships in tests, but was rejected by both France and England. Fulton was working on a steam-powered submarine that could carry a crew of one hundred when he died in 1815.

During the American Civil War, Horace L. Hunley financed the building of submarines for the Confederacy. The third of these vessels, the *H. L. Hunley*, attacked and sank the Union ship *Housatonic* on February 17, 1864, but was itself destroyed in the resulting explosion.

During the late nineteenth century, numerous submarines were built in the United States and Europe. Methods of moving the submarine evolved from hand-operated propellers to steam engines, gasoline engines, and electric motors. Submarines using diesel engines for surface travel and electric batteries for underwater travel were used successfully in World War I and II.

The development of nuclear power after World War II revolutionized submarine technology. Under the direction of Hyman Rickover, an engineer and officer in the U.S. Navy, American inventors Ross Gunn and Philip Abelson designed the *Nautilus*, the first nuclear submarine that was launched in 1954. By 1959, some nuclear submarines, known as strategic submarines,

*The first use of a nuclear submarine in active combat took place in 1982, when the British attack submarine* Conqueror *sank the Argentine ship* General Belgrano *during the conflict over the Falkland Islands.*

*Robert Fulton*

Robert Fulton, best known for his work in steamboat technology, was born in Little Britain, Pennsylvania, in 1765. As a child, Fulton enjoyed building mechanical devices. His interest turned to art as he matured, and although he managed to support himself through his sales of portraits and technical drawings, the general response his work received was disappointing and convinced him to concentrate on his engineering skills.

In 1797 while researching canals in Paris, France, Robert Fulton became fascinated with the notion of a "plunging boat," or submarine, and began designing one based on the ideas of American inventor David Bushnell. Fulton approached the French government, then at war with England, with the suggestion that his submarine could be used to place powder mines on the bottom of British warships. After some persuasion, the French agreed to fund the development of the boats and, in 1800, Fulton launched the first submarine, the *Nautilus*, at Rouen. The 24.5 ft (7.5 m) long, oval-shaped vessel sailed above the water like a normal ship, but the mast and sail could be laid flat against the deck when the craft was submerged to a depth of 25 ft (7.6 m) by filling its hollow metal keel with water. Fulton's plan was to hammer a spike from the metal conning tower into the bottom of a targeted ship. A time-released mine attached to the spike was designed to explode once the submarine was out of range. Although the system worked in the trials, British warships were much faster than the sloop used in the experiments and thus managed to elude the slower submarine. The French stopped funding the project after the failed battle attempt, but the British, who considered the technology promising, brought Fulton over to their side. Unfortunately, once again the submarine worked well in tests, but proved unsatisfactory on practical situations. After its failure in the Battle of Trafalgar (1805), the British too abandoned the project.

After these experiences, the undaunted Fulton turned to a new area of exploration—steam. Through his contacts in Paris, Fulton met Robert Livingston (1746-1813), the American foreign minister to France who also owned a 20-year monopoly on steam navigation in New York State. In 1802, the two decided to form a business partnership. The following year, they launched a steamboat on the Seine river that was based on the design of fellow American John Fitch.

Fulton returned to New York later in 1803 to continue developing his designs. After four years of work, Fulton launched the *Clermont*, a steam-powered vessel with a speed of nearly five miles per hour. The partnership between Fulton and Livingston thrived, and Fulton had at last achieved a recognized success.

Fulton's persistence and belief in his ideas helped steamboats become a major source of transportation on the rivers in the United States, and resulted in a significant reduction of domestic shipping costs.

were used to carry missiles with nuclear warheads. Other nuclear submarines, known as attack submarines, were designed to sink enemy ships and submarines. Strategic submarines and attack submarines became a vital part of naval forces worldwide. The first use of a nuclear submarine in active combat took place in 1982, when the British attack submarine *Conqueror* sank the Argentine ship *General Belgrano* during the conflict over the Falkland Islands.

## Raw Materials

The main material used in manufacturing a nuclear submarine is steel. Steel is used to make the inner hull that contains the crew and all the inner workings of the submarine, and the outer hull. Between the two hulls are the ballast tanks, which take in water to make the submarine sink and eject water to make the submarine rise.

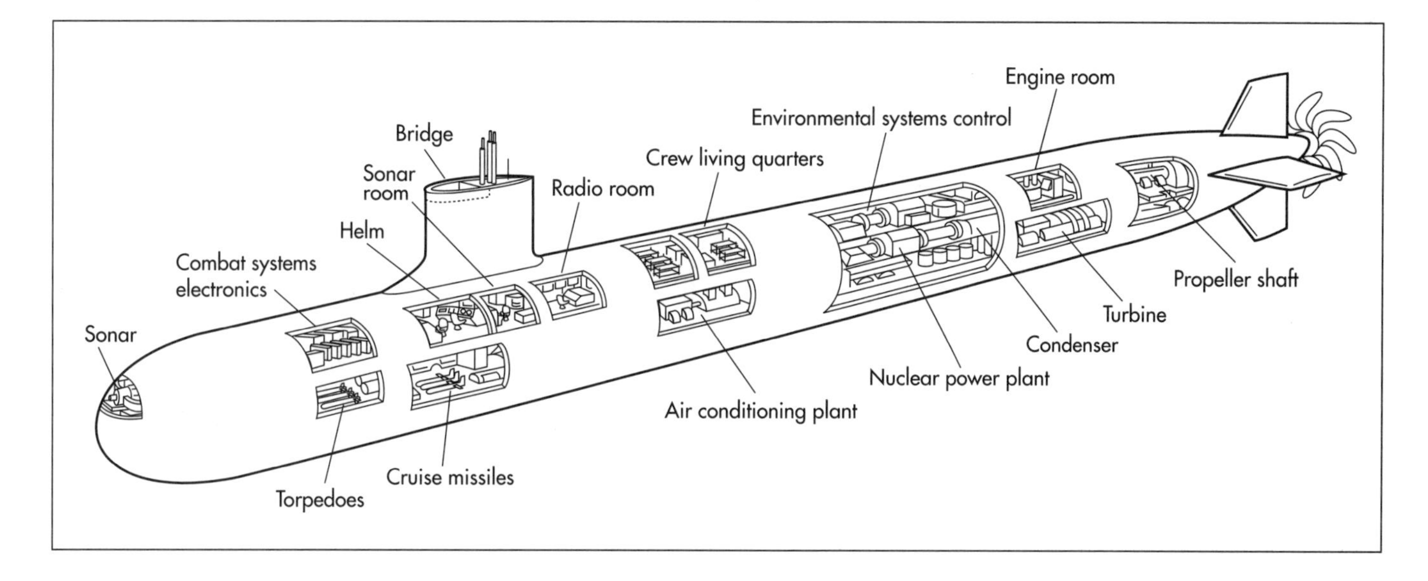


A typical submarine.

In addition to steel, various parts of a nuclear submarine are made from other metals, such as copper, aluminum, and brass. Other materials used to manufacture the thousands of components which make up a fully equipped nuclear submarine include glass and plastic. Electronic equipment includes semiconductors such as silicon and germanium. The nuclear reactor that powers the submarine depends on uranium or some other radioactive element as a source of energy.

## The Manufacturing Process

### *Preparing for manufacture*

1 Because nuclear submarines are only manufactured for military use, the decision to build them is made by a national government. In the United States, the Undersea Warfare Division of the Navy is responsible for requesting that a group of submarines, known as a flight, be manufactured.

2 The Navy accepts bids from thousands of companies to manufacture the many components which make up a nuclear submarine. The hull of the submarine is generally made by the Electric Boat Division of the General Dynamics Corporation. (The original Electric Boat Company made the first submarines used by the United States Navy in 1900.)

3 Funding for nuclear submarines is included in the defense budget presented by the President to Congress. If approved, the manufacturing process begins. The nuclear reactor is supplied by the government's Naval Reactor project. The methods used to manufacture these nuclear reactors are closely guarded and disclosure would be considered a breech of national security.

### *Making the hull*

4 Steel plates, approximately 2-3 in (5.1-7.6 cm) thick, are obtained from steel manufacturers. These plates are cut to the proper size with acetylene torches.

5 The cut steel plates are moved between large metal rollers under tons of pressure. The rollers, each about 28 in (71.1 cm) in diameter and about 15 ft (4.6 m) long, are set up so that one roller rests on two others. As the steel plate moves under the top roller and over the two bottom rollers it is bent into a curve. The plate is rolled back and forth until the desired curvature is obtained.

6 The curved steel plates are placed around a wooden template that outlines the shape of the hull. They are then welded together by hand to form a section of the hull. The section is lifted by a crane and placed next to another section. The two sections are rolled slowly under an automatic welder, which seals them together. The rotating sections move beneath the welder several times, resulting in an extremely strong seam.

7 The welded sections are strengthened by welding curved, T-shaped steel ribs around them. These are made by heating

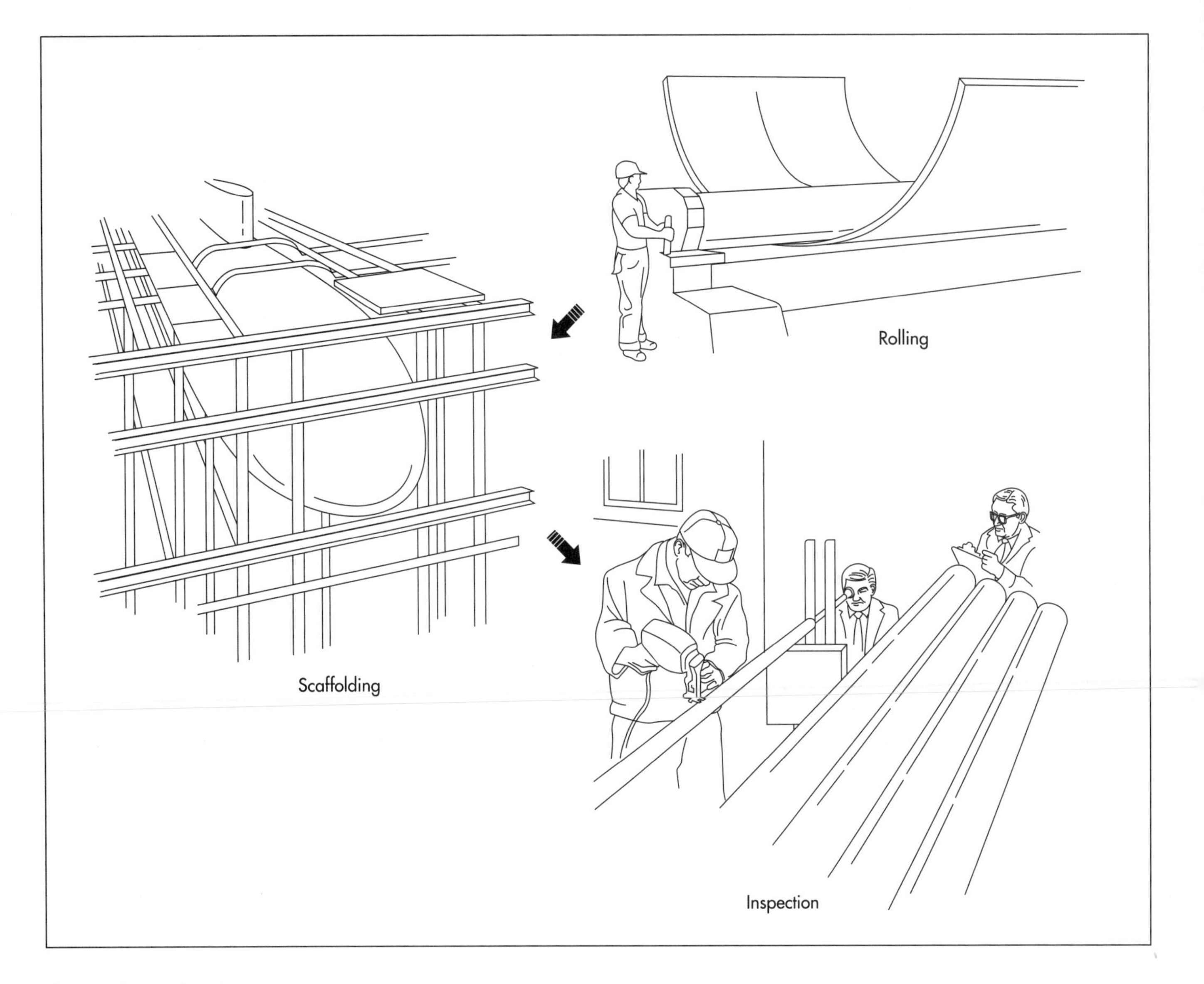


The manufacture of a submarine is highly complex and utilizes both manual and automated processes. Large sheets of steel are rolled and welded into the shape of the inner and outer hulls. Scaffolding is erected during manufacture so accessibility remains unencumbered. Every aspect of manufacture is checked by inspection and quality control measures. For example, welded steel components are inspected with x rays. Pipes are filled with helium in order to check for leaks. As a result, the Naval Reactors program is considered to have the best safety record of any nuclear power program.

bars of steel until they are soft enough to bend. Automatic hammers strike the ends of the bar, producing a curve matching the hull.

8 Welding several sections together produces an inner hull. The same process is repeated to form an outer hull. The inner hull is welded to steel ribs that are then welded to the outer hull. The steel ribs separate the two hulls, allowing space for the ballast tanks that control the depth of the submarine. The outer hull only extends as far as the bottom and sides of the inner hull, allowing the submarine to remain upright.

9 Meanwhile, steel plates are welded in place inside the inner hull in order to divide the submarine into several watertight compartments. Steel decks and bulkheads are also welded in place. Exterior welding seams are polished by high-speed grinding wheels, making them smooth. Not only does this improve the surface for painting, but it also provides the submarine with a streamlined surface that experiences little friction during travel. The hull is then painted with layers of protective coatings.

### Finishing the exterior

10 External components such as rudders and propellers are made using various metalworking techniques. One important method used for many metal components is sand casting. This process involves making a wood or plastic model of the desired part. The model is then surrounded by tightly packed, hardened sand held in a mold. The halves of the mold are separated, allowing the model to be removed. The shape of the

desired part remains as a cavity in the hardened sand. Molten metal is poured into the cavity and allowed to cool, resulting in the desired part.

11 The hull is surrounded by scaffolding, allowing workers to reach all parts of it. The external components are welded or otherwise attached. Certain components, such as sonar equipment, are attached to the hull then covered with smooth sheets of steel in order to reduce friction during underwater travel.

### *Finishing the interior*

12 Large equipment is placed within the inner hull as it is being built. Smaller equipment is brought into the inner hull after it is completed. The submarine is launched before much of the interior equipment is installed. After the launching ceremony, the submarine is towed into a fitting-out dock, where work on the interior continues. Vital components such as periscopes, snorkels, engines, and electronic equipment are installed. Equipment for the comfort of the crew, such as refrigerators, electric stoves, air conditioners, and washing machines are also installed at this time.

13 The nuclear reactor begins operating as the submarine begins its first sea trials. The crew is trained during an Atlantic Ocean cruise. Weapons are launched and tested, often in waters off Andros Island in the Bahamas. The submarine is officially commissioned in a ceremony which changes its designation from "Precommissioning Unit" (PCU) to "United States Ship" (USS). The submarine then undergoes a shakedown cruise before entering active service.

## Quality Control

The vital role it plays in national defense, the fact that the lives of its crew depend on its proper functioning, and the dangers inherent in its nuclear reactor ensure that quality control is more important for a nuclear submarine than for almost any other manufactured product. Before construction begins, the materials which will be used to build various components are inspected for any structural flaws. Previously when a new design for a nuclear submarine was proposed, a scale model was built to see if any improvements could be made. Scale drawings of the new design were made, then expanded into full-size paper patterns that allowed small details to be studied closely. A full-sized mockup of the interior was made in order to give builders a chance to adjust the location of components in order to save space or make them more readily accessible. Presently, design modeling, modification, and simulation are all enhanced by the use of computers.

When the steel plates are cut and rolled to form the hull, they are inspected to ensure that all dimensions are accurate to within one sixteenth of an inch (0.16 cm); smaller parts may need to be accurate to within one ten-thousandth of an inch (0.00025 cm) or less. Proper welding of all steel components is inspected with x rays. Pipes are inspected by filling them with helium and checking for leaks. Every instrument is tested to ensure it works properly. In particular, the nuclear reactor undergoes stringent tests to ensure that it is safe. As a result of these precautions, the Naval Reactors program is considered to have the best safety record of any nuclear power program.

After the submarine is commissioned, it undergoes a shakedown cruise to see how it would operate in wartime conditions. The speed and maneuverability of the submarine is tested to ensure that it meets the necessary requirements.

## Byproducts/Waste

The greatest concern dealing with wastes produced by nuclear submarines involves the radioactive waste produced by nuclear reactors. Although the waste produced by a nuclear submarine is much less than that produced by a larger nuclear power plant, similar problems of disposal exist. The Naval Reactors program has an excellent record of safely storing radioactive wastes. Some environmentalists, however, have expressed concern about the possibility of radioactive material being released if a nuclear submarine is sunk by accident or during military operations.

## The Future

Nuclear submarines are expected to remain a vital part of naval defense systems for many years to come. Future designs will feature new ways to improve the speed and depth capabilities of nuclear submarines. Research will also lead to an improved ability to detect enemy ships while remaining undetected. With the demise of the Soviet Union, leading to reduced defense budgets, the U.S. Navy faces the challenge of reducing the cost of nuclear submarines while retaining their effectiveness. With this goal in mind, the New Attack Submarine program was devised in the 1990s, with the goal of replacing large and expensive *Seawolf* attack submarines with smaller, less expensive, yet equally effective nuclear submarines.

## Where to Learn More

### Books

Clancy, Tom. *Submarine: A Guided Tour Inside a Nuclear Warship*. Berkley, 1993.

### Periodicals

George, Glenn R., and Lisa Megargle George. "The Naval Reactor Program: From Nautilus to the Millennium." *Nuclear News* (October 1998): 26-33.

Newman, Richard J. "Breaking the Surface." *U. S. News and World Report* (April 6, 1998): 28-31.

Wilson, Jim. "Run Silent, Run Deep." *Popular Mechanics* (January 1998): 62-66.

### Other

"Nuclear Submarines of the World." January 24, 1998. http://www.shima.demon.co.uk/sublist.htm/.

*—Rose Secrest*

# Nutcracker

## Background

A nutcracker is a device used to break open the shells of hard, dry fruits, known commonly as nuts, produced by certain species of trees. The edible material within the shell is known as the kernel. True nuts, including familiar foods such as pecans, hazelnuts, and walnuts, have shells which require nutcrackers. Other foods loosely called nuts include many which do not require a nutcracker, such as peanuts, almonds, and cashews, and those that do, such as the Brazil nut. Nutcrackers are also used to break open other hard foods, such as lobster.

A wide variety of nutcrackers are used in modern kitchens, ranging from simple tools that resemble pliers to more complex devices that rely on carefully controlled pressure to crack open shells without damaging kernels. Their continued use in an era of readily available shelled nuts is explained by the proclivity of gourmet cooks for freshly shelled nuts. Also, many people collect decorative nutcrackers, which are designed more for appearance than for daily use. These collectible nutcrackers, often in the shape of humans or animals, are usually carved from wood, although some are made of cast iron or other materials.

## History

Human beings have eaten nuts since prehistoric times, and have always faced the problem of breaking open the shells. The earliest nutcracker was probably a rock used to smash open the nut, resulting in pieces of kernel mixed with pieces of shell. This method was improved when simple tools were developed. Early nutcrackers were probably made by connecting two pieces of wood or metal with a hinge. By placing the nut between the two pieces and squeezing them together, it would have been possible to exert control over the pressure applied. In this way, the shell could be cracked with less damage to the kernel.

Although the exact details of the evolution of the nutcracker are lost to history, by the middle of the eighteenth century decorative wooden nutcrackers were carved by hand in many parts of Germany. These devices, also known as nut-biters, usually resembled a humorous human figure. The nut was placed in the figure's mouth. A set of handles were used to bring the figure's jaws together, cracking the nut. This type of nutcracker was so familiar in Germany in the early nineteenth century that the German writer E. T. A. Hoffman made one the hero of his fairy tale "Nussknacker und Mausekoönig" ("Nutcracker and Mouse King") in 1816. The Russian composer Peter Ilich Tchaikovsky adapted Hoffman's story into a ballet in 1891, and *The Nutcracker* has remained popular with audiences ever since, leading to an increased interest in collecting nutcrackers.

The part of the world most famous for collectible nutcrackers is Erzebrige, a mountainous region of Germany near the border of the Czech Republic. A productive mining region since the fourteenth century, Erzebrige also developed woodcarving as an important industry. The founder of nutcracker carving in Erzebrige was Friedrich Wilhelm Fuüchtner. Around the year 1870 Fuüchtner began using a lathe to carve nutcrackers in the form of simple human figures, such as soldiers and police officers. These nutcrackers were carved from woods such as pine,

*The part of the world most famous for collectible nutcrackers is Erzebrige, a mountainous region of Germany near the border of the Czech Republic.*

beech, and alder, then painted in bright colors. The village of Seiffen where Fuüchtner made his nutcrackers remains famous for its wooden nutcrackers and toys.

## Raw Materials

Although many decorative nutcrackers are still made by hand from wood, most modern nutcrackers in daily use are made from metal. Some nutcrackers of unusual design are made from various combinations of metal and wood or metal and hard plastic.

The most common metals used to make nutcrackers are steel and cast iron. Steel is an alloy of iron and a small amount of carbon. Cast iron is an alloy of iron and a somewhat larger amount of iron. The raw materials used to manufacture both of these materials are iron ore and coke. Coke is formed when coal is heated to a high temperature in the absence of air, resulting in a substance which is rich in carbon.

Chromium and nickel are often added to steel to form stainless steel. They may also be used to coat steel. Some steel nutcrackers, intended as impressive gifts, are coated with silver or gold.

## The Manufacturing Process

### *Making cast iron and steel*

1 Iron ore is dug out surface mines. The ore is obtained in pieces of various sizes, ranging from particles as small as 0.04 in (1 mm) in diameter to lumps as large as 40 in (1 m) in diameter. Large lumps are reduced in size by being crushed. The smaller pieces then pass through sieves to sort them by size. Small particles are melted together into larger pieces, a process known as sintering. To make cast iron and steel, the pieces of ore should be from 0.3-1 in (7-25 mm) in diameter.

2 A conveyor belt moves a mixture of iron ore and coke to the top of a blast furnace. A blast furnace is a tall vertical steel shaft lined with heat-resistant brick and graphite. The mixture of iron ore and coke, known as charge, falls through the blast furnace. Meanwhile, air heated to a temperature of 1,650-2,460° F (900-1,350° C) is blown into the blast furnace. The hot air burns the coke, releasing carbon monoxide and heat. The carbon monoxide reacts with the iron oxides present in the ore to form iron and carbon dioxide. The result of this process is molten pig iron, which consists of at least 90% iron, 3-5% carbon, and various impurities.

3 To make cast iron, the molten pig iron is allowed to cool into a solid. It is then mixed with scrap metal. The scrap metal is selected to result in a final product with the desired characteristics. The mixture of pig iron and scrap metal is moved by a conveyor belt to the top of a cupola. A cupola is similar to a blast furnace, but is somewhat smaller. The pig iron and scrap metal fall on a bed of hot coke through which air is blown. This process removes the impurities and some of the carbon, resulting in molten cast iron, which is poured into molds to form ingots as it cools.

4 To make steel, the pig iron must have most of its carbon removed. The most common method used to make steel is known as the basic oxygen process. This procedure uses a steel container lined with heat-resistant brick. Scrap metal of the proper type to produce steel with the desired characteristics is loaded into the container. Molten pig iron, usually transported directly from the blast furnace without cooling, is added to the scrap metal. Pure oxygen is blown into the mixture at extremely high speeds. The oxygen reacts with various impurities in the mixture to form a mixture of solids, known as slag, floating on the molten metal. The oxygen also removes most of the carbon in the form of carbon monoxide gas. Lime (calcium oxide) is also added to the mixture. The lime acts as a flux; that is, it causes the substances in the mixture to melt together at a lower temperature than they would without it. The lime also removes sulfur in the form of calcium sulfide, which forms part of the slag. The slag is removed from the molten steel, which is poured into molds to form ingots, cooled between rollers to form sheets, or otherwise solidified.

### *Shaping cast iron and steel*

5 Cast iron, as its name suggests, is usually shaped by casting. Casting involves pouring molten metal into a mold. The most common method used for cast iron is sand-

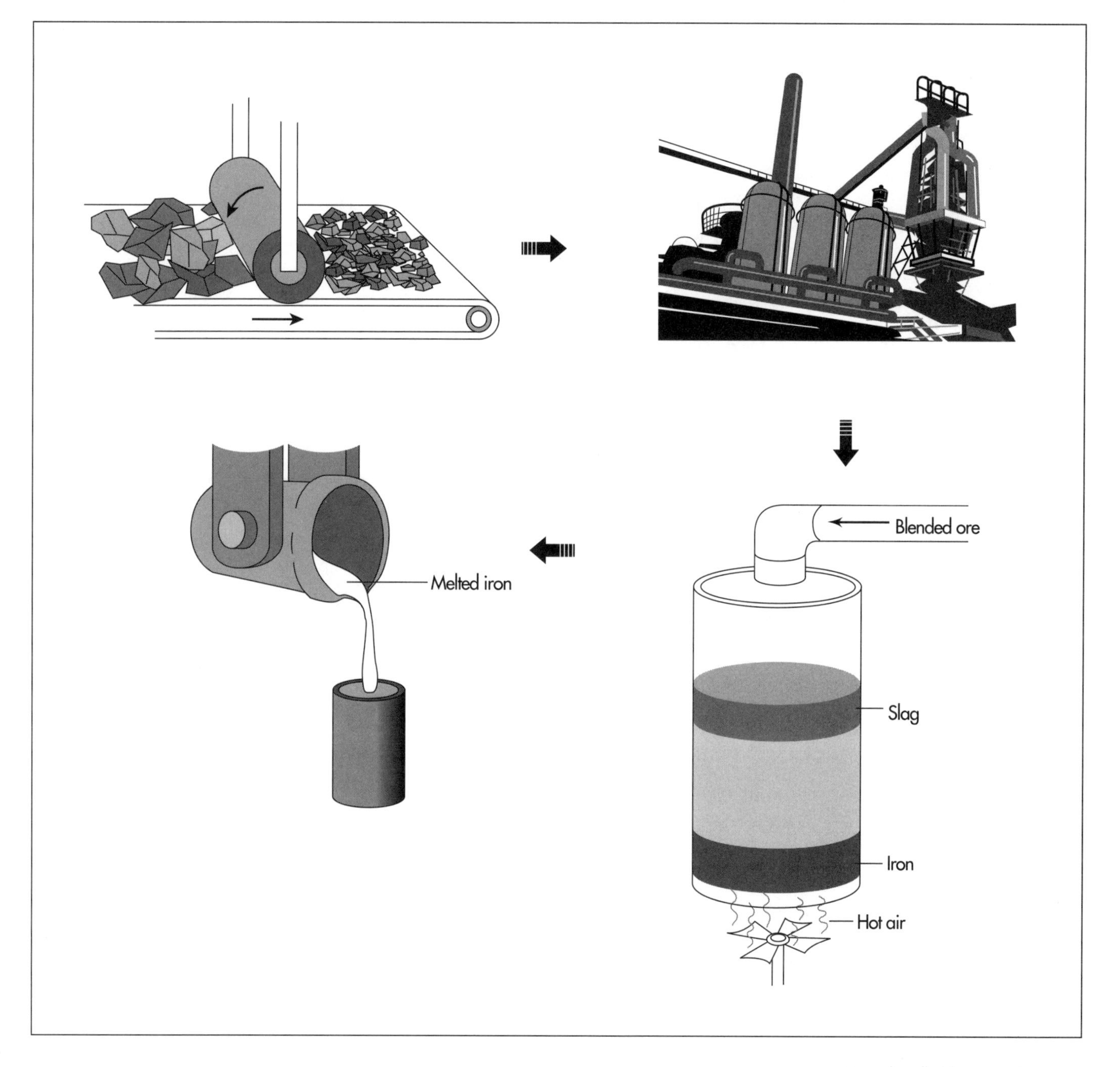


casting. A pattern in the shape of the desired product is shaped from wood, clay, metal, or plastic. The pattern is firmly packed into sand which is held together with various substances known as bonding agents. The sand mixture is hardened by various methods, which may involve heat, pressure, or chemicals. The pattern is removed from the hardened sand. Molten cast iron is poured into the resulting mold and allowed to cool into a solid.

6 Steel can be shaped in a variety of ways. Molten steel can be poured into molds, much like cast iron. It can also be shaped by cold stamping. This process involves pounding a sharp die in the shape of the desired product into a thin sheet of steel. The die cuts through the steel, forming the product. Steel can also be shaped by drop forging. This process also requires a shaped die, known as a drop hammer. The drop hammer is pounded into a red-hot bar of steel. The pressure of the die strengthens the steel at the same time it shapes the product.

### *Assembling the nutcracker*

7 Cast iron components must be cleaned and polished before being assembled

Most handheld nutcrackers are made of stainless steel. Steel is manufactured in a process that takes mined ore and refines it by melting the ore in a furnace. Once melted, the pure iron sinks to the bottom, while slag floats to the top. Chromium and nickel are added to create stainless steel.

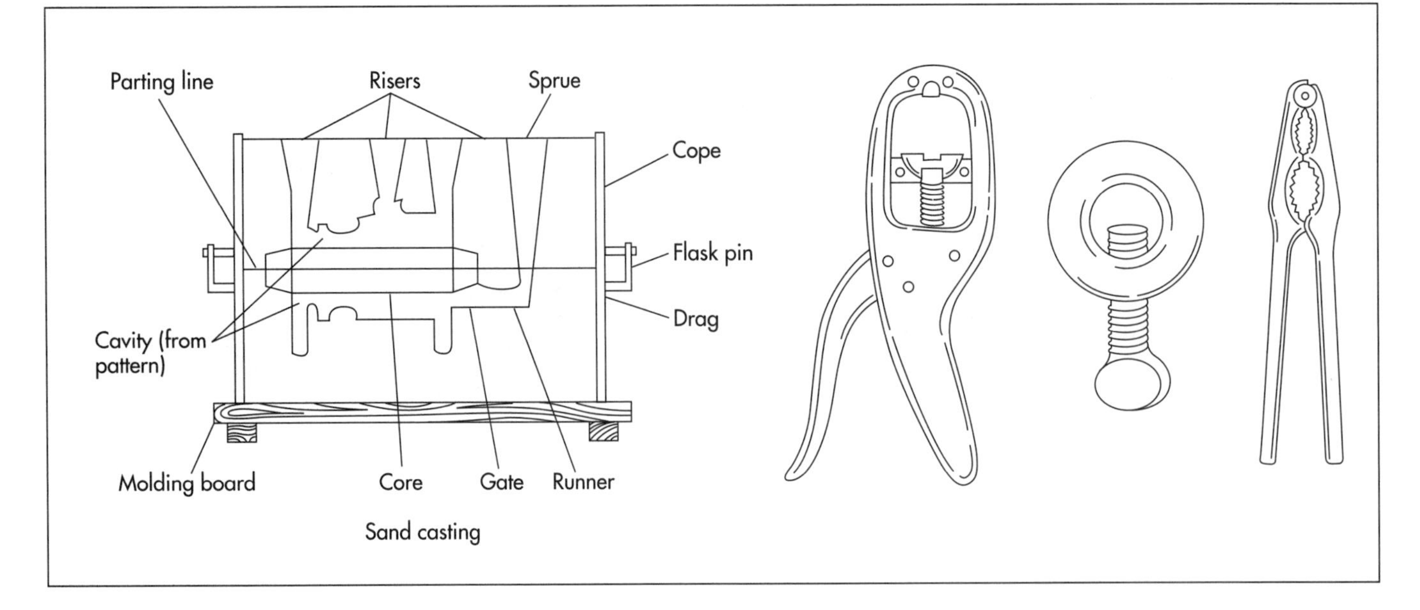


Sand casting

Sand casting is the most common method used to cast iron. A pattern is shaped from wood, clay, metal, or plastic and then firmly packed into sand held together with various substances known as bonding agents. The sand mixture is hardened into a mold using heat, pressure, or chemicals. Molten cast iron is poured into the resulting mold and allowed to cool into a solid.

into the nutcracker. Ball bearings strike the cast iron at high speed to remove any sand which clings to it. The cast iron is then polished with a surface grinder. A surface grinder is typically a rapidly spinning wheel covered with an abrasive substance. The abrasive grinds away a small portion of the surface of the cast iron, leaving it smooth.

8 Steel components are hardened by heating and cooling, a process known as tempering.

9 Depending on the exact structure of the nutcracker, assembly may be fairly simple or it may be quite complex. Precision drilling equipment it sued to form holes in the metal. The components are then brought together and rivets or screws pass through the holes and are tightened to hold the pieces of the nutcracker in place. Some nutcrackers may also require welding.

10 If the nutcracker is made form stainless steel, formed by adding chromium or nickel to the scrap metal and molten pig iron mixture, no protective coating is needed. Other metals are usually painted or covered with chromium or nickel. These metallic coatings are applied by giving a negative electrical charge to the nutcracker in a solution containing positively charged particles of chromium or nickel. The particles are attracted to the nutcracker, coating it evenly and smoothly. The coated nutcracker is removed from the solution, cleaned, and is ready to be packaged and shipped.

## Quality Control

After cast iron and steel parts are formed they are inspected to ensure that all the shapes are correct. Sharp edges or small irregularities produced during the shaping process are removed by hand using tools such as files.

During the assembly process, the tools used to drill holes may cause irregularities in the metal. Again, these can be removed using hand tools.

After being painted or plated, the nutcracker is inspected to ensure that the protective coating covers the entire surface. Any flaws found are corrected by repeating the procedure.

A sample nutcracker may be tested after it is manufactured to ensure that it operates correctly. In general, nutcrackers are extremely reliable tools, so frequent testing is not necessary.

## The Future

Although the nutcracker seems to be a simple device with little room for improvement, inventors are always working on better ways to break open shells without damaging kernels. Various designs have been proposed which are much different from the traditional plier-shaped nutcracker.

One simple but innovative device consists of a metal screw which rotates into a small plastic bowl. The nut in placed in the bowl and the screw is twisted until it cracks the shell. A more complex device which uses a similar mechanism consists of a small metal handle attached to a large handle which contains an opening. The nut is placed in the opening. Pulling the small handle toward the large handle causes a knob to be cranked up a notch toward the nut. Repeating the action increases the pressure until the nut cracks. Both these devices allow the amount of pressure placed on the nut to be carefully controlled; too little pressure fails to crack the shell, but too much pressure shatters the kernel.

A more radical design is found in the Texas Native Inertia Nutcracker. This unusual device, made from oak, steel, and aluminum, uses a rubber band to power a battering ram which cracks the nut. Imitations of this innovative device are usually made from hard plastic and might be shaped like a gun whose trigger pulls the rubber band. The future will continue to see changes in nutcracker design, while collectible wooden nutcrackers carved in the style of nineteenth century artisans will continue to be popular.

## Where to Learn More

### *Books*


Beard, James, et al., ed. *The Cooks' Catalogue*. Harper and Row, 1975.

Campbell, Susan. *Cooks' Tools*. Morrow, 1980.


### *Periodicals*


LaTorre, Bob. "Nutcrackers." *Mother Earth News* (November/December 1986): 54-56.

Smith, Louvinia T. "Nutcrackers." *Antiques and Collecting* (December 1996): 38-39.


### *Other*


"A Little History." August 11, 1997. http://members.aol.com/trcdesign/history.htm/ (June 29, 1999).

"The Story of the Nutcracker." December 28, 1992. http://www.serve.com/shea/germusa/nutcrack.htm/ (June 29, 1999).


*—Rose Secrest*

# Oatmeal

*When oatmeal was first introduced in the United States as a breakfast alternative, cartoons and editorials poked fun at the so-called oat-eaters, accusing them of robbing animals of their feeds and developing a whinny. Eventually, the medical profession found that the human consumption of oats was beneficial for the entire population, not just invalids and infants.*

Oatmeal is made from the ground or rolled seeds of oat grass (*Avena sativa*). It is cooked as cereal or used as an ingredient in baking.

## Background

Wild oats were eaten as early as the Neolithic and Bronze Ages. The Romans cultivated it. The Teutons and the Gauls used it to make gruel. Often growing weed-like amongst wheat and barley, oats thrive in cold climates with a short growing season, hence its prevalence in Scotland, Ireland, Scandinavia, and Germany. In those countries it became a basic foodstuff well into the nineteenth century. Haggis—a blend of oatmeal, goat organs, onions, and spices that are roasted in pieces of goat stomach—is still enjoyed as a traditional Scottish meal.

In the United States, however, oats were viewed as nothing more than horse feed until a German grocer named Ferdinand Schumacher emigrated to the United States in the mid-1800s and saw a ready market for oatmeal in the growing immigrant population. Up until that time, much of the United States was farmland. Breakfasts consisted of meats, eggs, breads, potatoes, fruits, and vegetables. The immigrants arriving in the growing urban areas had neither the means nor the resources to produce such morning meals.

The climate of the American Midwest was also conducive to growing oat grass and farmers were producing 150 bushels of the grain per year. In 1854, Schumacher began grinding oats in his Akron, Ohio, store, using a hand mill similar that used to grind coffee beans. Within two years he opened the German Mills American Oatmeal Company. Using a water wheel to generate power and rotate two large millstones, Schumacher's business was able to fill twenty 180 lb (81.7 kg) barrels with oatmeal daily.

This method for grinding oats was not revolutionary. It had been used for centuries. Oats were hulled by passing them between two stone wheels turning in opposite directions. The hulls and residue were sifted out of the oat grains, which are known as groats. The groats were crushed under a second set of millstones, producing a meal that could be eaten after it was cooked for three to four hours. But there was still enough powdery residue in the groats that the cooked oatmeal was pasty and lumpy.

One of Schumacher's employees, Asmus J. Ehrrichsen had the idea to replace the millstones with rotating knife blades. This substantially reduced the amount of residue and produced a meal of uniform taste and flakiness. However, one-quarter of Schumacher's oats were still ending up as residue, a product that he could not sell. Therefore, in 1878 Schumacher purchased a set of porcelain rollers from England. By rolling the groats, the residue was virtually eliminated and the cooking time was decreased to one hour.

In nearby Ravenna, another mill, owned by Henry Parsons Crowell, William Heston, and Henry Seymour, was also selling steel-cut oats. Instead of selling them in bulk out of an open barrel, Crowell measured out two pounds and placed them in clean paper boxes with the cooking directions printed on the outside. They called their product Quaker Oats.

George Cormack, an employee at Quaker Oats Company, designed a number of labor-saving devices. They included the first me-

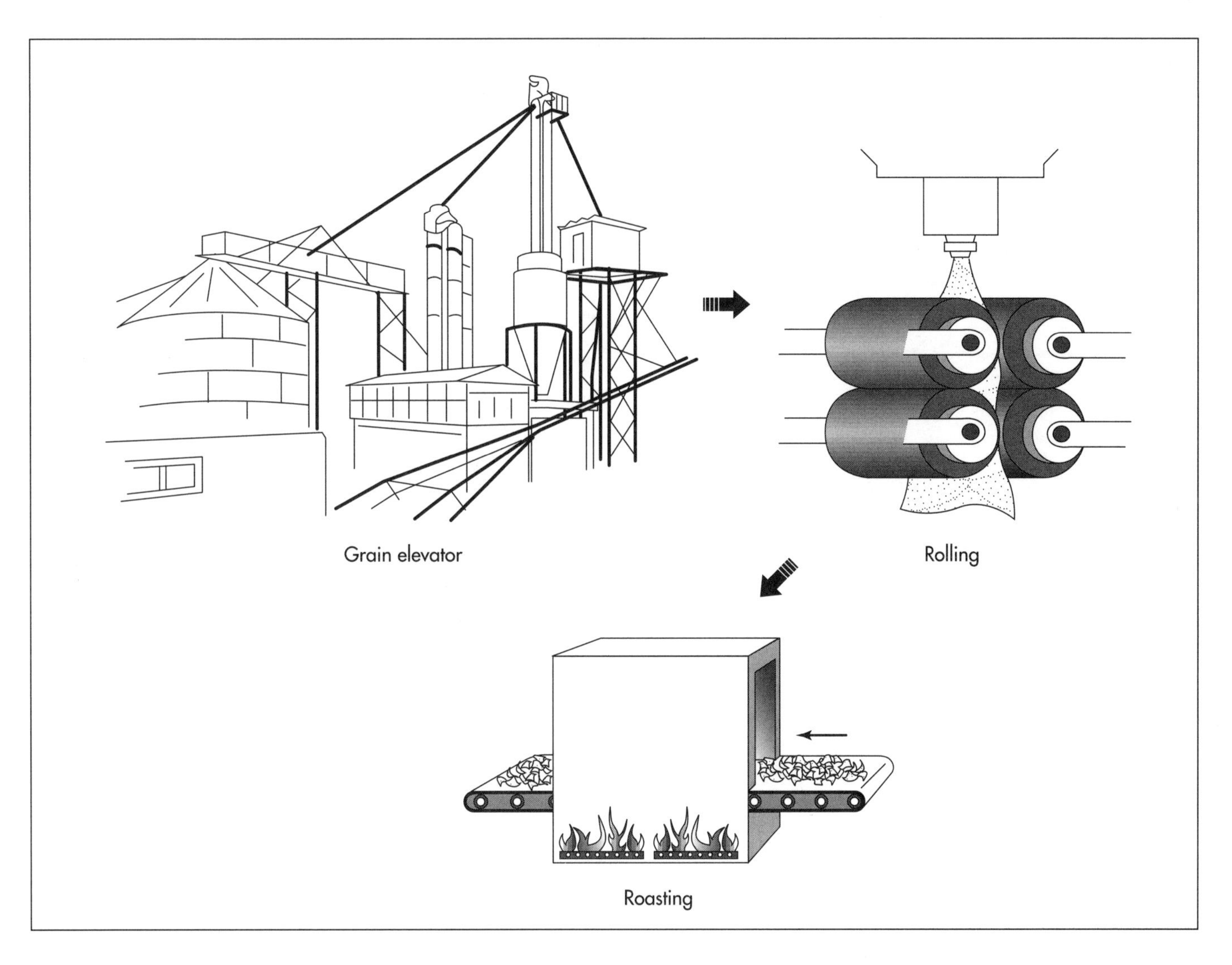


Oatmeal manufacture involves harvesting, washing, steaming, and hulling the oats. Standard oats are steel-cut, whereas quick-cooking oats are rolled between cylinders to produce a flatter flake. Once flaked, the oats are roasted and packaged.

chanical sorter that separated the oat grains by size instead of weight, ventilation systems that reduced spoilage, escalators and endless conveyer belts to move the product to and from the warehouses.

## Raw Materials

Oat grain is the only ingredient in oatmeal. The seeds of the *Avena* grasses are harvested in the fall. The thinner-skinned grains are preferable as they have a high protein content without being overly starchy. Additionally, the thin-skinned oats will yield 60% oatmeal, while the yield of thick-skinned oats is 50%. Oatmeal millers develop close relationships with their farmer-suppliers in order to insure that they receive the finest quality grain.

The grains should be milled as soon as possible thereafter to prevent spoilage or infestation of boll weevils. Rancidity is a major concern in the milling of oats. With a higher fat content than other cereal grains, the lipase enzyme can create a soapy taste. This is controlled by steaming the oats.

## The Manufacturing Process

### *Cleaning and sifting*

1 The oats are loaded onto moving trays and washed under a high-intensity water spray. Often the trays are perforated so that foreign material is discarded underneath.

### *Steaming*

2 The cleaned oats then move to a large steamer where they are subjected to moist heat for a predetermined length of time.

### Rolling or cutting

3 Standard oats are those that have been steel-cut. The oats are run through a machine with razor-sharp knife blades. Quick-cooking oats are rolled between cylinders to produce a flatter, lighter flake. These processes are usually repeated several times to produce the type of oat flake that is desired. In both processes, the hull is separated from the grain. The hulls are sifted out and used for other purposes.

### Roasting

4 The hulled oats are then placed into a roaster where they are toasted at a preset temperature for a pre-determined amount of time.

### Packaging

5 Pre-printed containers are filled with pre-measured amounts of oatmeal. A lid is vacuum-packed onto the top of the container. The containers are then loaded into cartons for shipment.

## Byproducts/Waste

The oat hulls that have been removed from the grain are often used for livestock feed and as fuel. The most common byproduct of the hulls is furfural, a liquid aldehyde (dehydrogenated alcohol) that is used as a phenolic resin or as a solvent. The list of products that contain furfural include nylon, synthetic rubber, lubricating oils, pharmaceuticals, antifreeze, charcoals, textiles, plastic bottle caps, buttons, glue, and antiseptics.

## The Future

Ironically, when Schumacher first started grinding oats in his Akron grocery store, the press had a field day with the idea of people eating horse food. Cartoons and editorials poked fun at the so-called oat-eaters, accusing them of robbing animals of their feeds and developing a whinny. Eventually, the medical profession found that the human consumption of oats was beneficial for the entire population, not just invalids and infants. By the end of the twentieth century, oatmeal was touted as one of the primary elements of a healthy diet. In 1997, the Federal Drug Administration ruled that manufacturers of foods made with the soluble fiber from whole oats could claim that when part of a diet low in saturated fat and cholesterol, these foods may reduce heart disease risk. Oatmeal was also being used by the food processing industry as an ingredient in meat substitutes.

## Where to Learn More

### Books

Marquette, Arthur F. *Brands, Trademarks and Good Will.* New York: McGraw-Hill, 1967.

### Other

Quaker Oats. http://www.quakeroats.com.

*—Mary McNulty*

# Olives

The olive tree boasts two prizes—the olive itself (called the table olive) and the precious oil pressed from the fruit's flesh. In fact, a third prize is the tree which has a twisted trunk full of character, grey-green leaves, and wood which can be used for carving and furniture-making. Fallen fruit looks edible, but it isn't. All olives, whether green or black, require processing before they can be eaten.

*In 1870, a bartender in California added an olive to a new concoction named the Martinez after his hometown. Today, the olive-ornamented cocktail is known as a martini.*

## Background

The olive tree has been given the Latin name *Olea europaea* and is from the botanical family called Oleaceae. It is an evergreen that typically grows from 10-40 ft (3-12 m) tall. The branches are fine and many, and the leathery leaves are spear-shaped and dark green on their tops and silver on their undersides.

The trees bloom in the late spring and produce clusters of small, white flowers. Olives grow erratically (unless the trees are cultivated and irrigated) and tend to either produce in alternate years or bear heavy crops and light ones alternately. Seedlings do not produce the best trees. Instead, seedlings are grafted to existing tree trunks or trees are grown from cuttings. Olives are first seen on trees within eight years, but the trees must grow for 15-20 years before they produce worthwhile crops, which they will do until they are about 80 years old. Once established, the trees are enduring and will live for several hundred years.

Olives mature on the tree and can be harvested for green table olives when the fruit is immature or left on the tree to ripen. The ripe olives are also harvested for processing as food but are left on the trees still longer if they are to be used for oil. Six to eight months after the flowers bloomed, the fruit will reach its greatest weight; and 20-30% of that weight (excluding the pit) is oil. Inside each olive, the pit contains one or two seeds; botanists call this kind of fruit with a seed-bearing stone a drupe; plums and peaches are other drupes.

Olives grow in subtropical climates in both the northern and southern hemispheres. Hundreds of varieties are grown; some produce only table olives, and others are cultivated for olive oil. Italy and Spain lead world production of olives; and Greece, Morocco, Tunisia, Portugal, Syria, and Turkey also consider the olive an important part of their economies. Europe produces three-fourths of the world's olives and also leads in consumption of both table olives and olive oil. California has also become a respected producer, especially since the health benefits of the olive have been widely recognized.

## History

Cultivation of the olive is as old as the civilizations that encircle the Mediterranean Sea. The indications that people had learned the secrets to making olives edible date from the isle of Crete in about 3,500 B.C. The Egyptians recorded their knowledge of the olive around 1,000 B.C., and the Phoenicians exported it to Greece, Libya, and Carthage. The Greeks further carried the olive to Sicily, Southern Italy, and Spain. The Romans also mastered olive cultivation. Around 600 B.C., they had a merchant marine and stock market just for the oil trade. Sardinia and the south of France became olive-growing regions, thanks to the Romans.

Olive branches, leaves, and wood gained sacred connotations in both Testaments of the Bible, like the dove's return to Noah's Ark with an olive leaf in its beak. In the Olympic Games in Greece, the victors were awarded crowns of olive branches and leaves. Oil figured in the anointing of athletes, rulers, and religious authorities and was used as lamp oil by most ancient civilizations on the Mediterranean rim. It was olive oil that burned on empty for eight days in the Hebrews' eternal flame during the miracle celebrated as Hanukkah. The olive's fragrant wood was reserved exclusively for altars to the gods, and all of these uses helped make the olive a symbol of peace.

In the 1500s, Spanish missionaries brought the both the grape and the olive to California. In South America, Italian immigrants planted the olive, and they were also responsible for plantings in Australia and southern Africa. The olive achieved new fame in California when, in 1870, an inventive bartender added the fruit to a new concoction named the Martinez for the town he lived in; the olive-ornamented cocktail is known today as the martini.

## Raw Materials

The olives themselves are the most important raw material. Depending on the curing method, pure water, caustic soda or lye, and coarse salt are used. Flavorings can be added to the brine. Among the favorites are red pepper or a variety of Mediterranean herbs for black olives and lemon or hot green peppers or chilies for green olives. Fennel, wine vinegar, or garlic can be used to add interest to any olive, but the time required for the olives to take on these flavors can range from a week for whole chilies to several months for a more subtle taste like the herb fennel.

Pitted green olives can be stuffed to add color, flavor, and texture. Almonds, pearl onions, sliced pimentos, mushrooms, anchovies, and pimento paste are the most common olive accessories.

## Design

"Design" of olives includes variety, color with green or ripe olives as the two basic differences, and method of curing. Kalamata olives from Greece are one of the best-known varieties and are distinguished by their purplish brown color and elongated shape with a sharp point. The green Manzanilla is the most famous Spanish olive and is now also cultivated in California. The Nicoise olive from France is famous for the tuna salad that requires the olive as an ingredient. Naturally cured olives can vary in color from a wonderful range of greens to purple, black, brown, and even the small Souri olive from Israel that is brownish pink.

The key to the flavor, color, and texture of the olive is the moment of harvest. Obviously, the fruit can be harvested when it is green and unripe, fully ripened to black or any stage in between. Older fruit can be salt-cured or dry cured to produce a salty, wrinkled product. Damaged fruit can still be used by pressing it into oil. It is the combination of the harvest, the cure, and any added flavors that yield the characteristics sought by the producer and consumer.

Until recently, most olives available in American grocery stores were artificially cured, meaning that they were treated with lye to remove their bitterness. This is still true for all canned black olives, many of the green olives imported from Spain and the black Nicoise from France, and other bottled versions; however, renewed appreciation of the olive has led to interest in naturally cured olives that are now generally available at deli counters and are bottled by some specialized manufacturers. Naturally cured olives are cured with either oil or brine and additives like wine vinegar for flavor.

Lye treatment is done to remove the bitterness of the olive. Olives contain oleuropein (after their botanical name *Olea europea*), and it is this substance (a compound called a glucoside) that makes them too bitter to eat directly from the tree. According to the purists, lye-cured olives are bland, either spongy or hard (but not crunchy), with most of the flavor gone. Lye-cured olives are also almost always pitted, and the most naturally flavorful part of the olive is adjacent to the pit. Curing with lye softens the olive so it can be picked when it is still hard, but olives to be naturally cured must be more ripe, handled carefully, and processed quickly.

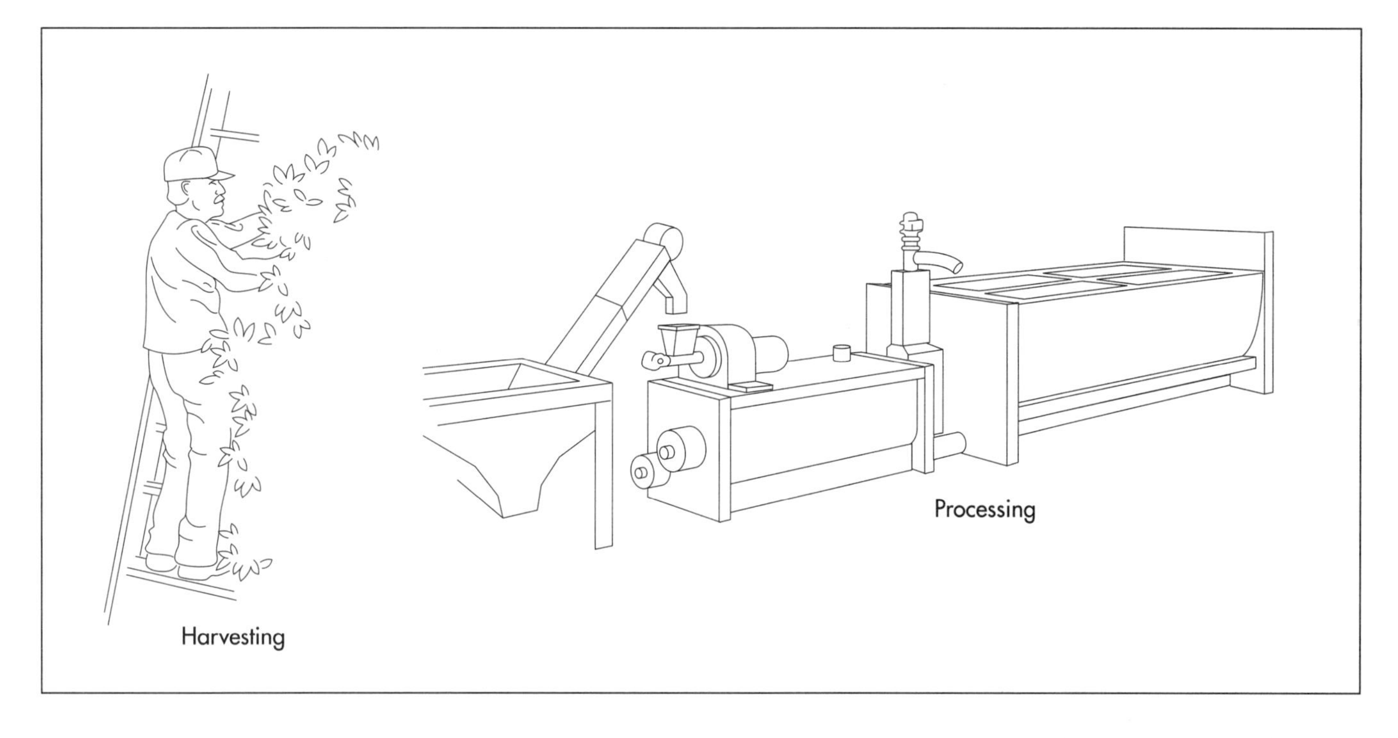


## The Manufacturing Process

### *In the field*

1 When olives are harvested by hand, sheets of netting or plastic are placed on the ground under the trees, and the harvesters climb ladders and comb the fruit from the branches. Long-handled rakes made of wood or plastic are used to pull the olives from the tree. There are other methods of harvesting including striking the branches with long canes or using shaped animal horns as combs to scrape the fruit from the branches. Pickers who use their fingers only employ a milking motion to strip the fruit from the trees. Hand picking is preferred by most growers, but it is also expensive.

Machine harvesting is a recent addition to the olive grower's arsenal. The machines were borrowed from the nut harvesters and are able to grasp the trunks of the trees and shake them. Each machine has a crew of six to nine men to operate the machine, shepherd the falling olives into the nets, and strike the branches to knock down the stubborn few by hand. The vibrations of the machine shake down about 80% of the tree's burden, and knocking at the branches with staves yields another 10% percent. About 1,100-1,800 lb (500-815 kg) of olives per day can be harvested in this manner. The trees are sensitive to such assaults by machines, however, and many purists prefer hand harvesting.

2 After harvesting of a tree's crop is completed, the nets filled with olives are emptied into baskets or crates, which are then transported to the processing plant.

### *In the processing plant*

3 At the processing plant, the harvest bags are emptied into 1,000 lb (450 kg) bins. From the bins, the olives are deposited onto conveyors and moved past a blower that blasts leaves and tree and dirt particles off the fruit. They are washed in pure water and placed in 55 gal (200 l) barrels.

4 For brine curing of green olives, 12-14% salt and water are added to the barrels filled with olives. One cup of live active brine is added to each barrel; the live active solution is previously used brine that contains airborne yeasts and sugars from the olives that fermented in the brine. The active ingredient transfers enough yeast to begin the curing process in the new batch of brine. If salt and water alone were added to the olives, fermentation (curing) would not begin on its own, so the live active brine is a starter. A salometer—a salinity meter or specific gravity meter—is used to measure

In order to produce edible olives, harvested olives are cleaned and then cured in a natural brine of salt, oil, and flavorings or artificially with lye. For green olives, the salinity is increased by 2% every two to three weeks from the initial salinity of 12-14%. Black olives begin their curing at 8-9% salinity; this is increased by 1-2% every two weeks until a maximum solution of 22-24% is reached. Taste of the final product depends upon variety, time of harvest, and curing solution.

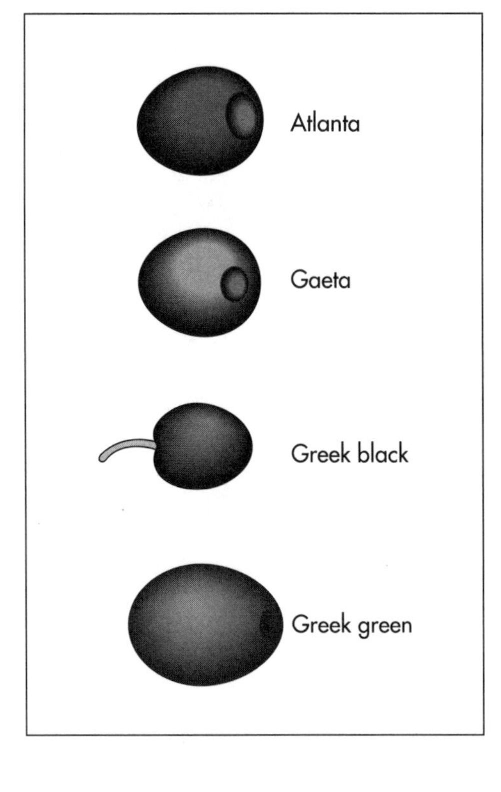


Different types of olives.

the percent of salt in solution in the barrels. For green olives, the salinity is increased by 2% every two to three weeks from the initial salinity of 12-14%. Black olives begin their curing at 8-9% salinity; this is increased by 1-2% every 2 weeks until a maximum solution of 22-24% is reached.

5 After curing is completed, the barrels of olives are emptied onto a shaker table and rinsed with clean water. The shaker table sorts the olives by size while inspectors watch and remove damaged fruit. The olives are moved to another station where they are pitted then stuffed. At filling stations, they are put in jars that are filled with an 8-11% saline solution. If the saline is flavored, herbs or other flavorings are also added to the brine. The jars are then capped and sealed for safety.

### *Other curing and canning methods*

6 Processing plants may use other methods of curing. Lye curing is accomplished with a solution containing lye, an alkaline byproduct of wood ash. The olives soak in lye solutions for 24 hours (as opposed to the six to eight weeks required for salt brine curing). The lye draws out the oleuropein to remove the olive's natural bitterness and make it edible; unfortunately, lye curing also changes the color and texture of the olive and removes many of its nutrients.

7 Dry (or Greek-style) curing is a method in which plump black olives are layered in barrels with dry rock salt (no liquid is added). The salt breaks down the bitterness and leaches it out. The olives are stirred daily, and purplish liquid leached from them is drained from the bottoms of the barrels. After four to six weeks, the olives are rinsed to remove the salt and glycoside and lightly coated in oil; they are wrinkled and purple in color, and these qualities are unpleasant to some despite the excellent flavor and nutritional value of dry-cured olives.

8 Black olives can also be cured by air curing. The olives are stored in burlap bags that allow air to pass through and around the olives. Over a period of weeks, the olives will cure, although they tend to be stronger in flavor than olives cured by other methods.

9 Green or black olives can be cured in water alone. They should be rinsed once or twice daily and consumed in about two weeks when the curing is complete. Water- and air-cured olives are not stable and should be kept in jars in the refrigerator; brine-, lye-, and salt-cured (dry-cured) olives will keep in crocks almost indefinitely.

10 In 1910, discovery of a method of canning black olives made commercial processing possible. Until that time, processing had been unsuccessful because the olives tended to discolor. The canning method consists of air ripening or lye-curing green olives in an oxygenated solution until they turn black, and treating them with ferrous gluconate. The iron additive fixes the black color, but the whole process removes most of the nutritional value of the olive. The olives are then packed in mild brine and processed in canners using pressure and heat.

## Quality Control

The quality of olive processing is protected by many sets of hands and eyes. Steps from hand-picking in the grove to hand-culling of olives on the shaker table are monitored by touch. All other processes are watched carefully. Chemistry is regulated by relatively simple instruments, and taste tests help assure the crunch of cured olives and the blending of flavors.

## Byproducts/Waste

Olive producers usually manufacture olive oil as well. Another byproduct that is growing in popularity is processed olive leaves. They are made into tea, put in caplets as crushed leaves, and processed as an extract or in tablets; all forms are believed to aid blood flow and inhibit viruses and diabetes.

Waste from olive processing consists of the pits and damaged fruit. The pits are sold as food for pigs, and all other olive waste can be ground and used as organic fertilizer. Some manufacturers return it to their groves to fertilize the olive trees.

## The Future

A ripe future is predicted for the olive business thanks to three occurrences. Medical studies have shown that olives and olive oil are healthful foods that provide vitamins, minerals, and other nutrients. They may reduce the risk of heart attacks and breast cancer, among other diseases. In America, the influence of immigrants from Spain, Italy, and the North Coast of Africa who are accustomed to naturally cured fruit has led to an interest in flavorful olives; specialty growers are reaching this market with carefully crafted, flavored olives. Finally, the "discovery" of crunchy, tasty, nutritious, naturally cured olives by a growing public is leading to the decline of canned ripe olives, which may disappear from the marketplace by about 2010.

## Where to Learn More

### *Books*

Klein, Maggie Blyth. *The Feast of the Olive*. San Francisco: Chronicle Books, 1994.

Rosenblum, Mort. *Olives: The Life and Lore of a Noble Fruit*. New York: North Point Press, 1996.

### *Periodicals*

Clark, Melissa. "An Ode to the Olive." *Vegetarian Times* (October 1997): 136.

Hamblin, Dora Jane. "To Italy, Olive Oil is Green Gold." *Smithsonian* (March 1985): 98.

Johnson, Elaine. "Know Your Olive Options." *Sunset* (April 1995): 164.

Kummer, Corby. "Real Olives: In Praise of an Old World Treat, Pits and All." *The Atlantic* (June 1993): 115.

Wing, Lucy. "A Taste of Olives." *Country Living* (September 1994): 142.

### *Other*

Australian Olive Association http://www.australianolives.com.au/ .

Australian Olive Association and Information Center. http://pom44.ucdavis.edu/olive2.html/.

Naomi's Olive Page, "An Ode to the Olive" http://www.bayarea.net/~emerald/olive.html/ .

The Olive Oil Source. http://www.oliveoilsource.com/ .

Santa Barbara Olive Company. Http://www.sbolive.com/ .

*—Gillian S. Holmes*

# Paintbrush

*Before about 1830, nearly all quality brushes were imported, but shortly thereafter, a number of American companies were founded that could produce paintbrushes rather quickly without much machinery to assist them.*

## Background

A paintbrush is a handheld tool used to apply paint or sealers to paintable surfaces. The brush picks up paint with filament, includes a ferrule that is a metal band that holds the filament and handle together and gives the brush strength, a spacer plug within the ferrule which helps the filament sits tightly in the brush and creates a reservoir for paint, epoxy to lock the filament, and a handle which provides comfort and good balance. The paintbrush industry categorizes their products based on the user of the product. Thus, there are consumer grade paintbrushes made for the homeowner who is painting small projects, professional grade paintbrushes for the professional house painter who requires a high-quality, long-lasting brush, and artistic grade paintbrushes.

Paintbrushes vary tremendously based on the quality of components used and are specifically constructed for the application of different paints and varnishes upon certain surfaces. The filament may be either animal bristle or synthetic and the brush quality largely rests on the differences in these materials. Inexpensive animal hair brushes used in lower grade brushes are of unbleached hog bristle, however, the most expensive animal hair brushes are of sable and are used for delicate hand painting. These synthetics vary greatly in quality and may be used for cheap brushes as well as better-quality brushes. Handles are of wood or plastic; the rounder the brush the easier it is to manipulate the brush for intricate movement.

Most paintbrushes are manufactured in a factory. However, the more expensive professional-quality brushes may still be produced in a factory but may be assembled, at least in part, by hand-assembly methods. Those who require delicate brushes for fine oil or watercolor painting may make their own brushes or purchase them from a specialist who produces them to order. These handmade brushes can be very expensive.

## History

Very little is known about the invention of the paintbrush. Nineteenth century histories of manufactures indicate that brushes are of relatively recent development. Then, as now, sable brushes were the very best bristle for close hand painting. Prior to the development of synthetics in paintbrushes materials such as rattan, whalebone or even shavings of wood were used in place of bristle for painting jobs that did not require much elasticity within the brush. Before about 1830, nearly all quality brushes were imported but shortly thereafter a number of American companies were founded that could produce paintbrushes rather quickly but without much machinery to assist them. Bristle was cleaned and mixed by hand, brush heads were affixed to the spacer by hand-gluing. A source from 1870 notes that the packing, papering, labeling was all completed by boys and girls. While these factories could produce brushes quickly, the process was not yet mechanized. Specialized machines for mixing, finishing, tapering, gluing, handle-making and attaching brush head to handle over 50 years later. However, fine brushes are still individually made by hand with great care at great cost.

## Raw Materials

The filament may be either of animal hair and is most often of long-haired hog bristle, often referred to simply as bristle. Other natural animal hairs used in American brushes include squirrel, goat, ox, badger, and horsehair. The most expensive animal-hair brushes are hand-made of sable. Synthetic filament used in paintbrushes are produced by extrusion (in which liquid synthetic is pushed through a mold and thus formed) and may be acrylic, polyester, nylon or amalon which is a very inexpensive petroleum-based synthetic. Different synthetics perform better with different kinds of paint so a painter should know the filament material as he or she chooses brushes. Synthetic filament may be of three constructions: solid extrusion, "x-shaped," or hollow. Solid extrusion synthetic filament lasts the longest and cleans up the easiest. X-shaped filament gives good performance and is a bit cheaper than solid filament. The hollow filament wears out quickly and is difficult to clean but is quite inexpensive. Consumer-grade paintbrushes may be of hog bristle or synthetic filament; however, water-based paints, such as latex, perform better when synthetic filament is used.

Handles may be either of wood or plastic. Different painters like the "feel" of specific handle materials; generally, professional painters prefer wood handles whereas the "do-it-yourself-er" often prefers plastic. Epoxy, a two-part glue consisting of epoxy resin and the other part consisting of a catalyst and curing agent, is required to affix the bristles within a metal band called the ferrule. The ferrule, the metal band between the handle and the bristles, is always of metal and may be tin-coated steel or another inexpensive metal. The spacer plug, either of wood or cardboard, is inserted in the brush head in the middle of the bristles (pushed within the ferrule). This plug provides a well that allows the brush to hold a reservoir of paint after it is dipped within the paint. The paint flows from this well to the brush tips.

## The Manufacturing Process

This process will describe the manufacture of a consumer-grade brush made of hog bristle with a plastic handle.

### *Mixing the bristle*

1 First, the bristle (often imported) is brought into the plant in small bundles that can be held in the hand. Each bundle includes bristle of the same length and taper ratio. However, brushes must include bristle of various length and taper ratio. The bundles must be untied and mixed together. As each different size and taper of bristle is unbundled it is placed with all bristles aligned in the same direction on a mixing machine. This machine is a series of belts that move back and forth, folding the bristle in and shuffling them together. This occurs as the bristle drops off the belt and lays onto the top of another belt with that set of bristle, then falls onto another set of bristle, etc. until the bristle is completely mixed (but still aligned in the same direction). This mixing takes about ten minute.

### *Picking the bristle and adding a ferrule*

2 The mixed bristles are then put into a machine that pinches off the proper amount of bristle (determined by weight) to form the size of brush under production. Then, the machine takes the bristle for individual brushes and shoves it into a metal ferrule (an oval band that helps attach and hide the attachment of the bristle to the brush).

### *Adding the plug*

3 The bristle and ferrule combination is put on a conveyor belt in which devices for patting the bristle further into the ferrule. When the bristle is pushed halfway into the ferrule the pieces are sent to the plugging station. Here, a wooden or cardboard plug, cut to fit the size of ferrule for the brush width under construction, is automatically shoved into the "butt end" of the ferrule (the end that will be attached to the handle). The bristle and the plug are patted again to ensure the bristles and plug are against the top edge of the ferrule.

### *Epoxying the bristles*

4 The brushes are pulled off the line by hand, put into racks with the ferrule end sticking up, and sent to the gluing station. Here, a worker then injects each butt-end of the brush with an epoxy with a machine that injects a squirt of epoxy by the touch of a trigger. This is done brush by brush with a hand-operated pump. The brush head is essentially complete; it takes about two min-

Most paintbrushes are mass-produced, but more expensive, professional-quality brushes are often hand assembled. Brushes used by the artist are handmade. Mass-produced brushes are either squared or chiseled. Typically, chiseled bristles indicate a higher-quality professional brush. Square construction and trim are often less expensive or used to spread paint over large areas.

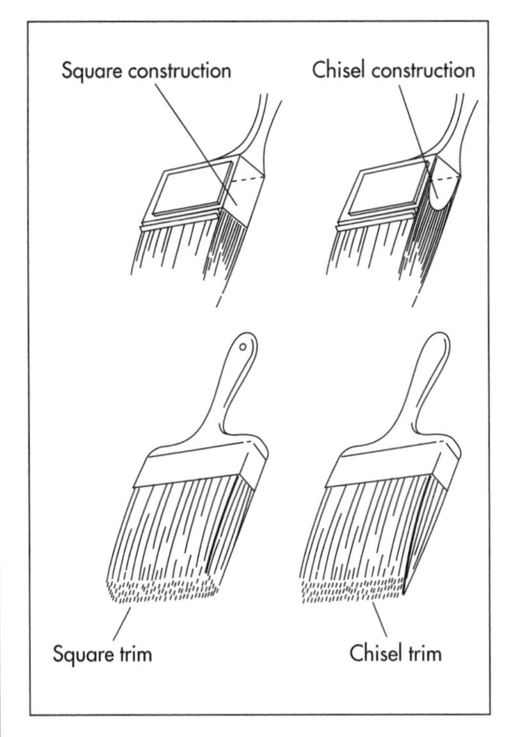


utes to pick the bristle, add the ferrule, put in the plug, and epoxy the bristles within the plug and ferrule. The brush head is now set aside to dry.

### Finishing the bristles

5 After the brush head is made and epoxied the manufacturer must "finish" the brush head. The head is then run through a series of equipment that clean out all loose hairs that escaped the epoxy. The brush head is also "tipped" meaning that the ends (that are dipped into paint) are slightly feathered or split so that they are finer and able to pick up paint more easily (the finer the bristles that fewer brush strokes the consumer will see when the paint has dried). The ends may also be tapered. A sanding wheel is used to feather and split the ends and clippers are often used for tapering. Now, the brush is set out to air dry overnight. The machinery and methods used to finish a brush is peculiar to each manufacturer and is part of the unique qualities of a brand-name brush.

### Making the handles

6 The handles are made earlier and may have come from another manufacturer. Some manufacturers produce their own handles elsewhere in the plant and send them to the brush-making department. Generally, consumer-quality brushes have plastic handles that are injection molded. To produce such a handle, a mold with two halves is clamped together and molten plastic is injected into the mold. The liquid plastic quickly hardens and the mold is opened. Many handles can be made in a series of molds that are connected. All the plastic handles are attached by a "stringer" or long, thin piece of plastic that must be broken to disconnect the handles. The handles do not require finishing.

### Putting on the handles

7 The brush heads are stacked up one on top of another after they are dried. The brush heads are taken, one at a time, and automatically inserted with the plastic molded handle which is forced against the ferrule. After insertion, the handles are nailed or riveted by machine and crimped to the ferrule so the brush head stays securely on the handle.

### Packaging

8 The same machine that inserted the handle into the ferrule also takes each finished brush and automatically packages the brushes individually. However, a number of paintbrushes come with minimal or no packaging and are sold in bins or cartons at the point-of-sale. Many brushes have minimal packaging that includes only small cardboard packaging that does not run the length of the brush.

## Quality Control

Brushes are extraordinarily varied in quality. Brush quality is determined by the use of materials and the methods of construction and the quality of a brush is generally well-marked on the packaging. Even if a brush is of lower-grade consumer quality, the materials are carefully monitored and chosen for their effectiveness as brush materials. Inferior brushes (and very cheap ones) are produced by using synthetic filament that is thick, untapered and unfeathered as the bristles show every brush stroke. Bristle that is used for consumer-grade brushes is often imported and is inspected once it arrives in the factory. The mixing process and particularly the finishing process ensures that ade-

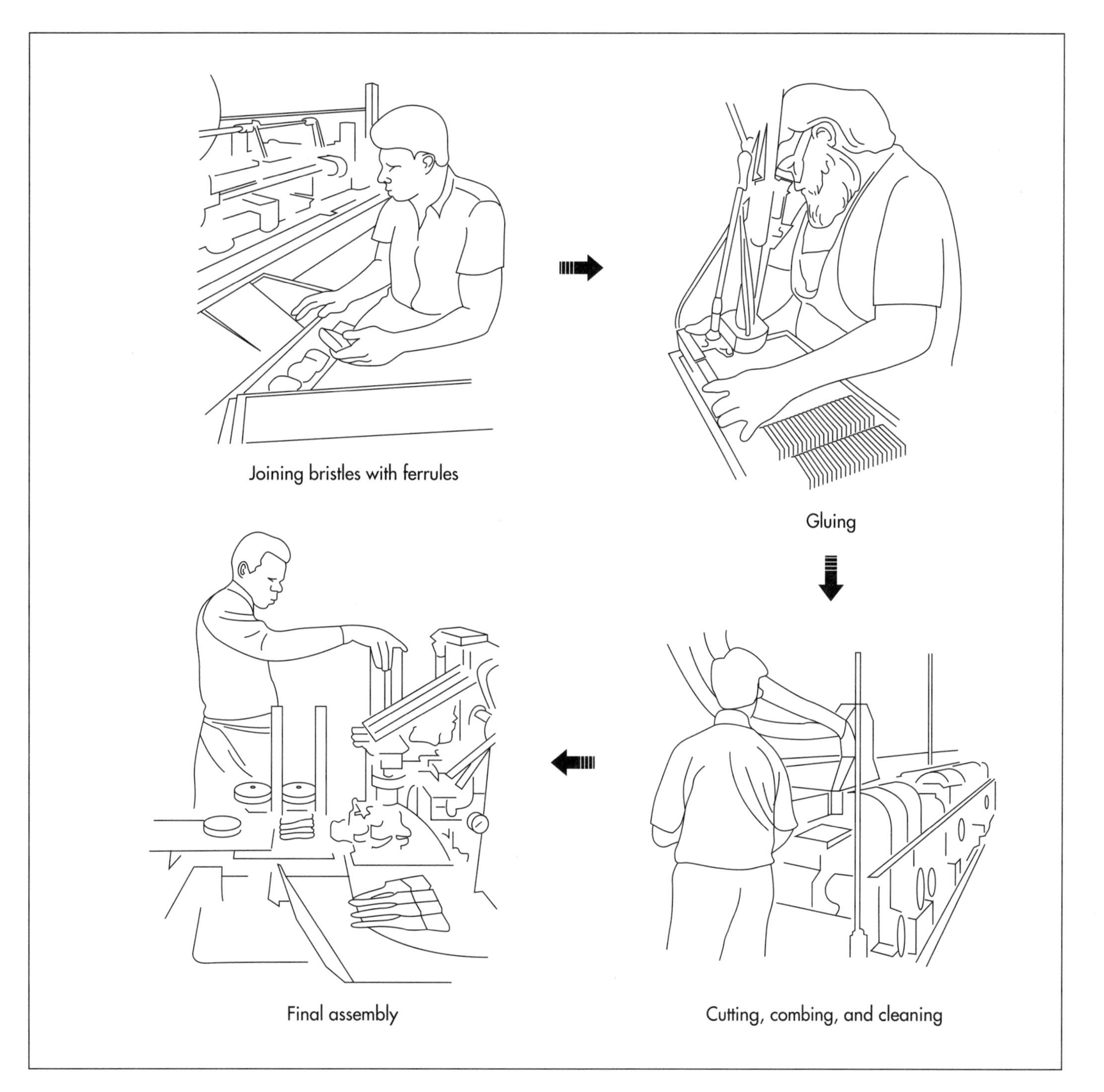
Joining bristles with ferrules

Gluing

Cutting, combing, and cleaning

Final assembly

quate bristle is processed enough to make good quality brushes.

Brush manufacturers employ brush inspectors who control quality by assessing the product at many stages of production. Furthermore, most American plants encourage the employees to visually monitor quality since so many of the processes described above are accomplished in plain sight and not within a "black box" of machinery. Employees are asked to pull pieces off the line when they believed the product is inferior.

## Byproducts/Waste

The primary byproducts of this manufacturing process are dust created from mixing filaments or bristle, handling plastic handles or ferrules, cutting out wooden or cardboard plugs, etc. Thus, most factories are vacuumed constantly using automatic systems. Epoxies used to secure the bristles within the ferrule and plug should not be inhaled extensively so the epoxies are ducted and filtered. Most of the parts of a paintbrush are recyclable (the ferrule perhaps is not). Plas-

In an automated process, bristles are separated into bundles that are joined with a metal ferrule, plugged, and glued. Once dry, the brush head is cleaned, combed, and trimmed. Handles are automatically inserted into the brush head and then nailed, riveted, or crimped to the ferrule.

tic handles can be recycled, bristles can be re-mixed. No harmful solvents are used in the manufacture of the paintbrush.

## Where to Learn More

### Books

Gottlieb, Leonard. *Factory Made: How Things are Manufactured.* Boston: Houghton Mifflin Co., 1978.

Greeley, Horace et al. *The Great Industries of the United States.* Hartford: J.B. Burr & Hyde, 1872.

Sloan, Annie and Kate Gwynn. *Classic Paints and Faux Finishes.* Pleasantville, NY: The Reader's Digest Association, Inc., 1993.

### Other

Osborn International. http://www.osborn.com.

Wooster Brush Company. "All About Paint Applicators: Information and Sales Tips." Wooster, OH: The Wooster Brush Company.

*—Nancy EV Bryk*

# Parachute

## Background

A parachute is a device used to slow the movement of a person or object as it falls or moves through the air. Used primarily for safe descent from high altitudes (e.g., a spacecraft reentering the atmosphere, a person or object dropped from an airplane), parachutes can also be used in horizontal configurations to slow objects like race cars that have finished their runs.

There are two basic types of parachutes. One is a dome canopy made of fabric in a shape that ranges from a hemisphere to a cone; the canopy traps air inside its envelope, creating a region of high pressure that retards movement in the direction opposite the entering air flow. The other is a rectangular parafoil, or ram-air canopy, consisting of a series of tubular cells; commonly used by sport jumpers, the parafoil acts as a wing, allowing the jumper to "fly" toward a target. Either type of parachute weighs less than 15 lb (7 kg) and costs from $1,200-$1,500.

In addition to the fabric canopy, a parachute designed to be used by a person must be equipped with a harness that is worn by the user. Attached to the harness is a container that holds the canopy; often this is a backpack, but it can also extend low enough for the user to sit on it. There is an actuation device that opens the container and releases the canopy for use; one of the most common actuation devices is a ripcord. When the container is opened, a small pilot chute about 3 ft (1 m) in diameter is pulled out, either by a spring mechanism or by hand. This pilot chute, in turn, pulls the main canopy from the container. Some type of deployment device, such as a fabric sleeve, is used to slow the opening of the canopy so that the suspension lines will have time to straighten. A gradual opening of the canopy also reduces the shock to the equipment and the user that a more sudden opening would cause.

## History

There is some evidence that rigid, umbrella-like parachutes were used for entertainment in China as early as the twelfth century, allowing people to jump from high places and float to the ground. The first recorded design for a parachute was drawn by Leonardo da Vinci in 1495. It consisted of a pyramid-shaped, linen canopy held open by a square, wooden frame. It was proposed as an escape device to allow people to jump from a burning building, but there is no evidence that it was ever tested.

Parachute development really began in the eighteenth century. In 1783 Louis-Sébastien Lenormand, a French physicist, jumped from a tree while holding two parasols. Two years later, J. P. Blanchard, another Frenchman, used silk to make the first parachute that was not held open by a rigid frame. There is some evidence that he used the device to jump from a hot air balloon.

There is extensive evidence that Andre Jacques Garnerin made numerous parachute jumps from hot air balloons, beginning in 1797. His first jump, in Paris, was from an altitude of at least 2,000 ft (600 m). In 1802, he jumped from an altitude of 8,000 ft (2,400 m); he rode in a basket attached to a wooden pole that extended downward from the apex (top) of the canopy, which was made of either silk or canvas. The parachute assembly weighed about 100 lb (45 kg). During the descent, the canopy oscillated so

*One parachute manufacturing plant lists its monthly materials use as exceeding 400,000 sq yd (330,000 $m^2$) of fabric, 500,000 yd (455 km) of tape and webbing, 2.3 million yd (2,000 km) of cord, and 3,000 lb (1,400 kg) of thread.*

wildly that Garnerin became airsick. In fact, he was once quoted as saying that he "usually experienced [painful vomiting] for several hours after a descent in a parachute." In 1804, French scientist Joseph Lelandes introduced the apex vent—a circular hole in the center of the canopy—and thus eliminated the troublesome oscillations.

Americans became involved in parachute development in 1901 when Charles Broadwick designed a parachute pack that was laced together with a cord. When the parachutist jumped, a line connecting the cord with the aircraft caused the cord to break, opening the pack and pulling out the parachute. In 1912, Captain Albert Berry of the U.S. Army accomplished the first parachute jump from a moving airplane. Parachutes did not become standard equipment for American military pilots until after World War I (German pilots used them during the final year of that war).

Parachutes were widely used during World War II, not only as life-saving devices for pilots, but also for troop deployment. In 1944, an American named Frank Derry patented a design that placed slots in the outer edge of the canopy to make a parachute steerable.

The world record for the highest parachute jump was set in 1960. Joe Kittinger, a test pilot for the U.S. Air Force's Project Excelsior ascended in a balloon to an altitude of 102,800 ft (31 km) and jumped. Using only a 6ft (1.8 m) parachute to keep him in a stable, vertical position, he experienced essentially free fall for four minutes and 38 seconds, reaching a speed of 714 mph (1,150 km/h). At an altitude of 17,500 ft (5.3 km), his 28-ft (8.5-m) parachute opened. In all, his fall lasted nearly 14 minutes.

## Raw Materials

Parachute canopies were first made of canvas. Silk proved to be more practical because it was thin, lightweight, strong, easy to pack, fire resistant, and springy. During World War II, the United States was unable to import silk from Japan, and parachute manufacturers began using nylon fabric. The material turned out to be superior to silk because it was more elastic, more resistant to mildew, and less expensive. Other fabrics, such as Dacron and Kevlar, have recently been used for parachute canopies, but nylon remains the most popular material. More specifically, parachutes are made of "ripstop" nylon that is woven with a double or extra-thick thread at regular intervals, creating a pattern of small squares. This structure keeps small tears from spreading.

Other fabric components such as reinforcing tape, harness straps, and suspension lines are also made of nylon. Metal connectors are made of forged steel that is plated with cadmium to prevent rusting. Ripcords are made from stainless steel cable.

One parachute manufacturing plant lists its monthly materials use as exceeding 400,000 sq yd (330,000 $m^2$) of fabric, 500,000 yd (455 km) of tape and webbing, 2.3 million yd (2,000 km) of cord, and 3,000 lb (1,400 kg) of thread.

## Design

A dome canopy may consist of a flat circle of fabric, or it may have a conical or parabolic shape that will not lie flat when spread out. It has a vent hole at the apex to allow some air to flow through the open canopy. Some designs also have a few mesh panels near the outer edge of the canopy to aid in steering the descent. Some designs use continuous suspension lines that run across the entire span of the canopy and extend to the harness on each end. Others—as described in "The Manufacturing Process"—use segments of suspension lines that are attached only to the outer edge of the canopy (and across the apex vent).

## The Manufacturing Process

### *Assembling*

1 Ripstop nylon cloth is spread on a long table and cut according to pattern pieces. The cutting may be done by a computer-guided mechanism or by a person using a round-bladed electric knife.

2 Four trapezoidal panels are sewn together to form a wedge-shaped "gore" about 13 ft (3.96 m) long. A two-needle industrial sewing machine stitches two parallel rows, maintaining consistent separation between

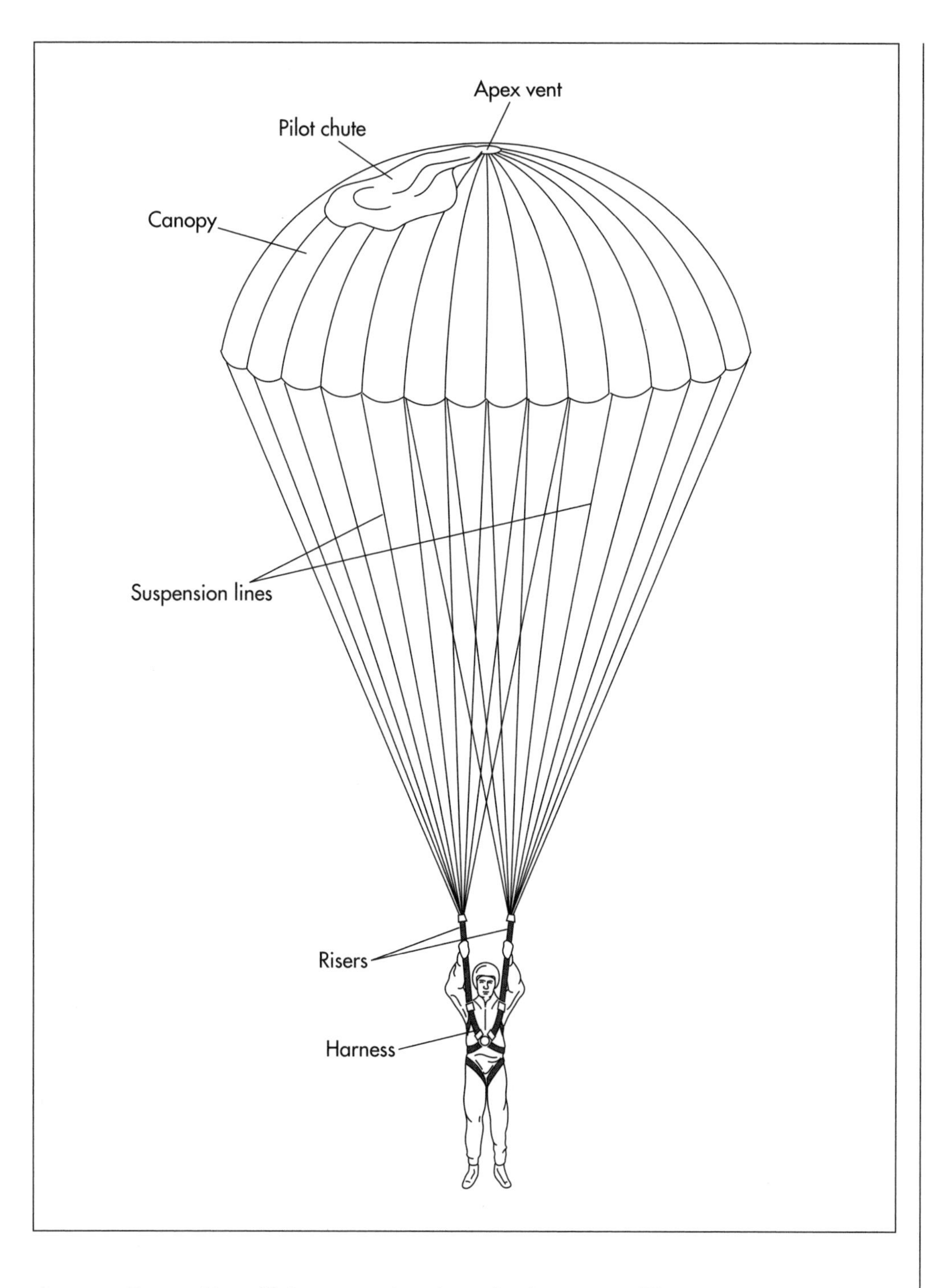


A typical dome canopy parachute.

the rows. To provide sufficient strength and enclose the raw fabric edges, a "French fell" seam is used; an attachment on the sewing machine folds the cloth edges as a highly skilled operator feeds the material through it. Depending on the parachute's specific design, a few of the gore sections may be sewn using mesh rather than ripstop nylon fabric for the largest panel.

3 A number of gores (typically 24) are sewn together, side by side, to form a circular canopy. The seams are sewn in the same manner as in Step 2.

4 Every panel and every seam is carefully inspected on a lighted inspection table to make certain that the seams are correctly folded and sewn and that there are no flaws in the cloth. If any weaving defects, sewn-in pleats, or an incorrect number of stitches per inch is found, the canopy is rejected. The problems are recorded on an inspection sheet, and they must be repaired before additional work is done.

A. French fell seam. B. Needle hem. C. V-tab. D. Outside view of stitched v-tab.

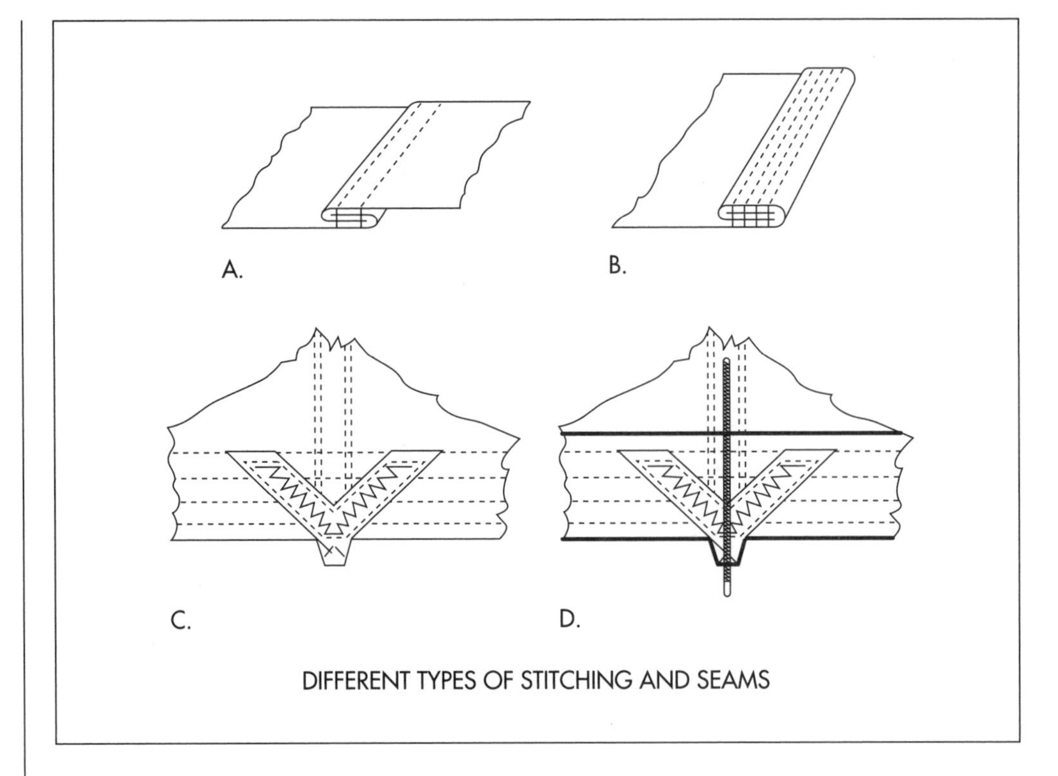


### Finishing

5 A tape the same width as the original seam is sewn on top of each radial seam using two more rows of stitching. This tape strengthens the canopy.

6 The top of each gore is a few inches (several centimeters) wide; after the gores are sewn together, their tops form a small open circle (the vent) at the center of the canopy. To reinforce the vent and to keep the cloth from fraying, the fabric is rolled around a piece of webbing and sewn with a four-needle sewing machine, which stitches four parallel rows at once.

7 The bottom of each gore is 2-3 ft (0.5-1 m) wide. Sewn together, these edges form the outer edge (the skirt) of the canopy. This edge is finished in the same manner as the vent, as in Step 6.

8 A short piece of reinforcing tape is sewn to the skirt at each radial tape. It is folded into a "V" pointing outward from the canopy. A specialized automatic sewing machine, designed for this specific operation, is used to sew precisely the same number of stitches in exactly the same pattern every time.

9 One end of a 20 ft (6 m) long suspension line is threaded through each V-shaped tab, which will distribute the load from the line to a section of the skirt hem. Using a special zigzag pattern that is both strong and elastic, the suspension cord is sewn to the canopy's hem tape and to the canopy seam for a length of 4-10 in (10-25 cm).

10 After the 24 suspension lines are sewn to the canopy, 12 1 ft (30 cm) long apex lines are similarly sewn to the central vent. One end of each line is stitched into a V-tab, then the line crosses the vent to the opposite seam where the other end is stitched into a V-tab.

### Rigging

11 The canopy is attached to the harness by tying the suspension lines to steel connector links on the harness. The lines must not be twisted or tangled if the parachute is to function properly. Attaching the lines to their correct sequential positions on the connecting links of the harness and making certain that the lines are straight is called rigging the parachute. The line end may be knotted at the harness link, or the end may be threaded back inside the line like a "Chinese fingertrap."

12 To keep the attaching knot or fingertrap from untying, the end of each sus-

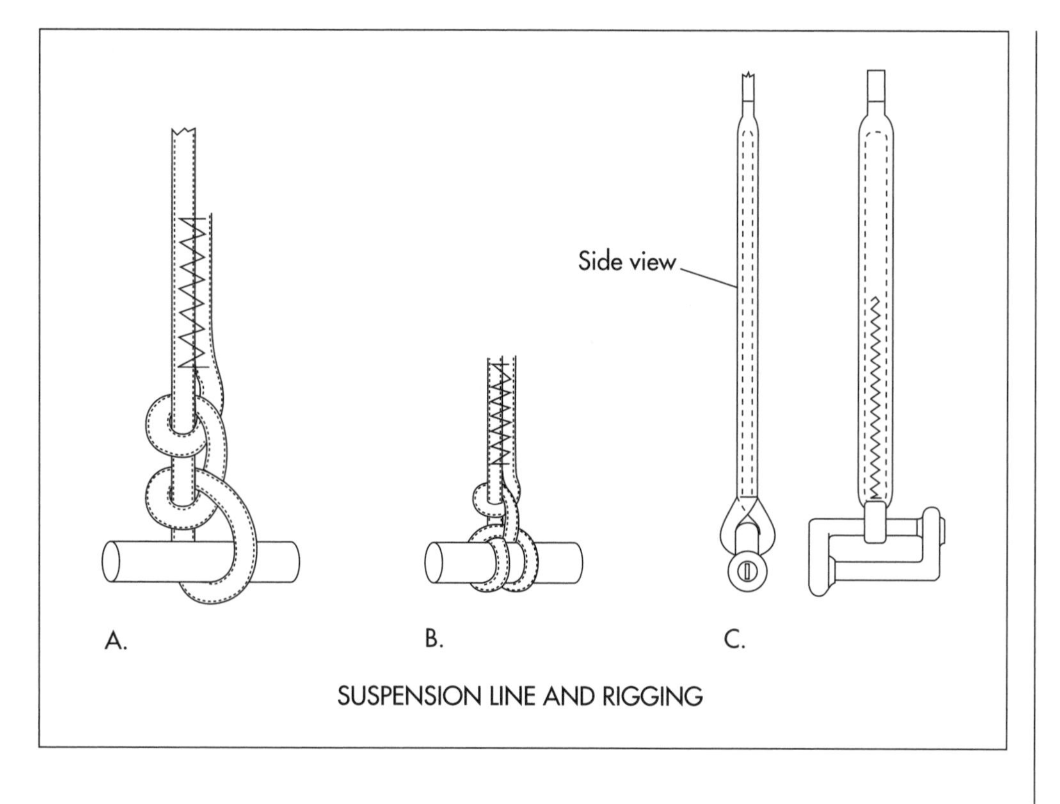


**A.** Two half hitches. **B.** Clove and half hitch. **C.** Braided suspension line.

pension line is zigzag stitched to the main section of the line.

13 Every assembly operation, every seam, even every stitch is reviewed for completion and correctness. When the parachute is approved, it is marked with a serial number, the date of manufacture, and a final inspection stamp.

14 A parachute rigger licensed by the Federal Aviation Administration (FAA) assembles the component parts (e.g., canopy, suspension lines, pilot chute) and carefully folds and arranges them in the pack, securing it with the appropriate activation device such as a ripcord.

## Quality Control

The quality control systems used by parachute manufacturers must meet the requirements for civil and/or military aviation equipment established by the federal government, under the supervision of the FAA. In addition to the lighted inspection tables mentioned, other types of testing equipment include tensile test machines (to measure strength of fabric and seams while being pulled), permeameters (to test the amount of air that can pass through the fabric), and basic measuring devices (e.g., to count stitches per inch).

## The Future

Like other manufacturers, parachute makers continually search for better materials and designs. Perhaps the most intriguing future development for parachutes, however, is their potential use to control the emergency descent of entire aircraft. At least one company, Ballistic Recovery Systems Inc. (BRS), is already manufacturing such General Aviation Recovery Devices (GARDs) for use on small airplanes.

Using an extremely low-porosity, strong, lightweight fabric for the canopy, the manufacturer bakes a 1,600-square-foot (150-$m^2$) canopy and vacuum-packs it into a 15x10x6-in (38x25x15-cm) bag weighing 25 lb (10 kg). The pack is installed inside the roof liner of the airplane near its center of gravity. To ensure that the parachute will deploy even in low-altitude emergencies, it is activated by a small rocket device.

By the late 1990s, more than 14,000 light and ultralight airplanes have already been equipped with GARDs costing $2,000-$4,000 each. As of June 1998, BRS had documented 121 lives saved by the devices. The FAA has approved a GARD system for two models of Cessna airplanes.

A system of five parafoils has been proposed for use on Boeing 747 commercial airliners. The complex system would allow the pilot to control the deployment of each canopy. Rather than dropping the airplane straight down, the system would establish a glide path that would allow the pilot to control and land the craft. The practicality of the proposed system has not yet been proven.

## Where to Learn More

### Books

Bates, Jim. *Parachuting: From Student to Skydiver.* Blue Ridge Summit, PA: TAB Books, 1990.

Poynter, Dan. *Parachuting: The Skydiver's Handbook.* Santa Barbara, CA: Para Publishing, 1989.

### Other

"Ballistic Recovery Systems." http://users.aol.com/BRSchute/BRS.HTML (February 1999).

"Strong Enterprises Company Overview." http://www.strongparachutes.com (February 1999).

*—Loretta Hall*

# Pepper

## Background

Pepper is often described as the "king of spices," and it shares a place on most dinner tables with salt. The word pepper originated from the Sanskrit word *pippali*, meaning berry. Pepper is now grown in Indonesia, Malaysia, Sri Lanka, Vietnam, and Kampuchea as well as the West coast of India, known as Malabar, where it originated. The United States is the largest importer of pepper. India is still the largest exporter of the spice, and Brazil may be among the newest exporter of pepper.

Both black and white pepper come from the shrub classified as *Piper nigrum. Piper nigrum* is one of about 1,000 species in the *Piper* genus that is part of the larger family of peppers called Piperaceae. The various species of Piper are grown mostly as woody shrubs, small trees, and vines in the tropical and subtropical regions of the world. The *Piper nigrum* is a climbing shrub that grows to about 30 ft (9 m) tall through a system of aerial roots, but is usually pruned to 12 ft (3.66 m) in cultivation. Its flowers are slender, dense spikes with about 50 blossoms each. The berry-like fruits it produces become peppercorns; each one is about 0.2 in (5 mm) in diameter and contains a single seed. It is indigenous to southern India and Sri Lanka, and has been cultivated in other countries with uniformly warm temperatures and with moist soil conditions. Because the plant also likes shade, it is sometimes grown interspersed within coffee and tea plantations. Each plant may produce berries for 40 years.

The hot taste sensation in pepper comes from a resin called chavicine in the peppercorns. Peppercorns also are the source of other heat-generating substances, including an alkaloid called piperine, which is used to add the pungent effect to brandy, and an oil that is distilled from the peppercorns for use in meat sauces.

As a natural medicinal agent, black pepper in tea form has been credited for relieving arthritis, nausea, fever, migraine headaches, poor digestion, strep throat, and even coma. It has also been used for non-medical applications as an insecticide. Of course, black pepper is a favorite spice of cooks because of its dark color and pungent aroma and flavor.

White pepper is also commonly used and is popular among chefs for its slightly milder flavor and the light color that compliments white sauces, mayonnaise, souffles, and other light-colored dishes. White pepper is also true pepper that is processed in the field differently than its black form.

A mixture of black and white peppercorns is called a mignonette. Ground pepper is also popular in mixes of spices. A French spice blend called *quartre epices* consists of white pepper, cloves, cinnamon, and either nutmeg or mace. Kitchen pepper is called for in some recipes for sauces and includes salt, white pepper, ginger, mace, cloves, and nutmeg. Pepper, therefore, proves itself to be a versatile and essential ingredient in combination with other spices, as well as in solitary glory in the pepper mill.

Other species of peppers, such as *P. longum*, *P. cubeba*, and *P. guineense*, produce peppercorns that are used locally for medicinal purposes, or are made into oleoresins, essential oils, or used as an adulterant of black pepper. Berries of pepper trees from the genus *Schinus*, family Euphorbiaceae, are

*Since 1950, consumption of pepper in the United States has risen from about 14,000-30,000 tons (12,700-27,200 metric tons) per year.*

not true peppers, but are often combined with true peppercorns for their color, rather than their flavor. *S. terebinthifolius* is the source of pink peppercorns, but must be used sparingly, because they are toxic if eaten in large quantity.

Betel leaf (*P. betel*) chewing, practiced by the Malays of Malaysia and Indonesia, is as popular as cigarette smoking in that region. Chewing the leaves aids digestion, decreases perspiration, and increases physical endurance.

Bell, cayenne, and chili peppers are not members of the *Piper* genus. They are classified within the family Solanacene, commonly known as nightshades. Comprised of over 2,000 species, the nightshade family is indigenous to Central and South America, although many species have been cultivated worldwide. Common nightshade species include potatoes, eggplant, tomatoes, tobacco, and petunia.

## History

Pepper was an important part of the spice trade between India and Europe as early as Greek and Roman times. Pepper remained largely unknown in Western Europe until the Middle Ages. During that time, the Genoese and Venetians monopolized sea trade routes and, therefore, also monopolized sale of pepper and other spices.

Knowledge of pepper truly flowered during the European period of exploration that began in the late fifteenth century. Pepper grows in hot, humid conditions near sea level, so many of the areas where pepper grows were simply unknown to Europeans until seafaring, exploring, and empire-building began. In addition, European tastes favored the "sweet pot," in which both sweet and savory ingredients were cooked in a single pot on the hearth. The spices used most often for this kind of cooking were nutmeg, cinnamon, mace, ginger, and cloves.

The pepper that was known in Europe from Roman times was the *Piper longum* (or long pepper) that is more aromatic and not so hot. Our familiar black pepper, or *Piper nigrum*, rose in popularity when the stove was introduced for cooking and sweet and savory foods could be prepared separately. Europeans valued pepper highly in the Middle Ages and the Renaissance, and pepper was often presented for gifts, rent, dowries, bribes, and to pay taxes.

Portuguese explorer Vasco da Gama reached India in 1497 and opened the trade route for pepper, among many other spices. Meanwhile, on the other side of the world, Christopher Columbus discovered the New World, and in the process, he made life complicated for pepper lovers. Columbus found a large and aromatic berry he dubbed "the Jamaican Pepper." This berry is extensively used as a ground spice today, but it is called allspice. His second peppery discovery was the capsicum. Its large, mild-flavored versions come in red, yellow, and green varieties; and it also includes these three colors in fiery hot chili peppers. The capsicum peppers are not related to the pepper found in shakers and mills. Cayenne pepper is ground from dried capsicums, so it also is not a variety of the dried berry.

To add further to the confusion Columbus unwittingly unleashed, the Spanish word for pepper is pimento; so the small slivers of red pimento found in olives are red pepper pieces, and allspice is also known as Jamaican pimento. Allspice, as this version of its name states, has a fragrance that suggests a mixture of cinnamon, nutmeg, cloves, and pepper. A few whole allspice berries added to the dinner-table pepper mill will spice up ground pepper.

## Raw Materials

Peppercorns are the only raw material for both black and white pepper in any form. If the manufacturer produces green peppercorns, brine consisting of pure water, salt, and preservatives is used. Green peppercorns are also packed in vinegar; the vinegar or brine should be washed off the berries before the peppercorns are used in cooking.

## The Manufacturing Process

### Cultivation

1 The pepper berries grow on bushes that are cultivated to heights of about 13 ft (4 m). If the berries were allowed to ripen fully, they would turn red; instead, they are har-

vested when they are green. Harvesting is done without any mechanical equipment. Women pick the unripened berries and transport them in large wicker baskets to drying platforms. The berries are spread on these large platforms to dry in the sun over a period of about a week and a half. In their dried state, the green berries blacken to become the peppercorns we use in pepper mills.

2 Alternatively, the pepper berries can be picked just as they begin to turn red. They are plunged into boiling water for approximately 10 minutes, and they turn black or dark brown in an hour. The peppercorns are spread in the sun to dry for three to four days before they are taken to the factory to be ground. This process is quicker than air-drying alone but requires the added step of the boiling water bath.

3 If white pepper is to be produced, the peppercorns are either stored in heaps after they have been boiled or they are harvested and packed in large sacks that are then lowered into running streams for seven to 15 days (depending on location). Bacterial action causes the outer husk of each peppercorn, called the pericarp, to break away from the remainder of the peppercorn. The berries are removed from the stream and placed in barrels partially immersed in water; workers trample the berries, much like stomping grapes, to agitate the peppercorns and remove any remaining husks. Some processors now use mechanical methods to grind off the outer coating to produce so-called decorticated pepper, but many exporters prefer the old-fashioned method.

### *In the factory*

4 Black and white pepper are processed in the factory by cleaning, grinding, and packaging. Blowers and gravity separators are used to remove dust, dirt clods, bits of twigs and stalk, and other impurities from the peppercorns after they are imported from the field. Sometimes, treatments are used to eliminate bacteria on the cleaned, dry peppercorns.

5 Grinding consists of using a series of rollers in a process called cold roll milling to crush the peppercorns. Cracked peppercorns are only crushed lightly to bruise the peppercorns and release their flavor. Further grinding steps crush peppercorns into coarse and fine grinds of pepper that are packaged separately. A sifter sorts the grains by size, and they are conveyed to packaging stations. Packaging varies widely among processors and includes bags, boxes, and canisters for large-volume commercial sales and smaller jars, cans, and mills for home use. Packing may also include the blending of pepper with other spices in a variety of spice mixes for preparing sauces, cajun recipes, Italian foods, seafood, and a range of other specialized blends.

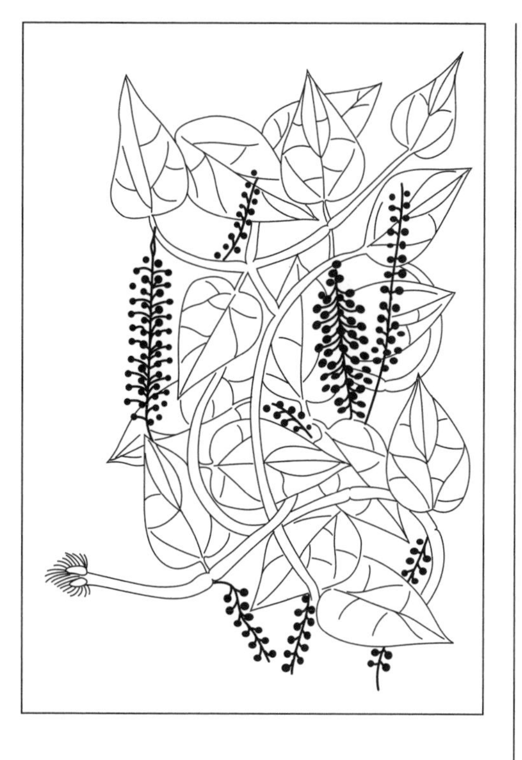

Black pepper plant *Piper nigrum.*

## Quality Control

Because pepper is harvested by hand, quality control begins in the field with the careful observations of the harvesters. Bulk importation of peppercorns is monitored, as with all agricultural products, by government inspectors. In the factory, machinery used to process pepper is simple, and the processing is observed throughout.

## The Future

The life of the spice called pepper seems guaranteed. Since 1950, consumption of pepper in the United States has risen from about 14,000-30,000 tons (12,700-27,200 metric tons) per year. Interest in gourmet

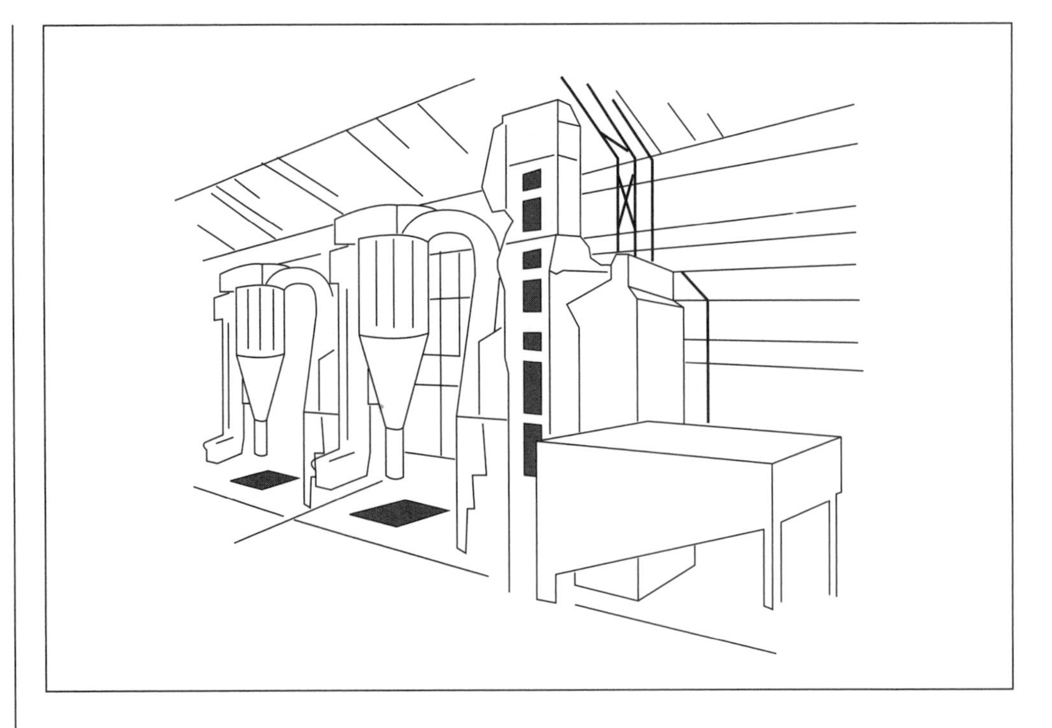

Cold roll milling machines crush the peppercorns by rolling them between a series of rollers. Cracked lightly, the peppercorns are able to release their flavor.

cooking, in types of cooking like cajun-style recipes that are spice-dependent, in restaurant dining, and in healthful food preparation have all sparked renewed enthusiasm for the flavor and goodness of pepper. Pepper will surely have an honored place at the table as long as there are cooks, kitchens, and taste-conscious consumers.

## Where to Learn More

### Books

Heinerman, John. *The Complete Book of Spices: Their Medical, Nutritional and Cooking Uses.* New Canaan, Connecticut: Keats Publishing Inc., 1983.

Norman, Jill. *The Complete Book of Spices: A Practical Guide to Spices and Aromatic Seeds.* New York: Viking Studio Books, 1990.

Stobart, Tom. *The International Wine and Food Society's Guide to Herbs, Spices and Flavorings.* New York: McGraw-Hill Book Co., 1970.

Walker, Jane. *Creative Cooking with Spices: Where They Come From & How to Use Them.* London: Quintet Publishing Ltd., 1985.

### Periodicals

Ee, Khoo Joo. "The life of spice; cloves, nutmeg, pepper, cinnamon." *UNESCO Courier* (June 1984): 20.

Wernick, Robert. "Men launched 1,000 ships in search of the dark condiment." *Smithsonian* (February 1984): 128.

### Other

Hela Spice Co. "The History of Spices." http://arcos.org/hela (March 24, 1999).

India Spice Board. http://www.indianspices.com (March 24, 1999).

McCormick & Company, Inc. March 24, 1999. http://www.mccormick.com (March 24, 1999).

*—Gillian S. Holmes*

# Pipe Organ

## Background

A pipe organ is a musical instrument that produces sound by blowing air through a series of hollow tubes controlled by keyboards. Pipe organs are distinguished from reed organs, in which air causes thin strips of metal to vibrate. They are also distinguished from electronic organs that use electrical devices to produce sounds similar to pipe organs. The large pipe organs used in public buildings are by far the biggest and most complicated musical instruments ever built.

A pipe organ consists of four basic parts. The console contains the keyboards, foot pedals, and stops. The pipes, which may be as short as 1 in (2.5 cm) or as long as 32 ft (10 m), produce the sound. The action is the complex mechanism which is operated by the console to control the flow of air to the pipes. The wind generator supplies air to the pipes.

A very small pipe organ may have a console with only one keyboard, with each key controlling the flow of air to one pipe. Most pipe organs, however, have consoles with two to five keyboards, a set of foot pedals, and a set of stops. Stops are controls which open or close the air supply to a group of pipes, known as a rank. In this way, each key can control the flow of air to several pipes.

The pipes exist in two basic forms. About four-fifths of the pipes in a typical pipe organ are flue pipes. A flue pipe consists of a hollow cylinder with an opening in the side of the pipe. The rest of the pipes are reed pipes. A reed pipe consists of a hollow cylinder, containing a vibrating strip of metal, connected to a hollow cone. The largest pipe organ in the world, located in Philadelphia, contains 28,500 pipes.

The action may be mechanical, pneumatic, electric, or electropneumatic. A mechanical action links the console to the valves which control the flow of air to the pipes with cranks, rollers, and levers. A pneumatic action uses air pressure, activated by the console, to control the valves. An electric action uses electromagnets, controlled by the console, to activate the valves. An electropneumatic action uses electromagnets, activated by the console, to control air pressure which activates the valves.

The wind generator of a modern pipe organ is usually a rotary blower, powered by an electric motor. Some small pipe organs use hand-pumped bellows as wind generators, as all pipe organs did until the beginning of the twentieth century.

The earliest known ancestor of the pipe organ was the hydraulus, invented by the Greek engineer Ctesibius in Alexandria, Egypt, in the third century B.C. This device contained a reservoir of air which was placed in a large container of water. Air was pumped into the reservoir, and the pressure of the water maintained a steady supply of air to the pipes. Pipe organs with bellows appeared about four hundred years later.

Medieval pipe organs had very large keys and could only play diatonic notes (the notes played by the white keys on modern keyboards). By the fourteenth century, keyboards could also play chromatic notes (the notes played by the black keys on modern keyboards). Keys were reduced in size by the end of the fifteenth century. By the year 1500, pipe organs in northern Germany had

*The largest pipe organ in the world, located in Philadelphia, Pennsylvania, contains 28,500 pipes.*

all the basic features found in modern instruments. Germany led the world in organ building for three hundred years.

Pipe organs fell out of favor during the eighteenth century, when orchestral music became popular. During the early nineteenth century, reed organs, which were smaller and less expensive than pipe organs, began to be used in small buildings and private homes. The increased availability of relatively inexpensive pianos in the early twentieth century, followed by the development of electronic organs in the middle of the century, led to the demise of reed organs in Europe and the United States. Small reed organs are still used in India.

Meanwhile, a renewed interest in pipe organs appeared in the middle of the nineteenth century, led by the French organ builder Aristide Cavaille-Coll and the British organ builder Henry Willis. These new pipe organs were better suited to playing orchestral music, greatly increasing their popularity.

The twentieth century brought the development of the electronic organ. The earliest ancestor of this device, known as the Telharmonium, was invented in the United States in 1904 by Thaddeus Cahill. This instrument weighed two tons (1800 kg) and was not a success. The first successful electronic organ was developed in France in 1928 by Edouard Coupleux and Armand Givelet. One of the most successful early electronic organs was the Hammond organ, invented by Laurens Hammond in 1934.

## Raw Materials

Pipe organs are primarily made of wood and metal. Wood used to make parts of the organ which are not visible, such as the action, may be made of plywood or soft woods such as poplar. Visible wooden parts, such as the console, are made from hard, decorative woods, such as mahogany or oak. Wood is also used to make some of the pipes. Woods used for pipes include poplar and mahogany.

Most pipes are made from metal. Metal pipes are most often made from alloys containing various amounts of tin and lead. Pipes may also be made from other metals, such as zinc and copper. The vibrating reeds inside reed pipes are usually made of brass.

Various small components, such as screws and bolts to hold parts of the action together, are made of steel. Other small components may be made of other materials, such as plastics and ceramics. Electronic organs require semiconducting materials, such as silicon and germanium, in order to manufacture the electrical circuits which produce the sound.

## Design

Every pipe organ must be individually crafted. Because only very small pipe organs are movable, the instrument must be able to produce the best possible sound in one particular location.

The organ builder inspects the site where the organ will be used. The acoustics of the location, as well as its physical dimensions, must be considered. The visual appearance of the pipe organ must be as beautiful as the sound it makes. Locations for pipes are selected with both factors in mind. Sometimes dummy pipes that do not actually produce sound are installed strictly to improve the appearance of the instrument.

The amount of money that a client is willing to spend for a pipe organ has an important influence on the design, such as the number of pipes that will be installed. Often a client will consider designs submitted by several organ builders, and will select the one that best supplies the desired characteristics within a specified budget.

## The Manufacturing Process

### *Making the pipes*

1 Lumber arrives at the pipe organ manufacturer and is inspected for flaws. It is then stored for about six months to allow it to adjust to the local climate. This avoids having the wood split or crack after it is shaped into pipes. Precision woodworking equipment is used to cut the wood to the proper size and shape.

2 Some metals, such as copper, zinc, and various alloys, may arrive at the pipe

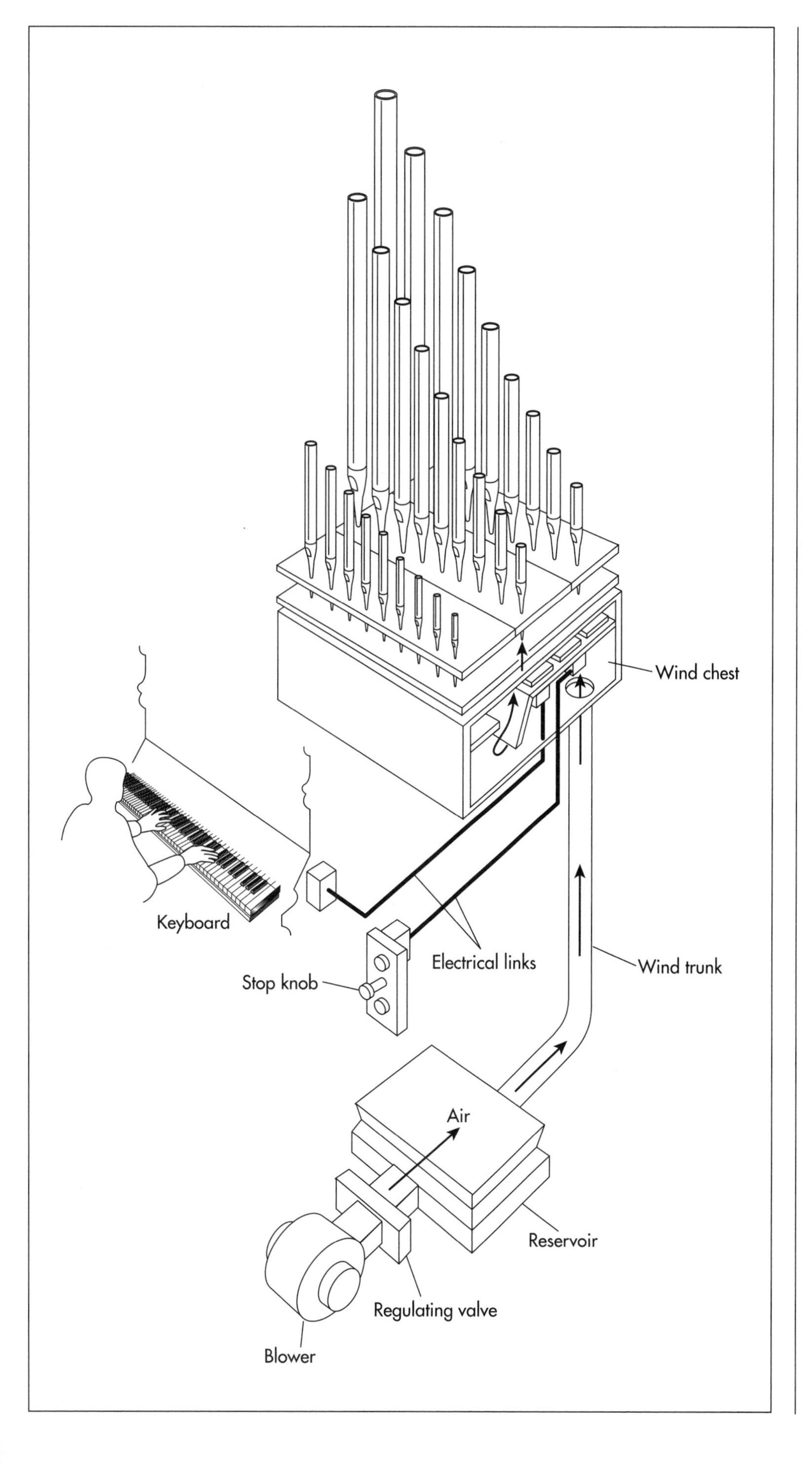


Powered by a rotary blower, a mechanical pipe organ creates sound by linking the console to the valves which control the flow of air to the pipes with cranks, rollers, and levers. The blower moves air through the wind trunk to the wind chest. Stop knobs open and close specific rows of pipes. When a row is open, the air can flow from the wind chest to the pipes as the keyboard is played, creating sound.

Pipe organs utilize two types of pipes. About four-fifths of the pipes in a typical pipe organ are flue pipes. A flue pipe consists of a hollow cylinder with an opening in the side of the pipe. The rest of the pipes are reed pipes. A reed pipe consists of a hollow cylinder, containing a vibrating strip of metal, connected to a hollow cone.

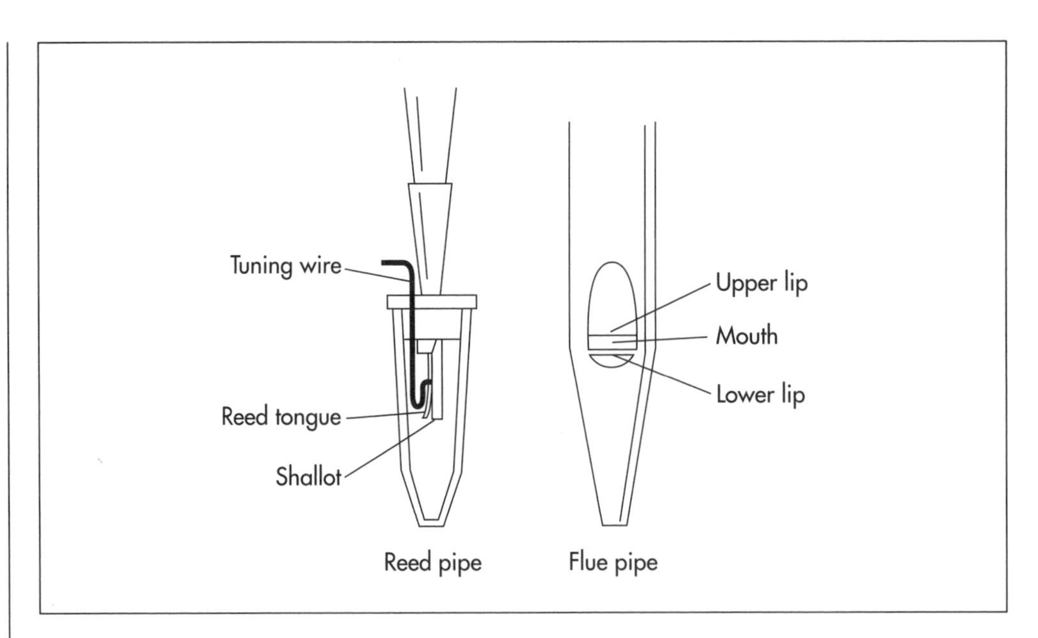


organ manufacturer in the form of sheets of the proper size and thickness. Sheets made of alloys of tin and lead are usually made by the pipe organ manufacturer. This is because small changes in the exact amounts of tin and lead present may cause large changes in the sound of the pipe. More tin tends to produce a brighter sound. More lead tends to produce a heavier sound.

3 Tin and lead are carefully weighed on accurate scales and mixed together in the desired amounts. The mixture is heated in an oven until it melts into a liquid. The molten alloy is poured into a large, shallow cooling tray. The metal cools into a sheet, which is removed from the tray. The exact amount of liquid that is poured into the tray determines the thickness of the sheet, which will affect the sound of the pipe.

4 The metal sheet is cut to the proper size. It is then bent around a wooden mandrel in the shape of the pipe being made. The sheet is hammered and rolled around the mandrel until it is shaped into a pipe. The seam, where one edge of the sheet meets the other edge, is smoothed and soldered. The pipe is then cut to the proper length.

5 For flue pipes, an opening of the proper size, shape, and location is cut into the side of the pipe. For reed pipes, a strip of brass is installed in the correct position. Small adjustments in the opening or the reed are often needed when the pipe organ is completed.

### *Assembling the organ*

6 Manufacturing the console, the action, and the wind chest is a long, slow, difficult process. Each of the thousands of wooden parts which make up a pipe organ is carefully carved using precision woodworking equipment, including joiners, sanders, and saws. Highly accurate drill presses are used to drill thousands of holes, each of which must be in the correct position to allow the pipe organ to operate. Wooden parts that will be visible are carefully polished smooth and given a protective, transparent finish, allowing the natural beauty of the wood to be seen. The keys, also carved from wood, are covered with layers of black or white plastic.

7 The various components, except the pipes, are assembled in order to test the action. A small pipe organ may be assembled where it is built, but most pipe organs must be assembled and tested in the place where they will be used. After the action is tested, the pipes are installed, tested, and adjusted.

## Quality Control

Raw materials are inspected before the process of building the pipe organ begins. Wood must be dry, evenly grained, and free from cracks or splits. Sheets of metal alloy must contain the proper amounts of each metal and must be of the proper thickness.

Constant visual inspection of all parts is necessary as they are built.

Before the pipes are installed, the console and action are tested to ensure that all mechanisms work properly. This procedure also blows dust out of the many holes drilled in the action that prevents it from causing problems later. Each pipe is tested for sound quality one at a time. The sound of each pipe is also compared to the sounds of its nearest neighbors. Then, an entire rank of pipes is tested and compared to neighboring ranks. Small adjustments are made as necessary, and the tests are repeated.

## The Future

Electronic organs will continue to reproduce the sound of pipe organs with greater accuracy. The most critical factor in this process will be improvements in sampling technology. Sampling involves converting sound into digital information, storing this information, then retrieving the information from memory and using it to reproduce the sound. Electronic organs will also make increasing use of Musical Instrument Digital Interface (MIDI) technology. MIDI technology allows electronic instruments of various kinds to work together and with computers. MIDI technology could allow electronic organs to reproduce the sound of almost any instrument, as well as producing sounds that have never been heard before.

On the other hand, many organ builders are showing more interest in creating instruments similar to those used prior to the nineteenth century. These pipe organs are better suited to playing music from the Baroque and Classical periods than organs built using designs developed during the Romantic period. Perhaps these two seemingly opposite trends will be combined to accurately reproduce sounds which have not been heard for hundreds of years.

## Where to Learn More

### Books

Sonnaillon, Bernard. *King of Instruments: A History of the Organ*. Rizzoli International Publications, 1985.

Williams, Peter. *A New History of the Organ: From the Greeks to the Present Day*. Indiana University Press, 1980.

### Periodicals

Webster, Donovan. "Pipe Dreams." *Smithsonian* (July 1997): 100-108.

### Other

"Pipe Organs and Music." http://www.orgel.com/home-e.html (September 3, 1998).

*—Rose Secrest*

# Pita Bread

*By the 1990s, the wholesale pita bread market in the United States was nearing $80 million in yearly sales.*

## Background

Nearly every civilization makes some type of bread. Prehistoric people of 10,000 years ago baked bread. The residents of Mesopotamia, what is now Iraq, were known to use stones to grind grain to which they added water and then cooked over an open fire.

Excavations of ancient Egyptian cities show that they grew wheat and barley and used them to bake flatbreads. It is believed that the Egyptians discovered leavened or raised bread accidentally when a mixture of grain and water was left in a warm place, releasing the naturally occurring yeast and producing a puffed-up dough.

Before a process for making yeast was developed, bakers would often set aside a piece of unbaked dough from each batch. By the time the next batch was made, the reserved dough had soured, or fermented, by airborne yeasts. It was then mixed with fresh dough to make it rise. In 1665, an enterprising baker thought to add brewer's yeast to his reserved dough.

At first, grain was manually ground by rubbing it between two stones. Then, a mechanical process was invented, in which a cattle-drive stone revolved on top of a lower, perpendicular, stationary stone. In time, the cattle were replaced by water mills or windmills. By the late eighteenth century, a Swiss miller had invented a steel roller mechanism that greatly simplified the grinding process.

Commercial bakeries first appeared in the Middle Ages, as towns and villages were established. In addition to baking bread for sale, these bakeries would set aside time for people who still wanted to mix their own dough and then bake it in the commercial ovens. These were large brick ovens heated by wood or coal. The loaves were moved in and out of the ovens with a long-handled wooden shovel called a peel.

It may have been the Bedouins who first made pita bread. After a long day in the sun, traversing the desert, they made camp and prepared a modest respite. Powdered grain was mixed with water to make dough which was formed into flat round loaves. The loaves were placed over the bottom of the mixing vessel and baked over an open fire. This bread was used as a utensil, as well as for food.

In remote Arab villages, bread is still baked in backyard stoves. Some Arab and Israeli communities have community ovens or bakeries that set aside special hours for families to bring in their homemade loaves.

When Middle Eastern immigrants began moving to the United States in large numbers in the 1970s, they introduced Americans to their cuisine. Pita bread became a popular bread choice, especially because the absence of shortening and the small amounts of sugar make it a low-fat food. By the 1990s, the wholesale pita bread market was nearing $80 million in yearly sales. Most of the pita is baked by specialty bakeries in the East, West and Midwest. Commerical pitas are typically baked with unbleached all-purpose flour or whole wheat flour. They range in size from 4-10 in (10.16-25.4 cm) in diameter.

## Raw Materials

Pita bread is made with grain flour, water, salt, and bakers' yeast. Harvested grain is

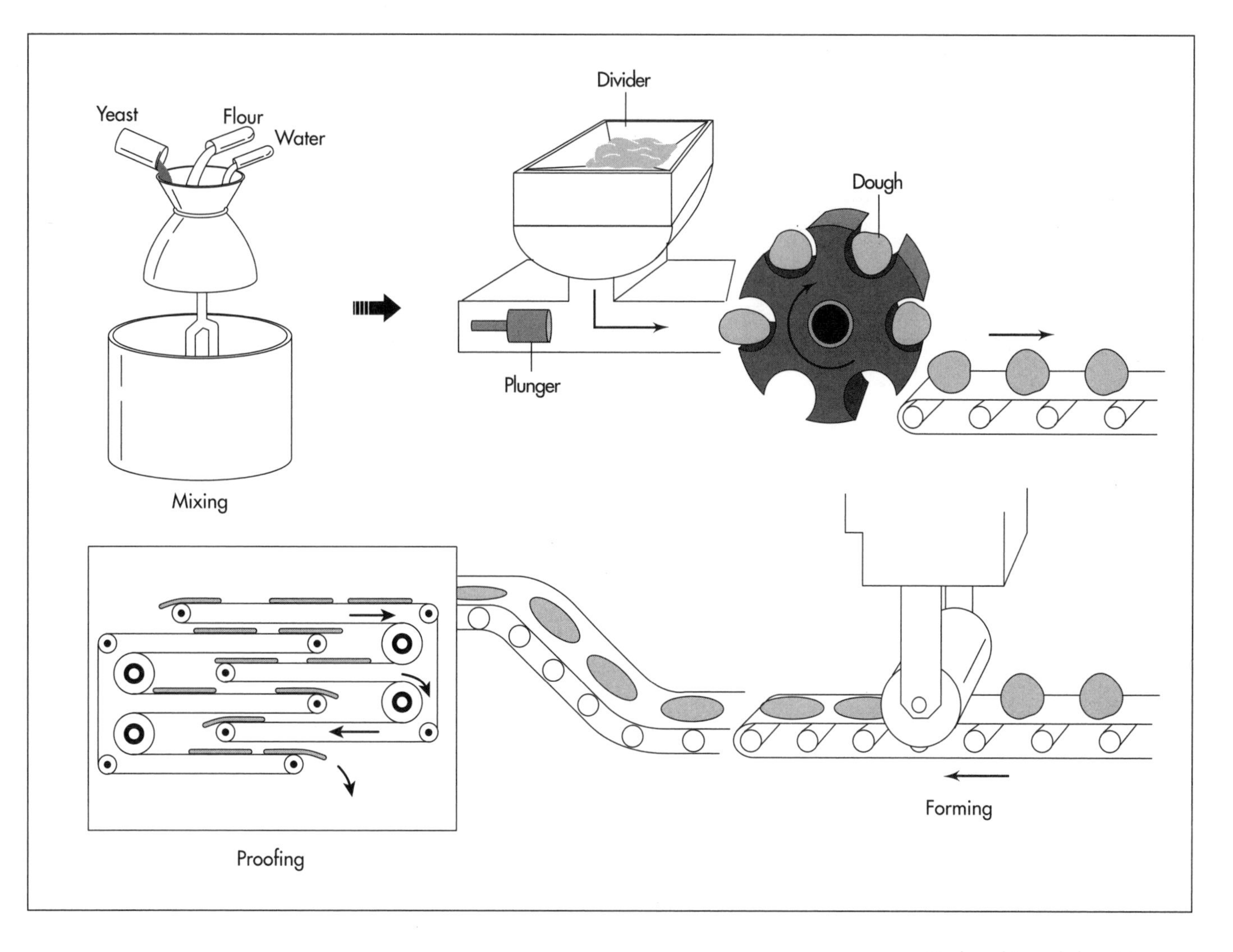


Pita bread is made by mixing premeasured amounts of flour, salt, water, and yeast. Once kneaded, the dough is fed through an extruder that forms the dough into tennis-ball sized portions. Next, the dough balls are allowed to rest and rise in a process called proofing. The dough balls pass under a series of rollers that press the dough into the desired-sized circles. The flattened dough is then passed under die-cuts that create circular pieces.

ground according to the type of bread being made. Grains are composed of three parts: bran (the hard outer layer), germ (the reproductive component), and endosperm (the soft inner core). Whole wheat flour is made from the grinding of all three parts. White flour is made solely from the endosperm. Because the nutrient-rich bran and germ have been removed in the processing of white flour, vitamins and minerals are often added. Grain mills grind the flour and then sell it in bulk to commercial bakeries. The bakeries store the flour in 100,000 lb (45,400 kg) bulk silos until ready for use.

Yeast is a single-celled fungus with enzymes that extract oxygen from starches or sugars present in food. This causes fermentation and leavening (rising). In commercial production, yeast strains are fed a solution of molasses, mineral salts and ammonia. After the fungus completes growth, the yeast is separated from the solution, washed and packaged. It is either combined with starch and compressed into cakes, or ground into powdered form and mixed with cornmeal. Bakeries purchase bakers' yeast in bulk from outside suppliers.

The water used must be of the purest quality, not only because it will be used for human consumption, but also because the hardness and pH affect the properties of the dough. Most processors filter the water so that it is of medium hardness (50-100 parts per million) with a neutral pH.

## The Manufacturing Process

### *Mixing the dough*

1 Premeasured amounts of flour, salt, water, and yeast are blended in commercial mixers in several hundred pound batches. Some bakeries may add a mold retardant

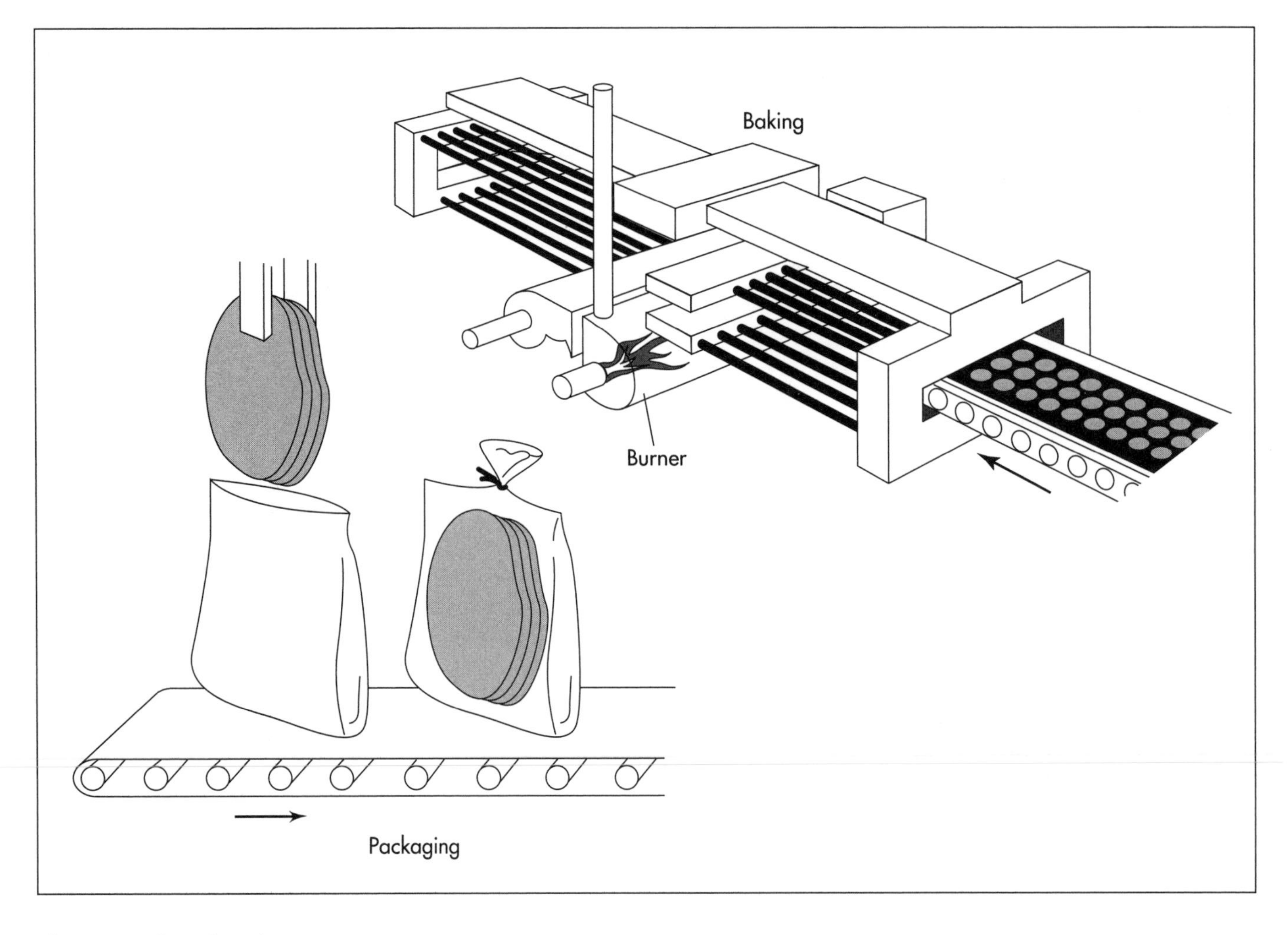


After a second proofing, the circle-shaped dough is baked quickly at a very high temperature so that the upper and lower crusts separate, forming a pocket.

such as calcium propionate. Large motorized arms in the mixer knead the dough to the desired elasticity.

### Extruding the dough

2 The dough is scooped out of the mixing bowl and fed into an extruder that forms the dough into tennis-ball sized portions. Each ball is then dropped into cups moving on a conveyor belt.

### First proofing

3 The dough balls are allowed to rest and rise in the cups for approximately 15 minutes. This process is called proofing. In some processing plants, the dough is allowed to proof in one continuous layer before it is cut into individual portions.

### Cutting and forming the pitas

4 The dough balls are turned out onto a sheeter that travels on a linear conveyer belt. The sheeter passes under a series of rollers that press the dough into the desired-sized circles. If the dough is still in one layer, the rollers press it to a thickness of about 0.125 in (0.3175 cm). The flattened dough is then passed under die-cuts that create circular pieces. The excess dough, about 10%, is recycled back into the extruder.

### Second proofing

5 The circular loaves move into the top shelf of a rotating proofer. As they slowly move down to the bottom of the proofer, the rises again. They exit the proofer and are conveyed into the oven.

### Baking the pitas

6 The ovens are kept at a very high temperature, between 800-900° F (426.6-482.2° C). The pita loaves move quickly through the ovens where they are exposed to the high heat for about one minute. The combination of the high heat and flash-baking causes the water in the dough to turn to

steam thus forming the pita pocket by separating the upper and lower crusts.

### Cooling and flattening the loaves

7 The baked pita loaves moves out of the oven and back and forth on a system of conveyers belt for about 20 minutes until they are cooled. Plant workers then manually flatten the puffed-up loaves. Burnt or undercooked loaves are discarded.

### Slicing the loaves

8 If the pita loaves are to be sliced in half, they are conveyed to slicing machines where rotating knife blades quickly slice the bread. The loaves can be cut individually or stacked in piles about six high and sliced.

### Packaging

9 Whether sliced or left whole, the pita are conveyed to the packaging area where they are stacked in a pre-determined amount and inserted into pre-printed plastic bags. Workers may close the bags manually with twist ties. Alternately, the bag openings may be fitted with a zipper tear-strip, in which case the bag is mechanically heat-sealed. The packaged pitas are loaded onto trays or into cartons for shipment. If the pitas are not going to be shipped immediately, they are flash-frozen and kept in industrial freezers that are regulated to a constant temperature of about 10° F (-12.2° C).

## Quality Control

As a foodstuff, pita bread is subject to stringent government food processing regulations, including, but not limited to the percent of additives allowed, sterilization of plant equipment, and cleanliness of plant workers. In addition to adhering to these regulations, processors control the quality of their products to meet consumer expectations by installing checkpoints are various stages of the processing. At each inspection station, the pita are tested for appearance, texture, and taste.

Because of its high moisture content, 38-40%, bread is particularly subject to bacteria growth. While the baking process destroys most of the bacteria, bread is still susceptible to re-inoculation of fungi after packages. There are a number of methods used to combat this including fungicides and ultraviolet lighting.

Labeling regulations stipulate the plant list baking date, ingredients and weight on packaging. If the pita bread is marketed as an organic product, its processing must adhere to the Organic Foods Production Act that enumerates various requirements. Those that pertain to bread processing include prohibitions against treating seeds with prohibited materials during the growing season and strict rules for commodities grown with fungi, such as yeast.

## Where to Learn More


### Books

Habeeb, Virginia T. *Pita the Great.* New York: Workman Publishing, 1986.

### Periodicals

Pacyniak, Bernard. "Kangaroo packs a wallop." *Bakery Production and Marketing* (April 24, 1990): 60.

"Pita sales zip along." *Packaging Digest* (September 1990): 42.

Sobel, Dava. "The Upper Crust." *Health* (December 1986): 45.

### Other

Leon International, division of Middle East Baking Co. http://leon-intl.com/pitabread.htm.


*—Mary McNulty*

# Plastic Doll

*Today, dolls of all materials represent a market valued at over $2 billion in shipments of the $15+ billion U.S. toy industry.*

## Background

Dolls have evolved over the centuries from religious symbols or idols in ceremonies to playthings by children, and are now also highly-prized collectibles. Doll collecting has become the second largest adult hobby in the United States, and many collectibles are made of plastic. One popular doll company, Alexander Doll Company, sells more than 500,000 dolls every year to both children and collectors.

Some of the earliest dolls date from around 2000 B.C. and were produced in Egypt. Made of wood into a simple paddle shape, these dolls were probably used in fertility rites. Other miniature figures were used in tombs. Wooden and clay dolls were also made by the early Greeks. Many of these ancient dolls were believed to have magical powers.

Doll-making was also an industry in the Roman era, and materials included wood, cloth, bone, and terra cotta. During this period, dolls were dedicated at puberty to the gods of Mercury, Jupiter, and Diana. Dolls were also often buried with their owners, especially since many died young. Other early cultures—including those found in India, Japan, and North America—used dolls to teach children their traditions. Materials included cloth, clay, buckskin, and corn husks. Dolls made from corn were often used in harvest celebrations.

As civilizations evolved, so did the doll, which became much more elaborate for those who could afford it. Wealthy children in Europe were given dolls of wax, wood, and composition during the late Medieval period (fourteenth century) and doll making became a commercial entity. Such dolls retained their popularity until the end of the nineteenth century. By the fifteenth century, Germany had established itself as an important center of doll-making, with France following suit. The making of luxury dolls continued into the sixteenth century with Paris an important center. Also during this period, the number of available doll accessories increased.

The next century continued to see dolls being played by both ordinary children and those born into nobility. Wax and wood were still popular, and *papier mache* was becoming more common. The doll by the end of this century became an accepted article of commerce, even as far as the colonies of America. American settlers also began to produce dolls of their own, usually made from wood.

Other materials became popular during the eighteenth century. Plaster of paris, which was easy to cast, was used despite it being so fragile. In Germany, hard paste porcelain was first made and was subsequently used to make heads. A common tradition during this period was for ladies to have wax dolls made as portraits of themselves and dressed in fashionable clothes. Another method of studying and examining costume that became popular was the use of paper figures. Paper was cheap and readily available; dresses were painted to fit the figure. Flat paper or card dolls were sold in sheet form in Europe at the end of the century.

During the nineteenth century, doll heads and dolls made of china were introduced and doll makers began looking for an unbreakable substance. Gutta percha, a rubber like substance from Mayala, came into regular use around 1840 in Europe. When the American firm Goodyear Rubber Company invented the vulcanizing process for hard rubber in 1851, this material was adapted by many doll makers in the United States and was still being used a hundred years later. Metal and celluloid were two other materials that were also used during this time in various countries.

Composition was still popular during the early twentieth century, and it was not until several years after World War II that the first true plastic doll was made in East Germany. Since then, plastic dolls abound in every toy store, with the most popular one named Barbie (with annual retail sales of almost $1 billion). Since 1959, more than one billion Barbie dolls and members of her family have been sold in more than 140 countries around the world. Every second, two Barbie dolls are sold somewhere in the world.

Today, dolls of all materials represent a market valued at over $2 billion in shipments of the $15+ billion U.S. toy industry. This represents a retail market of approximately $23 billion, a third of the global toy market.

## Design

Many toy designers utilize information from various sources, including parents, psychologists, educators, and other child-development specialists. This background provides valuable clues as to what consumers are looking for when they purchase toys, how children learn through play, and when youngsters are physically and cognitively ready for certain types of toys. Toys are also frequently tested by the children themselves in focus groups or at home to determine durability, age-appropriateness, play patterns, and marketability. Some toy manufacturers maintain in-house, year-round child care facilities for this purpose, while others establish relationships with universities and other research sites. A detailed evaluation of a product's safety is made upon completion of the very first prototype and updated as the toy nears production.

Doll manufacturers usually design a doll from the original concept to the finished product, which includes packaging and clothes. First, a simple sketch or drawing is made, followed by a color illustration. Next, a model is sculptured from wax. Once the design is approved, a master mold is made that is used to manufacture production molds.

## Raw Materials

Most plastic dolls are made from vinyl, otherwise known as polyvinyl chloride (PVC). Major resin producers supply PVC compounds, which is the world's second largest selling thermoplastic behind polyethylene. The basic building block of PVC is vinyl chloride, which is converted to PVC by a suspension process. All PVC must be compounded prior to use. Rigid compounds consist mostly of resin (85-90%), whereas flexible PVC contains 40-60% resin. Other additives include plasticizers, stabilizers, processing aids, lubricants, pigments, and fillers.

Some doll bodies are also made of polyethylene—a derivative of ethylene and a colorless, flammable gas. This gas is subjected to elevated temperatures and pressures in the presence of a catalyst, which converts the gas into a polymer. Other raw materials used in doll making include various paints to make facial features, nylon for the hair, and cloth and thread for the outfits.

## The Manufacturing Process

Two major plastic forming processes are used to make doll body parts.The heads and limbs are made by a process called rotational molding. Rotational molding is used for producing hollow, seamless products of all sizes and shapes with uniform wall thickness. Blow molding is sometimes used to make the torso if cost is an issue since it is a faster, more economical method.

### *Raw material preparation*

1 A separate compounding operation is required to convert the form of a resin, while also introducing any additives, into one suitable for the molding processes. Usually this step is done at the plastic manufacturer, though sometimes it is performed by the doll manufacturer if a special formulation is required.

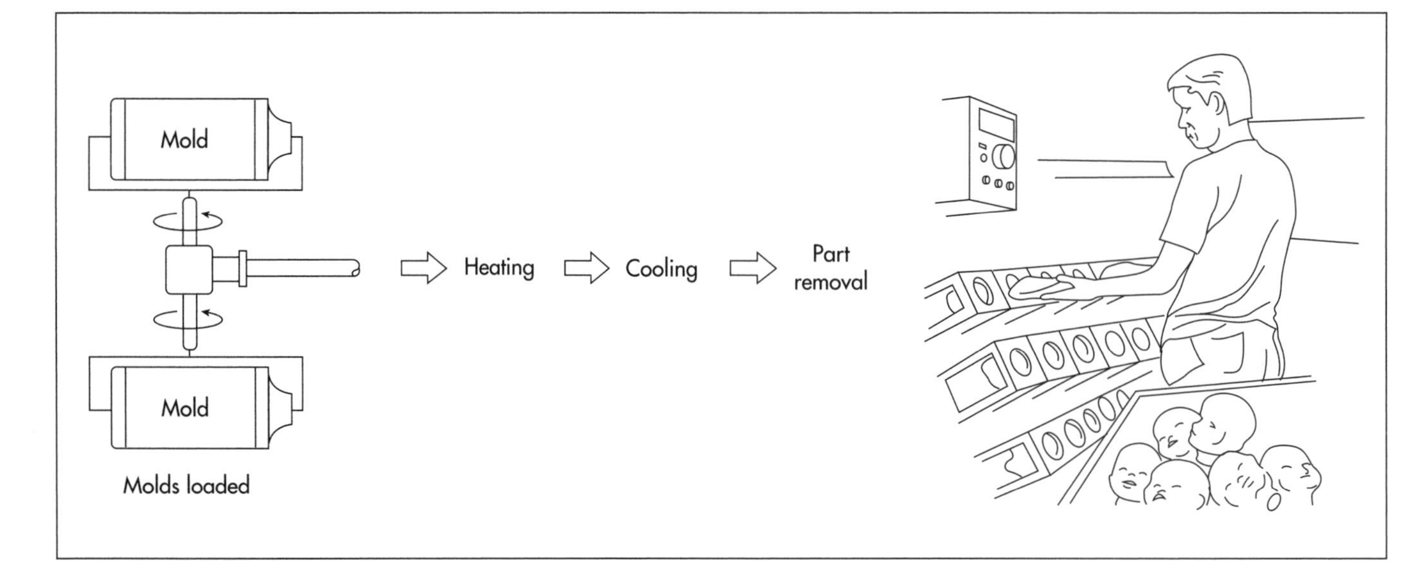


Plastic doll heads and limbs are made by a process called rotational molding. First, the mold cavities are filled, closed, and heated. While they are heating, the molds rotate biaxially in order to create a hollow, seamless shape.

### Molding

2 During rotational molding, the mold cavities are first filled with a predetermined amount of the compound, in liquid form. Each metal mold consists of multiple cavities and the quantity depends on the size of doll. For instance, as many as 60 heads can be made at once. After the molds are closed, they are placed in a heated oven and rotated biaxially. During the heating cycle, the resin melts, fuses, and then densifies into the shape of the mold cavity.

3 Next, the molds are slowly cooled inside a chamber using air and water. Once cooled, the molds are removed from the chamber, opened, and the finished part is removed.

4 During blow molding, a hollow tube called a parison is first formed out of molten plastic by extrusion through a tubular reservoir. This tube is then placed between two halves of a steel mold and forced to assume the shape of the mold cavity by use of air pressure. The air pressure, ranging from 80-120 psi, is introduced through the inside of the tube, forcing the plastic against the surface of the mold.

5 After either molding process, the part is trimmed by hand to remove the flange.

### Decorating the head

6 Before all the body parts are assembled to make the doll, the facial features and hair are applied. First, the eyes are inserted and then cheeks, lips, and sometimes lashes are spray painted using an air brushing method. This process can involve up to 15 steps.

7 If the hair isn't molded as part of the head, nylon is rooted into the vinyl using a special sewing machine that is operated by hand. The hair is then carefully trimmed, combed, and set. The head can now be attached to the body, along with the limbs.

### Dressing and packaging

8 The next step is to dress the doll and trim any hanging threads. Any special labels or tags are attached before packaging. Most dolls are packaged in boxes by hand since they often must be posed. Each doll is given a final check before the box is covered. The boxes are then packed in shipping cartons.

## Quality Control

In addition to spot checking the doll parts for defects during the molding process, plastic dolls must meet specific toy safety regulations first established by the United States Consumer Product Safety Commission (CPSC). These include requirements and test criteria for paint and other similar surface-coating materials; sharp edges and points; small parts which could be swallowed or inhaled; use-and-abuse testing; hazardous substances; flammability and toxicity; among others. All raw materials are tested by the supplier and are shipped with property data

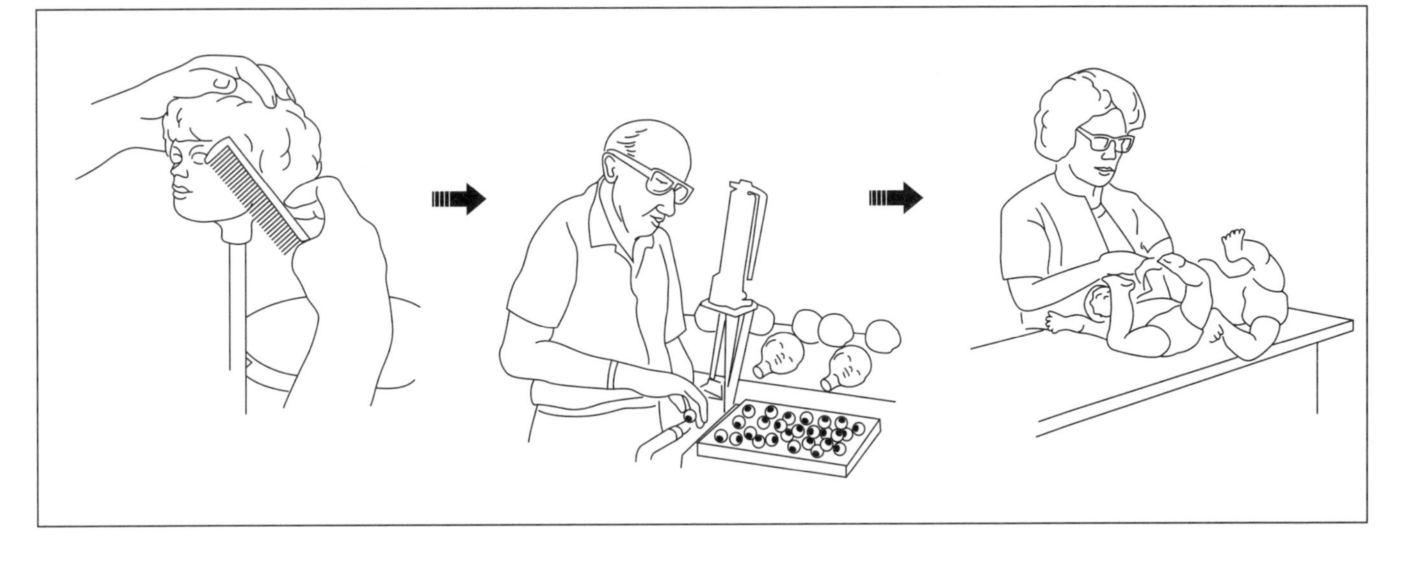

Before final assembly, the facial features and hair are applied to the doll head. If the hair isn't molded as part of the head, nylon is rooted into the vinyl using a special sewing machine that is operated by hand. The hair is then carefully trimmed, combed, and set.

sheets. Before the doll is packaged it is sent through a metal detector to make sure no metal contaminants are present.

The CPSC regulations are incorporated by reference in the toy industry's voluntary standard, ASTM F963, published by the American Society for Testing and Materials. ASTM F963 and federal regulations include more than 100 separate tests and design specifications to reduce or eliminate hazards that have the potential to cause injury under conditions of normal use or reasonably foreseeable abuse. These tests are conducted in-house or by independent testing laboratories.

Similar safety standards exist in Japan and Europe. An international standard is also under development that will serve as a set of requirements toward which all other countries can begin to harmonize.

## Byproducts/Waste

Any excess material produced during the molding process is usually discarded, since it cannot be recycled back into the process. The composition of the virgin material would be affected that would change the quality and color of the doll. Sometimes, the waste material is recycled for other purposes.

## The Future

The Internet and other computer technologies are having a profound effect on the toy industry and will continue to do so. Today, 40% of Americans aged nine to 14 have access to a computer and 22% actively surf the net, displacing some of the time that might otherwise be devoted to traditional play. Video games (now a market of almost $3 billion wholesale) and PC software also continue to gain in popularity, with double digit growth expected. However, the traditional toy market in the United States will grow by only a few percent and the basic doll market has shrunk from three to eight year olds to three- to five-year-olds. Overall, the toy and video game industry will grow by between 3-5%.

Consumers can expect to see many more examples of the use of technology in toys or to enhance the traditional version. For instance, the Barbie "Dress Designer" CD-ROM is an excellent example of creating new play dimensions for Barbie doll owners. As the cost of interactive microchip components continues to go down, more toy manufacturers will be able to incorporate interactive technology into toys at an affordable price for the consumer.

Though the basic doll market has become smaller, it is not expected to totally disappear. The demand is increasing for dolls that have more functions and are interactive. In the near future, plastic dolls will no doubt follow in the footsteps of the popular toy Furbie.

## Where to Learn More

### *Books*

King, Eileen. *The Collector's History of Dolls.* St. Martin's Press Inc., 1978.

### Periodicals

Vargas, Alexia. "A Pretty Population Explosion Sets Stage for Doll Wars." *Wall Street Journal* (December 16, 1997).

### Other

The Toy Manufacturers of America, Inc. 1115 Broadway, Suite 400, New York, NY 10010. http://www.toy-tma.com/ (June 29, 1999).

Uneeda Doll Co., Ltd. 200 Fifth Avenue, Ste. 556, New York, NY 10010. (212) 675-3313. Fax: (212) 929-6494.


*—Laurel Sheppard*

# Popcorn

## Background

Before about 1912, less than 19,000 acres (7,700 hectares) of farmland were dedicated to growing popcorn, but the electric popcorn machine and the microwave increased the demand for "prairie gold." Today, annual consumption of popcorn in America exceeds 1 billion lb (0.45 billion kg) or 71 quarts (67 liters) per person per year. The states of Indiana, Iowa, Illinois, and Ohio lead the field. Of the volume grown in the United States, 10% is used for seed and sold outside the United States; 30% is sold at ball games, movies, fairs, and circuses; and 60% is consumed in the home.

## History

Corn may have begun its long evolution as a kind of grass. In the Americas, corn varieties, including popcorn, were cultivated by the Aztecs and Mayans in Central America and Mexico and by the Incas in South America. The Aztecs decorated their Gods of Rain and Maize with strings of popcorn. North American Indians also strung the popped kernels on grass strings and used them for decorations and personal adornment. Archaeologists have found popped corn in dwelling caves in New Mexico, and the corn is estimated to be 5,600 years old. Scientists' best guesses for the age of popcorn and the place where it originated are 8,000 years and in Mexico. Curiously, popcorn was also common in parts of India, China, and Sumatra before the discovery of the Americas, but the paths and methods of its migration are unknown, as is the reason for its existence in these areas but not others. Part of the answer may be the hardiness of this type of corn over others or the change in climate conditions around the world over thousands of years.

Popcorn officially crossed into Western culture at the first Thanksgiving celebration. Popular legend has it that Quadequina, brother of the Indian chief Massosoit, brought a deerskin bag full of popped corn to that harvest celebration. The Indians' methods for popping corn varied from tribe to tribe. They probably discovered how to pop popcorn by accident because the hard kernel doesn't give any hint of the potential treat inside. The earliest poppers of corn may have thrown it into the fire and eaten the kernels when they popped and flew out of the flames. Our only historical evidence of early but more sophisticated popping methods is from the Incas whose ruins contain specially shaped clay pots with kernels of popped corn still inside them. The Incas apparently heated sand and placed it in these pots, then placed the corn on the sand. The pot was covered, and heat from the sand popped the kernels. The heavier sand stayed at the bottom of the pot, and the popped kernels rose above it where they could be reached.

Over 700 types of popcorn were being grown in the Americas by the time Columbus discovered these continents. French explorers in 1612 saw the Iroquois people popping corn in clay pots; and the Winnebago Indians who lived near the Great Lakes simply drove sticks into the cobs and held the cobs near the fire. Popcorn soup was a favorite method of using the grain among the Iroquois, and the Indians of Central America even made popcorn beer. Early explorers observed ornamental necklaces, bouquets, and headdresses made of popcorn.

*Annual consumption of popcorn in America exceeds 1 billion lb (0.45 billion kg) or 71 quarts (67 liters) per person per year.*

In early America, popcorn became a ritual part of many festivities including quilting bees and barn raisings. In cabins and homesteads, corn could be popped in the fireplace, seasoned with grease or butter, and shared by the family. Popped kernels were used as teeth in Halloween pumpkins and strung in long ropes to festoon Christmas trees. Popcorn was the accompaniment to banjo playing, singing, and the telling of ghost stories and folktales. In the 1700s, the first puffed cereal was created by pouring milk and sugar over popped corn; this breakfast dish was popular from Boston south to the Carolinas.

Popcorn was grown in family gardens or farms or bought from neighbors who grew more than they needed until about 1890 when it started to become recognized as a legitimate cash crop. The first automatic popcorn popper was a steam-powered machine invented by Charlie Cretors in 1885; before Cretors' invention, street vendors popped corn in wire baskets over open fires. By about 1890, the glass-sided popcorn machine with its gasoline burner became a popular feature of the circus, carnival, sideshow, local fair, and small town streets where popcorn vendors would sell bags of popcorn as dusk fell. The packaging of popcorn for use at home began in about 1914.

In 1893, Fred and Louis Rueckheim used the Chicago World's Fair to kick off their blend of popcorn, peanuts, and molasses. These German brothers made their name in America by manufacturing Cracker Jack, as this mixture came to be called, in a small kitchen and then at the World's Fair. In order to claim a prize, the consumer could mail in a coupon found in every box of Cracker Jack. After the Fair and until World War II, prizes were actually packed in the boxes, although this practice stopped during the War because the prizes were made in Japan. After the War, a bonus prize returned to every box.

When moving pictures became the rage and movie houses opened across the country, the street vendors of popcorn would rent space outside the theaters and sell bags of popcorn to movie ticket buyers. In 1925, Charles T. Manley perfected his electric popcorn machine, and popcorn vendors moved inside the theater where the trapped sounds and smells of popping corn often made more money than the feature film. During the Great Depression in the 1930s, vendors sold popcorn in five-cent bags, and popcorn became one of few affordable luxuries. Meanwhile, back in the theater, the paper bucket replaced the bag as the container for popcorn because the rustling bags made too much noise.

During World War II, popcorn was taken overseas as a treat for American servicemen and was adopted by other countries. In 1945, Percy Spencer applied microwave energy to popcorn and found that it popped; his discovery led to experiments with other foods and development of the microwave oven. Television brought popcorn into the home in the 1950s, when electric popcorn poppers and pre-packed corn for popping were developed and marketed. The 1970s and 1980s witnessed a boom in electric poppers, hot-air poppers, and microwave popcorn as the videotape industry brought movies and the desire for all the customs associated with movie-going into the home.

## Raw Materials

Selection of the best variety or hybrid of popcorn to be grown and processed for the kind of popcorn to be sold is critical to the raw materials comprising popcorn. In some forms of popcorn, the corn itself is the only raw material. For other methods of marketing popcorn such as microwave popcorn, soybean oil, salt, and flavoring are also needed.

### *Popcorn varieties and hybrids*

There are several commercial classifications of corn. Field corn (also called dent corn or cow corn) is fed to animals. Flour corn is mostly starchy center with a soft hull that allows it to be easily ground into flour. Sweet corn is the kind we eat at the dinner table. Flint corn is usually called Indian corn; its colorful kernels make it highly attractive, and it is used for decoration because it is tough and tasteless. Pod corn is also only used for decoration because each of its kernels has its own separate husk.

Popcorn, also a collection of varieties of *Zea mays*, is the only corn that pops; it is not dried kernels of sweet corn. There are sever-

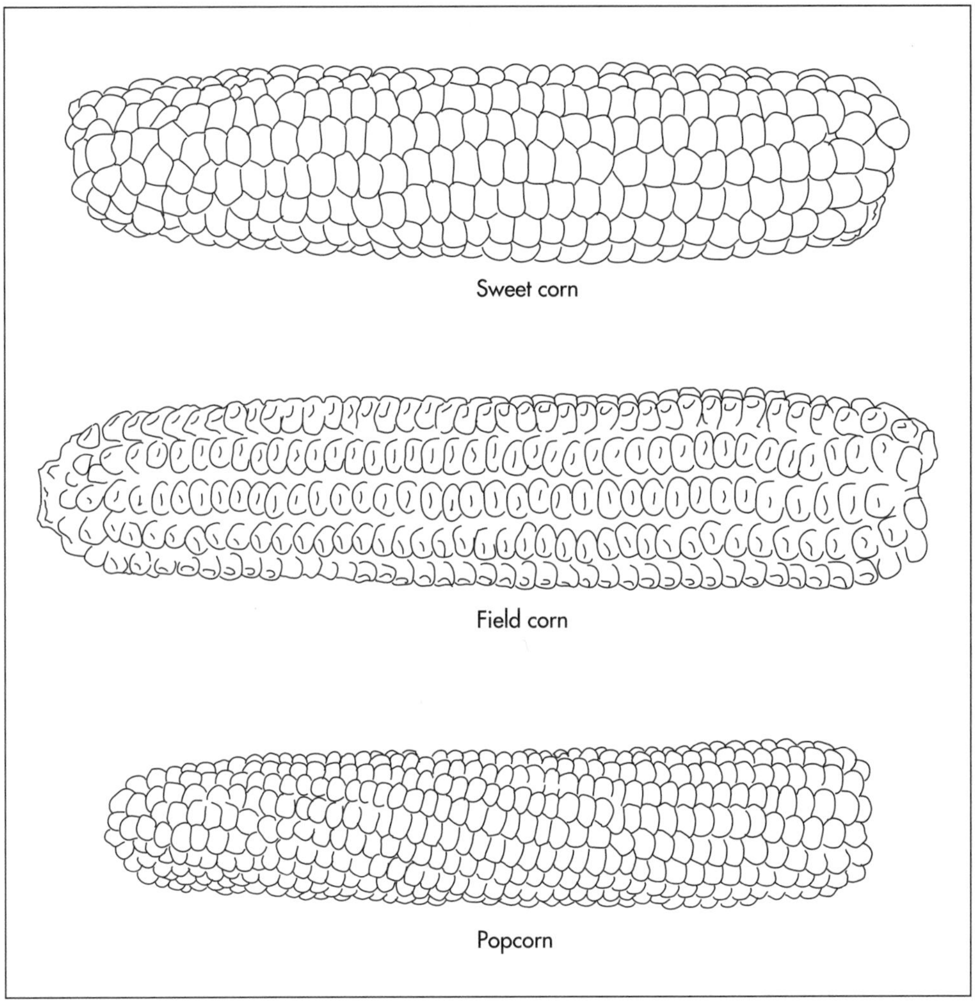

Sweet corn

Field corn

Popcorn

There are several commercial classifications of corn. Field corn (also called dent corn or cow corn) is fed to animals. Sweet corn is the kind we eat at the dinner table. Popcorn, also a collection of varieties of *Zea mays*, is the only corn that pops; it is not dried kernels of sweet corn.

al popular varieties of popcorn out of thousands of hybrids. White hull-less and yellow hull-less are the varieties sold most commonly and packaged in microwave bags. Rice popcorn is a variety with kernels that are pointed at both ends, and pearl popcorn produces round, compact kernels. Tiny red ears that are shaped like strawberries produce red kernels and are called strawberry popcorn. Black popcorn has black grains but pops as white kernels, and rainbow or calico corn has white, yellow, red, and blue kernels. Popcorn is also classified by the characteristics of its popped kernels, with the largest kernels called "Dynamite" and "Snow Puff."

The business of developing new hybrids and cultivating known, productive hybrids is key to the creation of popcorn. A hybrid is made by fertilizing one kind of popcorn plant with the pollen from another kind. The result is a seed that has characteristics of both plants. A major popcorn producer like Orville Redenbacher Popping Corn Company employs a team of scientists to pollinate its hybrid corn by hand. The kernels that are grown are used as seed to grow the popcorn that will be harvested and sold. As many as 30,000 new hybrids per year are created to try to improve the popcorn product. Producers also work with universities to develop ideal hybrids; millions of dollars are invested annually in this research.

Smaller growers like Snappy Popcorn rely on hybrids that are best suited to their location, climate, and type of product. When the hybrid is well matched to geography, it produces a greater yield. Hybrids are also chosen based on resistance to disease and damage from insects, stalk strength, how easily they grow, and how easily they can be pulled out of the ground. Types of kernels are important, so hybrids are chosen specifically to produce carmel corn, microwave popcorn,

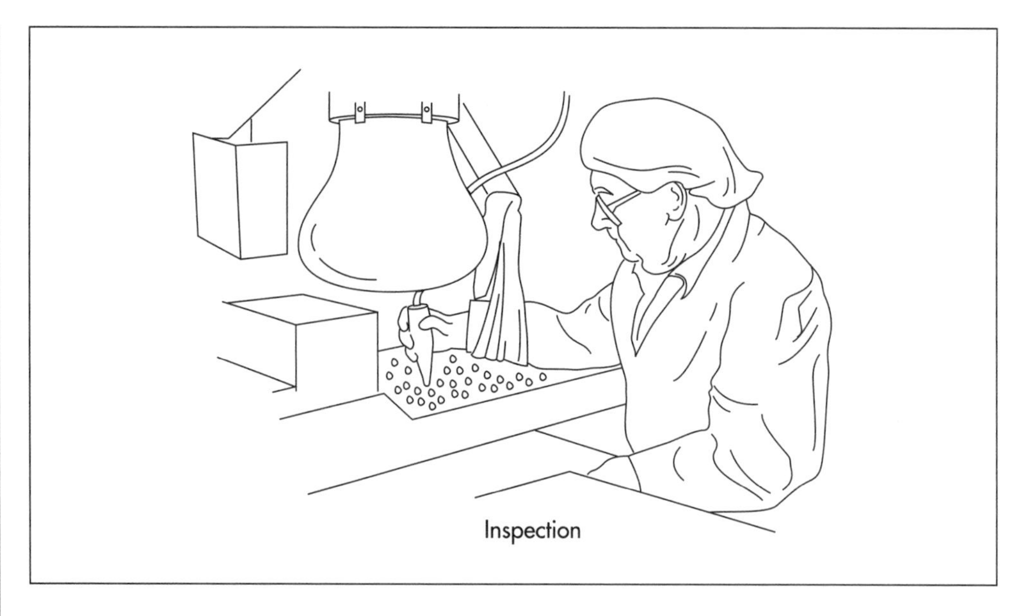


As a final step in the manufacturing process, quality-control inspectors observe the kernels as they move along a conveyor belt and suck out poor-quality kernels with a vacuum hose.

and movie theater popcorn. Movie theaters are interested in selling the greatest volume for the smallest investment, so high-expansion kernels are chosen for this market.

### *Popping methods*

Part of the "design" of popcorn is the method used to pop it. The dry method consists of putting the unpopped grain in a basket or wire cage, agitating it over a heat source like the campfire or coal stove, allowing the corn to pop, and seasoning it with butter and salt. In the wet-pop method, corn is placed in a container with a solid bottom. Oil is added (either before the corn or poured on top), and the oil helps to distribute the heat and cause more even and complete popping. Commercial popping machines use the wet-pop method, and coconut oil is used for its aroma and lightness. Microwave popcorn also uses the wet-pop method, although the moisture is present in a solidified form of oil, flavoring, and salt that melts when the microwaving process begins.

## The Manufacturing Process

### *Cultivation*

1 Popcorn grows best in rich soil. It is planted in checkrows, rows that intersect at right (90-degree) angles, so that it can be harvested by machine. Hybrid forms of popcorn have been perfected to produce the most grains per ear of corn, flavorful kernels, the correct internal moisture to insure that most of the corn pops, and other market-friendly characteristics. When the ears are ripe, the corn is harvested with either a picker that removes the ears and leaves the stalks temporarily or with a combine that crushes the corn stalks, mechanically removes the ears, and husks the corn. Combines tend to do more damage to the ears of corn. The ears are collected in the field in bins or boxes and moved into steel cribs using mechanical elevators or conveyors.

2 The ears are dried in cribs that are narrow and have open slots to minimize the time needed to dry them. A crib can be up to three stories high, as long as a city block, and with a capacity of up to 4 million lb (1.8 million kg) of corn. The ears are stored for eight to 12 months to allow them to dry, or in an alternative method, hot air is forced up into the cribs through holes in the bottoms of them to reduce the natural drying time. While in the cribs, the corn is carefully tended until the kernels reach a moisture content of 12.5-13.5% moisture, which is ideal for popping characteristics.

### *In the factory*

3 The dried ears of popcorn are then transferred by conveyor belt to the factory and a machine called a scalper. The scalper strips the kernels from the cobs. Simultaneously, a cleaner and de-stoner sort out the shuckings and any dirt or particles by passing it through a series of screens to separate

the kernels. They are cleaned and polished in another machine equipped with metal brushes that remove the chaff (sometimes called bee's wings). A gravity separator is then used to separate good kernels from bad; the kernels that have matured properly are lighter in weight, so the bad kernels drop through the bottom of the separator and are recycled for use as seed. The kernels near the two ends of the cob also tend to be either too small or too large to pop properly, and the gravity separator removes them as well.

4 Finally, in the portion of the factory called the fanning mill, fans blow dust and other fine material off the kernels, and the kernels are treated with a natural, inert fumigant to eliminate insects. Most manufacturers avoid pesticides altogether during the winter months when bugs are less common, and all must comply with government regulations regarding their use. Now completely processed, the popcorn kernels travel toward storage bins on a conveyor belt; quality-control personnel watch the passing flow and vacuum up bad kernels that may have escaped the previous sortings.

5 Types of popcorn with no other additives go directly to holding bins to await packaging. For microwave popcorn, measured amounts of salt, soybean oil, flavoring, and popcorn are pumped or dropped into the microwave bags. The bags are not vacuum-sealed, but they are air tight to prevent moisture in the air from affecting the contents.

6 In the packaging area, popcorn is conveyed from the holding bins to packing machines where it is placed in bags and then boxed for storage or shipment. Usually, the factory will bag a particular type of quantity, say 5 lb (2.27 kg) bags, until it has met its orders plus some for storage. Then the packing line is changed to accommodate different bags and quantities of popcorn.

## Quality Control

Quality control practices are essential in the field and factory. The process of pollinating the ears of corn correctly is essential to the production of any popcorn at all. In the factory, the cleaning processes are carefully monitored, and the series of screens and other devices are chosen to remove all stray materials and unwanted kernels. Even magnets are used to pull out bits of metal that may have been introduced by the farm machinery or storage bins. Finally, a team of quality-control inspectors simply observes the kernels as they move along a conveyor belt and removes poor-quality kernels with a vacuum hose.

## Byproducts/Waste

Cobs, husks, and stalks are sold for use as feed for cattle and other animals, so very little waste remains from popcorn cultivation and processing.

## The Future

Popcorn's future was assured in the 1980s when its nutritional benefits were widely publicized. Weight Watchers recommends popcorn as a snack for the weight-conscious, the American Dental Association endorses this sugar-free snack, and the American Cancer Society recognizes the benefits of the high fiber content of popcorn in possibly preventing several types of cancer. Popcorn's nutritional value is so high that doctors recommend it—even with oil—over many other snack foods.

Microwave packaging has also allowed popcorn manufacturers to enhance their product with flavorings that keep well and produce a range of good tastes when cooked. The competition to create the latest taste sensations (or borrow them from other trendy foods) is fierce in the popcorn trade, but this also helps assure the food's future. American popcorn makers compete among themselves for the best yield and novel flavors, but, increasingly, their competition is coming from growers in Argentina and South Africa.

## Where to Learn More

### Books

Russel, Solveig Paulson. *Peanuts, Popcorn, Ice Cream, Candy and Soda Pop and how they began* New York: Abingdon Press, 1970.

Woodside, Dave. *What Makes Popcorn Pop?* New York: Atheneum, 1980.

### *Periodicals*

"Exploding the Popcorn Myth." *Yankee* (February 1993): 27.

Hyatt, Joshua. "Surviving on Chaos." *Inc.* (May 1990): 60.

Kummer, Corby. "Hot popcorn: The First Popcorn was Made by Accident. Now There are Better Ways." *The Atlantic* (June 1988): 96.

### *Other*

Jolly Time Popcorn Company. http://www.jollytime.com.

The Popcorn Institute. http://www.popcorn.org.

Snappy Popcorn. http://www.netins.net/showcase/snappy/snappy.html.

Wabash Valley Farms. http://www.wabashvalleyfarms.com.


*—Gillian S. Holmes*

# Rice

## Background

As a main source of nourishment for over half the world's population, rice is by far one of the most important commercial food crops. Its annual yield worldwide is approximately 535 million tons. Fifty countries produce rice, with China and India supporting 50% of total production. Southeast Asian countries separately support an annual production rate of 9-23 million metric tons of which they export very little. Collectively, they are termed the Rice Bowl. Over 300 million acres of Asian land is used for growing rice. Rice production is so important to Asian cultures that oftentimes the word for rice in a particular Asian language also means food itself.

Rice is a member of the grass family (Gramineae). There are more that 10,000 species of grasses distributed among 600 genera. Grasses occur worldwide in a variety of habitats. They are dominant species in such ecosystems as prairies and steppes, and they are an important source of forage for herbivorous animals. Many grass species are also primary agricultural crops for humans. As well as rice, they include maize, wheat, sorghum, barley, oats, and sugar cane.

Typically, grass species are annual plants or are herbaceous perennials that die back to the ground at the end of the growing season and then regenerate the next season by shoots developing from underground root systems. Shoots generally are characterized by swollen nodes or bases. Leaves are long and narrow, varying in width from 0.28-0.79 in (7-20 mm). Flowers are small and are called florets. Grasses pollinate by using the wind to widely and opportunistically disperse grass pollen. The fruits are known as a caryopsis or grain, are one-seeded, and can contain a large concentration of starch.

Classified in the genus *Oryza*, there are two species of domesticated rice—*O. sativa* and *O. glaberrima*. *O. sativa* is the most common and often cultivated plant, occurring in Africa, America, Australia, China, New Guinea, and South Asia. The natural habitat of rice is tropical marshes, but it is now cultivated in a wide range of subtropical and tropical habitats. Unlike other agricultural crop grasses, rice plants thrive under extremely moist conditions and moderate temperatures. The ideal climate is roughly 75° F (24° C). Average plant height varies between 1.3-16.4 ft (0.4-5 m). Its growth cycle is between three to six months (agriculturally, this is broken down into three phases lasting approximately 120 days). Rice plants produce a variety of short- to long-grain rices, as well as aromatic grains.

There are three different types of rice: japonica, javanica, and indica. Japonica rice varieties are high yielding and tend to be resistant to disease. Javanica types of rice fall between japonica and indica varieties in terms of yield, use, and hardiness. Although quite hardy, indica yield less than japonica types and are most often grown in the tropics.

Because cultivation is so widespread, development of four distinct types of ecosystems has occurred. They are commonly referred to as irrigated, rainfed lowland, upland, and flood-prone agroecological zones. Irrigated ecosystems are the primary type found in East Asia. Irrigated ecosystems provide 75% of global rice production. Irrigated rice is grown in bunded (embanked), paddy fields. Rainfed lowland ecosystems only

*With one out of every three people on earth dependent on rice as a staple food in their diet and with 80-100 million new people to be fed annually, the importance of rice production to the worldwide human population is crucial.*

sustain one crop per growing season and fields are flooded as much as 19.7 in (50 cm) during part of the season. Rainfed lowland rice is grown in such areas as East India, Bangladesh, Indonesia, Philippines, and Thailand, and is 25% of total rice area used worldwide. Production is variable because of the lack of technology used in rice production. Rainfed lowland farmers are typically challenged by poor soil quality, drought/flood conditions, and erratic yields. Upland zones are found in Asia, Africa, and Latin America. It is the primary type of rice ecosystem in Latin America and West Africa. Upland rice fields are generally dry, unbunded, and directly seeded. Land utilized in upland rice production runs the gamut of descriptions. It can be low lying, drought-prone, rolling, or steep sloping. Usually, crops are either sown interspersed with another crop, intermittently with another crop, or the crop is shifted every few years to a new location. Lastly, flood-prone ecosystems are prevalent in South and Southeast Asia, and are characterized by periods of extreme flooding and drought. Yields are low and variable. Flooding occurs during the wet season from June to November, and rice varieties are chosen for their level of tolerance to submersion.

Rice is mostly eaten steamed or boiled, but it can also be dried and ground into a flour. Like most grains, rice can be used to make beer and liquors. Rice straw is used to make paper and can also be woven into mats, hats, and other products.

## History

Since it has been such an important grain worldwide, the domestication and cultivation of rice is one of the most important events in history that has had the greatest impact on the most people. When and where the domestication of rice took place is not specifically known, but new archaeological evidence points to an area along the Yangtze River in central China and dates back as far as 11,000 years. Researched by a team of Japanese and Chinese archaeologists and presented at the 1996 International Symposium on Agriculture and Civilizations in Nara, Japan, radiocarbon testing of 125 samples of rice grains and husks, as well as of rice impressions in pottery, from sites located along a specific portion of the Yangtze unanimously indicate a median age of over 11,000 years. Another discovery of possibly the oldest settlement found in China, which is located closely upstream from the other sites, gives credence to the new findings.

In any event, it wasn't until the development of puddling and transplanting of the rice plant that the spread of rice as an agricultural crop really began. Practiced in the wetlands of China, the concept of the rice paddy was adopted by Southeast Asia in roughly 2000 B.C. Wetland cultivation techniques migrated to Indonesia around 1500 B.C. and then to Japan by 100 B.C. To the West, rice was also an early important crop in India and Sri Lanka, dating as far back as 2500 B.C. and 1000 B.C. respectively.

The spread to Europe, Africa, and America occurred more slowly, first with the Moor's invasion of Spain in 700 A.D. and then later to the New World during the age of exploration and colonialism. Rice has been grown in the United States since the seventeenth century in such areas as the southeastern and southern states, as well as California.

## Raw Materials

The only raw material needed for commercial production of rice is the rice seed or seedlings. Additional use of herbicides, pesticides, and fertilizer can increase the likelihood of a larger yield.

## Design

Varieties of rice are selected and grown specifically for their end use. In the United States, long-grain rice is typically used for boiling, quick-cook products, and soup. Whereas, shorter- grain rice is used in cereal, baby food, and beer/liquors.

## The Manufacturing Process

### *Preparation*

1 Prior to planting, minimal soil manipulation is needed to prepare for cultivation. If the rice will be grown on a hilly terrain, the area must be leveled into terraces. Paddies are leveled and surrounded by dikes or levees with the aide of earth-moving equip-

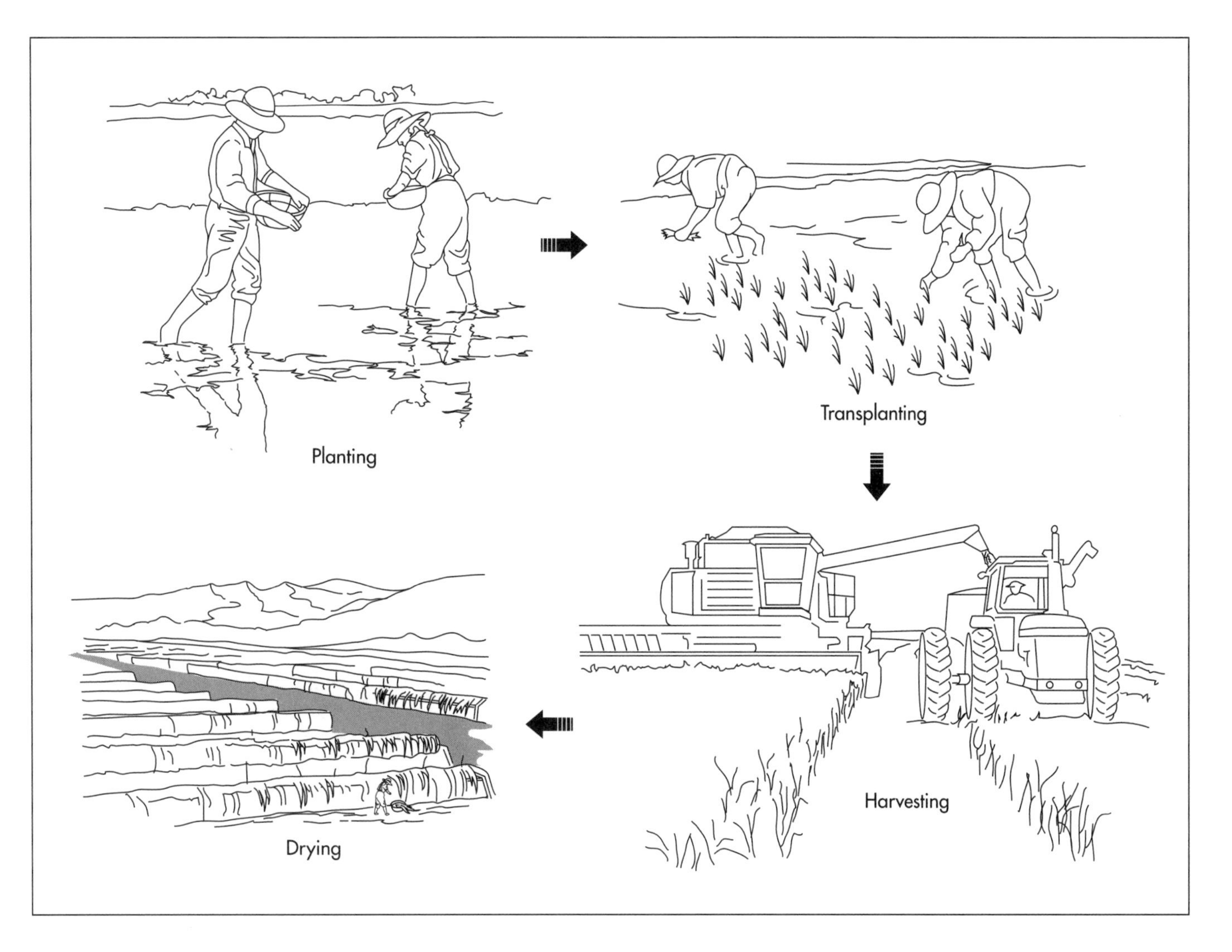


ment. Then, the fields are plowed before planting. In the United States, rice is most often planted on river deltas and plowing is accomplished with a disk plow, an off-set disk plow, or a chisel. Adequate irrigation of the terrace or river delta bed is required and accomplished by leveling and by controlling water with pumps, reservoirs, ditches, and streams.

### Planting

2 Rice seeds are soaked prior to planting.

3 Depending on the level of mechanization and the size of the planting, seeding occurs in three ways. In many Asian countries that haven't mechanized their farming practices, seeds are sown by hand. After 30-50 days of growth, the seedlings are transplanted in bunches from nursery beds to flooded paddies. Seeds can also be sown using a machine called a drill that places the seed in the ground. Larger enterprises often found in the United States sow rice seed by airplane. Low-flying planes distribute seed onto already flooded fields. An average distribution is 90-100 lb per acre (101-111 kg per hectare), creating roughly 15-30 seedlings per square foot.

### Harvesting

4 Once the plants have reached full growth (approximately three months after planting) and the grains begin to ripen—the tops begin to droop and the stem yellows—the water is drained from the fields. As the fields dry, the grains ripen further and harvesting is commenced.

5 Depending on the size of the operation and the amount of mechanization, rice is either harvested by hand or machine. By hand, rice stalks are cut by sharp knives or

The cultivation of rice begins by planting water-soaked seeds in a properly prepared bed. Oftentimes, the seedlings are transplanted to the paddy when they reach a certain size. When the grains begin to ripen, the water is drained from the fields. Harvesting begins when the grain yellow and the plants start to droop. Depending on the size of the operation and the amount of mechanization, rice is either harvested by hand or machine. Once harvested, the rice is usually dried in the fields with the help of sunshine.

At the processing plant, the rice is cleaned and hulled. At this point, brown rice needs no further processing. If white rice is desired, the brown rice is milled to remove the outer bran layers.

Milling/Shelling

sickles. This practice still occurs in many Asian countries. Rice can also be harvested by a mechanized hand harvester or by a tractor/horse-drawn machine that cuts and stacks the rice stalks. In the United States, most operations use large combines to harvest and thresh—separate the grain from the stalk—the rice stalks.

6 If the rice has been harvested by hand or by a semi-automated process, threshing is completed by flailing the stalks by hand or by using a mechanized thresher.

### Drying

7 Before milling, rice grains must be dried in order to decrease the moisture content to between 18-22%. This is done with artificially heated air or, more often, with the help of naturally occurring sunshine. Rice grains are left on racks in fields to dry out naturally. Once dried, the rice grain, now called rough rice, is ready for processing.

### Hulling

8 Hulling can be done by hand by rolling or grinding the rough rice between stones. However, more often it is processed at a mill with the help of automated processes. The rough rice is first cleaned by passing through a number of sieves that sift out the debris. Blown air removes top matter.

9 Once clean, the rice is hulled by a machine that mimics the action of the hand-held stones. The shelling machine loosens the hulls from the rice by rolling them between two sheets of metal coated with abrasives. 80-90% of the kernel hulls are removed during this process.

10 From the shelling machine, the grains and hulls are conveyed to a stone reel that aspirates the waste hulls and moves the kernels to a machine that separates the hulled from the unhulled grains. By shaking the kernels, the paddy machine forces the heavier unhulled grains to one side of the machine, while the lighter weight rice falls to the other end. The unhulled grains are then siphoned to another batch of shelling machines to complete the hulling process. Hulled rice grains are known as brown rice.

### Milling

Since it retains the outer bran layers of the rice grain, brown rice needs no other processing. However along with added vitamins and minerals, the bran layers also contain oil that makes brown rice spoil faster than milled white rice. That is one of the reasons why brown rice is milled further to create a more visually appealing white rice.

11 The brown rice runs through two huller machines that remove the outer bran layers from the grain. With the grains pressed against the inner wall of the huller and a spinning core, the bran layers are rubbed off. The core and inner wall move closer for the second hulling, ensuring removal of all bran layers.

12 The now light-colored grain is cooled and polished by a brush machine.

13 The smooth white rice is conveyed to a brewer's reel, where over a wire mesh screen broken kernels are sifted out. Oftentimes, the polished white rice is then coated with glucose to increase luster.

### Enriching

The milling process that produces white rice also removes much of the vitamins and minerals found primarily in the outer bran layers. Further processing is often done in order to restore the nutrients to the grain.

Once complete, the rice is called converted rice.

14 White rice is converted in one of two ways. Prior to milling, the rice is steeped under pressure in order to transfer all the vitamins and minerals from the bran layers to the kernel itself. Once done, the rice is steamed, dried, and then milled. Rice that has already been milled can be submersed in a vitamin and mineral bath that coats the grains. Once soaked, they are dried and mixed with unconverted rice.

## Quality Control

Quality control practices vary with the size and location of each farm. Large commercial rice farms in the United States more often than not apply the most effective combination of herbicides, fertilization, crop rotation, and newest farming equipment to optimize their yields. Smaller, less mechanized operations are more likely to be influenced by traditional cultural methods of farming rather than high technology. Certainly, there are benefits to both approaches and a union of the two is ideal. Rotating crops during consecutive years is a traditional practice that encourages large yield as is the planting of hardier seed varieties developed with the help of modern hybridization practices.

## Byproducts/Waste

Straw from the harvested rice plants is used as bedding for livestock. Oil extracted from discarded rice bran is used in livestock feed. Hulls are used to produce mulch that will eventually be used to recondition the farm soil.

The essential use of irrigation, flooding, and draining techniques in rice farming also produces runoff of pesticides, herbicides, and fertilizers into natural water systems. The extensive use of water in rice farming also increases its level of methane emissions. Rice farming is responsible for 14% of total global methane emissions.

## The Future

With one out of every three people on earth dependent on rice as a staple food in their diet and with 80-100 million new people to be fed annually, the importance of rice production to the worldwide human population is crucial. Scientists and farmers face the daunting task of increasing yield while minimizing rice farming's negative environmental effects. Organizations such as the International Rice Research Institute (IRRI) and the West African Rice Development Association (WARDA), and Centro Internacional de Agricultura Tropical (CIAT [International Center for Tropical Agriculture]) are conducting research that will eventually lead to more productive varieties of rice and rice hybrids, use of less water during the growing season, decrease in the use of fresh organic fertilizer that contributes to greenhouse effect, and crops more resistant to disease and pests.

## Where To Learn More

### Books

Huke, R.E. and E.H. *Rice: Then and Now*. International Rice Institute, 1990.

Johnson, Sylvia A. *Rice*. Minneapolis, MN: Lerner Publications Co., 1985.

### Periodicals

"Limiting Rice's Role in Global Warming." *Science News* (July 10, 1993): 30.

Normile, Dennis. "Yangtze Seen as Earliest Rice Site." *Science* (January 17, 1997): 309.

### Other

Riceweb. http://www.riceweb.org/ (June 29, 1999).

# Safety Razor

*Gillette employs 500 design engineers, who are constantly developing new shaving systems. Preliminary designs are developed into working prototypes that are tested by over 300 company employees taking part in Gillette's shave-at-work program.*

A safety razor is a device used to remove hair from areas of the body where it is undesirable such as the face for men and the legs and underarm regions for women. The modern blade razor consists of a specially designed blade mounted in a metal or plastic shell that is attached to a handle. This kind of razor can be designed as a refillable cartridge which can accept new blades or as a disposable unit which is intended to be thrown away after the blade becomes dull.

## History

Since primitive times, shaving has been an important cultural grooming practice. Cave painting show that even the prehistoric men practiced shaving by scraping hair off with crude implements such as stones, flint, clam shells, and other sharpened natural objects. With the advent of the Bronze Age, humans developed the ability forge simple metals and began to make razors from iron, bronze, and even gold. The ancient Egyptians began the custom of shaving their beards and heads, which was eventually adopted by the Greeks and Romans around 330 B.C. This practice was advantageous for soldiers because it prevented enemies from grasping their hair in hand-to-hand combat. The unshaven, unkempt tribes they fought became known as barbarians, meaning the unbarbered.

Until the nineteenth century, the most common razor was still a long handled open blade called a "cut-throat" razor which was difficult to use, required repeated sharpening, and was usually wielded by professional barbers. Credit for the first safety razor is generally given to a Frenchman, Jean-Jacques Perret, who modeled his design after a joiner's plane. He even wrote a book on the subject entitled *Pogonotomy or the Art of Learning to Shave Oneself.* As with the razors of today, Perret's design covered the blade on three sides to protect the user from nicks and cuts. However, it still required periodic sharpening to give a good shave. Similar inventions were introduced throughout the 1800s. Nonetheless, even as late as the early 1900s most men were still shaved periodically at the barber.

Shaving practices began to change dramatically around the turn of the century. In 1895, an American named King Camp Gillette had the idea of marketing a disposable blade that didn't require sharpening. Gillette designed a razor that had a separate handle and clamp unit that allowed the user to easily replace the blade when it became dull. However, metal working technology took another two years before it was able to make the paper thin steel blades required by Gillette's design. Even though he filed patents in 1901, Gillette could not market his disposable blades until 1903 when he produced a total of 51 razors and 168 blades. By 1905, sales rose to 90,000 razors and 2.5 million blades. Sales continued to grow over the next several years, reaching 0.3 million razors and 14 million blades in 1908. After Gillette's initial success, other manufacturers soon followed suit with their own designs, and an entire industry was born. Over the last 90 years, a variety of products have been introduced including tiny safety razors for women, long-life stainless steel blades, twin-blade safety razors, the completely disposable, one-piece plastic razor introduced by Bic, and the state of the art Sensor and Mach 3 shaving systems by Gillette.

## Design

Razor designs vary depending on the style. Some razors, such as the single piece disposables, are relatively simple. They consist of a hollow plastic handle, a blade, and a head assembly to keep the blade in place. They are primarily designed to be simple, economical, and disposable. The refillable cartridge style is more complicated. They are designed to give a more premium shaving experience with options like multiple blades, pivoting heads, and lubricating strips. For example, Gillette's Mach 3 razor, which was introduced in 1998, features a skin guard comprised of flexible microfins, a soft grip handle, water-activated moisturizers, a flow-through cartridge, optimal blade positioning, and other innovative features. The engineering behind some of these advancements is quite impressive. Gillette employs 500 design engineers, who are constantly developing new shaving systems. Preliminary designs are developed into working prototypes that are tested by over 300 company employees, who take part in Gillette's shave-at-work program. The company has 20 booths set up where employees use unmarked razors on different sides of their faces. They then rate performance attributes of each razor with the aid of a computer program. Engineers use this feedback to adjust their designs and create improved prototypes for further evaluation.

## Raw Materials

### Blades

Razor blades are periodically exposed to high levels of moisture and therefore must be made from a special corrosion resistant steel alloy. Furthermore, the grade of steel must be hard enough to allow the blade to hold its shape, yet malleable enough to allow it to be processed. The preferred type of steel is called carbide steel because it is made using a tungsten-carbon compound. One patented combination of elements used in stainless steel blade construction includes carbon (0.45-0.55%), silicon (0.4-1%); manganese (0.5-1.0%); chromium (12-14%) and molybdenum (1.0-1.6%); with the remainder being iron.

### Plastic parts

The plastic portions of a safety razor include the handle and blade cartridge, or portions thereof, depending on the razor design. These parts are typically molded from a number of different plastic resins including polystyrene, polypropylene, and phenylene-oxide based resins as well as elastomeric compounds. These resins are taken in pellet form and are melted and molded into the razor components through a combination of extrusion and injection molding techniques. For example, in making the handle for their advanced shaving systems, Gillette uses a coextrusion process which simultaneously molds an elastomer molded over polypropylene to create a surface that is easy to grip.

### Other components

Razors may contain a variety of miscellaneous parts which help hold the blade in place, guards which cover the blade during shipping, or springs or other release mechanisms which facilitate changing of the blades. These pieces are molded by similar processes. The more sophisticated brands include a lubricating strip made of polyurethane, or other similar materials, that is impregnated with acrylic polymers. These strips are mounted on the head of the razor, in front of the blades. The polymer film absorbs water and becomes very slippery, thus creating a lubricating surface that helps the blade glide across the surface of the face without snagging or cutting the skin.

## The Manufacturing Process

### Cutting blade formation

1 Blade manufacturing processes involve mixing and melting of the components in the steel. This mixture undergoes a process known as annealing, which makes the blades stronger. The steel is heated to temperatures of 1,967-2,048°F (1,075-1,120°C), then quenched in water to a temperature between -76- -112° F (-60- -80° C) to harden it. The next step is to temper the steel at a temperature of (482-752°F (250- 400°C).

2 The blades are then die stamped at a rate of 800-1,200 strokes a minute to form the appropriate cutting edge shape. The actual cutting edge of modern cartridge style razor blade is deceptively small. The entire cutting surface is only about 1.5 in (3.81cm) wide by 1 mm deep. This is compared to tra-

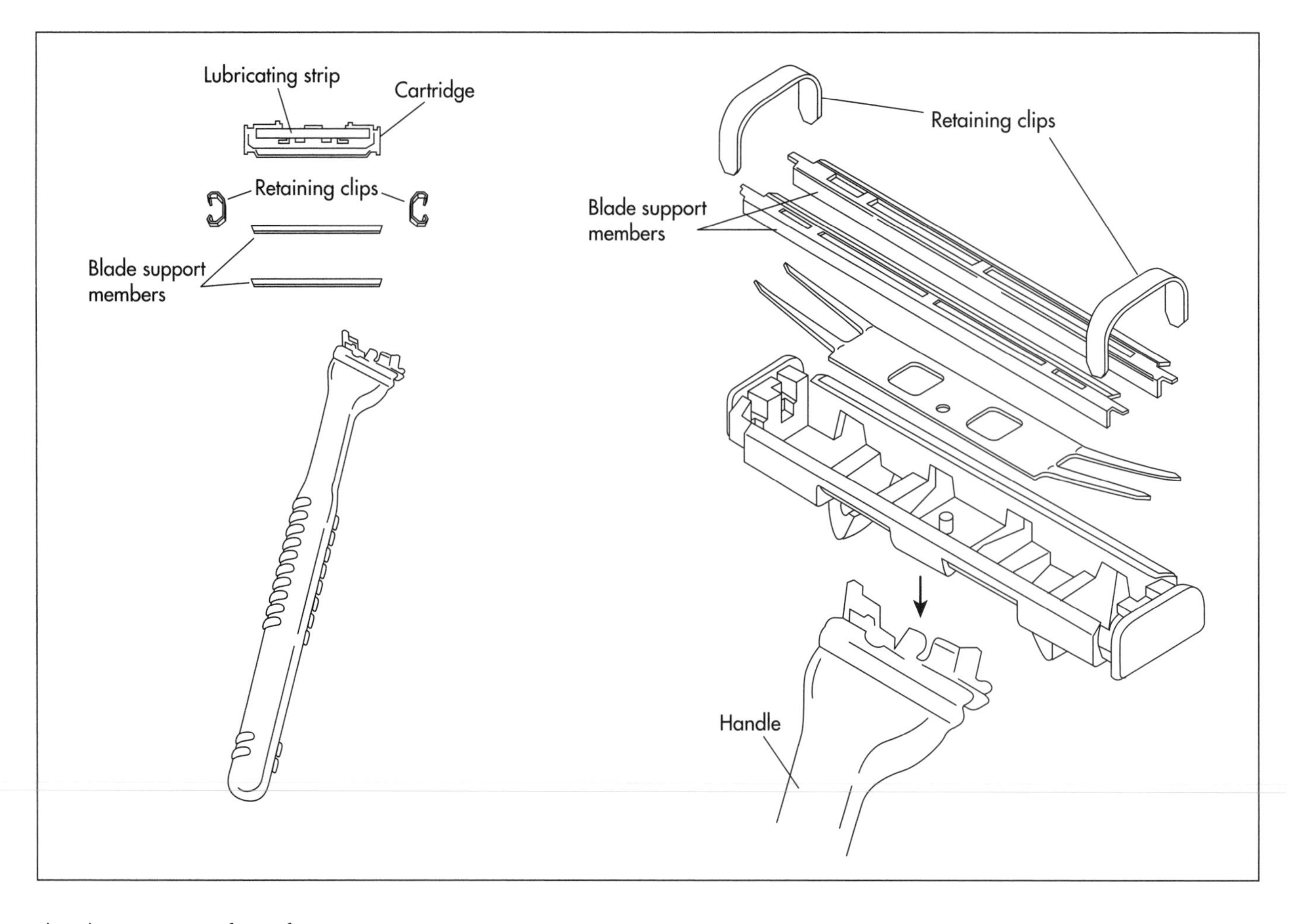


The plastic portions of a safety razor include the handle and blade cartridge. These parts are typically molded from a number of different plastic resins, including polystyrene, polypropylene, and phenylene-oxide based resins as well as elastomeric compounds. Razor blades are made from a special corrosion resistant blend of steel called carbide steel because it is made using a tungsten-carbon compound.

ditional razor blades which are almost 20 times wider and several times thicker. This design creates efficiencies in manufacturing by allowing the creation of a durable cutting surface using very little metal. Because the blade is so small, a special support structure is required to hold it inside the cartridge.

### *Support member formation*

3 At a separate work station, another sheet of metal passes through a die and cutter device to form a series of L-shaped support members. These support members are formed in a line with two edge runners connected to each side.

4 The row of supports, still connected to the edge runners is rolled onto a coil and transported to the next station. There the support pieces are severed from the edge runners which are collected in a waste bin. The support members are dropped into a funnel-like device equipped with a vibrating unit which deposits individual support members onto a conveyor belt. The belt transfers the members in a single file fashion the third work station where they are welded onto the cutting blade. The finished blade assembly is then ready for mounting in the cartridge. Because the entire process is automated, waste from broken or bent cutting blades and support members is minimized.

### *Plastic component molding*

5 Concurrent with the blade-making operations, the plastic components are molded and readied for assembly. The plastic resins are mixed with the plasticizers, colorants, antioxidants, stabilizers, and fillers. The powders are mixed together and melted in a special heated screw feeder. The resultant mixture is cut into pellets which can be used in subsequent molding operations.

6 Plastic razor parts are typically extrusion molded. In this process, molten plastic is shaped by being forced through the opening of a die. The parts can also be manufactured by injection molding, where plastic resin and other additives are mixed

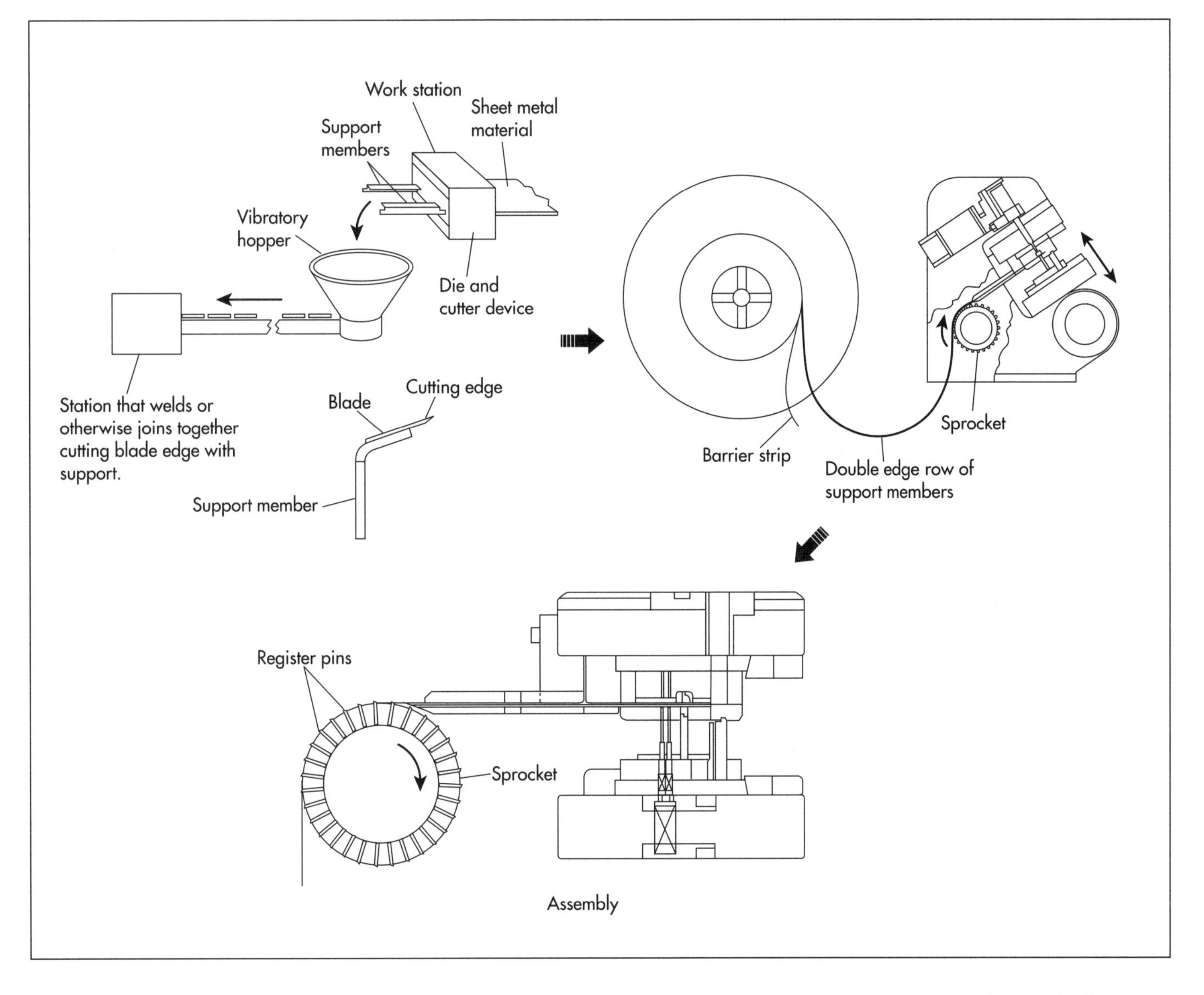


together, melted, and injected into a two piece mold under pressure. After the plastic has cooled, the mold is opened and the plastic parts are ejected. Major manufacturers have extremely efficient molding operations with cycle times for molded plastic parts routinely below 10 seconds. These processes are so efficient that the thermoplastic runners and other scrap from the molding process are reground, remelted, and reused.

### *Assembly of components*

7 The molded plastic components are fed to various work stations where the blade assembly is inserted into the cartridge. The work surfaces in these stations are equipped with vacuum lines to orient and hold the small blade parts in place during transport and insertion. Spring loaded arms push the blades into place and secure them in the cartridge slots. The finished cartridge may be attached to the razor handle during subsequent operations or they may be packaged separately. This step may include insertion of springs and other parts in the handle to allow ejection of the cartridge.

### *Packaging*

8 Razors are routinely packaged in clear plastic blister packs with a cardboard backing sheet that allow display of the razors design. Refill blade cartridges can be packaged in boxes, although most current designs require the cartridges to be held in a plastic tray that helps insert them into the handle.

The manufacture of safety razors involves first making the blade cartridge by die stamping the carbide steel and then welding the blades to steel support members. Simultaneously, the plastic components are extruded or injection molded and readied for assembly. The blade cartridge and plastic parts are automatically assembled at workstations that use vacuum lines to orient and hold the small blade parts in place during transport and insertion. Spring loaded arms push the blades into place and secure them in the cartridge slots. The finished cartridge may be attached to the razor handle during subsequent operations or they may be packaged separately.

## Quality Control

All finished razor components must conform to tight specifications before they are released. For example, blades must meet a designated hardness rating and contain a certain amount of steel. Gillette blades must meet a standard knows as Vickers hardness of at least 620 and a carbide density of 10-45 particles per 100 square microns to avoid rejection. The equipment itself operates so precisely that Gillette measures its reject rate in parts per million. Similarly, molded plastic parts are closely inspected by operators with lighted magnifying glasses to check for loose flashing or rough edges; they alert technicians when problems are discovered. In addition, razor components are checked by a computerized vision system which compares a critical dimension to a reference.

## The Future

Razor manufacturers like Gillette are constantly designing new and improved shaving systems. Their commitment to improved materials science continues to produce blades of increased hardness that are capable of sustaining sharp edge for more shaves. Advanced head design allows the blades to contact the face without cuts or nicks. They are also constantly updating their manufacturing equipment. The future manufacturing techniques also improve efficiency in molding and stamping operations. Gillette claims they are twice as fast as they were 10 years ago and have fewer defects.

## Where to Learn More

### Books

Panati, Charles. *Extraordinary Origins of Everyday Things.* New York: Perennial Library, 1987.

### Other

Gillette Company. 1999. http://www.gillette.com/ (April 5, 1999).

*—Randy Schueller*

# Sheet Music

## Background

Sheet music is a magic carpet. It is a printed page that, like a book, tells an original story created by the talent, imagination, training—and sometimes genius—of a writer. In the case of music, the writer is a composer or songwriter who uses a long-established set of notes and other symbols as well as lyrics (words that are sung to the music) and other words that instruct the singer or instrumentalist on the dynamics (loudness) and other characteristics of the piece. When the musician reads and performs the music, magic happens as the songwriter's composition is interpreted for the pleasure of the audience.

## History

Some of the earliest sheet music was laboriously written by scribes in the monasteries of medieval Europe. These beautiful examples were carefully inked on parchment and are prized today not only as music history but as artistic masterpieces. With the invention of the printing press, Johann Gutenberg and his followers developed methods of printing music, as well as words, during the fifteenth century. The printing of music was limited in quality and quantity for several hundred years, but the industry traveled to America with the founding of the Colonies.

The first music published in North America was *The Bay Psalm Book* printed in 1640 by Harvard College Press. The book contained only text because the congregations of churches were assumed to know the songs by heart. Publishing of music, complete with notation, became an industry by about 1800 when a number of firms in both America and Europe rolled out their presses to print both serious and popular music. This explosion was probably a direct result of the Industrial Revolution that gave rise to the middle class and allowed individuals more leisure time and money to spend on pianos for their homes, instruments for the town band, and attendance at the symphony. Composers were motivated to create when, during the nineteenth century, musicians began to pay for the privilege of performing the writer's music.

The growth of many styles of popular music that are considered American in character, including jazz, country-western, bluegrass, spirituals, and musical theater, is attributable not only to talented composers and artists but to the publishers who made it possible to imitate their music on Dad's banjo at home. By 1890, many department stores had counters for the sale of sheet music, and its popularity forced the price down. By 1910, Woolworth sold sheet music for 10 cents a copy.

The musicians of Tin Pan Alley in New York City were made famous early in the 1900s by the swift availability of their tunes in sheet music form; George Gershwin's "Rhapsody in Blue" (1924) is an excellent example. Composers Aaron Copeland, Charles Ives, and Virgil Thompson established their own publishing house and gave the American public its own contemporary, classical music. When Charles Lindberg made his solo flight across the Atlantic Ocean in 1927, 100 songs commemorating the event were printed in sheet music form within a year.

In 1892, a European firm established itself in the United States as G. Schirmer, Inc.; they publish a vast library of classical music recognizable by its yellow covers. The Eu-

*By 1910, Woolworth's department store sold sheet music for 10 cents a copy.*

ropean influence was felt even more strongly after World War I (1914-1919) and the rise of Hitler in Germany (1933) forced the immigration of Béla Bartòk, Arnold Schoenberg, and Igor Stravinsky, composers of international repute who imported music editors whose skills brought a classical tradition. Between the two World Wars, the phonograph and radio further popularized a wide range of music; and, after, World War II, television and technological improvements in the sound business accelerated popular interest in sheet music.

The sheet music industry experienced another boost in 1914 when the first performance rights society was established. The American Society of Composers, Authors and Publishers, Inc., (ASCAP) was followed by The Society of European Stage Authors and Composers (SESAC, Inc. 1931) and Broadcast Music, Inc. (BMI 1940). These organizations are essential to the orderly administration of performance data and distribution of royalties for music in copyright. These organizations also play an important role in funding the first efforts of young composers and songwriters.

## Raw Materials

The songwriter's or composer's music may be the most important "raw material" in the production of sheet music. When the writer has completed a piece of music, the business of publishing the music must be negotiated among the writer, his agent, and the publisher. A well-known songwriter may receive advances from a publishing company, but, typically, the contract between the writer and publishing house involves negotiation and distribution of royalties, where royalties are fees that are paid to the publisher and writer based on how often the piece of music is used.

## Design

Music in print appears in several major formats. Sheet music usually includes arrangements for voice and piano or guitar. Sometimes chord diagrams for other instruments are also shown. Sheet music that is collected under a songwriter's or performer's name is called a personality folio. Frank Sinatra and Barbra Streisand, for example, have popularized many pieces of music that wouldn't be known by the songwriters' names. However, popular writers like Henry Mancini or Herb Alpert have had songbooks of their own music published. Matching folios are collections of music that match all the songs on or in a compact disc, movie soundtrack, or musical. Mixed folios are similar, except they combine music by multiple writers under a theme cover like *Great Country Music Hits, Best Songs of the Century,* or *Music for the Tuba.*

Publishing of educational music is also a large part of the sheet music business. Schools, marching bands, drum corps, and choruses purchase or rent large quantities of copies of all categories of music that have been arranged specifically for voice or orchestra, for example, or for players at particular skill levels. Finally, most publishers also develop and sell MIDI sequences, which are electronically sequenced and recorded versions of songs; some are complete, professional arrangements and performances that are ready for use in performance.

## The Manufacturing Process

1 The process of making sheet music begins in the imagination of composer. The composer or songwriter may write a piece of music on his own accord, or he may receive a commission from an organization or individual like an opera company, film production company, or a jazz singer. The composer may have an established working relationship with a music publisher, or his or her agent may market the piece to publishers specializing in particular types of music.

2 When an interested publisher is found, the publisher's editorial department reviews the composition for its quality, market appeal, and the practicality of publishing it. Each publisher has an editorial policy that governs the range of types of music the firm publishes and the level or quality. For example, a music publisher that specializes in sheet music for students is not likely to be interested in a ballet score. A large music publishing house may have a number of divisions, however; following a pre-screening, both the beginning-level music and the ballet score will be forwarded to the editorial departments in the respective divisions, and both may be published by the same house.

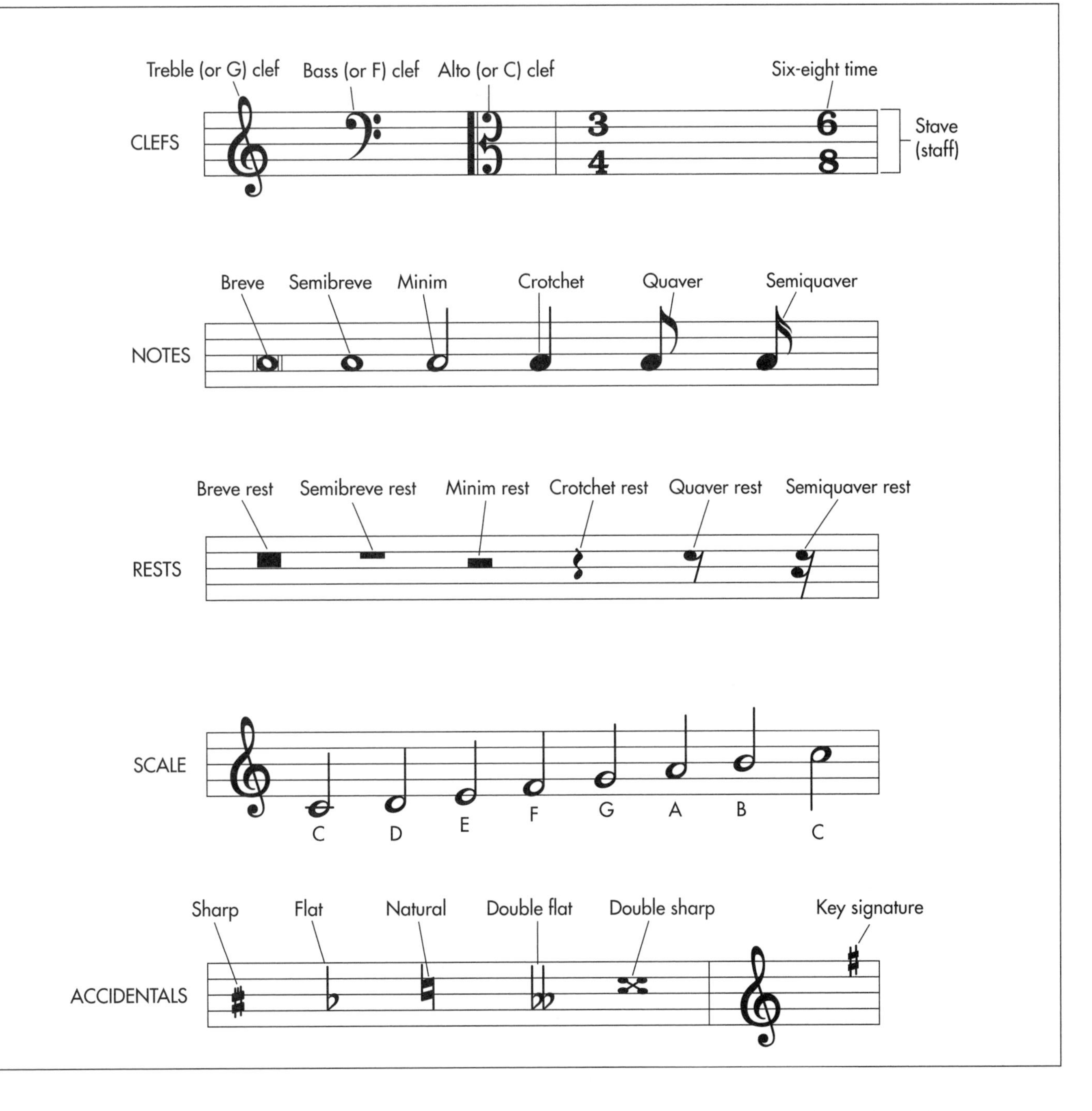


Types of musical notation.

3 The music publisher also reviews the composer's work in the Legal Division or Rights Clearance Division. The legal staff performs a number of functions. It negotiates a contract between the publishing house and the songwriter or composer. The new piece is registered with the U.S. Copyright Office. The publisher's legal staff works with the major performance rights organizations (ASCAP, BMI, and SESAC) so that royalties from performances of copyrighted music can be collected and negotiates other royalties from performing arts groups (ballet companies, musical theater groups, orchestras, opera houses, and others) and recording, and film, and television companies. The Rights Clearance Division collects and distributes royalties annually.

4 After the writer and the composition have cleared the hurdles of the editorial department, the U.S. Copyright Office, and the legal department, the piece of music is ready for publication. Traditionally, the composer's hand-drafted music is sent to an engraver who uses sets of type that include

musical staffs, notation, and text to etch the music into a metal plate. Engraving has several disadvantages; it is very expensive, and it has been outdated not only by publishing methods but by styles of notation. More recently, composers use specialized music software that writes out notation exactly as it will be printed. Music software, like Finale, produces sheet music that duplicates the quality of music produced through traditional engraving methods. The computer notations can be made through a standard keyboard; often, the publisher or music notation service has a piano-style keyboard linked to the computer, and the notation is entered by playing the piece. The music publisher not only enters the notation and expressive indications but adjusts the size, spacing, and layout by computer. The full score is completed and then the publisher extracts the parts for each instrument, places them in separate sheet music, and repeats the process of adjusting sizing and layout. The writer reviews printed drafts of proofs (from either the traditional engraving process or computer methods). When the music is finalized, the composer is given a hard copy and the piece on diskette to take to the agent and publishing houses.

5 If an engraving process is used, the music from the engraver is set up on pages to make camera-ready copy. The copy is photographed to make a negative, the negative is burned onto a metal plate using chemicals, and the plate is inserted in a printing press. The areas of the plate that are to be printed attract ink that is pressed onto a page in a process called offset lithography. The pages are collated, folded, or bound (depending on the length and style of the piece of music). A piece of popular sheet music will also bear a color photograph or design on the cover, and all published music includes dedications, copyright information, page numbers and headings, and instructions (if necessary to a beginning instrumentalist, for example). If the piece of music was written using computer software, that same software can be used to instruct a laser printer to print the music using fonts, notation styles, and layouts that are input as data into the program. The cover illustration or photograph can also be input using digital techniques.

6 The printed music is bundled and shrink-wrapped or boxed in quantity and shipped to the publisher's warehouse where it is stored by edition number for sale. Many types of sheet music aren't "sold" at all; instead, the publisher operates a rental department that rents the material to performance groups like orchestras or ensembles and to other organizations like the libraries of music schools and universities. The major publishers may carry over 20,000 titles, and they attend trade shows where music professionals everyone from opera company music directors to marching band leaders to performing artists shop for arrangements of sheet music and bargains in the rental department. The publisher's rental department also coordinates performance dates so that the top ten symphonies will not be performing the same works in the same season; they are also able to recommend complimentary pieces so performance groups can present an evening of innovative jazz or an afternoon in old Vienna.

7 After a customer buys or rents a piece of music, the order is processed by the publisher's trade department. Shipment is as simple as sending a box of 1,000 copies of Celine Dion's latest hit to a music store or as complex as assembling the thick scores of a Richard Wagner opera for the 100-plus different instruments in the orchestra, 10 principal singers, 60 chorus members, and the conductor, prompter, various directors, and other musical staff. The shipment is documented and invoiced by computer, and the music is sent to the customer. Finally, the composer's dream will be brought to life by artists who interpret that dream from the written page.

## Quality Control

The songwriter/composer has a say in the quality control of his or her music in print form. When proof copies of the music are provided by the publisher, the writer reviews the notations, chords and so forth to verify that they are as written or that they accurately represent the creation. The divisions of a large publishing house also bear responsibility for observing high standards and the legalities of the trade.

## The Future

Sheet music from bygone days has become a valuable collectible. Cover art greatly interests collectors who seek out the Art Deco designs of the 1920s and African-American songs published as early as 1835, for example. Photos of singers, band leaders, and Broadway productions as well as autographs by the songwriters, lyricists, and performers whet the collectors' appetites.

Sheet music shows every sign of maintaining its popularity as long as performers from the top of the charts to the beginning piano student at home want to play the latest tunes and the greatest classics. Software like Finale and Overture makes it easy for the youngest musician to experiment with composition and for the most experienced tunesmith to produce data and hardcopy versions of his or her latest song. The issue of immediacy somewhat hampers the music publisher who makes a considerable investment in copyrights and the physical production of sheet music. Despite the advent of many other technical diversions, music is among our most popular entertainments, and sheet music allows us to own copies of Mozart's genius and the joy of a simple Christmas carol.

## Where to Learn More

### Books

Braheny, John. *The Craft and Business of Song Writing*. Cincinnati: Writer's Digest Books, 1988.

Priest, Daniel B. *American Sheet Music with Prices: A Guide to Collecting Sheet Music from 1775 to 1975*. Des Moines, IA: Wallace-Homestead Book Co., 1978.

Sachs, Carolyn, ed. *An Introduction to Music Publishing*. New York: C. F. Peters, 1981.

Shemel, Sidney and M. William Krasilovsky. *This Business of Music: A Practical Guide to the Music Industry for Publishers, Writers, Record Companies, Producers, Artists, Agents*. Billboard Books, 1990.

### Periodicals

Mike, Dennis. "Classroom maestros: professional music software that's a boon to the classroom."*Electronics Learning* ( September 1994): 62.

Pogue, David. "Overture 2.0."*Macworld* (September 1997): 74.

### Other

Lewis Music Library. Massachusetts Institute of Technology. December 9, 1997. http://libraries.mit.edu/music/sheetmusic/ (June 29, 1999).

Special Collections Library for historic American sheet music at Duke University. 1999. http://scriptorium.lib.duke.edu/sheet-music/ (June 29, 1999).

*—Gillian S. Holmes*

# Silly Putty

*In 1968, the* Apollo 8 *astronauts carried Silly Putty into space in a specially designed sterling silver egg to alleviate boredom and help fasten down tools in the weightless environment.*

## Background

In 1943, Silly Putty was accidentally invented by James Wright, an engineer in General Electric's New Haven laboratory, which was under a government contract to create an inexpensive substitute for synthetic rubber for the war effort. By combining boric acid with silicone oil, a material resulted that would stretch and bounce farther than rubber, even at extreme temperatures. In addition, the substance would copy any newspaper or comic-book print that it touched.

There is some debate on who received the first patent. Corning Glass Works, who was also developing a substitute for rubber, applied for a patent in 1943 and received it in 1947 for treating dimethyl silicone polymer with boric oxide. Wright applied for his patent in 1944. In any event, Wright is still officially credited with the invention.

By 1945, General Electric (GE) had shared this discovery with scientists around the world, only to find that none of them, including those at the U.S. War Production Board, found it more practical than the synthetic rubber already then being produced. Several years later, an unemployed copywriter named Peter Hodgson recognized its marketing potential as a children's toy, after first seeing it advertised at a local toy store as an adult gift. Hodgson bought the production rights from GE and renamed it Silly Putty, packaging it in plastic eggs because Easter was on the way.

Though Hodgson introduced Silly Putty at the International Toy Fair in New York in February of 1950, it was not until several months later when an article appeared in *The New Yorker* magazine that sales took off. Initially, its market as a novelty item was 80% adult. However, by 1955 Silly Putty was most popular with kids ages six to 12 years old. Six years later, Silly Putty was introduced to the Soviet Union, followed by Europe, where it was a hit in Germany, Switzerland, the Netherlands, and Italy. By the time Hodgson died in 1976, Silly Putty had made him a multi-millionaire.

It was only after its success as a toy that practical uses were also found for Silly Putty. It picks up dirt, lint, and pet hair, and can stabilize wobbly furniture. It has also been used in stress-reduction and physical therapy (athletes have used it to strengthen their grip), and in other medical and scientific situations (like smoking cessation programs). In 1968, the *Apollo 8* astronauts carried Silly Putty into space in a specially designed sterling silver egg to alleviate boredom and help fasten down tools in the weightless environment. The Columbus Zoo in Ohio has even used it to make casts of the hands and feet of gorillas for educational purposes.

The eight million units produced in 1998 is four times what was produced in 1987. Binney & Smith, the maker of Crayola products who has manufactured Silly Putty since 1977, added four fluorescent colors in 1990—magenta, orange, green, and yellow. A market study at this time showed that nearly 70% of American households had purchased Silly Putty at some time.

In 1991, "Glow in the Dark" was introduced, though classic Silly Putty has remained the best seller. Most Silly Putty is still packaged in plastic eggs. Each egg contains 0.47 oz (13.5 g) and sells for about $1.00. Binney & Smith produces more than

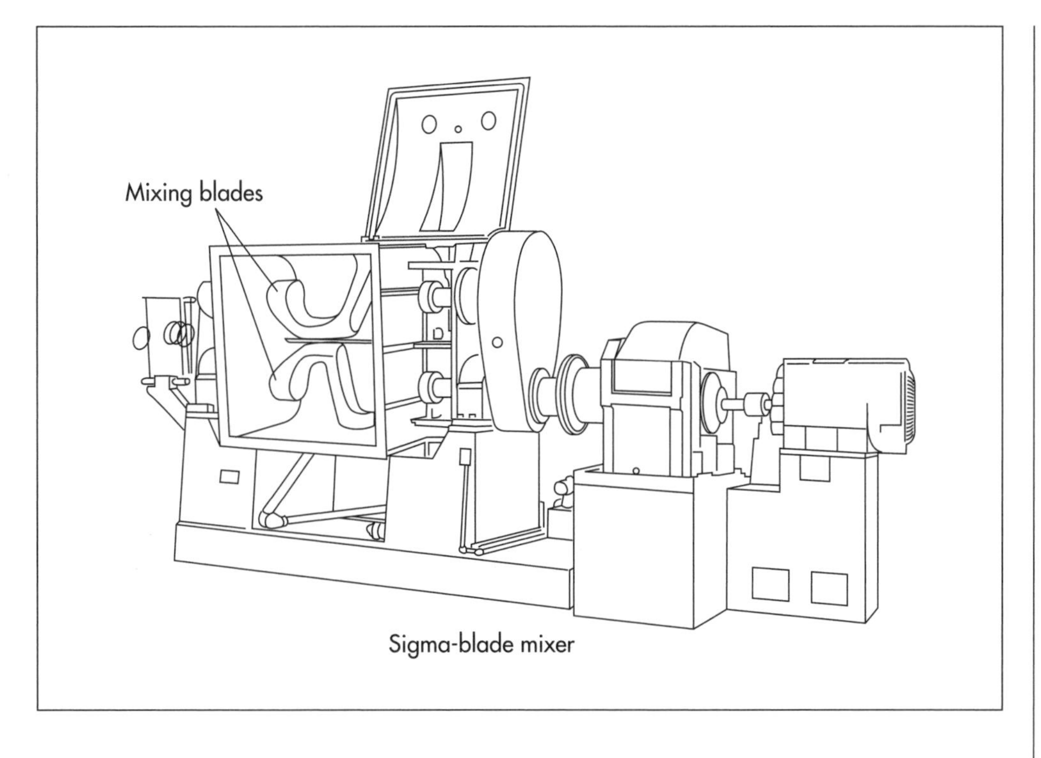


A sigma-blade mixer is used to manufacture Silly Putty. Raw materials are placed into the mixing bowl and blended together for half an hour. Once mixed, the machine operator tilts the mixing bowl and removes the material onto a cart. From there, the Silly Putty is cut and packaged.

12,000 eggs or 300 lb (136.2 kg) each day. More than 300 million eggs—or 4,000 tons of Silly Putty—have been sold since 1950, which is enough to stretch around the earth nearly three times.

## Raw Materials

Silly Putty is made from a mixture of silicone polymers (about 70 wt%) and other chemicals, including boric acid. Powdered fillers (clay and calcium carbonate) and dry pigments (to produce color or glitter) are also added. A homemade recipe can be made from mixing together water, white glue, and borax solution.

## Design

Silly putty was a serendipitous design that resulted from the combination of boric acid and silicone oil. The original design has not been significantly changed. Even the plastic eggs in which the silly putty is packaged has remained for the original marketing campaign.

## The Manufacturing Process

The process for making Silly Putty is relatively simple and involves only a few steps. After the raw materials are checked to make sure they meet specifications, they are weighed in the appropriate amounts to make up a batch.

### Mixing

1 The ingredients are placed into a large machine called a sigma blade mixer and blended together for half an hour. Once the batch is mixed, the machine operator tilts the machine and removes the sticky material onto a cart. From there, it is transported to the cutting operation.

### Cutting

2 The mixed material is first cut into basketball-sized pieces by hand. These pieces are then fed into the hoppers of a machine traditionally used for pulling taffy by the candy industry that extrudes and cuts the Silly Putty into smaller pieces similar in size to a golf ball. As the pieces come out of the machine, the pieces fall onto a conveyor belt that transports them to the packaging area.

### Packaging

3 There are two packaging steps. First, production workers place each piece into the plastic egg. The plastic eggs are then conveyed to a special packaging machine called a blister carding machine. This machine automatically places the egg, one per card, onto a special piece of cardboard called

a blister card. Next, the machine encases the egg into a vacuum formed plastic tray.

## Quality Control

A safety evaluation is conducted before manufacturing begins. At this point, the product ingredients are certified in a toxicological evaluation by the Art & Creative Materials Institute's consulting toxicologists. This certifies that the product contains no harmful or toxic materials and is then granted the AP (approved product) seal. Once this certification takes place, Silly Putty is labeled in accordance with ASTM (American Society for Testing and Materials) labeling standards.

The weight of each Silly Putty egg is controlled by sampling the extrusion process. Statistical process control is also used to maintain the correct weight. In addition to being visually checked for color, sample eggs are tested for bounce, stretch, and other performance properties.

## Byproducts/Waste

What little waste is produced is recycled back into the mixing process. Since it must be safe for child's play, the material used to make Silly Putty is nontoxic.

## The Future

Though Silly Putty is particularly popular with children, college students have recently taken a renewed interest. New uses will continue to be created and the artist community may find Silly Putty a medium they can't resist. At least one sculptor already is selling his creations for several thousand dollars a piece.

Silly Putty turns 50 in 2000 and a special commemorative product will be introduced. Silly Putty remains a classic toy that appeals to any age group and will continue to provide fun for many throughout the twenty-first century and beyond.

## Where to Learn More

### *Periodicals*

Chase, R. "Silly Putty's Turning 50."*Minneapolis Star Tribune*, September 16, 1998

"Silly putty: It's almost 50! It's still silly! And it still sells! "*The Courant*, August 2, 1995.

### *Other*

Binney & Smith. 1100 Church Lane Easton, PA 18044. (610) 253-6271.

*—Laurel Sheppard*

# Soccer Ball

## Background

People have played games similar to modern soccer around the world since ancient times. The oldest recorded soccer-like game is the Chinese game of *tsu-chu*, allegedly invented by the emperor Huang-Ti in 1697 B.C. Records from Huang-Ti's time describe a game played with a leather ball stuffed with animal hair and cork. Two teams vied to kick it through goal posts. The Japanese played a similar game called *kemari* in the same era. A North African game from the seventh century B.C. was also evidently similar to soccer, though it was a ceremonial game played as part of a fertility rite.

The ancient Greeks participated in a game involving kicking and throwing a ball on a marked field. It was called *espiskyros*. The Romans later had similar games, probably adapted from the Greek. One of the Roman games was called *follis*, and used a large light ball filled with hair. In follis, the players tried to keep the ball in the air with their hands. Another version was called *harpastum*. In this rougher game, players tried to tackle the person with the ball. Harpastum was popular among Roman soldiers, and it spread throughout Italy, and then across the Roman Empire. The game was brought to England, and from there its history becomes more narrowly British.

The balls used in early Britain were probably made from inflated animal bladders, though there are persistent rumors of games played with human heads. The skulls of either Roman or Viking oppressors were said to have been batted about at various Dark Age revelries. The British also may have had a soccer-like game played with the head of an animal, used in pre-Christian times as a fertility rite. This evolved in the Middle Ages into a game played on Shrove Tuesday. Teams competed to kick the head of an animal around their agricultural fields. The winning team got to bury the head on its ground, supposedly guaranteeing a good harvest.

By the 1300s, soccer (at that point called football) had evolved into a rough street game, where opposing mobs injured each other and crashed through houses and stores in their struggle to get the ball through the goal posts. King Edward II outlawed the game in 1314, and subsequent rulers had to renew his decree. As with the Roman soldiers, football was popular with British army men, and they apparently neglected their military training in order to play.

The game continued to be played in the streets of England at least through the eighteenth century, but in the nineteenth century, it became an upper-crust game, played at British public schools and colleges. Each school had its own rules, and the first standardized soccer rules were published in 1862, so that graduates of the different schools could play harmoniously. The English Football Association was founded in 1863. The term soccer dates to that time, when in British college student slang the term ruggers was the game played at Rugby, and soccer was the game played according to the Football Association rules.

*The English Football Association was founded in 1863. The term soccer dates to that time, when in British college student slang the term ruggers was the game played at Rugby, and soccer was the game played according to the Football Association rules.*

When the English Football Association was founded and it's official rules (based on Eton rules) of the game soccer drafted, it excluded certain game nuances that had developed at the Rugby school. Running with the ball and hacking (violently scrap-

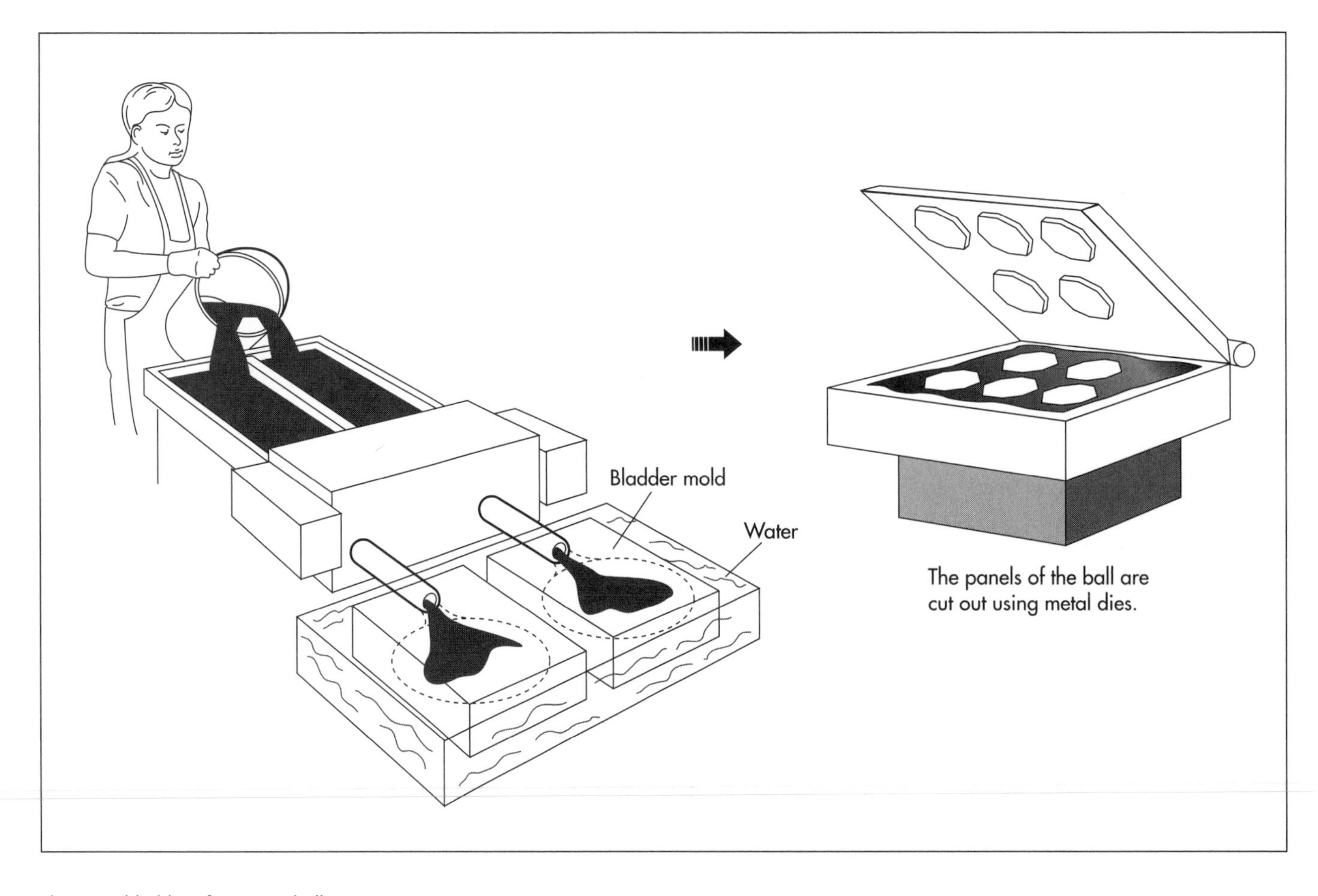


The panels of the ball are cut out using metal dies.

The inner bladder of a soccer ball can be made from either natural or synthetic rubber that is gently heated and forced into a mold. The outer panels are produced from sheets of synthetic leather backed with several layers of cloth, to strengthen the material. The sheets of synthetic leather are passed through a die-cutting machine that cuts the hexagonal panels and also punches the stitch holes. The panels are then silkscreened and imprinted with the manufacturer's logo.

ping at the ball with the boot to get it away from the player or tripping the player running with the ball) were not adopted by the Cambridge rules, and so Rugbeian teams refused to join the Football Association. In 1871, eight years after the organization of the English Football Association, Rugby teams formed the Rugby Football Union, which drafted its own official rules of the game rugby. With so much strong debate over hacking, the Rugby Football Union also discarded the practice. Over the years, rugby rules have been modified, but still the game and its equipment remains distinct from soccer.

Worldwide spread of soccer came with the expansion of the British empire. British laborers working on the Argentine and Uruguayan railroads took the game to those countries. International codification of the rules came in 1904, with the founding of the International Federation of Association Football (known by its acronym in French, FIFA). By the late 1990s, soccer was the most popular game on Earth, with millions of fans and a burgeoning equipment industry.

## Raw Materials

The standard soccer ball is made of synthetic leather, usually polyurethane or polyvinyl chloride, stitched around an inflated rubber or rubber-like bladder. Older balls were made of genuine leather and held shut with cotton laces. Modern balls have a valve. The synthetic leather panels are backed with cloth, usually polyester or a poly-cotton blend. The backing is held on with a latex adhesive. The ball is spherical, and for standard play must be no bigger than 28 in (71.12 cm) around, and no smaller than 27 in (68.6 cm). Its weight is specified at no less than 14 oz (0.392 kg) and no more than 16 oz (0.448 kg), filled to a pressure of 15 lb per sq in (6.8 kg per sq cm).

## The Manufacturing Process

There are two main elements of the soccer ball. One is the inner bladder, the second is the outer covering. Sometimes the whole soccer ball is made under one roof.

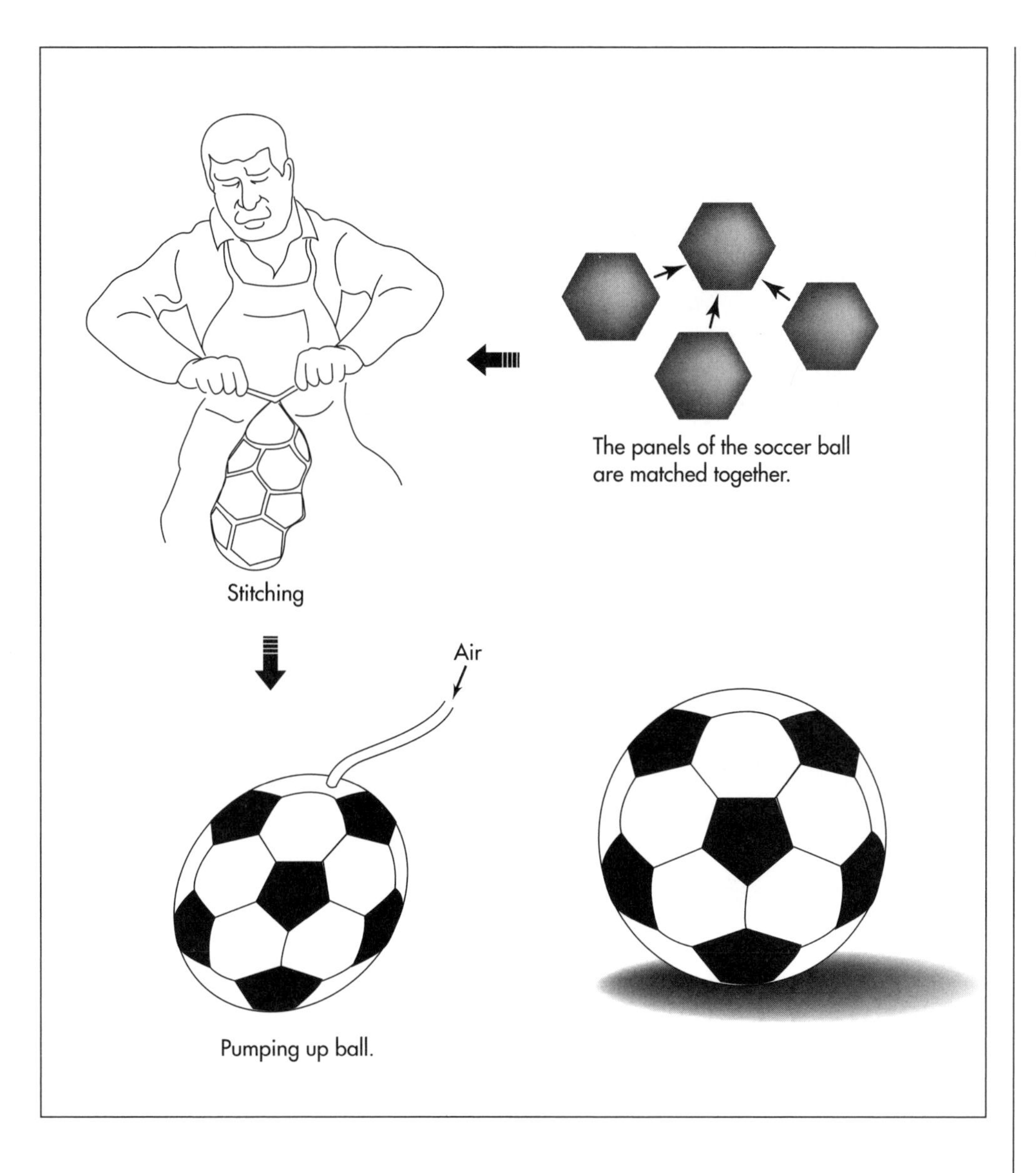


The 32 panels are sewn together by hand. The stitcher inserts the bladder and sews the final seams. In order to ensure that the bladder has not been punctured by a needle during stitching, a worker pumps up the ball. Then, the finished ball is deflated again and packed for shipping.

In many cases, the bladders are manufactured and the covering is die-cut and imprinted with logos at a central factory, and the stitching is done by hand elsewhere. The ball covering consists of 32 hexagonal panels, and the hand stitching of these is tedious and exacting work. Over 80% of all soccer balls are manufactured in Pakistan.

### *Lining the cover material*

1 Sheets of synthetic leather are backed with several layers of cloth, to strengthen the material. The sheets are fed through a press that applies a latex adhesive and attaches the cloth. Material for a cheaper ball will receive a two-layer backing. A sturdier, more expensive ball will have four layers of cloth attached.

### *Cutting the panels*

2 After the adhesive is fully dried, the sheets of synthetic leather are passed to a die-cutting machine. Workers load the cloth into the die cutter, which cuts the hexagonal panels and also punches the stitch holes.

### *Imprinting the panels*

3 Some of the cut panels are next taken to a screening area. Workers silkscreen the panels individually, imprinting them with the manufacturer's logo. This is done with a specially fast-drying paint.

### *Making the bladder*

4 The bladder can be made from either natural or synthetic rubber. The raw material is gently heated and forced into a

mold, where it forms a balloon. As the material cools, it wrinkles. Next, workers remove the bladders and partially inflate them to smooth them out.

### Stitching and final assembly

5 Now the balls are ready to be assembled. The 32 panels and the bladder are packed in a kit for the stitcher. The stitcher uses a pattern to guide him or her in assembling the panels in the proper order. The stitcher sews the panels together by hand.

6 When the cover is sewed, the stitcher inserts the bladder and sews the final seams. The ball is complete. In order to ensure that the bladder has not been punctured by a needle during stitching, a worker pumps up the ball. The ball may also be weighed and measured at this point. Then, the finished ball is deflated again and packed for shipping.

## Quality Control

The soccer ball is inspected at many points in its assembly. The bladders are checked as they are removed from the molds. The covering material is checked after it is lined with the backing material. Another important quality checkpoint is the silkscreen printing. The printed panels are visually inspected, and any faulty ones removed. After the ball is stitched together, an inspector looks it over carefully to see that no stitches have been missed. The bladder is inflated and the ball weighed and measured before the ball is passed on to the shipping area.

## The Future

Soccer is the world's most popular sport, and ball manufacturers are vying to create new sensations in soccer equipment. However, the shape, size, and weight of the ball is defined by international rules, and in a traditional sport, there is not too much room for innovation. Manufacturers are testing new synthetics, such as butyl for the inner bladder. For the covering, the aim is to create a softer, more pliable ball. While polyurethane and polyvinyl chloride are the synthetic leathers of choice, manufacturers are investigating new materials or new ways of treating these materials in order to come up with a better ball.

In the manufacturing process itself, the biggest change will hopefully lie in the eradication of unfair working conditions. Public outcry in the mid-1990s forced manufacturers to address the issue of child and prison labor. American consumers were confounded in 1994 when it was widely publicized that child laborers, living in appalling conditions of poverty and enslavement, sewed the majority soccer balls. Over 80% of all soccer balls are manufactured in Pakistan, where the stitching was routinely outsourced to remote villages where children did the work. In the mid-1990s, it was also revealed that some balls manufactured in China had been stitched by prison inmates. Under a barrage of negative publicity, reputable soccer ball manufacturers vowed to centralize their manufacturing under one roof, and not sub-contract to outside stitchers. Many of the largest manufacturers instituted human rights departments in order to guarantee that their balls were not being assembled by child or prison labor. Some athletes refused to endorse soccer products unless they had a guarantee that balls were produced under humane conditions. With high consumer awareness of the child labor problem, it is in the best interest of ball makers to monitor working conditions at their foreign plants. Some manufacturers are investing in technology that will eliminate hand stitching altogether. In the late 1990s, automatic stitching machines were able to produce low quality balls, suitable for non-professional play. Manufacturers hope to improve the stitching technology so that all grades of balls can be made by machine in the future.

## Where to Learn More

### Books

LaBlanc, Michael L. and Richard Henshaw, *The World Encyclopedia of Soccer*. Detroit, Michigan: Visible Ink, 1994.

### Periodicals

Cove, Tom. "The Last Word on Soccer Balls and Child Labor." *Sporting Goods Business* (February 4, 1998).

Newell, Kevin. "Goal Oriented: Soccer Companies with an Eye on Growing U.S. Sales Are Getting a Leg Up on Innovation." *Sporting Goods Business* (April 15, 1998).

Schanberg, Sydney H. "Six Cents an Hour."*Life* (June 1996).

*—Angela Woodward*

# Soy Milk

*Commercial manufacturers of soy milk are able to produce as many as 18,000 packages of soy milk per production hour.*

## Background

Soy milk is a high protein, iron-rich milky liquid produced from pressing ground, cooked soybeans. Creamy white soy milk resembles cow's milk but in fact differs from its dairy counterpart in a number of ways. Not only is it higher in protein and iron content, but it is cholesterol-free, low fat, and low sodium. It is, however, lower in calcium and must be fortified with calcium when given to growing children. Those who are allergic to cow's milk or are unable to digest lactose, the natural sugar found in cow's milk, find soy milk easy to digest since it is lactose-free. Those who are calorie-conscious can purchase reduced fat soy milk (called lite soy milk) but this is often lower in protein as well. Some do not enjoy the taste of original soy milk, so manufacturers now offer flavored soy milk. Soy milk can be substituted for milk in nearly any recipe. Those who merely want to boost protein intake often add powdered soy milk to other beverages; others find it economical to purchase it in powder form and then make soy milk when they add water to the powder. Children under one year of age should be given a formula of soy milk specifically developed with their nutritional needs in mind. Soy milk that is intentionally curdled is known as tofu.

## History

The soybean (*Glycine max*) is the world's foremost provider of protein and oil. The Chinese have been cultivating soybeans for thousands of years. The first written record of Chinese cultivation dates to the third centuryB.C. Many believe that the Chinese have been making soy milk for centuries—it has been sold in cafes and from street vendors for generations. So important to the Chinese are soybeans for the production of soy milk and tofu that soybeans are considered one of the five sacred grains along with rice, wheat, barley, and millet. Soybeans made their way to Japan by the sixth century and to Europe by the seventeenth century.

The beans came to the United States on ships from Asia and were used as ballast and often discarded once the ships docked. But soldiers during the Civil War substituted soy beans for coffee beans and were thus making their own form of soy beverage. By the nineteenth century, soy beverages were available in Europe as well.

However, the popularity of soybean products, including soy milk, came slowly to the United States. African-American agriculturist George Washington Carver began studying the soybean and touting its nutritive value in the early twentieth century. Shortly thereafter, doctors became intrigued with their use for its nutritional value, particularly for children unable to drink cow's milk. Soybean production has increased in the United States throughout the twentieth century and is a staple crop for many midwestern farmers, allowing soy milk producers a steady supply of the main ingredient. Soybeans are grown in 29 states and are our second largest cash crop.

Until the 1950s, soy milk was made in small quantities at home or in small shops and was not produced on a mass scale in this country. At this time, soy milk was bottled like soft drinks. Much of the technology now used in the production of soy milk was developed by the Japanese who use soy beverages (and other soy products) in tremendous quantities.

In the 1970s, when interest in soy and other non-dairy products soared, manufacturers began adding flavors to the bland soy milk. Shortly thereafter, the development of aseptic packaging (in which the milk is packaged in such a way that no air is introduced which can contain harmful bacteria) brought the beverage into the modern era.

## Raw Materials

Soy milk requires only soybeans and water (and steam) for its creation. Soy milk is nearly always fortified with calcium, vitamins D, and certain B vitamins. Highly concentrated flavorings, such as vanilla, carob, chocolate, and almond are often added to the finished product. Many companies add sugar and salt to the drink to make it more palatable to the consumer.

## The Manufacturing Process

The soybean is a low acid food and as such, is a good host for the breeding of harmful bacteria. Thus, the manufacturing process is "aseptic," meaning that at a certain point in its production, the soy milk is sealed off from any air because it might introduce dangerous bacteria into the product. The development of successful, affordable aseptic production of soy milk has been of tremendous importance in the mass production of this beverage. The initial phases of the production of soy milk do not have to be sealed off to air; only later does this happen.

### *Procuring the raw materials*

1 Soy milk manufacturers very often work directly with farmers so that the kind of soy bean that produces good soy milk is grown (one manufacturer gives the farmers the seeds for the soybeans they require). Generally soy milk producers seek large soybeans called clear hylem.

Once the soybeans are harvested and brought to the plant, the beans are cleaned in a grain elevator or bin on or off premises. The process may begin with the blending together of four to six tons of soybeans at one time. Some factories have two or more production lines running at one time, and thus use several tons of soybeans in a day.

### *De-hulling*

2 The soybeans are steamed and split in half. This loosens the hull on the bean. A vacuum sucks off the hulls.

### *Invalidating the indigestible enzyme*

3 Next, soybeans must be cooked in order to invalidate, or counteract, a specific enzyme which makes them indigestible to humans. This cooking occurs in the Enzyme Invalidator, in which the de-hulled soybeans are cooked using high pressure, Water, and high temperature (creating very hot live steam) to invalidate that enzyme.

### *Rough grinding*

4 The cooked soybeans then fall into the first rough grinder or mill. Water is added to the machine and the bean pieces are roughly ground in this first milling.

### *Finer grinding*

5 Although they have been ground once, the cooked soybeans are still rather coarse. Thus, the fine grinder further pulverizes the bean pellets into small particles. The hot slurry is white in color with minuscule particles of insoluble soybean particles.

### *Extracting*

6 A large centrifuge is then used to extract the tiny bits of soybean that are insoluble and cannot be included in the finished product. These particles are separated from the soy milk slurry using a centrifuge. A rubber roller presses the soy milk slurry against the surface of a drum within the centrifuge, forcing the liquid inside the drum while the fibers remain on the outside of the drum. The drum is then scraped of these fibers.

These soybean fibers are physically removed from the production process at this time. This waste soy fiber is called okara and it resembles mashed potatoes. A separate process dries the okara for use other than human consumption. The fiber-less soy liquid is raw soy milk at this point and is referred to in the industry as jun.

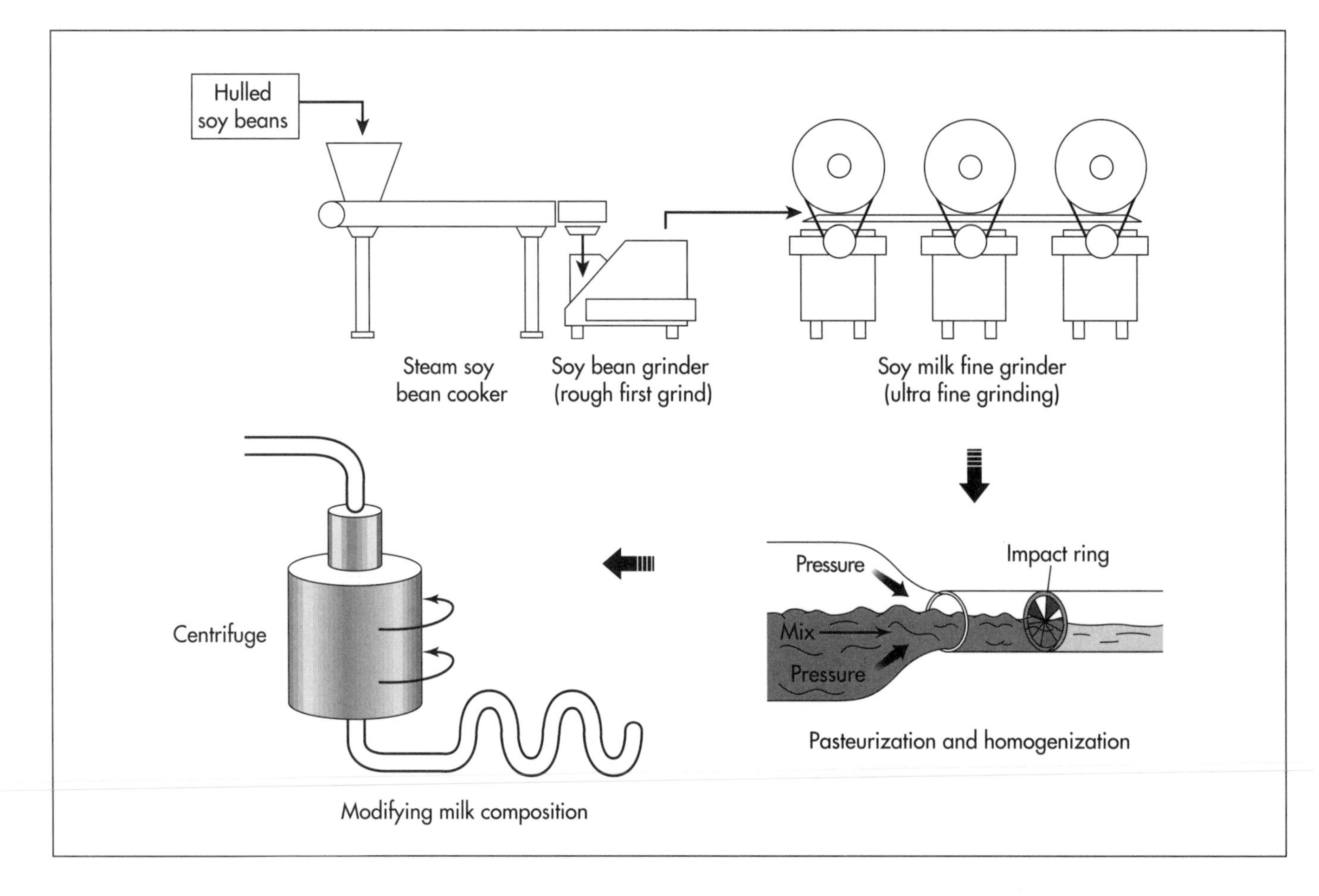


Good quality soybeans are harvested, cleaned, hulled, and pressure cooked. Next, the cooked soybeans are ground by a number of grinders that transform the beans into a milky slurry. The slurry is placed in a centrifuge that extracts any insoluble bits of bean. The separated soy liquid called jun is blended with vitamins, flavorings, and sugar and then sterilized and homogenized. The hot milk is cooled and packaged in such a way that it is never exposed to air.

### Blending

7 The jun is injected into large tanks and flavorings, sugar, and vitamins are mixed separately in smaller tanks. Ingredients of the smaller tank are infused into the larger tanks, thus blending the flavors with the raw milk.

### Aseptic sterilizing

8 At this point, it is essential that the jun be sealed within the equipment until the end of the manufacturing process (including packaging) in order to keep out air and ambient bacteria and germs that can grow in low-acid soy milk. Sterilization occurs with pressure and very hot temperatures within a vacuum for a short period of time.

### Homogenizing

9 From the sterilizer, the hot milk is sent to the homogenizer. This breaks down the fat particles and prevents them from separating from the rest of the mixture. In the homogenizer, which is essentially a high-pressure piston pump, the is blended as it is drawn into the pump cylinder and then forced back out in a repetitive motion.

### Cooling

10 Next, the hot milk is piped to the cooling tank. Here, the hot milk passes next to cold plates that lower the temperature of the soy milk to room temperature.

### Storing

11 The cooled milk is sent to the aseptic (sealed) tanks and held here in preparation for packaging. Here, the soy milk is refrigerated, pressurized, and sealed to ensure no bacteria thrives in the milk.

### Packaging

12 A very important part of the production is the aseptic packaging of the product. Packaging machines have been developed for this product that are able to mechanically package the product without exposing it to air. The cooled milk is sent to

this packaging machine which has a ribbon of flat packaging (cardboard) threaded into it. As the milk runs through the machine, the packaging surrounds the milk and a cutter cuts through the cardboard packaging and the milk, simultaneously folding the package and sealing the milk within it. A machine glues a plastic spout onto the sealed package. From here, the product is sent to an automatic sorter that packs a case and places it on a pallet. A modern factory is able to produce as many as 18,000 packages of soy milk in an hour.

## Quality Control

Quality control begins with acquiring high quality soybean for the production of soy milk. The beans considered most desirable for the process are called clear hylem, with a white (or colorless) hylem on the body of the bean. While the soybean is generally bland, the clear hylem variety is considered more flavorful. A number of soy milk producers market their product as organic and beans purchased from farmers for soy milk must be certified organic in order to be utilized.

The production of soy milk must be meticulously monitored to ensure that no bacteria grows in the low acid medium. Thus, many factories include over 200 quality control checkpoints in this production. Temperatures of water, steam, and the monitoring of pressure is essential in this process. In addition, the product is constantly analyzed as a sample of the product is taken off the line every 10 minutes and checked for pH, temperature, and bacterial growth (many samples are cultured). Because the product is sealed off from the workers for much of the production, visual checks occur primarily as the product comes off the line. Here, workers check to ensure packages are properly sealed.

## Byproducts/Waste

Until recently, the unusable okara was a significant waste problem for many soy milk production plants. Okara, the insoluble fiber that is removed from the raw soy Now, soy milk producers send the okara to a drying machine which takes the moisture out of the okara, transforming it into a high-fiber, high-protein animal feed. The dried okara is now sold to farmers for feed, thus eliminating a storage and waste problem at most soy milk plants. It has proven to be invaluable to farmers who raise organically fed animals because many soy milk producers only take in organically grown soybeans. Thus, the dried okara feed produced from these beans is considered organic and acceptable for feed.

## Where to Learn More

### Books

Herbst, Sharon Tyler.*Food Lover's Companion.* Barron's Educational Services, Inc.

### Other

Eden Foods. "History of Soy Beverages." Clinton, MI: Eden Foods, 1997.

Eden Foods. "Soy Beverage Manufacturing Process." Clinton, MI: Eden Foods, 1997.

Guelph Food Technology Centre. http://www.gftc.ca/tech/equip/soymilk.htm/ (June 29, 1999).

U.S. Soyfoods Directory. http://www.soyfoods.com (June 29, 1999).

*—Nancy EV Bryk*

# Spacesuit

*Presently, NASA has 17 completed Extravehicular Mobility Units (EMUs), each of which cost over $10.4 million to make.*

A spacesuit is a pressurized garment worn by astronauts during space flights. It is designed to protect them from the potentially damaging conditions experienced in space. Spacesuits are also known as Extravehicular Mobility Units (EMUs) to reflect the fact that they are also used as mobility aides when an astronaut takes a space walk outside of an orbiting spacecraft. They are composed of numerous tailor-made components that are produced by a variety of manufacturers and assembled by the National Aeronautics Space Agency (NASA) at their headquarters in Houston. The first spacesuits were introduced during the 1950s when space exploration began. They have evolved overtime becoming more functional and complicated. Today, NASA has 17 completed EMUs, each of which cost over $10.4 million to make.

## Background

On Earth, our atmosphere provides us with the environmental conditions we need to survive. We take for granted the things it provides such as air for breathing, protection from solar radiation, temperature regulation and consistent pressure. In space, none of these protective characteristics are present. For example, an environment without consistent pressure doesn't contain breathable oxygen. Also, the temperature in space is as cold as -459.4° F (-273° C). For humans to survive in space, these protective conditions had to be synthesized.

A spacesuit is designed to re-create the environmental conditions of Earth's atmosphere. It provides the basic necessities for life support such as oxygen, temperature control, pressurized enclosure, carbon dioxide removal, and protection from sunlight, solar radiation and tiny micrometeoroids. It is a life-support system for astronauts working outside Earth's atmosphere. Spacesuits have been used for many important tasks in space. These include aiding in payload deployment, retrieval and servicing of orbiting equipment, external inspection and repair of the orbiter, and taking stunning photographs.

## History

Spacesuits have evolved naturally as technological improvements have been made in areas of materials, electronics and fibers. During the early years of the space program, spacesuits were tailor made for each astronaut. These were much less complex than today's suits. In fact, the suit worn by Alan Shepard on the first U.S. suborbital was little more than a pressure suit adapted from the U.S. Navy high-altitude jet aircraft pressure suit. This suit had only two layers and it was difficult for the pilot to move his arms or legs.

The next generation spacesuit was designed to protect against depressurization while the astronauts were in an orbiting spacecraft. However, space walks in these suits were not possible because they did not protect against the harsh environment of space. These suits were made up of five layers. The layer closest to the body was a white cotton underwear that had attachments for biomedical devices. A blue nylon layer that provided comfort was next. On top of the blue nylon layer was a pressurized, black, neoprene-coated nylon layer. This provided oxygen in the event that cabin pressure failed. A Teflon layer was next to hold the suit's shape when pressurized, and the final layer was a white nylon

Sally Ride

Sally Ride is best known as the first American woman sent into outer space. Both scientist and professor, she has served as a fellow at the Stanford University Center for International Security and Arms Control, a member of the board of directors at Apple Computer Inc., and a space institute director and physics professor at the University of California at San Diego. Ride has chosen to write primarily for children about space travel and exploration.

Sally Kristen Ride is the older daughter of Dale Burdell and Carol Joyce (Anderson) Ride of Encino, California, and was born May 26, 1951. As author Karen O'Connor describes tomboy Ride in her young reader's book, *Sally Ride and the New Astronauts,* Sally would race her dad for the sports section of the newspaper when she was only five years old. An active, adventurous, yet also scholarly family, the Rides traveled throughout Europe for a year when Sally was nine and her sister Karen was seven. While Karen was inspired to become a minister, in the spirit of her parents, who were elders in their Presbyterian church, Ride's own developing taste for exploration would eventually lead her to apply to the space program almost on a whim. "I don't know why I wanted to do it," she confessed to *Newsweek* prior to embarking on her first spaceflight.

The opportunity was serendipitous, since the year she began job-hunting marked the first time NASA had opened its space program to applicants since the late 1960s, and the very first time women would not be excluded from consideration. Ride became one of thirty-five chosen from an original field of applicants numbering eight thousand for the spaceflight training of 1978. "Why I was selected remains a complete mystery," she later admitted to John Grossmann in a 1985 interview in *Health.* "None of us has ever been told."

Ride would subsequently become, at thirty-one, the youngest person sent into orbit as well as the first American woman in space, the first American woman to make two spaceflights, and, coincidentally, the first astronaut to marry another astronaut in active duty.

Ride left NASA in 1987 for Stanford's Center for International Security and Arms Control, and two years later she became director of the California Space Institute and physics professor at the University of California at San Diego.

material that reflected sunlight and guarded against accidental damage.

For the first space walks that occurred during the *Gemini* missions in 1965, a seven layer suit was used for extra protection. The extra layers were composed of aluminized Mylar, which provided more thermal protection and protection from micrometeoroids. These suits had a total weight of 33 lb (15 kg). While they were adequate, there were certain problems associated with them. For example, the face mask on the helmet quickly fogged so vision was hampered. Also, the gas cooling system was not adequate because it could not remove excessive heat and moisture quickly enough.

The Apollo missions utilized more complicated suits that solved some of these problems. For moon walks, the astronauts wore a seven layer garment with a life-support backpack. The total weight was about 57 lb (26 kg). For the Space Shuttle missions, NASA introduced the Extravehicular Mobility Unit (EMU). This was a spacesuit designed for space walks that did not require a connection to the orbiter. One primary difference in these suits was that they were designed for multiple astronaut use instead of being custom made like the previous spacesuits. Over the last 20 years, the EMUs have undergone steady improvements however, they still look the same as they did when the shuttle program began in 1981. Currently, the EMU has 14 layers of protection and weighs over 275 lb (125 kg).

## Raw Materials

Numerous raw materials are used for constructing a spacesuit. Fabric materials include a variety of different synthetic polymers. The innermost layer is made up of a Nylon tricot material. Another layer is composed of spandex, an elastic wearable polymer. There is also a layer of urethane-coated nylon, which is involved in pressurization. Dacron—a type of polyester—is used for a pressure-restraining layer. Other synthetic fabrics used include Neoprene that is a type of sponge rubber, aluminized Mylar, Gortex, Kevlar, and Nomex.

Beyond synthetic fibers other raw materials have important roles. Fiberglass is the primary material for the hard upper torso segment. Lithium hydroxide is used in making the filter which removes carbon dioxide and water vapor during a space walk. A silver zinc blend comprises the battery that powers the suit. Plastic tubing is woven into the fabric to transport cooling water throughout the suit. A polycarbonate material is used for constructing the shell of the helmet. Various other components are used to make up the electronic circuitry and suit controls.

## Design

A single EMU spacesuit is constructed from various tailor-made components produced by over 80 companies. The size of the parts vary ranging from one-eighth-inch washers to a 30 inch (76.2 cm) long water tank. The EMU consists of 18 separate items. Some of the major components are outlined below.

The primary life support system is a self-contained backpack that is fitted with an oxygen supply, carbon-dioxide removal filters, electrical power, ventilating fan and communication equipment. It provides the astronaut with most of the things needed to survive such as oxygen, air purification, temperature control and communication. As much as seven hours worth of oxygen can be stored in the suit's tank. A secondary oxygen pack is also found on the suit. This provides an additional 30 minutes of emergency oxygen.

The helmet is a large plastic, pressurized bubble that has a neck ring and a ventilation distribution pad. It also has a purge valve, which is used with a secondary oxygen pack. In the helmet, there is a straw to a drink bag in case the astronaut gets thirsty, a visor which shields rays from the bright sun, and a camera which records extra vehicular activities. Since space walks can last over seven hours at a time, the suit is fitted with a urine collection system to allow for bathroom breaks. The MSOR assembly attaches to the outside of the helmet. This device (also known as a "Snoopy Cap") snaps into place with a chin strap. It consists of headphones and a microphone for two way communication. It also has four small "head lamps" which shine extra light where needed. The visor is manually adjusted to shield the astronaut's eyes.

To maintain temperature, a liquid cooling and ventilation garment is worn under the outer garment. It is composed of cooling tubes, which have fluid flowing through them. The undergarment is a mesh one-piece suit composed of spandex. It has a zipper to allow for front entry. It has over 300 ft of plastic tubing intertwined within which it circulates cool water. Normally, the circulating water is maintained from 40-50° F (4.4-9.9° C). The temperature is controlled by a valve on the display control panel. The lower garment weighs 8.4 lb (3.8 kg) when loaded with water.

The lower torso assembly is made up of the pants, boots, "brief" unit, knee and ankle joints and the waist connection. It is composed of a pressure bladder of urethane-coated nylon. A restraining layer of Dacron and an outer thermal garment composed of Neoprene-coated nylon. It also has five layers of aluminized Mylar and a fabric surface layer composed of Teflon, Kevlar, and Nomex. This part of the suit can be made shorter or longer by adjusting the sizing rings in the thigh and leg section. The boots have an insulated toe cap to improve heat retention. Thermal socks are also worn. The urine storage device is also located in this section of the suit. Old models could hold up to 950 milliliters of liquid. Currently, a disposable diaper type garment is used.

The arm assembly is adjustable just like the lower torso assembly. The gloves contain

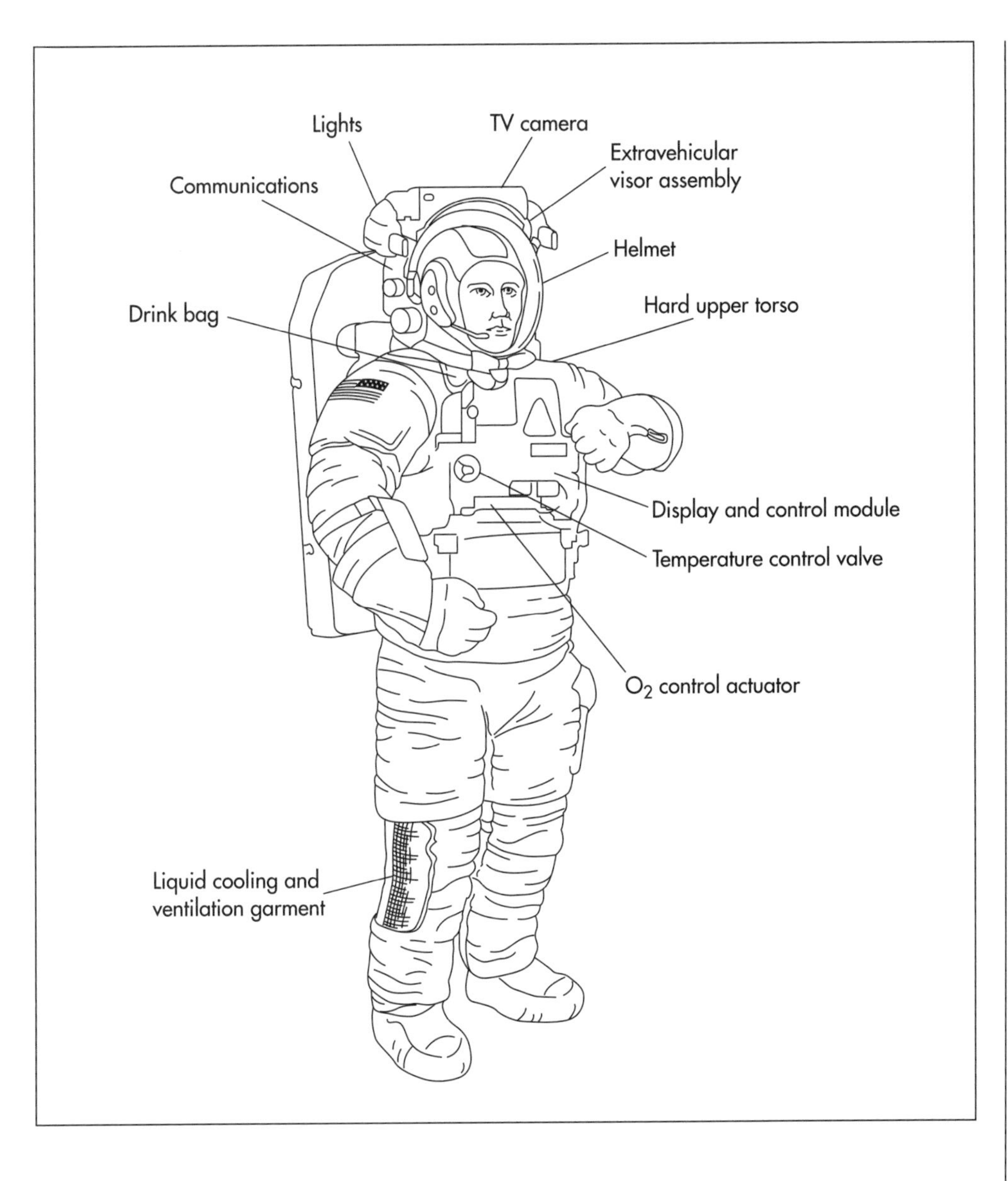


An Extravehicular Mobility Unit (EMU).

miniature battery-powered heaters in each finger. The rest of the unit is covered by padding and an additional protective outer layer.

The hard upper torso is constructed with fiberglass and metal. It is where most of the suit pieces attach including the helmet, arms, life support system display, control module and lower torso. It includes oxygen bottles, water storage tanks, a sublimator, a contaminant control cartridge, regulators, sensors, valves, and a communications system. Oxygen, carbon dioxide and water vapor leave the suit through the ventilation garment near the astronaut's feet and elbows. A drinkbag in the upper torso can hold as much as 32 oz (907.2 g) of water. The astronaut can take a drink through the mouthpiece that extends into the helmet.

Chest mounted control module lets the astronaut monitor the suit's status and connect to external sources of fluids and electricity. It contains all the mechanical and electrical operating controls and also a visual display panel. A silver zinc, rechargeable battery which operates at 17 volts is used to power the suit. This control module is integrated with the warning system found in the hard upper torso to ensure that the astronaut knows the status of the suit's environment. The suit connects to the orbiter through an umbilical line. It is disconnected prior to leaving the airlock.

The white suit weighs about 275 lb (124.8 kg) on earth and has a product life expectancy of about 15 years. It is pressurized to 4.3 lb (1.95 kg) per square inch and can be recharged by hooking up directly to the orbiter. The exist-

The primary life support system is a self-contained backpack that is fitted with an oxygen supply, carbon-dioxide removal filters, electrical power, ventilating fan, and communication equipment.

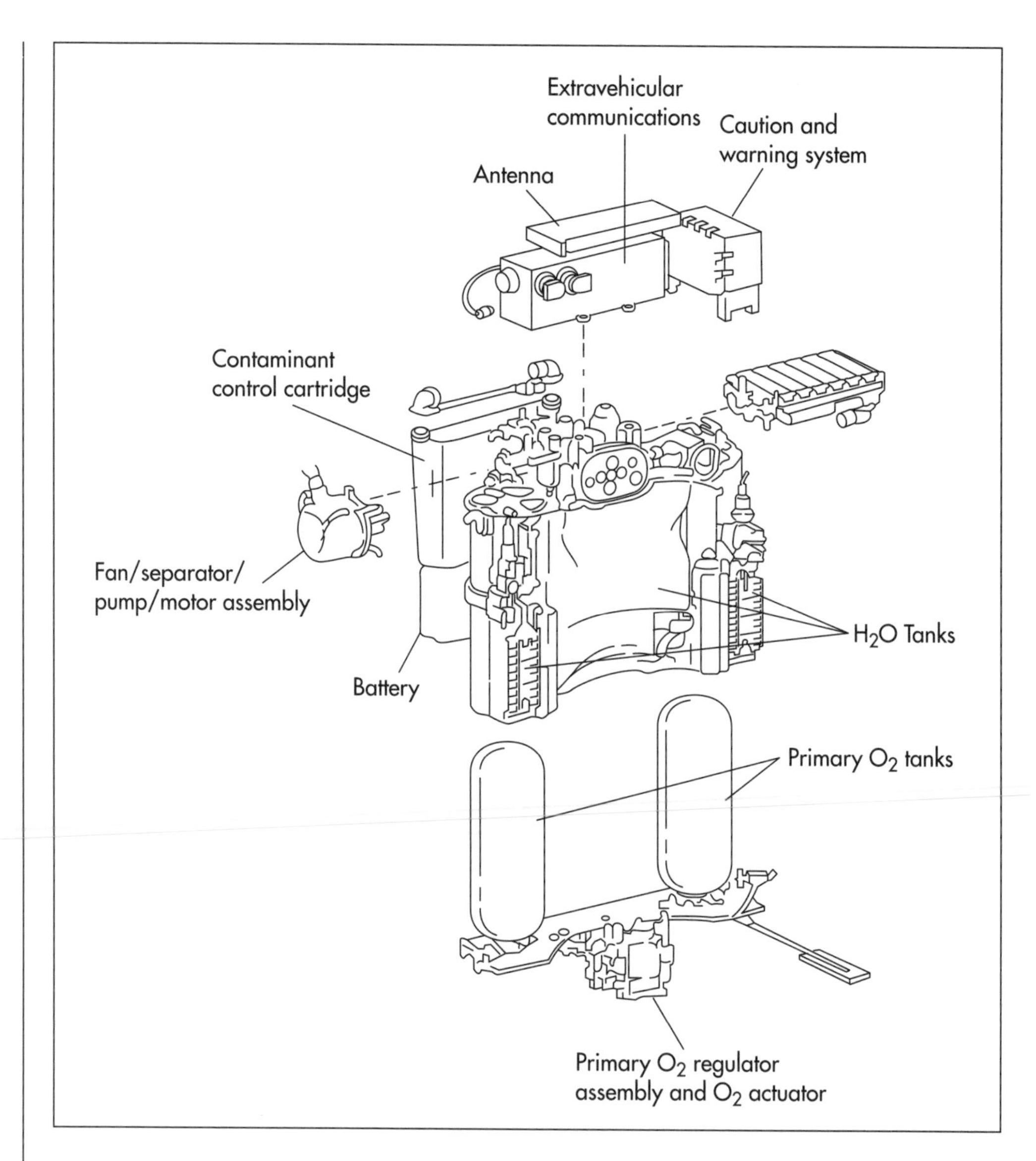


ing spacesuits are modular so they can be shared by multiple astronauts. The four basic interchangeable sections include the helmet, the hard upper torso, the arms and the lower torso assembly. These parts are adjustable and can be resized to fit over 95% of all astronauts. Each set of arms and legs comes in different sizes which can be fine- tuned to fit the specific astronaut. The arms allow for as much as a one inch adjustment. The legs allow for up to a three inch adjustment.

It takes about 15 minutes to put on the spacesuit. To put the spacesuit on the astronaut first puts on the lower garment that contains the liquid cooling and ventilation system. The lower torso assembly is put on next with the boots being attached. Next, the astronaut slides into the upper torso unit which is mounted with the life-support backpack on a special connector in the airlock chamber. The waste rings are connected and then the gloves and helmet are put on.

## The Manufacturing Process

The manufacture of a spacesuit is a complicated process. It can be broken down into two phases of production. First the individual components are constructed. Then the parts are brought together in a primary manufacture location, such as NASA headquarters in Houston, and assembled. The general process is outline as follows.

### *Helmet and visor assembly*

1 The helmet and visor may be constructed using traditional blow molding tech-

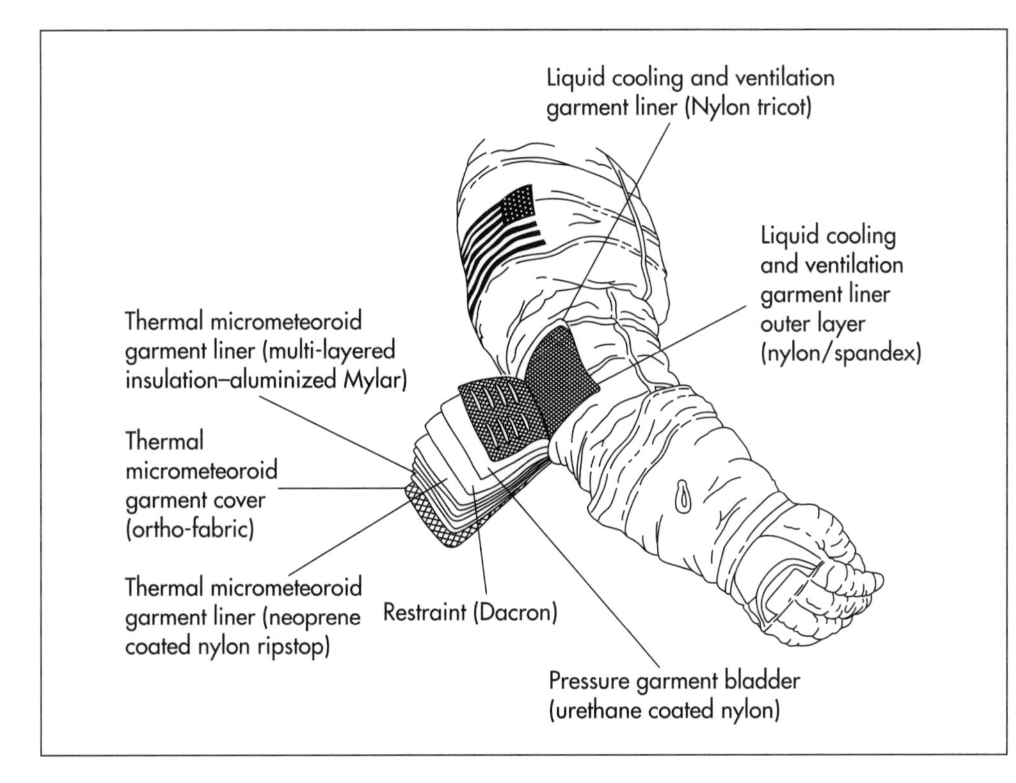


An EMU is made of 14 protective layers. Fabric materials include a variety of different synthetic polymers. The innermost layer is a Nylon tricot material. Another layer is composed of spandex, an elastic wearable polymer. There is also a layer of urethane-coated nylon, which is involved in pressurization. Dacron—a type of polyester—is used for a pressure-restraining layer. Other synthetic fabrics used include Neoprene that is a type of sponge rubber, aluminized Mylar, Gortex, Kevlar, and Nomex.

niques. Pellets of polycarbonate are loaded into a injection-molding machine. They are melted and forced into a cavity which as the approximate size and shape of the helmet. When the cavity is opened, the primary piece of the helmet is constructed. A connecting device is added at the open end so the helmet can be fastened to the hard upper torso. The ventilation distribution pad is added along with purge valves before the helmet is packaged and shipped. The visor assembly is similarly fitted with "head lamps" and communication equipment.

### *Life-support systems*

2 The life support systems are put together in a number of steps. All the pieces are fitted to the outer backpack housing. First, the pressurized oxygen tanks are filled, capped, and put into the housing. The carbon dioxide removal equipment is put together. This typically involves a filter canister that is filled with lithium hydroxide which gets attached to a hose. The backpack is then fitted with a ventilating fan system, electrical power, a radio, a warning system, and the water cooling equipment. When completely assembled, the life support system can attach directly to the hard upper torso.

### *Control module*

3 The key components of the control module are built in separate units and then assembled. This modular approach allows key parts to be easily serviced if necessary. The chest mounted control module contains all of the electronic controls, a digital display and other electronic interfaces. The primary purge valve is also added to this part.

### *Cooling garment*

4 The cooling garment is worn inside the pressure layers. It is made out of a combination of nylon, spandex fibers and liquid cooling tubes. The nylon tricot is first cut into a long underwear-like shape. Meanwhile, the spandex fibers are woven into a sheet of fabric and cut into the same shape. The spandex is then fitted with a series of cooling tubes and then sewn together with the nylon layer. A front zipper is then attached as well as connectors for attachment to the life support system.

### *Upper and lower torso*

5 The lower torso, arm assembly, and gloves are made in a similar manner. The various layers of synthetic fibers are woven together and then cut into the appro-

priate shape. Connection rings are attached at the ends and the various segments are attached. The gloves are fitted with miniature heaters in every finger and covered with insulation padding.

6 The hard upper torso is forged using a combination of fiberglass and metal. It has four openings where the lower torso assembly, the two arms, and the helmet attach. Additionally, adapters are added where the life support pack and the control module can be attached.

### *Final assembly*

7 All the parts are shipped to NASA to be assembled. This is done on the ground where the suit can be tested prior to use in space.

## Quality Control

The individual suppliers conduct quality control tests at each step of the production process. This ensures that every part is made to exacting standards and will function in the extreme environment of space. NASA also conducts extensive tests on the completely assembled suit. They check for things such as air leakage, depressurization, or nonfunctional life support systems. The quality control testing is crucial because a single malfunction could have dire consequences for an astronaut.

## The Future

The current EMU design is the result of many years of research and development. While they are a powerful tool for orbital operations, many improvements are possible. It has been suggested that the spacesuit of the future may look dramatically different than the current suit. One area that can be improved is the development of suits that can operate at higher pressures than the current EMU. This would have the advantage of reducing time currently required for prebreathing prior to a space walk. To make higher pressure suits improvements will have to be made in the connecting joints on each part of the suit. Another improvement can be in the resizing of the suit in orbit. Currently, it takes a significant amount of time to remove or add extending inserts in the leg and arm areas. One other possible improvement is in the electronic controls of the suit. What now requires complex command codes will be done with the push of a single button in the future.

## Where to Learn More

### *Books*

*Suited for Spacewalking.* NASA, 1998.

### *Other*

Hamilton-Standard Company. http://www.hamilton-standard.com/.

*—Perry Romanowski*

# Sponge

## Background

There are many different varieties of sea sponges, and these come in widely varying shapes and sizes. They can be very large, and grow in elaborate branched formations, or be round and small, or grow flat or in a tube shape. Some are brilliantly colored, though they fade when they are harvested. Sea sponges are thought to have evolved at least 700 million years ago. They are among the simplest animal organisms, having no specialized organs such as heart and lungs, and no locomotion. Sponges live attached to rocks on the sea bed. Their bodies consist of skeletons made of a soft material called spongin, and a leathery skin broken by pores. The sponge eats by pumping seawater in through its pores. It filters microscopic plants from the water, and expels the excess water through one or more large holes called oscula. It also absorbs oxygen directly from seawater. Sponges are slow-growing, taking several years to reach full size, and some live for hundreds of years.

Sea sponges were used since ancient times in the Mediterranean region where they are most common. Roman soldiers each carried a personal sponge, which served the purpose of modern toilet paper, and they were certainly used for other purposes as well. Artificial sponges were first developed by the Du Pont company—a leader in synthetic materials manufacturing industry that also invented nylon—in the 1940s. Three DuPont engineers patented the cellulose sponge process, and DuPont held onto the secret until 1952, when it sold its sponge technology to General Mills. In the second half of the twentieth century, cellulose sponges rapidly replaced the natural sponge for most common household uses.

## Raw Materials

Many different types of sponge are harvested and dried for human use, but the most common one is the *Spongia oficinalis*, also known as the glove sponge. Another common type used commercially is the sheep's wool sponge, or *Hippospongia canaliculata*. Synthetic sponges are made of three basic ingredients: cellulose derived from wood pulp, sodium sulphate, and hemp fiber. Other materials needed are chemical softeners, which break the cellulose down into the proper consistency, bleach, and dye.

## Harvesting Sea Sponges

To gather natural sponges, specially trained divers descend into sponge-growing waters with a large two-pronged hook and a string bag. Traditional sponge divers in Greece used no special breathing equipment. The men of seaside villages were trained from childhood and were expert deep water divers. The sponge industry in the United States centers around Tarpon Springs, Florida, a community that was founded by Greek immigrant divers. Today's sponge divers use modern diving equipment such as wet suits and oxygen tanks. The divers pry sponges off the rocks or reefs where they grow, and bring them up in their string bags. The divers pile the sponges on the deck of their boat and cover them with wet cloths. The animals die on the boat, and their skins rot off. After the skins have decayed, the harvesters wash the sponges and string them on a long, thin rope to dry in the sun. After they have dried completely, the harvesters wash the sponges several more times. This is all the preparation the sponges need to be ready for sale.

*Sea sponges are thought to have evolved at least 700 million years ago. They are among the simplest animal organisms, having no specialized organs such as heart and lungs, and no locomotion.*

Natural sponges are the skeletons of a kind of simple sea animal. They grow in warm, shallow waters, and are particularly plentiful in the eastern Mediterranean and off the western coast of Florida. Artificial sponges have largely replaced natural ones in the United States, where at least 80% of the sponges in use are manmade.

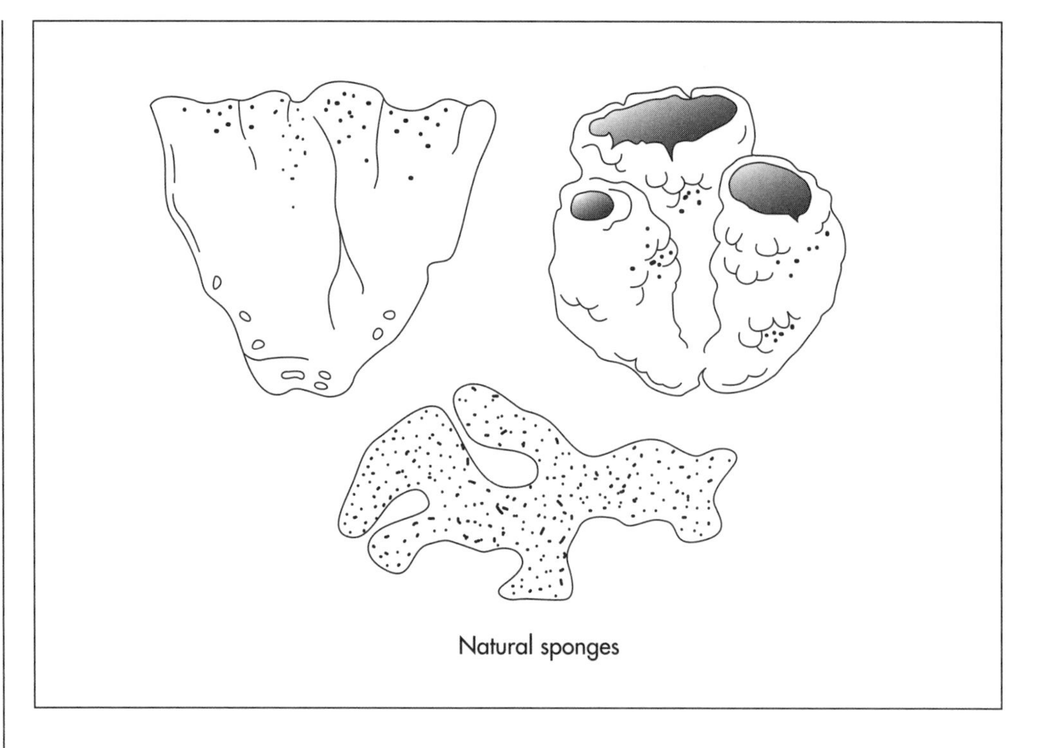

Natural sponges

## The Manufacturing Process

The steps necessary in the manufacture of synthetic sponge is discussed below.

1 The cellulose used for sponges arrives at the sponge factory in large, stiff sheets. Workers take the sheets and soak them in a vat of water mixed with certain chemical softeners. The cellulose becomes soft and jelly-like. Then workers load the cellulose into a revolving mixer, which is a large rotating metal drum. Workers add the sodium sulphate crystals, cut hemp fibers, and dye, and close the mixer. The mixer is set to rotate, and it churns the ingredients so that they are thoroughly amalgamated.

2 From the mixer, workers pour the material into a large rectangular mold that may be 2 ft (61 cm) high, 2 ft (61 cm) wide, and 6 ft (1.8 m) long. The mold is heated, and the cellulose mixture cooks. As it cooks, the sodium sulphate crystals melt, and drain away through openings in the bottom of the mold. It is their melting that leaves the characteristic pores in the finished sponge. The size of the pores is determined by the size of the sodium sulphate crystals. A rough sponge used for washing a car, for instance, is made with coarse crystals, while a fine sponge of the type used for applying make-up is made with very fine crystals. As the celluolose mix cooks, then cools, it becomes a hard, porous block.

3 The sponge block is then soaked in a vat of bleach. This removes dirt and impurities, and also brightens the color. Next the sponge is cleaned in water. Additional washings alter the texture, making the sponge more pliable. The sponge is left to dry, to prepare it for cutting.

4 Some manufacturers make the sponge and cut and package it themselves. Others produce the raw blocks of sponge, and then sell them to a company known as a converter. The converter cuts the sponges according to its customers needs, and takes care of the packaging and distribution. Whether at the first manufacturing facility or at the converter, workers cut the sponges on an automatic cutter. They load each big rectangle of sponge into a machine that slices it into the desired size. Because the sponge block is rectangular, it can be cut into many smaller rectangles with little or no waste.

5 Many household sponges have a textured plastic scouring pad attached to one side. This is attached in a process called laminating, after the sponge is cut. The scouring pad, which is cut to the same size as the sponge, is affixed to the sponge in a laminat-

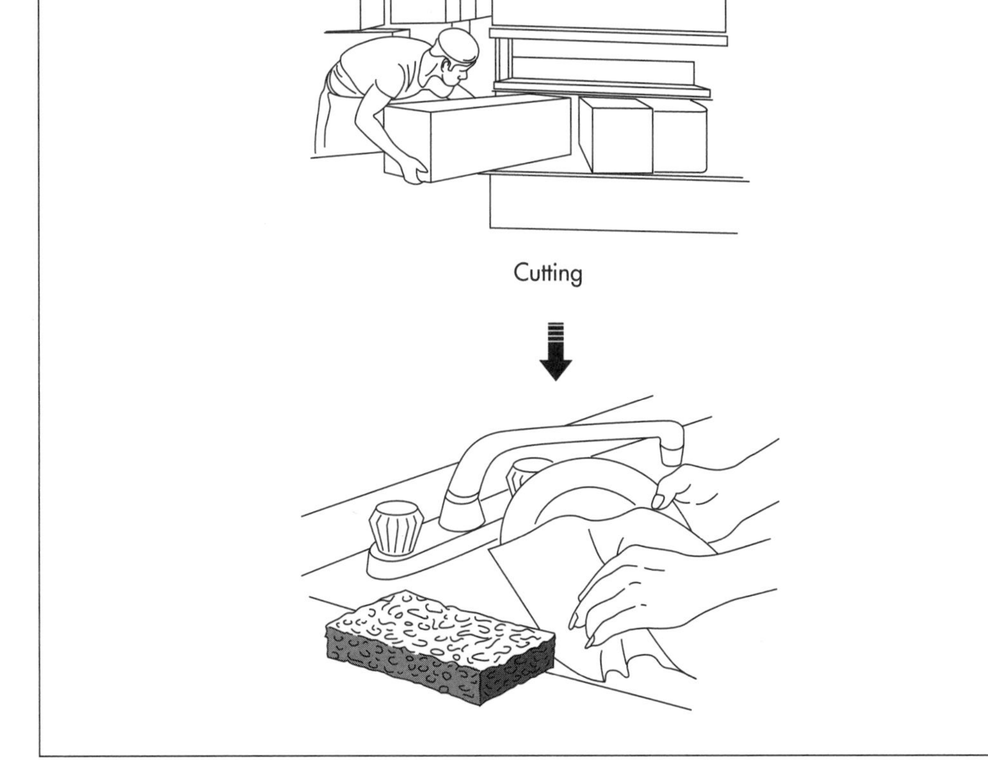


Softened cellulose is mixed with sodium sulphate crystals, cut hemp fibers, and dye in a large, revolving metal drum. Once blended, the material is poured into a large rectangular mold, which may be 2 ft ( 61 cm) high, 2 ft (61 cm) wide, and 6 ft (182.9 cm) long. As the mold cooks, the sodium sulphate crystals melt, and drain away through openings in the bottom of the mold. It is their melting that leaves the characteristic pores in the finished sponge.

ing machine that uses a specialized sponge glue made of moisture-cured polyurethane. Next, the sponges move to a packaging area where they are sealed in plastic. The packaged sponges are boxed, and the boxes sent to a warehouse for further distribution.

## Quality Control

A sponge manufacturer typically checks the product for quality at many steps along the manufacturing process. The raw ingredients are analyzed when they come into the plant to make sure they conform to standards. In a modern facility, most of the machinery is monitored by computers, that maintain the proper proportions in the mix, for example, and control the temperature of the mold during the cooking process. The finished sponges are checked for tenacity, that is, how easily they tear. An inspector takes a random sample from the batch and puts it in a specially built machine. The machine measures the force needed to tear the sponge. Another test is of color. In this case, a sample sponge is examined under a spectrograph.

## Byproducts/Waste

Sponge manufacturing produces no harmful byproducts and little waste. Sponge material that is lost in trimming, such as when an uneven end is cut off the large block, is ground up and recycled. It can be thrown in the mixer at the beginning of the process, and become part of a new sponge.

## Where to Learn More

### Books

Esbensen, Barbara Juster. *Sponges Are Skeletons* (New York: Harper Collins, 1993).

### Periodicals

Sookdeo, Richard. "Ex-sponging Bacteria." *Fortune* (October 31, 1994).

*—Angela Woodward*

# Statuary

## Background

Sculpture is three-dimensional art, and statuary is affordable sculpture for everyone. Statuary encompasses the sublime to the ridiculous it is as familiar as red- and-green lawn gnomes and as exotic the *Winged Victory*, an ancient Greek sculpture displayed in the Louvre Museum in Paris, France. The availability of free time to spend in our yards and gardens has boosted interest in ornamenting quiet corners of flower gardens with cherubs, birdbaths, and gargoyles. The same statuary can be used indoors as paperweights, decorations for the mantelpiece, wall display, or other ornamentation. It provides a relatively inexpensive means of echoing a design or artistic theme in a room, establishing atmosphere, and providing interest in an empty space.

Cast sculpture is the parent of statuary, but the process of making both is essentially the same. The major differences are that sculpture is produced in very limited quantities, materials like statuary bronze are more expensive, sculptures are often large in size (prohibitively so for the average home), and all these factors make sculpture expensive or, indeed, priceless. Statuary is cast using molds and is made of cement, plaster, or resin; but sculpture can be made of almost any material or many materials from marble and bronze to feathers and hubcaps. Any method or material that adds dimension to artwork has potential value to the sculptor. The finish of statuary, like sculpture, consists of either an applied finish like color or a natural finish that allows the true color and beauty of the material to show or to be enhanced by sanding or polishing.

## History

Sculpture has been an important art form in all cultures that have evolved since Paleolithic times. Its three dimensions have permitted artists to interpret mythical characters or pay homage to kings, and construction materials like metal and stone are more durable and adaptable than canvas, paper, and paint. Artistic trends in sculpture have always kept pace and similarity with other styles, although its development has also followed architecture, the filling of large spaces, and the idea of freeing the personality trapped in stone.

The sculpture of all the great civilizations has been copied in making statuary. Statuary is just as old as sculpture, and these same civilizations have prized statuary for display in private and public gardens, patios and piazzas, and interior decoration. The eruption of Mount Vesuvius in 79A.D. perfectly preserved the lifestyles of the inhabitants of Pompeii and Herculaneum, and their homes and gardens housed ornamental statuary.

The rise of civilizations in Europe and Asia produced new styles of statuary, but, periodically, interest in ancient forms and styles was reborn, so statuary has helped preserve and personalize the marble friezes of the Parthenon and clay statues of early Chinese dynasties. In modern America, private homes can feature French Gothic decor, thanks to statuary manufacturers who reproduce the great works of the period in affordable formats.

The extensive gardens of royal palaces and parks in many cultures have included the display of fountains and sculpture. Repro-

*Cast sculpture is the parent of statuary, but the process of making both is essentially the same. The major differences are that sculpture is produced in very limited quantities, materials like statuary bronze are more expensive, sculptures are often large in, and all these factors make sculpture expensive or, indeed, priceless.*

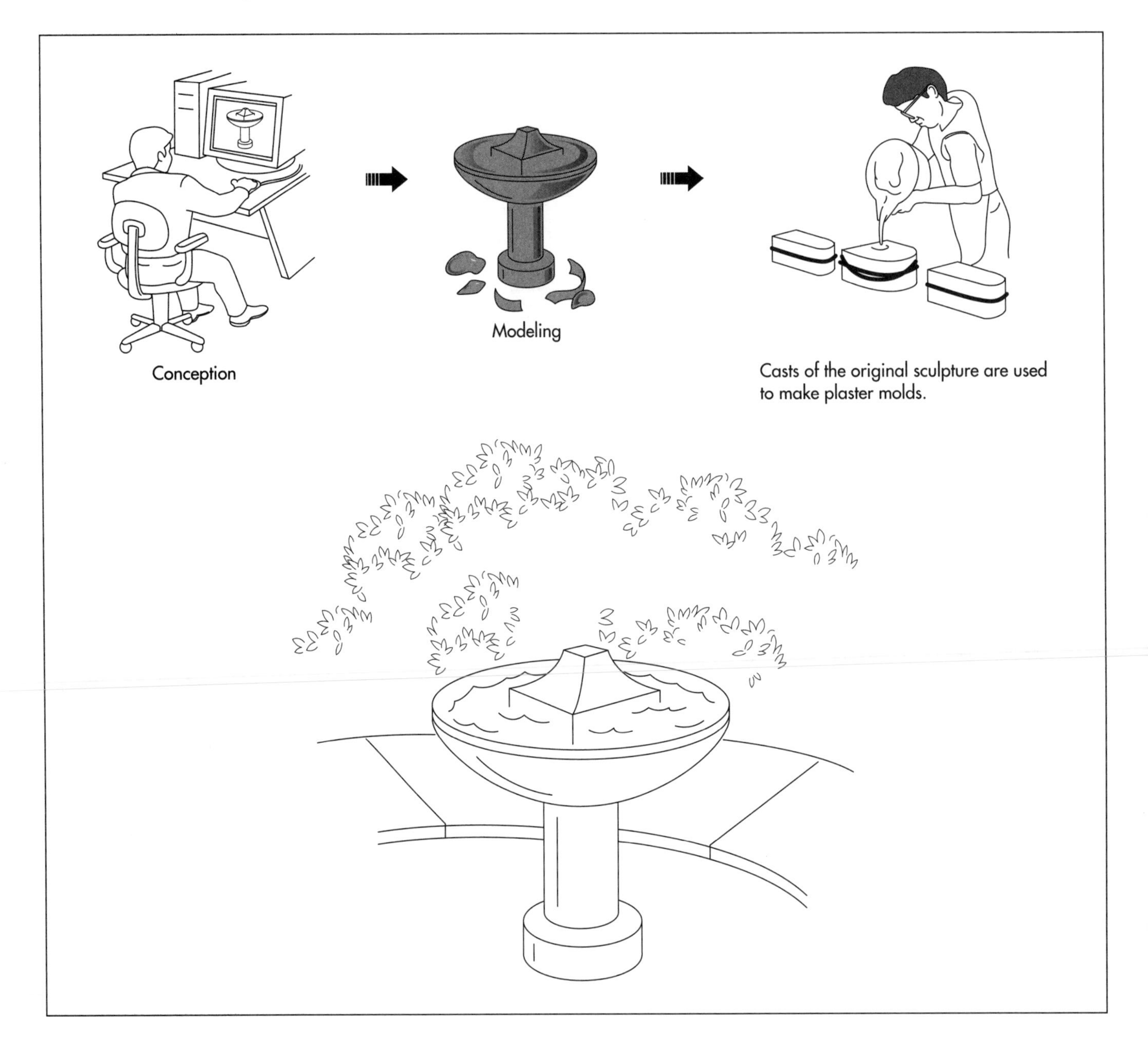


Statuary is cast using molds and is made of cement, plaster, or resin.

duction through statuary allows small spaces to be filled with similar artwork that adds personality, whimsy, and even scholarly knowledge on a more reasonable scale. The renaissance of interest in gardening has helped spur enthusiasm for relief statuary that can be mounted on walls or inset among paving stones as well as statuary birds, rabbits, fairies, and other real and imaginary woodland inhabitants.

## Raw Materials

Materials for home statuary consist of those for making molds and those for the actual sculpture. Flexible molds are used to cast statuary, as opposed to waste molds (that are used once and wasted) or piece molds (that are constructed in sections) for making sculpture. Molds are made from either rubber or silicone, depending on the material that will be cast. Cement and plaster statues are created from rubber molds, and resin statues require silicone molds. Both materials can be molded with different surface finishes, but they can be removed more cleanly from their respective statue types.

The statue-making cement, plaster, and resin are all mixed into liquids by adding clean water to powder components supplied by specialized manufacturers. The grade, dried color, and fineness of the powders will affect the appearance of the final product. They are chosen carefully and may be very different from one style of statue to another. Plaster is used to make statues for the interior only, but cement and resin can be made for indoor or outdoor use. Additives are mixed with the synthetic or polyester resin particularly to make it durable, give it the properties to withstand temperature extremes, and make it non-porous so moisture won't penetrate and cause damage during frosts and freezing. Sometimes, fiberglass are mixed in to add strength; the resin is then termed glass-fiber reinforced polyester resin or GRP. Statues made of enriched cement sometimes include marble dust and are left unfinished so the statue has the natural texture of the cement.

## Design

Designs for home statuary originate from three sources. Old molds made for original sculpture and for statuary makers that have gone out of business are sometimes available for purchase. Statuary makers hire artists to design statues based on existing art; these designers are basically copyists, but they are experts in the process of sketching, modeling clay pieces from their sketches, and making molds from those models. This type of design requires skills in proportioning and interpreting the original art and in reproducing it. Reproduction designers also look at ancient sources with modern eyes and often adapt an older work for modern tastes. Finally, manufacturers of statuary commission designers to create original works. Like reproduction specialists, these designers are also knowledgeable about the processes of modeling and molding casts for statuary, and they use antique sources or themes and new trends in home and garden decorating to create original works.

## The Manufacturing Process

1 For original artwork and some reproductions, the process may begin with a sketch from which the artist makes a clay or plaster original or model. Usually, this step is completed in the design process so the manufacturer can see the three-dimensional piece, review and approve it, and commit to full-scale manufacture. When the work is reviewed, modified (if necessary), and approved, a mold is made from the sculpted model.

2 Using the mold made especially from an original design or an existing mold that may be several generations removed from its original, liquid resin, plaster, or cement is poured into the mold. Although each statue is made individually, many molds for the same design are used to reproduce a number of statues at the same time. The molds are stored in temperature- and humidity-controlled rooms for 24-72 hours, depending on the configuration of the piece. Heat is not applied, so this is called "cold cure" statuary.

3 When the statue has dried partially, the mold is carefully removed. This is possible because the flexible mold can be removed from complex detailing and undercut areas of the statue without damaging either the statue or the mold. The mold can be cleaned and reused. The statue is stored again in a drying room to be completed cured or dehydrated. This step also takes 24-72 hours depending on the material, size, and complexity of design of the piece. The dried statues are inspected before the finishing processes begin.

4 Finishing involves two phases of sanding for most of the statue designs. Enriched cement statues and other pieces with rough appearances aren't sanded at all or are sanded once before priming. Statuary with the finest finish is sanded then painted with primer and sanded a second time. The statues are inspected before they are painted because the quality of the sanded finish will affect quality of the paint job and the final appearance.

5 Statues of the same color are painted in batches in the paint room. Spray guns are used to apply the most popular colors. Most statuary makers will custom finish or paint statues at the customer's request; in this case, the statues are painted individually in small work stations. The painted statues are left to dry overnight.

6 The statues are moved to the packing and shipping department where small ones are boxed in cardboard and larger models are crated in wood. They are stored pending shipment.

## Quality Control

Statuary making is a hand manufacturing process, so opportunities for inspecting the product for quality occur throughout production. Quality control is practiced first in choosing the artists who will design the statues then in accepting the designs for production. Initials casts, prepared molds, and dried statuary are all checked. Sanded and primed statues are also inspected carefully because these preparations prior to painting are critical to the look of the finished statues. Painting is observed during the process and inspected after the paint has dried.

## Byproducts/Waste

Statue manufacturers produce many varieties of statuary for interior or exterior use and in many artistic styles. Waste is minimal. The quality of the plaster, resin, and cement used for the statuary prevents air bubbles and other flaws, so few of the statues have to be destroyed because of errors discovered after they have been cast. Rubber and silicone molds are long lasting. Careful observations throughout manufacturing also limit waste.

## The Future

Statuary's strong future is founded on its past. The museums of the world hold a range of art work that can be reproduced at full size, in smaller versions, and in modernized adaptations to suit every taste. Other sources of sculpted inspiration, like the gargoyles on the great cathedrals and public buildings of Europe, further multiply the possibilities. Interest in sculpture is fueled by the always evolving trends in interior decoration, the popularity of gardening as a hobby, and fascination with art. Mass production of statuary for the home and garden creates affordable art that can stand on the desktop or add personality to a stone walkway or lawn.

## Where to Learn More

### Books

Dawson, Robert. *Practical Sculpture: Creating with plastic media.* New York: The Viking Press, 1970.

Johnson, Lillian.*Sculpture: The Basic Methods and Materials.* New York: David McKay Co., Inc., 1960.

### Periodicals

Adams, William Howard. "The Power of Bare Suggestion." *House & Garden* (April 1985): 234+.

Greenberg, Cara. "Heroics Come Home: Raise High the Roof Beams garden Statuary Is Moving Indoors." *Metropolitan Home* (July 1991): 32+.

### Other

Design Toscano. http://www.aaweb.com/toscano/ (June 29, 1999).

Eaglemount Statuary. http://www.olympus.net/eaglemount/ (June 29, 1999).

Gargoyles Statuary. http://www.gargoylestatuary.com/ (June 29, 1999).

Orlandi Statuary. http://www.statue.com/ (June 29, 1999).

*—Gillian S. Holmes*

# Steel Pipe

Steel pipes are long, hollow tubes that are used for a variety of purposes. They are produced by two distinct methods which result in either a welded or seamless pipe. In both methods, raw steel is first cast into a more workable starting form. It is then made into a pipe by stretching the steel out into a seamless tube or forcing the edges together and sealing them with a weld. The first methods for producing steel pipe were introduced in the early 1800s, and they have steadily evolved into the modern processes we use today. Each year, millions of tons of steel pipe are produced. Its versatility makes it the most often used product produced by the steel industry.

Steel pipes are found in a variety of places. Since they are strong, they are used underground for transporting water and gas throughout cities and towns. They are also employed in construction to protect electrical wires. While steel pipes are strong, they can also be lightweight. This makes them perfect for use in bicycle frame manufacture. Other places they find utility is in automobiles, refrigeration units, heating and plumbing systems, flagpoles, street lamps, and medicine to name a few.

## History

People have used pipes for thousands of years. Perhaps the first use was by ancient agriculturalists who diverted water from streams and rivers into their fields. Archeological evidence suggests that the Chinese used reed pipe for transporting water to desired locations as early as 2000 B.C. Clay tubes that were used by other ancient civilizations have been discovered. During the first century A.D., the first lead pipes were constructed in Europe. In tropical countries, bamboo tubes were used to transport water. Colonial Americans used wood for a similar purpose. In 1652, the first waterworks was made in Boston using hollow logs.

Development of the modern day welded steel pipe can be traced back to the early 1800s. In 1815, William Murdock invented a coal burning lamp system. To fit the entire city of London with these lights, Murdock joined together the barrels from discarded muskets. He used this continuous pipeline to transport the coal gas. When his lighting system proved successful a greater demand was created for long metal tubes. To produce enough tubes to meet this demand, a variety of inventors set to work on developing new pipe making processes.

An early notable method for producing metal tubes quickly and inexpensively was patented by James Russell in 1824. In his method, tubes were created by joining together opposite edges of a flat iron strip. The metal was first heated until it was malleable. Using a drop hammer, the edges folded together and welded. The pipe was finished by passing it through a groove and rolling mill.

Russell's method was not used long because in the next year, Cornelius Whitehouse developed a better method for making metal tubes. This process, called the butt-weld process is the basis for our current pipe-making procedures. In his method, thin sheets of iron were heated and drawn through a cone-shaped opening. As the metal went through the opening, its edges curled up and created a pipe shape. The two ends were welded together to finish the pipe. The first manufacturing plant to use

*Archeological evidence suggests that the Chinese used reed pipe for transporting water to desired locations as early as 2000 B.C.*

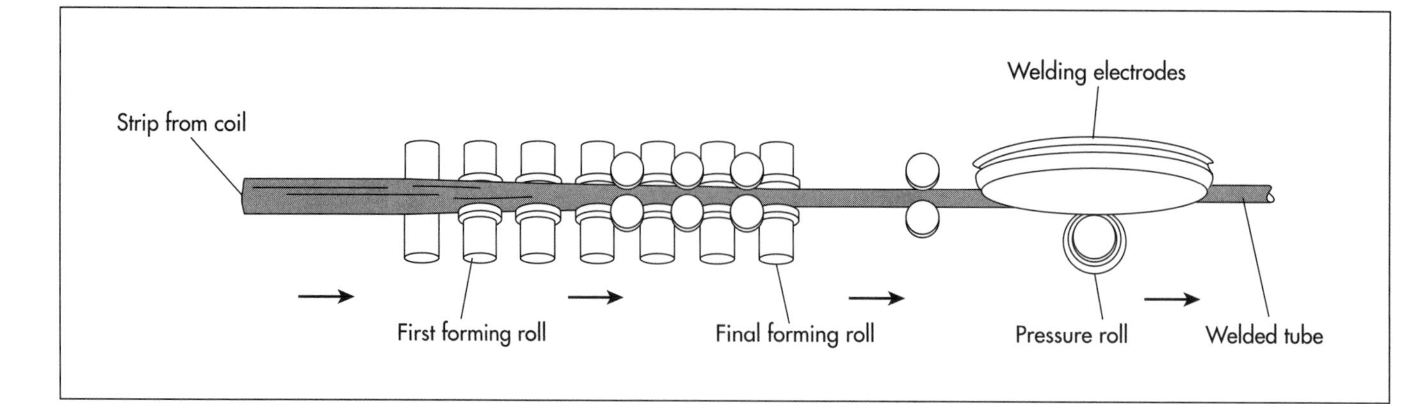


Welded pipe is formed by rolling steel strips through a series of grooved rollers that mold the material into a circular shape. Next, the unwelded pipe passes by welding electrodes. These devices seal the two ends of the pipe together.

this process in the United States was opened in 1832 in Philadelphia.

Gradually, improvements were made in the Whitehouse method. One of the most important innovations was introduced by John Moon in 1911. He suggested the continuous process method in which a manufacturing plant could produce pipe in an unending stream. He built machinery for this specific purpose and many pipe manufacturing facilities adopted it.

While the welded tube processes were being developed, a need for seamless metal pipes arouse. Seamless pipes are those which do not have a welded seam. They were first made by drilling a hole through the center of a solid cylinder. This method was developed during the late 1800s. These types of pipes were perfect for bicycle frames because they have thin walls, are lightweight but are strong. In 1895, the first plant to produce seamless tubes was built. As bicycle manufacturing gave way to auto manufacturing, seamless tubes were still needed for gasoline and oil lines. This demand was made even greater as larger oil deposits were found.

As early as 1840, ironworkers could already produce seamless tubes. In one method, a hole was drilled through a solid metal, round billet. The billet was then heated and drawn through a series of dies which elongated it to form a pipe. This method was inefficient because it was difficult to drill the hole in the center. This resulted in an uneven pipe with one side being thicker than the other. In 1888, an improved method was awarded a patent. In this process the solid billed was cast around a fireproof brick core. When it was cooled, the brick was removed leaving a hole in the middle. Since then new roller techniques have replaced these methods.

## Design

There are two types of steel pipe, one is seamless and another has a single welded seam along its length. Both have different uses. Seamless tubes are typically more light weight, and have thinner walls. They are used for bicycles and transporting liquids. Seamed tubes are heavier and more rigid. The have a better consistency and are typically straighter. They are used for things such as gas transportation, electrical conduit and plumbing. Typically, they are used in instances when the pipe is not put under a high degree of stress.

Certain pipe characteristics can be controlled during production. For example, the diameter of the pipe is often modified depending how it will be used. The diameter can range from tiny pipes used to make hypodermic needles, to large pipes used to transport gas throughout a city. The wall thickness of the pipe can also be controlled. Often the type of steel will also have an impact on pipe's the strength and flexibility. Other controllable characteristics include length, coating material, and end finish.

## Raw Materials

The primary raw material in pipe production is steel. Steel is made up of primarily iron. Other metals that may be present in the alloy include aluminum, manganese, titanium, tungsten, vanadium, and zirconium. Some finishing materials are sometimes used during production. For example, paint may be

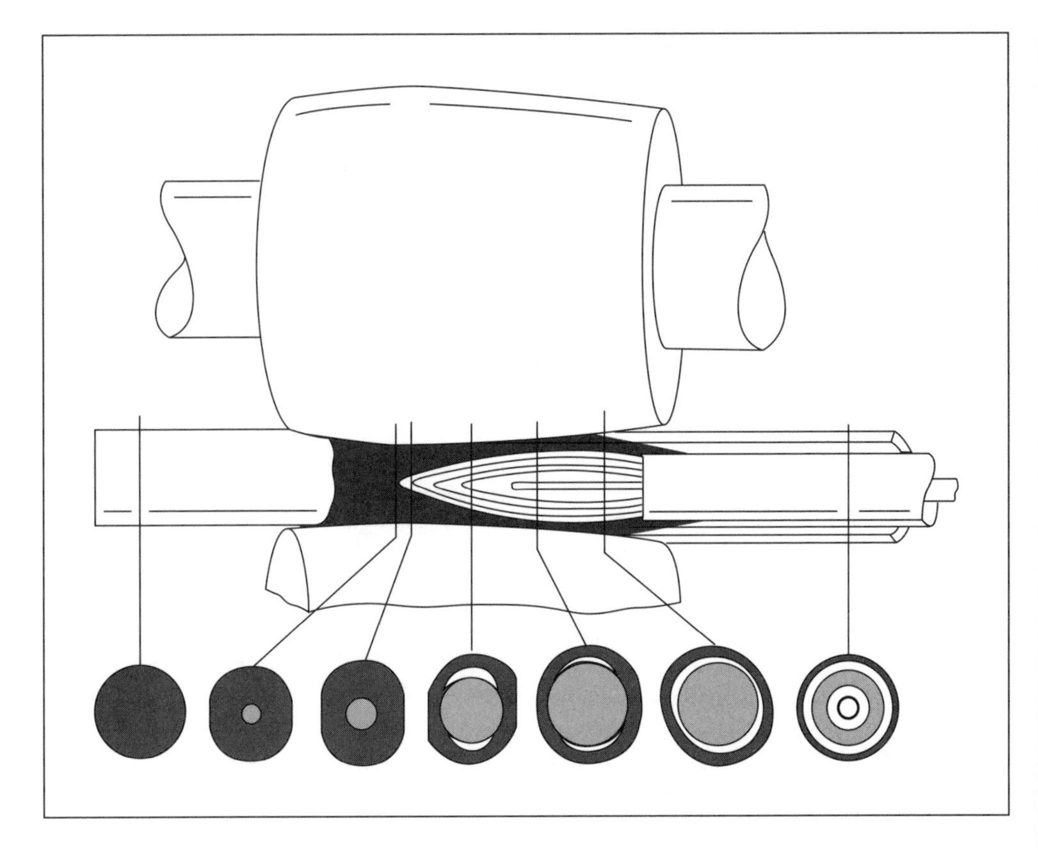

Seamless pipe is manufactured using a process that heats and molds a solid billet into a cylindrical shape and then rolls it until it is stretched and hollowed. Since the hollowed center is irregularly shaped, a bullet-shaped piercer point is pushed through the middle of the billet as it is being rolled.

used if the pipe is coated. Typically, a light amount of oil is applied to steel pipes at the end of the production line. This helps protect the pipe. While it is not actually a part of the finished product, sulfuric acid is used in one manufacturing step to clean the pipe.

## The Manufacturing Process

Steel pipes are made by two different processes. The overall production method for both processes involves three steps. First, raw steel is converted into a more workable form. Next, the pipe is formed on a continuous or semicontinuous production line. Finally, the pipe is cut and modified to meet the customer's needs.

### *Ingot production*

1 Molten steel is made by melting iron ore and coke (a carbon-rich substance that results when coal is heated in the absence of air) in a furnace, then removing most of the carbon by blasting oxygen into the liquid. The molten steel is then poured into large, thick-walled iron molds, where it cools into ingots.

2 In order to form flat products such as plates and sheets, or long products such as bars and rods, ingots are shaped between large rollers under enormous pressure.

### *Producing blooms and slabs*

3 To produce a bloom, the ingot is passed through a pair of grooved steel rollers that are stacked. These types of rollers are called "two-high mills." In some cases, three rollers are used. The rollers are mounted so that their grooves coincide, and they move in opposite directions. This action causes the steel to be squeezed and stretched into thinner, longer pieces. When the rollers are reversed by the human operator, the steel is pulled back through making it thinner and longer. This process is repeated until the steel achieves the desired shape. During this process, machines called manipulators flip the steel so that each side is processed evenly.

4 Ingots may also be rolled into slabs in a process that is similar to the bloom making process. The steel is passed through a pair of stacked rollers which stretch it. However, there are also rollers mounted on the side to control the width of the slabs. When

the steel acquires the desired shape, the uneven ends are cut off and the slabs or blooms are cut into shorter pieces.

### Further processing

5 Blooms are typically processed further before they are made into pipes. Blooms are converted into billets by putting them through more rolling devices which make them longer and more narrow. The billets are cut by devices known as flying shears. These are a pair of synchronized shears that race along with the moving billet and cut it. This allows efficient cuts without stopping the manufacturing process. These billets are stacked and will eventually become seamless pipe.

6 Slabs are also reworked. To make them malleable, they are first heated to 2,200° F (1,204° C). This causes an oxide coating to form on the surface of the slab. This coating is broken off with a scale breaker and high pressure water spray. The slabs are then sent through a series of rollers on a hot mill and made into thin narrow strips of steel called skelp. This mill can be as long as a half mile. As the slabs pass through the rollers, they become thinner and longer. In the course of about three minutes a single slab can be converted from a 6 in (15.2 cm) thick piece of steel to a thin steel ribbon that can be a quarter mile long.

7 After stretching, the steel is pickled. This process involves running it through a series of tanks that contain sulfuric acid to clean the metal. To finish, it is rinsed with cold and hot water, dried and then rolled up on large spools and packaged for transport to a pipe making facility.

### Pipe making

8 Both skelp and billets are used to make pipes. Skelp is made into welded pipe. It is first placed on an unwinding machine. As the spool of steel is unwound, it is heated. The steel is then passed through a series of grooved rollers. As it passes by, the rollers cause the edges of the skelp to curl together. This forms an unwelded pipe.

9 The steel next passes by welding electrodes. These devices seal the two ends of the pipe together. The welded seam is then passed through a high pressure roller which helps create a tight weld. The pipe is then cut to a desired length and stacked for further processing. Welded steel pipe is a continuous process and depending on the size of the pipe, it can be made as fast as 1,100 ft (335.3 m) per minute.

10 When seamless pipe is needed, square billets are used for production. They are heated and molded to form a cylinder shape, also called a round. The round is then put in a furnace where it is heated white-hot. The heated round is then rolled with great pressure. This high pressure rolling causes the billet to stretch out and a hole to form in the center. Since this hole is irregularly shaped, a bullet shaped piercer point is pushed through the middle of the billet as it is being rolled. After the piercing stage, the pipe may still be of irregular thickness and shape. To correct this it is passed through another series of rolling mills.

### Final processing

11 After either type of pipe is made, they may be put through a straightening machine. They may also be fitted with joints so two or more pieces of pipe can be connected. The most common type of joint for pipes with smaller diameters is threading—tight grooves that are cut into the end of the pipe. The pipes are also sent through a measuring machine. This information along with other quality control data is automatically stenciled on the pipe. The pipe is then sprayed with a light coating of protective oil. Most pipe is typically treated to prevent it from rusting. This is done by galvanizing it or giving it a coating of zinc. Depending on the use of the pipe, other paints or coatings may be used.

## Quality Control

A variety of measures are taken to ensure that the finished steel pipe meets specifications. For example, x-ray gauges are used to regulate the thickness of the steel. The gauges work by utilizing two x rays. One ray is directed at a steel of known thickness. The other is directed at the passing steel on the production line. If there is any variance between the two rays, the gauge will automatically trigger a resizing of the rollers to compensate.

Pipes are also inspected for defects at the end of the process. One method of testing a pipe is by using a special machine. This machine fills the pipe with water and then increases the pressure to see if it holds. Defective pipes are returned for scrap.

## Where to Learn More

### Books

*Pipe Characteristic Handbook.* Williams Natural Gas Company Engineering Group. Pennwell Publishing. 1996.

*Kirk Othmer Encyclopedia of Chemical Technology.* John Wiley & Sons. New York: 1992.

*Steel Pipe: A Guide for Design and Installation.* American Water Works Association. 1989.

*—Perry Romanowski*

# Sunflower Seed

*The sunflower plant originated in western North America. It is thought to have been domesticated around 1000 B.C. by Native Americans.*

Sunflower seeds have become a popular snack food. The sunflower plant is an annual herb that has large yellow flowers, broad leaves and can grow from 3-15 ft (0.91-4.6 m) high. To make the finished product enjoyed by millions annually, the seeds are harvested after about 120 days, dried, roasted, salted and packaged.

## Background

The sunflower plant originated in western North America. It is thought to have been domesticated around 1000 B.C. by Native Americans. Spanish explorers brought the sunflower to Europe in 1510. However, it was not until the late 1800s when the flower was introduced to Russia that the sunflower became a food crop. In 1860, Russian farmers made significant improvements in the way that the sunflower was cultivated. During this time, they became the world's largest producer of sunflower seeds. Today, they remain a world leader along with Europe, Argentina, and the United States. Production in the United States has emphasized oil producing varieties, but snack food producing sunflowers have steadily increased.

Sunflowers are technically classified as *Helianthus annuus*. They are a large plant and are grown throughout the world because of their relatively short growing season. In the United States, some varieties reach maturity from 90-100 days after planting. Domesticated sunflowers typically have a single stalk topped by a large flower. This is significantly different from the smaller, multiply branched wild sunflower. Sunflowers have large yellow, ray flower petals on the outer edge that do not produce seeds. The sunflower head is composed of 1,000-2,000 tiny little flowers joined together at the base. These flowers are disk-shaped and can be brown, yellow or purple. During the growing season, the individual flowers are each pollinated. Seed development then begins moving from the outer rim of the flower toward the center. It generally takes 30 days after the last flower is pollinated for the plant to mature.

The sunflower plants reach various heights, but most are from 5-7 ft (1.52-2.1 m) tall. The width of the flower heads is relatively large, typically between 3-6 in (7.62-15.24 cm), although some can reach more than a foot. An exception is the dwarf varieties that are only 3-4 ft (0.91-1.22 m) high and have smaller flower heads. A common characteristic of sunflowers is a tendency for their flowering heads to follow the movement of the sun during the day. This phenomenon, called heliotropism, has the benefit of reducing bird damage and disease development.

Most sunflower plants grown in the United States are used for the oil production. The plants have been bred over time and have steadily improved in quality and consistency. Many options are now available including dwarf varieties and high oil types. Certain hybrids have higher yields and a reduced oil content. The modern sunflower crop is self pollinating so insect vectors are not necessary. Other traits of the crops that have been controlled are disease resistance, speed of maturity, and seed size.

## Design

While most sunflower seeds are used for their oil, sunflower seeds are also sold as a food product. These products are available in many different flavors. Most are sold salted and in their shell. They can be coated and sold as barbeque, sour cream, or ranch. Certain varieties of sunflower seed are made unsalted or with reduced sodium and fat for more health conscious consumers. For smaller sunflower seeds, the shells are removed and only the kernel is sold. These variants are generally easier to eat. Seeds are packaged in plastic bags or glass jars, which are also available in various sizes.

## Raw Materials

The seed is the primary ingredient in all sunflower seed products. They are four sided and flat. They are generally a quarter inch long and an eighth of an inch wide. They have a black seed coat with dark or grey stripes. The coat, or hull, surrounds a small kernel which is composed of about 20% protein and 30% lipids. Additionally, it contains a high level of iron and dietary fiber. The high linoleic acid content of the kernel makes it prone to rancidity and thus gives it a limited shelf life.

While plain sunflower seeds are sold as a snack food, most varieties are soaked or coated with ingredients to improve the seed's characteristics. Flavor enhancers are often added to increase appeal and differentiate product types. Salt is the most common flavor enhancer. It can provide a subtle taste effect that removes the "off" flavor inherent in raw sunflower seeds. A small amount of sugar or dry corn syrup can be added to impart a sweet flavor. Spices and herbs such as garlic, onion powder, or paprika can also have a unique effect on how a sunflower seed tastes. Both natural and artificial flavors can be included.

Beyond flavoring ingredients, manufacturers include texture and appearance modifiers, antioxidants and preservatives with sunflower seed recipes. Texture modifiers such as maltodextrin or cornstarch help control the feel of the sunflowers when they are put in the mouth. Color modifiers are used to change the seed's appearance. Typically, natural coloring materials are used. These modifiers are useful for sunflower flavors such as barbeque or sour cream. Antioxidants are sometimes added to improve the sunflower seed's shelf life by inhibiting natural rancidity reactions. Salt has the added benefit of also preserving the seeds.

To grow the best sunflower plants, the soil must be adequately treated. Sunflower growth is dependant on the amount of nitrogen available more than any other nutrient. For this reason, about 100 lb (45.4 kg) of nitrogen fertilizer per acre is used. Phosphorous and potassium are also included in the fertilizer. To protect the crop from damage, herbicides are added to the soil. Insecticides are also used, but only to a limited extent to prevent the killing of beneficial pollinating insects.

## The Manufacturing Process

Producing sunflower seeds involves the basic process of growing and harvesting the plants, separating the seeds, roasting them and then packaging.

### *Planting and growing*

1 Sunflower seed production begins in early spring when the fields are prepared and the seeds are planted. The seedbeds are tilled and the soil is kept moist. Having adequate moisture content in the soil is the most important planting requirement. The seeding rate (number of seeds planted per acre) is particularly important for sunflower seeds sold as snack foods because a high seeding rate results in smaller seeds which are less desirable. Roughly 17,000 seeds are planted per acre for snack food sunflower seeds. Row spacing is typically between 20-30 in (50.8-76.2 cm). It is thought that the best orientation of the plants is in a North-South direction. A common planting method is to use a corn planter fitted with a special sunflower seed plate. The seeds are typically treated with a fungicide prior to planting.

2 Sunflowers are ready to harvest when the black part of their heads turns brown. In the United States this is generally in late September or October. The seeds typically mature earlier than this but the heads must first dry out to make harvesting efficient. To minimize losses to birds and disease, harvesting is done in a timely manner. A modified grain head is put on the front of the harvesting combine to reduce seed loss. This special device collects the sunflower heads

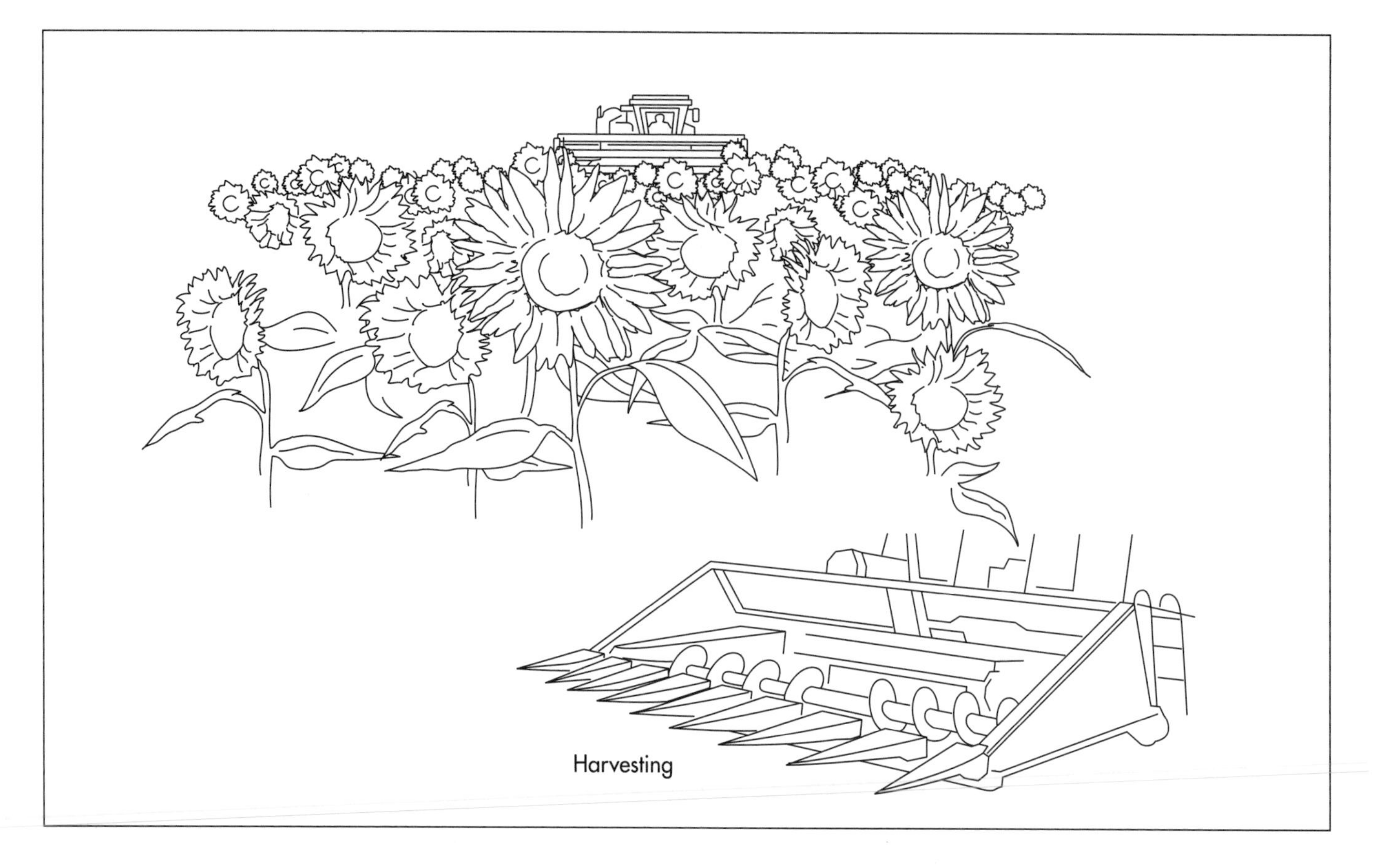


Sunflowers are harvested in the United States in late September or October. A modified grain head is put on the front of the harvesting combine to reduce seed loss. This special device collects the sunflower heads, while minimizing the amount of stem material.

while minimizing the amount of stem material. Harvesting speed is reduced to prevent dehulling of the seeds.

3 The seeds are rapidly dried to under 10% moisture content. Growers typically have facilities on their farms to store seeds until they can be transferred to the local grain elevators. From here, the seeds are transported by truck or train to the processing sites.

### *Processing*

4 When the sunflower seeds arrive at the processing plant, they are emptied onto wire screens and shaken to remove dirt and unwanted debris. They are also inspected to ensure they meet previously determined specifications. Factors such as moisture content, appearance and taste are used to evaluate the shipment. They are next transferred to a large bin and further cleaned.

5 The seeds are then passed on to sizing screens which separate them by size. These screens have holes that allow smaller sized seeds to fall through. The largest seeds will be further processed as snack foods. Medium sized seeds are destined for use in toppings for cookies, salad, or ice cream. The smallest seeds are sold as bird and pet feed.

6 Snack food sunflower seeds are transferred through large ovens. Here, they are dry roasted, reducing the moisture level in the seed further. The medium sized seeds are first sent through de-hulling machines which remove their shells. They are then roasted in oil.

7 Both types of seeds can be flavored as desired. There are a variety of ways in which this can be accomplished. In one procedure, the warm seeds are transferred out of the roasters and put into a large, rotating container. As they are moved around, they are combined with the flavoring ingredients. Oils are sometimes used to make the ingredients stick better.

### *Packaging*

8 From the flavoring stations, the seeds are transferred to packaging machines. Here, the sunflower seeds are weighed and placed in packaging. Typically, these are

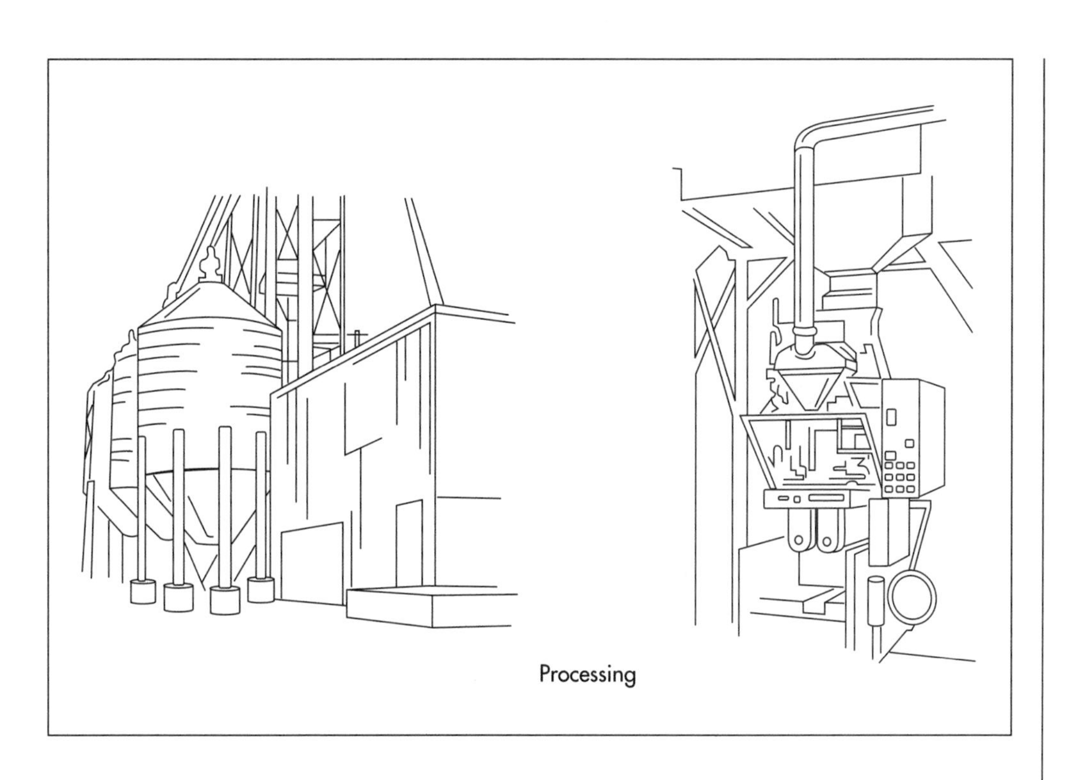
Processing

When the sunflower seeds arrive at the processing plant, they are cleaned and inspected Next, the seeds are passed on to sizing screens that separate them by size. These screens have holes that allow smaller sized seeds to fall through. The largest seeds will be further processed as snack foods. Medium sized seeds are destined for use in toppings for cookies, salad, or ice cream. The smallest seeds are sold as bird and pet feed. Snack food sunflower seeds are dry roasted, reducing the moisture level in the seed further. The medium sized seeds are hulled and then roasted in oil. Once flavored, the seeds are packaged in air tight containers to prevent spoilage.

sealed plastic bags. It is important that this packaging be air tight to prevent the uptake of moisture by the product because moisture can cause them to become rancid. From the packaging station, the sunflower seeds are transferred to boxes, put on pallets, and shipped to stores.

## Quality Control

To ensure that a consistent and quality product is made, sunflower seeds are examined during each step of production. While the plants are growing, they are frequently checked to make sure they are growing properly and free from disease. The seeds are also inspected when they are received at the manufacturing plant. They are subjected to a variety of laboratory tests to ensure that only high quality seeds are used. The finished products are also tested. This may include a chemical analysis or a consumer taste test. Packaging components may also be inspected.

## Byproducts/Waste

The hulls are the primary byproduct of sunflower seed production. This material is high in fiber and is often used as a feed additive for cows. The hulls have also been sold as poultry litter, fireplace logs and fillers for plastics. These markets have been limited and the hulls are often just burned by the factories for energy.

## The Future

Current sunflower research promises to produce improved crops. These plants will be designed to produce a greater yield per acre, grow faster, be better tasting, and more tolerant to pests and diseases. Marketers will also develop new recipes with flavors designed to attract more customers.

## Where to Learn More

### Books

Booth, Gordon. *Snack Food.* New York: Van Nostrand Reinhold Co., 1990.

Hoseney, Carl. *Principles of Cereal Science and Technology.* St. Paul, MN: American Association of Cereal Chemists, 1994.

Macrae, R. et al., ed. *Encyclopedia of Food Science, Food Technology, and Nutrition.* San Diego: Academic Press, 1993.

Salunkhe, D.K., J. Chavan, R. Adsule, and S. Kadam. *World Oilseeds-Chemistry, Technology and Utilization.* New York: Van Nostrand Reinhold Co., 1992.

—*Perry Romanowski*

# Suspension Bridge

*Completed in 1998 with a span of 6,529 ft (1,991 m), the Akashi Kaikyo Bridge took ten years to build, cost $3.6 billion, and involved only six injuries (no deaths). A century earlier, construction of the Brooklyn Bridge, with a span of 1,600 ft (490 m) took 14 years and resulted in the loss of 27 lives.*

In a suspension bridge, the traffic-carrying deck is supported by a series of wire ropes that hang from massive cables draped between tall towers. The Brooklyn Bridge in New York City and the Golden Gate Bridge in San Francisco are two of the most famous suspension bridges. The Akashi Kaikyo Bridge in Japan, which was completed in 1998, contains the world's longest suspension span (distance between support towers)—6,529 ft (1,991 m); the entire bridge, including the portions between the towers and the shores, totals nearly 2.5 mi (4 km). Construction of the Akashi Kaikyo Bridge took ten years, cost $3.6 billion, and involved only six injuries (no deaths). A century earlier, construction of the Brooklyn Bridge, with a span of 1,600 ft (490 m) took 14 years and resulted in the loss of 27 lives.

## Background

Suspension bridges are one of the earliest types devised by man. The most primitive version is a vine rope linking two sides of a chasm; a person travels across by hanging from the rope and pulling himself along, hand over hand. Such primitive bridges—some as long as 660 ft (200 m)—are still being used in areas such as rural India. Somewhat more sophisticated designs incorporate a flat surface on which a person can walk, sometimes with the assistance of vine handrails.

By the eighth century, Chinese bridge builders were constructing suspension bridges by laying planks between pairs of iron chains, essentially providing a flexible deck resting on cables. Similar bridges were built in various parts of the world during subsequent centuries. But the modern era of suspension bridges did not begin until 1808 when an American named James Finley patented a system for suspending a rigid deck from a bridge's cables.

Although Finley built more than a dozen small bridges, the first major bridge that incorporated his technique was built by Thomas Telford over the Menai Straits in England. Completed in 1825, it had stone towers 153 ft (47 m) tall, was 1,710 ft (521 m) long, and boasted a span of 580 ft (177 m). The roadway, which was 30 ft (9 m) wide, was built on a rigid platform suspended from iron chain cables. The bridge is still in use, although the iron chains were replaced with steel bar links in 1939.

Another American, John Roebling, developed two major improvements to suspension bridge design during the mid-1800s. One was to stiffen the rigid deck platform with trusses (arrays of horizontal and vertical girders that are cross-braced with diagonal beams). Experience had shown that wind or rhythmic traffic loads could send insufficiently stiffened decks into vibrations that could grow out of control and literally rip a bridge apart.

Roebling's other important innovation involved construction of the bridge's supporting cables. Around 1830, French engineers had shown that cables consisting of many strands of wire worked better than chains to suspend bridges. Roebling developed a method for "spinning," or constructing, the cables in place on the bridge rather than transporting ungainly prefabricated cables and laboring them into position. His method is still commonly (though not exclusively) used on new bridges.

The history of suspension bridges is liberally sprinkled with examples of successful

*A steel worker tying cable strands for the suspension cable of the new Tacoma, Washington, Narrows Bridge on October 21, 1949.*

The Tacoma Narrows Bridge was the third largest suspension span bridge in the world and only five months old when it collapsed on Saturday, November 7, 1940. The center span, measuring 2,800 ft (853.4 m), stretched between two 425 ft (129.5 m) high towers, while the side spans were each 1,100 ft (304.8 m) long. The suspension cables hung from the towers and were anchored 1,000 ft (304.8 m) back towards the river banks. The designer, Leon Moisseiff, was one of the world's foremost bridge engineers.

Moisseiff's intent was to produce a very slender deck span arching gently between the tall towers. His design combined the principles of cable suspension with a girder design of steel plate stiffeners—running along the side of the roadway—that had been streamlined to only 8 ft (2.4 m) deep.

The $6.4 million bridge was nicknamed "Galloping Gertie" by people who experienced its strange behavior. Forced to endure undulations that pitched and rolled the deck, workmen complained of seasickness. After the opening, it became a challenging sporting event for motorists to cross even during light winds, and complaints about seasickness became common.

State and Toll Bridge Authority engineers were more than a little nervous about the behavior of the slender two-lane span, which was only 39 ft (11.9 m) wide. Its shallow depth in relation to the length of the span (8-2,800 ft [2.4-853.4 m]) resulted in a ratio of 1:350, nearly three times more flexible than the Golden Gate or George Washington bridges. Engineers tried several methods to stabilize the oscillations, but none worked.

Witnesses included Kenneth Arkin, chairman of the Toll Bridge Authority, and Professor Farquharson. By 10:00, Arkin saw that the wind velocity had risen from 38-42 mi (61.1-67.6 km) per hour while the deck rose and fell 3 ft (0.9 m) 38 times in one minute. He and Farquharson halted traffic.

Leonard Coatsworth, a newspaperman, had abandoned his car in the middle of the bridge when he could drive no further because of the undulations. He turned back briefly, remembering that his daughter's pet dog was in the car, but was thrown to his hands and knees. By 10:30, suspender ropes began to tear, breaking the deck and hurling Coatsworth's car into the water. Within a half hour, the rest of the deck fell section by section.

Engineers looking into the problem of the twisting bridges were able to explain that winds do not hit the bridge at the same angle, with the same intensity, all the time. For instance, wind coming from below lifts one edge, pushing down the opposite. The deck, trying to straighten itself, twists back. Repeated twists grow in amplitude, causing the bridge to oscillate in different directions. The study of wind behavior grew into an entire engineering discipline called aerodynamics. Eventually no bridge, building, or other exposed structure was designed without testing a model in a wind tunnel. With the development of graphic capabilities, some of this testing is now done on computers.

bridges that were widely believed to be impossible when proposed by a visionary engineer. One example was a railway bridge Roebling constructed between 1851-1855 across the Niagra River gorge. The first truss-stiffened suspension bridge, it was supported by four 10 in (250 cm) diameter cables strung between stone towers. Forty years after completion, the bridge was successfully carrying traffic 2.5 times as heavy as it was designed for; at that point it was retired and dismantled.

In 1869, Roebling died in an accident while surveying the site for the Brooklyn Bridge, which he had designed. His son, Washington Roebling, spent the next 14 years building the famous structure. This was the first suspension bridge to use cables made of steel rather than wrought iron (a relatively soft type of iron that, while hot, can be shaped by machines or formed by hammering). Each of the four 16 in (40 cm) diameter cables consists of more than 5,000 parallel strands of steel wire. More than a century after its completion, the Brooklyn Bridge carries heavy loads of modern traffic.

Another landmark suspension bridge was built across the Golden Gate—the mouth of San Francisco Bay—from 1933-1937 by Joseph Strauss. The Golden Gate Bridge is 6,450 ft (1,966 m) long, with a main span of 4,200 ft (1,280 m). Its two towers are 746 ft (227 m) tall; they support two 7,125-ton (6.5 million kg) cables that contain a total of 80,000 mi (129,000 km) of steel wire. Despite rigorous safety precautions, 11 workers died; 19 were saved by a safety net hanging below the deck during construction—an innovation that became standard on later bridge projects.

One of the most famous bridge failures in America was the 1940 collapse of the Tacoma Narrows Bridge on Puget Sound in the state of Washington. Then the third-longest suspension bridge in the world, it had been designed to be exceptionally sleek. Only wide enough for two traffic lanes and sidewalks, the span was 2,800 ft (853 m) long. Rather than being stiffened with trusses, the deck was reinforced by two steel girders that were only 8 ft (2.4 m) high, with some cross-bracing connecting them. This design not only provided less rigidity than trusses, but it also allowed the wind to exert strong forces on the structure rather than passing harmlessly through an open truss arrangement. Four months after it was completed, the bridge was set into a pattern of increasing oscillations by 42 mph (68 km/h) winds and tore itself apart. The replacement bridge, built a decade later, was designed with a deck stiffened with a steel truss 33 ft (10 m) thick.

## Raw Materials

Many of the components of a suspension bridge are made of steel. The girders used to make the deck rigid are one example. Steel is also used for the saddles, or open channels, on which the cables rest atop a suspension bridge's towers.

When steel is drawn (stretched) into wires, its strength increases; consequently, a relatively flexible bundle of steel wires is stronger than a solid steel bar of the same diameter. This is the reason steel cable is used to support suspension bridges. For the Akashi Kaikyo Bridge, a new low-alloy steel strengthened with silicon was developed; its tensile strength (resistance against pulling forces) is 12% greater than any previous steel wire formulation. On some suspension bridges, the steel wires forming the cables have been galvanized (coated with zinc).

The towers of most suspension bridges are made of steel, although a few have been built of steel-reinforced concrete.

## Design

Each suspension bridge must be designed individually to take into account many factors. For example, the geology of the site provides a foundation for the towers and cable anchorages, and may be susceptible to earthquakes. The depth and nature of the water being bridged (e.g., fresh or saltwater, and strength of currents) may affect both the physical design and the choice of materials like protective coatings for the steel. In navigable waters, it may be necessary to protect a tower from possible ship collisions by building up an artificial island at its base.

Since the Tacoma Narrows Bridge disaster, all new bridge designs have been tested by placing scale models in wind tunnels, as the Golden Gate Bridge's design had been. For the Akashi Kaikyo Bridge, for example, the world's largest wind tunnel was constructed to test 1/100th-scale models of bridge sections.

In very long bridges, it may be necessary to take the earth's curvature into account when designing the towers. For example, in the New York's Verrazano Narrows Bridge, the towers, which are 700 ft (215 m) tall and stand 4,260 ft (298 m) apart, are about 1.75 in (4.5 cm) farther apart at the top than they are at the bottom.

## The Manufacturing Process

Construction of a suspension bridge involves sequential construction of the three

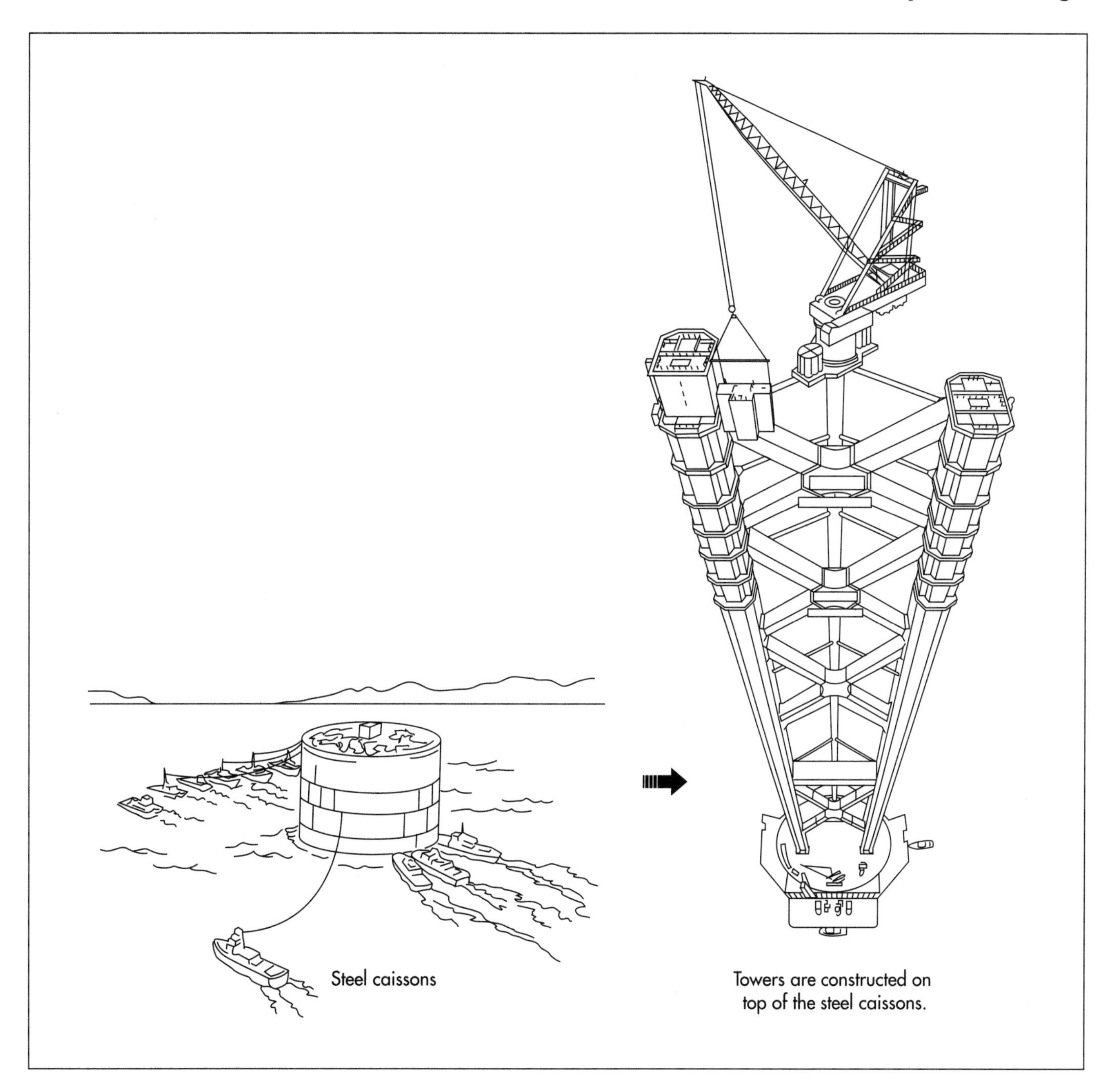


major components: the towers and cable anchorages, the support cable itself, and the deck structure.

### *Tower construction*

1 Tower foundations are prepared by digging down to a sufficiently firm rock formation. Some bridges are designed so that their towers are built on dry land, which makes construction easier. If a tower will stand in water, its construction begins with lowering a caisson (a steel and concrete cylinder that acts as a circular dam) to the ground beneath the water; removing the water from the caisson's interior allows workers to excavate a foundation without actually working in water. When the excavation is complete, a concrete tower foundation is formed and poured.

2 Construction details vary with each unique bridge. As an example, consider the Akashi Kaikyo Bridge. Each of its two steel towers consists of two columns. Each column is composed of 30 vertical blocks (or layers), each of which is 33 ft (10 m)

Tower constructions that will stand in water begin with caissons (a steel and concrete cylinder that acts as a circular dam) that are lowered to the ground beneath the water, emptied of water, and filled with concrete in preparation for the actual towers.

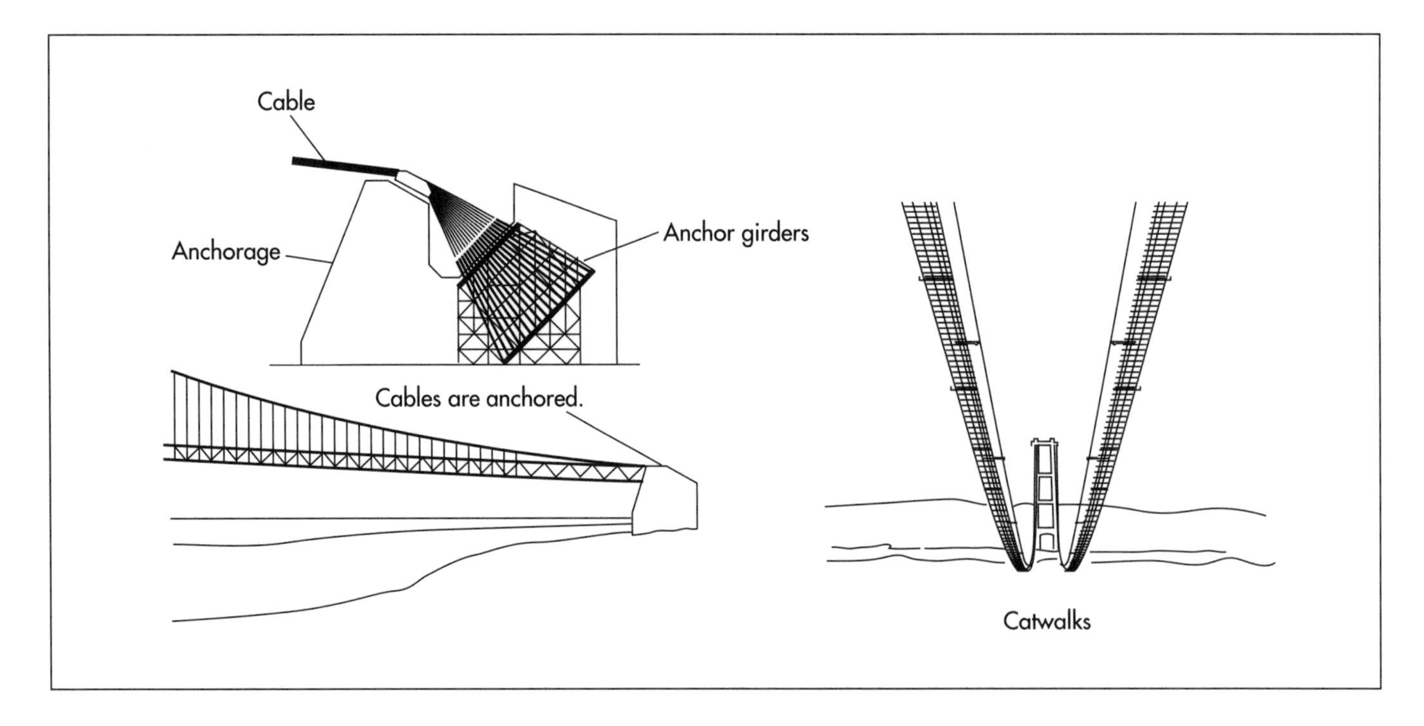


Anchorages—structures that support the bridge's cables—are massive concrete blocks securely attached to strong rock formations. When the towers and anchorages have been completed, a pilot line must be strung along the cable's eventual path, from one anchorage across the towers to the other anchorage.

tall; each of these blocks, in turn, consists of three horizontal sections. A crane positioned between the columns lifted three sections into place on each column, completing a layer. After completing a block on each column, the"bootstrapping" crane was jacked up to the next level, where it lifted the sections of the next layer into place. At appropriate intervals, diagonal bracing was added between the columns.

### Anchorage construction

3 Anchorages are the structures to which the ends of the bridge's cables are secured. They are massive concrete blocks securely attached to strong rock formations. During construction of the anchorages, strong eyebars (steel bars with a circular hole at one end) are embedded in the concrete. Mounted in front of the anchorage is a spray saddle, which will support the cable at the point where its individual wire bundles (see Step 5) fan out—each wire bundle will be secured to one of the anchorage's eyebars.

### Cable construction

4 When the towers and anchorages have been completed, a pilot line must be strung along the cable's eventual path, from one anchorage across the towers to the other anchorage. Various methods can been used to position the pilot line. For the Niagra River bridge, for example, Roebling offered a reward of $10 to the first youngster who could fly a kite with a pilot line attached across the gorge to make the connection. Today, a helicopter might be used. Or the line might be taken across the expanse by boat and then lifted into position. When the pilot line is in place, a catwalk is constructed for the bridge's entire length, about 3 ft (1 m) below the pilot line, so workers can attend to the cable formation.

5 To begin spinning the cable, a large spool of wire is positioned at the anchorage. The free end of the wire is looped around a strand shoe (a steel channel anchored to an eyebar). Between the spool and the strand shoe, the wire is looped around a spinning wheel that is mounted on the pilot line. This wheel carries the wire across the bridge's path, and the wire is looped around a strand shoe at the other anchorage; the wheel then returns to the first anchorage, laying another strand in place. The process is repeated until a bundle of the desired number of wire strands is formed (this varies from about 125 strands to more than 400). During the spinning, workers standing on the catwalk make sure the wire unwinds smoothly, freeing any kinks. As spools are exhausted, the end of the wire is spliced to the wire from a new spool, forming a continuous strand. When the bundle is thick enough, tape or wire straps are applied at in-

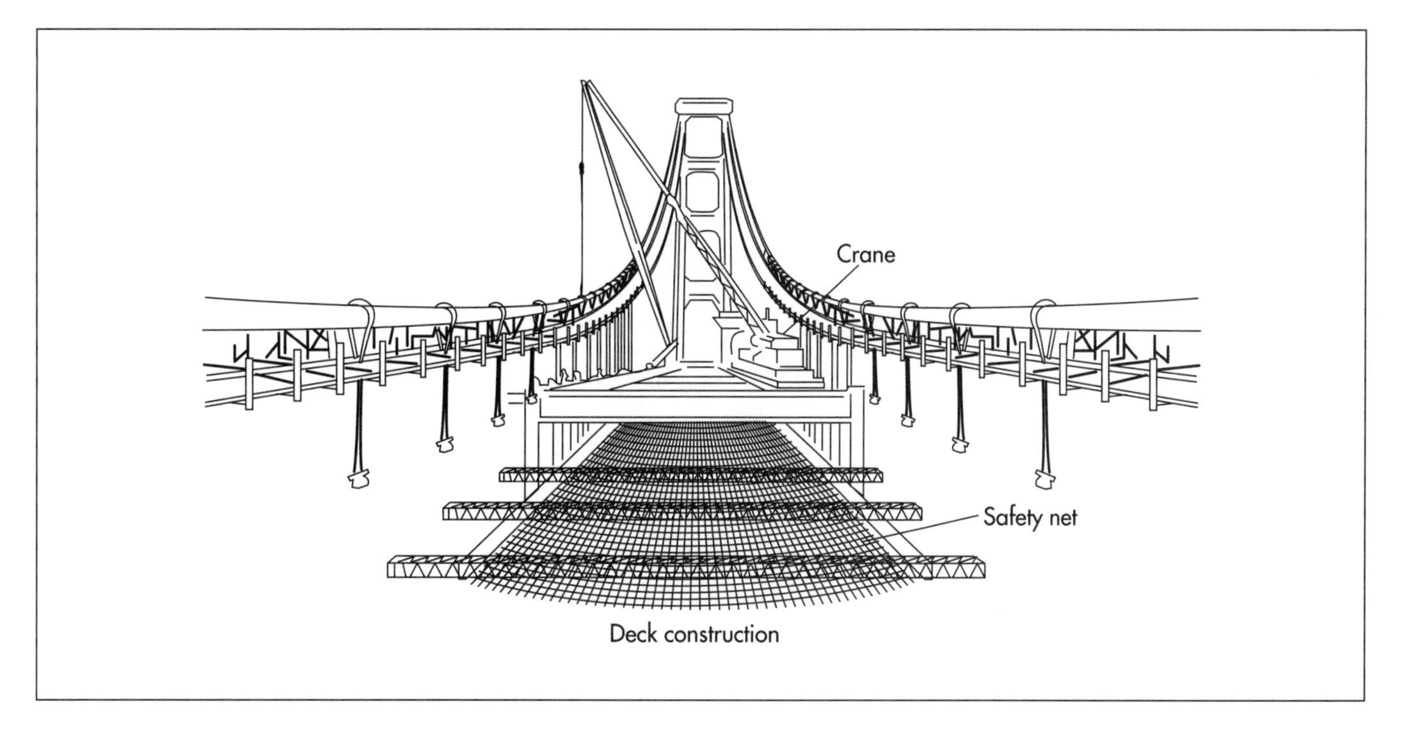

Deck construction

tervals to keep the wires together. The wire coming off the spool is cut and secured to the anchorage. Then the process begins again for the next bundle.

The number of bundles needed for a complete cable varies; on the Golden Gate Bridge it is 61, and on the Akashi Kaikyo Bridge it is 290. When the proper number have been spun, a special arrangement of radially positioned jacks is used to compress the bundles into a compact cable, and steel wire is wrapped around it. Steel clamps are mounted around the cable at predetermined intervals to serve as anchoring points for the vertical cables that will connect the decking to the support cable.

### *Deck construction*

6 After vertical cables are attached to the main support cable, the deck structure can be started. The structure must be built in both directions from the support towers at the correct rate in order to keep the forces on the towers balanced at all times. In one technique, a moving crane that rolls atop the main suspension cable lifts deck sections into place, where workers attach them to previously placed sections and to the vertical cables that hang from the main suspension cables, extending the completed length. Alternatively, the crane may rest directly on the deck and move forward as each section is placed.

### *Finishing*

7 When the deck structure is complete, it is covered with a base layer (e.g., steel plates) and paved over. Painting the steel surfaces and installing electric lines for lighting are examples of other finishing steps. In addition, ongoing maintenance procedures begin. For example, a permanent staff of 17 ironworkers and 38 painters continue to work daily on the Golden Gate Bridge, replacing corroding rivets and other steel components and touching up the paint that protects the bridge.

## The Future

Each suspension bridge is designed uniquely, with attention given to both function and aesthetics. New materials may be used, or even developed, to make the bridge less bulky and more efficient. And innovative designers sometimes create unusual solutions to their challenges. For example, the design approved in 1998 to replace the east span of the San Francisco-Oakland Bay Bridge that was severely damaged by a 1989 earthquake is a suspension bridge supported by only one tower. Its main cables are anchored, not in the massive anchorages de-

Once the vertical cables are attached to the main support cable, the deck structure must be built in both directions from the support towers at the correct rate in order to keep the forces on the towers balanced at all times. A moving crane lifts deck sections into place, where workers attach them to previously placed sections and to the vertical cables that hang from the main suspension cables.

scribed above, but in the deck support structure of the bridge itself.

Perhaps the most ambitious plans currently under development for a new suspension bridge are those for connecting Sicily to the Italian mainland. Because the support towers will have to be built on opposite shores of the Messina Strait, the main span will be 9,500-10,800 ft (2,900-3,300 m) long. One proposed design uses towers that are 1,312 ft (400 m) tall. Developers hope to build the bridge by 2006.

## Where to Learn More

### Books

Brown, David J. *Bridges.* New York: Macmillan, 1993.

Cassady, Stephen. *Spanning the Gate.* Mill Valley, CA: Squarebooks, 1979.

Kingston, Jeremy. *How Bridges are Made.* New York: Facts on File, 1985.

Troitsky, M.S. *Planning and Design of Bridges.*New York: John Wiley & Sons, 1994.

### Periodicals

Kashima, Satoshi and Makoto Kitagawa. "The Longest Suspension Bridge." *Scientific American* (December 1997): 88ff.

Kosowatz, John J."Building a New Gateway to China." *Scientific American* (December 1997): 106ff.

*—Loretta Hall*

# Sword

## Background

The development of the sword was not possible until ancient civilizations discovered how to mine and work metal. Thus, the first swords were probably made of the oldest worked metal, pure copper. The earliest copper mines were in Egypt around 3700B.C., and in Anatolia (in what is now Turkey) around the same time. By about 1900B.C., copper working had spread across Europe, and presumably copper swords were made during this era. Copper alloyed with tin produces bronze, and this metal made stronger weapons than pure copper. The earliest bronze swords were made by the Egyptians in about 2500B.C. They made blades by heating bronze ingots or by casting molten metal in clay molds. Bronze swords were used throughout the ancient world, until bronze was replaced by iron as the metal used to make weapons. The Hittites knew how to smelt iron as early as 3000B.C., but an efficient method of forming the iron into blades was not discovered until somewhere around 1400 B.C. The Hittites were the first to harden iron for blades by heating it with carbon, hammering it into shape, and then quenching it in water. They kept their methods secret for as long as they could, but gradually ironworking spread across the ancient world. The Romans used iron swords with double blades, a weapon for hand-to-hand fighting. A bigger sword, which could be used to fight from horseback, came into vogue in Western Europe by the third century. Both the Vikings and Saxons were renowned swordsmiths. They used sophisticated ironworking techniques both in forming and decorating their blades.

During the Middle Ages in Europe, the sword was the preferred weapon of the knight in armor. The medieval sword was made of steel, and so sharp and heavy that it could easily cut a man in half. The quality of the sword depended to a great extent on the quality of the metal. Production of swords was specialized in certain towns or areas where skilled ironworkers had access to good metal and knew how to work it. From the sixth century, the lower Rhine in Germany was a center of sword manufacturing, and later swords were exported from Milan, Brescia, and Passau. Toledo, in Spain, was renowned for its swords. A test of the Toledo sword's sharpness was to throw a silk scarf into the air so that it floated down onto the sword blade. The edge was so sharp that the silk would rend on impact.

Perhaps the strongest swords ever made were the weapons of the samurai in Japan. As far back as the eighth century until the end of the feudal period in the nineteenth century, Japanese smiths made blades of exceptional hardness by welding strips of iron and steel together, then folding the resulting sandwich over on itself and pounding it flat again. This process was repeated from 12-28 times. Old blades were passed down in families, and some were still in use in World War II. These swords were so sharp and strong they could cut through a machine gun barrel.

During the sixteenth century, the sword evolved from a slashing weapon into a more refined thrusting rapier. The rapier had a long, thin blade sometimes reaching 6 ft (1.83 m) in length. When carried at the waist, the longest of rapiers would inconveniently hit the street. By the end of the century, the rapier became more lightweight

*With swordplay arose the art of the duel, a privilege primarily reserved for the upper class. From 1600-1789, 40,000 aristocrats lost their lives in duels.*

and its length was shortened to 3 ft (0.91 m). These adjustments gave birth to swordplay and expertise.

With swordplay arose the art of the duel, a privilege primarily reserved for the upper class. From 1600-1789, 40,000 aristocrats lost their lives in duels. Since Germans preferred heavier swords, dueling was often violent and resulted in injury and death. It was tolerated by the ruling monarchs because of its rigid exclusion of the lower classes. In Germany, dueling as an aristocratic sport unified the upper classes and distinguished them from the masses. In France, dueling was more of an art that did not necessarily have to end in injury or death. With the French Revolution and the abolition of aristocracy, dueling was considered a sport for all. The French used lighter weight épées—a sword with no cutting edge that tapers to a point—and duels were usually fought until the first blood was drawn. By the end of the nineteenth century, Frenchmen averaged 400-500 duels per year with a nonexistent death rate. The English banned dueling in 1844.

Swords declined in utility after the introduction of firearms, though they persisted for a surprisingly long time. The British army was still perfecting its sword design in the first decade of the twentieth century, and its last change in design was in 1920. The cutlass, a wide sword used in the British Navy, was not withdrawn from service until 1936. Swords made today are for the most part ceremonial. They are still part of some military dress uniforms. The only place where swords are actually still employed as weapons seems to be Japan, where they are said to be a choice murder weapon of underworld gangsters and far-right political assassins.

## Fencing as Sport

With the refinement of sword design and the popularity of dueling came the sport of fencing. During the eighteenth century Domenico Angelo, an Italian that studied swordsmanship in Paris, moved to London and gained a reputation as an expert duelist. Challenged by Ireland's master swordsman Dr. Keys, Angelo quickly out-maneuvered Dr. Keys' slashing techniques with his own fencing moves. His victory made Angelo popular with the upper class as a teacher of dueling. He opened a school and fencing as sport was established.

Modern fencing is done with blunted foils, épées, and sabers. A typical uniform is equipped with a padded jacket, gauntlets, and wire-mesh helmets. It is an official Olympic sport and it's popularity has been sustained by the romantic swashbuckling of early Hollywood films and recent epics like *Star Wars* and *Braveheart.*

## Raw Materials

The swords commonly in use in Europe in the Middle Ages were made of steel. Steel is an alloy of iron and carbon, and iron heated properly over a charcoal fire becomes steel. But the theory behind the process was not understood until the nineteenth century, and not many communities knew how to make good steel. Iron smelters roasted ore in charcoal fires, and produced wrought iron, cast iron and carbon steel, depending on the heat and makeup of the ore. Cast iron contains more than 2.2% carbon. It is too hard to work, and until the fourteenth century in Europe, it was considered a waste product. Wrought iron contains less than 0.3% carbon. It is a soft, workable metal most used for tools. But wrought iron swords bent in use, and so were inferior to steel ones. Steel suitable for swords contains from 0.3-2.2% carbon, and it is both soft and workable and can be hardened by heating it to red hot and then quenching it in water.

Until the fourteenth century, when the mechanical bellows was invented and iron production became more organized in Europe, production of steel was haphazard, and primitive furnaces produced steel more by luck than design. The invention of the bellows and the blast furnace in the fourteenth century allowed smelters to heat ore to higher temperatures, producing wrought iron that could be converted to steel. A common kind of steel available in Medieval Europe was called blister steel. It was made from thin rods of wrought iron. The iron rods were packed in charcoal dust and set inside a tight iron box or small furnace. The iron was heated in the furnace and blown with the bellows. When the iron reached white hot, it began to absorb carbon from the charcoal, and turn to steel. Rods of small diame-

Sword hilt assembly.

ter could be transformed into steel in about 24 hours, and bigger rods took longer.

The finest steel was imported from India, called Wootz steel. Indian metallurgy was renowned from the time of the Roman Empire, and blades made in the Persian Empire and across the east were usually made from imported Wootz. European Crusaders encountered Wootz steel in the superior weapons of their eastern enemies. Crusading knights began bringing Wootz steel back to Europe in the eleventh century, but the secret of making it remained in India until the nineteenth century. Blades made from Wootz showed a grainy pattern in the metal, formed by the fibrous layout of crystals in the steel. The appearance has been compared to watered silk, or damask fabric. The swordsmith usually emphasized the pattern by etching the blade with acid. The most skilled smiths could make the crystalline pattern appear in regular formations along the blade. This ancient art is now lost. Eastern blades with patterned metal are called Damascus swords, named after the city that was a major east-west trading point. To confuse matters, some European swords are also called "Damascus." In this case, European smiths tried to copy the eastern swords by marking blades and inlaying the metal. But in true Damascus blades, the patterning is inherent in the steel itself, and not imposed on it.

Indian metallurgists had several ways of preparing Wootz steel. In one method, wrought iron plates were immersed in a crucible filled with molten cast iron. Cast iron has a high carbon content, and when heated, the carbon leached from the cast iron to the wrought plates. The resulting metal was a mixture of soft iron and hard carbon steel, dispersed in granules throughout the ingot. Another method was to crush iron ore and

The blade core is formed from two or more thin iron rods that have been heated, forged, and twisted with a pair of tongs. Next, the twisted rods are drawn out and a seam along the edge of the blade is opened with a heated tool. A thin piece of steel that has been roughened or "scarfed" along one edge, is then set into the groove. The smith then heats the metal so that both the iron and steel are molten and join. The blade is tempered—transformed from soft, workable metal into a hard blade—by holding the blade over a fire and then quenching the blade in a vat of oil or brine. The blade is polished and decorated.

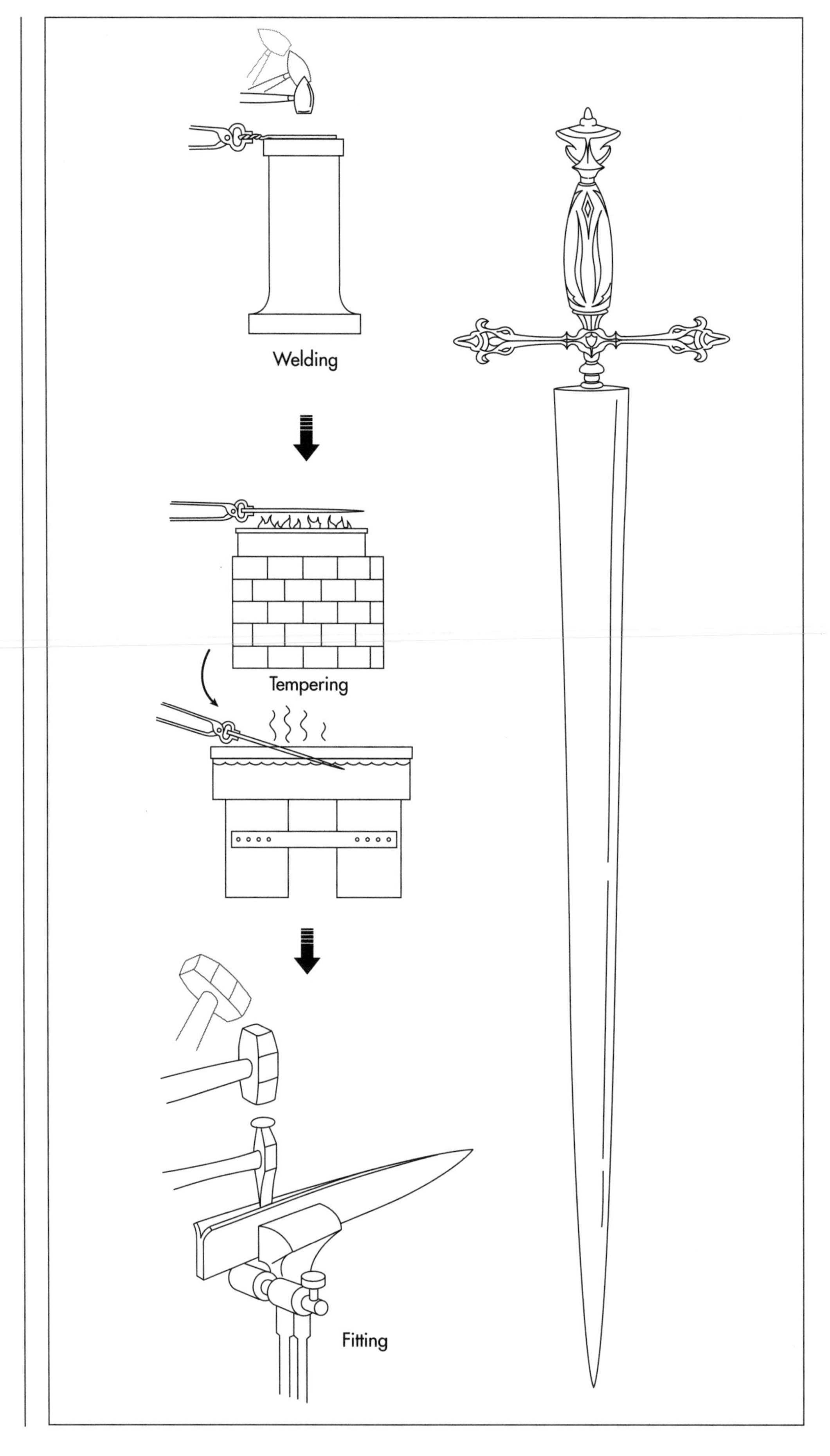

wash it repeatedly, in the panning process used by goldminers. This refined ore was then dried and placed in a small clay crucible. The smelter added charcoal and other plant matter, sealed the crucible, and fired it in a charcoal fire for one to two days. Then, the sealed crucible was cooled for another period of days. The clay was broken open, and the ingot was then packed in clay mixed with iron filings. Next, the smelter reheated this mixture to red heat. At this point, the metal was soft enough to work, and could be successfully forged into weapons.

## The Manufacturing Process

Different metal workers made swords in many different ways, and most of the techniques of swordmaking were never written down. In general, specialized smiths made swords. The finishing of the blade, which often involved elaborate inlay work, was done separately by a jeweler. Then the blade was sometimes sent to a cutler, who did the final assembling of the blade in the grip. What follows is a general process for a type of sword made with an iron core and steel blades. A Roman sword was presumably made this way, as were the swords of Toledo.

### *Forming the core*

1 Some swords were made with a core of wrought iron, and fitted with steel blades. The softer center made the weapon more flexible and resilient. The core is formed from twisted rods of iron. The smith takes two or more thin iron rods and heats them in the forge until they are white-hot. Then, the smith fastens one end of the rods in a vice, and twists them all together using a pair of tongs. One long narrow piece is left to protrude from the center of the bundle. This forms the tang that holds the sword to its hilt.

### *Drawing out*

2 The twisted rods are then "drawn out"—a smith's term for making the iron thinner. The metal is heated to an orange-red, then placed on the anvil. The smith strikes the metal with measured blows that stretch the body and make it long and sword-shaped.

### *Fitting the blades*

3 Next, the smith fixes the iron into a vice and opens a seam along the edge with a heated tool. A thin piece of steel that has been roughened or "scarfed" along one edge, is then set into the groove. The smith then heats the metal so that both the iron and steel are molten on the surface. This was evidently quite an art, as the metals had different melting temperatures. The smith closes the seam by hammering it deftly, and the molten metals join. For a two-edged sword, the process is then repeated on the other side.

### *Packing the edge*

4 Now, the smith heats the blade so it barely turns red. The smith often held the blade in the shadow of a box or barrel, to see the color in the dark. When the whole length of the blade is the right color, the smith sets it on the anvil and strikes quick blows with a small hammer all along the steel edges. This masses the steel fibers, and makes a stronger weapon that keeps its sharpness longer.

### *Tempering*

5 Now the blade is tempered—transformed from soft, workable metal into a hard blade. The smith holds the blade over a fire that may be a long fire built specially to fit swords. The difficulty is in getting an even heat all along the length of the metal. When every part is glowing an even color, the smith quenches the blade in a vat of oil or brine. For this first quenching, the blade is placed in the vat with the blade held flat, parallel to the liquid's surface. After it cools, the smith cleans off the metal scale that collects on the blade's surface. Then, the smith heats the sword again, in a slightly different way. The smith heats a long iron bar to orange-red, and lays the sword on it. When the sword heats to a blue or purple color, the smith lifts it with tongs and quenches it again, this time edge down (perpendicular to the first quenching).

### *Filing and grinding*

6 The blade is next polished with a series of fine files. The edges are ground sharp on a grindstone, a rotating wheel of textured stone.

### Decorating

7 Many blades were elaborately decorated with inlaid patterns. Usually the sword was sent to a jeweler for this step. The jeweler engraved a pattern on the metal, and then often etched it out with acid.

### Assembling

8 For the final step, the blade is attached to a hilt. The smith had made the blade with a narrow piece called the tang protruding from the end of the sword opposite the tip. The smith prepares a crosspiece with a hole punched through the center. The simplest grip was usually made of wood. It was carved as a solid piece, and then the smith, (or cutler, if a specialist did the finishing) bored a hole through it from end to end. A third piece is called the pommel. It is the rounded end of the grip. It would also be carved in one piece, and drilled end to end. Then, the smith heats the tang, and fits the pieces over it. The hot metal bores out and fills the holes in the pieces, and effectively joins them. The tang is long enough that a bit of it still protrudes through the pommel. This is folded over and tacked down.

## Byproducts/Waste

The production of iron and steel for swords required massive amounts of charcoal. Charcoal is made from slowly charred wood. The amount of trees needed to provide charcoal was so enormous that Queen Elizabeth I of England had to put a limit on how much timber could be felled, fearing her country would run out. An ironworks in colonial America that produced 15 tons of iron a week used up about four square miles of forest each year. So iron production on a massive scale ran into the danger of deforestation.

Iron itself was usually assiduously recycled. Old nails and horseshoes made excellent sword cores, and smiths usually kept a scrap heap of broken or useless tools and parts that could be melted down and re-used. An unskilled smith, however, could waste steel if he burnt the narrow edge of a blade. If heated too high, the steel became brittle and useless. And in this condition, it was not recyclable.

## Where to Learn More

### Books

Bealer, Alex W. *The Art of Blacksmithing.* New York: Funk & Wagnalls, 1969.

Evangelista, Nick. *Encyclopedia of the Sword.* Greenwood Publishing Group, 1995.

Figiel, Leo S. *On Damascus Steel.* New York: Atlantis Arts Press, 1991.

Oakeshott, Ewart. *A Knight and His Weapons.* Chester Springs, PA: Dufour Editions, 1997.

Wilkinson-Latham, Robert. *Swords in Color.* New York: Arco Publishing Company, 1978.

### Periodicals

"Murder a la Mode." *The Economist* (April 29, 1995).

*—Angela Woodward*

# Telephone

## History

Throughout history, people have devised methods for communicating over long distances. The earliest methods involved crude systems such as drum beating or smoke signaling. These systems evolved into optical telegraphy, and by the early 1800s, electric telegraphy. The first simple telephones, which were comprised of a long string and two cans, were known in the early eighteenth century.

A working electrical voice-transmission system was first demonstrated by Johann Philipp Reis in 1863. His machine consisted of a vibrating membrane that opened or closed an electric circuit. While Reis only used his machine to demonstrate the nature of sound, other inventors tried to find more practical applications of this technology. They were found by Alexander Graham Bell in 1876 when he was awarded a patent for the first operational telephone. This invention proved to revolutionize the way people communicate throughout the world.

Bell's interest in telephony was primarily derived from his background in vocal physiology and his speech instruction to the deaf. His breakthrough experiment occurred on June 2, 1875. He and his assistant, Thomas Watson, were working on a harmonic telegraph. When a reed stuck on Watson's transmitter an intermittent current was converted to a continuous current. Bell was able to hear the sound on his receiver confirming his belief that sound could be transmitted and reconverted through an electric wire by using a continuous electric current.

The original telephone design that Bell patented was much different than the phone we know today. In a real sense, it was just a modified version of a telegraph. The primary difference was that it could transmit true sound. Bell continued to improve upon his design. After two years, he created a magnetic telephone which was the precursor to modern phones. This design consisted of a transmitter, receiver, and a magnet. The transmitter and receiver each contained a diaphragm, which is a metal disk. During a phone call, the vibrations of the caller's voice caused the diaphragm in the transmitter to move. This motion was transferred along the phone line to the receiver. The receiving diaphragm began vibrating thereby producing sound and completing the call.

While the magnetic phone was an important breakthrough, it had significant drawbacks. For example, callers had to shout to overcome noise and voice distortions. Additionally, there was a time lapse in the transmission which resulted in nearly incoherent conversations. These problems were eventually solved as the telephone underwent numerous design changes. The first phones made available to consumers used a single microphone. This required the user to speak into it and then put it to the ear to listen. Thomas Edison introduced a model that had a moveable listening earpiece and stationary speaking tube. When placing a call, the receiver was lifted and the user was connected directly to an operator who would then switch wires manually to transmit. In 1878, the first manual telephone exchange was opened. It served 21 customers in New Haven, Connecticut. Use of the telephone spread rapidly and in 1891, the first automatic number calling mechanism was introduced.

Long-distance service was first made available in 1881. However, the transmission rates

*Long-distance telephone service was first made available in 1881.*

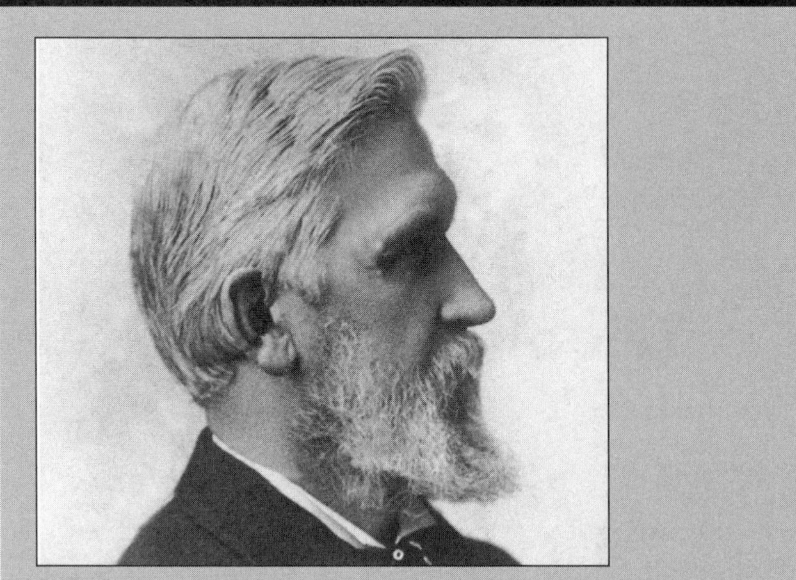

*Elisha Gray*

Elisha Gray was Alexander Graham Bell's principle rival, first for invention of the harmonic telegraph and then of the telephone. He was a prolific inventor, granted some 70 patents during his lifetime. Born in Barnesville, Ohio, on August 2, 1935, and brought up on a farm, Gray had to leave school early when his father died but later continued his studies at Oberlin College, where he concentrated on physical sciences, especially electricity, and supported himself as a carpenter.

After leaving Oberlin, Gray continued his electrical experiments, concentrating on telegraphy. In 1867, he patented an improved telegraph relay, and later, a telegraph switch, an "annunciator" for hotels and large business offices, a telegraphic repeater, and a telegraph line printer. He also experimented with ways to transmit multiple, separate messages simultaneously across a single wire, a subject that was also engaging the efforts of Bell. Gray prevailed, filing his harmonic telegraph patent application in February 1875, two days before Bell's similar application.

Gray now began investigating ways to transmit voice messages, soon developing a telephone design that featured a liquid transmitter and variable resistance. In one of the most remarkable coincidences in the history of invention, Gray filed notice of his intent to patent his device on February 14, 1876—just two hours after Bell had filed his own telephone patent at the same office. Western Union Telegraph Company purchased the rights to Gray's telephone and went into the telephone business; the Bell Telephone Company launched a bitter lawsuit in return.

Meanwhile, Gray had been a founding partner in 1869 of Gray and Barton, an electric-equipment shop in Cleveland, Ohio. This became Western Electric Manufacturing of Chicago in 1872, which evolved into Western Electric Company, which, ironically, became the largest single component of Bell Telephone in 1881.

were not good and it was difficult to hear. In 1900, two workers at Bell System designed loading coils that could minimize distortions. In 1912, the vacuum tube was adapted to the phone as an amplifier. This made it possible to have a transcontinental phone line, first demonstrated in 1915. In 1956, a submarine cable was laid across the Atlantic to allow transatlantic telephone communication. The telecommunication industry was revolutionized in 1962 when orbiting communication satellites were utilized. In 1980, a fiber-optic system was introduced, again revolutionizing the industry.

## Background

Telephones still operate on the same basic principles that Bell introduced over one hundred years ago. If a person wishes to make a call, they pick up the handset. This causes the phone to be connected to a routing network. When the numbers are pressed on a touch-tone keypad, signals are sent down the phone line to the routing station. Here, each digit is recognized as a combination of tone frequencies. The specific number combination causes a signal to be sent to another phone causing it to ring. When that phone is picked up, a connection between the two phones is initiated.

The mouthpiece acts as a microphone. Sound waves from the user's voice cause a thin, plastic disk inside the phone to vibrate. This changes the distance between the plastic disk and another metal disk. The intensity of an electric field between the two disks is changed as a result and a varying electric current is sent down the phone line. The receiver on the other phone picks up this current. As it enters the receiver, it passes through a set of electromagnets. These magnets cause a metal diaphragm to vibrate. This vibration reproduces the voice that initiated the current. An amplifier in the receiver makes it easier to hear. When one of the phones is hung up the electric current is broken, causing all of the routing connections to be released.

The system of transmission presented describes what happens during a local call. It varies slightly for other types of calls such as long distance or cellular. Long distance calls are not always connected directly through wires. In some cases, the signal is converted

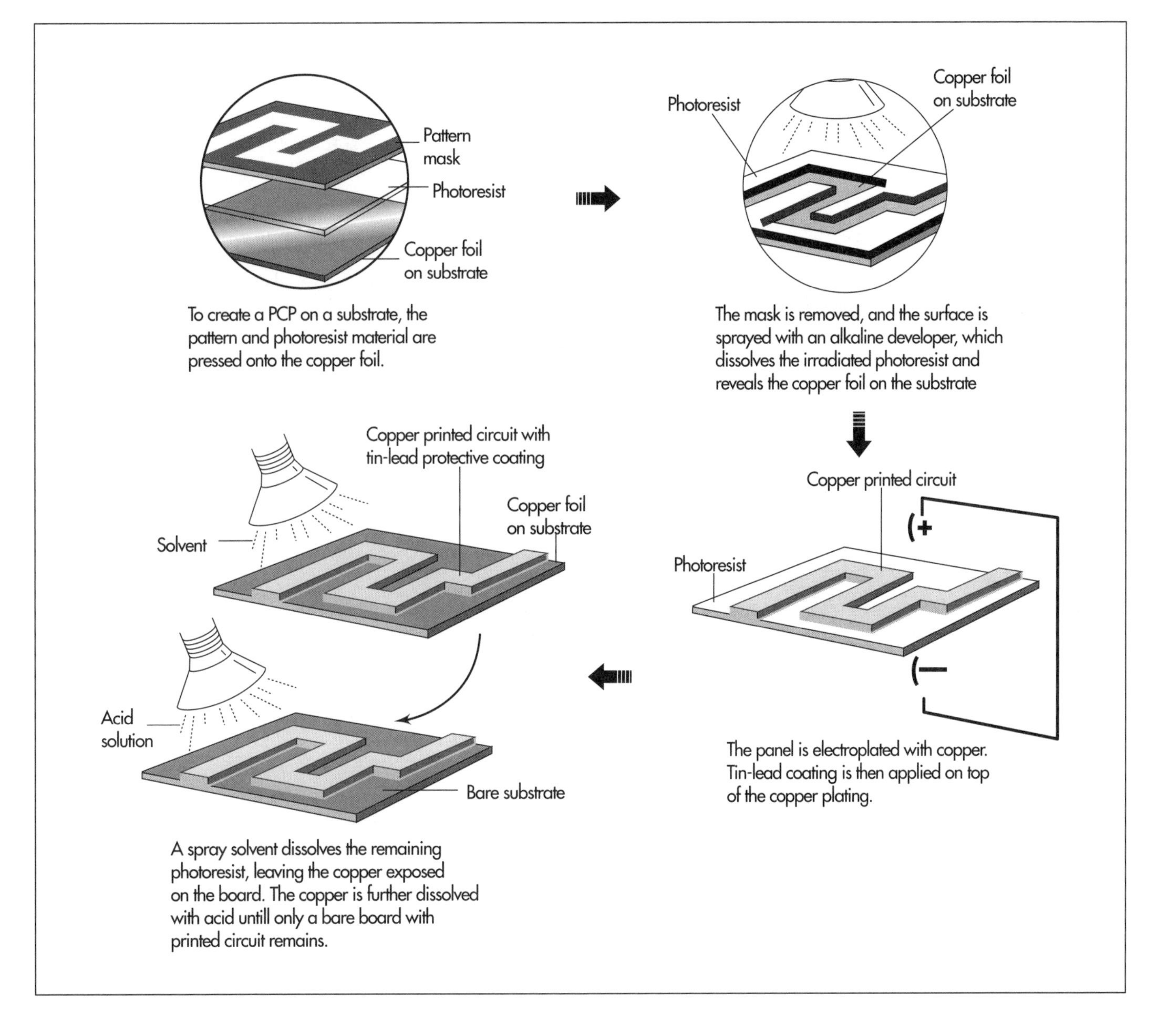


to a satellite dish signal and transmitted via a satellite. For cellular phones, the signal is sent to a cellular antenna. Here, it is sent via radio waves to the appropriate cell phone.

## Raw Materials

A variety of raw materials are used for making telephones. Materials range from glass, ceramics, paper, metals, rubber and plastics. The primary components on the circuit board are made from silicon. The outer housing of the phone is typically made of a strong, high-impact resistant polymer. To modify the characteristics of this polymer, various fillers and colorants are used. The speakers require magnetic materials.

## Design

Modern telephones come in many shapes and sizes, but they all have the same general features. They consist of a single handset which contains both the transmitter and receiver. The handset rests on the base when the phone is not in use. They also have a dialing system which is either a rotary dial or a touch-tone keypad. Recently, rotary phones have been phased out in favor of the more useful keypad. To alert the consumer that they have an incoming call, phones are equipped with ringers. A wide variety of specialized phones are also produced. Speaker phones are made to allow the consumer to carry on a telephone conversation

The electronic components of the telephone are sophisticated and use the latest in electronic processing technology. The circuit board is produced the same way that boards are made for other types of electronic equipment. The preprinted, nonconductive board is passed through a series of machines that place the appropriate chips, diodes, capacitors, and other electronic parts in the appropriate places. To affix the electronic parts to the board, a wave soldering machine is used.

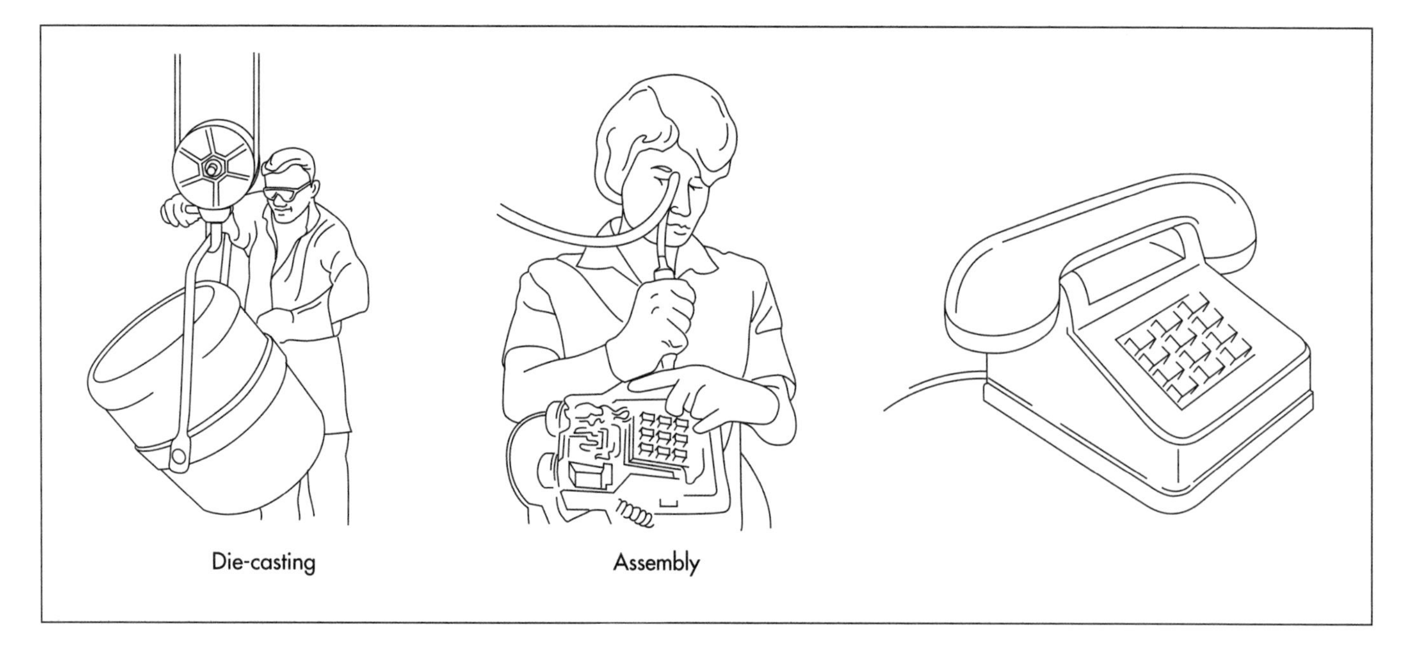


The individual parts of the telephone are assembled both automatically and manually. The transmitter and receiver are put together by machines. These parts are then fed onto the main assembly line and inserted into the molded headset. Similarly, the internal electronics, including the touch-tone pad, are inserted into the main housing and attached with screws.

without holding the handset. Cordless phones are also available. These models do not require direct connection of the handset to the base. Instead the user's voice is converted to radio waves and then sent to the phone base. This, in turn, gets converted to an electric signal and sent down the phone lines. Another type of common phone is the cellular phone. These phones use radio waves and an antenna system to communicate between phones.

## The Manufacturing Process

Since there are so many different parts that go into making a telephone, the components are typically produced by different companies and then assembled by the phone manufacturer. The main components include the internal electronics, the handset, and the various plastic parts.

### *Plastic parts*

1 To produce the plastic parts like the base, handset casing, and push buttons, injection molding is typically done. In this process, pellets of plastic polymer are put into the hopper of an injection molding machine. They then pass through a hydraulically controlled screw and are melted. While the screw rotates, the melted plastic is moved through a nozzle and injected into a mold. Just prior to this injection, the two halves of a mold are brought together to form a cavity which matches the shape of the telephone part. While inside the mold, the plastic is held under pressure for a set amount of time and then allowed to cool. As it cools, it hardens and forms into the shape of the part. This mold is coated with chromium to create a shiny surface.

2 After a short while, the mold halves are opened and the part is ejected. The mold then closes again and the process begins again. At this point in process, many of the parts are manually inspected to ensure that no significantly damaged parts are used. If there are damaged parts, they are set aside to be remelted and reformed into new parts.

### *Internal electronics*

3 The electronic components of the telephone are sophisticated and use the latest in electronic processing technology. The circuit board is produced the same way that boards are made for other types of electronic equipment. The process begins with a board made of non-conducting material that has the electronic configuration printed on it using a conducting material. This board is then passed through a series of machines that place the appropriate chips, diodes, capacitors and other electronic parts in the appropriate places. To prevent damage caused by dust, the process is completed in a specially cleaned room. When completed, it is sent to the next step for soldering.

4 To affix the electronic parts to the board, a wave soldering machine is used. Before being put into the machine, the board is washed to remove contaminants. Upon entering, the board is heated using infrared heat. The underside of the board is passed over a wave of molten solder and through capillary action, all of the necessary spots are filled. As the board is allowed to cool, the solder hardens and the pieces stay in place. This creates an electrical connection between the printed circuits and the components.

### Assembly and Packaging

5 The individual parts are assembled both automatically and manually. The transmitter and receiver are put together by machines. These parts are then fed onto the main assembly line and inserted into the molded headset. Similarly, the internal electronics, including the touch-tone pad, are inserted into the main housing and attached with screws. The headset is then put on the phone base and the phone chord may also be put on.

6 After all of the phone pieces are assembled, the completed phones are put in final packaging. Typically, they are wrapped in plastic and put in boxes. A packaging material such as polystyrene is also included to protect the device from damage during shipping. An owner's manual or other literature is included and the box is sealed with tape. The boxes are stacked on pallets, shipped to distributors and finally, customers.

## Quality Control

To ensure the quality of each telephone, visual and electrical inspections are done throughout the entire production process and most flaws are detected. Additionally, the each completed phone is tested to make sure it works. Often these tests are done under different environmental conditions such as excessive heat and humidity to simulate the extremes that are experienced in a real life setting. Plastic parts are given torture tests to ensure they will still function even after a level of consumer abuse. For example, the numbers on a touch tone pad are put under a rubber-finger that taps the buttons enough times that it equals forty years of dialing. Since many of the parts that make up the telephone are produced by subcontractors, telephone manufacturers rely heavily on these suppliers for good quality. To ensure consistent manufacturing, most telephone makers set quality specifications for individual parts that the suppliers must meet.

## The Future

Telephone technology is improving rapidly. In the future, cordless phones will be designed to be smaller and lighter. They will have wider transmission and better reception ranges. The sound quality will also be improved. Other technologies that integrate the phone with computers and cable television will be commonplace. Dialing systems will also be improved. In a recent demonstration, one company showed a technology that accepted voice command to dial a phone number.

## Where to Learn More

### Books

Bigelow, Stephen. *Understanding Telephone Electronics* New York: Butterworth-Heinemann, 1997.

Noll, A. *Introduction to Telephones and Telephone Systems*. New York: Artech House Publishers, 1998.

Grosvenor, E and M. Wesson. *Alexander Graham Bell: The Life and Times of the Man Who Invented the Telephone*. Harry N. Abrams Inc., 1997.

### Periodicals

Woolley, S. "Dial tones? No, Web tones."*Forbes* (January 26, 1998).

*—Perry Romanowski*

# Thread

*Ninety-five percent of all sewing thread that is manufactured is used in commercial and industrial sewing.*

## Background

Thread is a tightly twisted strand of two or more plys of yarn that are circular when cut in cross section. It is used for hand sewing and in home sewing machines. Ninety-five percent of all sewing thread that is manufactured is used in commercial and industrial sewing. Sewing thread is distinguished from yarn by the fact that thread is used to sew together garments or other products, but yarn is the collection of fibers used to weave or knit into a textile fabric. The terms are confusing and are often used interchangeably; thread can be made of yarn, but yarn is not made of threads.

Thread is wound on spools or large cones that are marked on their ends with the size or fineness of the thread. Thread for handwork and machines (both home and commercial machines) has to be smooth and friction-free. It should be easy to thread through needles, and it should move easily when tension is applied to it. Strength to hold stitches when garments are being worn and during laundering is a requirement, as is elasticity during stitching and wear.

The three basic types of thread are based on their origin and are animal, plant, or synthetic. Silk thread is best for wools and silks (fabrics of animal origin). It is strong, very elastic, and fine in diameter. Silk is also used for tailoring, to finish the edges of buttonholes, to sew on buttons, and for decoration. Buttonhole twist is about three times the diameter of sewing silk and shiny or lustrous. It is strong and can be permanently stretched.

Cotton threads are made of the cellulose from plants and are used to stitch fabrics like linen, rayon, and cotton that also have plant origins. Plant-based fabrics may shrink, and cotton thread has the same shrinkage characteristics. Also, cotton thread does not stretch, so it is useful for woven fabrics but not stretchy knits. Cotton is also used to make basting thread that is used to hold pieces of a garment together temporarily until the garment is sewn. Cotton basting thread is inexpensive and is loosely twisted so it can be broken easily and pulled out when permanent stitching is in place.

Nylon and polyester threads are preferred for synthetics and stretch knits. Both types of synthetic threads have the same characteristics including no shrinkage, high strength, and excellent abilities to stretch and recover that make them suitable for knits, preshrunk fabrics, and sheers. Nylon and polyester are the only threads that can be made from a single yarn or single ply. Nylon thread was popular in the 1960s and 1970s because it is clear and so matches any fabric; however, nylon's stretching properties and relative brittleness also work as disadvantages, and this thread has become less popular. Synthetic threads are marketed under a variety of names including polypropylene, Kevlar, Teflon, and Nomex. Production of Kevlar and Nomex is reviewed and approved by Underwriters Laboratories because these threads are used to stitch fire-resistant and -retardant products such as suits for firefighters and motor racing drivers as well as sleepers and crib linens for infants.

Specialized thread is also treated after spinning and depending on use. Garments made of fabric treated with water repellant are also sewn with treated thread. Metallic thread is used for decoration, but it is sensitive to heat and steam and must be handled with care. Mending thread is made in the same types as those sold on spools, but it comes in short lengths so it is economical to buy a small amount. Upholstery thread is reinforced for strength so it can be used to upholster furniture. Its opposite is invisible thread that is very fine and made from nylon to be strong. Invisible thread is used to sew light- to medium-weight synthetic fabrics. Other specialized threads are made for decorative sewing like needlepoint. These include soft cotton embroidery thread, cotton perle that is twisted and has a lustrous or pearl-like finish, stranded cotton with six fine strands that can be used together or separated for finer stitching, tapestry wool for decorative work, and mending yarn for reinforcing and darning.

Thread can also be made of combinations of fiber. Thread with a core of polyester and an outer wrap of cotton, spun polyester, or rayon combines the characteristics of both fibers, with the outer fiber creating the appearance or finish of the thread.

## History

Thread has been essential to humans since the first garments were made for warmth and protection. Early sewing thread consisted of thin strips of animal hide that were used to stitch together larger pieces of hide and fur. The advance of civilizations brought many refinements in clothing and adornments, including the spinning and dyeing of thread. The Egyptians were skilled at making thread from plant fibers and in using the wool and hair from domestic animals in spinning. They and the Phoenicians also pioneered the use of berries and plant matter in the manufacture of colorful and longlasting dyes. The Chinese and Japanese discovered the beauties of silk fibers spun as thread and made as cloth.

Sewing thread had at least two historical heydays. In the Middle Ages, improvements in shipping, wool production and processing, and the opening of the Silk Road to Asia provided fertile ground for the flowering of woven tapestries and needlework. Tapestries were an art form that allowed large spaces to be filled with colorful scenes that were also portable. Tapestries were woven of wool yarn primarily, but linen, cotton, silk, and gold and silver were also used in weaving these magnificent paintings in fiber. Seamstresses and needleworkers were inspired to use the same materials in smaller works stitched with sewing or embroidery thread.

During the Industrial Revolution in the seventeenth and eighteenth centuries, production of thread moved out of the cottages and into factories equipped with high-speed machines. Machine manufacture generated more uniform thread with fewer flaws, and producers could devote more time to maximizing the characteristics of the types of fibers being used. Stronger thread, truer dye colors, and production of a wider variety of thread for different applications were among the direct results.

## Raw Materials

### *Silk thread*

Silk worms produce cocoons from which silk is made. The female silk worms feed on mulberry leaves until the worms mature and begin to spin cocoons. The worm secretes the silk thread from specialized glands below its mouth. It completes the cocoon around itself; over a two- to three-week period, it becomes a pupa then a moth. The cocoons can be harvested when the pupas are still inside so that so-called nett silk can be produced; cocoons are also collected after the moths have hatched, and silk from these is termed Schappe silk.

### *Cotton thread*

Cotton is picked from cotton plants in the field and compressed into large bales. The bales often contain dirt, broken pieces of cotton boll, seeds, and other impurities, so cleaning is a first step at the mill. The bale is broken, the fibers are opened by a comb-like device, mixed together, and cleaned. The cleaned cotton fibers are termed laps. The laps are fed into a carding machine that separates the fibers. Further cleaning, combing, and sorting readies the fibers for processing into thread.

Cotton thread is produced from large cotton bales that are cleaned, combed, carded, twisted, and spun. Polyester thread is derived from petroleum that has gone through a cracking process. During the cracking process, crude oil is broken down into a number of components that will be processed into a range of products from gasoline to plastics including polyester. Xylene, a hydrocarbon compound, is generated during cracking. Nitric acid and glycol are added to modify the xylene by a series of chemical reactions. The fluid is heated and condensed in an autoclave, and the molecules align to form long molecules called polyester. The resulting mass is extruded, cooled with water, and cut into chips. These chips are shipped from the refinery to the thread manufacturer for spinning.

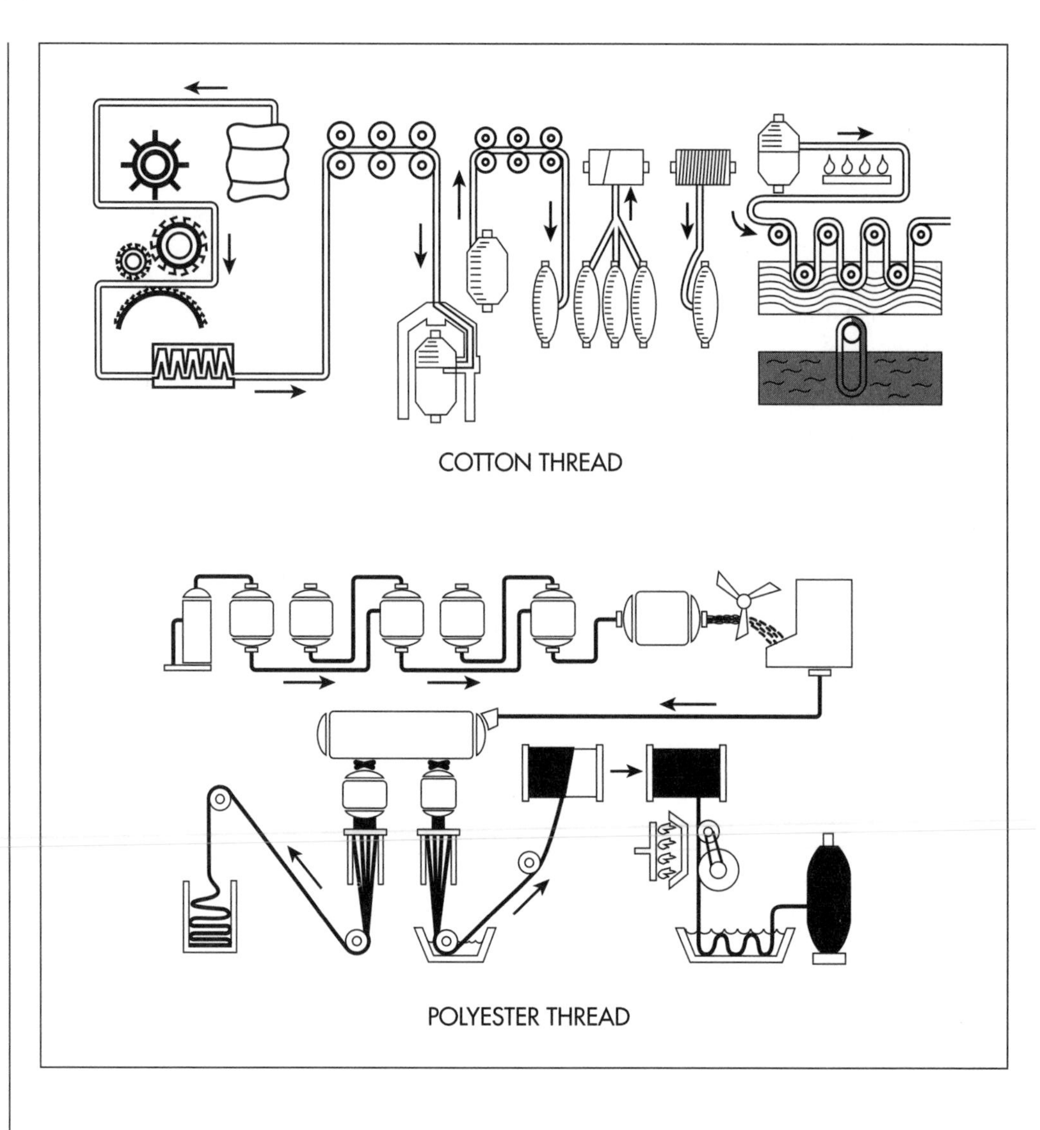


### *Polyester thread*

Polyester is a petroleum product. During the cracking process, crude oil is broken down into a number of components that will be processed into a range of products from gasoline to plastics including polyester. Xylene, a hydrocarbon compound, is generated during cracking. Nitric acid and glycol are added to modify the xylene by a series of chemical reactions. The fluid is heated and condensed in an autoclave, and the molecules align to form long molecules called polyester. The resulting mass is extruded, cooled with water, and cut into chips. These chips are shipped from the refinery to the thread manufacturer for spinning.

## Design

Engineers who design sewing thread are called seam engineers. They are experienced in the practical aspects of sewing, sewing machine operation, and clothing manufacture. When a new sewing thread is designed, the needs of the specific market are analyzed carefully, and a prototype thread is produced and tested under actual use conditions. The best thread products are needed to meet the demands of new and everchanging markets.

## The Manufacturing Process

### *Nett silk*

1 Cocoons containing the pupas are heated with air to keep the pupas inside.

2 The cocoons are immersed in water and a brushing motion unwraps the silk thread; a single thread can be up to 1,000 yd (3,000 m) long.

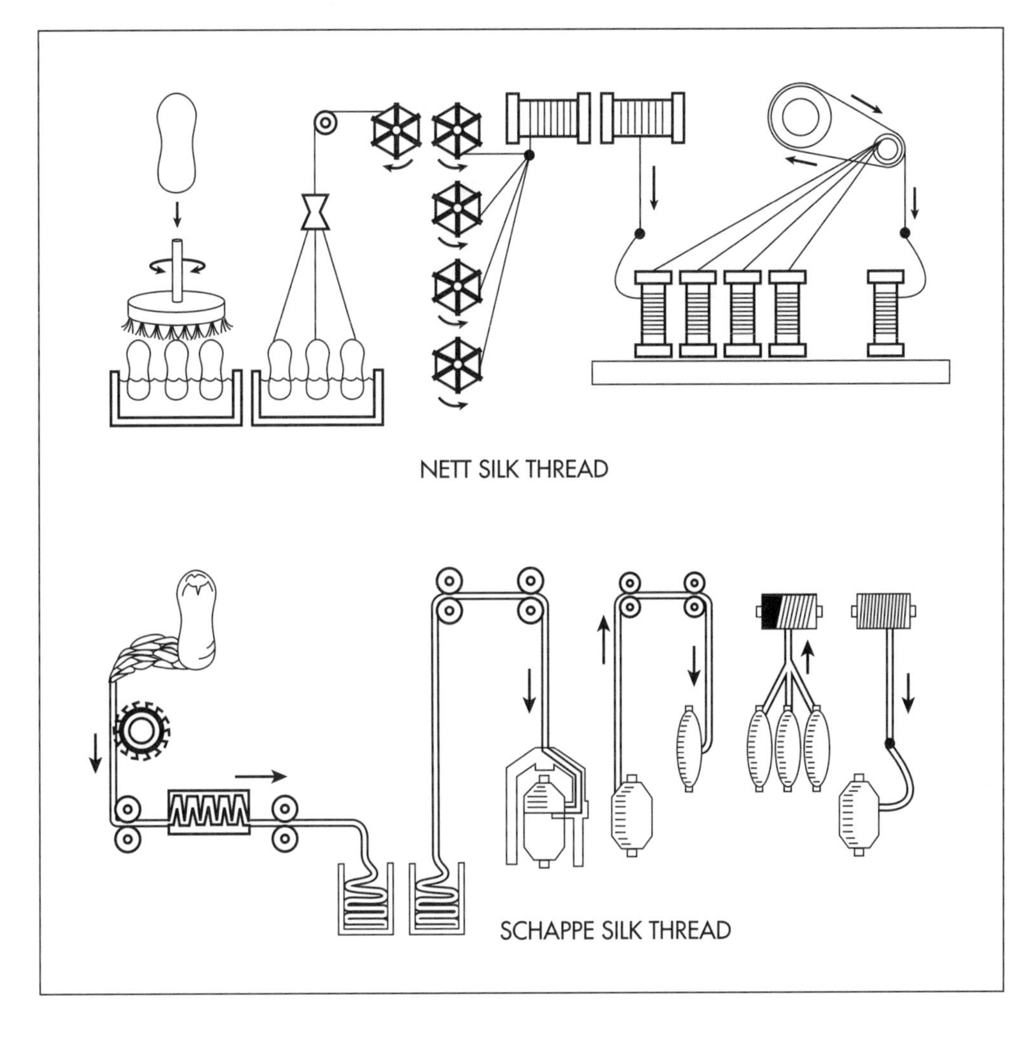


There are two types of silk manufacture. Both use the cocoon of the silk worm to produce silk thread, however, one processes the cocoons that still contain the pupa, producing a soft silk called nett silk. The other utilizes cocoons that have already hatched to produce schappe silk. To make nett silk hot air is used to inhibit the growth of the pupa inside the cocoon, the cocoon is unraveled, and several yarns are wound together to produce a thread. Since it is derived from the older, hatched cocoon, schappe silk manufacture begins with the soaking, softening, washing, and drying of the cocoons. Next, the cocoons pass through steel rollers with combs to produce bundles of long, straight fibers of silk, which are combed, twisted, and spun into thread.

3 Several threads from the same cocoon are wound together in long loops or hanks to produce a single yarn. Three of these single yarns are twisted to produce nett silk thread. More than three single yarns can be used depending on the planned use of the thread.

### *Schappe silk*

4 Schappe silk is made from cocoons from which the silk moths have hatched. This matured material is tougher than nett silk cocoons and must be softened, washed, and dried.

5 The cocoons are passed through steel rollers with combs to produce bundles of long, straight fibers of silk called combed top or peigné.

6 Several bundles are grouped together to form a band that is also combed into a narrower band and fed through rotating rollers. The emerging yarn is slightly twisted and is called the roving.

7 The roving is spun to form a single thread, and several single threads are wound together, twisted, and fed onto a bobbin or cone.

### *Cotton thread*

8 Cleaned, combed, and sorted cotton is fed through a series of rollers in a process called drawing that generates a narrow band of cotton fiber.

9 The fiber is slightly twisted to form roving, and the roving is drawn and twisted again.

10 It is spun to form a single thread that is wound and twisted with others to form the thread.

11 Cotton thread is singed over an open flame and mercerized by immersion in

caustic soda. These processes strengthen the thread and give it a lustrous finish. The treated cotton thread is wound on bobbins or cones.

### *Spun polyester thread*

12 The chips of polyester received from the refinery are spun into long filaments that are banded together to form polyester tow. Tow contains more than 170,000 continuous, fine, parallel filaments in a long band.

13 The tow is stretched to the breaking point so that weaknesses in the filaments are exposed. The weak points are cut out, and fibers that are 2.5-4.75 in (60-120 mm) are produced.

14 These resulting strong fibers are banded together again in parallel, and several bands are combined until a narrow, even, high-quality band of parallel fibers results.

15 These bands are further stretched and slightly twisted into roving, and the roving is spun and twisted into single thread. The drawing and twisting process stretches the fibers up to 10-20 times their length in the tow stage, and the spindles on the spinning machines that do this work turn at 12,000 revolutions per minute.

### *Dyeing and packaging finished thread*

All types of thread may be finished in similar ways.

16 After manufacture, the thread is dyed. Dye is mixed in large vats; several hundred colors can be produced, and dye mixing is controlled by computer. Large cones or bobbins of manufactured thread are lowered into pressurized vessels, and the dye (and other additives) are released into the vessels by computer. After it is removed from the dyeing vessel, the thread is dried for 24 hours. About 6.6 tons (6 metric tons) of thread can be dyed in one day; this equals 66 million yd (200 million m) of sewing thread or enough to circle the equator five times.

17 The dyed thread is wound on smaller spools for industrial or home use, and the spools are packed into boxes for shipment. Optical devices monitor the spool winding and packing processes.

## Quality Control

Thread must emerge from the factory with a great combination of performance characteristics. During its lifetime, thread will be passed through a number of metal guides on a sewing machine, held in tension by the sewing machine needle and subjected to its action, bent into position from the bobbin to interlock with the thread from the needle, stretched and abraded when the garment is worn, and laundered or dry cleaned many times. Thread functions not only to clothe but to protect us. We count on the thread in products like shoes and automobile safety belts to protect us throughout the life of the product.

Product manufacturers place considerable demand on thread makers to improve the quality of thread and to adapt to demands of the workplace. For example, a lubricating finish is applied to thread that is used to stitch thermoplastic materials so the sewing machine needle is constantly lubricated to prevent its frictional heat from rising and fusing the material.

Production of almost all types of threads is fully automated. A device called an Autoconer monitors the thread quality, cuts out imperfections, and rejoins the ends to produce an even thread without knots, joins, or weak areas.

## Byproducts/Waste

Cotton thread production results in more waste than the manufacture of other types of thread because of the parts of the cotton that cannot be used. The volume is still relatively minor but must be disposed. Imperfect fibers and fiber dust are generated in the early stages of production and, again, must be disposed.

## The Future

In the future, the manufacture of thread must continue to adapt to new types of fabric and clothing design and production. Futurists may predict that clothing and its varieties will become obsolete; however, the reality is that the consumer wants to express personality through a closet filled with clothing for a variety of activities. And this clothing is sewn together with specialized threads.

## Where to Learn More

### Books

Cream, Penelope, ed. *The Complete Book of Sewing*. New York: DK Publishing, Inc., 1996.

Hollen, Norma and Jane Saddler. *Textiles*. New York: The Macmillan Company, 1968.

Tortora, Phyllis G. *Understanding Textiles*. New York: The Macmillan Company, 1992.

### Other

Coats & Clarke, North America. http://www.coats.com.

Coats Viyella PLC, Thread Division. http://www.coats-viyella.co.uk.

Gütermann of America, Inc., and Gütermann AG, Germany. http:// www.guetermann. com.

Saunders Thread Company. http://www.saunders-thread.com/sewing.html.

Synthetic Thread Company. http://www.syntheticthread.com.

*—Gillian S. Holmes*

# Toilet

*The Old Testament contains several references to toilets, from laws about how to cover waste out of doors to mention of King Eglon of Moab's indoor privy chamber.*

## Background

A system for dealing with excrement is necessary in every human community, and the need becomes more pressing the more densely populated the area. Though simple pit latrines are still common in many rural areas today, more complex lavatory designs date back thousands of years. The Old Testament contains several references to toilets, from laws about how to cover waste out of doors to mention of King Eglon of Moab's indoor privy chamber. Some kind of lavatory flushed with water is believed to have been used by residents of the Indus Valley by around 2000 B.C. Even earlier, in about 2750 B.C., the ancient Indian city of Mohendro Daro was equipped with toilets connected to a drain. Dating back to approximately 4000 B.C., the neolithic stone huts of the Scara Brae settlement in the Orkney Islands seem to have had indoor lavatory provisions. Apparently used as toilets, stone chairs have also been unearthed from the site of the Sumerian city of Ashnunnack, dating to around 4000 B.C. The palace of King Minos of Crete, from about 2000 B.C., had elaborate indoor plumbing, including marble toilets that were flushed with water dumped from a vase in an adjoining room.

The remains of Roman lavatories are still extant in many places. Some private Roman houses had their own toilets, which were in most cases a seat located over a drain or a cesspit. Roman public lavatories were more impressive. They were often built next to or as part of public baths. Rows of stone or marble seats in pairs, divided by armrests, stood over a trench. Excess water from the baths flowed into the trench, and washed the waste into a main sewer. A smaller trench filled with fresh water flowed past the base of the stone toilets. This water was used for rinsing. Roman forts, which housed hundreds of soldiers, also boasted impressive toilet facilities. The builders of Housesteads, a Roman fort in northern England dating to 122 A.D., diverted a river to flow underneath the latrine and carry waste out of the fort. The latrine itself was a large room with benches built around three walls. The benches had about 20 holes with no dividers for privacy. Roman cities also took care of the needs of travelers by erecting huge vases along the roadways for people to urinate into, thus keeping waste off the public streets.

During the Middle Ages, lavatories drained with running water were common in British abbeys, which housed large groups of monks. Similar to the Roman forts, abbey latrines were usually meant for many people to use at once, and drained over a river or stone drain. Stone castles were often designed with vertical shafts for the emptying of waste. The waste flowed into a trench leading in most cases to the moat. Indoor toilets consisted of wooden closets or cupboards, which concealed a seat over a chamber pot. Servants emptied the pot into the moat.

In Medieval European cities, common practice was to empty indoor chamber pots directly into the streets, a foul practice that bred disease. Something akin to the modern flushing toilet first came into use in England at the end of the sixteenth century. A water-operated "water closet" was invented in 1596 by Sir John Harrington. Queen Elizabeth I had Harrington's device installed in her palace, setting the vogue among the nobility. However, flushing toilets did not catch on with the bulk of the population until much later. The first British patent for a water closet was awarded to Alexander Cumming in 1775. His device used a pan with a sliding door. The pan contained a few inches of water. When finished, the user would pull a lever that opened the pan, letting the contents slide out into a drain, and at the same time opening a valve that let fresh water into the pan. The Bramah water closet, patented by Joseph Bramah in 1778, used a similar but more complex flushing device that kept the water running for about 15 seconds. By about 1815, water closets of this type had become common in London households. A modern sewer system was completed in London in 1853, and a large-scale toilet manufacturing industry dates to around this time.

## Raw Materials

Toilet bowls and tanks are made from a special clay called vitreous china. Vitreous china is a mix of several kinds of clay, called ball clay and china clay, silica, and a fluxing agent. Clays are hardened by first drying in air, then being fired (baked) in a very hot oven called a kiln. Usually a shiny, waterproof coating called a glaze is applied only after a first firing, and the clay is fired a second time. Vitreous china is an exception, in that clay and glaze can be fired together. The whole clay body vitrifies, or turns glassy, so the toilet is actually waterproof and stainproof through its entire thickness.

Toilet seats are generally made from one of two materials. Plastic toilet seats are made from a type of thermoplastic called polystyrene. The less expensive and more common type of toilet seat is made from a blend of wood and plastic. The wood is hardwood, usually maple or birch, which has been ground up into the consistency of flour. This wood flour is blended with a powdered plastic resin called melamine. Zinc stearate is a third ingredient in wooden toilet seats. This prevents the wood-resin mix from sticking to the mold in the manufacturing process. The metal tank fixtures are made of stainless steel or copper, and the joints that hold the seat to the bowl are usually a rubber-like plastic.

*A chamber pot.*

Some Victorians couldn't abide the thought of indoor toilets because they reviled at the notion of odor and unclean gases associated with them. Today, it is difficult to imagine life without indoor plumbing. How awful to have to scurry to the outhouse in cold weather or to stumble to the privy late at night when duty called.

One did not always have to walk to the privy on these occasions, however. Instead, one could use a ceramic chamber pot. It functioned like an indoor toilet that did not flush—one perched upon it for defecation or used it as a urinal and then the "slop jar" was emptied into the outhouse. Some chamber pots were decorated with lacy covers along the edge of the bowl called silencers and presumably muffled the noise of clanking of the top upon the bowl at night so that others weren't awakened by its use.

The chamber pot in the photo is part of a large set of ceramics used for personal hygiene in the days before indoor plumbing. Many bedrooms had a pitcher for fresh water, a basin to hold the water for cleansing, a soap dish and a chamber pot. These ceramics were always fashionably decorated, so that the bedroom could be attractively appointed even for these disagreeable tasks.

## The Manufacturing Process

### *Plastic seat*

1 Plastic seats begin as pellets of polystyrene. A worker feeds the pellets into a hopper attached to an injection molding machine. From the hopper, a precisely measured amount of pellets flows into a

A plastic toilet seat is made by a process called injection molding, where plastic pellets are melted and injected into a mold. A wooden toilet seat is produced from a mixture of wood powder and melamine mixture that is heated to 300° F (149° C). Once both types of seats are molded, they are hung on an overhead conveyor rack that moves them along to the finishing area.

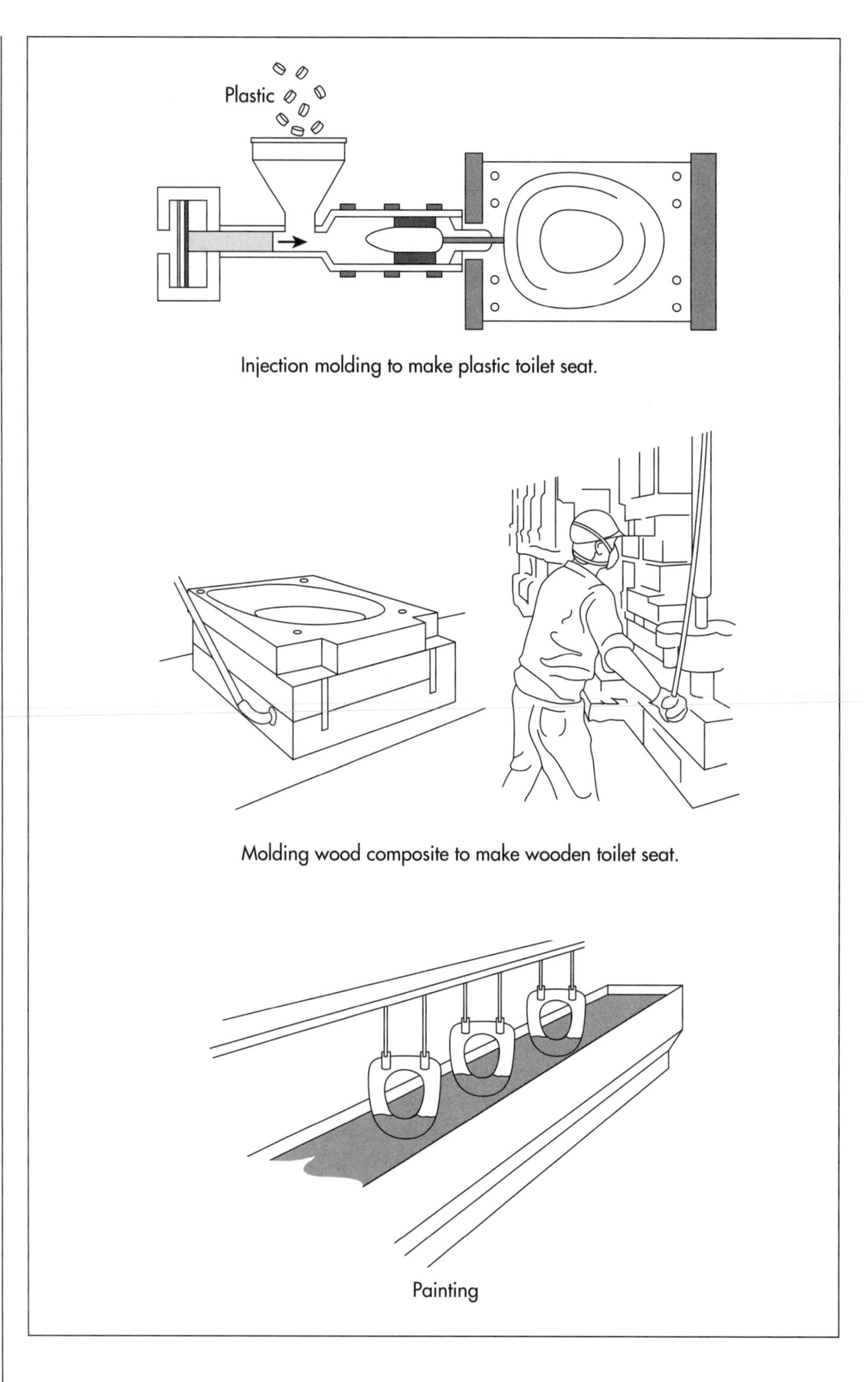


Injection molding to make plastic toilet seat.

Molding wood composite to make wooden toilet seat.

Painting

container that heats the material until it melts. Then the liquid polystyrene flows through a small hole in the center of a two-part mold. The mold is made of chrome-plated machined die steel. Its two halves are hollowed in the shape of the toilet seat and cover. When the mold is full, it is clamped together by a huge hydraulic press. This exerts 10,000 lb per sq in (4,540 kg per sq cm) of pressure on the mold, and heats the polystyrene to 400° F (204° C).

2 The plastic in the mold begins to solidify. Then cool water is pumped through a channel system around the mold to bring the temperature down. A worker releases the hydraulic clamp and separates the two halves of the mold. The worker removes the seat and cover from the mold, breaking off the extra plastic that formed in the water channel. Then, the worker places the seat and cover into a water bath.

3 After the seat and cover have cooled in the bath, a worker takes them to a finishing area for the final steps. Here holes are drilled for the hinges. Then, a worker smooths the rough edges at a sanding machine. The sander is a rotating wheel covered with an abrasive material. The worker passes the seat or cover along the wheel until any plastic fragments from the drilling or from the mold are sanded off. A similar machine with a softer surface may next be used to give a final polish.

### *Wooden seat*

4 For wooden toilet seats, the first step is to mix the wood flour and the plastic resin. Workers wearing protective masks slit open bags of wood flour and empty them into a mix box. Then, the worker adds the powdered plastic resin that makes up 15% of the formula. Last, a small amount of zinc stearate is added. The mixture is passed to an attrition mill, which grinds the particles down further. After milling, the powdered mixture may be measured into boxes for loading into the molding press. Or it may be set aside, and later measured and scooped by hand into the press.

5 The processed wood and melamine mixture is next emptied into the bottom half of a mold. A worker makes sure the mix fills the mold evenly and smooths the surface. Then the worker lowers the top half of the mold and begins to heat the whole thing to 300° F (149° C). While it heats, the mold is clamped at 150 tons of force. After 6.5 minutes, the wood flour and melamine have fused together and hardened. Then, the worker opens the mold and hangs the seat and cover on an overhead conveyor rack, which moves it along to the finishing area.

6 Wooden seats are finished in the same way as plastic seats. First, they are drilled, then sanded. Then, they are hung again on an overhead conveyor and taken to the painting area. The conveyor automatically lowers the seats into a tank of paint. Then the conveyor pulls them up and passes them into an enclosed room called a vapor chamber. A paint solvent is released as a vapor, and this carries off any excess paint without leaving drip marks. Next, the painted seats pass along the conveyor into a drying oven. The paint-vapor-drying process is repeated four times. The first two coats are a primer, and the second two are an enamel paint that produces a smooth, hard, plastic-like surface.

7 Both plastic and wooden seats are assembled and packaged the same way. The seats and covers are screwed together and packed with the necessary mounting hardware. Then, they are boxed and moved to a warehouse or distribution center.

### *Bowl and tank*

8 The toilet bowl and tank are made at a type of factory known as a pottery. The pottery receives huge amounts of vitreous china in a liquid form called slurry slip. Workers at the pottery first thin the slurry slip to a watery consistency. Then, they feed it through very fine screens in order to sieve out any impurities. The purified slip is thickened again, and pumped into storage tanks in preparation for use in casting.

9 Next, the slip is carried through hoses and pumps into the casting shop. Workers fill plaster of Paris molds with the slip. The molds are in the shape of the desired piece, except they are about 12% bigger, to allow for shrinkage. The workers fill the molds completely with the slip, and let it sit for about an hour. Then, the workers drain out any excess slip. This is recycled for later use. The clay sits in the mold for another few hours. The plaster of Paris absorbs water from the clay, and the clay dries to the point where the mold can be safely removed. At this point, the casting is semisolid, and is called greenware. Workers use hand tools and sponges to smooth the edges of the casting and to make holes for drains and fittings.

10 The greenware castings are left to dry in the open air for several days. Then they are put into a dryer for 20 hours. The

Toilet bases are cast from a slurry of vitreous china and molded into the base shape. Once molded, the greenware, as it is called, goes through a series of drying, glazing, and firing steps until it reaches final inspection.

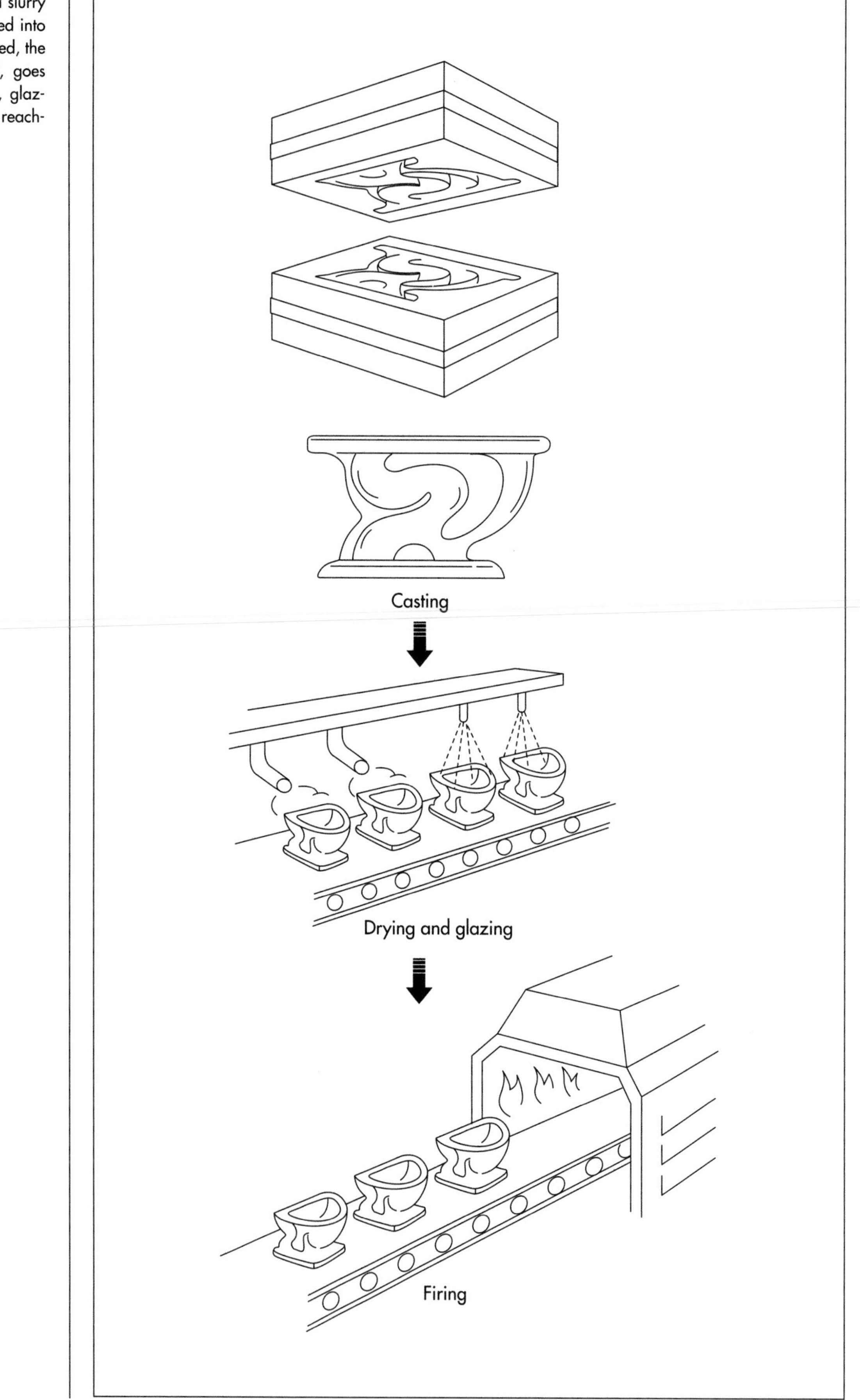

dryer is set to 200° F (93° C). After the castings come out of the dryer, they have lost all but about 0.5% of their moisture. At this point workers spray the greenware castings with glaze. Now, the pieces are ready for the kiln.

11 The kilns at a large industrial pottery are warehouse-sized tunnels, and the pieces move through the kiln on a conveyance called a car. Each car is loaded with a number of pieces, and then it moves automatically through the hot kiln at a very slow pace. Because rapid changes in temperature will cause the clay to crack, the cars move leisurely through graduated temperature zones: the first zone is about 400° F (204° C), and it increases in the middle of the kiln to over 2,200° F (1,204° C) degrees. The temperature gradually decreases from there, so that the final temperature is only about 200° F (93° C). The whole firing process takes approximately 40 hours.

12 When the pieces are removed from the kiln and fully cool, they are ready for inspection. After inspection, the flushing mechanism is installed. This is either manufactured at the plumbing fixture company or bought from a contractor. The seat too may be installed at this time, or the parts may be sold separately and assembled by a plumbing distributor.

## Quality Control

As with any industrial process, quality checks are taken at several points in the manufacturing of toilets. The clay is sieved and purified before it is pumped into the factory's tanks. Workers doing the manual finishing of the castings check the pieces for cracks or deformities. After firing, each toilet is tested individually. Random sample checks are not a good enough gauge of quality: each piece must be inspected for cracks. There are several ways to do this. One test is to bounce a hard rubber ball against the piece. It should emit a clear, bell-like ringing sound. A cracked piece will give off a dull sound, indicating a crack that might not have been visually obvious.

## Byproducts/Waste

The pottery is able to recycle much of its clay. As long as it has not been fired, all the clay is reusable. Even the air-dried greenware can be scrapped, softened and reprocessed into the watery slip of the first step of the process.

## Where to Learn More

### *Books*

Barlow, Ronald S. *The Vanishing American Outhouse.* El Cajon, California: Windmill Publishing Company, 1989.

Hart-Davis, Adam. *Thunder, Flush and Thomas Crapper.* North Pomfret, Vermont: Trafalgar Square Publishing, 1997.

Reyburn, Wallace. *Flushed with Pride: The Story of Thomas Crapper.* Englewood Cliffs, New Jersey: Prentice-Hall, 1971.

*—Angela Woodward*

# Tuba

*During the 1820s, various instrument manufacturers produced the ophicleide, a precursor to the tuba. It was a type of keyed-bugle instrument that could reportedly produce a pitch as low as F.*

A tuba is a brass instrument characterized by its large size and deep sound. It consists of vertically coiled tubing, three or four valves, a wide conical bore, flared bell, and a cup-shaped mouthpiece. The different segments of the tuba are shaped and bent from machinable brass using standard techniques. The pieces are then soldered together to complete the instrument. First built in the early nineteenth century, the tuba was invented to provide a bass sound in brass ensembles.

## Background

Tubas belong to a class of instruments known as brass wind instruments. The sounds that these instruments make are produced by the buzzing of the musicians lips against the mouthpiece. This causes the column of air inside the instrument to vibrate. As it exits the instrument through the bell, audible sound is created. The tone that is heard is directly related to the length of tubing through which the air traveled. By engaging a valve, the tubing is shortened or lengthened and the tone is changed. In an orchestra, the tuba is used to produce the lowest notes. Depending on the type of tuba it can reach notes as low as bottom D of the piano. It is often used to play quick staccato solos but can also play sustained melodies.

## History

The development of the tuba began during the early nineteenth century. During the 1820s, various instrument manufacturers produced the ophicleide, a precursor to the tuba. It was a type of keyed-bugle instrument that could reportedly produce a pitch as low as F. However, these instruments were difficult to play and were never truly embraced by the music community. To satisfy the desire of bandleaders for a valved brass instrument that was capable of playing low notes the tuba was invented.

In 1835, the first bass tuba was patented by Johann Gottfried Moritz (an instrument maker) and Wilheim Wieprecht (a bandmaster) It was completely distinct from the keyed ophicleide in that it was equipped with a new type of short piston valve called the Berliner-Pumpe. Their prototype tuba looked different than the modern tuba but it shared some key characteristics. For example, it was pitched in F, the standard pitch of orchestral tubas. It had five valves that could lower the pitch of the instrument. It was also made from brass with silver fittings.

Next, Moritz and Wieprecht designed a wider-bored instrument that they called a bombardon. It had the same general shape as the ophicleide but also had valves. Their design was adopted by other instrument makers who modified them slightly and began producing the bombardon in different sizes and shapes. Within a few years, rotary valves were used instead of the Berliner-Pumpe valves. In 1845, Adolfe Sax patented a family of sax-horns that ranged from sopranino to contrabass. The lowest member of these instruments closely resembled the modern tuba.

In 1849, the hellicon bass was introduced. This was an instrument with a circular coil that was designed to rest on a player's shoulder. It first gained popularity in Great Britain brass bands because of its ease of portability. In the 1860s the famous American bandleader, John Philip Sousa, commissioned C. G. Conn to produce a more audience-friendly version of the hellicon.

The instrument was later known as the sousaphone.

When the tuba was first introduced, it rapidly replaced the ophicleide in bands and orchestras in Germany. It was more slowly accepted by other countries, particularly Britain and France. By 1880, the tuba had become the standard brass bass instrument and the ophicleide was obsolete.

## Design

The general design of the tuba has changed little since the early days of its development. It is a massive instrument with a large bell of anywhere from 14-30 in (35.56-76.2 cm) in diameter. The main tubing is shaped in an elliptical coil with the bell pointing up. There are numerous variations on this shape however. For example, the sousaphone is a circular shaped instrument with the bell pointing forward. Certain upright tubas have also been made with a forward pointing bell. Other tubas have left or right facing bells. The position of the bell is important because it affects the overall tone quality the instrument produces.

Most tubas have four valves that are of the piston variety. However, certain instruments have anywhere from two to six valves. Rotary valves are also used for certain brands of tubas. These valves are faster than piston valves but less durable. A large, cup-shaped mouthpiece is used. Tubas are available in five different keys including double B flat, double C, E flat, F, and the double G contrabass bugle.

## Raw Materials

The primary raw material used for making standard tubas is brass. Brass is an alloy composed of copper and zinc. Other metals that may also be added to modify the characteristics include tin and nickel. The most common type used in the construction of instruments is yellow brass that contains a 70/30 blend of copper and zinc. This material is yellow in color and malleable. Other types of brass are also used when different effects are desired. For example, red-gold brass, which is composed of 80% copper and 20% zinc, provides a more gold color and a slightly different sound. Silver brass that includes copper, zinc, and nickel is also used.

The zinc in brass makes the alloy workable at lower temperatures. Some custom manufacturers use special blends of brass for different parts of the tuba. For example, a material such as Ambronze which is composed of 85% copper, 13% zinc, and 2% tin, may be used to make the bell. This material gives the instrument a unique sound.

Beyond brass, only a few other materials are used to make the tuba. Most of the screws are composed of stainless steel. In some cases, certain surfaces on the instrument such as the valves or the sliding pipes are coated with chromium or a nickel alloy. This reduces friction and helps the pieces move more freely. During production, flux and solder are used to connect the various tubes. To reduce the possibility of damage, valves are lined with felt where the valve button meets the instrument body. For decoration, a material such as mother of pearl is used on the valve button. Cork may also be used to line the water key. To reduce weight, the bulk of the sousaphone is composed of fiberglass.

## The Manufacturing Process

The demand for tubas is much less than for other brass instruments such as trumpets, bugles, or trombones. Therefore, production is not typically done in a highly automated fashion. The three general steps of production include piece forming, assembly and final finishing.

### *Piece making*

1 The main body, valves, bell, and mouthpiece are produced separately during manufacture. In some cases, these different parts may be made by various contractors who ship them to the tuba manufacturer for assembly. The main body is made by first putting a tube of brass on a pole-shaped mandrel. Lubrication is applied, and then a doughnut shaped die is drawn down the mandrel to reshape the brass giving it a consistent shape and thickness. The tube is then heated to make it more workable. It is then soaked in sulfuric or nitric acid to remove the oxide that is formed by the heating process.

A tuba is made up of a main body, valves, bell, and mouthpiece. All parts are manufactured from brass in various drawing, hammering, and bending operations.

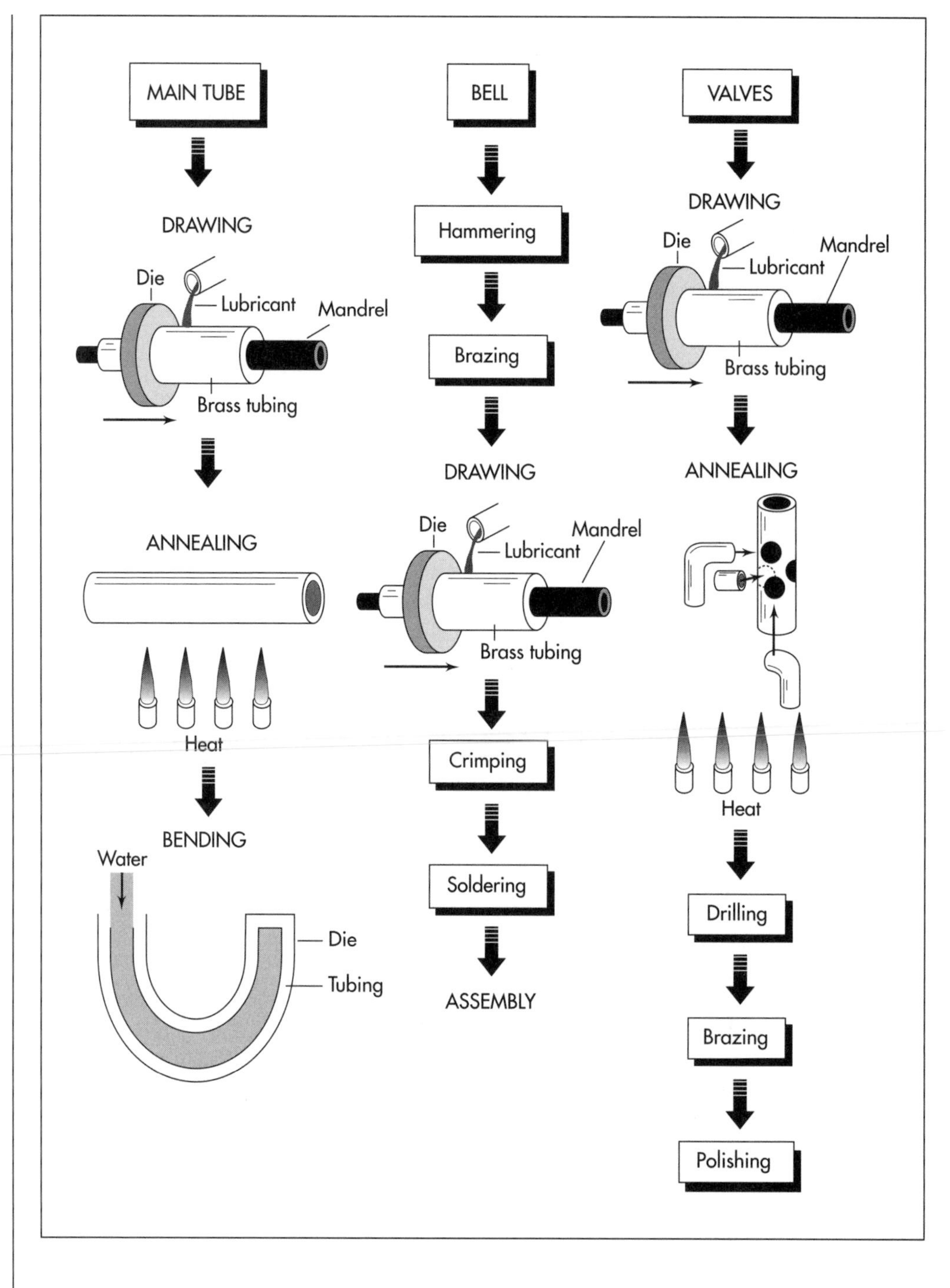


2 The modified tubes are then moved to another area where they are bent to form the appropriate curves and angles necessary for the correct tone. Various bending methods are employed. In one procedure, the tube is first placed in a die that matches the desired curve. This requires a small bit of mechanical bending. Water is then pumped through the tube at high pressures, causing the metal to expand and automatically take on the shape of the die. Another bending technique employs a ball bearing that is sent through the pipe. Here, the ball bearing works much like the water forcing the walls of the tube to conform to the die.

3 The valve system is made using many of the same techniques as the main tube. The casing is first cut from a long piece of tubing. It is threaded on the ends to allow the valves to be taken apart more easily. Using computerized systems, holes are drilled into the casing which match up exactly to the holes in the pistons. The pistons are also constructed similarly. The knuckles, which are pieces of tubing that

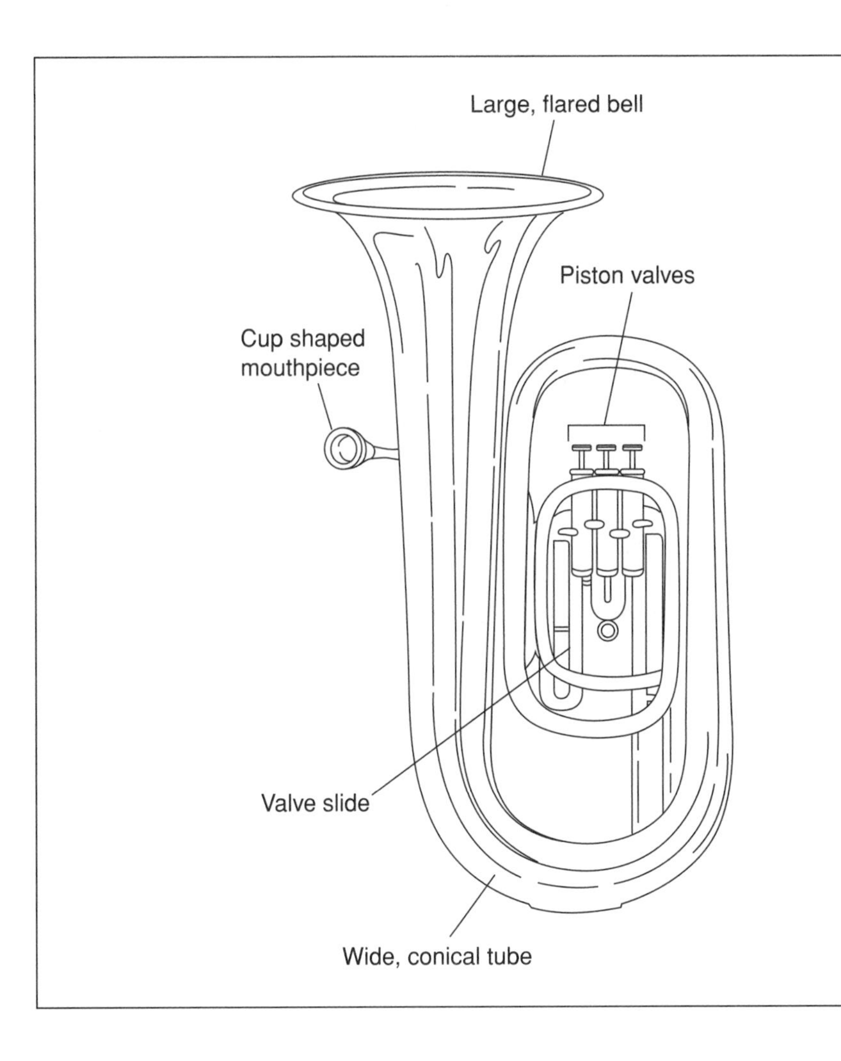


A tuba.

connect the main tube to the valves, are bent into the different angles as required for the instrument. They are then heated, or annealed, and washed in an acid bath to remove oxides. The whole assembly is held together in jigs and the joints are soldered together with a blow torch. It is then sent for another acid wash to remove excess solder and oxide, and then to a polishing machine.

4 The bell is produced from a large brass sheet that is cut to a flat dress shaped pattern. It is then wrapped around a sufficiently wide pole and pounded into shape. At the narrow end of the bell, the ends are soldered together in a butt joint. Where the bell widens, the ends are designed to overlap. They are connected using a lap joint which is then brazed with high temperatures to ensure the proper seal. The narrow end is then drawn on a mandrel to make it the same size as the end of the main tube. The wider end is then hammered further to give it a more of a bell shape. A small wire is attached to the rim of the bell and the metal is crimped around it.

5 While the mouthpiece may be made using various metals, brass is often used. It is made by pouring molten metal into a two-piece die. The die is brought together and held under pressure for a set amount of time. It is then allowed to cool causing the metal to harden. After a short time, the mold halves are opened and the mouthpiece is ejected.

### *Assembly*

6 All the parts of the tuba can then be assembled. The valve assembly is joined to the main tube and soldered. The bell is also soldered to the main tube. The pistons are inserted into the valve casings and then screwed onto the main tubing. Other pieces such as the water key or thumb rings are also soldered on at this point.

### Final finishing

7 The tuba is then cleaned and polished. It may also be electroplated with a gold of silver alloy to change its appearance. The company name is engraved on the metal using automatic or manual techniques. The instrument may then be put in a lightweight, felt-lined case along with the mouthpiece and other accessories. These are further packed in heavy boxes and shipped to customers.

## Quality Control

The quality of each tuba part is checked during the various phases of manufacture. The primary test method is visual inspection. Inspectors check for things such as deformed parts, inadequate welds, and other variations. In addition to visual checks, more rigorous measurements can also be performed. Measuring devices such as a vernier caliper or micrometer as used to check the length, width and thickness of each part.

The sound quality of the instrument is also checked prior to shipment. Manufacturers may employ professional musicians who can verify that the tone and pitch are within standards set for the specific model of instrument. Depending on the thoroughness of these tests, the instrument may also be checked under different acoustical settings

## Where to Learn More

### Books

Whitener, Scott. *A Complete Guide to Brass.* Schirmer Books, 1990.

Phillips, H and W. Winkle. *The Art of Tuba and Euphonium.* Summy-Birc, 1992.

Othmer, Kirk. *Encyclopedia of Chemical Technology*, 1992.

### Other

United States Patent 5,644,095. 1997.

United States Patent 5,012,714. 1991.

*—Perry Romanowski*

# TV Dinner

TV dinners are frozen trays of pre-cooked food. Also known as frozen dinners, they are assembled automatically on a conveyor system. In this process, the food is initially prepared and cooked. It is then placed on the trays and rapidly frozen. The concept of a frozen dinner was first made popular in the 1950s. With the increased use of microwave cooking, frozen dinners have become a large part of the grocery market. They generate over $4.5 billion in sales each year and that number continues to grow.

*TV dinners generate over $4.5 billion in sales each year.*

## History

The development of the TV dinner traces its history back to the origin of the technology for freezing food for later use. The practice of freezing food has been known for centuries. No doubt, this technology was discovered accidently by people living in cold climates such as the Arctic. However, it was not until the nineteenth century that any real commercial use of frozen food technology was known. The earliest commercial attempts at producing frozen food were centered on meats. One of the first patents related to freezing food was issued to H. Benjamin in 1842. Later in 1861, a U.S. patent was issued to Enoch Piper for a method of freezing fish. The incidence of frozen food became much more widespread later in the century with the advent of mechanical refrigerators. In 1861, the first meat freezing plant was established in Sydney, Australia. One of the first successful shipments of frozen meats occurred in 1869.

Success in the frozen beef market prompted food manufacturers to develop freezing methods for other food types. One method was the "cold-pack" process that was used around 1905. This early technology was based on a process called slow freezing. In this method, food was processed and then put into large containers. The containers were put in low-temperature storage rooms and allowed to stay there until frozen solid. This could take anywhere from one to three days. Unfortunately, this technique had two significant drawbacks. First, for some products like vegetables, freezing was too slow. The vegetable's center would start to spoil before it was frozen. Second, during freezing large ice crystals would be produced throughout the food. This lead to a break down in the food structure, and when it was thawed, the taste and appearance became undesirable.

Clarence Birdseye improved on this process when he developed a quick-freezing method. During the early 1900s, Birdseye worked for the U.S. government as a naturalist. Stationed in the Arctic, he had the opportunity to see how native Americans preserved their food during the winter. They used a combination of ice, low temperatures, and wind to instantly and thoroughly freeze fish. When this fish was thawed, it looked and tasted as good as if it were fresh. Birdseye returned from the Arctic and adapted this technology for commercial use. By using his method, Birdseye was able to reduce the time it took to freeze food from three days to a few minutes. He perfected the method and in 1924 began the Birdseye Seafoods company.

The product was a success and he turned his attention to methods for freezing different types of foods. In 1930, after years of development, he patented a flash-freezing system that packed meat, fish or vegetables in waxed-cardboard containers. He helped get these products in the grocery stores by co-developing refrigerated grocery display cases in 1934. Since freezers were not widely available to consumers, this product did not succeed immediately. However, in 1945 airlines began to serve frozen meals. In the early 1950s freezer technology had advanced to the point that people could afford to have them in their houses. This led to the introduction of TV dinners in 1954. Since this time, they have been a convenient alternative to homemade meals.

## Background

TV dinners represent a unique adaptation of frozen food technology. Most foods will spoil over time depending on storage conditions. This degradation is the result of natural chemical reactions and microbial growth. People discovered that food could be made to last longer was by freezing it. When food is frozen, the food-spoiling chemical reactions like oxidation by enzymes are slowed. Also, the growth of microorganisms such as bacteria and mold is stopped because these organisms cannot flourish in the cold temperatures. Since the process does not kill all microorganisms, those that survive a reactivated when the food is thawed.

While frozen foods resemble fresh food more closely than food preserved by other techniques, they do undergo some changes. The freezing process causes ice crystals to form throughout the product. These crystals cause a certain amount of degradation in texture and taste by disrupting the cell structure of the food. This problem was significantly reduced by the development of the quick-freezing method which produced much smaller ice crystals.

Not all foods are suitable to be frozen, particularly vegetables. For example, of the thousands of different types of peas that are available, only a few varieties produce a good tasting frozen product. A large amount of research has been done to determine exactly the types of food that are usable. It has been found that most meats, fish, and poultry can be frozen. However, certain meats and fish that are high in fat content tend to breakdown slowly even when frozen. This limits the shelf life.

## Design

TV dinners are popular for a variety of reasons such as convenience, quality, and ease of preparation. One of the greatest appeal of frozen dinners is that they are so easy to prepare. In fact, people who are not good cooks can enjoy nearly any type of dinner they want. Typically, all that is necessary is for them to heat up the product in the oven or microwave. These products require little preparation. Today, there are thousands of different types of frozen dinner products on the market, and more products are being introduced each day. The earliest TV dinners included a meat product, potatoes and a vegetable and a dessert. This has been expanded to include pasta dinners, oriental dinners, ethnic and specialty plate dinners. There are also special dinners for people who are watching their weight.

The types of food sold in TV dinners has become quite varied. Different types of meats include beef, chicken, turkey, and even sausage. Any number of vegetable dishes can include peas, corn, broccoli, and cauliflower. Mashed, whipped, and baked potatoes can be included. Pasta dishes, such as lasagne, spaghetti, linguini, or fettuccini, can make up the whole TV dinner. Typically, desserts like apple strudel or cranberry sauce are also included.

A distinguishing characteristic of a TV dinner is the partitioned plate container in which it is sold. The first TV dinners used aluminum trays covered with cardboard. While they are still used, these types of trays have given way to plastic and paper trays which are more compatible with the microwave. The food is arranged in the different compartments to keep everything separate. Dinners that are designed for home consumption are generally sold in sizes ranging from 10 oz to 1 lb (0.28 - 0.45 kg).

Preparation can be done either in a microwave or conventional oven. The disadvantage to microwave cooking is that the meats do not get the baked texture. Everything tends to be a bit soggy. However, ovens take much longer to cook than microwaves.

## Raw Materials

The primary raw materials used in the production of TV dinners are the food ingredi-

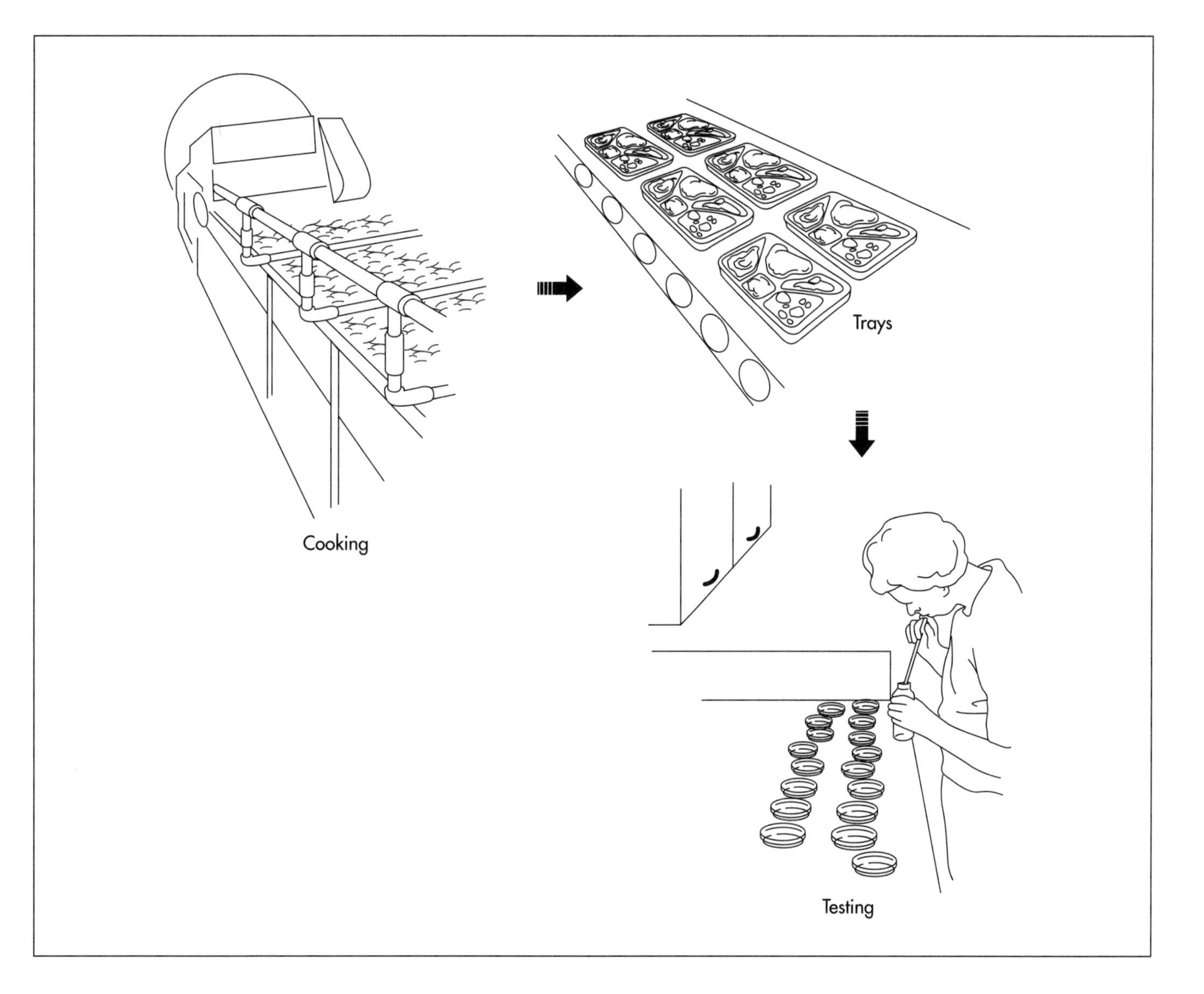


ents. To ensure good quality TV dinners, only high quality food is used. Depending on the variety of the TV dinner, this generally includes meats, potatoes, vegetables, fruits, or pasta. Since tv dinners are prepared foods other ingredients are also needed. This includes such materials as flour, water, and eggs. Flavoring ingredients such as salt, sugar, onion powder, pepper and various spices help improve taste. Artificial colorants are used to improve appearance. A preservative like sodium benzoate is also added to maintain quality during storage.

Since TV dinners are a frozen product, it is imperative that the raw materials are available at the appropriate time. For certain manufacturers, harvesting is scheduled to take place at the same time so the maximum amount of food raw materials can be utilized in the minimum amount of time. Most frozen vegetables and fruits are prepared and frozen within four hours after harvesting.

For making the cartons, various materials such as aluminum, cardboard, paperboard, and plastic is used. These are typically provided prefabricated to the TV dinner manufacturer. They are made by typical molding processes. The carton also contains the printed labels and directions. This is also typically done by contract manufacturers and shipped to the TV dinner maker.

## The Manufacturing Process

The process for producing TV dinners is highly automated. It can be broken down into three stages. First the food is processed

A unique adaptation of frozen food technology, the process for producing TV dinners is highly automated and can can be broken down into three stages. First the food is processed and prepared. Next, it is loaded into the packaging and then frozen.

For health and safety reasons, the government sets strict guidelines for minimum food quality. Raw ingredients are checked to make sure characteristics such as pH, odor, taste, moisture content, and appearance are within accepted standards. While the food is processed, it is tasted and analyzed to make sure the ingredients are put in at the proper proportions.

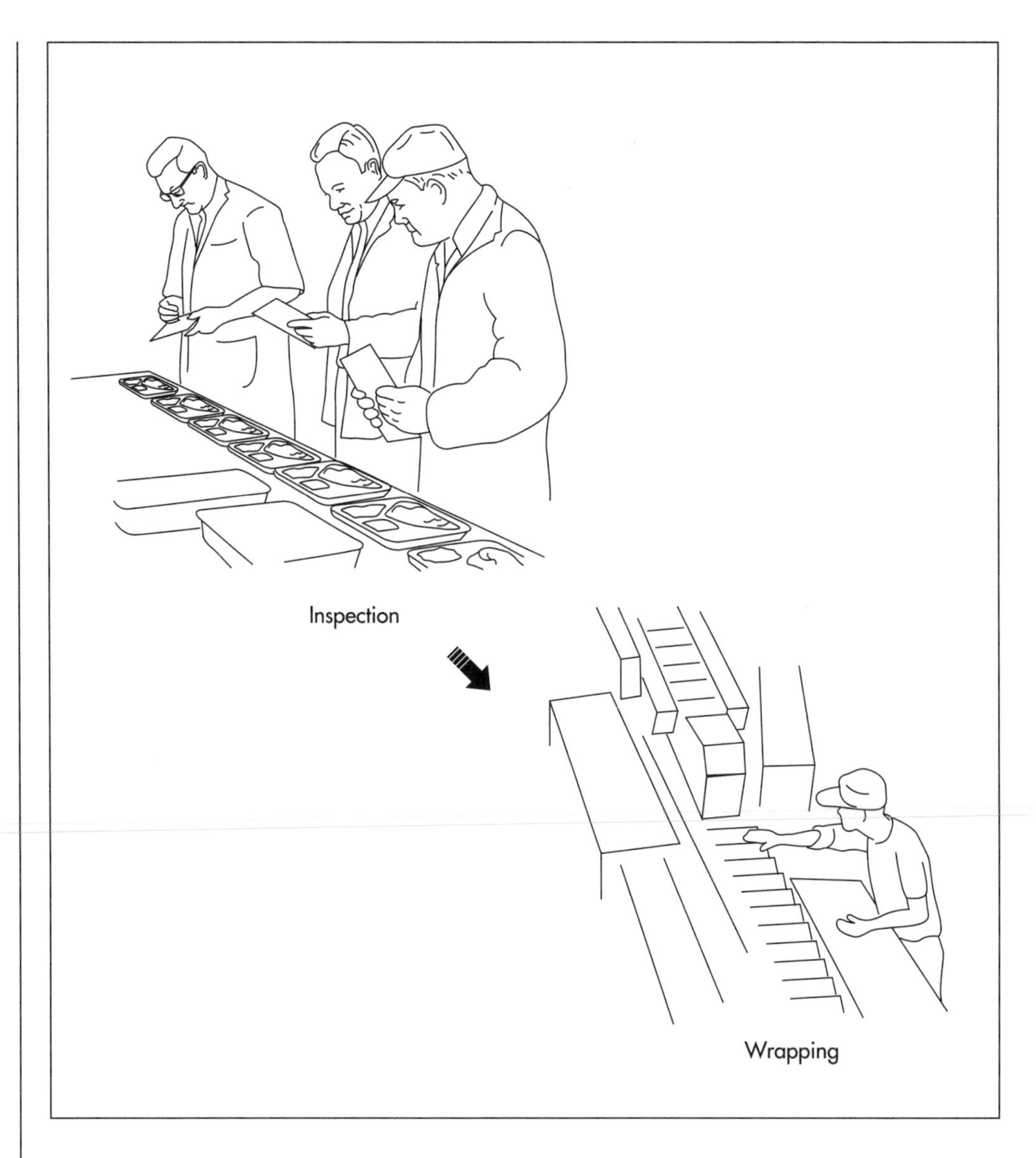


and prepared. Next, it is loaded into the packaging and then frozen.

### Food preparation

1 The first step in the process involves the processing of the incoming food. Depending on the type of food, this will mean different things. For vegetables and fruits, they are placed on a movable conveyor belt and washed. They are then put into a container and steamed or boiled for one to three minutes. This process, known as blanching, helps destroy enzymes in the food that can cause the chemical changes that negatively affect flavor and color. Before meats can be cooked, they are trimmed of fat and cut into appropriate sizes. Fish is cleaned, scaled and cut into fillets. Poultry is thoroughly washed and dressed.

2 Next, the different food dishes are prepared in large quantities. Meat is flavored, seasoned, and put on trays. It is then cooked in an oven for a predetermined amount of time. Vegetables may be further steamed or boiled, and mixed with flavoring ingredients. Potatoes are typically whipped in large, stainless steel containers. Other ingredients called for in the recipe are added and then they are cooked. When this step is complete, all of the food is cooked and ready to be packaged. It is then sent to the filling lines.

### Tray loading

3 Trays are put at the beginning of the filling line and the conveyor system is started. As the trays pass under various filling machines, food is placed in the compart-

ments. The amount of food placed on the tray is strictly regulated by the filling devices. This ensures that every TV dinner gets exactly the same amount of food.

4 After a tray is completely filled with food, it next moves to the packaging station. Here, dinners are covered with either aluminum foil, paper or some other topping. The food is packed tightly and a partial vacuum is created to ensure that the container is airtight and no evaporation takes place. Evaporation can have a variety of negative effects. For example, it can cause food to dry out. It can also allow package ice to form which can result in freezer burn giving food a dull, dried-out appearance.

### *Freezing and shipping*

5 From the tray loading station the dinners are moved to the freezing units. Today, most manufacturers use one of three rapid freezing methods: cold air blast, direct liquid immersion and indirect contact with refrigerated plates. For TV dinners, the cold air blast method is most often used. In this freezing technique, the trays are carried into a freeze tunnel and passed through a series of refrigerated coils. Fans inside the freeze tunnel blow cold air around the trays. Since the temperature can get as low as -75° F (-59° C) the dinners freeze instantly.

6 As the dinners exit the freezers, they are stacked and put into cardboard cases. These cases are put onto pallets and placed in a refrigerated storage facility. They are then transported in a refrigerated truck and stored in the grocers freezer. Good quality food that is prepared properly and frozen will remain in nearly perfect condition if it is kept at 0° F (-18° C) during shipping and storage.

## Quality Control

In the United States, quality control is a highly regulated and important aspect of every food processing facility. For health and safety reasons the government sets strict guidelines for minimum food quality. Meat is particularly well regulated because there can represent a significant health risk if poor quality meat is used. Quality control begins with the receipt of raw materials. They are checked to make sure characteristics such as pH, odor, taste, moisture content and appearance are within accepted standards. Next, the processing equipment is sterilized and checked for microorganisms before manufacture begins. While the food is processed, it is tasted and analyzed to make sure the ingredients are put in at the proper proportions. During the filling process, quality control workers are stationed at various points of the production line. At the filling section, they ensure that each compartment is filled correctly. At the end of the filling line, workers watch to make sure that each tray is set before it is covered.

## Future

Future improvements in TV dinner manufacture will focus on improving quality, speeding production, and increasing sales. A recent development has been the application of cryogenic freezing methods. This is a super fast freezing method that has allowed the utilization of foods that had previously been unsuitable for freezing. This method is also thought to produce a better tasting product. In addition to new freezing methods, new packaging materials will be used. Manufacturers are constantly trying to solve the problems associated with microwave heating. They have introduced special trays that give meat a baked texture. There may also be trays that allow some components to be heated while other remain cool.

## Where to Learn More

### *Books*

Bald, W.B. and A. Robards, ed. *Food Freezing Today and Tomorrow.* New York: Springer-Verlag, 1991.

Erickson, M. and Y. Hung. *Quality in Frozen Food.* New York: Chapman and Hall, 1997.

Macrae, R., et al., ed. *Encyclopedia of Food Science, Food Technology and Nutrition.* San Diego: Academic Press, 1993.

Mallett, C.P. *Frozen Food Technology.* New York: Routledge, 1993.

*—Perry Romanowski*

# Vegetarian Burger

*By the close of the twentieth century, vegetarianism was enjoying its strongest popularity, with an estimated 15 million practitioners in the United States alone.*

A vegetarian burger is a meatless patty made of ground grains or soybean curd, and vegetables. It is often referred to as a veggieburger.

## Background

Americans' love affair with hamburgers began sometime in the 1850s when German immigrants introduced the Hamburg steak to their new country. Made with a mixture of ground beef and seasonings and served on a roll, it quickly became the quintessential American meal. In fact, hamburgers were the foundation for the proliferation of fast-food chain restaurants in the United States and eventually around the world. A fat content of 15-30% supplies the juicy taste that consumers love, but has also been linked to health problems. This, and the rise in popularity of a vegetarian diet, led food processors to develop a meatless burger.

Although the term vegetarian did not exist until the 1800s, the theory or practice of following a meatless diet can be traced as far back as the first millennium. The Buddhist religion forbade the killing of animals for food. Buddhist priests who had spent time in China were responsible for introducing tofu, a white cheeselike substance that results from the soaking or boiling of soybeans, to Japan in the eighth century. The sixth century B.C. Greek philosopher and mathematician Pythagoras advocated a kinship between humans and animals, and his followers often adhered to a vegetarian diet. Plato, Epicurus, and Plutarch were other early vegetarians.

In the Christian religion, the avoidance of meat has often been viewed as a penance. Some monastic orders forbid the consumption of meat. For centuries, Catholics were instructed to forgo meat on Fridays and even now avoid it during the season of Lent. In the 1800s, the Bible Christians sect was created when a group separated from the Church of England, citing the Bible's prohibition of meat consumption as one reason for the split. William Metcalf, a Bible Christian minister, and 41 followers, arrived in the United States in 1817. One of those followers was Sylvester Graham who traveled the country extolling the virtues of vegetarianism. One of his particular favorite foodstuffs was whole grain flour, and it is from him that we got Graham crackers.

By 1847, British Bible Christians had established the Vegetarian Society of Great Britain. The American Vegetarian Society followed in 1850. Up until this time, the primary impetus for following a vegetarian diet was a concern for animal life . In the twentieth century, the healthful benefits of a meatless diet became another, equally compelling force. Once again, this came from within the religious community: the Church of Seventh Day Adventists, which claims that 50% of its members and nearly 100% of its clergy are practicing vegetarians.

One of its most famous members was John Harvey Kellogg of corn flake fame. Kellogg was physician-in-chief of the Adventist-run Western Health Reform Institute in Battle Creek, Michigan. Kellogg believed that meat consumption was ruinous to the human colon and thus the Institute's kitchen was strictly vegetarian. Kellogg and his wife developed the first meat substitute, a seasoned peanut and flour mixture called nuttose. Worthington Foods, the country's oldest vegetarian foods company, was established in 1939. Its initial target market was members of the Church of Seventh Day Adventists. Today, the company produces a veggieburger under its Morningstar Farms brand.

By the 1960s, vegetarian restaurants were cropping up throughout the United States. In 1971, *Diet for a Small Planet*, by Frances Moore Lappe was published. Although Lappe's purpose was to alert the public to the negative effects of animal farming on the environment and people rather than to write a treatise on vegetarianism, her book convinced many to drop meat from their diet. Equally influential was the burgeoning animal rights movement, buoyed by the publication in 1975 of *Animal Liberation* by Peter Singer, and the founding of People for the Ethical Treatment of Animals (PETA) in 1980.

By the close of the twentieth century, vegetarianism was enjoying its strongest popularity, with an estimated 15 million practitioners in the United States alone. An entire industry devoted to the processing of high-protein vegetable foods to simulate the taste of meat has evolved.

One successful meatless burger company is Gardenburger, Inc., founded by Paul Wenner. Wenner became interested in the correlation between nutrition and health in the 1960s. Chronically ill most of his life, Wenner experimented with various food combinations and ultimately became a vegetarian. After working as a cooking teacher for a number of years, he opened the Gardenhouse restaurant and Gourmet Cooking School in Gresham, Oregon. It was here that the original Gardenburger, a mixture of mushrooms, brown rice, onions, oats, and low fat cheeses, was created. In 1985, he was forced to close the restaurant. Undaunted, he established Wholesome and Hearth Foods, Inc., and began to distribute his meatless burger nationwide.

Other major brands include Boca Burger and Harvest Burger. Although some of the smaller firms produce their veggieburgers by hand, most companies employ modern food-processing machinery.

## Raw Materials

Veggieburgers are created with a variety of ingredients including, but not limited to soybeans, rice, whole wheat, black beans, corn, lentils, mushrooms, carrots, and zucchini. Some companies add stabilizers such as tapioca starch and vegetable gum. These ingredients are purchased from outside suppliers and then processed in-house. When the grains and vegetables arrive at the plant, they are examined for quality. Rotted specimens are discarded.

## The Manufacturing Process

### Washing

1 Grains and vegetables are loaded into separate machines for thorough cleansing to remove dirt, bacteria created by spoilage, chemical residue, and any other foreign materials that may exist. Some factories have conveyer belts that move the food products under high-pressure sprayers. Others use hollow drums that tumble the food while water is sprayed on it.

### Cooking the grains

2 The base grain, whether it be whole wheat, rice, or beans, is cooked in large vats of water until softened. The resulting puree is strained, separating the product from excess water, and any remaining foreign matter.

### Dicing the vegetables

3 The vegetables are diced into tiny pieces. In some factories, this is done by a machine that is calibrated to slice the vegetables into uniform sizes. Other, smaller companies, still do this by hand.

### Combining the grains and vegetables

4 Pre-measured amounts of the grain puree and the diced vegetables are combined into an industrial mixing bowl that blends the ingredients thoroughly.

### Forming the patties

5 The mixture is then loaded into an automatic patty-making machine, or press. The press is a cylindrical device with several stacks of round molds topped by a plunger. When the plunger is depressed, the ground mixture is formed into patties.

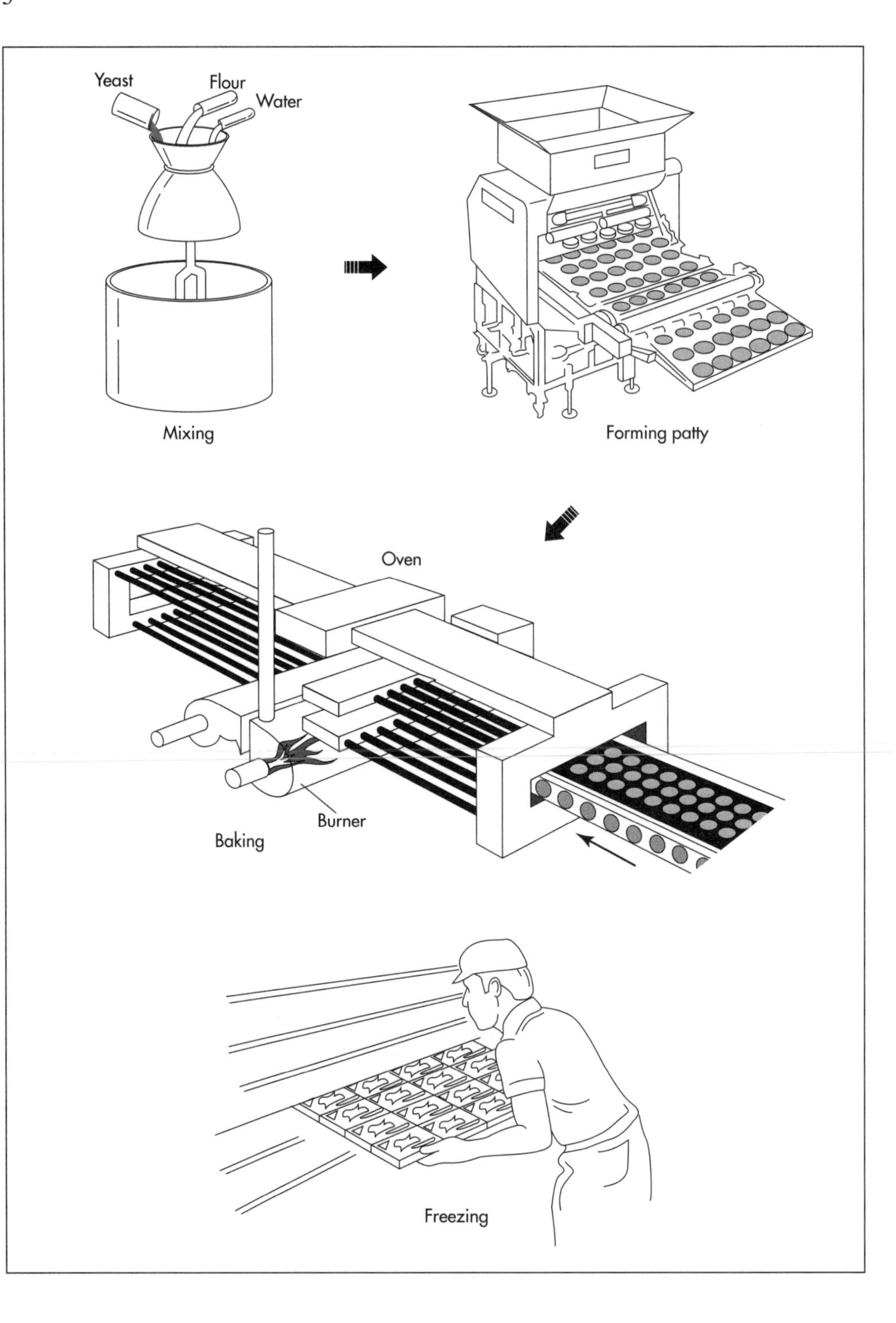


Once washed, grains and diced vegetables are cooked separately and then mixed together. The mixture is loaded into an automatic patty-making machine, or press. The patties are loaded onto perforated baking trays and baked. Next, the trays are loaded into a freezing chamber, where the burgers are quick frozen to retain consistency and taste.

### *Baking the patties*

6 The patties are loaded onto perforated baking trays, then placed in an oven for about an hour and a half at a preset temperature.

### *Patties are quick-frozen*

7 The trays are loaded into a freezing chamber in which the temperature is below the freezing point of 32° F (0° C). The goal is to freeze the patties in 30 minutes or less. Because vegetables contain a jelly-like protoplasm, the speedy processes promotes the formation of ice crystals through the tissues. When the patties are cooked, the water is reabsorbed as the ice crystals melt.

### *Patties are vacuum-packed and packaged*

8 The patties are conveyed to a vacuum-packing machine which envelopes the

patties in pre-measured plastic sleeves, drawing out the excess air and sealing each end. Then, they are loaded into pre-printed cardboard packages, usually four patties to a package. The frozen varieties are kept in temperature-controlled refrigerated compartments before and during shipment.

## Quality Control

The Food and Drug Administration issues strict standards for the commercial processing of food. These regulations include sterilization of factory equipment, quality of ingredients, and storage safeguards. Raw materials are tasted and judged visually upon their arrival at the plant. Tasters also sample the product at various points along the processing line.

## The Future

While the trend toward a more healthful diet is expected to continue, it is not readily apparent that the veggieburger will become an integral part of that diet. The primary challenge facing companies that produce meatless burgers is to create a patty that pleases palates accustomed to beef and the fat that gives it its flavor.

On the positive side, the products received major media coverage toward the end of the 1990s. Boca Burgers were served at the White House and in the Senate. Gardenburger's consumer market share jumped from 24-51% after it purchased advertising time on "Seinfeld," a popular television program. However, most industry analysts think that the real breakthrough will only occur if one of the major hamburger chains, such as Burger King or McDonald's, puts a veggieburger on the menu.

## Where to Learn More


### Books

Messina, Virginia and Mark Messina. *The Vegetarian Way*. New York: Harmony Books, 1996.

Wenner, Paul. *GardenCuisine: Heal Yourself*. New York: Simon & Schuster, 1997.

### Periodicals

"Veggieburgers Are Ringing Up Meaty Sales." *Chicago Tribune*, 6 December 1998, sec. 5, p. 1.

### Other

Boca Burger. http://www.bocaburger.com (March 4, 1999).

Gardenburger, Inc. 1997. http://www.gardenburger.com (March 4, 1999).

Worthington Foods. http://www.morningstarfarms.com (March 4, 1999).


*—Mary McNulty*

# Video Game

*Recent statistics show that 70% of all children in the United States have home video game systems.*

## Background

Video games are played at the arcade, at home on a television or personal computer, and as a handheld portable game. They are packaged in large consoles, game paks that can only be played on the same manufacturer's hardware (i.e. Nintendo, Sega Genesis, and Sony Playstation), and as CD-ROMs. Made up of a program that instructs the computer to display specific visual and audio effects, video games utilize cutting-edge technology in order to provide fast paced entertainment. Recent statistics show that 70% of all children in the United States have home video game systems. Over four billion dollars is spent on arcade video games annually.

## History

A precursor to the video game, pinball machines were introduced during the 1930s and remained popular through the 1970s. In 1971, a video arcade game was produced called Computer Space. Invented by Nolan Bushnell and Ted Dabney, Computer Space was the first real coin-operated video game, but for various reasons, it never became popular. It did however, lay the groundwork for the next video game that Bushnell and Dabney introduced: the phenomenally successful arcade game Pong. Modeled after the game of ping pong, it was an electronic game in which players tried to hit a flashing dot passed their opponent's video paddle. With the success of Pong, Bushnell and Dabney started the Atari Company, and in 1975, they introduced a home version of Pong. In 1976, Warner Communication purchased Atari for $28 million and expanded its home line of video game cartridges.

At the same time Bushnell and Dabney were developing Pong, Ralph Baer, who was working for Sanders Associates, was designing a home video game system called The Odyssey. Developed in 1969, Baer's system was finally manufactured and distributed by Magnavox in 1972. The Odyssey was a package of 12 different plug-in games that were housed on circuit cards. Each game came with plastic overlays that, when placed over the television screen, simulated the appropriate background. For example, a plastic overlay of a hockey rink was included with the hockey game. The Odyssey also offered an electronic shooting gallery with four additional games and an electronic rifle. Eighty-five thousand systems were sold.

Rapid advances in electronics technology during the 1970s led to the development of more complicated games, such as Space Invaders and Pac-Man. Introduced in 1983 as a joint venture between the Namco Company of Japan and Midway of the United States, Pac-Man has sold hundreds of thousands of games and remains one of the most popular video games.

When personal computers became available, computer games were created. Many of these games were adaptations of arcade or home video game systems, however unique games were also developed. The computer game industry grew swiftly during the 1980s powered by various companies, especially the Nintendo Corporation. In the late 1980s, the CD-ROM was introduced. These disks could hold more information on them, and allowed the development of more sophisticated, interactive games. In 1995, digital video disks (DVDs) were first produced for home computers. Since they have a storage capacity over twenty times greater than CD-ROMs, they promise to revolutionize computer games.

## Design

Design is the key aspect of making all video games. It is typically done by a team of skilled computer programmers, writers, artists, and other game designers. During this phase of development, they generate the game's specifications, which includes game type, objective, and graphics.

While creating a video game is rarely a step by step process, there are a variety of tasks that must be accomplished during the development phase. In the beginning, the type and objective of the game is determined. In general, games fall within six categories, or genres, including fighting, shooting, strategy, simulations, adventure, and run, jump and avoid (RJA). Fighting games require the players to battle with each other or the computer. Presently, they are the most popular and encompass such titles as Mortal Kombat and Street Fighter. Shooting games involve battles in which the player tries to destroy enemy tanks, ships, or planes. Strategy games include such classics as chess, bridge or checkers. Simulations are games, which reproduce real life situations such as flying or driving. Adventure games are computerized versions of role-playing fantasy games. The RJA games are those like the Super Mario games in which a character tries to reach a goal while being impeded by various obstacles.

The action of the game is dependent upon its intended venue. An arcade game must have immediate action. A home version usually includes various skill levels in order to keep the player interested over a longer period of time. A handheld version of a video game is simplified to be played in miniature.

## Raw Materials

Although the most important raw material in creating a video game is imagination, a number of supplies are necessary to bring that imagination to life. Once the story has been created, characters and background are drawn on storyboards, then transferred to electronic format directly by an artist or via digitization. Lifelike action is captured on film and sound is recorded on digital audio tape (DAT).

Once design is complete, a variety of raw materials are used to produce video games. This includes the materials that go into making the storage medium, the accessories, and the packaging. The most common storage mediums are floppy disks and CDs. These are made with hard plastics such as polycarbonates. CDs have a thin layer of aluminum or silver coating. Additionally, they are coated with a protective, clear acrylic coating. Floppy disks are made with a thin plastic that is coated with a magnetic material. Plastics are also used to make the accessory pieces that some computer games require. In each of the plastics used, a variety of fillers and colorants are incorporated to modify its characteristics. Typically, computer games are packaged in highly decorated cardboard boxes.

## The Manufacturing Process

Creating a video game is a long, multifaceted process that can take up to one year to complete one game.

### *Creating the story*

1 Typically, writers are responsible for creating a game's story complete with a setting, characters, and plot. This gives the game a purpose and makes it more enjoyable for the player. It also provides an objective for the player and a guideline for the rules of the game. This information is then used to create the game's manual. Often, the inspiration for a story is derived from popular movies or books. The story is transferred to storyboards, where preliminary drawings are also added. Storyboards are a series of one-panel sketches pinned to a board. Each sketch is accompanied by story dialogue and/or character description.

### *Capturing action with art*

After the type of game and story are outlined, the game's format can be determined. The format refers to what the player sees when playing the game. A variety of formats exist including platform, top-down, scrolling, isometric, three dimension (3D),

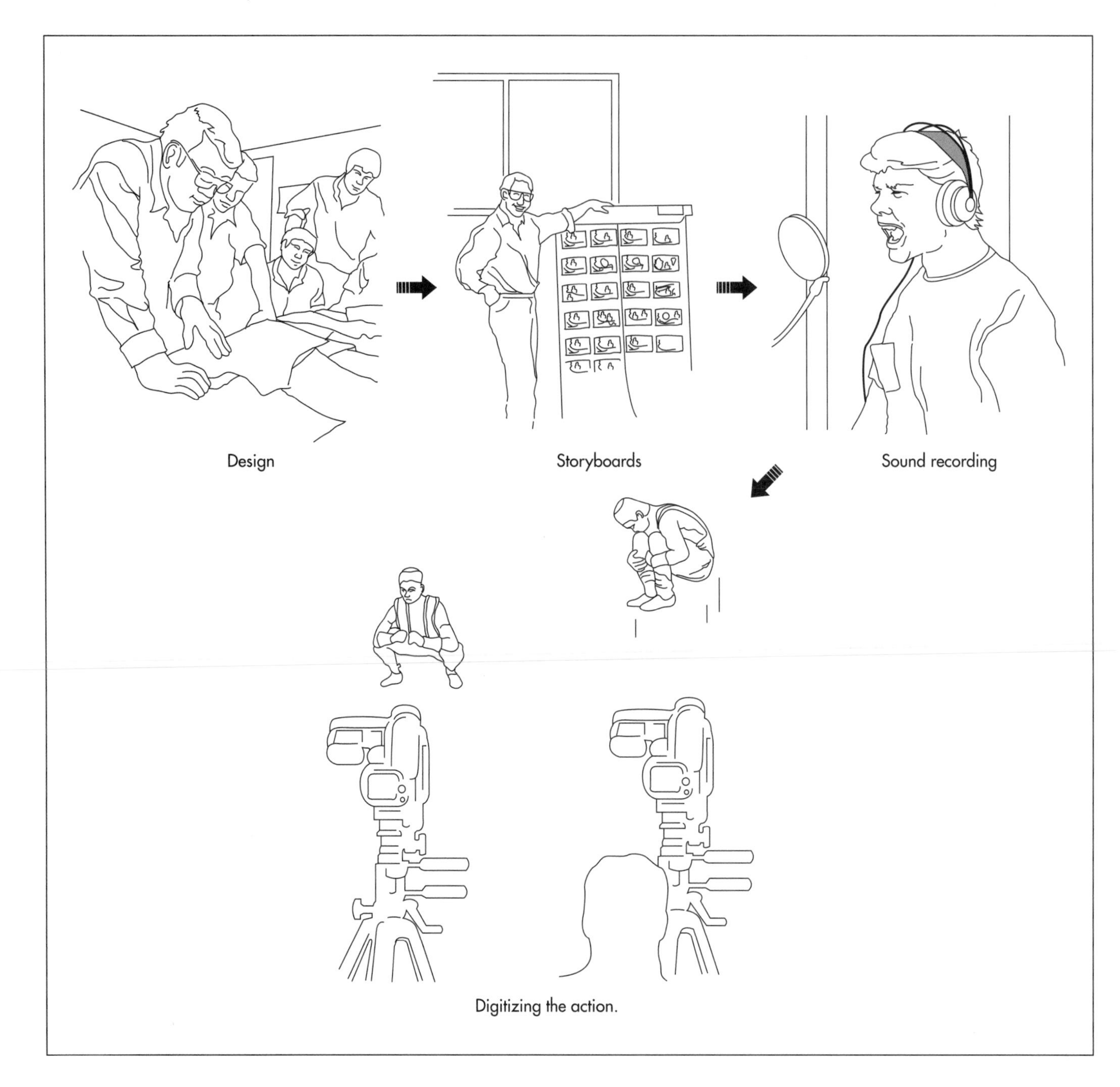


Digitizing the action.

Creating a video game is a long, multifaceted process. A team of designers, artists, and programmers work together to create the final product. After the type of game and story are outlined, the game's format can be determined. The format refers to what the player sees when playing the game and a game may utilize one or more of these formats. The artist adds drawings to storyboards, including character descriptions and arrows showing how the characters will move. Sound and life-like action is recorded.

and text. Platform games are those that feature a side view of the player's character. Top-down games give a bird's eye view of the player's character. They are often used for war games. The isometric format is a top-down game, which uses perspective tricks to give the illusion of 3D. True 3D games are just now becoming a reality with the introduction of CDs and DVDs. These represent the future of computer game formats. Text game formats have limited graphics and are only used for interactive fiction. In general, all games may use one or more of these formats.

2 The artist adds drawings to storyboards, including character descriptions and arrows showing how the characters will move.

3 Final pictures are created in two ways. An artist can render the final picture by creating converted graphics. Converted graphics are images that have been drawn by the artist using a computer. More life-like images are created by filming the action with an actor and then electronically digitizing the image. The artist reviews the videotaped action on a RGB monitor that produces brighter images and chooses each

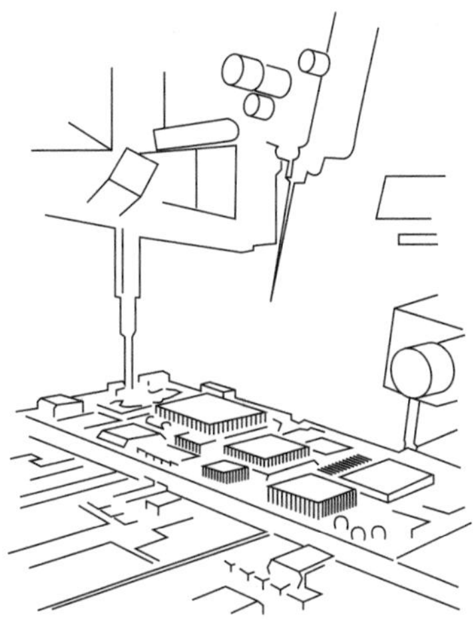

Computer chip board

After the programmer has written the program for the game and the recorded sound and action has been digitized, the game code is transferred to a master compact disk. This disk will be used to mass produce thousands of copies. The disk is put into a laser-cutting machine. While the disk is spun, the binary code from the computer game's program sends a signal to the laser. The laser will then cut pits in the photo resistive coating corresponding to the program. The disk is then chemically etched and given a metal coating.

frame that will be used to create an action. Typically, six to 10 frames are used to show a character running/walking one step. Four to five frames are needed to illustrate a punch or a kick. The background is also created by the artist using both converted and digitized images. Though seemingly cohesive to the viewer, the background is really a collage of many different images.

### *Recording dialogue and sound effects*

4 Dialogue and sound effects are recorded in a sound studio using various audio techniques. Once recorded on digital audio tape (DAT), the sounds are computerized by a synthesizer—a computer that specifically alters and translates sound into data.

### *Writing the program*

5 When all of the preliminary design elements are determined, programming, or coding, can begin. The first step in this process is drawing a flowchart, which shows the logical progression of the computer program. A variety of programming languages are used such as C++, Java, or visual BASIC. The code is typically produced by a team of programmers, each working on a different phase of the game, and can take up to seven months to produce. To speed the coding process, previously developed algorithms are often modified and adapted to the new game. This is more efficient because it eliminates the need to continually rewrite similar programs and reduces the chances of serious errors. Each action can require many individual instructions written by the programmer, and roughly 250,000 individual commands are written to create a video game program. Sound and graphics must also be programmed separately.

### *Testing*

6 The testing phase of game development helps reveal fundamental design and programming problems. Testing can be completed in a number of ways. The programmers can run the game and try to discover gross problems themselves. Additionally, professional playtesters are used. These are people who are specifically trained to play the games and look for subtle errors. They are typically game designers themselves, and have experience with many types of games. Beyond finding errors, playtesters also give criticisms and suggestions to make the game better. In some cases, computer game developers use people from the general population to test games. This gives them information about consumer acceptance. The information obtained from the testing phase is reviewed. Reprogramming is then done until the game is appropriately tweeked.

### *Burning the disks*

7 When the programming is completed, the game code is transferred to a master compact disk. This disk will be used to mass

produce the thousands of copies needed. The master disk is composed of a smoothly polished glass coated with an adhesive and a photo resistive material. This disk is put into a laser-cutting machine. While the disk is spun, the binary code from the computer game's program sends a signal to the laser. The laser will then cut pits in the photo resistive coating corresponding to the program. The disk is then chemically etched and given a metal coating. At this point, it is an exact replica of a finished disk.

8 A metal inverse copy of the master disk is then made through a process known as electroforming. Using an electrical current, metal is deposited on the master disk surface. This inverse copy is then used to make multiple metal masters through a similar process. These metal masters are used to make multiple inverse stamping copies.

9 To mass produce plastic compact disks, the stamping copies are put into a die in an injection molding machine. In this machine, polycarbonate pellets are placed in a hopper and drawn through a long screw device. While it is drawn, the plastic becomes molten. It is then injected into the die with the stamping copy. It is held under pressure for a few moments and then allowed to cool. As it cools, it hardens giving it exactly the same pattern of pits as the master copy. It is then released from the die and inspected for flaws. If any flaws are found, the disk is rejected.

10 The disk is then transferred to a machine, which punches a hole in its center. It is then coated with a thin film of either aluminum or silver using a vacuum deposition technique. Next, a clear acrylic plastic is applied. Finally, the disk is transported to a silk screen assembly and a label is printed on it.

### *Packaging the game*

11 All the components of the game are transported to the packaging line. Here, each part of the game is placed in a preprinted, cardboard box. This process may either be done by hand or by automated machinery. The game boxes are then conveyored to a shrink-wrap machine that seals them in plastic. They are cased and ready for shipment.

## Quality Control

The process of transferring the computer game program to a compact disk, or DVD, must be done in a clean, dust-free environment. This is because dust particles are much larger than the pits carved in a disk, and a single particle can ruin a disk. Therefore, strict quality control measures are taken to control the environment around the disk-making process. Other visual inspections are done at various points during the disk manufacture. Random samples of finished disks are also tested to make sure the program is working properly. Beyond the checks involved in disk manufacture, the other components of the game are also checked to ensure they meet the required specifications. At the end of the manufacturing process, random samples of the finished product are checked to make sure it includes all of the necessary components.

## The Future

Computer game programming continues to become more sophisticated as the available hardware improves. The most important recent advancement that promises to revolutionize gaming is the development of DVD technology. This will allow a much greater amount of information to be included in the game's program. This should improve many aspects of the game such as the artificial intelligence routines, the graphics, and the special effects. Things such as video clips will be included to make the games more interactive.

## Where To Learn More

### *Books*

Gruber, Diana. *Action Arcade Adventure Set.* Coriolis Group, 1994.

Katz, Arnie and Laurie Yates. *Inside Electronic Game Design.* Prima Publications, 1997.

Sawyer, Ben. *The Ultimate Game Developers Sourcebook.* Coriolis Group, 1997.

### *Periodicals*

Bunn, Austin. "Joystick City." *The Village Voice* (December 30, 1997).

Wright, Maury. "DVD: Breathtaking Sight and Sound, Significant Challenges."*EDN* (August 15, 1997).

*—Perry Romanowski*

# Vodka

*The concentrated vapors, or fine spirits, contain 95-100% alcohol. This translates to 190 proof. In order to make it drinkable, water is added to the spirits to decrease the alcohol percentage to 40, and the proof to 80.*

Vodka is an alcoholic beverage distilled at a high proof from a fermented vegetable or grain mash. Proof is a measurement of the alcohol content. Each degree of proof equals a half percent of alcohol. Thus, 100 proof is that which contains 50% alcohol, 90 proof contains 45%, and so on. Because distilled vodka can have a proof as high as 145, all taste and odor has been eliminated, making vodka a neutral spirit. Water is added to bring the proof down to a range between 80 and 100.

## Background

The practice of allowing certain grains, fruits, and sugars to ferment so that they produce an intoxicating beverage has been around since ancient times. Fermentation is a chemical change brought about by the yeast, bacteria, and mold in an animal or vegetable organism. In the production of alcoholic beverages, yeast enzymes act on the sugars in the mash (usually dextrose and maltose) and convert them to ethyl alcohol.

It was in the tenth century writings of an Arabian alchemist named Albukassen that the first written account of distillation was found. Distillation was also said mentioned among the writings of the thirteenth century Majorcan mystic Ramon Llull. Distillation is a heating and condensing process that drives gas or vapor from liquids or solids to form a new substance. Distilled spirits are also known as ardent (Latin for burn) spirits.

There is disagreement among Russians and Poles as to which country was the first to distill vodka. Most historical references credit Russia. In any event, the drinking of vodka has been documented since the fourth century in eastern and northern Europe. In those regions, it was common to distill alcoholic beverages to a very high proof, eliminating any aroma or flavor.

Vodka remained primarily an eastern and northern European preference for centuries. It was not until the 1930s that it began to gain popularity in Western Europe and North America. A 1930 British publication, the *Savoy Cocktail Book*, was the first to include recipes for vodka drinks. The "Blue Monday" combined vodka with Cointreau and blue vegetable juice. A "Russian Cocktail" called for the addition of creme de cacao and dry gin to the neutral spirit.

One the primary vodka producers, the Smirnoff family distillery began business in 1818 in Moscow. A century later the distillery was churning out one million bottles daily. However, after the 1917 Bolshevik revolution, the family lost control of the business. In 1934, a Russian immigrant named Rudolph Kunitt, bought the American rights to the Smirnoff name. Kunitt opened a distillery in Bethel, Connecticut, and struggled along for five years, at best only producing 20 cases a day. He sold his business to the Heublein Company in 1939.

Heublein executive John C. Martin found that vodka was especially popular in the California film industry and he cultivated those customers. In 1946, he met the owner of a Los Angeles restaurant, the Cock 'n' Bull, who was trying to unload an overstock of ginger beer. Since one of vodka's attributes is its ability to mix with almost anything, the two men experimented with a vodka and ginger beer concoction. They added a slice of lime, called their invention the "Moscow Mule," and had an instant success on their hands.

Louis Pasteur

Louis Pasteur was one of the most extraordinary scientists in history, leaving a legacy of scientific contributions which include an understanding of how microorganisms carry on the biochemical process of fermentation, the establishment of the causal relationship between microorganisms and disease, and the concept of destroying microorganisms to halt the transmission of communicable disease. These achievements led him to be called the founder of microbiology.

After his early education Pasteur went to Paris, studied at the Sorbonne, then began teaching chemistry while still a student. After being appointed chemistry professor at a new university in Lille, France, Pasteur began work on yeast cells and showed how they produce alcohol and carbon dioxide from sugar during the process of fermentation. Fermentation is a form of cellular respiration carried on by yeast cells, a way of getting energy for cells when there is no oxygen present. He found that fermentation would take place only when living yeast cells were present.

Establishing himself as a serious, hard-working chemist, Pasteur was called upon to tackle some of the problems plaguing the French beverage industry at the time. Of special concern was the spoiling of wine and beer, which caused great economic loss and tarnished France's reputation for fine vintage wines. Vintners wanted to know the cause of l'amer, a condition that was destroying the best burgundies. Pasteur looked at wine under the microscope and noticed that when aged properly the liquid contained little spherical yeast cells. But when the wine turned sour, there was a proliferation of bacterial cells which were producing lactic acid. Pasteur suggested that heating the wine gently at about 120°F would kill the bacteria that produced lactic acid and let the wine age properly. Pasteur's book*Etudes sur le Vin*, published in 1866 was a testament to two of his great passions—the scientific method and his love of wine. It caused another French Revolution—one in wine-making, as Pasteur suggested that greater cleanliness was need to eliminate bacteria and that this could be done with heat . Some wine-makers were aghast at the thought but doing so solved the industry's problem.

The idea of heating to kill microorganisms was applied to other perishable fluids like milk and the idea of pasteurization was born. Several decades later in the United States the pasteurization of milk was championed by American bacteriologist Alice Catherine Evans who linked bacteria in milk with the disease brucellosis, a type of fever found in different variations in many countries.

By the 1950s, New Yorkers were drinking vodka too. From 40,000 cases sold in the United States in 1950, vodka sales jumped to just over one million in 1954. The following year, 4.5 million cases were sold. By the mid-1960s, vodka nudged out gin; by 1976, it surpassed whiskey. By the end of the decade, the martini was more likely to be made with vodka than with it original ingredient, gin. At the close of the twentieth century, vodka accounted for 25% of the distilled spirits market.

Until the middle of the eighteenth century, vodka production was essentially a home-based, one-pot operation called batching. Heating potatoes or grains until the starch was released and converted to sugar made a mash. The resulting liquid matter was allowed to ferment, and then heating it at a high temperature to release the intoxicating vapors distilled the liquid.

It was soon discovered that multiple distillations produced a spirit of a higher proof and of greater purity. In 1826, Robert Stein invented the continuous still that allowed for repeated recycling of steam and alcohol until all of the spirit has been extracted. Aeneas Coffey improved Stein's design. Mod-

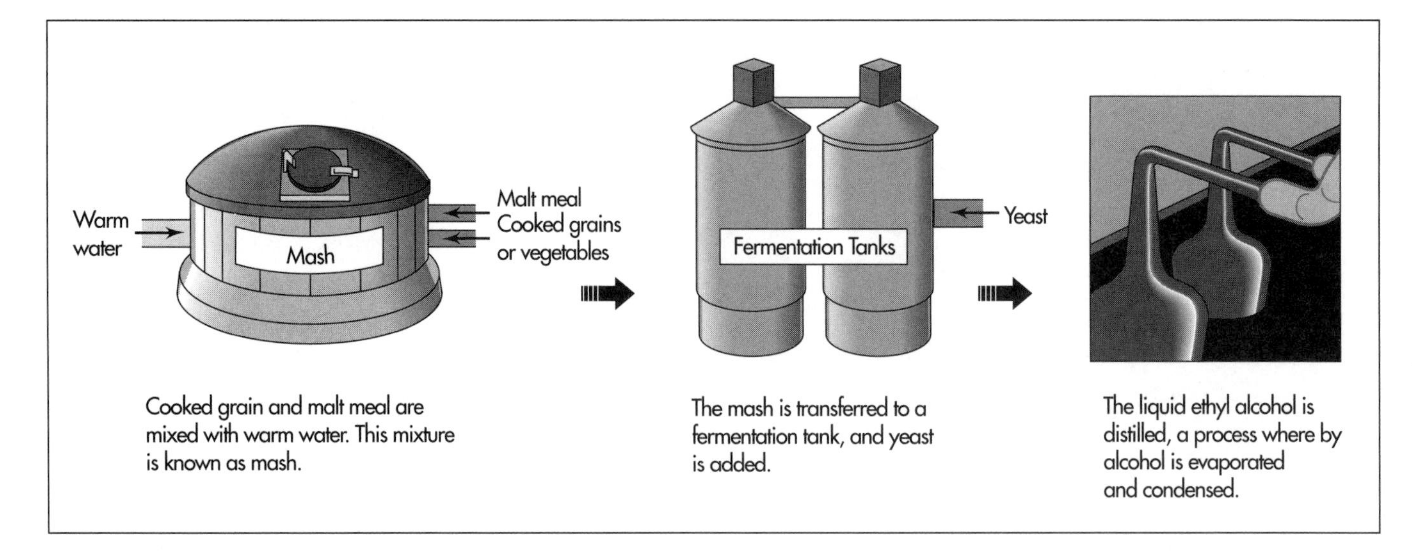


Cooked grain and malt meal are mixed with warm water. This mixture is known as mash.

The mash is transferred to a fermentation tank, and yeast is added.

The liquid ethyl alcohol is distilled, a process where by alcohol is evaporated and condensed.

ern continuous stills usually contain three primary sections: still heads (where the vapors are collected), fractionating columns (where the ethyl alcohol is broken down), and condensers (where the vapors are reconverted to liquid).

Louis Pasteur's development of pasteurization began when a French distiller asked him for advice on fermentation. Pasteur's research led him to the discovery of lactic acid and its role in fermentation. Today, lactic acid is used as an inoculation against bacteria in the production of vodka.

At first, charcoal filtration was the universal procedure used to purify the vodka. Then at the beginning of the twentieth century, the process of rectification was developed. In rectification, the spirits are passed through several purifying cylinders designed to eliminate dangerous imperfections such as solvents, fusil oil, and methanol.

## Raw Materials

### *Vegetables or grains*

Because it is a neutral spirit, devoid of color and odor, vodka can be distilled from virtually any fermentable ingredients. Originally, it was made from potatoes. Although some eastern European vodkas are still made from potatoes and corn, most of the high quality imports and all vodka made in the United States are distilled from cereal grains, such as wheat. Distillers either purchase the grain from suppliers, or grow it in company-owned fields.

### *Water*

Water is added at the end of the distillation process to decrease the alcohol content. This is either purchased from outside suppliers or brought in from company-owned wells.

### *Malt meal*

Because vegetables and grains contain starches rather than sugars, an active ingredient must be added to the mash to facilitate the conversion of starch to sugar. These particular converted sugars, maltose, and dextrin respond most effectively to the enzyme diastase that is found in malt. Therefore, malt grains are soaked in water and allowed to germinate. Then, they are coarsely ground into a meal and added during the mash process.

### *Yeast*

A microscopic single-celled fungus, yeast contains enzymes that allow food cells to extract oxygen from starches or sugars, producing alcohol. In the manufacturing of alcoholic beverages, the yeast species *Sacchasomyces cereviseal* is used. It is purchased from outside suppliers.

### *Flavorings*

In the latter part of the twentieth century, flavored vodkas became popular. Thus, herbs, grasses, spices, and fruit essences may be added to the vodka after distillation. These are usually purchased from an outside supplier.

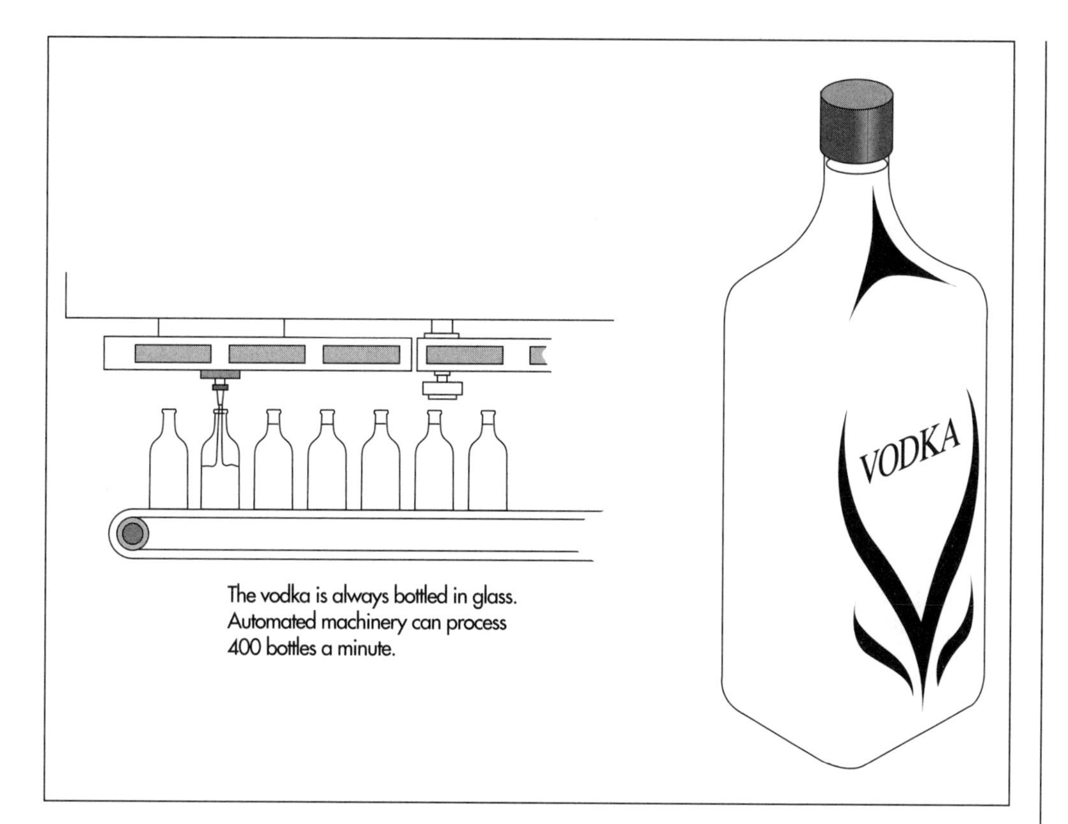

The vodka is always bottled in glass. Automated machinery can process 400 bottles a minute.

## The Manufacturing Process

### *Mash preparation*

1 The grain or vegetables are loaded into an automatic mash tub. Much like a washing machine, the tub is fitted with agitators that break down the grain as the tub rotates. A ground malt meal is added to promote the conversion of starches to sugar.

### *Sterilization and inoculation*

2 Preventing the growth of bacteria is very important in the manufacture of distilled spirits. First, the mash is sterilized by heating it to the boiling point. Then, it is injected with lactic-acid bacteria to raise the acidity level needed for fermentation. When the desired acidity level is reached, the mash is inoculated once again.

### *Fermentation*

3 The mash is poured into large stainless-steel vats. Yeast is added and the vats are closed. Over the next two to four days, enzymes in the yeast convert the sugars in the mash to ethyl alcohol.

### *Distillation and rectification*

4 The liquid ethyl alcohol is pumped to stills, stainless steel columns made up of vaporization chambers stacked on top of each other. The alcohol is continuously cycled up and down, and heated with steam, until the vapors are released and condensed. This process also removes impurities. The vapors rise into the upper chambers (still heads) where they are concentrated. The extracted materials flow into the lower chambers and are discarded. Some of the grain residue may be sold as livestock feed.

### *Water added*

5 The concentrated vapors, or fine spirits, contain 95-100% alcohol. This translates to 190 proof. In order to make it drinkable, water is added to the spirits to decrease the alcohol percentage to 40, and the proof to 80.

### *Bottling*

6 Alcoholic beverages are stored in glass bottles because glass is non-reactive. Other receptacles, such as plastic, would cause a chemical change in the beverage. The bottling procedure is highly mecha-

nized as the bottles are cleaned, filled, capped, sealed, labeled, and loaded into cartons. This can be done at rates as high as 400 bottles per minute.

## Quality Control

Although tasters draw off quantities of vodka for sampling throughout the distilling process, most of the controls on vodka quality come from local, state, and federal governments. At the federal level, the Bureau of Alcohol, Tobacco and Firearms issues strict guidelines for production, labeling, importation, advertising, and even plant security. For example, charcoal-filtered vodka imports are not permitted. Flavored vodkas must list the predominant flavor (pepper, lemon, peach, etc.) on the label. The relationships between suppliers and producers are strictly regulated as well.

## Where to Learn More

### Books

Grimes, William. *Straight Up or On the Rocks: A Cultural History of American Drink.* New York: Simon & Schuster, 1993.

Grossman, Harold J. *Grossman's Guide to Wines, Beers and Spirits.* 7th edition, revised by Harriet Lembeck. New York: Charles Scribner's Sons, 1983.

### Other

Lancut Distillery. http://www.lancutdistillery.euro-index.com.pl/spiritproduction. htm (March 9, 1999).

*—Mary McNulty*

# Voting Machine

The term voting machine encompasses a variety of mechanical devices used for recording votes. The voting machines commonly used in the United States may operate using punch cards, mechanical levers, optical scanning, or direct electronic recording.

## Background

In colonial America, voters recorded their choices by placing balls, coins, bullets or beans in a container. Paper ballots did not come into use until around 1800. At first, paper ballots were usually supplied by the political parties, and each party used a different color. The voter would typically hold his ballot up in the air for all to see which color he chose. Municipalities also used voice votes, where the voter simply spoke the name of his candidate or answered "yea" or "nay." Some communities in the United States continued to hold elections based on voice votes until after the Civil War. A new system of paper ballots was first used in the Australian state of Victoria in 1856. The ballot was made from a uniform size, weight and color of paper, and it listed all the candidates from any parties. Voters marked their choices by checking a box next to their preferred candidate's name. Then, they dropped the ballot into a sealed ballot box. This became known as the Australian ballot, and it was adopted in the United States in the 1870s. The Australian ballot obviously offered confidentiality that the earlier voting systems lacked.

The famed inventor Thomas Edison took out a patent on a mechanical voting machine in 1869, but his device was never used. Another inventor, Jacob H. Myers, developed a similar machine, which became the first mechanical voting device ever used in a U.S. election when it was adopted by the city of Lockport, New York, in 1892. Several companies developed and marketed similar voting machines, which were known as mechanical lever devices. They were widespread across the United States by 1930. More than half the nation voted on lever machines by the 1960s. The voter enabled the machine by pulling a lever that also closed the privacy curtain around the booth. To mark candidate choices, the voter pulled down pointers next to his or her selections. The pointers moved counter wheels inside the machine. The machine re-set every time the curtain lever was raised, and at the end of the day, its counter wheels revealed how many votes had been cast for each candidate on the ballot. These machines were thought to reduce the possibility of ballot tampering, though apparently those inclined to rig elections still found ways to do so.

Two more voting systems were developed by entrepreneurs in the 1960s. The first person to develop a voting machine using computer punch cards was a Mr. Martin A. Coyle, who marketed his machine in Ohio beginning in 1960. A University of California political science professor, Joseph P. Harris, took out a patent for a punch card voting system called Votomatic in 1963. Computer giant International Business Machines bought the rights to manufacture Votomatic machines in 1965. Close to 40% of all votes in the 1996 presidential election were cast on punch card machines. A second method of recording votes, using optical scan equipment, was also developed in the 1960s. In the late 1990s, almost a quarter of all registered voters used some kind of optical scanning machine. These are similar

*Prior to the paper ballot, voters in colonial America recorded their choices by placing balls, coins, bullets, or beans in a container.*

to the readers used for grading standardized tests. The voter marks his or her vote by filling in a dot or completing an arrow with a number two pencil, and the machine reads and tallies the pencil marks.

A smaller, but growing segment of the population votes on machines known as Direct Recording Electronic (DRE). This is essentially the electronic version of the mechanical lever system. By touching keys or buttons, voters enter their choices into an electronic memory in the machine. It tallies its votes automatically, and a district using these machines can have its election results in as little as half an hour. Many DRE machines use buttons next to a printed ballot. Others use a liquid crystal display to show the ballot, and the voter can push buttons or touch the display. Using these machines is similar to using an automatic teller machine.

Whatever the system the device uses, all voting machines have similar constraints. They are used, for the most part, only twice a year, and in the meantime they sit in storage. They need to be simple to operate, and also simple to maintain and repair. Relatively untrained and unskilled poll workers or election administrators need to be able to set them up and keep them working during polling, and yet they need to be highly reliable. In general, large municipalities buy more advanced equipment first, and the technological gains trickle down to smaller cities as they buy used equipment from their bigger neighbors.

## Standards

During the 1970s, many small companies were making voting machines, some with dubious reliability. A document from the Federal Election Commission states that at this time "nearly anybody could cobble together a 'voting machine' in their garage (and some of them looked like they had been)...." Purchase of voting equipment was done by municipalities or counties, usually with no state standards to guide them. The National Bureau of Standards first issued a report on computers and vote tallying in 1975, and concluded that there were rampant problems with the technology, and few people running local elections were trained to handle them. Unscrupulous election officials could use technological problems to manipulate election outcomes. For example, when Jesse Jackson became the first black presidential candidate to run in a primary in New York in 1984, outraged supporters claimed that Jackson lost thousands of votes because voting machines in poor repair had been selected for neighborhoods with strong Jackson followings. The National Bureau of Standards (now called the National Institute of Standards and Technology) worked with the Federal Election Commission (FEC) to study the feasibility of implementing national voting machine standards. A report was released in 1982, and in 1990 the FEC issued its first voluntary standards for punch-card, optical scanning and DRE voting systems. As of 1998, 31 states had adopted the standards, meaning the remaining states either had their own standards or used none. However, voting system reliability is thought to have improved substantially since the implementation of the voluntary standards, which ask manufacturers to submit their devices to independent testing authorities.

## Raw Materials

The raw materials for the direct recording electronic voting machine are light metals and plastics. It uses printed computer circuit boards, control panels, printers, lights, and memory cartridges made of the materials most often used for these components in other electronics such as computers or video games.

## The Manufacturing Process

The various components for the voting machine are assembled separately at different work stations or assembly lines. Then the components are assembled and the device is tested. Typically, the manufacturer makes many of the components under one roof, and buys others from sub-contractors according to specifications.

### *Electronics assembly*

1 The manufacturer maintains one area of the factory for assembling the electronic components of the voting machine. These are the printer module, the circuit boards, and the memory cartridges. Workers at tables or put together the small electronic parts according to directions, snapping and

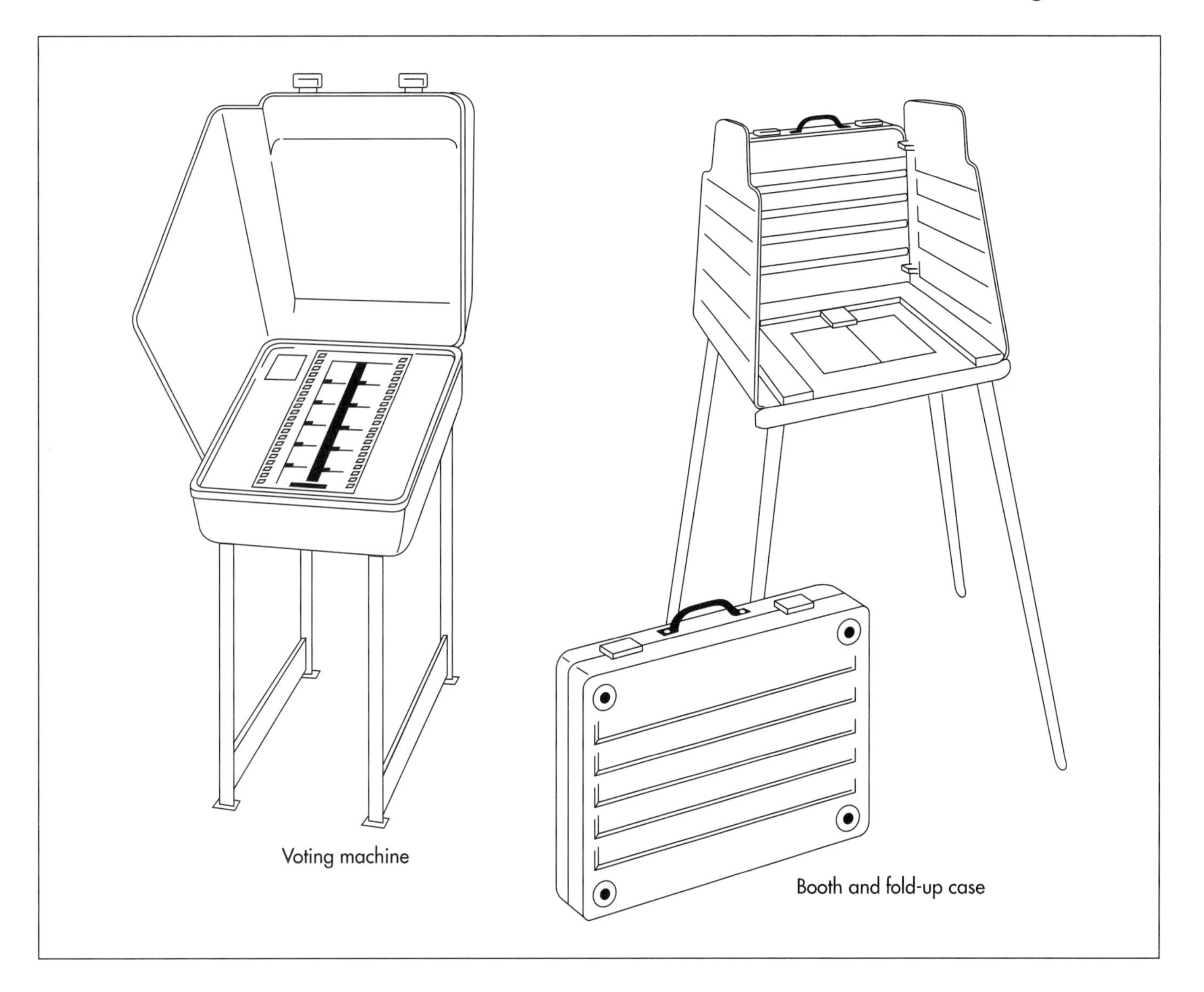


A portable voting machine system.

soldering pieces together. All of the electronics components are tested individually as they are completed.

### Other components

2 Workers assemble other components at separate stations. These include the ballot frame, which is the panel that holds the buttons voters push to select candidates, and the control panel used by the election judge to set the machine. When all the interior parts are assembled, they are sent to a conveyor belt that winds through a final assembly area.

### The case

3 The case that houses the workings of the voting machine is made from lightweight plastic, and is about the size of a large suitcase. A typical manufacturer subcontracts this to another company that specializes in this type of part. The cases arrive from the sub-contractor made according to the original design, with holes for screws and bolts. Workers on the production line drop the components into the case and affix them with screws or rivets. Other hardware is attached to the case at this point, including the privacy panel or curtain, and light fixtures.

### The chassis

4 Once everything is assembled inside and on the case, workers take the case to another assembly line. Here, the case is screwed onto the chassis, which is the table-like stand that holds the machine up. Now, the machine

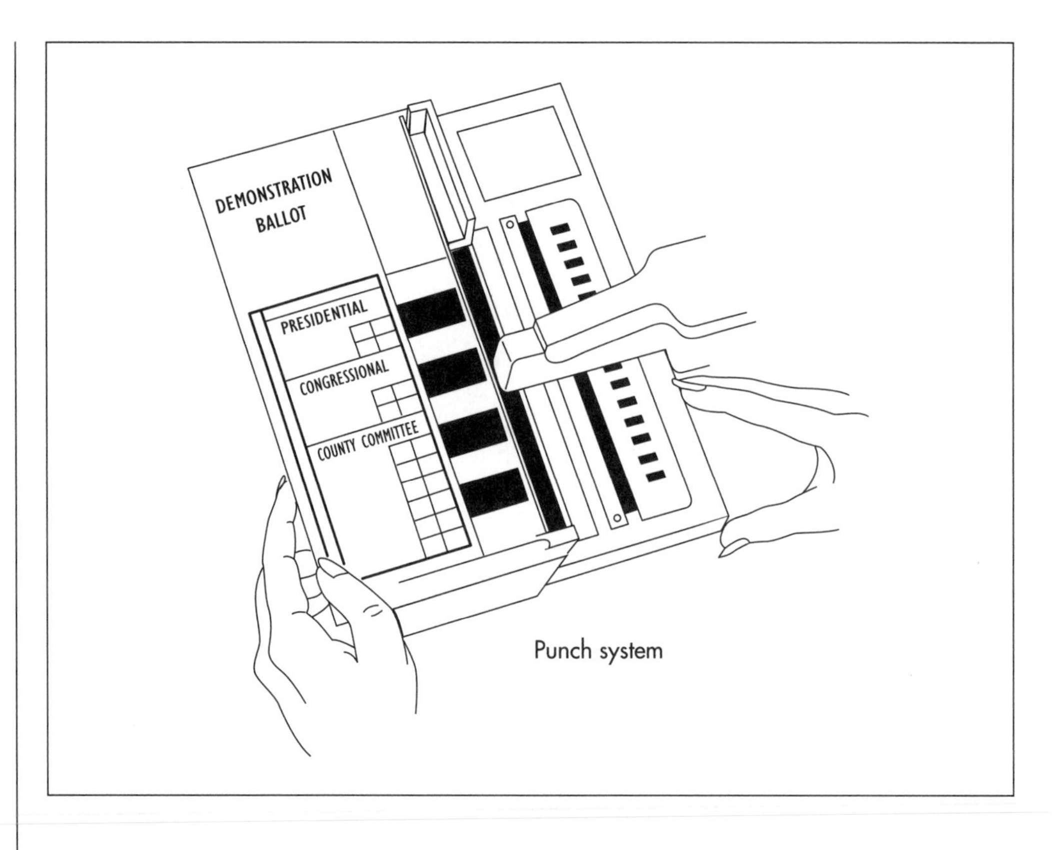


A typical data punch system used to cast votes in an election.

is fully assembled. It goes through a diagnostic and testing step before it is packed up ready for storage or distribution.

### *Diagnostics*

5 It is extremely important for voting machines to be able to work after long periods of sitting on a shelf or in a warehouse. DRE machines have a built-in self-diagnostic system designed to make sure the machine is ready to work every time it is taken out for an election. After the machine is assembled at the plant, workers start the self-diagnostic system, and the machine checks itself to make sure its electronics are functioning. Then, workers actually run through a sample election on the machine. They fit it with a dummy ballot, and vote on it from ten to a hundred times. If it passes this test, the machine can be stored for later sale. Before the machine is shipped to a customer, it goes through one more diagnostic and setting step. The manufacturer plugs the machine into a laptop computer and sets options on it, such as whether lights will blink or not, what county it is going to, and entering any passwords the voting district wants to use. The computer also does a further diagnostic, testing the internal electronics such as the setting of the clock.

## Quality Control

Since the Federal Election Commission issued its voluntary voting systems standards in 1990, manufacturers usually work with national guidelines for what the machine should be able to do. The guidelines don't specify how the machine should work, but ask that it meets technical standards for its hardware and software, that it proves it is secure from tampering, and that it provides adequate documentation in cases of challenged election results. Though not every state has adopted these standards, and they are voluntary, it behooves the manufacturer to be able to state that its machines meet or exceed FEC standards. Independent testing authorities run tests on voting machines submitted by manufacturers, to determine whether they conform to the FEC's standards. The testing authority examines and evaluates the machine's software and looks at its documentation (how it shows what happened, in case of a contested vote). The testing lab also simulates the conditions the machine is subjected to in storage, transport, and in operation, to make sure it works or is easy to repair in varying circumstances. The testing authority documents how the machine holds up under "normal and abnormal conditions," meaning that it subjects the machine to a variety of stress tests. The testing authority also works with manufacturers to see that they learn to perform similar tests themselves. However, the independent testing authority does not certify entire voting machine companies. It only evaluates the particular model. It may provide guidance to the company so that it can run its own tests and make sure its internal quality control system is up to par.

## The Future

Voting machine technology lags behind that of many other industries. For example, optical scanning was used in many areas long before it was applied to voting machines in the 1960s. And the most modern, electronically advanced machines actually make up the smallest portion of the voting machines in use at the turn of the twenty-first century. Voting machines must be simple, durable, and easy to use by even the most unsophisticated voter, and municipalities are for the most part reluctant to change and upgrade expensive machines that are used only twice a year. People in the industry have sometimes suggested sweeping innovations, such as systems that allow voters to call in their votes by telephone. However, this raises daunting security issues. In the late 1990s, there has been speculation that voting on the Internet might be a convenient and technologically feasible future development. Yet given the history of the voting machine in the United States, it seems unlikely that any new system would gain a foothold quickly. Even in the face of a proven better technology, the older systems will likely persist, with smaller and more rural districts lagging behind urban centers.

## Where to Learn More

### Books

Moloney, Martha A. *Mechanized Vote Recording: A Survey.* Frankfort, Kentucky: Legislative Research Commission, 1975.

National Clearinghouse of Election Administration. *Voting System Standards.* Washington, DC: National Clearinghouse of Election Administration, 1990.

### Periodicals

Baquet, Dean, with Martin Gottlieb. "Politics of Voting Machines: A 6-Year Fight for $50 Million." *New York Times.* (October 20, 1990.)

*—Angela Woodward*

# Wheelbarrow

*The modern wheelbarrow was invented in China, possibly as early as 100 B.C.*

## Background

A wheelbarrow is a carrier, usually having only one wheel, consisting of a tray bolted to two handles and two legs. While known mostly as a device for carrying small loads for the household gardener, a wheelbarrow is often also used in construction and industry for carrying larger loads.

The birthplace of the modern wheelbarrow was China, possibly as early as 100 B.C. One early version consisted of a large single wheel at or near the front of a platform. The load would be placed behind the wheel, and the operator would lift the heavy end and push the load. Sometimes a small basket would be used to carry the load, and if it was heavy, a second operator could pull from the front. Another type, probably invented by Chuko Liang in 300 A.D., had a carrying surface that was basically a large wheel housing that encased the top half of the wheel. A flat platform projected out from both sides at axle level could then carry large loads, people and goods in pannier fashion while the operator steered it. Loads were higher off the ground than those carried in modern wheelbarrows. Chinese wheelbarrows were hence similar to rickshaws in that the goal was to carry heavy loads long distances. Tseng Min-Hsung in 1200 A.D. boasted that "ways which are as winding as the bowels of a sheep will not defeat it."

Unlike Chinese wheelbarrows, European wheelbarrows were designed to carry small loads over short distances. It is difficult to trace the progression of the wheelbarrow from China to Europe. Possibly, Arab traders brought it to the Middle East and Europeans learned of it during the Crusades. Ancient Greeks might have used the wheelbarrow for construction, while the Romans might have adapted it for agriculture. After Rome fell, the wheelbarrow could have remained in use in Byzantium until the Crusaders learned of it during their journeys. However, it is most likely that it was an independent invention of the late middle ages, created by putting a wheel on the two-person handbarrow already in use for carrying such items as stones or sheaves of corn.

The first use of the wheelbarrow in Europe was probably in agriculture; then, it spread to construction, transportation of goods, mining, and brick making. Some wheelbarrows had a wooden, box-like body with feet. Others had a flat slatted or wicker framework with feet. French, Flemish, and Dutch wheelbarrows usually had a rack to rest the load or basket against. Swiss wheelbarrows had solid wheels, while English ones had four feet and spoked wheels. Bohemians of the time also used spoked wheels, but no feet were used to help lower the load. To assist when moving loads, European wheelbarrows of around 1200 A.D. had leather straps that wound around the handles. Operators would slip the straps around their necks.

The wheelbarrow's advantages were that loads could be lifted and carried close to the ground, as opposed to two-person handbarrows that required carrying to be done at waist level. A wheelbarrow carrying a basket of goods could be unloaded quickly and put back into action, although it was too unwieldy at this date to be emptied by simply tilting and twisting it. One person using a wheelbarrow cuts labor costs in half, and it's easier than two people coordinating their movements as they carry a load. Wheelbarrows quickly became items crafted by carpenters to be sold to construction workers. In 1222, records for the king of England's works show that eight wheelbarrows were purchased from the town of Canterbury for construction at Dover.

The modern wheelbarrow can have one, two, or four wheels. These wheels can be either in front or beneath the load. Other conveniences include storage space compartments or side clips to carry tools. Wheelbarrows are considered a necessity for the backyard gardener as well as industry, for they are considered simple, unmotorized, yet effective ways for one person to carry a heavy load.

## Raw Materials

A wheelbarrow consists of a tray or bed composed of steel, wood, or plastic. A steel brace attaches this bed to steel support legs and to a steel or plastic wheel, with a rubber tire around it. In two- or four-wheeled models, the wheels may be similar to bicycle tires, complete with inner tubes. Some handles are metal with foam or hard plastic grips, while some wheelbarrows intended for the home gardener have solid wooden handles with no grips.

## The Manufacturing Process

Because wheelbarrows exist in a variety of forms, made from many different materials, manufacturing procedures vary widely depending on the exact kind of wheelbarrow being made. Some wooden wheelbarrows are simple enough to be made with hand tools used by weekend hobbyists with modest carpentry skills. Some heavy-duty wheelbarrows, intended for industrial use, are manufactured with heavy machinery which shapes thick steel sheets. In order to discuss several different techniques used for making wheelbarrows, the following outline will describe the manufacturing of a typical garden wheelbarrow, with wooden handles, steel legs and braces, a rubber tire, and a plastic tray.

### *Making the tray*

1 A typical plastic used to make wheelbarrow trays is polyethylene. Polyethylene is formed by linking molecules of ethylene into a long chain. Ethylene is a hydrocarbon which is found in petroleum and natural gas. For industrial purposes, however, ethylene is usually obtained by heating heavier hydrocarbons, obtained from petroleum, in the presence of a catalyst. This process, known as cracking, transforms larger molecules into smaller molecules, such as ethylene.

2 The process of combining numerous small molecules into a long chain is known as polymerization. Ethylene can be polymerized into two types of polyethylene. Low density polyethylene, made by subjecting ethylene to very high pressure, is generally used in the form of a thin, flexible film, as in plastic bags. High density polyethylene is much stiffer and is used to make wheelbarrow trays and other rigid items. High density polyethylene is polymerized by subjecting ethylene to a pressure of about 250-500 lb per sq in (18-35 kg per sq cm) in the presence of a catalyst.

3 High density polyethylene can be formed into a wide variety of shapes using various techniques. For wheelbarrow trays, it is generally produced in the form of a thick sheet. This can be done by extrusion. This process involves melting granules of polyethylene into a liquid and forcing it through a nozzle. The opening of the nozzle has the width and thickness of the desired sheet. As the liquid polyethylene emerges from the nozzle it cools into a sheet that can be cut to the desired length.

4 Sheets of polyethylene are shipped to the wheelbarrow manufacturer and inspected. They are then shaped into trays using a technique known as vacuum forming or thermo forming. This process involves heating the sheet until it is soft. The soft plastic is then placed on top of an open box. The box contains a mold in the shape of the wheelbarrow tray. Air is removed from the box, and the resulting vacuum causes the air pressure outside the box to force the sheet against the mold. The plastic cools into the wheelbarrow tray and is removed.

### *Making the legs and braces*

5 The support legs and braces are usually made of steel. Steel is made by combining iron ore and coke (a carbon-rich substance made by heating coal in the absence

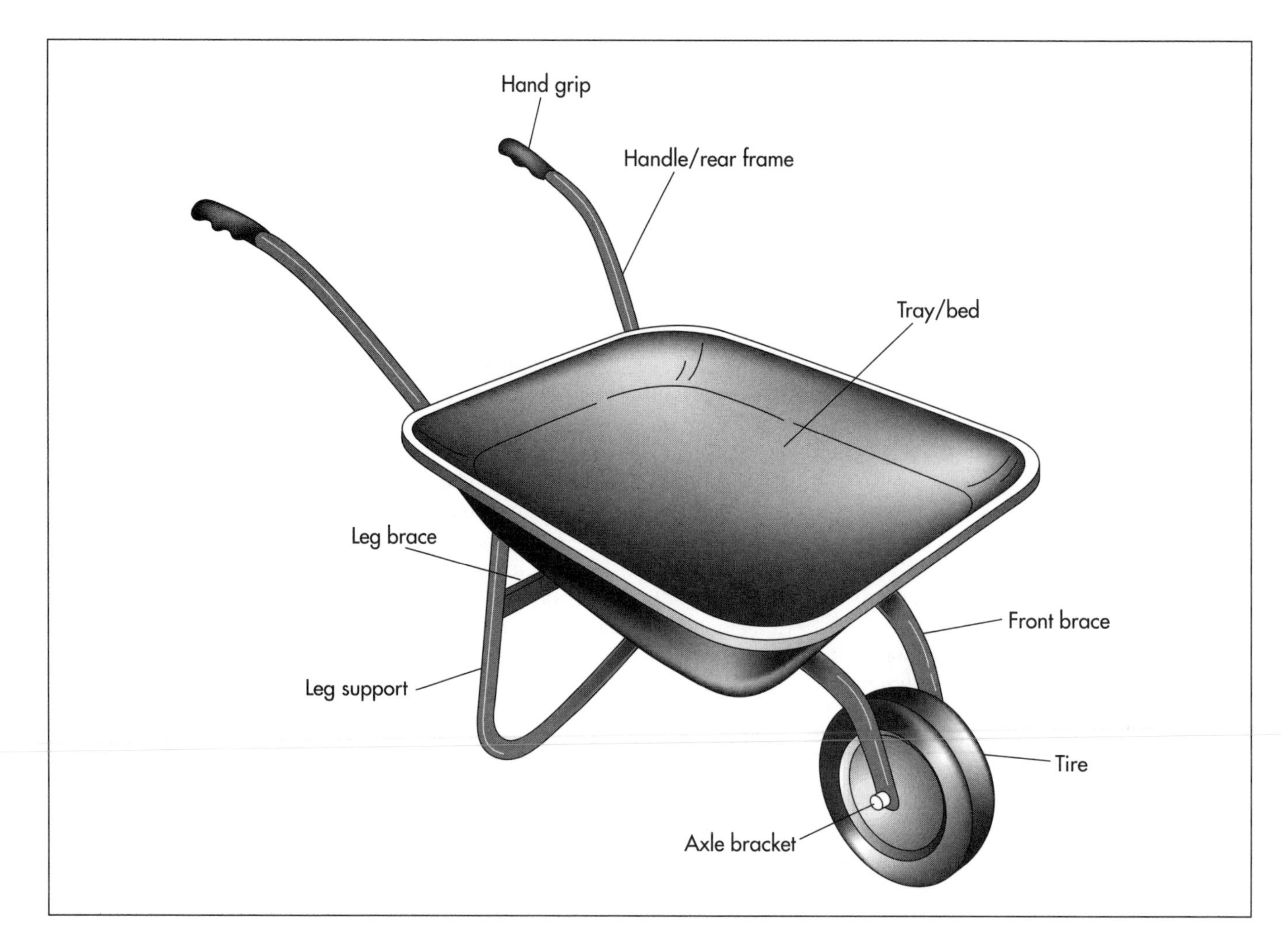


A typical wheelbarrow.

of air) and heating them with very hot air in a blast furnace. The resulting mixture of iron and carbon is known as pig iron. Oxygen is blasted into molten pig iron to remove most of the carbon. The resulting molten steel is cooled into a variety of shapes.

6 Strips of steel arrive at the wheelbarrow manufacturer and are inspected. Sharp dies are used to punch various components out of the steel strips using large, powerful presses. Holes are punched into these components to allow them to be bolted into place.

7 The steel components are usually painted for protection. A degreasing solution is used to clean the components, which are then dried. Paint is applied using an electrostatic process. The steel and the paint are given opposite electrical charges. The opposite charges attract, causing the paint to cover the steel evenly and completely. The paint is then baked onto the steel in an oven.

### *Making the wheel and tire*

8 The wheel may be made of steel or plastic. It is made using methods similar to those used to make other steel parts or the plastic tray.

9 The tire is usually made of natural or synthetic rubber. Natural rubber is made from latex, a substance secreted by rubber trees. The latex is filtered to remove foreign substances, and water is added to make the liquid latex thinner. A dilute acid is added to cause solid rubber to be deposited out of the liquid mixture. Synthetic rubber can be made from a wide variety of chemicals. The most common form of synthetic rubber is styrene-butadiene rubber. This substance is made by mixing styrene and butadiene, both derived from petroleum, in an emulsion of

soap and water. The two chemicals combine to form synthetic rubber.

10 The natural or synthetic rubber is heated until it melts into a liquid. It is then poured into a mold in the shape of the tire. The rubber cools and the tire is removed from the mold. The tire is shipped to the wheelbarrow manufacturer and is placed around the wheel.

### *Making the handles*

11 Lumber arrives at the wheelbarrow manufacturer and is inspected. The wood is stored and allowed to dry. It is then cut and shaped by woodworking equipment such as saws and lathes. Holes are drilled to allow the handles to be bolted into place. The wooden handles are then coated with varnish for protection.

### *Packaging and shipping*

12 Wheelbarrows are generally not assembled until they reach the consumer. Instead, the components are packaged in strong cardboard boxes and shipped to retailers. By not being fully assembled, the wheelbarrows take up much less space during shipping and storage.

## Quality Control

The most important factor in the quality control of wheelbarrows is ensuring that all the components fit together correctly. During the shaping of steel components, each part is compared to a standard part of the same type that is known to be correct. In a similar way, the plastic tray is compared to a standard tray. Wooden handles are compared to a drawing or to a sample handle. The most critical part of this comparison process is ensuring that holes are drilled in the correct positions to allow bolts to hold the wheelbarrow together.

Because a wheelbarrow is generally assembled into its final form by the consumer, it is critical that all necessary components are included in the shipping container. Instructions for assembly must also be included. Large components and instruction sheets are inspected individually to ensure that they are all present. Instead of counting all the bolts to ensure that there are enough included to assemble the wheelbarrow, the manufacturer can save time by weighing a batch of bolts on an accurate scale. A batch of bolts of the correct weight will contain the correct number of bolts.

A sample wheelbarrow can be tested by assembling it, filling it with a load, and moving it around. Factors to be considered in order to produce a wheelbarrow that will satisfy consumers include ease of assembly, ease of motion, stability, ability to be steered accurately, and ability to carry a heavy load with minimal effort on the part of the user.

## The Future

The wheelbarrow has changed over thousands of years from handheld carriers with no wheels to modern lawn carts with four wheels. The increasing popularity of gardening as a hobby, combined with the aging of the American population, suggests that wheelbarrows will continue to evolve in ways that will make them more versatile and easier to use.

## Where to Learn More

### *Periodicals*

Lewis, M. J. T. "The Origins of the Wheelbarrow." *Technology and Culture* (July 1994): 453-475.

Matthies, Andrea L. "The Medieval Wheelbarrow." *Technology and Culture* (April 1991): 356-364.

Randolph, Jeff. "Wheelbarrows." *Flower and Garden* (April/May 1995): 26-28.

Shelton, Will. "Will's Indestructible Weekend Wheelbarrow." *Mother Earth News* (December 1997/January 1998): 76-77.

### *Other*

Erie Products. http://www.new-form.com/erie.html/ (15 August 1998).

*—Rose Secrest*

# Wind Chime

*Bells were first invented in prehistoric times, and the earliest evidence of them dates from 3000 B.C. and was found at archaeological sites in Southeast Asia.*

## Background

The wind chime is a musical instrument that harnesses the wind as its player and composer. The wind chime is also called a wind bell, and it is made in three basic configurations:

- a cluster of similar objects (metal tubes, seashells, glass pieces, lengths of bamboo, or fragments of pottery) that hang from a central support and tinkle when moved by the wind;
- a cluster of objects including a clapper suspended from the support that chimes as the breezes blow against a flat plate or wind catcher at the end of the clapper;
- a bell with a long clapper attached to a wind catcher that plays as the wind strikes the clapper against the bell.

## History

The wind chime is a product of many cultures and many times. Bells were first invented in prehistoric times, and the earliest evidence of them dates from 3000 B.C. and was found at archaeological sites in Southeast Asia. By 2000 B.C., bells had been developed independently along the shores of the Mediterranean and were being cast from bronze by the Egyptians.

The ancient civilizations of Asia can claim the musical and artistic refinements of the wind chime as theirs. The Chinese were casting bells by 1100 B.C., and they created the wind bell that they call *feng-ling*. Wind chimes were considered religious objects that were believed to attract kind spirits and drive away evil ones. The Buddhists especially revered wind bells and hung them by the hundreds from the eaves of shrines, pagodas, temples, and in caves. In Japan, China, Tibet, and Bali, the religious custom was adopted in the secular world, and many homes were similarly decked with the distinctive chimes. In Japan, wind chimes are called *fürin*.

Where bamboo grows easily, wind chimes made of lengths of bamboo may be even more ancient. In Bali especially, large sets of bamboo chimes are hung in the rice fields where they perform the practical services of startling birds and animals, as well as charming the spirits.

Metal sets of tuned, tubular bells were perfected for use in orchestras in the nineteenth century. Precisely tuned, high-quality wind chimes seemed to follow the development of instrumental chimes and have been aided by technology that allows precision cutting and tuning to both ancient and modern musical scales.

Wind bells and chimes became a decorative art as well that used a wide variety of materials and ranged from simple to extremely elaborate in size, construction, and ornamentation. The double attraction of sound and appearance helped the wind chime spread to the Western world during the nineteenth and twentieth centuries.

## Raw Materials

The traditional or percussion wind chime is of the simplest in appearance but the most precise in design. Its material requirements include wood pieces for the top or support and the wind catcher, metal pipes or tubes that are usually made of aluminum, a top ring for suspending the whole chime, metal

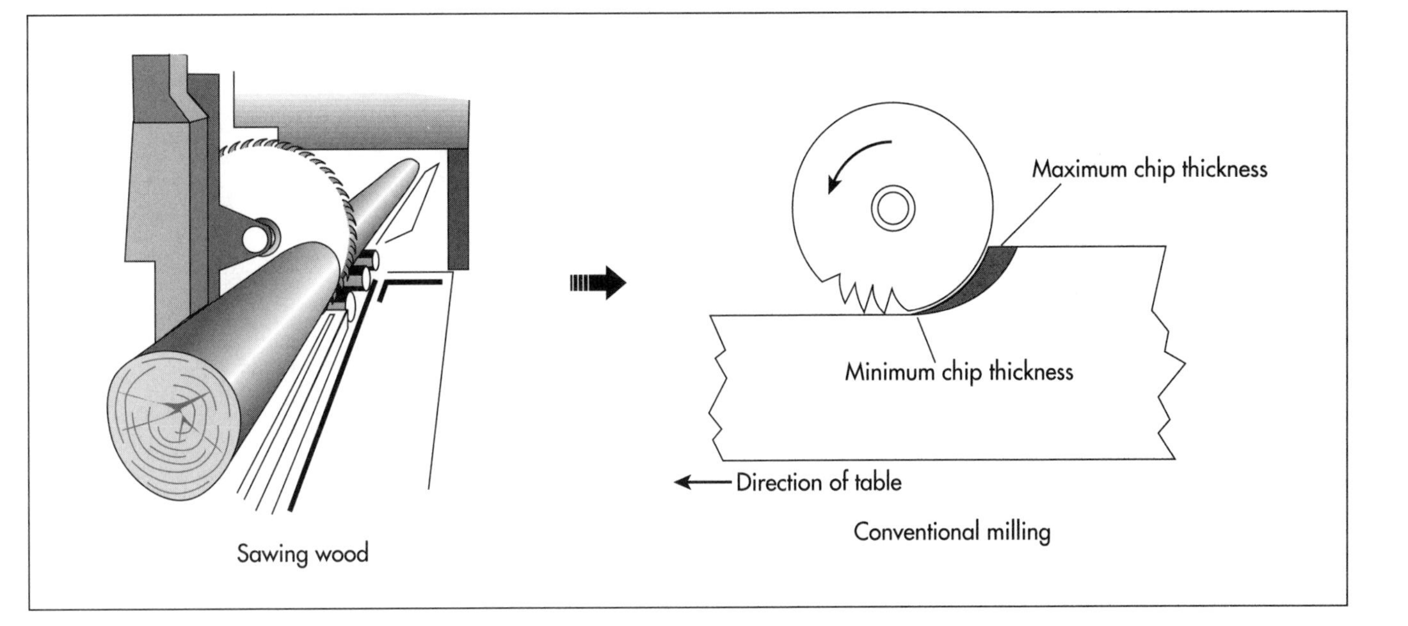


Wood tops are sawed, shaped, sanded, and weatherproofed. Holes are drilled through the wood to accommodate support strings. Aluminum tubing is cut using conventional milling process.

fasteners or nails, and braided nylon string for stitching the pieces together. The wood can be of many types, and it is treated with oil as the only weatherproofing.

Novelty wind chimes are made with brass, bronze, copper, or bamboo tubes. In place of tubes, they may make music with suspended shells, clay pottery shapes, die-cut metal pieces, glass, or other tinkling objects. Their supports and clappers may be equally diverse. Driftwood, metal shapes, and resin ornaments are among hundreds of available choices. Paint, glue, many varieties of string, and hangers of different styles and materials are used for this school of wind chime design.

## Design

Wind chimes come in every size and shape. The tiniest can be worn as earrings, and another mini model fits in the palm of the hand and has three pipes. The largest has tubes that are 58 in (147 cm) long and resonates like the bells in Big Ben's clock tower. Some are mounted on magnets or plant stakes, and most are hung from ornamental shapes or objects.

The emphasis among many designers and manufacturers today is the quality of the sound produced by the chimes. Woodstock Percussion, Inc., began a design revolution in wind chimes in 1979 when the firm began to produce chimes with distinctive sounds and scales that are musically tuned by computer. They looked to other cultures to find historical scales on which to base wind chime melodies and designed chimes with the feel and tone of the chimes of Bali in Indonesia or the Gregorian chant mode.

Durability is also an issue because most chimes are designed for outdoor display, so they must withstand not only the wind but sun, temperature extremes, and precipitation.

## The Manufacturing Process

The type of chime described here is a traditional, percussion wind chime made of aluminum tubes, any of a variety of types and shapes of wood as the top, braided nylon string that is used to attach the tubes to their top and the top to the hanger, and metal fasteners or nails to clip the strings to the wood.

1 The wood top is fabricated by cutting it into the selected shape, machining and sanding the cut piece, and treating it with oil to protect it against the elements. Holes are drilled through the wood to allow for the supporting strings to pass through.

2 The manufacturer buys long lengths of aluminum tubing from a specialty supplier. The tubes are 8-12 ft (2.4-3.7 m) long, depending on the diameter of the tube chosen to suit the overall size of the design. The tubes are precisely cut to initial or working

Four to eight tubes are used in a typical wind chime. Each tube is then tuned with the help of a computer that has been programmed to compare the sound when the tube is struck to the pitches from the original design. If necessary, the tubes are trimmed slightly to adjust the sound.

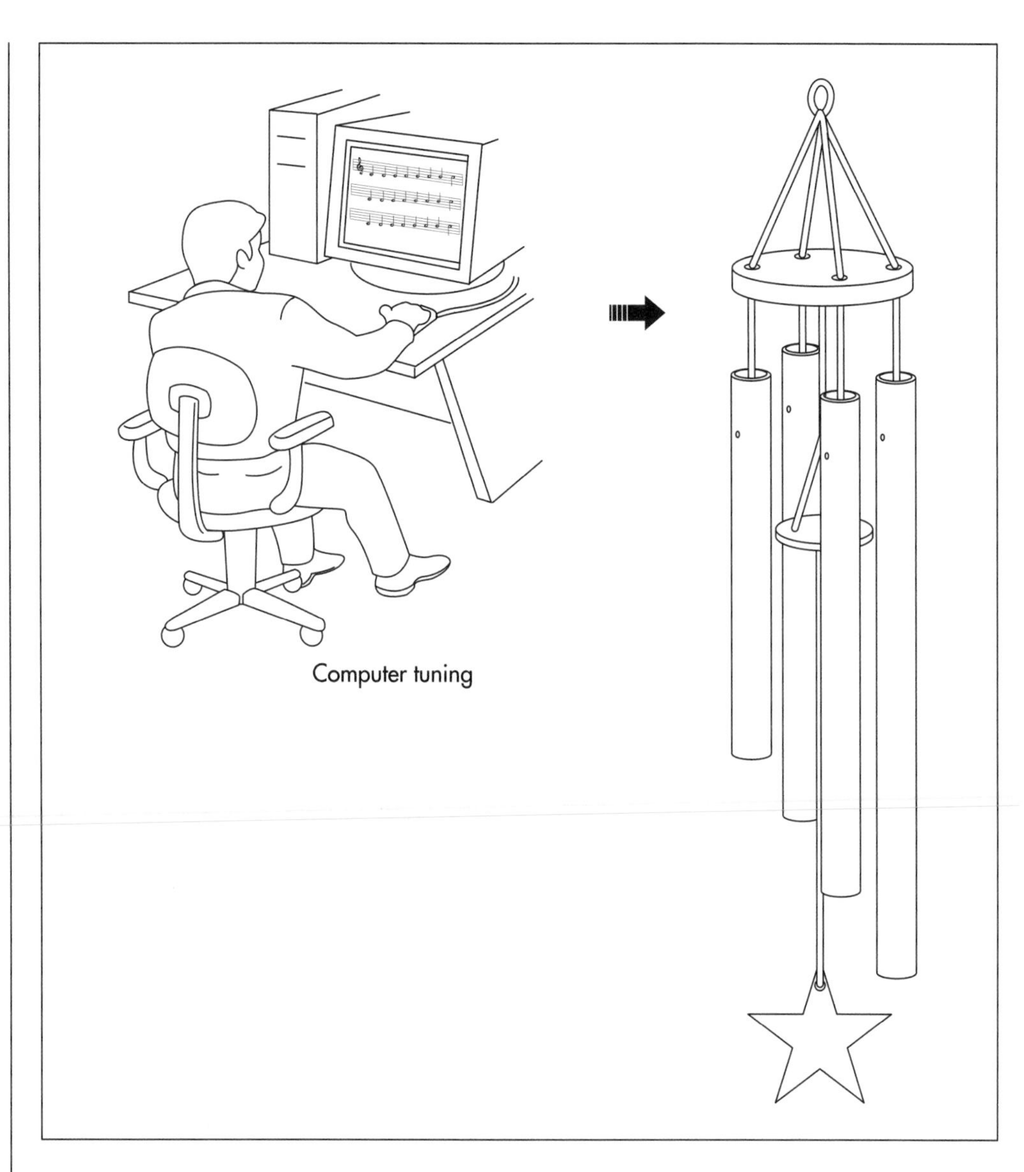
Computer tuning

lengths by using a lathe or a milling machine. Four to eight tubes are used in a typical wind chime. Each tube in a precision instrument is then computer tuned; the computer has been programmed to know the material and compare the sound when the tube is struck to exact pitches for the musical scale that, again, is part of the chosen design. If necessary, the tubes are trimmed slightly to adjust the sound. Holes are then drilled near the tops of the cut pipes so they can be suspended; these holes are countersunk so there are no sharp, external edges.

3 In the assembly area, the wind chime is put together by hand in what is essentially a sewing operation. The braided nylon line is strung through the tubes and the top. The top is linked to the hanger, and the center string assembly is added. At the bottom end of the center string, the clapper is attached; some manufacturers call the clapper a feather or wind catcher. A hang tag is tied to the completed wind chime.

4 The finished product is given a final visual inspection and packaged for sale or shipment.

## Quality Control

Quality inspections are performed throughout the manufacturing process. Skilled wood and metal workers handle these materials and reject any that have cuts or nicks. In the assembly area, the assembly crew is also responsible for rejecting materials and workmanship that are substandard. The final quality inspection occurs just before the complete wind chime is packed.

## Byproducts/Waste

Wind chime manufacturers do not make byproducts, per se, but they often produce many varieties of wind chime or versions of their designs. They also generate very little waste. Aluminum waste is minimized by careful machining techniques. The small amount of waste aluminum that does result is recycled. Hazards associated with wind chime production are also extremely limited.

## The Future

Like many products with a long history, the wind chime seems to have prospects for an equally long future. Gardening is the number one hobby in the United States, and wind chimes are pleasing accessories for gardens, decks, and patios. The wind chime also experienced a renaissance in interest starting in about 1979 when computer technology began to make it possible to produce an in-tune and truly musical wind chime. The instrument's ties to music and many cultures make it universally loved.

## Where to Learn More

### Books

Sadie, Stanley. *The New Grove Dictionary of Musical Instruments, 1984.*

### Periodicals

Zetlin, Melinda. "A ringing success." *Nation's Business*(February 1996): 12+.

### Other

Woodstock Percussion, Inc. http://www.wodstockchimes.com/ (June 29, 1999).

*—Gillian S. Holmes*

# Index

## B

## D

## E

## G

## H

## J



## K



## L

## P

## T

## U



## V

## W



## Y



## Z